MONOGRAPHS ON
STATISTICS AND APPLIED PROBABILITY

General Editors

D. R. Cox, D. V. Hinkley, D. Rubin and B. W. Silverman

Probability, Statistics and Time
M. S. Bartlett

The Statistical Analysis of Spatial Pattern
M. S. Bartlett

Stochastic Population Models in Ecology and Epidemiology
M. S. Bartlett

Risk Theory
R. E. Beard, T. Pentikäinen and E. Pesonen

Residuals and Influence in Regression
R. D. Cook and S. Weisberg

Point Processes
D. R. Cox and V. Isham

Analysis of Binary Data
D. R. Cox

The Statistical Analysis of Series of Events
D. R. Cox and P. A. W. Lewis

Analysis of Survival Data
D. R. Cox and D. Oakes

Queues
D. R. Cox and W. L. Smith

Stochastic Modelling and Control
M. H. A. Davis and R. Vinter

Stochastic Abundance Models
S. Engen

The Analysis of Contingency Tables
B. S. Everitt

An Introduction to Latent Variable Models
B. S. Everitt

(Full details concerning this series are available from the Publishers)

Bandit problems

SEQUENTIAL ALLOCATION OF EXPERIMENTS

DONALD A. BERRY
BERT FRISTEDT

Department of Theoretical Statistics
School of Mathematics
University of Minnesota

LONDON   NEW YORK

CHAPMAN AND HALL

First published in 1985 by
Chapman and Hall Ltd
11 New Fetter Lane, London EC4P 4EE
Published in the USA by
Chapman and Hall
29 West 35th Street, New York NY 10001

© *1985 D. A. Berry and B. Fristedt*

Printed in Great Britain by J. W. Arrowsmith Ltd, Bristol

ISBN 0 412 24810 7

British Library Cataloguing in Publication Data

Berry, Donald A.
  Bandit problems: sequential allocation of
  experiments. —— (Monographs on statistics and
  applied probability)
  1. Mathematical statistics
  I. Title   II. Fristedt, Bert   III. Series
  519.5   QA276

  ISBN 0-412-24810-7

Library of Congress Cataloging in Publication Data

Berry, Donald A.
  Bandit problems.
  (Monographs on statistics and applied probability)
  Bibliography: p.
  Includes index.
  1. Experimental design. I. Fristedt, Bert,
  1937–   . II. Title. III. Series.
QA279, B47 1985   001.4'34   85-9696
ISBN 0-412-24810-7

Contents

Preface

Our purpose in writing this monograph is to give a comprehensive treatment of the subject. We define bandit problems and give the necessary foundations in Chapter 2. Many of the important results that have appeared in the literature are presented in later chapters; these are interspersed with new results. We give proofs unless they are very easy or the result is not used in the sequel. We have simplified a number of arguments so many of the proofs given tend to be conceptual rather than calculational. All results given have been incorporated into our style and notation.

The exposition is aimed at a variety of types of readers. Bandit problems and the associated mathematical and technical issues are developed from first principles. Since we have tried to be comprehensive the mathematical level is sometimes advanced; for example, we use measure-theoretic notions freely in Chapter 2. But the mathematically uninitiated reader can easily sidestep such discussion when it occurs in Chapter 2 and elsewhere. We have tried to appeal to graduate students and professionals in engineering, biometry, economics, management science, and operations research, as well as those in mathematics and statistics. The monograph could serve as a reference for professionals or as a test in a semester or year-long graduate level course.

A uniform treatment of the numerous papers that deal with bandit problems is not possible. We have tried to compensate with the inclusion of an Annotated Bibliography. In it we list about 200 papers and books that have appeared on the subject and comment on those available to us. We say what problems the authors address, indicate when we think the paper contains mistakes, and occasionally criticize the approach taken.

The individual chapters stand as much on their own as seems reasonable, but all readers should learn our notation as given in

Section 2.2. Chapter 4 should be read before Chapters 5, 6 or 7. Chapters 5 and 6 form a natural pair. Chapter 3 should be read or skimmed early, and either of Chapters 8 or 9 can be read without having learned the content of the earlier chapters.

Results of other authors are appropriately credited, so those given without reference are new. As an example of the latter, in Chapter 6 we prove a converse of the famous Gittins–Jones theorem. Most of the examples (of which there are many) and most of the tables and figures are new. The tables and figures given in Section 5.4 were made with the programming assistance of K. Samaranayake. The tables in Section 5.6 are due to M. K. Clayton. Information for Figures 8.2 and 8.3 was supplied by A. J. Petkau.

We have benefited greatly from conversations with many people concerning the preparation of this monograph. These include J. A. Bather, M. K. Clayton, D. C. Heath, N. C. Jain, R. E. McCulloch, S. Orey, A. J. Petkau, and M. Schäl. The contributions of S. G. Eick and W. D. Sudderth in this regard have been especially numerous and deep. We thank T. S. Ferguson for commenting on the manuscript. We also thank our typists, P. Linman and A. M. Ruggles, for their diligence through our numerous revisions. Finally, we thank our wives, Donna and Shirin, for their patience and support.

Minneapolis, Minnesota   D. A. Berry
October 1984   B. Fristedt

CHAPTER 1

Introduction

Suppose two treatments are available for a certain disease. Patients arrive at a clinic one at a time and one of the treatments must be used on each. Information as to the effectiveness of the treatments accrues as they are used. The overall objective is to treat as many patients as effectively as possible. This seemingly innocent but important problem is surprisingly difficult, even when the responses are dichotomous, either success or failure. It is an example of a two-armed bandit problem.

A bandit problem in statistical decision theory involves sequential selections from $k \geqslant 2$ stochastic processes (or 'arms', machines, treatments, etc.). Time may be discrete or continuous and the processes themselves may be discrete or continuous. The processes are characterized by parameters which are typically unknown. The process selected for observation at any time depends on the previous selections and results. A decision procedure (or strategy) specifies which process to select at any time for every history of previous selections and observations. A utility is defined on the space of all histories. This provides a definition for the utility of a strategy in the usual way, by averaging over all possible histories resulting from the strategy.

Most of the literature, and most of this monograph, deals with discrete time. In such a setting, each of the k arms generates an infinite sequence of random variables. An observation on a particular sequence is made by selecting the corresponding arm. The mth member of a sequence is observed if the corresponding arm is 'selected' at stage m. The classical objective in bandit problems is to maximize the expected value of the payoff, $\Sigma_1^\infty \alpha_m Z_m$, where Z_m is the variable observed at stage m and the α_m are nonnegative numbers. Though our approach will be somewhat more general (see Section 3.2), the *discount factors* α_m are usually assumed to be known with $0 < \Sigma_1^\infty \alpha_m < \infty$ and,

1

sometimes, $\alpha_m \geqslant \alpha_{m+1}$. $\mathbf{A} = (\alpha_1, \alpha_2, \ldots)$ is called a *discount sequence*. A strategy is *optimal* if it yields the maximal expected payoff. An arm is said to be *optimal* if it is the first selection when following some optimal strategy.

The discount sequences which are most frequently considered in the literature are:

(i) finite horizon uniform: $(1, \ldots, 1, 0, \ldots)$,
(ii) geometric: $(1, \alpha, \alpha^2, \ldots), 0 < \alpha < 1$.

For (i) the objective is to maximize the sum of the first n observations (where n is the horizon). If the mth observation is made at time m and is paid in inflated monetary units, then (ii) may be appropriate where $\alpha^{-1} - 1$ is the inflation rate. Further discussion of discount sequences and some motivation for general discounting are given in Chapter 3.

There are two benefits derived from selecting an arm: (1) immediate payoff, and (2) information that can make for better later selections and greater future payoff. When an arm is selected, available information concerning the arms is modified and the discount sequence changes: $(\alpha_1, \alpha_2, \alpha_3, \ldots)$ becomes $(\alpha_2, \alpha_3, \ldots)$. For example, in (i) above, the horizon is decreased by 1. In (ii) above, the new sequence is proportional to the original sequence. The decision problem is unchanged if the discount sequence is multiplied by a constant, so for geometric discounting (and only for this case) the discount sequence is effectively the same throughout the trial. This characteristic can make bandit problems with geometric discounting more tractable than other problems (see Chapter 6).

1.1 Bayesian and other approaches

Consider the Bernoulli case for ease of discussion. Given a discount sequence $\mathbf{A}$, the utility of a particular strategy can be calculated as a function of $\mathbf{A}$ and of the Bernoulli parameters θ_1 and θ_2. It is generally too much to expect that a strategy will exist that is best for all pairs (θ_1, θ_2).

The vast majority of the bandit literature takes one of two approaches. In the Bayesian approach, the utility of a strategy is averaged over (θ_1, θ_2) with respect to some measure; papers by Bradt, Johnson, and Karlin (1956), Bellman (1956), and Feldman (1962) are early examples. This measure represents information that is present about the various processes separate from the current experiment. Many adherents to the Bayesian approach regard this measure as being subjective (Savage, 1954; Barnett, 1982) and quantifying the knowledge of the experimenter concerning the two arms.

The second approach taken in the literature is to consider particular strategies and compare their utilities as a function of (θ_1, θ_2). Papers by Robbins (1952) and Isbell (1959) are early examples. When the utility of one strategy uniformly dominates that of the others then, of course, it is best in the class of strategies under consideration. When one is not uniformly best, the various strategies can be compared using tables (e.g. Wahrenberger, Antle, and Klimko, 1977).

A third alternative is the minimax approach in which nature is regarded as an opponent in a two-person, zero-sum game (e.g. Vogel, 1960a, b). Nature chooses (θ_1, θ_2) in the unit square, or in a subset of it, according to some *a priori* restriction. The decision maker's goal is to minimize the expected difference between what is achieved and what could be achieved were (θ_1, θ_2) known. Nature's goal is to maximize this expected difference. The minimax approach is discussed in Chapter 9.

Thompson (1933, 1935) posed the first bandit problem. He considered two Bernoulli processes, uniform discounting ($\alpha_1 = \ldots = \alpha_n = 1, \alpha_{n+1} = \ldots = 0$), and took a Bayesian point of view. In this setting the objective is to maximize the expected number of successes in the first n trials. Thompson regarded the two processes as independent with their parameters having beta distributions (cf. Chapter 7).

After Thompson (1933, 1935), the bandit problem received little attention until it was studied by Robbins (1952, 1956), who also considered two arms but took the second approach mentioned above. Robbins (1952) suggested a selection strategy that depends on the history only through the last selection and the result of that selection; namely, the same arm is selected after a success and the other after a failure. Robbins's objective was to maximize the long-run proportion of successes. He showed that the 'stay on a winner, switch on a loser' strategy uniformly dominates random selection. This originated an approach called 'finite memory': the decision maker's choice at any stage can depend only on the selections and results in the previous r stages; Isbell (1959) and Smith and Pyke (1965) are examples of this approach.

Bradt, Johnson, and Karlin (1956) took a Bayesian approach for the finite horizon, uniform case. They characterized optimal strategies when one parameter is known *a priori*.

With a Bayesian approach, a strategy requires (and 'remembers') only the sufficient statistics: the numbers of successes and failures on the two arms. Much of the recent bandit literature – and most of this monograph – takes the Bayesian approach. It is not that researchers in bandit problems tend to be 'Bayesians'; rather, Bayes's theorem provides a convenient mathematical formalism that allows for adaptive learning, and so is an ideal tool in sequential decision problems.

With a Bayesian approach, a bandit is a typical problem in dynamic programming. When the horizon is finite ($\alpha_{n+1} = \alpha_{n+2} = \ldots = 0$ for some n), backwards induction can be used to determine optimal strategies (cf. Sections 2.3 and 2.4). One first finds the maximal conditional expected payoff (together with the arm or arms that give it) at the very last stage for every possible $(n-1)$-history (sequence of selections and results), optimal and otherwise. Here, 'conditional' refers to the particular history. Proceeding to the penultimate stage, one maximizes the conditional expected payoff from the last two observations for every possible $(n-2)$-history. Continuing backwards, while remembering the optimal arms at each partial history, gives all optimal strategies. The problem is four-dimensional in the Bernoulli setting since that is the dimension of a minimal sufficient statistic. But a computer program requiring on the order of $n^3/6$ storage locations can be devised.

1.2 Myopic strategies

Feldman (1962) solved the Bernoulli two-armed bandit problem with uniform discounting for a deceptively difficult special case: both probabilities of success are known, but not which goes with which arm. Feldman showed that *myopic* strategies are optimal: at every stage, select the arm with greater expected immediate gain (the unconditional probability of success with arm j is the current mean of the Bernoulli parameter θ_j).

It is important to recognize that myopic strategies are not optimal – or even good – in general. The following is a case in point.

Example 1.2.1 Suppose **A** is uniform with horizon n. Assume θ_2 is known to be $1/2$ (selecting arm 2 is like tossing a fair coin). And θ_1 is either 1 or 0 (the other coin is either two-headed or two-tailed); let r be the initial probability that $\theta_1 = 1$ so r is the prior mean of θ_1. The fact that a single selection of arm 1 reveals complete information makes the analysis of this problem rather easy. If $r < 1/2$ then a myopic strategy indicates selections of arm 2 indefinitely, and has utility $n/2$. On the other hand, selecting arm 1 initially and then indefinitely if it is successful and never again if it is not, results in n successes with probability r and an average of $(n-1)/2$ successes with probability $1 - r$. The advantage of this strategy over the myopic is

$$rn + (1-r)(n-1)/2 - n/2 = [r(n+1)-1]/2,$$

which is positive for $r > 1/(n+1)$.

All we have shown is that the indicated strategy is better than the myopic when $r > 1/(n+1)$, but, as a consequence of Theorem 5.2.2, it is optimal. □

In this example, and in bandit problems generally, it may be wise to sacrifice some potential early payoff for the prospect of gaining information that will allow for more informed choices later. This aspect prompted Whittle (1982, p. 210) to claim that a bandit problem 'embodies in essential form a conflict evident in all human action'. The 'information versus immediate payoff' question makes the general problem difficult; the issue is seldom as clear as it is in Example 1.2.1.

In the clinical trial setting, sacrificing early payoff for information means that patients arriving later are likely to be treated better because they benefit from the responses of early patients (cf. Corollary 5.2.3). This also characterizes most designs actually used in clinical trials – but in the extreme. Patients are assigned treatments randomly in a clinical trial to gain information about the various treatments; the accumulating information is seldom used in treatment assignment during the trial. After the trial, patients can be assigned the treatment found to be most effective. While a sequential design with the exclusive purpose of gathering information is a special case of our approach (see Section 3.6), our general framework allows for weighting of future and present patients. This is done by appropriate choice of the discount sequence. An advantage of the geometric discount sequence in this regard is that it is 'democratic': the current patient is always weighted the same when compared with future patients.

1.3 Preview

Chapter 2 lays various foundations for the rest of the monograph. Firstly, it introduces the basic ideas of bandit problems to the uninitiated. In particular, Sections 2.1 and 2.4 contain many easy examples which embody the flavour of bandit problems. Secondly, the chapter constructs a technical framework for the chapters that follow. The fundamental equation of dynamic programming is developed and optimal strategies are shown to exist (Sections 2.3 and 2.5). Readers are urged not to get bogged down in the technical considerations in Sections 2.2, 2.3, and 2.5 at the expense of the various interesting aspects of bandits appearing in the rest of the chapter and in later chapters. We have written the other chapters to stand on their own as much as possible; in particular, only occasional reference to Chapter 2 is made.

Chapter 3 deals with the role of the discount sequence. Our approach to discounting is more general than that of other authors. A purpose of Chapter 3 is to motivate our approach, and to explain its applicability. In so doing we present a theory of random discounting. We also describe a setting in which the stages of a bandit problem occur in real time, and randomly. This setting will be discussed again in Chapters 5 and 8.

Most of this monograph treats independent arms. Chapter 4 develops some basic results for the case in which the arms are independent and also Bernoulli. Many of these results are applied in Chapters 5, 6, and 7.

Chapter 5 deals with the special case of two arms when the characteristics of one arm are completely known. Such bandits have been treated as stopping problems in the literature, and they are correctly regarded as stopping problems for some discount sequences. We characterize such sequences and give a variety of results for the case in which the problem is to decide when to stop selecting the unknown arm. For ease of presentation, in Sections 5.1 to 5.4 we assume that the unknown arm is Bernoulli. This assumption is dropped in Section 5.5 and a rather comprehensive setting is given in Section 5.6. In view of the main result in Chapter 6, the results of Chapter 5 take on new meaning when the discount sequence is geometric.

In Chapter 6 we give the celebrated result of Gittins and Jones (1974). This result shows that when the discounting is geometric, a bandit problem involving k independent arms can be solved by solving k different two-armed bandits, each involving one known and one unknown arm. We prove the converse of this result by showing that a bandit involving two or more unknown arms can be solved in this way only when the discount sequence is geometric.

Chapter 7 treats the case of two independent Bernoulli arms when the horizon is finite and discounting is uniform. We give sufficient conditions for optimality, some depending on the horizon and some not.

In Chapter 8 we turn to the setting of continuous time in which both payoff and information accrue continuously. We treat Weiner processes and Lévy processes, mostly by example. We show, partly by example, the difficulties involved in treating continuous-time bandits. In particular, we argue that a satisfactory treatment of the foundations of continuous-time bandits is essentially an open problem.

Chapter 9 takes a different approach from that of the rest of the monograph. Instead of averaging over the unknown characteristics of the arms, the decision maker takes the worst-case point of view – a minimax approach.

References

Barnett, V. (1982) *Comparative Statistical Inference* (2nd edn), Wiley, New York.

Bellman, R. (1956) A problem in the sequential design of experiments. *Sankhyā A* **16**: 221–229.

Bradt, R. N., Johnson, S. M. and Karlin, S. (1956) On sequential designs for maximizing the sum of n observations. *Ann. Math. Statist.* **27**: 1060–1074.

Feldman, D. (1962) Contributions to the 'two-armed bandit' problem. *Ann. Math. Statist.* **33**: 847–856.

Gittins, J. C. and Jones, D. M. (1974) A dynamic allocation index for the sequential design of experiments. In *Progress in Statistics* (eds J. Gani et al.), pp. 241–266, North-Holland, Amsterdam.

Isbell, J. R. (1959) On a problem of Robbins. *Ann. Math. Statist.* **30**: 606–610.

Robbins, H. (1952) Some aspects of the sequential design of experiments. *Bull. Amer. Math. Soc.* **58**: 527–535.

Robbins, H. (1956) A sequential decision problem with finite memory. *Proc. Nat. Acad. Sci. U.S.A.* **42**: 920–923.

Savage, L. J. (1954) *The Foundations of Statistics*, Wiley, New York.

Smith, C. V. and Pyke, R. (1965) The Robbins–Isbell two-armed bandit problem with finite memory. *Ann. Math. Statist.* **36**: 1375–1386.

Thompson, W. R. (1933) On the likelihood that one unknown probability exceeds another in view of the evidence of two samples. *Biometrika* **25**: 275–294.

Thompson, W. R. (1935) On the theory of apportionment. *Am. J. Math.* **57**: 450–456.

Vogel, W. (1960a) A sequential design for the two-armed bandit. *Ann. Math. Statist.* **31**: 430–443.

Vogel, W. (1960b) An asymptotic minimax theorem for the two-armed bandit problem. *Ann. Math. Statist.* **31**: 444–451.

Wahrenberger, D. L., Antle, C. E. and Klimko, L. A. (1977) Bayesian rules for the two-armed bandit problem. *Biometrika* **64**: 172–174.

Whittle, P. (1982) *Optimization Over Time: Dynamic Programming and Stochastic Control*, Vol. I, Wiley, New York.

CHAPTER 2

Notation and preliminaries

Bandit problems are described formally in this chapter. One purpose of the chapter is to develop some necessary notation. Another is to define strategies rigorously. It is shown that optimal strategies exist and that the 'fundamental equation of dynamic programming' is satisfied.

Some of the more important notions are developed in a simple example in Section 2.1. The general setting is presented in Section 2.2. In Section 2.3 and 2.5 the existence of optimal strategies is proved; the method of dynamic programming is used in the finite horizon case in Section 2.3. A variety of finite horizon examples is given in Section 2.4. In Section 2.6 an approximate method for the infinite horizon case is discussed and an illustrative example given.

Much of the literature that deals with bandit problems, and other stochastic decision problems, implicitly assumes that everything works the way one would like it to work. The reader will see that a careful development is quite technical. Some readers may want to skim the technical Sections 2.2, 2.3, and 2.5; others will skip them entirely.

2.1 An example: maximizing the sum of two observations

Suppose there are two opportunities to receive payoff and the objective is to maximize the sum of the two observations, so the discount sequence is $A = (1, 1, 0, \dots)$. At the first stage the decision maker has two choices: one is a random variable which may be regarded as being produced by a mechanism, arm 1, and the other is the known constant λ which, with future considerations in mind, we will regard as being produced by arm 2. Let Z_1 denote the observation at stage 1; so $Z_1 = \lambda$ in case the decision maker selects arm 2, and Z_1 is random in case arm 1 is chosen.

Taking into account the result and selection at stage 1, the decision

9

maker faces the same choice at stage 2. The setting is now different for two reasons. First, there is but one remaining observation. Second, there may now be different information available to the decision maker, depending on whether arm 1 was selected initially and on the resulting observation.

The notion of a strategy has been defined implicitly above. More formally, a *strategy* is a function that assigns to each (partial) history of observations the integer 1 or 2 indicating the arm to be observed at the next stage. Thus, a strategy τ assigns to the empty history the integer indicating the arm to be observed initially, and $\tau(z_1)$ indicates the arm to be observed at stage 2 when z_1 is observed at stage 1. We require

$$\{z_1 : \tau(z_1) = i\}$$

to be a Borel set for $i = 1$ and therefore also for $i = 2$; that is, we require the function $z_1 \mapsto \tau(z_1)$ to be Borel measurable. The decision maker's objective is to choose τ to maximize $E_\tau(Z_1 + Z_2)$, where Z_m indicates the observation at stage m and the subscript τ indicates the dependence on the strategy τ. The use of expectation implies the existence of an underlying probability structure; we now turn our attention to this structure.

The characteristics of arm 2 are known to the decision maker: whenever arm 2 is selected a known constant λ is observed. The characteristics of arm 1 are not completely known. Let X_m denote the outcome from arm 1 at stage m; it is useful for notational reasons to assume there is an outcome on arm 1 at each stage whether or not arm 1 is in fact selected.

We assume that the X_m are normally distributed random variables having mean θ and variance 1. The parameter θ is unknown and is assumed to have an initial distribution which is normal with mean μ and variance $\rho^2 > 0$, both of which are known. Conditional on θ, X_1 and X_2 are independent. The value of X_1 is known to the decision maker at stage 2 if and only if arm 1 was selected initially. Accordingly, the full strategy τ can be specified prior to stage 1, for a robot can carry out a decision maker's wishes by evaluating τ at the value of Z_1.

The *utility* or *worth* of a strategy τ is $E_\tau(Z_1 + Z_2)$. For this to be well-defined, Z_2 must be a random variable; that is, it must be measurable. That it is measurable follows from the measurability of $z_1 \mapsto \tau(z_1)$ and the identity

$$Z_2 = X_2 \mathbf{1}_{\{z_1 : \tau(z_1) = 1\}}(Z_1) + \lambda \mathbf{1}_{\{z_1 : \tau(z_1) = 2\}}(Z_1). \qquad (2.1.1)$$

We introduce a notation that will be adapted more generally. The worth of a strategy depends on the discount sequence and the available information concerning the arms. In the case at hand the information about arm 1 is indexed by (μ, ρ), arm 2 is specified by the constant λ, and the discount sequence is $\mathbf{A} = (\alpha_1, \alpha_2, \ldots)$ $= (1, 1, 0, 0, 0, \ldots)$; the problem is called the $((\mu, \rho), \lambda; \mathbf{A})$-bandit. The worth of a strategy τ is

$$W((\mu, \rho), \lambda; \mathbf{A}; \tau) = E_\tau \sum_{m=1}^\infty \alpha_m Z_m$$

$$= E_\tau (Z_1 + Z_2).$$

The *value* of the $((\mu, \rho), \lambda; (1, 1, 0, \ldots))$-bandit is defined to be

$$V((\mu, \rho), \lambda; (1, 1, 0, \ldots)) = \sup_\tau W((\mu, \rho), \lambda; (1, 1, 0, \ldots); \tau).$$

Any τ for which the supremum is attained is an *optimal strategy*.

We shall calculate $V((\mu, \rho), \lambda; (1, 1, 0, \ldots))$ and find an optimal strategy. We begin by calculating a strategy (call it τ_2) that is best among those that begin with arm 2 at stage 1. Since

$$E(X_2) = E(X_1) = E(E(X_1 | \theta)) = E(\theta) = \mu, \tag{2.1.2}$$

we can take

$$\tau_2(z_1) = \begin{cases} 2 \text{ if } \mu \leqslant \lambda \\ 1 \text{ if } \mu > \lambda. \end{cases}$$

Moreover, with '$\vee$' denoting maximum,

$$W((\mu, \rho), \lambda; (1, 1, 0, \ldots); \tau_2) = \lambda + (\lambda \vee \mu). \tag{2.1.3}$$

The next step will be to find a strategy, say τ_1, that is best among those that begin with arm 1. Accordingly, $Z_1 = X_1$ in (2.1.1) and

$$E(Z_2 | X_1) = \mathbf{1}_{\{z_1 : \tau_1(z_1) = 1\}}(X_1) E(X_2 | X_1) + \mathbf{1}_{\{z_1 : \tau_1(z_1) = 2\}}(X_1) \lambda.$$

This is maximized by τ_1 defined by $\tau_1(z_1) = 2$ if $E(X_2 | X_1 = z_1) \leqslant \lambda$ and $\tau_1(z_1) = 1$ otherwise. A well-known formula (DeGroot 1970, p. 248) for the posterior mean of a normal distribution with a normally distributed mean gives

$$E(X_2 | X_1) = \frac{\mu + \rho^2 X_1}{1 + \rho^2}.$$

So we can take

$$\tau_1(z_1) = \begin{cases} 2 & \text{if } \dfrac{\mu + \rho^2 z_1}{1 + \rho^2} \leqslant \lambda \\[3mm] 1 & \text{if } \dfrac{\mu + \rho^2 z_1}{1 + \rho^2} > \lambda, \end{cases}$$

and

$$W((\mu, \rho), \lambda; (1, 1, 0, \ldots); \tau_1) = \mu + E\left(\frac{\mu + \rho^2 X_1}{1 + \rho^2} \vee \lambda\right). \quad (2.1.4)$$

From (2.1.3) and (2.1.4) we obtain

$$V((\mu, \rho), \lambda; (1, 1, 0, \ldots))$$

$$= W((\mu, \rho), \lambda; (1, 1, 0, \ldots); \tau_1) \vee W((\mu, \rho), \lambda; (1, 1, 0, \ldots); \tau_2)$$

$$= \left[\mu + E\left(\frac{\mu + \rho^2 X_1}{1 + \rho^2} \vee \lambda\right)\right] \vee [\lambda + (\lambda \vee \mu)]$$

$$= \left[\mu + E\left(\frac{\mu + \rho^2 X_1}{1 + \rho^2} \vee \lambda\right)\right] \vee 2\lambda \vee [\lambda + \mu]$$

$$= \left[\mu + E\left(\frac{\mu + \rho^2 X_1}{1 + \rho^2} \vee \lambda\right)\right] \vee 2\lambda.$$

Therefore, optimal strategies depend on the sign of

$$\mu - \lambda + E\left[\left(\frac{\mu + \rho^2 X_1}{1 + \rho^2} - \lambda\right) \vee 0\right] = \frac{\rho^2}{\sqrt{(1 + \rho^2)}}[\Psi(t) - t],$$

where $t = (\lambda - \mu)\rho^{-2}\sqrt{(1 + \rho^2)}$, the transform

$$\Psi(t) = \int_t^\infty (x - t)\varphi(x)\,dx$$

$$= \varphi(t) - t[1 - \Phi(t)]$$

(cf. DeGroot 1970, p. 247), and φ and Φ are the standard normal density and distribution functions, respectively. The function Ψ is positive, continuous, and strictly decreasing. So there is a unique t_0 satisfying $\Psi(t_0) = t_0$; numerically, $t_0 \approx 0.2760$.

If $t \geqslant t_0$ then τ is optimal if $\tau(\varnothing) = \tau(z_1) = 2$; that is, arm 2 is optimal at both stages. If $t \leqslant t_0$ then τ is optimal if $\tau(\varnothing) = 1$ and

$$\tau(z_1) = \begin{cases} 1 \text{ for } z_1 \geqslant \lambda + (\lambda - \mu)/\rho^2 \\ 2 \text{ for } z_1 \leqslant \lambda + (\lambda - \mu)/\rho^2. \end{cases}$$

So arm 1 is optimal initially if $(\lambda - \mu)\rho^{-2}\sqrt{(1 + \rho^2)} \leqslant t_0$ and it is optimal at stage 2 if its current mean is greater than λ.

The optimal initial selection has various properties which can aid in understanding bandit problems more generally. If the mean μ of an observation on the unknown arm is at least as large as the sure-thing λ, then the unknown arm is optimal. If, on the other hand, $\mu < \lambda$ then whether arm 1 is optimal depends on ρ. Since $\rho^{-2}\sqrt{(1 + \rho^2)}$ decreases to 0 as $\rho \to \infty$, arm 1 is optimal for sufficiently large ρ for any fixed μ and λ! So if the characteristics of arm 1 are sufficiently uncertain (that is, if ρ is sufficiently large), then any expected loss will be tolerated on the initial selection for the possibility of a large gain on the second selection. (Many statisticians who take the Bayesian approach to inference recommend using improper priors and taking $\rho \to \infty$ for convenience in the case of normal sampling. This may not be grossly incorrect in a problem in which some sampling will occur in any case. But it can be very misleading and lead to inane strategies when the issue is to sample or not!)

One purpose of this example is to give a concrete setting for some of the abstract notions to be developed in later sections of this chapter. An important quantity in the example is $E(X_2 | X_1)$. When the distribution of an arm is normal with unknown mean and known variance, its distribution conditional on observations from the arm is easy to calculate. This is true in large part because it has a single unknown one-dimensional parameter. No such simple parametrization exists in general.

In the general setting of the following section, an arm will be characterized by a probability measure F on the Borel field of subsets of $\mathscr{D}$, the space of probability distributions on $\mathbb{R}$ with the topology of convergence in distribution. Considering $F(\cdot | X_1)$ for a general F, there is no problem in defining, up to a set of probability 0, $F(\mathscr{C} | X_1)$ for each fixed Borel subset $\mathscr{C}$ of $\mathscr{D}$. But we also want, for each fixed ω in the underlying probability space, that the function

$$\mathscr{C} \mapsto F(\mathscr{C} | X_1)(\omega)$$

be a probability measure on $\mathscr{D}$. By a result of Parthasarathy (1967, Theorem V.8.1), we can have this and more. Namely, there exists a function $f(\mathscr{C}, x)$ such that for each x the function $\mathscr{C} \mapsto f(\mathscr{C}, x)$ is a probability measure on $\mathscr{D}$, for each fixed $\mathscr{C}$ the function $x \mapsto f(\mathscr{C}, x)$ is a Borel measurable function on $\mathbb{R}$, and, almost surely,

$$F(\mathscr{C} \mid X_1)(\omega) = f(\mathscr{C}, X_1(\omega)). \tag{2.1.5}$$

In the example of this section, $f(\cdot, x)$ assigns probability one to the set of normal distributions having variance one; the mean of this distribution is itself normally distributed with mean $(\mu + \rho^2 x)/(1 + \rho^2)$ and variance $\rho^2/(1 + \rho^2)$.

2.2   General setting

The development in this section depends on the preceding section for some notation and concepts. However, the setting is general and calculations involving normal distributions will not play a role. There is an arbitrary finite number k of arms.

Recall from Section 2.1 that $\mathscr{D}$ denotes the space of probability distributions on $\mathbb{R}$. We use the topology of convergence in distribution on $\mathscr{D}$. Thus, with Q_l, $Q \in \mathscr{D}$ viewed as cumulative distribution functions, $Q_l \to Q$ as $l \to \infty$ if and only if $Q_l(x) \to Q(x)$ for every x at which Q is continuous. From the measure-theoretic viewpoint, $Q_l \to Q$ as $l \to \infty$ if and only if $Q_l(B) \to Q(B)$ for every Borel subset B of $\mathbb{R}$ for which $Q(\text{boundary of } B) = 0$, or equivalently, $\int_{\mathbb{R}} h \, dQ_l \to \int_{\mathbb{R}} h \, dQ$ as $l \to \infty$ for every bounded continuous function h on $\mathbb{R}$. It is shown in Parthasarathy (1967, Section II.6) that $\mathscr{D}$ inherits many properties from $\mathbb{R}$: $\mathscr{D}$ is separable, locally compact, metrizable to be complete, and its topology is determined by convergent sequences and the limits of such sequences. The Borel field of subsets of $\mathscr{D}$ is the smallest σ-field containing the open sets; it is this σ-field of subsets of $\mathscr{D}$ that is used throughout.

The space $\mathscr{D}^k$ of ordered k-tuples of members of $\mathscr{D}$ will be considered to have the product topology arising from the above-defined topology on $\mathscr{D}$. The Borel field generated by this product topology is the only σ-field of subsets of $\mathscr{D}^k$ that will be considered; it is the product σ-field of k copies of the Borel σ-field of $\mathscr{D}$. The component Q_i of $(Q_1, \ldots, Q_k) \in \mathscr{D}^k$ governs observations on arm i. Since $(Q_1, \ldots, Q_k)$ is random, the probability distribution G of $(Q_1, \ldots, Q_k)$ plays a central role.

The space $\mathscr{D}(\mathscr{D}^k)$ of probability distributions on $\mathscr{D}^k$ will therefore also play a central role. A member G of $\mathscr{D}(\mathscr{D}^k)$ represents the decision maker's prior information concerning the k arms. The member of $\mathscr{D}(\mathscr{D}^2)$ that represents the prior information for the example of Section 2.1 is supported by a rather small subset of $\mathscr{D}^2$ – namely (normal distributions with variance 1) × (a constant). We use the topology of convergence in distribution on $\mathscr{D}(\mathscr{D}^k)$. Thus, for members G_l and G of $\mathscr{D}(\mathscr{D}^k)$, $G_l \to G$ if and only if

$$\int_{\mathscr{D}^k} h \, dG_l \to \int_{\mathscr{D}^k} h \, dG$$

for every bounded continuous h on $\mathscr{D}^k$. According to Parthasarathy (1967, Section II.6) the space $\mathscr{D}(\mathscr{D}^k)$ inherits from $\mathscr{D}^k$ the properties of being separable, locally compact, metrizable to be complete, and having its topology determined by the convergent sequences and their limits. The σ-field of subsets of $\mathscr{D}(\mathscr{D}^k)$ that will be used is the Borel field.

We turn to the construction of a probability space Ω with a natural structure rich enough to reflect randomness in the structure of the arms as well as in the observations resulting from selecting arms according to an arbitrary strategy (yet to be defined). Let Ω be the product space obtained from $\mathscr{D}^k$ and infinitely many copies of the open unit interval, one for each pair (i, m), $1 \leqslant i \leqslant k$, $m = 1, 2, \ldots$; that is,

$$\Omega = \mathscr{D}^k \times \mathop{\mathsf{X}}_{i=1}^{k} \mathop{\mathsf{X}}_{m=1}^{\infty} (0, 1).$$

The probability measure P on Ω is the product of G on $\mathscr{D}^k$ and Lebesgue measure on each unit interval, and it is defined on the product σ-field $\mathscr{F}$ of the various Borel fields. A member of Ω can be written in the form

$$\omega = (Q_i, 1 \leqslant i \leqslant k; \omega_{im}, 1 \leqslant i \leqslant k, m = 1, 2, \ldots) \qquad (2.2.1)$$

where each $Q_i \in \mathscr{D}$ and each $\omega_{im} \in (0, 1)$. Denoting by Q_i^{-1} the right-continuous 'inverse' function of Q_i, regarded as a cumulative distribution function, we set

$$X_{im}(\omega) = Q_i^{-1}(\omega_{im})$$

for ω given by (2.2.1). We leave it to the reader to check that each X_{im} is measurable and thus a random variable and that, conditional on $(Q_1, \ldots, Q_k)$,

$$\{X_{im}: 1 \leqslant i \leqslant k, m = 1, 2, \ldots\}$$

is an independent family with the conditional distribution of X_{im} being Q_i. The value of X_{im} is the outcome on arm i at stage m (whether or not X_{im} is observed by the decision maker, as defined below).

A *strategy* τ assigns to each (partial) history of observations an integer from 1 to k indicating the arm to be selected at the next stage. Thus, $\tau(\varnothing)$ indicates the arm to be selected initially when following τ, $\tau(z_1)$ indicates the arm to be selected at the stage 2 given that z_1 is observed at stage 1, $\tau(z_1, z_2)$ is the arm to be selected at stage 3, etc. The value of τ indicates the arm to be selected at a particular stage. The outcome on that arm at that stage is an *observation*. Using all previous observations, the decision maker can evaluate τ for the next stage. For τ to be a strategy, we require that the set of observations for which arm i is indicated at stage m,

$$\{(z_1, \ldots, z_{m-1}): \tau(z_1, \ldots, z_{m-1}) = i\},$$

be a Borel subset of $\mathbb{R}^{m-1}$.

According to the description of a strategy in Chapter 1, the arm selected at any stage may depend on the history of observations *and* arms selected. Our notation is consistent with that definition but the dependence of the current selection on the prior selections is not as clear from the notation as is its dependence on the previous observations. To see that our notation is consistent with the earlier description, consider, for example, $\tau(z_1, z_2, z_3)$ for particular values of z_1, z_2, and z_3. Given τ, the arms associated with the three observations are determined: z_1 resulted from arm $\tau(\varnothing)$, z_2 from arm $\tau(z_1)$, and z_3 from arm $\tau(z_1, z_2)$.

The sequence of observed random variables is recursively defined via

$$Z_1 = X_{\tau(\varnothing), 1},$$
$$Z_m = X_{\tau(z_1, \ldots, z_{m-1}), m}, \quad m > 1.$$

That each Z_m is in fact measurable (and thus a random variable) follows by induction.

Since the general problem will be to choose τ to maximize the expected value of

$$\sum_{m=1}^{\infty} \alpha_m Z_m,$$

it is natural for us to introduce the following requirements on G and the discount sequence $\mathbf{A} = (\alpha_1, \alpha_2, \ldots)$. We require that each $\alpha_m \geqslant 0$

and $\sum \alpha_m < \infty$. We also require that each component Q_i of $(Q_1, \ldots, Q_k) \in \mathcal{D}^k$ has finite first absolute moment with G-probability one and, moreover, that this moment has finite G-expectation. We use $\mathcal{D}^*(\mathcal{D}^k)$ to denote the subspace of $\mathcal{D}(\mathcal{D}^k)$ consisting of those G's satisfying this condition. Since we sometimes write expectations as integrals, the above condition can be written in a variety of ways:

$$E(|X_{i1}|) = E(E(|X_{i1}| \big| Q_i)) = \int_{\mathcal{D}} \int_{\mathbb{R}} |x| Q_i(\mathrm{d}x) G(\mathrm{d}(Q_1, \ldots, Q_k))$$

$$= \int_{\mathcal{D}} \left(\int_{\mathbb{R}} |x| Q_i(\mathrm{d}x) \right) F_i(\mathrm{d}Q_i) < \infty$$

where F_i denotes the distribution of Q_i, that is, the ith marginal distribution of G. Since

$$|Z_m| \leqslant \bigvee_{i=1}^{k} |X_{im}| \leqslant \sum_{i=1}^{k} |X_{im}|,$$

Z_m is integrable and for any strategy τ,

$$\left| E_\tau \left(\sum_{m=1}^{\infty} \alpha_m Z_m \right) \right| \leqslant \left[\sum_{i=1}^{k} E(|X_{im}|) \right] \sum_{m=1}^{\infty} \alpha_m < \infty,$$

where the subscript τ indicates the dependence of the expectation on the strategy τ. The quantity $E_\tau(\sum_{m=1}^{\infty} \alpha_m Z_m)$, called the *worth* of the strategy τ, will be denoted by $W(G; \mathbf{A}; \tau)$. The *value* of the $(G; \mathbf{A})$-*bandit* is the maximal worth:

$$V(G; \mathbf{A}) = \sup_\tau W(G; \mathbf{A}; \tau) = \sup_\tau E_\tau \left(\sum_{m=1}^{\infty} \alpha_m Z_m \right). \qquad (2.2.2)$$

A strategy for which $V(G; \mathbf{A})$ is attained is *optimal*. In Sections 2.3 and 2.5 it will be shown that there is an optimal strategy for every G and $\mathbf{A}$.

At the first stage of any bandit problem, the decision maker is faced with an initial distribution and a discount sequence. At the second stage (after the initial selection and observation) the decision maker is faced with a new distribution and a new discount sequence, that is, with a new bandit problem. So the second stage selection can be viewed as the initial selection in this new bandit; and similar statements apply at every future stage. Therefore, if an optimal initial selection were known for all possible bandits – that is, all possible pairs $(G; \mathbf{A})$ – then an optimal strategy would be known for every bandit. This gives the initial selection special significance. However, as the subsequent development shows, an optimal initial selection cannot be made without considering future selections.

The worth of selecting an arm can be separated into the sum of the expected payoff at stage 1 and the expected value of the best that can be subsequently achieved. Partitioning the set of strategies according to the arm selected initially (as in Section 2.1), the supremum in (2.2.2) can be represented as the maximum of k suprema:

$$
\begin{aligned}
V(G; \mathbf{A}) &= \bigvee_{i=1}^{k} \sup_{\tau(\varnothing)=i} W(G; \mathbf{A}; \tau) \\
&= \bigvee_{i=1}^{k} \left[\alpha_1 E(X_{i1}) + \sup_{\tau(\varnothing)=i} E_\tau \left(\sum_{m=2}^{\infty} \alpha_m Z_m \right) \right].
\end{aligned}
\tag{2.2.3}
$$

We proceed to introduce notation that will aid in the study of (2.2.3) and which reflects the fact that the decision maker is faced with a new bandit at every stage. For a discount sequence $\mathbf{A} = (\alpha_1, \alpha_2, \alpha_3, \ldots)$, $\mathbf{A}^{(1)}$ denotes the discount sequence $(\alpha_2, \alpha_3, \ldots)$. More generally, $\mathbf{A}^{(m)}$ denotes the discount sequence $(\alpha_{m+1}, \alpha_{m+2}, \ldots)$.

The discount sequence $\mathbf{A}^{(1)}$ and the posterior distribution of $(Q_1, \ldots, Q_k)$ after stage 1 characterize the bandit confronting the decision maker at stage 2. To indicate the dependence of this posterior distribution on the observation we use $(x)_i G$ to denote a version of the conditional distribution of $(Q_1, \ldots, Q_k)$ given observation x on arm i. Lemma 2.2.1 below asserts that $(x)_i G$ can be chosen to depend measurably on (x, G). No suitable reference for this lemma was found in the literature. (This lemma, or a variation, is fundamental in statistical contexts involving random prior distributions, but such considerations are seldom made explicit.) For the lemma recall that

$$
\Omega = \mathscr{D}^k \times \mathop{\mathsf{X}}_{i=1}^{k} \mathop{\mathsf{X}}_{m=1}^{\infty} (0, 1).
$$

Lemma 2.2.1 For each $j = 1, \ldots, k$ there exists a measurable function $(x, G) \mapsto (x)_j G$ from $\mathbb{R} \times \mathscr{D}(\mathscr{D}^k)$ into $\mathscr{D}(\mathscr{D}^k)$ such that for every Borel subset C of $\mathscr{D}^k$,

$$
\left((X_{jm}(\omega))_j G \right)(C) = P\left(C \times \mathop{\mathsf{X}}_{i=1}^{k} \mathop{\mathsf{X}}_{m=1}^{\infty} (0, 1) \,\middle|\, X_{jm} \right)(\omega) \qquad \text{a.e. } \omega.
\tag{2.2.4}
$$

Remarks The following proof is essentially due to Varadhan (1983) who treats the joint measurability problem, but Varadhan was not concerned with our special product space structure. Rhenius (1977) proves a similar result using similar methods.   □

Before proving the lemma we introduce notation and two other lemmas. We use $\mathscr{D}(\Omega)$ to denote the space of probability distributions on $(\Omega, \mathscr{F})$. Again we use the topology of convergence in distribution: $P_l \to P$ if and only if

$$\int_\Omega h \, dP_l \to \int_\Omega h \, dP$$

for every bounded continuous h on Ω. The measurable subsets of $\mathscr{D}(\Omega)$ are the Borel sets. As in similar contexts discussed previously, $\mathscr{D}(\Omega)$ is separable, locally compact, metrizable to be complete, and has a topology determined by the convergent sequences and their limits. We omit the easy proof of the following result.

Lemma 2.2.2 The function which maps $P \in \mathscr{D}(\Omega)$ to its marginal on $\mathscr{D}^k$ is continuous. So is the function which maps $G \in \mathscr{D}(\mathscr{D}^k)$ to the member of $\mathscr{D}(\Omega)$ that is the product of G and Lebesgue measure on each $(0, 1)$.

Lemma 2.2.3 If Y is a bounded random variable, then $P \mapsto \int_\Omega Y \, dP$ is a measurable function from $\mathscr{D}(\Omega)$ into $\mathbb{R}$ and $P \mapsto Y \, dP$ is a measurable function from $\mathscr{D}(\Omega)$ into $\mathscr{D}(\Omega)$.

Proof (cf. Dubins and Freedman, 1966, 3.1) Suppose first that Y is continuous. The function $P \mapsto \int_\Omega Y \, dP$ is then continuous by definition. Suppose that $P_n \to P$ as $n \to \infty$ and that h is a bounded continuous function on Ω. Then hY is bounded and continuous, so

$$\lim_{n \to \infty} \int_\Omega h \,(Y \, dP_n) \to \int_\Omega h \,(Y \, dP).$$

Thus, by definition $Y \, dP_n \to Y \, dP$ as $n \to \infty$. Therefore $P \to Y \, dP$ is a continuous function from $\mathscr{D}(\Omega)$ into $\mathscr{D}(\Omega)$.

Next suppose that $Y = \mathbf{1}_B$ for some closed $B \subset \Omega$. Let

$$Y_q(\omega) = 0 \vee [1 - (\text{distance from } B \text{ to } \omega)q].$$

Since Y_q is continuous, it follows that the functions $P \mapsto \int_\Omega Y_q \, dP$ and $P \mapsto Y_q \, dP$ are continuous. For each P, $\int_\Omega Y_q \, dP \to \int_\Omega \mathbf{1}_B \, dP$ as $q \to \infty$ and, for each bounded continuous h, $\int_\Omega h \,(Y_q \, dP) \to \int_\Omega h \,(\mathbf{1}_B \, dP)$ as

$q \to \infty$, by Lebesgue's dominated convergence theorem. Therefore, the functions $P \mapsto \int_\Omega 1_B \, dP$ and $P \mapsto 1_B \, dP$ are the pointwise limits of sequences of continuous functions, and so are measurable.

The class of measurable B's for which the functions $P \mapsto \int_\Omega 1_B \, dP$ and $P \mapsto 1_B \, dP$ are measurable contains Ω and is closed under disjoint unions and proper differences (by addition and subtraction of measurable functions). It is also closed under increasing limits – by the monotone (or dominated) convergence theorem in the case of the first function, and by the dominated convergence theorem applied when the integrand is multiplied by an arbitrary bounded continuous h in the case of the second function. Therefore (Chow and Teicher, 1978, Theorem 1.3.2, for instance), the class of B's for which $P \mapsto \int_\Omega 1_B \, dP$ and $P \mapsto 1_B \, dP$ contains the σ-field generated by the closed sets.

For any bounded random variable Y we conclude that the functions $P \mapsto \int_B Y \, dP$ and $P \mapsto Y \, dP$ are measurable by taking limits of the two sequences of functions obtained when Y is replaced by a sequence of simple functions approaching Y everywhere.   □

Remark Eick (1984) helped us with the proof of Lemma 2.2.3 and with formulating the lemma in a manner most useful for the proof of Lemma 2.2.1.   □

Proof of Lemma 2.2.1 For $x \in [r2^{-l}, (r+1)2^{-l})$, define the probability measure $((x)_j G)_l$ by

$$((x)_j G)_l(C) = \begin{cases} G(C) & \text{if } P(r2^{-l} \leq X_{jn} < (r+1)2^{-l}) = 0 \\ P(C \times \overset{k}{\underset{i=1}{\mathsf{X}}} \overset{\infty}{\underset{m=1}{\mathsf{X}}} (0,1) \mid r2^{-l} \leq X_{jn} < (r+1)2^{-l}) \end{cases}$$

$$\text{otherwise}$$

for C a measurable subset of $\mathscr{D}^k$. Notice that, although n indexes the random variable, the definition is independent of n. Because of the rather simple dependence on x, the measurability of $(x, G) \mapsto ((x)_j G)_l$ will follow from the measurability of $G \mapsto ((x)_j G)_l$ for each fixed x. This measurability is an immediate consequence of Lemmas 2.2.2 and 2.2.3 with the indicator function of $\{\omega : r2^{-l} \leq X_{jn}(\omega) < (r+1)2^{-l}\}$ being used for Y in Lemma 2.2.3. Let

$$(x)_j G = \begin{cases} \lim_{l \to \infty} ((x)_j G)_l & \text{if the limit exists} \\ G & \text{otherwise.} \end{cases} \quad (2.2.5)$$

Clearly, the function $(x, G) \mapsto (x)_j G$ is measurable.

It remains to prove (2.2.4) for each measurable $C \subset \mathscr{D}^k$. By the martingale convergence theorem,

$$\left((X_{jn}(\omega))_j G \right)_l (C) \to P(C \times \underset{i=1}{\overset{k}{\mathsf{X}}} \underset{m=1}{\overset{\infty}{\mathsf{X}}} (0,1) \,|\, X_{jn})(\omega) \qquad \text{a.e. } \omega$$

as $l + \infty$. If ω does not belong to the exceptional set and is such that $X_{jn}(\omega)$ equals an x for which the limit in (2.2.5) exists, then

$$((X_{jn}(\omega))_j G)(C) = P\left(C \times \underset{i=1}{\overset{k}{\mathsf{X}}} \underset{m=1}{\overset{\infty}{\mathsf{X}}} (0,1) \,\middle|\, X_{jn} \right)(\omega).$$

We shall complete the proof by showing that

$$\lim_{l \to \infty} ((X_{jn}(\omega))_j G)_l \qquad (2.2.6)$$

exists for almost every ω. To obtain a candidate for the limit we first observe that the range of X_{jn} equals $\mathbb{R}$; any $x \in \mathbb{R}$ equals $Q_j^{-1}(\omega_{jn})$ for some Q_j and ω_{jn}. In view of this fact and Theorem V.2.2 of Parthasarathy (1967), Theorem V.8.1 of Parthasarathy (1967) is applicable. This latter theorem asserts the existence of a regular conditional probability distribution P_ω on Ω, one property of which is

$$E(h|X_{jn})(\omega) = \int_\Omega h \, dP_\omega \qquad \text{a.e. } \omega, \qquad (2.2.7)$$

for every bounded measurable h on Ω. In case h depends on $(Q_1, \ldots, Q_k)$, (2.2.7) can be rewritten

$$E(h|X_{jn})(\omega) = \int_{\mathscr{D}^k} h \, dG_\omega \qquad \text{a.e. } \omega, \qquad (2.2.8)$$

where G_ω denotes the projection of P_ω on $\mathscr{D}^k$. The measure G_ω is the candidate for the almost sure limit at (2.2.6).

For a bounded measurable h that depends only on $(Q_1, \ldots, Q_k)$, the martingale convergence theorem and (2.2.8) give

$$\lim_{l \to \infty} \int_{\mathscr{D}^k} h \, d((X_{jn}(\omega))_j G) = \int_{\mathscr{D}^k} h \, dG_\omega \qquad \text{a.e. } \omega; \qquad (2.2.9)$$

the exceptional null set may depend on h. If we restrict to a countable determining set of continuous h's, the null set may be chosen independently of h. The equality in (2.2.9) for such h's is sufficient for concluding that the limit at (2.2.6) exists and equals G_ω. $\qquad \square$

The existence of a regular conditional probability distribution was used in the proof of Lemma 2.2.1; there is little significance in the reverse implication. Nevertheless, the following corollary of Lemma 2.2.1 sheds further light on that lemma.

Corollary 2.2.4 For each measurable subset $C \subset \mathscr{D}^k$, each $G \in \mathscr{D}(\mathscr{D}^k)$, and each $i = 1, \ldots, k$, the function $x \mapsto ((x)_i G)(C)$ is measurable.

Proof The function of interest is the composition of $x \mapsto (x)_i G$ and $G \mapsto G(C)$. The first is measurable by Lemma 2.2.1 and the second by Lemma 2.2.3. $\qquad\qquad\qquad\qquad\qquad\qquad\qquad\qquad\qquad\qquad$ $\square$

We now describe two distinct ways of viewing a bandit problem. The previous development makes it clear that they are equivalent.

Suppose a decision maker can give instructions to an aide who is to carry out the decision maker's wishes. One way is to give a strategy τ to the aide. As indicated earlier in this section, the aide will evaluate $\tau(\varnothing)$, select that arm at stage 1, observe a consequent number z_1, evaluate $\tau(z_1)$, select that arm at stage 2, observe z_2, evaluate $\tau(z_1, z_2)$, select that arm at stage 3, etc.

Another way is for the decision maker to consider all pairs of distributions in $\mathscr{D}^*(\mathscr{D}^k)$ and discount sequences, $(G, \mathbf{A})$, and to specify $\tau_{G,\mathbf{A}}(\varnothing)$, the arm to be selected initially for each $(G, \mathbf{A})$-bandit. When faced with a *particular* G and $\mathbf{A}$, the aide selects $\tau_{G,\mathbf{A}}(\varnothing)$ and observes the result z_1. The aide will then (with ease) calculate $\mathbf{A}^{(1)}$ and (possibly with difficulty) calculate $(z_1)_j G$ with $j = \tau_{G,\mathbf{A}}(\varnothing)$, which distribution the aide might call $G^{(1)}$, the current information about the arms after stage 1. At stage 2 the aide will select arm $\tau_{G^{(1)}, \mathbf{A}^{(1)}}(\varnothing)$ and observe z_2. Then the aide will calculate $\mathbf{A}^{(2)}$ and $G^{(2)}$ and select arm $\tau_{G^{(2)}, \mathbf{A}^{(2)}}(\varnothing)$ at stage 3. And so on.

There are a number of things worth mentioning concerning the latter approach. First, since any particular outcome may have probability 0, the aide cannot be permitted to calculate conditional distributions after making observations but must choose versions in advance. Secondly, it is important for the evaluation of expectations and therefore for the assessment of strategies, that, for instance, $\{z_1 : \tau(z_1) = i\}$ be measurable. Now $\tau(z_1)$ is defined implicitly:

$$\tau(z_1) = \tau_{G^{(1)}, \mathbf{A}^{(1)}}(\varnothing).$$

Lemma 2.2.1 helps by assuring that $G^{(1)}$ depends measurably on z_1 and the decision maker must guarantee that $\tau_{G,A}(\varnothing)$ depends measurably on G.

The notation $G^{(1)}$ (and more generally, $G^{(m)}$) for the conditional distribution on $\mathscr{D}^k$ after stage 1 (and m) is useful for avoiding subscripted subscripts and notations such as $(z_2)_j(z_1)_i G$. It should be emphasized that $G^{(1)}$, say, can be regarded as either a function $(z_1)_{\tau(\varnothing)} G$ of the initial observation, or as a function $(Z_1(\omega))_{\tau(\varnothing)} G$ of ω. The context will distinguish between the two interpretations. For instance, $G^{(1)}$ is a random distribution in the expression $E_\tau V(G^{(1)}; A^{(1)})$.

Many discussions may involve several G's simultaneously. Rather than using corresponding subscripts and superscripts on P and E, we will use P and E without subscripts and a G to the right of a conditioning bar to indicate the G from which P is constructed. Accordingly,

$$P_\tau(Z_2 > 7 | G)$$

is the probability that the second observation is larger than 7 when strategy τ is followed in a $(G; A)$-bandit. Also,

$$E_\tau(X_{12} | G, Z_1) = E_\tau(X_{12} | G^{(1)})$$

is a random variable which depends on τ, while no subscript on E is necessary when writing $E(X_{12} | G, X_{31})$; this last expression cannot be simplified using the notation $G^{(1)}$.

Consistent with the above comments we can write

$$E(|X_{j2}| \,\big|\, G) = E_\tau\left(E(|X_{j1}| \,\big|\, G^{(1)}) \,\Big|\, G\right). \qquad (2.2.10)$$

When we speak of the $(G; A)$-bandit we assume implicitly that $G \in \mathscr{D}^*(\mathscr{D}^k)$ and not just $G \in \mathscr{D}(\mathscr{D}^k)$. From (2.2.10) we see that $G \in \mathscr{D}^*(\mathscr{D}^k)$ implies $G^{(1)} \in \mathscr{D}^*(\mathscr{D}^k)$ with probability one; so that it makes sense to speak of the $(G^{(1)}; A^{(1)})$-bandit.

In case $G = F_1 \times \ldots \times F_k$, $(x)_i$ can be applied directly to F_i with the subscript dropped; so

$$(x)_i(F_1 \times \ldots \times F_k) = (F_1 \times \ldots \times (x)F_i \times \ldots \times F_k)$$

where $(x)F_i$ denotes the conditional distribution of Q_i given an observation x on arm i. In the example of Section 2.1, $(x)F_1$ is supported by normal distributions having variance 1 and distributes the mean of such a normal distribution normally with mean $(\mu + \rho^2 x)/(1 + \rho^2)$ and variance $\rho^2/(1 + \rho^2)$ (F_1 is denoted by F in Section 2.1 since arm 1 is the only unknown arm in that section).

In the case of Bernoulli arms, where all outcomes are either 1 or 0, we frequently refer to the outcomes as 'success' and 'failure'. In this case we replace $(1)_i$ by σ_i and $(0)_i$ by φ_i. Accordingly, $\sigma_1 \varphi_1 \sigma_2 G$ is the conditional distribution of $(Q_1, \ldots, Q_k)$ given a success and a failure on arm 1 and a success on arm 2, and, with exponents in lieu of repetitions, $\sigma_1 \varphi_4^3 G$ is the conditional distribution of $(Q_1, \ldots, Q_k)$ given a success on arm 1 and three failures on arm 4.

At various places, we will need a topology on the space of discount sequences as well as on the other spaces we have discussed, such as $\mathscr{D} (\mathscr{D}^k)$. We will use the l_1-metric topology arising from the l_1-norm. The norm of a discount sequence $\mathbf{A} = (\alpha_1, \alpha_2, \ldots)$ is

$$|\mathbf{A}|_1 = \sum_{m=1}^{\infty} \alpha_m$$

and the distance between it and a discount sequence $\mathbf{B} = (\beta_1, \beta_2, \ldots)$ is

$$|\mathbf{A} - \mathbf{B}|_1 = \sum_{m=1}^{\infty} |\alpha_m - \beta_m|.$$

We will use $\mathscr{A}$ to denote this metric space of discount sequences.

Notice that

$$|\mathbf{A}^{(m)}|_1 = \sum_{p=m+1}^{\infty} \alpha_p$$

for which we will also use the notation γ_{m+1} in Chapter 5 in order to facilitate the algebraic manipulations there. In particular, $\gamma_1 = |\mathbf{A}|_1$.

2.3   Dynamic programming for finite horizons

The *horizon* of a discount sequence $\mathbf{A} = (\alpha_1, \alpha_2, \ldots)$, or of a bandit with that discount sequence, equals

$$\inf \{n : \alpha_m = 0 \quad \text{for } m > n].$$

The horizon may be infinite, and it equals 0 if $\alpha_m = 0$ for all m. The main feature of this section is a proof of the existence of an optimal strategy for each bandit having finite horizon. Also included is the beginning of our study of the dependence of the value function V and optimal strategies on G and $\mathbf{A}$.

The following lemma shows the existence of an optimal strategy when the horizon is finite. It will be extended to general discount sequences in Section 2.5. Recall from Section 2.2 that $\mathscr{A}$ denotes the space of all discount sequences and $\mathscr{D}^*(\mathscr{D}^k)$ denotes the space of all G such that $E(|X_{i1}| \,|\, G) < \infty$ for all i.

Lemma 2.3.1 There exists a strategy $\tau_{G,\mathbf{A}}$ for each $\mathbf{A} \in \mathscr{A}$ having finite horizon and each $G \in \mathscr{D}^*(\mathscr{D}^k)$ such that

$$\{(G;\mathbf{A};z_1,\ldots,z_{m-1}): \tau_{G,\mathbf{A}}(z_1,\ldots,z_{m-1}) = i\}$$

is a measurable subset of $\mathscr{D}^*(\mathscr{D}^k) \times \mathscr{A} \times (-\infty,\infty)^{m-1}$ for each $i = 1,\ldots,k$ and $m = 1, 2, \ldots$, and $\tau_{G,\mathbf{A}}$ is, for each G and $\mathbf{A}$, optimal for the $(G;\mathbf{A})$-bandit. Restricted to discount sequences having finite horizon, the function $(G,\mathbf{A}) \mapsto V(G;\mathbf{A})$ is measurable and satisfies

$$V(G;\mathbf{A}) = \bigvee_{i=1}^{k} E\left(\alpha_1 X_{i1} + V((X_{i1})_i G;\mathbf{A}^{(1)}) \,\middle|\, G\right). \qquad (2.3.1)$$

Remark Recall that $(X_{i1})_i G$ denotes the measure obtained by conditioning G on the observation X_{i1} on arm i and $\mathbf{A}^{(1)}$ denotes the sequence obtained by dropping the first member from $\mathbf{A}$. Thus, (2.3.1) gives V recursively for finite horizon bandits, but the recursion proceeds backwards. *Dynamic programming* and *backwards induction* are terms often used for such a backward recursion. □

Proof of Lemma 2.3.1 The set of discount sequences having a particular horizon is measurable. Accordingly, we may proceed by induction on the horizon. The only discount sequence having horizon 0 is $\mathbf{0} = (0,0,0,\ldots)$. Clearly $V(G;\mathbf{0}) = 0$ for every G and all strategies are optimal. To be specific we set

$$\tau_{G,\mathbf{0}}(z_1,\ldots,z_{m-1}) = 1$$

or all $m = 1, 2, \ldots$ and $z_1, z_2, \ldots, z_{m-1}$. Equation (2.3.1) obviously holds with $\mathbf{A} = \mathbf{0}$.

Suppose $\tau_{G,\mathbf{A}}$ has been defined and the assertions of the lemma established for all discount sequences $\mathbf{A}$ having horizon less than n. We proceed to define $\tau_{G,\mathbf{A}}$ for $\mathbf{A}$ having horizon n. Define a strategy $\tau_{G,\mathbf{A}}^i$

that selects arm i initially and then proceeds optimally:

$$\tau_{G,A}^i(\varnothing) = i$$

$$\tau_{G,A}^i(z_1, \ldots, z_{m-1}) = \tau_{(z_1)_i G, A^{(1)}}(z_2, \ldots, z_{m-1}) \quad \text{if } m > 1.$$

Since the functions $A \mapsto A^{(1)}$ and $(z_1, G) \mapsto (z_1)_i G$ are measurable, the induction hypothesis yields the measurability of

$$\{(G; A; z_1, \ldots, z_{m-1}): \tau_{G,A}^i(z_1, \ldots, z_{m-1}) = j\}$$

for each i and m and, in particular, that each $\tau_{G,A}^i$ is a strategy.

To prove (2.3.1) we consider the right-hand side of (2.2.3):

$$\sup_{\tau(\varnothing) = i} E_\tau \left(\sum_{m=2}^\infty \alpha_m Z_m \Big| G \right) = \sup_{\tau(\varnothing) = i} E_\tau \left(E_\tau \left(\sum_{m=2}^\infty \alpha_m Z_m \Big| X_{i1} \right) \Big| G \right)$$

$$\leq E_\tau \left(\sup_{\tau(\varnothing) = i} \left(E_\tau \left(\sum_{m=2}^\infty \alpha_m Z_m \Big| X_{i1} \right) \Big| G \right) \right).$$

It is clear that each supremum on the right-hand side is attained for $\tau_{G,A}^i$ and the corresponding supremum equals $V((X_{i1})_i G; A^{(1)})$, which, by the induction hypothesis and the measurability of $(x, G) \mapsto (x)_i G$, is a measurable function of (ω, G, A). So, (2.3.1) follows from (2.2.3); and from (2.3.1) and the induction hypothesis we obtain the measurability of $(G, A) \mapsto V(G; A)$.

We complete the proof by setting $\tau_{G,A}$ equal to the $\tau_{G,A}^i$ having the smallest i for which the maximum in (2.3.1) occurs.   □

For calculational purposes, the following rewritings of (2.3.1) may be useful:

$$V(G; A) = \bigvee_{i=1}^k \int_{\mathscr{D}^k} \int_R \left[\alpha_1 x + V((x)_i G; A^{(1)}) \right]$$

$$\cdot Q_i(dx) G(d(Q_1, \ldots, Q_k))$$

$$= \bigvee_{i=1}^k \int_{\mathscr{D}} \int_R \left[\alpha_1 x + V((x)_i G; A^{(1)}) \right] Q_i(dx) F_i(dQ_i). \qquad (2.3.2)$$

We now show how to use (2.3.2) to find optimal strategies. Suppose the $(G; A)$-bandit has horizon 1. Since the horizon of $A^{(1)}$ is zero,

$V(G^{(1)}; \mathbf{A}^{(1)}) = 0$ for any $G^{(1)}$ and so (2.3.2) simplifies to

$$
\begin{aligned}
V(G;(\alpha_1, 0, 0, \ldots)) &= \alpha_1 \bigvee_{i=1}^{k} \int_{\mathscr{D}^k} \int_{\mathbb{R}} x\, Q_i(dx)\, G(d(Q_1, \ldots, Q_k)) \\
&= \alpha_1 \bigvee_{i=1}^{k} \int_{\mathscr{D}} \int_{\mathbb{R}} x\, Q_i(dx)\, F_i(dQ_i).
\end{aligned}
\tag{2.3.3}
$$

An optimal initial selection is any arm i for which the maximum in (2.3.3) is attained.

Suppose the $(G; \mathbf{A})$-bandit has horizon 2. To use (2.3.2) we require the values of bandits having $\mathbf{A}^{(1)}$, which has horizon 1, as the discount sequence. Not all distributions need be considered in conjunction with $\mathbf{A}^{(1)}$, only those of the form $(x)_i G$ for some possible observation x on some arm i. The desired values $V((x)_i G; \mathbf{A}^{(1)})$ can be obtained from (2.3.3). When (2.3.2) is used to find $V(G; \mathbf{A})$, the optimal initial selections are found as those indicated by the i's for which the maximum in (2.3.2) is attained.

Now that we know (in principle) how to find $V(G; \mathbf{A})$ for any $(G; \mathbf{A})$ with a horizon of 2, we can (again, in principle) find $V(G; \mathbf{A})$ when the horizon is 3 by applying (2.3.2). In general, a bandit with horizon n can be solved (in principle) by first solving many bandits with horizon $n-1$, etc.

This process can be carried out to solve an arbitrary $(G; \mathbf{A})$-bandit in which the horizon of $\mathbf{A}$ is $n < \infty$. The first step is to calculate, for each $m = 1, \ldots, n-1$ and each allocation of m selections on the k arms, the possible conditional distributions $G^{(m)}$ of $(Q_1, \ldots, Q_k)$ given these observations. Since $\mathbf{A}^{(n-1)}$ has horizon 1, each $V(G^{(n-1)}; \mathbf{A}^{(n-1)})$ is obtained from (2.3.3). Then each $V(G^{(n-2)}; \mathbf{A}^{(n-2)})$ is calculated from (2.3.2). This process of backward induction or dynamic programming continues until $V(G; \mathbf{A})$ is obtained. For each $(G^{(m)}; \mathbf{A}^{(m)})$-bandit that can arise starting with the $(G; \mathbf{A})$-bandit, the optimal initial selections are indicated by those i's for which the maximum in (2.3.2) is attained. These selections can be pieced together to give all optimal strategies for the $(G; \mathbf{A})$-bandit.

The method described above is illustrated with several examples in the next section.

Remark Equations such as (2.3.2) are common in the literature; the term 'fundamental equation of dynamic programming' is used. In various probabilistic contexts, posterior distributions have to be identified with the 'states' in the basic dynamic programming formulations. We do not think it is a trivial matter to do so and then to complete the appropriate arguments; for us, Lemma 2.2.1 has played a central role. We feel that the literature contains some oversights in this regard.   □

2.4 Examples having finite horizons

Various examples are provided here to illustrate the technique described in Section 2.3 and to give some flavour of the variety of issues that can arise in bandit problems. Detailed calculations are given in the first two examples. Only the most interesting aspects of the remaining examples are given. The interested reader can perform the requisite backward induction to verify our statements in these latter examples.

Only Examples 2.4.2 and 2.4.5 have more than two arms. Only Example 2.4.6 has an arm that is not Bernoulli (except that there is an arbitrary but known arm in Example 2.4.7). Example 2.4.7 is the only example concerned with an infinite subset of $\mathscr{D}(\mathscr{D}^k)$. Example 2.4.1 is the only one having dependent arms.

The first example is due to Bradt, Johnson, and Karlin (1956), who gave it as a counterexample to the 'stay-with-a-winner rule'. The setting is quite simple but the arms are dependent.

Example 2.4.1 Suppose $k = 2$ and the arms are Bernoulli with parameters θ_1 and θ_2 so that

$$P(X_{im} = 1 | \theta_i) = \theta_i = 1 - P(X_{im} = 0 | \theta_i). \qquad (2.4.1)$$

The distributions Q_1 and Q_2 do not appear explicitly in (2.4.1); they are represented by the random parameters θ_1 and θ_2, and a probability distribution G on $\mathscr{D}^2$ can be represented by a distribution, that we also call G, on the unit square.

The parameters θ_1 and θ_2 are known to be either both small or both large. More precisely, G is a two-point probability measure:

$$G = \frac{4}{5}\delta_{(1/10,0)} + \frac{1}{5}\delta_{(9/10,1)},$$

where $\delta_{(x,y)}$ is the delta measure at (x, y). As in Section 2.1, assume **A** $= (1, 1, 0, 0, \ldots)$.

The reader can easily list all possible strategies, calculate $V(G; \mathbf{A}) = 53/100$, and conclude that the only optimal strategy τ is given by $\tau(\emptyset) = 1$, $\tau(1) = 2$, $\tau(0) = 1$. Accordingly, arm 1 should be selected initially and a switch to arm 2 should be made if and only if a success is observed at stage 1. This fact gave the example its original importance. Our goal is to illustrate the dynamic programming method described in the preceding section and in so doing we will reach the above conclusions in a rather laborious manner.

As indicated in Section 2.3, we begin by listing all the $G^{(m)}$, $m = 1, \ldots, n-1$, that can arise, where n is the horizon. In this example $n = 2$ and so 1 is the only relevant value of m. The distributions $G^{(1)}$ that can arise are four in number and are given by (see Section 2.2 for notation):

$$\sigma_1 G = \frac{4}{13} \delta_{(1/10,0)} + \frac{9}{13} \delta_{(9/10,1)},$$

$$\phi_1 G = \frac{36}{37} \delta_{(1/10,0)} + \frac{1}{37} \delta_{(9/10,1)},$$

$$\sigma_2 G = \qquad\qquad\quad \delta_{(9/10,1)},$$

$$\phi_2 G = \delta_{(1/10,0)}.$$

From (2.3.3) we obtain:

$$V(\sigma_1 G; \mathbf{A}^{(1)}) = \left(\frac{4}{13} \cdot \frac{1}{10} + \frac{9}{13} \cdot \frac{9}{10} \right) \vee \left(\frac{4}{13} \cdot 0 + \frac{9}{13} \cdot 1 \right)$$

$$= \frac{17}{26} \vee \frac{9}{13} = \frac{9}{13}, \tag{2.4.2}$$

$$V(\varphi_1 G; \mathbf{A}^{(1)}) = \left(\frac{36}{37} \cdot \frac{1}{10} + \frac{1}{37} \cdot \frac{9}{10} \right) \vee \left(\frac{36}{37} \cdot 0 + \frac{1}{37} \cdot 1 \right)$$

$$= \frac{9}{74} \vee \frac{1}{37} = \frac{9}{74}, \tag{2.4.3}$$

$$V(\sigma_2 G; \mathbf{A}^{(1)}) = \left(1 \cdot \frac{9}{10} \right) \vee (1 \cdot 1) = \frac{9}{10} \vee 1 = 1, \tag{2.4.4}$$

$$V(\varphi_2 G; \mathbf{A}^{(1)}) = \left(1 \cdot \frac{1}{10} \right) \vee (1 \cdot 0) = \frac{1}{10} \vee 0 = \frac{1}{10}. \tag{2.4.5}$$

To make the optimal selections clear, we have written the maxima in detail and in the order of the arm number. For instance, arm 2 is the optimal selection in the $(\sigma_1 G; (1, 0, 0, \ldots))$-bandit since 9/13 is the maximum and is second in the expression $(17/26) \vee (9/13)$.

For the second and, in this example, last step in the recursion, we use (2.3.2) to obtain

$$V(G; \mathbf{A}) = \left(\frac{4}{5} \left[\frac{1}{10} \left(1 + \frac{9}{13} \right) + \frac{9}{10} \left(0 + \frac{9}{74} \right) \right] + \frac{1}{5} \left[\frac{9}{10} \left(1 + \frac{9}{13} \right) \right. \right.$$

$$\left. \left. + \frac{1}{10} \left(0 + \frac{9}{74} \right) \right] \right) \vee \left(\frac{4}{5} \left[0(1 + 1) + 1 \left(0 + \frac{1}{10} \right) \right] \right.$$

$$\left. + \frac{1}{5} \left[1(1 + 1) + 0 \left(0 + \frac{1}{10} \right) \right] \right)$$

$$= \frac{53}{100} \vee \frac{12}{25} = \frac{53}{100};$$

and so the only optimal first selection is arm 1. From (2.4.2) and (2.4.3) we see that, at stage 2, the decision maker should select arm 2 after a success with arm 1 and arm 1 after a failure with arm 1. Calculations (2.4.4) and (2.4.5) are necessary to determine that arm 1 is optimal initially, but once that determination is made they are irrelevant.

□

Although we were able to make a complete list of all distributions $G^{(1)}$ in the above example, this is not necessary for an explicit solution using dynamic programming. In particular, the example in Section 2.1 was solved with dynamic programming, although that terminology was not used.

While every possible strategy is evaluated in the previous example, this does not happen using dynamic programming when the horizon is larger than 2. The next example (in which the horizon is 3) makes this clear.

Example 2.4.2 Suppose $k = 3$ and the arms are Bernoulli with parameters θ_1, θ_2, θ_3. The distribution G can be regarded as a distribution on the cube $\{(u_1, u_2, u_3): 0 \leqslant u_i \leqslant 1\}$ and the distributions F_i can be regarded as the corresponding marginal distributions of the θ_i. We suppose the distribution is supported by the plane $u_3 = 8/15$ on which it has a (two-dimensional) uniform density. So the three parameters are independent: $G = F_1 \times F_2 \times F_3$ where F_1 and F_2 are uniform distributions on $(0, 1)$, written $U(0, 1)$, and

$F_3 = \delta_{8/15}$. Take $\mathbf{A} = (1,1,1,0,0,\ldots)$; so the objective is to maximize the expected sum of the first three observations.

The first step in the dynamic program is to calculate all possible $G^{(m)}$ for $m = 1, 2$. Parameters θ_1 and θ_2 have beta densities and θ_3 remains known; on $(0,1) \times (0,1) \times \{8/15\}$:

$$d(\sigma_1{}^{s_1}\varphi_1{}^{f_1}\sigma_2{}^{s_2}\varphi_2{}^{f_2}\sigma_3{}^{s_3}\varphi_3{}^{f_3}G(u_1,u_2,u_3))$$
$$\propto u_1{}^{s_1}(1-u_1)^{f_1}u_2{}^{s_2}(1-u_2)^{f_2}\,du_1\,du_2. \qquad (2.4.6)$$

The possible $G^{(1)}$ are obtained by letting $s_1 + f_1 + s_2 + f_2 + s_3 + f_3 = 1$, namely: $\sigma_1 G$, $\varphi_1 G$, $\sigma_2 G$, $\varphi_2 G$, and G. The possible $G^{(2)}$ are obtained by letting $s_1 + f_1 + s_2 + f_2 + s_3 + f_3 = 2$. There are fifteen such distributions; these include the five mentioned above which now correspond to $s_3 + f_3 \geq 1$.

The fifteen values $V(G^{(2)}; \mathbf{A}^{(2)})$ were easily obtained using (2.3.3), multiplied by 180, and entered into Table 2.1(i). The maxima have been written explicitly in the order corresponding to the arm numbers so as to facilitate the eventual identification of optimal strategies.

For the next recursive step, (2.3.2) is to be used five times to calculate $V(\sigma_1 G; \mathbf{A}^{(1)})$, $V(\varphi_1 G; \mathbf{A}^{(1)})$, $V(\sigma_2 G; \mathbf{A}^{(1)})$, $V(\varphi_2 G; \mathbf{A}^{(1)})$, and $V(G; \mathbf{A}^{(1)})$; these are given in Table 2.1(ii). Let us, for instance, calculate $V(\sigma_1 G; \mathbf{A}^{(1)})$. To do this we need the normalizing constant in (2.4.6): it equals 2 for $s_1 = 1, f_1 = s_2 = f_2 = 0$. We may regard $\mathcal{D}$ to be $[0,1]$; so from (2.3.2) and Table 2.1(i) we obtain

$$180V(\sigma_1 G; \mathbf{A}^{(1)}) = \left(\int_0^1 [u_1(180+135)+(1-u_1)(0+96)]2u_1\,du_1 \right)$$

$$\vee \left(\int_0^1 [u_2(180+120)+(1-u_2)(0+120)]\,du_2 \right)$$

$$\vee \left(\frac{8}{15}(180+120) + \frac{7}{15}(0+120) \right)$$

$$= 242 \vee 210 \vee 216$$

which is entered in the appropriate place in Table 2.1(ii). Only one use of (2.3.2) is required at the last step in the recursion. The result of that calculation is entered in Table 2.1(iii).

From Table 2.1(iii) we immediately obtain that $V(G; \mathbf{A}) = 310/180$. In addition, we see that both arms 1 and 2 are optimal initially (they are obviously equally good by symmetry). Correspondingly, there are

Table 2.1 180 $V(G^{(m)}; \mathbf{A}^{(m)})$ *for m observations and various possible* $G^{(m)}$

(i) $m = 2$: $\mathbf{A}^{(2)} = (1,0,0,\ldots)$

		φ_1	σ_1	φ_1^2	$\varphi_1\sigma_1$	σ_1^2
	90 v 90 v 96	60 v 90 v 96	120 v 90 v 96	45 v 90 v 96	90 v 90 v 96	135 v 90 v 96
φ_2	90 v 60 v 96	60 v 60 v 96	120 v 60 v 96			
σ_2	90 v 120 v 96	60 v 120 v 96	120 v 120 v 96			
φ_2^2	90 v 45 v 96					
$\varphi_2\sigma_2$	90 v 90 v 96					
σ_2^2	90 v 135 v 96					

(ii) $m = 1$: $\mathbf{A}^{(1)} = (1,1,0,0,\ldots)$

		φ_1	σ_1
	198 v 198 v 192	156 v 198 v 192	242 v 210 v 216
φ_2	198 v 156 v 192		
σ_2	210 v 242 v 216		

(iii) $m = 0$: $\mathbf{A} = (1,1,1,0,0,\ldots)$

310 v 310 v 294

at least two optimal strategies; let τ be one with $\tau(\varnothing) = 1$. From the right-hand portion of Table 2.1(ii) we obtain $\tau(1) = 1$ and $\tau(0) = 2$. From the rightmost two entries in Table 2.1(i) we obtain $\tau(1,1) = 1$ and $\tau(1,0) = 3$. From the lower two entries in the second column of Table 2.1(i) we obtain $\tau(0,1) = 2$ and $\tau(0,0) = 3$, which, since selections after stage 3 are irrelevant, completes the description of τ. From Table 2.1 there is just one other optimal strategy and it is obtained by interchanging the roles of arms 1 and 2 in the above description.

There are a number of locations in Table 2.1 (for example, $\sigma_1\sigma_2 G$) that play no role in the eventual specification of an optimal strategy since they cannot be reached when following some optimal strategy.

But every value in the table is required to calculate $V(G; \mathbf{A})$ and, therefore, for specifying optimal strategies.

Another interesting aspect of the table is that many strategies are not evaluated; an example is the myopic strategy. There is a unique myopic strategy in this example; it selects arm 3 at all three stages and has worth $3(8/15) = 288/180$. Using backwards induction, strategies are ruled out if they do not perform well after the stage under consideration; so the myopic strategy was ruled out at stage 2. □

Table 2.2.

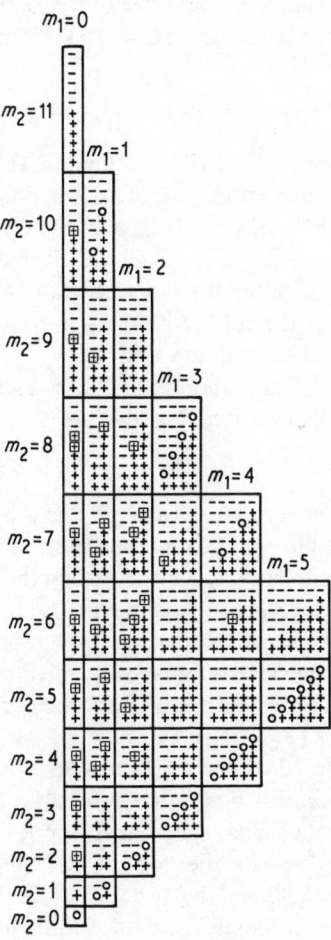

The next example is somewhat more interesting than the last in that the horizon is larger. It supposes there are two independent arms (arms 1 and 2 of the previous example) and a horizon of 12.

Example 2.4.3 Suppose $k = 2$ and arms 1 and 2 are independent Bernoulli distributions with uniformly distributed parameters: regarding F_1 and F_2 as distributions of θ_1 and θ_2, $F_i = U(0, 1)$. Take $\alpha_1 = \ldots = \alpha_{12} = 1, \alpha_{13} = \ldots = 0$, so the objective is to maximize the expected number of successes in the first 12 trials.

Every optimal strategy can be found from Table 2.2. Optimal initial selections are indicated in the table for every $(G^{(m)}; \mathbf{A}^{(m)})$ up to $m = 11$ that is possible when starting from $G = F_1 \times F_2$ and $\mathbf{A}$. The table gives the sign of

$$\Delta(G^{(m)}; \mathbf{A}^{(m)}) = W(G^{(m)}; \mathbf{A}^{(m)}; \tau_1) - W(G^{(m)}; \mathbf{A}^{(m)}; \tau_2), \quad (2.4.7)$$

where τ_i is a strategy for the $(G^{(m)}; \mathbf{A}^{(m)})$-bandit that begins with arm i and proceeds optimally thereafter. (The function Δ is used here for the various $(G^{(m)}; \mathbf{A}^{(m)})$ merely to display the optimal strategy. It is discussed in detail in Section 4.2 and used subsequently to show various properties of optimal strategies.) So (2.4.7) is positive when arm 1 is optimal for the $(G^{(m)}; \mathbf{A}^{(m)})$-bandit, negative when arm 2 is optimal, and zero when both are optimal.

Table 2.2, which indicates the sign of Δ, uses a scheme similar to that of Table 2.1. The distributions possible are

$$G^{(m)} = F_1^{(m_1)} \times F_2^{(m_2)} = \sigma^{s_1} \varphi^{f_1} F_1 \times \sigma^{s_2} \varphi^{f_2} F_2$$

where $s_i + f_i = m_i$, $m = m_1 + m_2$, and $F_i^{(m_i)}$ is a beta distribution; cf. (2.4.6). Within each (m_1, m_2) box, s_2 runs from 0 in the bottom row to m_2 in the top row; similarly, s_1 runs from 0 in the leftmost column to m_1 in the rightmost. Because of symmetry, the part of the table with $m_2 < m_1$ has been omitted. A square circumscribing a '+' in the table indicates a state in which the optimal selection disagrees with a myopic selection; that is, a state $(G^{(m)}; \mathbf{A}^{(m)})$ for which arm 1 is optimal but $E(X_{1, m+1} | G^{(m)}) < E(X_{2, m+1} | G^{(m)})$.

To see how Table 2.2 can be used, consider the first few selections. The 'O' in the lowermost box indicates that both arms are optimal initially (this is obvious from symmetry). Suppose arm 2 is used and is successful. The '−' in the top of the $(m_1 = 0, m_2 = 1)$ box indicates that arm 2 should be selected again. Suppose it gives a second success, then it should be used a third time. If it now gives a failure, arm 1 should be used at stage 4 even though the probability of success on arm 2 is $(s_2 + 1)/(m_2 + 2) = 3/5$ while on arm 1 it is $(s_1 + 1)/(m_1 + 2) = 1/2$.

Except for considerations of symmetry, which are available occasionally, there is no easy calculation which indicates an optimal arm at any stage – there is no alternative to an extensive description such as that in Table 2.2. Still, there are some attributes of optimal strategies that are evinced by Table 2.2 and that generalize to uniform discounting with independent arms. These will be discussed in Chapter 7.   □

The next example has the same arms as the previous one, but the discount sequence is very different.

Example 2.4.4 As in the previous example, suppose $k = 2$ with $F_i = U(0, 1)$, $i = 1, 2$. Let $\mathbf{A} = (0, 0, 0, 0, 1, 0, 0, \ldots)$; so the objective is to maximize the expected value of the fifth observation. There are four learning observations in preparation for the only observation whose payoff counts.

The value of this bandit is 37/60. It is obvious that the fifth observation will be taken on the arm with greater current expected value. One might guess in view of symmetry that any nonsequential strategy (one that ignores the observations) that selects both arms twice is optimal. There are many optimal strategies, and some are nonsequential, but this strategy is not one of them. Any choices whatever can be made at stages 1, 2, and 3. For example, arm 1 can be selected the first three times; arm 2 must then be chosen at stage 4 if $z_1 + z_2 + z_3$ is either 1 or 2 and both arms are optimal at stage 4 if this sum is 0 or 3. This makes it clear that the set of optimal strategies includes those nonsequential strategies that select, in any order, one of the arms three times and the other once. There is no optimal strategy, sequential or not, that with probability one indicates two selections on each arm. Pearson (1980, Theorem 4.2.1 and its proof) shows that if an even number n of learning observations on these two arms are available, then when restricted to nonsequential strategies, it is uniquely optimal to allocate $(n/2) - 1$ to either arm and $(n/2) + 1$ to the other.   □

Bandit problems frequently involve a trade-off between immediate payoff and gaining information. Since more information is available later in the experiment than earlier, one might expect a greater contribution to the value of a bandit from later stages than from earlier stages; at least when following an optimal strategy (cf. Section 1.2). While this is true in a variety of settings (see Corollary 5.2.3, for example), it is not true in general, as the next example shows.

Example 2.4.5 Suppose there are two independent Bernoulli arms with parameters θ_1 and θ_2. Where F_1 and F_2 are viewed as distributions of θ_1 and θ_2, let $F_1 = \frac{2}{9}\delta_1 + \frac{7}{9}\delta_0$ and $F_2 = \frac{1}{12}\delta_1 + \frac{3}{4}\delta_{1/2} + \frac{1}{6}\delta_0$. The discount sequence is $\mathbf{A} = (1,1,1,0,0,\ldots)$. An optimal strategy τ (uniquely optimal through stage 3) is as follows:

$$\tau(\varnothing) = 2,$$
$$\tau(1) = 2, \ \tau(0) = 1,$$
$$\tau(1,1) = \tau(1,0) = 2, \ \tau(0,1) = 1, \ \tau(0,0) = 2.$$

The worth of τ and the value of this bandit equals $E_\tau(Z_1 + Z_2 + Z_3 | G)$ where

$$E_\tau(Z_1 | G) = E(\theta_2 | G) = \frac{198}{432}$$

$$E_\tau(Z_2 | G) = E(\theta_2^2 + (1 - \theta_2)\theta_1 | G) = \frac{169}{432}$$

$$E_\tau(Z_3 | G) = E(\theta_2^3 + \theta_2^2(1 - \theta_2) + (1 - \theta_2)\theta_1^2$$
$$+ (1 - \theta_2)(1 - \theta_1)\theta_2 | G) = \frac{232}{432}.$$

The total is 599/432; but the interesting aspect of these calculations is that the expected contribution from the second stage is less than that from the first! □

The above example indicates that $E_\tau(Z_m | G)$ is not necessarily increasing in m when τ is an optimal strategy. However, the next result shows that the last observation (at stage n) is at least as large as every other observation in expectation. Its proof is due to Eick (1984).

Theorem 2.4.1 If the horizon of $\mathbf{A}$ is n then for any G, $\mathbf{A}$, optimal strategy τ, and $m < n$,

$$E_\tau(Z_n | G) \geqslant E_\tau(Z_m | G).$$

Proof Suppose τ is an optimal strategy with

$$E_\tau(Z_n|G) < E_\tau(Z_m|G)$$

for some $m < n$. Let τ^* be a modification of τ that imitates τ up to stage n and at stage n selects the arm indicated by τ at stage m. That is,

$$\tau^*(z_1, \ldots, z_j) = \tau(z_1, \ldots, z_j)$$

for all $(z_1, \ldots, z_j)$ with $j = 0, 1, \ldots, n-2$, and

$$\tau^*(z_1, \ldots, z_{m-1}, \ldots, z_{n-1}) = \tau(z_1, \ldots, z_{m-1})$$

for all $(z_1, \ldots, z_{n-1})$. (Continuations beyond stage n are of course irrelevant.) Then

$$E_{\tau^*}(Z_j|G) = E_\tau(Z_j|G)$$

for $j = 1, 2, \ldots, n-1$, and

$$E_{\tau^*}(Z_n|G) = E_\tau(Z_m|G) > E_\tau(Z_n|G).$$

So τ cannot be optimal.   □

The next example illustrates how the result on one arm can affect the choice among the other arms, even when the arms are independent.

Example 2.4.6 Suppose there are 3 independent arms, so $G = F_1 \times F_2 \times F_3$, and $\mathbf{A} = (1, 1, 1, 0, 0, \ldots)$. Arms 1 and 2 are Bernoulli arms (with two-point prior distributions on the parameters). Arm 3 yields only 0's, only $\frac{1}{2}$'s, or only 1's. Since there is a non-Bernoulli arm, we will retreat from the convention used in previous examples of regarding F_i to be a probability distribution on $[0, 1]$. The distribution $G = F_1 \times F_2 \times F_3$ on $\mathscr{D}^3$ is given by:

$$F_1(\{\delta_1\}) = 17/57, \qquad F_1\left(\left\{\frac{19}{20}\delta_0 + \frac{1}{20}\delta_1\right\}\right) = 40/57,$$

$$F_2(\{\delta_1\}) = 107/470, \qquad F_2\left(\left\{\frac{47}{66}\delta_0 + \frac{19}{66}\delta_1\right\}\right) = 363/470,$$

$$F_3(\{\delta_1\}) = 9/20, \qquad F_3(\{\delta_{1/2}\}) = 1/10, \qquad F_3(\{\delta_0\}) = 9/20.$$

An optimal strategy τ, uniquely optimal in the first three stages, is as follows:

$$\tau(\varnothing) = 3$$
$$\tau(1) = 3, \ \tau(\tfrac{1}{2}) = 2, \ \tau(0) = 1,$$
$$\tau(1, 1) = 3, \ \tau(\tfrac{1}{2}, 1) = 2, \ \tau(\tfrac{1}{2}, 0) = 3, \ \tau(0, 1) = 1, \ \tau(0, 0) = 2;$$

and $V(G; \mathbf{A}) = 1153/600$. The interesting feature of this optimal strategy is that the choice between arms 1 and 2 at stage 2 depends on the result on arm 3 at stage 1, even though the arms are independent.

$\square$

The next example generalizes Example 1.2.1 in a number of ways. The setting in which only one arm has unknown characteristics is generalized in Chapter 5.

Example 2.4.7 Suppose that there are two independent arms so that $G = F_1 \times F_2$. Further, suppose as in Section 2.1 that, for some constant λ,

$$F_2 \{Q_2 : \int_{\mathbb{R}} x Q_2 (\mathrm{d}x) = \lambda\} = 1,$$

that is, that the random distribution on arm 2 has mean λ with probability one: we frequently label such an arm as 'known'. With no loss we may assume that each observation on arm 2 equals λ. We suppose that arm 1 is Bernoulli and thus consider F_1 to be the distribution of the Bernoulli parameter θ_1. We assume $0 < \lambda < 1$; if not, the problem is trivial. We assume that F_1 is supported by $[0, \lambda] \cup \{1\}$.

Recall that $|A^{(m)}|_1 = \sum_{p = m + 1}^{\infty} \alpha_p$. We assume

$$\frac{|\mathbf{A}^{(m + 1)}|_1}{|\mathbf{A}^{(m)}|_1} \leqslant \frac{|\mathbf{A}^{(m)}|_1}{|\mathbf{A}^{(m - 1)}|_1} \tag{2.4.8}$$

whenever $|\mathbf{A}^{(m)}|_1 \neq 0$.

The assumption in this section that the horizon is finite is not necessary for this example.

Since φF_1 is supported by $[0, \lambda]$, selections of arm 2 are always optimal following a failure on arm 1. What is not so obvious is that it is always optimal to continue selecting arm 1 following a success if the selection of arm 1 was optimal. Also, it is always optimal to continue selecting arm 2 once it has been selected if that selection was optimal. These assertions may be incorrect if (2.4.8) is not satisfied; they are verified for general F_1 in Theorem 5.2.2 in case (2.4.8) holds.

It is easy to calculate the value and find optimal strategies, for there are only two strategies that satisfy the above assertions – namely, τ_2: select arm 2 indefinitely, and τ_1: select arm 1 until (if ever) a failure is observed and select arm 2 thereafter. From these assertions,

$$V(F_1, \lambda; \mathbf{A}) = W(F_1, \lambda; \mathbf{A}; \tau_1) \vee W(F_1, \lambda; \mathbf{A}; \tau_2)$$

$$= \left(\sum_{m=1}^{\infty} \alpha_m \int_{[0,1]} [u_1^m + (1 - u_1^{m-1})\lambda] F_1(du_1) \right) \vee \left(|\mathbf{A}|_1 \lambda \right)$$

Furthermore, arm 1 is uniquely optimal if

$$\lambda < \frac{\sum\limits_{m=1}^{\infty} \alpha_m E(\theta_1^m | F_1)}{\sum\limits_{m=1}^{\infty} \alpha_m E(\theta_1^{m-1} | F_1)}, \tag{2.4.9}$$

arm 2 is uniquely optimal if the reverse inequality holds, and both are optimal if (and only if) equality holds.

Table 2.3 gives the right-hand side of (2.4.9) for two particular distributions and for a family of discount sequences. Namely, assume $F_1 = pU(0, 1/2) + (1 - p)\delta_1$, where $p = 1/2$ or $9/10$, $U(0, 1/2)$ is the uniform distribution on $(0, 1/2)$, and $\mathbf{A}$ is the n-horizon uniform sequence. For large n, this quantity is approximately $1 - p/[(1 - p)n]$.

Table 2.3 *Right-hand side of (2.4.9)*

n	1	2	5	10	20	50	100	200	500	1000
$p = 1/2$	0.625	0.718	0.844	0.912	0.953	0.981	0.990	0.995	0.998	0.999
$p = 9/10$	0.325	0.377	0.485	0.600	0.723	0.856	0.920	0.958	0.982	0.991

□

When the horizon is finite, the proof of the existence of optimal strategies, given in Section 2.3, was explicit and allowed us to calculate optimal strategies in the examples of this section. The general existence proof given in the next section is not so constructive.

2.5 Existence of an optimal strategy

The purpose of this section is to extend Lemma 2.3.1 by showing that there exists an optimal strategy for all G and any discount sequence $\mathbf{A}$. We begin with a theorem that asserts continuity in $\mathbf{A}$. This theorem is useful when approximating arbitrary discount sequences by those having finite horizons. (Recall that $|\mathbf{A}|_1 = \sum\limits_{m=1}^{\infty} \alpha_m$.)

Theorem 2.5.1 The function $(G, \mathbf{A}) \mapsto V(G; \mathbf{A})$ is measurable and satisfies

$$-\infty < \bigvee_{i=1}^{k} E(X_{i1}|G)|\mathbf{A}|_1 \leqslant V(G; \mathbf{A})$$

$$\leqslant E\left(\bigvee_{i=1}^{k} E(X_{i1}|Q_i) \Big| G_i \right)|\mathbf{A}|_1 \leqslant E\left(\bigvee_{i=1}^{k} X_{i1} \Big| G \right)|\mathbf{A}|_1 < \infty \qquad (2.5.1)$$

and

$$V(G; \mathbf{A}) = \bigvee_{i=1}^{k} V^{(i)}(G; \mathbf{A}), \qquad (2.5.2)$$

where

$$V^{(i)}(G; \mathbf{A}) = E(\alpha_1 X_{i1} + V((X_{i1})_i G; \mathbf{A}^{(1)})|G).$$

For each $G \in \mathscr{D}^*(\mathscr{D}^k)$, the function $V(G; \cdot)$ is uniformly continuous.

Proof The finiteness inequalities in (2.5.1) follow from the fact that $G \in \mathscr{D}^*(D^k)$ and from the inequality

$$E\left(\bigvee_{i=1}^{k} X_{i1} \Big| G \right) \leqslant E\left(\sum_{i=1}^{k} |X_{i1}| \Big| G \right) = \sum_{i=1}^{k} E(|X_{i1}||G).$$

The second inequality follows from

$$W(G; \mathbf{A}; \tau_i) = E(X_{i1}|G)|\mathbf{A}|_1,$$

where τ_i denotes the strategy: always select arm i. To prove the third inequality in (2.5.1) we proceed as follows:

$$
\begin{aligned}
E_\tau(Z_m|G) &= E(X_{\tau(Z_1,\ldots,Z_{m-1}),m}|G) \\
&= \sum_{i=1}^k E\left(1_{\{\tau(Z_1,\ldots,Z_{m-1})=i\}} X_{im}\Big|G\right) \\
&= \sum_{i=1}^k E\left(E(1_{\{\tau(Z_1,\ldots,Z_{m-1})=i\}} X_{im}|Q_1,\ldots,Q_k, \right. \\
&\qquad\qquad\qquad\qquad\qquad\qquad\qquad\left. Z_1,\ldots,Z_{m-1})\Big|G\right) \\
&= \sum_{i=1}^k E\left(1_{\{\tau(Z_1,\ldots,Z_{m-1})=i\}} E(X_{im}|Q_1,\ldots,Q_k, \right. \\
&\qquad\qquad\qquad\qquad\qquad\qquad\qquad\left. Z_1,\ldots,Z_{m-1})\Big|G\right) \\
&\leqslant \sum_{i=1}^k E\left(1_{\{\tau(Z_1,\ldots,Z_{m-1})=i\}} \bigvee_{j=1}^k E(X_{jm}|Q_1,\ldots,Q_k)\Big|G\right) \\
&= E\left(\bigvee_{j=1}^k E(X_{jm}|Q_j)\Big|G\right) = E\left(\bigvee_{j=1}^k E(X_{j1}|Q_j)\Big|G\right).
\end{aligned}
$$

(2.5.3)

The fourth inequality in (2.5.1) follows from

$$
\begin{aligned}
\bigvee_{i=1}^k E(X_{i1}|Q_i) &= \bigvee_{i=1}^k E(X_{i1}|Q_1,\ldots,Q_k) \\
&\leqslant E\left(\bigvee_{i=1}^k X_{i1}\Big|Q_1,\ldots,Q_k\right).
\end{aligned}
$$

To show uniform continuity in $\mathbf{A}$, fix G and let $\varepsilon > 0$. Let $\mathbf{A} = (\alpha_1,\alpha_2,\ldots)$ and $\mathbf{B} = (\beta_1,\beta_2,\ldots)$ be discount sequences satisfying

$$
|\mathbf{A}-\mathbf{B}|_1 < \varepsilon / E\left(\bigvee_{i=1}^k |X_{i1}|\Big|G\right).
$$

For any strategy τ,

$$
\begin{aligned}
|W(G;\mathbf{A};\tau) - W(G;\mathbf{B};\tau)| &= \left|\sum_{m=1}^\infty (\alpha_m-\beta_m)E_\tau(Z_m|G)\right| \\
&\leqslant \sum_{m=1}^\infty |\alpha_m-\beta_m| E_\tau(|Z_m||G) \leqslant |\mathbf{A}-\mathbf{B}|_1 E\left(\bigvee_{i=1}^k |X_{i1}|\Big|G\right) < \varepsilon.
\end{aligned}
$$

Hence, for fixed G, $W(G; \mathbf{A}; \tau)$ depends continuously on $\mathbf{A}$, uniformly in $\mathbf{A}$ and τ. The uniform continuity of $V(G; \cdot)$ then follows from its definition (2.2.2).

The function $(G, \mathbf{A}) \mapsto (G, \mathbf{A}_n)$, where $\mathbf{A}_n = (\alpha_1, \alpha_2, \ldots, \alpha_n, 0, 0, 0, \ldots)$ when $\mathbf{A} = (\alpha_1, \alpha_2, \ldots)$, is measurable (in fact, continuous), so, by Lemma 2.3.1, $(G, \mathbf{A}) \mapsto V(G; \mathbf{A}_n)$ is measurable. The pointwise limit as $n \to \infty$ of these functions is the function $(G, \mathbf{A}) \mapsto V(G; \mathbf{A})$ which is, therefore, measurable. Equality (2.5.2) follows from its finite horizon version (2.3.1) and Lebesgue's dominated convergence theorem, the applicability of which follows from (2.5.1). □

Remarks Suppose strategies were permitted to depend on $Q_1, \ldots, Q_k$; that is, that the decision maker is told the distributions $Q_1, \ldots, Q_k$ before stage 1. The best such 'strategy' would be the one that always selects the arm that maximizes $E(X_{i1} | Q_i)$ and its worth would equal

$$E\left(\bigvee_{i=1}^{k} E(X_{i1} | Q_i) \Big| G \right) |\mathbf{A}|_1 \quad \text{(cf. (2.5.1))}.$$

As in Section 2.3, (2.5.2) can be rewritten as (2.3.2). □

The following example shows that V is not continuous in G.

Example 2.5.1 Let $\mathbf{A} = (1, 1, 0, 0, 0, \ldots)$, $k = 2$, and for $l = 2, 3, \ldots$, $G_l = F_{1l} \times F_{2l}$ where, under F_{2l}, observations on arm 2 always equal 0 and F_{1l} has atoms of size $1/2$ at each of the following two Q_1's:

$$Q_1(\{-l^{-1}\}) = Q_1(\{-1\}) = 1/2,$$
$$Q_1(\{l^{-1}\}) = Q_1(\{1\}) = 1/2.$$

Clearly an optimal strategy is to select arm 1 at stage 1 and then, at stage 2, select arm 1 if a positive result was observed and select arm 2 otherwise. An easy calculation gives

$$V(G_l; \mathbf{A}) = (1 + l^{-1})/4 \to 1/4 \text{ as } l \to \infty.$$

As $l \to \infty$, $G_l \to G = F_1 \times F_2$ where, under F_2, observations on arm 2 always equal 0 and F_1 has atoms of size $1/2$ at each of the following two Q_1's:

$$Q_1(\{0\}) = Q_1(\{-1\}) = 1/2,$$
$$Q_1(\{0\}) = Q_1(\{+1\}) = 1/2.$$

Clearly an optimal strategy is to select arm 1 at stage 1 and then, at stage 2, select arm 1 if $+1$ was observed at stage 1 and otherwise select arm 2. An easy calculation gives $V(G; \mathbf{A}) = 1/8 < 1/4$. $\qquad\square$

The reason for this lack of continuity in the preceding example is that the limit G hides information much better than do the G_l, even for large l. It is difficult to imagine an example where 'better' could be replaced by 'worse'. Accordingly, for fixed $\mathbf{A}$ we conjecture that $V(\cdot; \mathbf{A})$ is a lower semicontinuous function on $\{G : P(X_{im} < -c|G) = 0$ for some $c < \infty$ and each $i\}$.

The next theorem asserts the existence of an optimal strategy. This has already been proved (Lemma 2.3.1) in case the horizon is finite. The proof was accomplished by constructing optimal strategies recursively: first for bandits having horizon 0, then for those having horizon 1, then for those having horizon 2, etc. Then (2.5.2) was obtained for the finite horizon case. By going to the limit we have obtained (in Theorem 2.5.1) the same formula for V in the infinite horizon case. In the infinite horizon case this formula cannot be viewed as a recursion on the horizon since both $\mathbf{A}$ and $\mathbf{A}^{(1)}$ have infinite horizon, but it can be viewed as being recursive on the stages. Accordingly, we will use (2.5.2) in the proof of the next theorem to define optimal strategies stage-by-stage beginning from stage 1. This definition will not be very constructive since to carry it out in a specific example one needs to know $V(G; \mathbf{A})$ for a wide variety of bandits. This stage-by-stage definition will give an optimal strategy for every bandit; in the finite horizon case it will be the same strategy that was constructed in the proof of Lemma 2.3.1.

Theorem 2.5.2 There exists a strategy $\tau_{G, \mathbf{A}}$ for each $\mathbf{A} \in \mathscr{A}$ and each $G \in \mathscr{D}^*(\mathscr{D}^k)$ such that

$$\{(G; \mathbf{A}; z_1, \ldots, z_{m-1}) : \tau_{G, \mathbf{A}}(z_1, \ldots, z_{m-1}) = i\} \qquad (2.5.4)$$

is a measurable subset of $\mathscr{D}^*(\mathscr{D}^{k} \times \mathscr{A} \times (-\infty, \infty))^{m-1}$ for each $i = 1, \ldots, k$ and $m = 1, 2, \ldots$, and $\tau_{G, \mathbf{A}}$ is optimal for the $(G; \mathbf{A})$-bandit.

Proof Define $\tau_{G, \mathbf{A}}$ recursively as follows: $\tau_{G, \mathbf{A}}(\varnothing)$ is the smallest i for which the maximum in (2.5.2) is attained and, for $m > 1$,

$$\tau_{G, \mathbf{A}}(z_1, \ldots, z_{m-1}) = \tau_{G^{(1)}, \mathbf{A}^{(1)}}(z_2, \ldots, z_{m-1}) \qquad (2.5.5)$$

where

$$G^{(1)} = (z_1)_{\tau_{G, \mathbf{A}}(\varnothing)} G.$$

The measurability of (2.5.4) follows immediately by an induction argument on m using Lemma 2.2.1 to deal with $G^{(1)}$ in (2.5.5).

For the remainder of the proof we write τ for $\tau_{G,\,A}$. From (2.5.2) and the definition of τ we conclude that

$$V(G; \mathbf{A}) = E_\tau(\alpha_1 Z_1 + V(G^{(1)}; \mathbf{A}^{(1)})|G)$$

and, by induction,

$$V(G; \mathbf{A}) = E_\tau\left(\sum_{m=1}^{n} \alpha_m Z_m + V(G^{(n)}; \mathbf{A}^{(n)})\bigg|G\right).$$

Since

$$W(G; \mathbf{A}; \tau) = E_\tau\left(\sum_{m=1}^{\infty} \alpha_m Z_m\bigg|G\right),$$

we can complete the proof by showing

$$\lim_{n \to \infty} E_\tau\left(\sum_{m=n+1}^{\infty} \alpha_m Z_m - V(G^{(n)}; \mathbf{A}^{(n)})\bigg|G\right) = 0. \qquad (2.5.6)$$

We have

$$\left|E_\tau\left(\sum_{m=n+1}^{\infty} \alpha_m Z_m\bigg|G\right)\right| \leqslant \sum_{m=n+1}^{\infty} \alpha_m E_\tau\left(|Z_m|\bigg|G\right)$$

$$\leqslant \sum_{m=n+1}^{\infty} \alpha_m E\left(\bigvee_{i=1}^{k} |X_{im}|\bigg|G\right)$$

$$= E\left(\bigvee_{i=1}^{k} |X_{i1}|\bigg|G\right)|\mathbf{A}^{(n)}|_1 \to 0.$$

From (2.5.1),

$$|V(G^{(n)}; \mathbf{A}^{(n)})| \leqslant E\left(\bigvee_{i=1}^{k} |X_{i1}|\bigg|G^{(n)}\right)|\mathbf{A}^{(n)}|_1$$

and, hence

$$E_\tau(|V(G^{(n)}; \mathbf{A}^{(n)})|\big|G) \leqslant E\left(\bigvee_{i=1}^{k} |X_{i1}|\bigg|G\right)|\mathbf{A}^{(n)}|_1 \to 0.$$

Therefore, (2.5.6) holds.   □

We immediately obtain

Corollary 2.5.3 Any i for which the maximum in (2.5.2) is attained is an optimal initial selection in the $(G; \mathbf{A})$-bandit, and conversely.

Remark The requirement in the preceding proof that $\tau_{G, \mathbf{A}}(\varnothing)$ be chosen to yield the maximum in (2.5.2) gives $\tau_{G, \mathbf{A}}$ the property (used in gambling theory literature; e.g. Dubins and Savage (1976)) of being *conserving* at stage 1. The recursive relation (2.5.5) makes $\tau_{G, \mathbf{A}}$ conserving at every stage – that is, *thrifty*. We could have relied on a general theorem to conclude that for our context any thrifty strategy is optimal. Instead we adapted arguments used in the proofs of such general theorems, which, incidentally, have hypotheses in addition to thriftiness. □

Although optimal strategies and V are the main objects of study, the behaviour of the function W may be of some interest.

Theorem 2.5.4 For each fixed τ, the function $(G, \mathbf{A}) \mapsto W(G; \mathbf{A}; \tau)$ is measurable. For each fixed G, the function $\mathbf{A} \mapsto W(G; \mathbf{A}; \tau)$ is continuous, uniformly in $\mathbf{A}$ and τ.

Proof The proof of uniform continuity is contained in the proof of Theorem 2.5.1.

To prove measurability it suffices to prove measurability of the function $G \mapsto E_\tau(Z_m | G)$ for each m. But this measurability follows easily from

$$Z_m = \sum_{i=1}^{k} X_{im} \mathbf{1}_{\{\tau(Z_1, \ldots, Z_{m-1}) = i\}}$$

and an application of Lemma 2.2.3 using approximations of each summand by bounded random variables. □

While optimal strategies exist generally, finding them can be difficult when the horizon is infinite. One route is to find V; we turn to approximating V in the infinite horizon case.

2.6 Approximating value functions for infinite horizons

A method for finding an upper or lower bound for the function $V(\cdot; \mathbf{A})$ is to replace $V(\cdot; \mathbf{A}^{(1)})$ with an upper or lower bound in (2.2.5). The same statement applies as well for $V(\cdot; \mathbf{A}^{(m)})$ and $V(\cdot; \mathbf{A}^{(m+1)})$ for all integers $m \geqslant 0$. Therefore, upper and lower bounds for $V(\cdot; \mathbf{A}^{(n)})$ can be used via dynamic programming to find bounds for $V(\cdot; \mathbf{A})$.

Two bounds are given by Theorem 2.5.1. Letting $(G^{(n)}; \mathbf{A}^{(n)})$ play the role of $(G; \mathbf{A})$, we have

$$\bigvee_{i=1}^{k} E(X_{i1}\Big|G^{(n)})|\mathbf{A}^{(n)}|_1 \leqslant V(G^{(n)}; \mathbf{A}^{(n)})$$

$$\leqslant E\left(\bigvee_{i=1}^{k} E(X_{i1}|Q_i)\Big|G^{(n)}\right)|\mathbf{A}^{(n)}|_1 \qquad (2.6.1)$$

Let $L_n(G; \mathbf{A})$ denote the approximation for $V(G; \mathbf{A})$ obtained by dynamic programming using the left side of (2.6.1) to begin the backward induction; thus, $L_n(G; \mathbf{A})$ equals $V(G; \mathbf{A}^*)$ where $\mathbf{A}^* = (\alpha_1, \alpha_2, \ldots, \alpha_n, |\mathbf{A}^{(n)}|_1, 0, 0, \ldots)$. Similarly, let $U_n(G; \mathbf{A})$ denote the approximation obtained using the right side of (2.6.1) as starting values. The next theorem indicates that V is bounded by each L_n and each U_n; in addition, it gives a very crude bound for the difference between L_n and U_n.

Theorem 2.6.1 For all $G \in \mathcal{D}(\mathcal{D}^k)$ and all discount sequences $\mathbf{A}$,

$$L_n(G; \mathbf{A}) \leqslant V(G; \mathbf{A}) \leqslant U_n(G; \mathbf{A}) \qquad (2.6.2)$$

and

$$U_n(G; \mathbf{A}) - L_n(G; \mathbf{A})$$

$$\leqslant \left[E\left(\bigvee_{i=1}^{k} E(X_{i1}|Q_i)\Big|G\right) - \bigvee_{i=1}^{k} E(X_{i1}|G)\right]|\mathbf{A}^{(n)}|_1. \qquad (2.6.3)$$

In particular,

$$\lim_{n \to \infty} [U_n(G; \mathbf{A}) - L_n(G; \mathbf{A})] = 0.$$

Proof The inequalities in (2.6.2) hold as indicated previously. The method of dynamic programming and analogues of (2.3.1) are appropriate for the following three modifications of the $(G; \mathbf{A})$-bandit. In the first, the experiment terminates at stage n and a (random) amount

$$\sum_{m=n+1}^{\infty} \left(\alpha_m \bigvee_{i=1}^{k} E(X_{im}|Q_i)\right) \qquad (2.6.4)$$

is added to $\Sigma_{m=1}^{n} \alpha_m Z_m$. The value for the modification is $U_n(G; \mathbf{A})$.

In the second, after n stages the decision maker must select one arm to be used for all stages $m > n$. The value for this modification is clearly $L_n(G; \mathbf{A})$.

Finally, consider a third modification whose value is obviously no larger than $L_n(G; \mathbf{A})$. The decision maker must, before stage 1, select one arm to be used for all stages $m > n$. Of course, that choice will be a j for which

$$E(X_{j1}|G) = \bigvee_{i=1}^{k} E(X_{i1}|G).$$

Let V_i denote the value for the ith modification. From (2.6.4) we see that

$$V_1(G; \mathbf{A}) \leqslant V_3(G; \mathbf{A})$$

$$+ |\mathbf{A}^{(n)}|_1 \left[E\left(\bigvee_{i=1}^{k} E(X_{i1}|Q_i) \middle| G \right) - \bigvee_{i=1}^{k} E(X_{i1}|G) \right]. \quad \Box \quad (2.6.5)$$

Remarks This theorem can be used in two ways to decide on a truncation stage in order to approximate V with any desired accuracy. First, n can be chosen so that the right side of (2.6.3) is sufficiently small. Second, $U_n - L_n$ can be calculated for various values of n until the desired accuracy is attained. While the second can involve a number of dynamic programs, the total calculation time is usually less because the bound provided by (2.6.3) is so crude. $\quad \Box$

Example 2.6.1 Let $\mathbf{A}$ be geometric: $\mathbf{A} = (1, 0.9, (0.9)^2, \ldots)$. Suppose there are two Bernoulli arms. Assume θ_1 is uniformly distributed on $[0, 1]$ and $\theta_2 = 0.6$.

To use (2.6.3) we calculate

$$E\left(E(X_{i1}|\theta_1) \vee E(X_{i2}|\theta_2) \right) = \int_0^1 (u_1 \vee 0.6) \, du_1 = 0.68,$$

$$E(X_{11}|G) \vee E(X_{21}|G) = 0.5 \vee 0.6 = 0.6,$$

and

$$|\mathbf{A}^{(n)}|_1 = \sum_{m=n+1}^{\infty} (0.9)^{m-1} = 10(0.9)^n.$$

Suppose we want to approximate $V(G; \mathbf{A})$ with an error no greater than 0.005. Accordingly, we require the right-hand side of (2.6.3) to be no larger than 0.01:

$$(0.08)(10)(0.9)^n \leqslant 0.01.$$

The smallest n that suffices is 42. Dynamic programming gives $L_{42}(G; \mathbf{A}) \approx 6.38827$ and $U_{42}(G; \mathbf{A}) \approx 6.39020$. The average, 6.38923, is an approximation as desired and, in fact, is in error by no more than 0.001. In carrying out the two dynamic programs only one strategy is obtained: the one derived in calculating $L_{42}(G; \mathbf{A})$. In view of (2.6.3) we know that its worth is within 0.01 of the value. Having calculated $U_{42}(G; \mathbf{A})$ we know that $V - L_{42} \leqslant 0.002$. In view of the results described below it turns out that $V - L_{42} \approx 4.3 \times 10^{-6}$.

Proceeding without reference to (2.6.3) we can generate selected L_n and U_n as indicated in Table 2.4, stopping when desired accuracy is achieved. For example, we would stop at $n = 28$ if we want our estimate to be in error by less than 0.005.

Table 2.4

n	$L_n(G; \mathbf{A})$	$U_n(g; \mathbf{A})$	$U_n - L_n$
1	6.2	6.72	0.52
2	6.335	6.648	0.313
3	6.335	6.5832	0.2482
4	6.36524	6.53183	0.16659
5	6.3736756	6.514334	0.1406584
10	6.3855284	6.4536415	0.0681131
15	6.3874732	6.4249362	0.0374630
20	6.3880147	6.4095730	0.0215583
28	6.3882272	6.3972135	0.0089863
50	6.3882686	6.3890906	0.0008220
100	6.3882698	6.3882740	0.0000042
150	6.3882698	6.3882698	0.0000000

It is evident that neither of these approaches is as good as using L_n for modest n. For example, L_7 is about as accurate as $(L_{50} + U_{50})/2$. However, we do not have a way of assessing error using the sequence $\{L_n\}$ alone.   □

References

Bradt, R. N., Johnson, S. M. and Karlin, S. (1956) On sequential designs for maximizing the sum of n observations. *Ann. Math. Statist.* **27**: 1060–1074.

Chow, Y. S. and Teicher, H. (1978) *Probability Theory*, Springer-Verlag, New York.

DeGroot, M. H. (1970) *Optimal Statistical Decisions*, McGraw-Hill, New York.

Dubins, L. and Freedman, D. (1964) Measurable sets of measures, *Pac. J. Math.* **14**: 1211–1222.

Dubins, L. E. and Savage, L. J. (1976) *Inequalities for Stochastic Processes: How to Gamble If You Must*, Dover, New York.

Eick, S. G. (1984) Personal communication.

Parthasarathy, K. R. (1967) *Probability Measures on Metric Spaces*, Academic Press, New York.

Pearson, L. M. (1980) Treatment allocation for clinical trials in stages. Ph.D. thesis, Univ. of Minnesota, USA.

Rhenius, D. (1977) Faktorisierung von übergangswahrscheinlichkeiten in Markoffschen Lernmodellen. *Arbeiten aus den Psychologischen Instituten der Univ. Hamburg Nr.* 44.

Varadhan, S. R. S. (1983) Personal communication via N. C. Jain.

CHAPTER 3

The discount sequence

The particular discount sequence plays a critical role in any bandit or other decision problem. Various interpretations of discount sequences are discussed in this chapter. One purpose of the discussion is to aid a user in choosing an appropriate sequence; another is to motivate interest in the generality of discounting allowed in this monograph.

As indicated in Chapter 1, much of the bandit literature concerns maximizing the sum of n observations. In our notation the corresponding discount sequence is the n-horizon uniform sequence: $\alpha_m = 1$ for $m \leqslant n$ and $\alpha_m = 0$ for $m > n$ (see Chapter 7 for an extensive treatment of this case). The other important discount sequence in the literature is the geometric: $\alpha_m = \alpha^{m-1}$ for some $\alpha \in (0, 1)$ (see Chapter 6). There has been very little discussion in the literature of other cases, and the question arises as to whether our more general approach has any practical relevance. This chapter gives an affirmative answer. Some of the ideas in this chapter are considered in Berry (1983).

The possibility that the discount sequence is unknown is considered in the first four sections of this chapter. Random uniform discount sequences are discussed in Section 3.1. The next three sections are devoted to more general random discount sequences. An important consideration is whether or not learning about the unknown discount sequence is possible as the experiment develops. The two cases of observing and not observing the discount factors are both considered. The orientation throughout the first four sections is to identify hypotheses under which random discount factors may, without loss, be replaced by their expectations.

Section 3.5 deals with nonrandom real-time discount sequences. Time is continuous and times at which trials (that is, stages) occur are random.

Section 3.6 motivates discount sequences that are not monotone. The important special case $(0, 0, \ldots, 0, 1, 0, 0, \ldots)$ is discussed.

3.1 Mixtures of uniform sequences

Consider Bernoulli trials and suppose a success at any stage is worth 1, as in uniform discounting. Also suppose the experiment may terminate at any stage for reasons outside the decision maker's control; let η_m be the probability of termination with the mth trial, $m = 1, 2, \ldots$. For example, in a clinical trial the number of patients in the population to be treated may not be precisely known. Or, a new arm that is clearly better than any in the trial may be discovered at any time. The true discount sequence may be an n-horizon uniform, all successes being equally valued, but n is not known.

This situation can be placed in the framework of Chapter 2 as follows. As in Chapter 2 we assume that there is an infinite sequence of trials. We account for the fact that the experiment may have terminated before a given stage by specifying that a success at that stage is worth only the probability that the experiment has not yet terminated; that is, the appropriate discount sequence $(\alpha_1, \alpha_2, \ldots)$ is given by

$$\alpha_m = \sum_{l=m}^{\infty} \eta_l. \qquad (3.1.1)$$

Any nonincreasing discount sequence can arise in this manner by normalizing to $\alpha_1 = 1$. However, it may happen that $(\alpha_1, \alpha_2, \ldots)$ given by (3.1.1) is not a discount sequence because $\Sigma \alpha_m = \Sigma l\eta_l = \infty$.

Assume the length of the trial is independent of the strategy followed and responses obtained. Then the setting is precisely the one described in the first two chapters. In particular, (2.2.2) applies where α_m is defined in (3.1.1). Though the discount sequence is random since it is a mixture of uniforms, the problem is identical to one in which the discount sequence is deterministic – namely, an average of uniforms weighted by the η_m's.

For example, assume there is a constant probability of terminating at each stage given that that stage has been reached, so that

$$\eta_m = (1 - \alpha)\alpha^{m-1} \qquad \text{for some } \alpha \in (0, 1).$$

Then $\alpha_m = \alpha^{m-1}$, $m = 1, 2, \ldots$, and the appropriate discount sequence is geometric.

3.2 Random discount sequences

The previous section dealt with mixtures of uniform discount sequences. The generalization to mixtures of arbitrary sequences is considered in this section.

Results in general are considerably more complicated than in the uniform setting for an important reason: the character of the decision problem depends upon whether or not discount factors are observed at each stage. Mixtures of uniform sequences are special in this regard: conditioning on previous discount factors offers no advantage. A discount factor is 0 when it is not 1 in the uniform case, so it can be assumed to be 1 without loss – if it is 0 then any selection (and continuation) is of no consequence.

In the next section we discuss bandit problems in which the discount factors are not observed and, hence, strategies cannot depend on them. In Section 3.4, selections are allowed to depend on all previous discount factors. (A third possibility, one not discussed here, is that $\alpha_m Z_m$ is observed, but not α_m and Z_m individually.) In this section we shall discuss some facets common to both settings.

The decision maker's prior knowledge concerning the random discount sequence consists of a probability distribution $\mathbf{H}$ on the space $\mathscr{A}$ of all discount sequences. We assume

$$E\left(\sum_{m=1}^{\infty} \alpha_m \middle| \mathbf{H} \right) < \infty,$$

where $\mathbf{H}$ appears in the notation to make explicit the dependence of expected values on the distribution of the random discount sequence. Some expectations depend on both G and $\mathbf{H}$ – for instance, we write $E(\alpha_m X_{im} | G, \mathbf{H})$. We always assume the arms to be independent of the discount sequence, so we deal only with the product measure $G \times \mathbf{H}$. Accordingly,

$$E(\alpha_m X_{im} | G, \mathbf{H}) = E(\alpha_m | \mathbf{H}) E(X_{im} | G).$$

However, in Section 3.4, $E_\tau(\alpha_m Z_m | G, \mathbf{H})$ may not factor; for τ and therefore Z_m may depend on $\alpha_1, \ldots, \alpha_{m-1}$.

For the purposes of this chapter we extend previous terminology and notation by speaking of the $(G; \mathbf{H})$-bandit and using $W(G; \mathbf{H}; \tau)$ and $V(G; \mathbf{H})$ instead of $W(G; \mathbf{A}; \tau)$ and $V(G; \mathbf{A})$. The value $V(G; \mathbf{H})$ depends on the class of strategies being considered. Therefore $V(G; \mathbf{H})$ will be no smaller in the setting of Section 3.4 than in that of

Section 3.3. For either of the two cases, the important equation (2.5.2) is valid when modified appropriately:

$$V(G; \mathbf{H}) = \bigvee_{i=1}^{k} E(\alpha_1 X_{i1} + V((X_{i1})_i G; \mathbf{H}^{(1)})|G, \mathbf{H}) \quad (3.2.1)$$

where $\mathbf{H}^{(1)}$ denotes the conditional distribution of the discount sequence $\mathbf{A}^{(1)} = (\alpha_2, \alpha_3, \ldots)$ given what has been observed at stage 1. Formula (3.2.1) means different things depending on context: in Section 3.4 we condition on α_1 to obtain $\mathbf{H}^{(1)}$ but in Section 3.3 we will only condition on being at the second stage. In general, (3.2.1) can be rewritten:

$$V(G; \mathbf{H}) = \bigvee_{i=1}^{k} [E(\alpha_1|\mathbf{H})E(X_{i1}|G) + E((V(X_{i1})_i G; \mathbf{H}^{(1)})|G, \mathbf{H})];$$

$$(3.2.2)$$

although, as will be seen in the next section, in the nonobservable case the simpler relation (2.5.2) applies with α_m replaced by its expected value under $\mathbf{H}$.

The topology we use on the space of all $\mathbf{H}$'s is the topology of convergence in distribution; the measurable sets of $\mathbf{H}$'s are the Borel sets.

3.3 Nonobservable discount factors

In this section strategies τ are defined as in Chapter 2. As such they are measurable with respect to the σ-field generated by $\{X_{im}: i = 1, \ldots, k; m = 1, 2, \ldots\}$. (This requirement on strategies will be expressed by saying that the discount factors are 'nonobservable'.) Hence, each Z_m is measurable with respect to this σ-field. On the other hand, the random discount sequence is independent of it. Therefore, for each τ,

$$W(G; \mathbf{H}; \tau) = \sum_{m=1}^{\infty} E_\tau(\alpha_m Z_m|G, \mathbf{H})$$

$$= \sum_{m=1}^{\infty} E(\alpha_m|\mathbf{H})E_\tau(Z_m|G)$$

$$= E_\tau\left(\sum_{m=1}^{\infty} E(\alpha_m|\mathbf{H})Z_m|G\right)$$

which leads to (2.2.2) with $E(\alpha_m | \mathbf{H})$ playing the role of α_m. Thus we have proved the following result.

Theorem 3.3.1 Suppose a random discount sequence $\mathbf{A}$ governed by a distribution $\mathbf{H}$ is independent of $\{X_{im} : i = 1, \ldots, k; m = 1, 2, \ldots\}$ and the discount factors are not observable. Then, for any G, the $(G; \mathbf{H})$-bandit is equivalent to the $(G; E(\mathbf{A} | \mathbf{H}))$-bandit; that is,

$$W(G; \mathbf{H}; \tau) = W(G; E(\mathbf{A} | \mathbf{H}); \tau)$$

for all τ. In particular, the two bandits have the same value and the same set of optimal strategies. In addition, $V(G; \mathbf{H})$ and $W(G; \mathbf{H}; \tau)$ depend continuously on $\mathbf{H}$ and measurably on the pair $(G, \mathbf{H})$.

Example 3.3.1 Suppose $E(\mathbf{A} | \mathbf{H}) = (1, \ldots, 1, 0, \ldots)$. Then the discount sequence relevant for choosing a strategy is the finite horizon uniform. Chapter 7 and Examples 2.4.1 to 2.4.3 and 2.4.5 to 2.4.7 apply.   □

Example 3.3.2 Suppose $E(\alpha_m | \mathbf{H}) = \alpha^{m-1}$ for known $\alpha > 0$. The relevant discount sequence is geometric. Chapter 6 applies.   □

Example 3.3.3 Suppose $\mathbf{H}$ assigns probability $\frac{1}{2}$ to each of two geometric discount sequences: $(3/4, (3/4)^2, \ldots)$ and $(1/4, (1/4)^2, \ldots)$. Suppose that $k = 2$ and that, under G, both arms are Bernoulli with arm 2 having a known probability λ of success and arm 1 having probability 1 or 0 of success, each with probability $1/2$. By Theorem 3.3.1 the $(G; \mathbf{H})$-bandit is equivalent to the $(G; \mathbf{A})$-bandit where

$$\mathbf{A} = \left(\frac{1}{2}, \frac{5}{16}, \ldots, \frac{1}{2}\left(\frac{3}{4}\right)^m + \frac{1}{2}\left(\frac{1}{4}\right)^m, \ldots \right).$$

Notice that (2.4.8) is not satisfied for this sequence for any $m > 1$ even though each component geometric sequence does satisfy (2.4.8).

Complete information is obtained with a single observation of arm 1. Accordingly, whenever arm 1 is selected it is to be used indefinitely thereafter if successful, and never again otherwise. So the only uncertainty in describing an optimal strategy is specifying when to select arm 1. Straightforward calculations show that there is a sequence $\{\lambda_i\}$ with $0.7273 \approx 8/11 = \lambda_0 < \lambda_1 < \lambda_2 < \ldots < \lambda_\infty = \lim \lambda_r = 4/5$, with the following interpretation. If $\lambda \in [0, \lambda_0]$ then arm 1 is optimal initially. If $\lambda \in [\lambda_0, \lambda_1]$ then it is optimal to select arm 2 initially, followed by arm 1 at stage 2. Generally, if $\lambda \in [\lambda_{r-1}, \lambda_r]$ then it is optimal to select arm 2 at stages 1 through r, followed by arm 1 at stage $r + 1$. If $\lambda \in [\lambda_\infty, 1]$ then arm 2 is optimal indefinitely and arm 1 should never be used.

When a nonrandom discount sequence is geometric, Theorem 5.2.2 applies to show that there is an optimal strategy that either begins with arm 1 or else never uses arm 1. This example shows that for a mixture of geometrics, such simplicity may be lost – after observing the known arm for some time, the decision maker may want to switch to the unknown arm despite having received no additional information. □

3.4 Observable discount factors

In this and the following section, and in these two sections only, a strategy τ depends on previous discount factors as well as on previous observations. Thus $\tau(z_1, z_2; \alpha_1, \alpha_2)$ indicates the arm to be observed at stage 3 if z_1 and z_2 are the outcomes on arms $\tau(\varnothing)$ and $\tau(z_1; \alpha_1)$, respectively, at stages 1 and 2 and α_1 and α_2 are the first two discount factors.

Example 3.4.1 We use the same G and $\mathbf{H}$ as in Example 3.3.3. As in that example, if arm 1 is selected initially then optimal continuations are clear. So suppose arm 2 is selected initially. At stage 2 the discount sequence becomes known – one of two geometrics. Since (2.4.8) is satisfied for geometric sequences, Example 2.4.7 applies for the bandit problem starting at stage 2.

Accordingly, one of the following three strategies is optimal:

τ_1: select arm 1 initially, then proceed optimally;

τ_2: select arm 2 always;

τ_{2*}: select arm 2 initially; if $\alpha_1 = 1/4$, observe arm 2 thereafter, if $\alpha_1 = 3/4$, observe arm 1 and thereafter proceed optimally.

Easy calculations show:

$$W(G; \mathbf{H}; \tau_1) = 7\lambda/12 + 5/6,$$
$$W(G; \mathbf{H}; \tau_2) = 5\lambda/3,$$
$$W(G; \mathbf{H}; \tau_{2*}) = 185\lambda/192 + 9/16.$$

Therefore τ_1 is optimal if $\lambda \leqslant 52/73 \approx 0.7123$, τ_{2*} is optimal if $52/73 \leqslant \lambda \leqslant 4/5$, and τ_2 is optimal for $\lambda \geqslant 4/5$.

If $52/73 < \lambda < 4/5$, then the value is larger than that for the same bandit in the nonobservable setting. If λ is not in this interval then the values $\lambda \neq 52/73$ and $\lambda \neq 4/5$ are the same; if in addition then optimal strategies are the same.   □

The $(G; H)$-bandit treated in Examples 3.3.3 and 3.4.1 is rather unusual in that the observable version is easier to solve than is the nonobservable version. The observable version is generally very difficult. The only general results we give for the observable case are the lemma and two theorems of this section.

Minor changes in the methods leading to Lemma 2.3.1, Theorem 2.5.1, and Theorem 2.5.2 lead to the following theorem and lemma.

Theorem 3.4.1 In the observable case, (3.2.1) holds. An optimal initial selection for an arbitrary $(G; H)$-bandit is given by the smallest i for which the maximum in (3.2.1) is attained. These optimal initial selections fit together to constitute optimal strategies $\tau_{G,H}$ whose dependence on (G, H) is measurable. The value function $(G, H) \mapsto V(G; H)$ is measurable.

For the lemma we need a notation. As in Section 2.5 let A_n denote the discount sequence obtained from A by replacing all terms after the nth term by zeros. The mapping $A \mapsto A_n$ induces a mapping, say $H \mapsto H_n$, of probability measures on the space of discount sequences.

Lemma 3.4.2 For H_n defined as above, $V(G; H_n) \mapsto V(G; H)$ as $n \to \infty$.

Despite the preceding lemma, V does not depend continuously on H, as the following example shows.

Example 3.4.2 Let H assign probability $\frac{1}{2}$ to each of the discount sequences $(1, 0, 2, 0, 0, 0, \ldots)$ and $(1, 2, 0, 0, 0, 0, \ldots)$ and let H_n^* assign probability $\frac{1}{2}$ to each of the discount sequences $(1 - n^{-1}, 0, 2, 0, 0, 0, \ldots)$ and $(1 + n^{-1}, 2, 0, 0, 0, 0, \ldots)$. So $H_n^* \to H$ as $n \to \infty$. Suppose that under G, there are two independent Bernoulli arms having parameters θ_1 and θ_2 where θ_1 is uniformly distributed on $(0, 1)$ and $\theta_2 = 2/5$ with probability one.

Consider the $(G; \mathbf{H})$-bandit. No information concerning the discount sequence is available at stage 1. The fact that the decision maker learns the discount sequence at stage 2 is irrelevant since, in any case, an optimal selection at stage 3 is the arm with the higher current mean. So the $(G; \mathbf{H})$-bandit is equivalent to one with nonrandom discounting: $(G; (1, 1, 1, 0, 0, \ldots))$. The first three selections of an optimal strategy τ are as follows:

$$\tau(\varnothing) = 1,$$
$$\tau(1; 1) = 1,$$
$$\tau(1, 1; 1, 0) = \tau(1, 1; 1, 2) = \tau(1, 0; 1, 0) = \tau(1, 0; 1, 2) = 1,$$
$$\tau(0; 1) = 2,$$
$$\tau(0, 1; 1, 0) = \tau(0, 1; 1, 2) = \tau(0, 0; 1, 0) = \tau(0, 0; 1, 2) = 2.$$

An easy calculation gives $V(G; \mathbf{H}) = 94/60$.

Now consider the $(G; \mathbf{H}_n^*)$-bandit. Complete information concerning the discount sequence is present at stage 1. It is easy to see that for all n, an optimal strategy τ^* satisfies:

$$\tau^*(\varnothing) = 1,$$
$$\tau^*(1; 1 - n^{-1}) = \tau(1; 1 + n^{-1}) = 1,$$
$$\tau^*(1, 1; 1 - n^{-1}, 0) = \tau^*(1, 0; 1 - n^{-1}, 0) = 1,$$
$$\tau^*(0; 1 - n^{-1}) = 1, \qquad \tau^*(0; 1 + n^{-1}) = 2,$$
$$\tau^*(0, 1; 1 - n^{-1}, 0) = 1, \qquad \tau^* = (0, 0; 1 - n^{-1}, 0) = 2;$$

and $V(G; \mathbf{H}_n^*) = 95/60$.
So $V(G; \mathbf{H}_n^*) \nrightarrow V(G; \mathbf{H})$. $\qquad \qquad \square$

When the discount sequence is observable, one typically cannot replace a random discount sequence by a nonrandom sequence without changing the problem in an essential way. The next result gives a special situation where such a replacement is possible.

Theorem 3.4.3 Suppose an observable random discount sequence $\mathbf{A} = (\alpha_1, \alpha_2, \ldots)$ is independent of $\{X_{im}: i = 1, \ldots, k; m = 1, 2, \ldots\}$ and is given by

$$\alpha_m = \prod_{l=1}^{m} U_l, \qquad (3.4.1)$$

where, under the distribution $\mathbf{H}$, $\{U_l: l = 1, 2, \ldots\}$ is an independent sequence. Then, for all G,

$$V(G; \mathbf{H}) = V(G; E(\mathbf{A}|\mathbf{H}))$$

and each optimal strategy for the $(G; E(\mathbf{A}|\mathbf{H}))$-bandit is optimal for the $(G; \mathbf{H})$-bandit.

Proof The proof is by induction followed by a limiting argument. Let

$$\mathscr{H}_n = \{\mathbf{H}: P(U_{n+1} = 0|\mathbf{H}) = 1, \{U_l: l \geq 1\} \text{ is independent under } \mathbf{H}\}.$$

The conclusion of the theorem obviously holds for $\mathbf{H} \in \mathscr{H}_0$. Assume it holds for all $\mathbf{H} \in \mathscr{H}_{n-1}$ and fix $\mathbf{H} \in \mathscr{H}_{\bar{n}}$.

The distribution of $\{U_l: l = 1, 2, \ldots\}$ is governed by $\mathbf{H}$ and, hence, $\mathbf{H}$ determines the distribution $\mathbf{H}_1$ of the random discount sequence

$$(U_2(\omega), \, U_2(\omega)U_3(\omega), \ldots, \prod_{l=2}^{m} U_l(\omega), \ldots).$$

For a set S of discount sequences let S/u, for $u > 0$, denote the set of discount sequences obtained by dividing each member of each sequence in S by u. It is clear that the random distribution $\mathbf{H}^{(1)}$ used in (3.2.2) almost surely belongs to $\mathscr{H}_{n-1}$ and satisfies

$$\mathbf{H}^{(1)}(S, \omega) = \mathbf{H}_1(S/U_1(\omega))$$

if $U_1(\omega) \neq 0$ and, when $U_1(\omega) = 0$,

$$\mathbf{H}^{(1)}(\{0, 0, 0, \ldots)\}, \omega) = 1.$$

In the random setting, as in the nonrandom, multiplication of the discount sequence by a constant multiplies the value by that constant. Hence,

$$V(G^{(1)}(\cdot, \omega); \mathbf{H}^{(1)}(\cdot, \omega)) = U_1(\omega) V(G^{(1)}(\cdot, \omega); \mathbf{H}_1).$$

Since $\mathbf{H}_1$ is not random,

$$E_\tau(U_1(\omega) V(G^{(1)}(\cdot, \omega); \mathbf{H}_1)|G, \mathbf{H})$$
$$= E(U_1(\omega)|\mathbf{H}) E_\tau(V(G^{(1)}(\cdot, \omega); \mathbf{H}_1)|G, \mathbf{H}),$$

which, by the induction hypothesis, equals

$$E(U_1|\mathbf{H}) E_\tau(V(G^{(1)}; (E(U_2|\mathbf{H}), E(U_2 U_3|\mathbf{H}), \ldots))|G).$$

On moving the constant $E(U_1|\mathbf{H})$ through the expectations and through V, this expression becomes

$$E_\tau(V(G_i^{(1)}; (E(U_1U_2|\mathbf{H}), E(U_1U_2U_3|\mathbf{H}), \dots))|G).$$

By comparing (3.2.2) with (2.5.2) we see that the $(G;\mathbf{H})$-bandit has the same value and optimal initial selections as does the $(G; E(\mathbf{A}|\mathbf{H}))$-bandit. This completes the induction step.

If $\mathbf{H}$ is such that $P(U_n > 0|\mathbf{H}) > 0$ for all n, a limiting argument using Lebesgue's dominated convergence theorem and Lemma 3.4.2 easily completes the proof. $\qquad\square$

Remark Suppose the hypothesis and, therefore, the conclusion of Theorem 3.4.3 hold. Since the class of strategies is much richer when the discount sequence is random and discount factors are observable, there may be optimal strategies for the $(G;\mathbf{H})$-bandit that are meaningless for the $(G; E(\mathbf{A}|\mathbf{H}))$-bandit; for, when two or more selections are equally good for the $(G; E(\mathbf{A}|\mathbf{H}))$-bandit, the choice from among these selections in the $(G;\mathbf{H})$-bandit can depend on preceding values of the discount sequence – all resulting strategies are, of course, equally good. $\qquad\square$

Formula (3.4.1), with a highly dependent sequence $\{U_l: l = 1, 2, \dots \}$, is appropriate for Example 3.4.1. In that example, $U_l = 1/4$ or $3/4$ each with probability $1/2$ and $U_l = U_1$ for each l. This is an instance of a certain type of dependence of the sequence $\{U_l: l = 1, 2, \dots \}$: a random distribution R on $\mathbb{R}$ is chosen and then $\{U_l: l = 1, 2, \dots \}$ is a conditionally independent sequence in which the conditional distribution of each U is R. In Example 3.4.1, R is a one-point distribution concentrated at either $1/4$ or $3/4$, each with probability $1/2$. It would be interesting to pursue the study of observable sequences in this manner for an arbitrary member of $\mathscr{D}(\mathbb{R})$ governing R.

When (3.4.1) holds, an interpretation of U_l is that the payoff at stage l is discounted by the random factor U_l as compared with the previous stage: $(1 - U_l)/U_l$ is the random inflation rate.

The next section deals with the situation in which stages occur in real time and discounting is also in terms of real time.

3.5 Real-time discounting

Consider a clinical trial in which patients arrive at haphazard times. These times are not predictable in advance, and, indeed, the number of patients to arrive in any fixed time period is unknown. Suppose that the response of a patient who arrives and is treated at time $t \geq 0$ is weighted by the factor $\exp(-\beta t)$, $0 < \beta < \infty$. Were the patients' arrival times known in advance to be $t_1, t_2, t_3, \ldots$, then the appropriate discount sequence would be nonrandom:

$$\mathbf{A} = (\exp(-\beta t_1), \exp(-\beta t_2), \exp(-\beta t_3), \ldots).$$

Since arrival times are not known, we let T_m denote the random time at which the mth trial (or stage, or patient) occurs. Let $Y_1 = T_1$ and $Y_m = T_m - T_{m-1}$, $m = 2, 3, \ldots$, denote the interarrival times. The appropriate discount sequence is random:

$$(\exp(-\beta T_1), \exp(-\beta T_2), \ldots, \exp(-\beta T_m), \ldots)$$

$$= (\exp(-\beta Y_1), \exp(-\beta(Y_1 + Y_2)), \ldots,$$

$$\exp\left(-\beta \sum_{l=1}^{m} Y_l\right), \ldots)$$

$$= (U_1, U_1 U_2, \ldots, \prod_{l=1}^{m} U_l, \ldots)$$

where $U_l = \exp(-\beta Y_l)$, $l = 1, 2, \ldots$.

The ability to observe discount factors means in the current context that the decision maker observes $Y_1, \ldots, Y_m$ and uses this information when selecting an arm at the mth stage. If $\{Y_l : l = 1, 2, \ldots\}$ is an independent sequence and is independent of $\{X_{im} : i = 1, \ldots, k; m = 1, 2, \ldots\}$, then the same is true for $\{U_l : l = 1, 2, \ldots\}$. Theorem 3.4.3 does not apply directly because the selection at stage m in the current setting may depend on U_m as well as on $U_1, \ldots, U_{m-1}$. Nevertheless, an argument similar to the proof of Theorem 3.4.3 applies to show that the discount factors may, without loss, be replaced by their expected values. This argument uses the fact that the discounting is exponential.

The situation is more complicated for real-time *discount functions* that are not of the form $\exp(-\beta t)$. Suppose, for example, that

$$\alpha_t = \begin{cases} 1, & t \in [0, 1] \\ 0, & t \in (1, \infty); \end{cases}$$

in a clinical trial the objective is to maximize the sum of the responses of all patients arriving in $[0, 1]$. If the first stage occurs at time 0.01 the arm selected may be very different than if it occurs at time 0.99; a 'risky' arm may be appropriate in the former and an arm with large mean may be a clear choice in the latter.

Allowing strategies that depend on real time t does not fit into the framework we have developed. But, when $\{Y_l : l = 1, 2, \ldots\}$, as defined above, is independent of $\{X_{im} : i = 1, \ldots, k; m = 1, 2, \ldots\}$ and is a sequence of independent exponentially distributed random variables having mean $\kappa > 0$, a simple artifice gives an arbitrarily good approximation using a bandit as described in Chapter 2.

We let $\varepsilon > 0$ and construct a bandit whose mth stage occurs at real time $m\varepsilon$. If ε is small then we are introducing too many stages by having them occur at each integral multiple of ε. We would like to correct for this by concealing observations so that the lengths of intervals between successive unconcealed observations are independent random variables having mean κ and distributions that approach the exponential as $\varepsilon \downarrow 0$.

Concealing observations is not a possibility reckoned with in Chapter 2, so we need to obtain the effect of concealment in another manner. Instead of concealing an observation at a particular stage, we suppose that a constant c_ε is observed that gives no information about the character of the various arms. For this purpose we consider a number c_ε such that $P(X_{im} = c_\varepsilon | G) = 0$ for $i = 1, \ldots, k$. To allow for the observation c_ε on any arm we need to replace $G \in \mathscr{D}^*(\mathscr{D}^k)$ by an appropriate $G_\varepsilon \in \mathscr{D}^*(\mathscr{D}^k)$. Define $\psi_\varepsilon : \mathscr{D}^k \to \mathscr{D}^k$ by

$$\psi_\varepsilon(Q_1, \ldots, Q_k) = \left(\frac{\varepsilon}{\kappa}Q_1 + \left(1 - \frac{\varepsilon}{\kappa}\right)\delta_{c_\varepsilon}, \ldots, \frac{\varepsilon}{\kappa}Q_k + \left(1 - \frac{\varepsilon}{\kappa}\right)\delta_{c_\varepsilon}\right)$$

and let $G_\varepsilon = G \circ \psi_\varepsilon^{-1}$. For each i, $P(X_{im} = c_\varepsilon | G_\varepsilon) = 1 - \varepsilon/\kappa$ as desired. In addition, the conditional distribution $(c_\varepsilon)_i G_\varepsilon$ equals G_ε. This last fact is needed since we want the 'concealed observations' to contain no information about the arms.

The unwanted observations of c_ε, while not being relevant for strategies, do contribute the quantity

$$\sum_{m=1}^{\infty} c_\varepsilon (1 - \varepsilon/\kappa)\alpha_{\varepsilon m} = \varepsilon^{-1} c_\varepsilon (1 - \varepsilon/\kappa) \sum_{m=1}^{\infty} \alpha_{\varepsilon m}\varepsilon \qquad (3.5.1)$$

to the worth of any strategy. We want (3.5.1) to approach 0 as $\varepsilon \downarrow 0$ so we suppose, as is natural, that $\int_0^\infty \alpha_t \, dt < \infty$ and we choose c_ε so that $\varepsilon^{-1} c_\varepsilon \to 0$.

In (3.5.1), and implicitly elsewhere, we have been using the discount sequence $\mathbf{A}_\varepsilon = (\alpha_\varepsilon, \alpha_{2\varepsilon}, \ldots)$ when the distribution is G_ε. Let τ_ε denote an optimal strategy for the $(G_\varepsilon; \mathbf{A}_\varepsilon)$-bandit. We would like to let $\varepsilon \downarrow 0$ to obtain an optimal strategy for the original setting, which we call the $(G; \alpha_t, \kappa)$-bandit. To accomplish this we make the following definition for the $(G; \alpha_t, \kappa)$-bandit. A *strategy* is a function that assigns an integer indicating the arm to be selected at the mth stage to each sequence $t_1 < \ldots < t_m$ of observation times and each (partial) history $z_1, \ldots, z_{m-1}$ of observations. Thus $\tau(\varnothing; t_1)$ indicates the arm to be selected at stage 1 if stage 1 occurs at time t_1; $\tau(z_1; t_1, t_2)$ indicates the arm to be selected at stage 2 if stage 2 occurs at time t_2 and z_1 was observed at time t_1 (necessarily on arm $\tau(\varnothing; t_1)$); etc. The definition of the $(G; \alpha_t, \kappa)$-bandit is now complete; we denote its value by $V(G; \alpha_t, \kappa)$.

Since the optimal strategy τ_ε for the $(G_\varepsilon; \mathbf{A}_\varepsilon)$-bandit is not a strategy for the $(G; \alpha_t, \kappa)$-bandit according to the preceding definition, we define a strategy $\tau_{\varepsilon,\kappa}$ for the $(G; \alpha_t, \kappa)$-bandit that is closely related to τ_ε. The observations of c_ε for the $(G_\varepsilon; \mathbf{A}_\varepsilon)$-bandit must not be used in the definition of $\tau_{\varepsilon,\kappa}$ except as timekeepers. Set

$$\tau_{\varepsilon,\kappa}(z_1, \ldots, z_{m-1}; t_1, \ldots, t_m)$$
$$= \tau_\varepsilon(c_\varepsilon, \ldots, c_\varepsilon, z_1, c_\varepsilon, \ldots, c_\varepsilon, z_2, \ldots, z_{m-1}, c_\varepsilon, \ldots, c_\varepsilon)$$

where $z_1, \ldots, z_{m-1}$ occur at positions $[t_1/\varepsilon], \ldots, [t_{m-1}/\varepsilon]$ and the number of positions is $[t_m/\varepsilon] - 1$; here $[t]$ denotes the greatest integer no larger than t.

The preceding discussion leads to the following theorem, the formal proof of which we will omit.

Theorem 3.5.1 Suppose the discount function α_t is nonnegative, of bounded variation, and that it satisfies $\int_0^\infty \alpha_t \, dt < \infty$. For $\varepsilon > 0$ let c_ε satisfy $P(X_{il} = c_\varepsilon | G) = 0$ for $i = 1, \ldots, k$, and suppose that $\varepsilon^{-1} c_\varepsilon \to 0$ as $\varepsilon \downarrow 0$. Then

$$V(G; \alpha_t, \kappa) = \lim_{\varepsilon \downarrow 0} V(G_\varepsilon; \mathbf{A}_\varepsilon).$$

In addition, τ_κ is an optimal strategy for the $(G; \alpha_t, \kappa)$-bandit, where τ_κ is defined by

$$\tau_\kappa(z_1, \ldots, z_{m-1}; t_1, \ldots, t_m) \qquad (3.5.2)$$
$$= \limsup \tau_{\kappa,\varepsilon}(z_1, \ldots, z_{m-1}; t_1, \ldots, t_m)$$
$$(\varepsilon \downarrow 0 \text{ through a fixed sequence}).$$

Remarks If $P(X_{im} = 0|G) = 0$ for each i, then c_ε can be chosen to be 0 for each ε. The lim sup in (3.5.2) can be replaced by lim inf or any other scheme that chooses an i for which $i = \tau_{\varepsilon,\kappa}(z_1, \ldots, z_{m-1}; t_1, \ldots, t_m)$ for infinitely many ε in the sequence. $\qquad\qquad\square$

3.6 Nonmonotone discount sequences

In many potential applications of bandit problems, future observations are worth less than the current one. The corresponding discount sequence is decreasing.

In other applications certain future observations are worth more than the current one. Future investments may necessarily be larger than they are now. Or in a clinical trial, the frequency of patient arrivals may be increasing, necessitating future treatment in larger batches.

Most of the bandit literature – and the sequential decision theory literature generally – concerns monotone discount sequences. The principal exception is the sequence given by $\alpha_n = 1, \alpha_m = 0, m \neq n$ (cf. Example 2.4.4): the only observation with payoff is the one at stage n and the first $n-1$ observations are made sequentially with the sole objective of obtaining information to aid in the nth selection. While information and payoff separate nicely in this problem, it is still quite difficult. When $\mathbf{A} = (0, \ldots, 0, 1, 0, 0, \ldots)$, the problem is similar to a conventional problem in decision theory: there is a sampling period followed by a terminal decision. The terminal decision is simply to choose the best arm. The sampling is sequential with a restriction on the total number of observations. Some appropriate references are Lindley and Barnett (1965), Ray (1965), Pratt (1966), and Clayton (1983).

References

Berry, D. A. (1983) Bandit problems with random discounting. *Mathematical Learning Models – Theory and Algorithms* (eds U. Herkenrath, D. Kalin and W. Vogel), pp. 12–25, Springer-Verlag, New York.

Clayton, M. K. (1983) Bayes sequential sampling for choosing the better of two populations. Ph.D. thesis, Univ. of Minnesota, USA.

Lindley, D. V. and Barnett, B. N. (1965) Sequential sampling: two decision problems with linear losses for binomial and normal random variables. *Biometrika* **52**: 507–532.

Pratt, J. W. (1966) The outer needle of some Bayes sequential continuation regions. *Biometrika* **53**: 455–467.

Ray, S. N. (2965) Bounds on the maximal sample size of a Bayes sequential procedure. *Ann. Math. Statist.* **36**: 859–878.

CHAPTER 4

Independent Bernoulli arms

Many of the examples in the first three chapters assume independent arms: $G = F_1 \times \ldots \times F_k$. In this chapter we consider independent Bernoulli arms. As has been our convention in the Bernoulli case, we regard F_i as a distribution on the Bernoulli parameter $\theta_i \in [0, 1]$ rather than on $Q_i \in \mathscr{D}$; and consistent with an earlier modification of notation, we write the conditional distribution of $(\theta_1, \theta_2, \ldots, \theta_k)$ given success on arm 1, say, as

$$\sigma_1(F_1, F_2, \ldots, F_k) = (\sigma F_1, F_2, \ldots, F_k),$$

and given a failure on arm 1 and as

$$\varphi_1(F_1, F_2, \ldots, F_k) = (\varphi F_1, F_2, \ldots, F_k).$$

Some of the definitions and results in this chapter also apply when the F_i are arbitrary, with, perhaps, some modification. These will be given without announcing such generality; various extensions will be developed as needed in later chapters. Also, some of the results can be extended to certain kinds of dependence among the arms, but most do not apply more generally. One result (Corollary 4.3.10) is given for a particular class of distributions for which the arms are dependent.

Some results in this chapter have intrinsic importance; notable examples are Theorems 4.1.6, 4.3.6, 4.3.8, and 4.3.9. But the primary purpose of the chapter is to set the stage and develop technical foundations for the following three chapters, which deal exclusively with settings in which the arms are independent. In Chapter 5, **A** is arbitrary, $k = 2$, and F_2 associates all its mass with a constant. In Chapter 6, **A** is geometric and k is arbitrary. In Chapter 7, **A** is finite horizon uniform, $k = 2$, and the arms are Bernoulli.

The most efficient way to read this chapter will depend on the reader. Though we give examples during the discourse, the primary motivation is frequently provided by results in Chapters 5, 6, and 7. So the reader may choose to skim this chapter first, paying special attention to the theorems mentioned above. Such a reader can refer back later for more serious but selective perusal.

4.1 Monotonicity of the value function

The purpose of this section is to describe the behaviour of V in the independent case. We shall repeat some of the notation and terminology of Chapter 2 as it applies to the current setting. The worth of strategy τ depends on the initial state of information $(F_1, \ldots, F_k; \mathbf{A})$ and is written $W(F_1, \ldots, F_k; \mathbf{A}; \tau)$. Recall that

$$V(F_1, \ldots, F_k; \mathbf{A}) = \sup_{\tau} W(F_1, \ldots, F_k; \mathbf{A}; \tau).$$

We saw in Example 2.5.1 that V is not always continuous in the distribution G. However, it is continuous on the restricted domain of the present setting, as we prove in the following theorem.

Theorem 4.1.1 In the independent Bernoulli case, the value V is a continuous function of $(F_1, \ldots, F_k; \mathbf{A})$.

Proof We must compare two bandits, say $(F_1, \ldots, F_k; \mathbf{A})$ and $(F_1^*, \ldots, F_k^*; \mathbf{A}^*)$, and then take the suprema of their worths over τ. We omit the details, noting only the method of proof. For any fixed $\varepsilon > 0$, n, τ and history of n observations, the probabilities that the two bandits yield this history for the first n trials (when strategy τ is used) differ by less than ε if $(F_1^*, \ldots, F_k^*)$ is sufficiently close to $(F_1, \ldots, F_k)$. (This also gives uniform continuity assuming one of the usual metrics for convergence in distribution.) $\square$

The nonnegativity of each X_{im} immediately gives the next result.

Theorem 4.1.2 The value V is an increasing function of $\mathbf{A}$.

The following example illustrates that monotonicity of V in $(F_1, \ldots, F_k)$ can be tricky.

Example 4.1.1 Let $k = 2$ and $\mathbf{A} = (1, \alpha, \alpha^2, \ldots)$, where $\alpha = 0.95$. Suppose θ_2 is known to be 0.7, that is, $F_2 = \delta_{0.7}$. Consider two different distributions for arm 1: F_1^* assigns probability $1/2$ to each of

1/4 and 3/4, that is, $F_1^* = \frac{1}{2}\delta_{1/4} + \frac{1}{2}\delta_{3/4}$; whereas $F_1 = \frac{1}{2}\delta_0 + \frac{1}{2}\delta_{3/4}$. From Example 5.4.1 it will follow that arm 2 is always optimal for the $(F_1^*, F_2; \mathbf{A})$-bandit and hence, that

$$V(F_1^*, F_2; \mathbf{A}) = 0.7 \sum_{m=1}^{\infty} \alpha^{m-1} = 0.7/(1 - 0.95) = 14.$$

Define strategy τ as follows: $\tau(\emptyset) = 1$ and, for $m > 2$, $\tau(z_1, \ldots, z_{m-1}) = 1$ if $z_1 = 1$ and $\tau(z_1, \ldots, z_{m-1}) = 2$ if $z_1 = 0$. We have

$$W(F_1, F_2; \mathbf{A}; \tau) = \frac{449}{32} > 14$$

and, hence, $V(F_1, F_2; \mathbf{A}) > 14$. (It happens that τ is optimal but this is incidental to our purpose.)

This is a rather surprising conclusion for it indicates that the $(F_1, F_2; \mathbf{A})$-bandit is preferable to the $(F_1^*, F_2; \mathbf{A})$-bandit even though θ_1 is stochastically larger under F_1^* than under F_1. □

The first of the following two definitions is equivalent to stochastic ordering of random variables and the second is motivated by the preceding example. The importance of the second definition will be made clear in Chapter 5. The second definition will be extended in Section 4.3; this extension will play an important role in Chapter 7.

Definition 4.1.1 A one-dimensional distribution function F^* is *to the right of* a one-dimensional distribution function F if $F^*(x) \leqslant F(x)$ for every x; that is, in terms of measures, $F^*[x, \infty) \geqslant F[x, \infty)$ for every x. If, in addition, $F^* \neq F$, this relationship is *strict*.

Recall from Chapter 2 that $\sigma^s \varphi^f F$ is the current distribution of an arm that has yielded s successes and f failures and whose prior distribution is F.

Definition 4.1.2 A one-dimensional distribution F^* on $[0, 1]$ is *strongly to the right of* a one-dimensional distribution F on $[0, 1]$ if $\sigma^s \varphi^f F^*$ is to the right of $\sigma^s \varphi^f F$ for every pair (s, f) of nonnegative integers for which both $\sigma^s \varphi^f F^*$ and $\sigma^s \varphi^f F$ are defined. If, in addition, $F^* \neq F$, this relationship is *strict*.

In Example 4.1.1 the distribution F_1^* is to the right of F_1, but it is not strongly to the right:

$$\sigma F_1(\{3/4\}) = 1 > \sigma F_1^*[3/4, 1] = 3/4.$$

The following three easy lemmas are given without proof. The first says that the probability of success is at least as large under F^* as under F when F^* is to the right of F. The second says that 'strongly to the right' is preserved under identical experimental results. The third asserts that successes on arm 1 move the distribution of θ_1 strongly to the right and that the opposite happens with failures.

As a natural adjustment of notation we write $E(\theta_1|F_1)$, for example, in lieu of $E(\theta_1|G)$ when $G = (F_1, \ldots, F_k)$, for this expectation depends on G only through F_1.

Lemma 4.1.3 If F_1^* is to the right of F_1 then $E(\theta_1|F_1^*) \geqslant E(\theta_1|F_1)$ with equality if and only if $F_1^* = F_1$.

For any one-dimensional distribution F, if $F = \delta_0$ then σF is not defined, and the same is true for φF when $F = \delta_1$. The next two lemmas do not cover these possibilities for F or F^*.

Lemma 4.1.4 If F^* is strongly to the right of F then σF^* and φF^* are, respectively, strongly to the right of σF and φF (when all these distributions exist).

Lemma 4.1.5 For any F, σF is strongly to the right of F, which is strongly to the right of φF (when these distributions exist). The relationships are strict if and only if F is not concentrated at one point.

While Example 4.1.1 rules out monotonicity of V in F_i with respect to 'to the right', the following theorem gives monotonicity of V with respect to 'strongly to the right'. In view of Lemma 4.1.5 this implies that the value is not decreased if the number of successes on an arm were to be increased or the number of failures decreased. The result has inherent significance but is not of great importance in the sequel; a number of papers discussed in the Annotated Bibliography prove special cases.

Theorem 4.1.6 Suppose distribution F_i^* is strongly to the right of F_i for $i = 1, \ldots, k$. Then for any $\mathbf{A}$,

$$V(F_1^*, \ldots, F_k^*; \mathbf{A}) \geqslant V(F_1, \ldots, F_k; \mathbf{A}).$$

Remarks The following proof of this result exploits Lemma 4.1.4, which says that 'strongly to the right' is preserved in the two bandits from one stage to the next. It uses truncation and induction on the horizon of the truncated sequence.   □

Proof of Theorem 4.1.6 For an inductive proof, assume first that the horizon n of $\mathbf{A}$ is 0. Then $\mathbf{A} = (0, 0, \ldots)$ and hence, $V(F_1^*, \ldots, F_k^*; \mathbf{A}) = 0 = V(F_1, \ldots, F_k; \mathbf{A})$.

Take $n \geqslant 1$ and assume that the conclusion of the theorem holds for all discount sequences having horizon less than n. Suppose the horizon of $\mathbf{A}$ is n and let τ denote an optimal strategy for the $(F_1, \ldots, F_k; \mathbf{A})$-bandit. Define τ^* to be a strategy for the $(F_1^*, \ldots, F_k^*; \mathbf{A})$-bandit that specifies the same first selection as τ and is optimal for the new bandit presenting itself at stage 2. We need only show

$$W(F_1^*, \ldots, F_k^*; \mathbf{A}; \tau^*) \geqslant V(F_1, \ldots, F_k; \mathbf{A}).$$

For notational ease assume $\tau(\varnothing) = 1$, and therefore $\tau^*(\varnothing) = 1$. Then from (2.5.2) we have

$$
\begin{aligned}
W(F_1^*, \ldots, F_k^*; \mathbf{A}; \tau^*) = {} & \alpha_1 E(\theta_1 | F_1^*) \\
& + E(\theta_1 | F_1^*) \, V(\sigma F_1^*, F_2^*, \ldots, F_k^*; \mathbf{A}^{(1)}) \\
& + E(1 - \theta_1 | F_1^*) \, V(\varphi F_1^*, F_2^*, \ldots, F_k^*; \mathbf{A}^{(1)}).
\end{aligned}
$$

Since τ is optimal in the $(F_1, \ldots, F_k; \mathbf{A})$-bandit,

$$
\begin{aligned}
V(F_1, \ldots, F_k; \mathbf{A}) = {} & \alpha_1 E(\theta_1 | F_1) \\
& + E(\theta_1 | F_1) \, V(\sigma F_1, F_2, \ldots, F_k; \mathbf{A}^{(1)}) \\
& + E(1 - \theta_1 | F_1) \, V(\varphi F_1, F_2, \ldots, F_k; \mathbf{A}^{(1)}).
\end{aligned}
$$

We want to show that the following is nonnegative:

$$
\begin{aligned}
W(F_1^*, \ldots, & F_k^*; \mathbf{A}; \tau^*) - V(F_1, \ldots, F_k; \mathbf{A}) \\
= {} & \alpha_1 [E(\theta_1 | F_1^*) - E(\theta_1 | F_1)] \\
& + E(\theta_1 | F_1) [V(\sigma F_1^*, F_2^*, \ldots, F_k^*; \mathbf{A}^{(1)}) \\
& \qquad\qquad\qquad - V(\sigma F_1, F_2, \ldots, F_k; \mathbf{A}^{(1)})]
\end{aligned}
$$

$$+ E(1 - \theta_1 | F_1^*)[V(\varphi F_1^*, F_2^*, \ldots, F_k^*; \mathbf{A}^{(1)})$$
$$- V(\varphi F_1, F_2, \ldots, F_k; \mathbf{A}^{(1)})]$$
$$+ [E(\theta_1 | F_1^*) - E(\theta_1 | F_1)][V(\sigma F_1^*, F_2^*, \ldots, F_k^*; \mathbf{A}^{(1)})$$
$$- V(\varphi F_1, F_2, \ldots, F_k; \mathbf{A}^{(1)})]. \qquad (4.1.1)$$

The first term on the right-hand side of (4.1.1) is nonnegative by Lemma 4.1.3. The next two terms are nonnegative by Lemma 4.1.4 and the induction hypothesis. That the last term is nonnegative follows from Lemmas 4.1.3, 4.1.4, and 4.1.5, and the induction hypothesis.

The result for a discount sequence with infinite horizon now follows from the continuity of V in $\mathbf{A}$ (Theorem 4.1.1).   □

In the next section we assume $k = 2$ and introduce a notation for the difference in worths between the two initial selections. We give a recursion for this difference that is useful in later demonstrations concerning the nature of optimal strategies.

4.2   The advantage of one arm over another

In this section we specialize to $k = 2$ and define a function of $(F_1, F_2; \mathbf{A})$ which represents the difference between selecting arm 1 followed by an optimal continuation and arm 2 followed by an optimal continuation. Analogous functions can be defined when k is arbitrary – see Quisel (1965) – but we have had little success with these generalizations.

There are two possible initial selections: arm 1 and arm 2. As in Theorem 2.5.1, let $V^{(i)}$ denote the worth of selecting arm i initially and then continuing with an optimal strategy:

$$V^{(i)}(F_1, F_2; \mathbf{A}) = \sup_{\tau(\varnothing) = i} W(F_1, F_2; \mathbf{A}; \tau).$$

From (2.5.2), the value $V(F_1, F_2; \mathbf{A})$ is the maximum of these two numbers. We have

$$V^{(1)}(F_1, F_2; \mathbf{A}) = \alpha_1 E(\theta_1 | F_1) + E(\theta_1 | F_1) V(\sigma F_1, F_2; \mathbf{A}^{(1)})$$
$$+ E(1 - \theta_1 | F_1) V(\varphi F_1, F_2; \mathbf{A}^{(1)}), \qquad (4.2.1)$$

$$V^{(2)}(F_1, F_2; \mathbf{A}) = \alpha_1 E(\theta_2 | F_2) + E(\theta_2 | F_2) V(F_1, \sigma F_2; \mathbf{A}^{(1)})$$
$$+ E(1 - \theta_2 | F_2) V(F_1, \varphi F_2; \mathbf{A}^{(1)}). \qquad (4.2.2)$$

Let $\Delta(F_1, F_2; \mathbf{A})$ denote the advantage of arm 1 over arm 2 assuming optimal continuations in both cases:

$$\Delta(F_1, F_2; \mathbf{A}) = V^{(1)}(F_1, F_2; \mathbf{A}) - V^{(2)}(F_1, F_2; \mathbf{A}).$$

The sign of Δ indicates the optimal initial selection: arm 1 if $\Delta \geqslant 0$ and arm 2 if $\Delta \leqslant 0$. In fact, the sign of Δ was used in Example 2.4.3 to display an optimal strategy. But as the following development makes clear, the sign of $\Delta(F_1, F_2; \mathbf{A})$ depends on the magnitude of $\Delta(\sigma F_1, F_2; \mathbf{A}^{(1)})$, for example, and not just its sign. Therefore, we need to consider the magnitude of the Δ function when we want to find optimal strategies.

For all $\mathbf{A}$ and any (F_1, F_2),

$$V^{(1)}(F_1, F_2; \mathbf{A}) = V(F_1, F_2; \mathbf{A}) - \Delta^-(F_1, F_2; \mathbf{A}), \qquad (4.2.3)$$

$$V^{(2)}(F_1, F_2; \mathbf{A}) = V(F_1, F_2; \mathbf{A}) - \Delta^+(F_1, F_2; \mathbf{A}), \qquad (4.2.4)$$

where $\Delta^+ = 0 \vee \Delta$ and $\Delta^- = 0 \vee (-\Delta)$. In view of (4.2.4) and (4.2.3), respectively, (4.2.1) and (4.2.2) become

$$
\begin{aligned}
V^{(1)}(F_1, F_2; \mathbf{A}) = {}& \alpha_1 E(\theta_1 | F_1) \\
& + E(\theta_1 | F_1)[V^{(2)}(\sigma F_1, F_2; \mathbf{A}^{(1)}) + \Delta^+(\sigma F_1, F_2; \mathbf{A}^{(1)})] \\
& + E(1 - \theta_1 | F_1)[V^{(2)}(\varphi F_1, F_2; \mathbf{A}^{(1)}) + \Delta^+(\varphi F_1, F_2; \mathbf{A}^{(1)})].
\end{aligned}
$$
$$(4.2.5)$$

$$
\begin{aligned}
V^{(2)}(F_1, F_2; \mathbf{A}) = {}& \alpha_1 E(\theta_2 | F_2) \\
& + E(\theta_2 | F_2)[V^{(1)}(F_1, \sigma F_2; \mathbf{A}^{(1)}) + \Delta^-(F_1, \sigma F_2; \mathbf{A}^{(1)})] \\
& + E(1 - \theta_2 | F_2)[V^{(1)}(F_1, \varphi F_2; \mathbf{A}^{(1)}) + \Delta^-(F_1, \varphi F_2; \mathbf{A}^{(1)})].
\end{aligned}
$$
$$(4.2.6)$$

The sum

$$
\begin{aligned}
& \alpha_1 E(\theta_1 | F_1) + E(\theta_1 | F_1) V^{(2)}(\sigma F_1, F_2; \mathbf{A}^{(1)}) \\
& + E(1 - \theta_1 | F_1) V^{(2)}(\varphi F_1, F_2; \mathbf{A}^{(1)})
\end{aligned}
$$
$$(4.2.7)$$

in (4.2.5) amounts to the worth of selecting arm 1 first and arm 2 second and then continuing optimally. Likewise,

$$
\begin{aligned}
& \alpha_1 E(\theta_2 | F_2) + E(\theta_2 | F_2) V^{(1)}(F_1, \sigma F_2; \mathbf{A}^{(1)}) \\
& + E(1 - \theta_2 | F_2) V^{(1)}(F_1, \varphi F_2; \mathbf{A}^{(1)})
\end{aligned}
$$
$$(4.2.8)$$

in (4.2.6) is the expected worth of selecting arm 2 first and arm 1 second and then continuing optimally. Since the selections are exchanged in these two interpretations, the difference between (4.2.7) and (4.2.8) is $(\alpha_1 - \alpha_2)[E(\theta_1 | F_1) - E(\theta_2 | F_2)]$. Therefore, subtracting (4.2.6) from (4.2.5) gives the following lemma.

Lemma 4.2.1 When there are two independent Bernoulli arms,

$$
\begin{aligned}
\Delta(F_1, F_2; \mathbf{A}) = {} & (\alpha_1 - \alpha_2)\left[E(\theta_1 \mid F_1) - E(\theta_2 \mid F_2)\right] \\
& + E(\theta_1 \mid F_1)\Delta^+(\sigma F_1, F_2; \mathbf{A}^{(1)}) \\
& + E(1 - \theta_1 \mid F_1)\Delta^+(\varphi F_1, F_2; \mathbf{A}^{(1)}) \\
& - E(\theta_2 \mid F_2)\Delta^-(F_1, \sigma F_2; \mathbf{A}^{(1)}) \\
& - E(1 - \theta_2 \mid F_2)\Delta^-(F_1, \varphi F_2; \mathbf{A}^{(1)}). \qquad (4.2.9)
\end{aligned}
$$

Remark In the repeated application of (4.2.9) certain Δ's will not be defined when F_1 or F_2 gives zero probability to the open interval (0,1). For example, $\varphi \sigma F_1$ is not defined when $F_1(\{1, 0\}) = 1$. But the multiplier of such a term is always 0, and our convention is that the product is 0.   □

A rather easy consequence of later development is that not all terms in (4.2.9) can be zero, except when F_1 and F_2 are the same one-point distribution. In addition, it seems reasonable to expect that the vanishing of particular terms in (4.2.9) implies the vanishing of other terms. In fact, when $\mathbf{A}$ is nonincreasing, it will follow from Theorem 4.3.6 that $\Delta(\sigma F_1, F_2; \mathbf{A}) > 0$ whenever $\Delta(\varphi F_1, F_2; \mathbf{A}) > 0$ and, symmetrically, $\Delta(F_1, \sigma F_2; \mathbf{A}) < 0$ whenever $\Delta(F_1, \varphi F_2; \mathbf{A}) < 0$. So an arm that is optimal after a failure is also optimal after a success.

When the horizon n is finite, (4.2.9) defines Δ recursively. The uniform case (in which the objective is to maximize the expected successes in the first n observations) is treated in Chapter 7. In this case the right-hand side of (4.2.9) reduces to the first term when $n = 1$; when $n \geqslant 2$ the first term is zero, and the other terms apply. A detailed discussion of the application of (4.2.9) when $n = 2$ is given in Section 7.2.

When the horizon is 1, (4.2.9) makes it clear that Δ depends on F_1 and F_2 only through their first moments. Applying induction to (4.2.9) on the horizon n of $\mathbf{A}$ shows that $\Delta(F_1, F_2; \mathbf{A})$ depends on F_1 and F_2 only through their first n moments.

Proposition 4.2.2 If the horizon of $\mathbf{A}$ is n and $E(\theta_1^r \mid F_1^*) = E(\theta_1^r \mid F_1)$ for $r = 1, 2, \ldots, n$, then $\Delta(F_1^*, F_2; \mathbf{A}) = \Delta(F_1, F_2; \mathbf{A})$.

4.3 Staying with a winner

Since the arms are Bernoulli, the only possible outcomes when an arm is selected are success and failure. When the arm selected has known probability of success, we are indifferent to these two outcomes, except for the difference in immediate income. But, in view of Theorem 4.1.6, the information contained in a success is distinctly preferred to that of failure when the arm is unknown since σF is strongly to the right of φF (Lemma 4.1.5). However, this preference may not translate into a desire to stay with the successful arm. Example 2.4.1 gives a setting with dependent arms in which it does not translate. As we will see (Examples 5.2.1 and 5.2.2), even when the arms are independent, it may not be optimal to stay with a winner.

A detailed analysis of this issue is carried out in Chapter 5 when only one arm is unknown. The current section gives a partial characterization of discount sequences for which staying with a winner is a property of optimal strategies when there are two independent Bernoulli arms. This result, and most of the other results in this section, assumes that $\mathbf{A}$ is nonincreasing, $\alpha_m \geq \alpha_{m+1}$ for $m \geq 1$. Examples 4.3.3 and 4.4.4 clarify the role of this assumption.

The 'stay-with-a-winner' rule, Theorem 4.3.8, generalizes Theorem 6.2 of Berry (1972) which is the corresponding result for finite horizon uniform discounting. The current proof is more direct and simpler. It uses the fact that $\Delta(F_1^*, F_2; \mathbf{A}) \geq \Delta(F_1, F_2; \mathbf{A})$ if F_1^* is strongly to the right of F_1, which is contained in Theorem 4.3.3. Because a stronger result is needed in Chapter 7, the forthcoming Theorem 4.3.3 uses a weaker notion of order among distribution measures; this notion is defined next. Since this definition applies to any arm, we drop the subscript from the Bernoulli parameter θ as well as from its distribution F.

Definition 4.3.1 Let $m \geq 0$. For one-dimensional distribution measures, $F^* \overset{m}{\succeq} F$ (read F^* *m-greater than F*) if

$$E(\theta|\sigma^s \varphi^f F^*) \geq E(\theta|\sigma^s \varphi^f F) \tag{4.3.1}$$

whenever $s + f \leq m$ and $\sigma^s \varphi^f F^*$ and $\sigma^s \varphi^f F$ are both defined. If, in addition, $E(\theta|F^*) > E(\theta|F)$, then $F^* \overset{m}{\succ} F$ (read F^* *strictly m-greater than F*).

The following two examples illustrate this notion. The first is trivial in that there is obviously no stochastic ordering even though the first several moments under F^* are larger than under F. The second has the flavour of Example 4.3.1 since one distribution measure is to the right of the other but it is not strongly to the right. Repeated failures eventually reverse the inequality between the means in the first example, while successes do so in the second.

Example 4.3.1 Suppose F^* has density

$$dF^*(u) \propto u^9(1-u)^4 \, du,$$

on $(0, 1)$ and $F = \delta_{1/2}$. Then

$$E(\theta | \sigma^s \varphi^f F^*) = \frac{s+10}{s+f+15}, \qquad E(\theta | \sigma^s \varphi^f F) = \frac{1}{2}.$$

So $F^* \overset{m}{\succsim} F$ for $m \leqslant 5$ and F^* and F are not comparable for $m > 5$. □

Example 4.3.2 Consider the family of two-point distributions F_x for $x \in (0, 1)$ where $F_x = \frac{1}{2}\delta_x + \frac{1}{2}\delta_1$. Then

$$E(\theta | \sigma^s \varphi^f F_x) = \begin{cases} x & \text{if } f > 0 \\ \dfrac{1+x^{s+1}}{1+x^s} & \text{if } f = 0. \end{cases}$$

So $F_y \overset{m}{\succsim} F_x$ provided y is sufficiently larger than x, but for fixed x and y this relation does not hold for arbitrarily large m. For example, $F_{0.9} \overset{m}{\succsim} F_x$ only for $x \in [x^*, 0.9]$ where x^* is given to two decimals in the following table:

m	0	1	2	3	4	10	∞
x^*	0	0.05	0.25	0.43	0.56	0.87	0.90

The reason small values of x are excluded here is the same reason the order of the V's in Example 4.1.1 is opposite from that of naive expectation: when a success is observed, θ is more likely to be 1 under F_x for smaller x (or more likely to be 3/4 in Example 4.1.1). □

The following two propositions clarify the strict version of Definition 4.3.1 and relate 'm-greater than' with 'strongly to the right' (Definition 4.1.2). The proof of the first is easy and is omitted.

Proposition 4.3.1 If $F^* \overset{m}{\underset{\sim}{\geq}} F$ and (4.3.1) is strict for some (s, f) with $s + f \leq m$, then $F^* \overset{m}{\underset{\sim}{>}} F$.

Proposition 4.3.2 A distribution F^* is strongly to the right of F if and only if $F^* \overset{m}{\underset{\sim}{\geq}} F$ for every m. The first relationship is strict if and only if the second is strict for every m.

In view of Proposition 4.3.2 we use '∞-greater than' synonymously with 'strongly to the right'. The proof of Proposition 4.3.2 is omitted. The 'only if' part is easy and relevant, for example, in connection with the subsequent Corollary 4.3.7 from which it follows that arm 1 is optimal if F_1 is strongly to the right of F_2. The 'if' part is more difficult to prove and will not be used in the sequel.

When the horizon n is finite it would not be surprising, in view of Propositions 4.2.2 and 4.3.2, that a general theorem with a hypothesis involving 'strongly to the right' could be strengthened by using '$(n-1)$-greater than' instead. For example, the next theorem strengthens Theorem 4.1.6 in this way. While it is true for $k \geq 2$, it is stated for $k = 2$ because that is the setting of this section. The proof mimics that of Theorem 4.1.6 in the obvious way and so is omitted.

Theorem 4.3.3 Suppose that **A** has horizon n and distribution F_i^* is $(n-1)$-greater than F_i for $i = 1, 2$. Then

$$V(F_1^*, F_2^*; \mathbf{A}) \geq V(F_1, F_2; \mathbf{A}).$$

We will require the following two lemmas that are analogous to Lemmas 4.1.4 and 4.1.5. We omit the proof of Lemma 4.3.4. Lemma 4.3.5 is a logical consequence of Lemma 4.1.5.

Lemma 4.3.4 For $m \geq 1$, if $F^* \overset{m}{\underset{\sim}{\geq}} F$ then $\sigma F^* \overset{m-1}{\underset{\sim}{\geq}} \sigma F$ and $\varphi F^* \overset{m-1}{\underset{\sim}{\geq}} \varphi F$ (whenever these distributions exist).

Lemma 4.3.5 For any F and $m \geq 0$, $\sigma F \overset{m}{\underset{\sim}{\geq}} F \overset{m}{\underset{\sim}{\geq}} \varphi F$ (whenever these distributions exist). These relationships are strict unless F is concentrated at one point.

It is convenient to have a terminology for functions of distribution measures that are monotonic with respect to 'm-greater than':

Definition 4.3.2 Let $m \in \{1, 2, \ldots, \infty\}$. A function h on the set of distribution measures on $[0, 1]$ is *m-increasing* if $h(F^*) \geqslant h(F)$ when $F^* \overset{m}{\underset{\sim}{\succ}} F$. It is *strictly m-increasing* if, in addition, $h(F^*) > h(F)$ whenever $F^* \overset{m}{\succ} F$. Similarly, h is *m-decreasing* or *strictly m-decreasing* according as $-h$ is *m*-increasing or strictly *m*-increasing.

Using this terminology, in view of Theorem 4.1.6 the conclusion of Theorem 4.3.3 is that V is $(n-1)$-increasing in either F_1 or F_2.

The most important step in proving the stay-with-a-winner rule is the next theorem.

Theorem 4.3.6 Assume $\mathbf{A}$ is nonincreasing and has horizon $n \in \{1, 2, \ldots, \infty\}$. For fixed F_2, $\Delta(F_1, F_2; \mathbf{A})$ is a strictly $(n-1)$-increasing function of F_1.

Remark By symmetry, $\Delta(F_1, F_2; \mathbf{A})$ is strictly $(n-1)$-decreasing in F_2. $\square$

Proof The parts of the proof are labelled from (i) to (iv). In (i) we use Lemma 4.2.1 to obtain an expression for $\Delta(F_1^*, F_2; \mathbf{A}) - \Delta(F_1, F_2; \mathbf{A})$ that we will apply when $F_1^* \overset{n-1}{\underset{\sim}{\succ}} F_1$. In (ii) we use induction on the horizon to show that

$$\Delta(F_1^*, F_2; \mathbf{A}) \geqslant \Delta(F_1, F_2; \mathbf{A}) \tag{4.3.2}$$

when $F_1^* \overset{n-1}{\underset{\sim}{\succ}} F_1$ and $n < \infty$. In (iii) we let the horizon approach ∞ to obtain (4.3.2) for all $\mathbf{A}$. Finally, in (iv) we use induction on the smallest m for which $\alpha_{m+1} < \alpha_1$ to prove $\Delta(F_1^*, F_2; \mathbf{A}) > \Delta(F_1, F_2; \mathbf{A})$ whenever $F_1^* \overset{n-1}{\succ} F_1$, that is, when $F_1^* \overset{n-1}{\underset{\sim}{\succ}} F_1$ and $E(\theta_1 | F_1^*) > E(\theta_1 | F_1)$. Throughout we will assume that neither F_1^* nor F_2 is supported by $\{0\}$ and that neither F_1 nor F_2 is supported by $\{1\}$. These cases are easily treated separately without induction.

(i) Lemma 4.2.1 gives

$$\begin{aligned}
\Delta(F_1^*, F_2; \mathbf{A}) - \Delta(F_1, F_2; \mathbf{A}) &= (\alpha_1 - \alpha_2)[E(\theta_1 | F_1^*) - E(\theta_1 | F_1)] \\
&+ E(\theta_1 | F_1^*)[\Delta^+(\sigma F_1^*, F_2; \mathbf{A}^{(1)}) - \Delta^+(\sigma F_1, F_2; \mathbf{A}^{(1)})] \\
&+ E(1 - \theta_1 | F_1^*)[\Delta^+(\varphi F_1^*, F_2; \mathbf{A}^{(1)}) - \Delta^+(\varphi F_1, F_2; \mathbf{A}^{(1)})] \\
&+ [E(\theta_1 | F_1^*) - E(\theta_1 | F_1)][\Delta^+(\sigma F_1, F_2; \mathbf{A}^{(1)}) \\
&\hspace{4cm} - \Delta^+(\varphi F_1, F_2; \mathbf{A}^{(1)})] \\
&+ E(\theta_2 | F_2)[\Delta^-(F_1, \sigma F_2; \mathbf{A}^{(1)}) - \Delta^-(F_1^*, \sigma F_2; \mathbf{A}^{(1)})] \\
&+ E(1 - \theta_2 | F_2)[\Delta^-(F_1, \varphi F_2; \mathbf{A}^{(1)}) - \Delta^-(F_1^*, \varphi F_2; \mathbf{A}^{(1)})].
\end{aligned}$$
$$\tag{4.3.3}$$

As indicated following Lemma 4.2.1, some of the quantities in (4.3.3) are not defined in case F_1^* is supported by $\{0\}$ or F_1 is supported by $\{1\}$. We interpret an undefined term multiplied by 0 to equal 0. We take $\Delta^+(\sigma F_1, F_2; \mathbf{A}^{(1)})$ to equal 0 when it is undefined and observe that this interpretation makes the second term on the right-hand side of (4.3.3) positive and the fourth term zero. So the remainder of the proof can proceed without further attention to this case.

(ii) In view of (4.2.9), $\Delta(F_1^*, F_2; \mathbf{A}) \geqslant \Delta(F_1, F_2; \mathbf{A})$ when $\mathbf{A}$ has horizon 1 and $F_1^* \overset{0}{\succsim} F_1$. Suppose $\mathbf{A}$ has finite horizon $n > 1$ and that $\Delta(F_1^*, F_2; \mathbf{A}^{(1)}) \geqslant \Delta(F_1, F_2; \mathbf{A}^{(1)})$ whenever $F_1^* \overset{n-2}{\succsim} F_1$. We now show that (4.3.3) is nonnegative when $F_1^* \overset{n-1}{\succsim} F_1$.

The first term on the right-hand side of (4.3.3) is nonnegative since (4.3.1) applies with $s = f = 0$ and since $\alpha_1 \geqslant \alpha_2$ by hypothesis. The next two and the last two terms are nonnegative for similar reasons; consider the second term for definiteness. Since $F_1^* \overset{n-1}{\succsim} F_1$ it follows immediately from Lemma 4.3.4 that $\sigma F_1^* \overset{n-2}{\succsim} \sigma F_1$. Therefore

$$\Delta(\sigma F_1^*, F_2; \mathbf{A}^{(1)}) \geqslant \Delta(\sigma F_1, F_2; \mathbf{A}^{(1)}) \tag{4.3.4}$$

by the inductive hypothesis since the horizon of $\mathbf{A}^{(1)}$ is $n - 1$. Since $\Delta^+ = \Delta \vee 0$, it follows that (4.3.4) holds as well with Δ^+ in place of Δ. (The proof that each of the last two terms in (4.3.3) is nonnegative is even easier. Lemma 4.3.4 need not be used; rather Definition 4.3.1 gives $F_1^* \overset{n-2}{\succsim} F_1$ as an immediate consequence of $F_1^* \overset{n-1}{\succsim} F_1$.)

The first factor in the fourth term is nonnegative as indicated in discussing the first term. The other factor in the fourth term is nonnegative by the inductive hypothesis since $\sigma F_1 \overset{n-2}{\succsim} \varphi F_1$ in view of Lemma 4.3.5.

Therefore,

$$\Delta(F_1^*, F_2; \mathbf{A}) - \Delta(F_1, F_2; \mathbf{A}) \geqslant 0.$$

(iii) Suppose $n = \infty$ and $F_1^* \overset{\infty}{\succsim} F_1$. Since $F_1^* \overset{n-1}{\succsim} F_1$ for each finite n, part (ii) gives

$$\Delta(F_1^*, F_2; \mathbf{A}_n) - \Delta(F_1, F_2; \mathbf{A}_n) \geqslant 0,$$

where $\mathbf{A}_n$ is the truncated version of $\mathbf{A}$ – it agrees with $\mathbf{A}$ through stage n and has only zeros thereafter. Let $n \to \infty$ to obtain

$$\Delta(F_1^*, F_2; \mathbf{A}) - \Delta(F_1, F_2; \mathbf{A}) \geqslant 0.$$

(iv) Let $m_* = \inf\{m: \alpha_{m+1} < \alpha_1\}$. Suppose $m_* = 1$ and that $F_1^* \overset{n-1}{>} F_1$, where n is the horizon $\mathbf{A}$. By parts (ii) and (iii), every term on the right-hand side of (4.3.3) is nonnegative and it is easily seen that the first term of (4.3.3) is positive. Hence,

$$\Delta(F_1^*, F_2; \mathbf{A}) > \Delta(F_1, F_2; \mathbf{A}).$$

Suppose m_* (necessarily finite) is larger than 1 and assume that $\Delta(F_1^*, F_2; \mathbf{A}^{(1)}) > \Delta(F_1, F_2; \mathbf{A}^{(1)})$ whenever $F_1^* \overset{n-2}{>} F_1$. By parts (ii) and (iii) every term on the right-hand side of (4.3.3) is nonnegative; it remains to show that at least one of them is positive.

If F_1 is not supported by one point, Lemma 4.3.5 implies that $\sigma F_1 \overset{n-2}{>} \varphi F_1$ and so, by the induction hypothesis, the fourth term on the right-hand side of (4.3.3) is positive. If F_1 is supported by one point, then, by the induction hypothesis, the second term or the penultimate term on the right-hand side of (4.3.3) is positive because, as a consequence of Lemma 4.3.5 and the induction hypothesis, $\Delta(\sigma F_1^*, F_2; \mathbf{A}^{(1)}) > 0$ or $\Delta(F_1^*, \sigma F_2; \mathbf{A}^{(1)}) \leq 0$.   □

Corollary 4.3.7 Suppose $\mathbf{A}$ is nonincreasing with horizon $n \in \{1, 2, \ldots, \infty\}$. If $F_1 \overset{n-1}{\geq} F_2$ then arm 1 is optimal initially for the $(F_1, F_2; \mathbf{A})$-bandit. Moreover, if $F_1 \overset{n-1}{>} F_2$ then arm 1 is uniquely optimal.

Proof By symmetry, $\Delta(F_2, F_2; \mathbf{A}) = 0$. The result now follows by applying Theorem 4.3.6.   □

The next result is a stay-with-a-winner rule. The bandit is trivial when both F_1 and F_2 are one-point distributions; this possibility is not considered in the theorem. The theorem says that an arm optimal initially continues to be optimal after a success when $\alpha_1 = \alpha_2$ or when the arm has smaller initial probability of success (that is, has smaller mean). So the result applies in the setting of Chapter 7 ($\alpha_1 = \ldots = \alpha_n = 1$, $\alpha_{n+1} = \ldots = 0$). It will be supplemented by results in Chapters 5 and 6.

Theorem 4.3.8 Suppose that $\mathbf{A}$ is an arbitrary nonincreasing sequence with horizon $n \in \{2, 3, \ldots, \infty\}$ and the support of either F_1 and F_2 consists of more than one point. Then $\Delta(F_1, F_2; \mathbf{A}) \geq 0$

implies $\Delta(\sigma F_1, F_2; \mathbf{A}^{(1)}) > 0$ provided

$$\text{(i)} \quad \alpha_1 = \alpha_2$$

or

$$\text{(ii)} \quad E(\theta_1 | F_1) \leqslant E(\theta_2 | F_2).$$

Proof From Lemma 4.2.1 and the hypotheses $\Delta(F_1, F_2; \mathbf{A}) \geqslant 0$ and (i) or (ii), we obtain

$$0 \leqslant E(\theta_1 | F_1)\Delta^+(\sigma F_1, F_2; \mathbf{A}^{(1)}) + E(1 - \theta_1 | F_1)\Delta^+(\varphi F_1, F_2; \mathbf{A}^{(1)})$$
$$- E(\theta_2 | F_2)\Delta^-(F_1, \sigma F_2; \mathbf{A}^{(1)}) - E(1 - \theta_2 | F_2)\Delta^-(F_1, \varphi F_2; \mathbf{A}^{(1)}).$$
$$(4.3.5)$$

Suppose, for a proof by contradiction, that $\Delta^+(\sigma F_1, F_2; \mathbf{A}^{(1)}) = 0$. Then $\Delta^+(\varphi F_1, F_2; \mathbf{A}^{(1)}) = 0$ and so each of the four terms in (4.3.5) equals 0. Since, by Lemma 4.3.5 and Theorem 4.3.6,

$$\Delta^-(F_1, \sigma F_2; \mathbf{A}^{(1)}) \geqslant -\Delta(F_1, \sigma F_2; \mathbf{A}^{(1)})$$
$$> -\Delta(\sigma F_1, F_2; \mathbf{A}^{(1)}) \geqslant -\Delta^+(\sigma F_1, F_2; \mathbf{A}^{(1)}) = 0,$$

it must be that $E(\theta_2 | F_2) = 0$ in order that the third term in (4.3.5) equals 0. Thus, the support of F_2 consists of the one point 0; so, F_1 has more than one point in its support and $\Delta(\sigma F_1, F_2; \mathbf{A}^{(1)}) > 0$, the desired contradiction. □

There is no switch-from-a-loser rule analogous to the stay-with-a-winner rule. One arm may be so much better *a priori* than the other that even many losses with it would not make the other arm worth considering. For example, F_1 may be 'wholly to the right' of F_2. The case in which one arm is known can be fruitful in understanding this asymmetry between success and failure, for in this case success and failure are equivalent except for the obvious difference in immediate income.

Robbins (1952) considered a particular stay-with-a-winner strategy: choose randomly initially and always switch on a loser. He calculated the asymptotic advantage of this 'play-the-winner rule' over random selection. This strategy has been considered extensively in the 'ranking and selection' literature (Sobel and Weiss, 1970; Nordbrock, 1976).

There are many examples where one should not select the arm having larger mean (cf. Example 1.2.1). However, myopic strategies are optimal in various special cases; one is given in the next result.

Theorem 4.3.9 Suppose that $\mathbf{A}$ is nonincreasing and F_1 and F_2 are supported by the same two points. Then an arm is optimal initially if and only if it has the larger mean. Specifically, arm 1 is optimal initially if and only if $E(\theta_1 | F_1) \geqslant E(\theta_2 | F_2)$.

Proof When both are supported by the same two points, it is easy to check that $F_1 \overset{n-1}{\underset{\sim}{\succ}} F_2$ for any $n \geqslant 1$ if and only if $E(\theta_1 | F_1) \geqslant E(\theta_2 | F_2)$. The result follows from Corollary 4.3.7.   □

The following example shows that the hypothesis that $\mathbf{A}$ is nonincreasing cannot be dropped from Theorem 4.3.9, nor from either Corollary 4.3.7 or Theorem 4.3.6.

Example 4.3.3 Let $k = 2$, $\mathbf{A} = (0, 1, 0, 0, 0, \ldots)$, and suppose that $F_i = (1 - p_i)\delta_0 + p_i\delta_{1/2}$ for $i = 1, 2$ where $1 > p_1 > p_2 > 0$. It is easy to check that the only optimal initial selection is arm 2 and that it should be followed by a selection of arm 2 if and only if a success is observed at stage 1. Since $E(\theta_1 | F_1) > E(\theta_2 | F_2)$, the conclusion of Theorem 4.3.9 fails. The advantage gained by selecting arm 2 is due to the fact that a success with it at stage 1 indicates that it is the appropriate selection at stage 2. If p_2 is sufficiently small an initial selection of arm 1 is worthless; whatever is observed at stage 1, arm 1 will be optimal at stage 2.   □

Since F_1 is strongly to the right of F_2 in the preceding example, it shows that the hypothesis on $\mathbf{A}$ cannot be dropped in either Corollary 4.3.7 or Theorem 4.3.6. The next example is a little more complicated and shows that the hypothesis on $\mathbf{A}$ cannot be dropped from Theorem 4.3.8.

Example 4.3.4 Let $k = 2$, $\mathbf{A} = (0, 0, 1, 0, 0, 0, \ldots)$, and suppose that $F_i = (1 - p_i)\delta_{1/4} + p_i\delta_{1/2}$ for $i = 1, 2$. Further suppose that $4p_1/(3 + p_1) > p_2 > p_1$. Then both (i) and (ii) of Theorem 4.3.8 are satisfied. By considering a small number of cases it is easy to check that $\Delta(F_1, F_2; \mathbf{A}) > 0$ and $\Delta(\sigma F_1, F_2; \mathbf{A}^{(1)}) < 0$; in fact, the only optimal selections (through stage 3) are given by $\tau(\varnothing) = 1$, $\tau(1) = 2$, $\tau(0) = 1$, $\tau(1, 1) = 2$, $\tau(1, 0) = 1$, $\tau(0, 1) = 1$, and $\tau(0, 0) = 2$.   □

One of the early papers dealing with a two-armed bandit is Feldman (1962); see also DeGroot (1970, Section 14.7). The discount sequence in the setting considered by Feldman is $\mathbf{A} = (1, \ldots, 1, 0, 0, \ldots)$. He considered two *dependent* arms. In particular, G is the following very special distribution: (θ_1, θ_2) is known to be either (a, b) or (b, a). So the two parameters are known but not which goes with which arm. While this problem is outside the setting of the current chapter, it is discussed here since Feldman's result is an easy consequence of Theorem 4.3.9; the basic idea for the following argument is due to Kadane (1969).

Assume that discount sequence is nonincreasing. Consider Feldman's prior and assume without loss that $a > b$. The initial probability that (θ_1, θ_2) equals (a, b) is $G(a, b)$, and $G(b, a) = 1 - G(a, b)$. Construct a new distribution G^* with support $\{(a, b), (b, a), (a, a), (b, b)\}$ so that θ_1 and θ_2 are independent and the (marginal) probability of $\{\theta_1 = a\}$ is $G(a, b)$:

$$G^*(a, b) = G^2(a, b), \qquad G^*(b, a) = G^2(b, a),$$
$$G^*(a, a) = G^*(b, b) = G(a, b)G(b, a).$$

So $\quad F_1 = G(a, b)\delta_a + G(b, a)\delta_b \quad$ and $\quad F_2 = G(b, a)\delta_a + G(a, b)\delta_b$. Theorem 4.3.9 applies to G^* to show that arm 1 is optimal if and only if $G(a, b) > G(b, a)$. But if it were known in advance that the arms were identical $(\theta_1 = \theta_2 = a$ or $\theta_1 = \theta_2 = b)$ then neither arm would be strictly preferred. The only possibilities that influence the preference for an arm have $\theta_1 \neq \theta_2$ (that is, either $(\theta_1, \theta_2) = (a, b)$ or (b, a)). Therefore, Theorem 4.3.9 applies to show that arm 1 is optimal when and only when it is *a priori* at least as likely that $(\theta_1, \theta_2) = (a, b)$ as that $(\theta_1, \theta_2) = (b, a)$. Therefore, we have the following as a generalization of Feldman's result.

Corollary 4.3.10 Let $0 \leqslant b \leqslant a \leqslant 1$. Suppose $k = 2$ and that the two arms are Bernoulli with parameters a for arm 1 and b for arm 2 with probability $G(a, b)$ and a for arm 2 and b for arm 1 with probability $G(b, a) = 1 - G(a, b)$. Suppose the discount sequence is nonincreasing. Then arm 1 is optimal initially if and only if $G(a, b) \geqslant 1/2$ and arm 2 is optimal initially if and only if $G(b, a) \geqslant 1/2$.

Feldman's result has been generalized in a number of other directions by Fabius and van Zwet (1970), Kelley (1974), Rodman (1978), and Zaborskis (1976). These directions are indicated in the Annotated Bibliography.

References

Berry, D. A. (1972) A Bernoulli two-armed bandit. *Ann. Math. Statist.* **43**: 871–897.

DeGroot, M. H. (1970) *Optimal Statistical Decisions*, McGraw-Hill, New York.

Fabius, J. and van Zwet, W. R. (1970) Some remarks on the two-armed bandit. *Ann. Math. Statist.* **41**: 1906–1916.

Feldman, D. (1962) Contributions to the 'two-armed bandit' problem. *Ann. Math. Statist.* **33**: 847–856.

Kadane, J. B. (1969) Personal communication.

Kelley, T. A. (1974) A note on the Bernoulli two-armed bandit problem. *Ann. Statist.* **2**: 1056–1062.

Nordbrock, E. (1976) An improved play-the-winner sampling procedure for selecting the better of two binomial populations. *J. Amer. Statist. Assoc.* **71**: 137–139.

Quisel, K. (1965) Extensions of the two-armed bandit and related processes with on-line experimentation. Tech. Rep. No. 137, Institute for Mathematical Studies in the Social Sciences, Stanford Univ., USA.

Robbins, H. (1952) Some aspects of the sequential design of experiments. *Bull. Amer. Math. Soc.* **58**: 527–535.

Rodman, L. (1978) On the many-armed bandit problem. *Ann. Prob.* **6**: 491–498.

Sobel, M. and Weiss, G. H. (1970) Play-the-winner sampling for selecting the better of two binomial populations. *Biometrika* **57**: 357–365.

Zaborskis, A. A. (1976) Sequential Bayesian plan for choosing the best method of medical treatment. *Avtomatika i Telemekhanika* **2**: 144–153.

CHAPTER 5

Two arms, one arm known

In this chapter we assume that there are two arms ($k = 2$) and that one arm, say arm 2 for definiteness, has known mean λ. The only uncertainty is embodied in F_1, now abbreviated to F, the distribution of the random measure Q_1. For arbitrary λ we can, without loss, assume that arm 2 always produces the known observation λ. Since G is given by the pair (F, λ), we now speak of the $(F, \lambda; \mathbf{A})$-bandit.

Depending on $\mathbf{A}$, continuous-time approximations may be available. For example, if $\mathbf{A}$ is n-horizon uniform with n large and F is Bernoulli, then the example discussed in Section 8.2 applies as an approximation.

There are two main types of results in this chapter. One compares bandits that differ in some respect – say that have different λ's. The other type of result gives properties of optimal strategies. The chapter is organized partly on the basis of various rather weak restrictions on $\mathbf{A}$. We give two easy and intuitive results in the present section that apply for arbitrary discount sequences, as well as arbitrary F.

Theorem 5.0.1 For all F and $\mathbf{A}$, the value $V(F, \lambda; \mathbf{A})$ is a continuous nondecreasing function of λ.

Remark This result is immediate in the Bernoulli case in view of Theorems 4.1.1 and 4.1.6 since a one-point distribution on $\lambda^* > \lambda$ is strongly to the right of a one-point distribution on λ. □

Proof of Theorem 5.0.1 Suppose $\lambda^* > \lambda$ and τ is an optimal strategy in the $(F, \lambda; \mathbf{A})$-bandit; such a strategy exists in view of Theorem 2.5.2. Suppose τ is followed in the $(F, \lambda^*; \mathbf{A})$-bandit. The only change in worth as compared with the $(F, \lambda; \mathbf{A})$-bandit is when arm 2 is selected;

the corresponding observation is λ^* as opposed to λ. Therefore

$$V(F, \lambda; \mathbf{A}) = W(F, \lambda; \mathbf{A}; \tau)$$
$$\leqslant W(F, \lambda^*; \mathbf{A}; \tau) \leqslant V(F, \lambda^*; \mathbf{A}).$$

To prove continuity let τ^* be optimal for the $(F, \lambda^*; \mathbf{A})$-bandit. Since the only difference between $W(F, \lambda; \mathbf{A}, \tau^*)$ and $W(F, \lambda^*; \mathbf{A}; \tau^*)$ is a result of the observations on arm 2,

$$V(F, \lambda^*; \mathbf{A}) = W(F, \lambda^*; \mathbf{A}; \tau^*)$$
$$\leqslant W(F, \lambda; \mathbf{A}; \tau^*) + (\lambda^* - \lambda)|\mathbf{A}|_1$$
$$\leqslant V(F, \lambda; \mathbf{A}) + (\lambda^* - \lambda)|\mathbf{A}|_1.$$

So, not only is V a continuous function of λ, but it is absolutely continuous with a derivative bounded by $|\mathbf{A}|_1$. $\qquad \square$

Since arm 2 is known, it is used only to achieve immediate payoff; using arm 1 can gain information as well. So arm 1 is optimal if it gives greater immediate payoff than does arm 2. (The analogous result in continuous time is illustrated in the example of Figure 8.2 by that fact that the boundary is less than 0.) This fundamental fact is intuitive and easy to show:

Theorem 5.0.2 If $E(X_{11}|F) \geqslant \lambda$ then arm 1 is optimal for any $\mathbf{A}$.

Proof Suppose τ is an optimal strategy in the $(F, \lambda; \mathbf{A}^{(1)})$-bandit. Let τ^* be the strategy in the $(F, \lambda; \mathbf{A})$-bandit which indicates arm 1 initially and then follows τ, thereby ignoring the initial result with arm 1. This strategy has worth

$$W(F, \lambda; \mathbf{A}; \tau^*) = \alpha_1 E(X_{11}|F) + V(F, \lambda; \mathbf{A}^{(1)})$$
$$\geqslant \alpha_1 \lambda + V(F, \lambda; \mathbf{A}^{(1)}) = W(F, \lambda; \mathbf{A}; \tau_2),$$

say, where τ_2 is a strategy which indicates arm 2 initially and then proceeds optimally (by following τ, for example). Since there is a strategy that starts with arm 1 and is at least as good as the best strategy that starts with arm 2, arm 1 is optimal. $\qquad \square$

For an example of how much larger the mean of arm 2 must be to compete with an unknown arm 1, see Example 5.5.3.

The first four sections of this chapter assume arm 1 to be Bernoulli with the random parameter θ_1. As has become our convention in such a setting, we regard F as the distribution of θ_1 (on $[0, 1]$) rather than as a distribution on $\mathscr{D}$.

In Section 5.1 we give various properties of $\Delta(F, \lambda; \mathbf{A})$ when $\mathbf{A}$ is monotone. In Section 5.2 we consider discount sequences which may not be monotone and give necessary and sufficient conditions on $\mathbf{A}$ for the $(F, \lambda; \mathbf{A})$-bandit to be a stopping problem for all (F, λ). So we answer the question: for which $\mathbf{A}$ does the problem always reduce to deciding when to stop selecting arm 1?

Section 5.3 is based on the main result of Section 5.2. It contains a rather complete characterization of optimal strategies when $\mathbf{A}$ is restricted to the class determined by that result. In addition, Section 5.3 gives separate demonstrations of the results of Section 5.1 for this class, which contains some monotone and some non-monotone sequences.

Section 5.4 gives Bernoulli examples in which optimal strategies are calculated using the results of Section 5.3. (Those examples in which $\mathbf{A}$ is geometric will be given greater relevance by the Gittins–Jones result discussed in Chapter 6.) We also find bounds which provide sufficient conditions for an arm to be optimal. These bounds are compared by example with some exact derivations.

Many of the results derived in Section 5.2 and 5.3 apply also outside the Bernoulli setting. Throughout Section 5.2 and in the initial part of Section 5.3 the Bernoulli assumption is unnecessary; it is made in these sections to facilitate the presentation. The results for which we will make no general claims will be so indicated. Most such results concern bandits that arise from another bandit after a success or a failure on arm 1. In Section 5.5 we discuss general distributions for arm 1. The most important results from the earlier sections and those to be used in the later development will be repeated. A rather comprehensive example is discussed in Section 5.6; the distribution F of Q_1 is a Dirichlet process as defined by Ferguson (1973).

In Section 5.7 we turn to discounting in real time as described in Section 3.5. We describe the extent to which the results of Sections 5.2 and 5.3 apply in this setting.

5.1 Monotone discount sequences

In this section we assume arm 1 is Bernoulli.

An important class of discount sequences has $\alpha_m \geqslant \alpha_{m+1}$ for $m = 1, 2, \ldots$. With this monotonicity assumption we show that there is a 'break-even value' $\Lambda \in [0, 1]$ for which arm 2 is optimal when $\lambda \geqslant \Lambda$ and arm 1 is optimal when $\lambda \leqslant \Lambda$ (Corollary 5.1.2). We also show that when arm 1 is optimal and yields a success it is optimal again (Theorem 5.1.3). When $\mathbf{A}$ is monotone we can apply Theorem 4.3.6 to show that arm 2 becomes more desirable when λ increases.

Corollary 5.1.1 The difference $\Delta(F, \lambda; \mathbf{A})$ is strictly decreasing in λ when $\mathbf{A}$ is nonincreasing with $\mathbf{A} \neq \mathbf{0}$.

Proof Let n denote the horizon of $\mathbf{A}$. According to Theorem 4.3.6, $\Delta(F, \cdot; \mathbf{A})$ is strictly $(n-1)$-decreasing. The result follows since a one-point distribution on $\lambda^* > \lambda$ is ∞-greater than a one-point distribution on λ. $\qquad\square$

The next result is a consequence of this fact. It states that there is a unique value of λ, say $\Lambda(F, \mathbf{A})$, such that both arms are optimal when $\lambda = \Lambda(F, \mathbf{A})$.

Corollary 5.1.2 For each nonincreasing discount sequence $\mathbf{A}$ with $\mathbf{A} \neq \mathbf{0}$ and each distribution F on $[0, 1]$, there exists a unique $\Lambda(F, \mathbf{A}) \in [0, 1]$ such that arm 1 is optimal initially in the $(F, \lambda; \mathbf{A})$-bandit if and only if $\lambda \leqslant \Lambda(F, \mathbf{A})$ and arm 2 is optimal if and only if $\lambda \geqslant \Lambda(F, \mathbf{A})$.

Remarks Corollary 5.1.2 generalizes the results of Bradt, Johnson, and Karlin (1956, Lemma 4.2) where uniform discounting was considered, and of Bellman (1956, Theorem 2) where geometric discounting was considered.

The break-even value of λ, $\Lambda(F, \mathbf{A})$, is called a 'dynamic allocation index' by Gittins and Jones (1974). It plays an especially important role in multi-armed bandits with geometric discounting (Theorem 6.1.1). $\qquad\square$

Proof From Corollary 5.1.1, there exists a unique $\Lambda(F, \mathbf{A}) \in [0, 1]$ such that arm 1 is uniquely optimal initially in the $(F, \lambda; \mathbf{A})$-bandit if $\lambda < \Lambda(F, \mathbf{A})$ and arm 2 is uniquely optimal initially if $\lambda > \Lambda(F, \mathbf{A})$. It remains to prove that both arms are optimal initially in the

$(F, \Lambda(F, \mathbf{A}); \mathbf{A})$-bandit. We assume $0 < \Lambda < 1$ and leave it to the reader to consider the easier cases $\Lambda = 0$ and $\Lambda = 1$.

For $\lambda < \Lambda(F, \mathbf{A})$,

$$V^{(1)}(F, \lambda; \mathbf{A}) > V^{(2)}(F, \lambda; \mathbf{A})$$

(quantities defined in Theorem 2.5.1); while the inequality is reversed for $\lambda < \Lambda(F, \mathbf{A})$. By Theorem 5.0.1, $V^{(1)}$ and $V^{(2)}$ are continuous. Therefore,

$$V^{(1)}(F, \Lambda(F, \mathbf{A}); \mathbf{A}) = V^{(2)}(F, \Lambda(F, \mathbf{A}); \mathbf{A}),$$

which is equivalent to both arms being optimal initially for the $(F, \Lambda(F, \mathbf{A}); \mathbf{A})$-bandit.

$\square$

It seems reasonable to expect the desirability of arm 2 to increase with λ whether or not $\mathbf{A}$ is monotone. However, the next example shows that the monotonicity hypothesis in Corollaries 5.1.1 and 5.1.2 cannot be dropped. Namely, for a particular discount sequence which is not monotonic the example shows that arm 2 can be optimal for one value of λ but not for a larger value of λ.

Example 5.1.1 Let $\mathbf{A} = (1, 0, a, 0, 0, 0, \ldots)$ and let

$$F = (5/8)\delta_0 + (1/4)\delta_{1/4} + (1/8)\delta_1;$$

so $E(\theta_1 | F) = 3/16$. We shall show that when a is sufficiently large, arm 2 is optimal initally when $\lambda = 1/4$ but not when $\lambda = 1/2$.

Let us first consider the $(F, 1/4; \mathbf{A})$-bandit. Irrespective of the selection and result at stage 1, arm 1 is clearly optimal at stage 2 since $\alpha_2 = 0$. At stage 3, arm 1 is optimal (not necessarily uniquely) if a success was obtained at stage 2 and arm 2 is optimal otherwise – this being the case irrespective of the arm selected and result obtained at stage 1. Accordingly, arm 2 is optimal at stage 1 for any $a \geqslant 0$ since $E(\theta_1 | F) = 3/16 < 1/4 = \lambda$.

Now consider the $(F, 1/2; \mathbf{A})$-bandit. Again, arm 1 can be selected without loss at stage 2. And as in the case $\lambda = 1/4$, arm 1 should not be selected at stage 3 if it has yielded a failure previously. Therefore, if arm 1 is selected initially there is only one continuation that need be considered: select arm 1 at stage 2 and again at stage 3 if and only if successes were obtained at both stages 1 and 2. The probability of the latter is $(1/8)(1)^2 + (1/4)(1/4)^2 = 9/64$. So the expected payoff using this strategy is

$$\frac{3}{16}(1) + \left[\left(\frac{9}{64}\right)\frac{11}{12} + \left(1 - \frac{9}{64}\right)\frac{1}{2}\right]a = \frac{3}{16} + \frac{143}{256}a.$$

The first term is the contribution from stage 1 and the second that of stage 3. If arm 2 is selected initially and arm 1 at stage 2 then the optimal selection at stage 3 is arm 1 in the case of a success at stage 2 $(3/4 > 1/2)$ and arm 2 in case of a failure at this stage. The expected payoff of this strategy is

$$\frac{1}{2}(1) + \left[\left(\frac{3}{16}\right)\frac{3}{4} + \left(1 - \frac{3}{16}\right)\frac{1}{2}\right]a = \frac{1}{2} + \frac{140}{256}a < \frac{3}{16} + \frac{143}{256}a$$

which is true when $a > 80/3$.

Hence, arm 2 is uniquely optimal when $\lambda = 1/4$ and, assuming $a > 80/3$, arm 1 is uniquely optimal when $\lambda = 1/2$!   □

As a corollary of Theorem 5.0.2, and of Theorem 4.3.8, which applies when **A** is nonincreasing, it follows that arm 1 continues to be optimal if it is successful.

Theorem 5.1.3 Suppose **A** is nonincreasing. Then $\Delta(F, \lambda; \mathbf{A}) \geqslant 0$ implies $\Delta(\sigma F, \lambda; \mathbf{A}^{(1)}) \geqslant 0$. (In other words, $\Lambda(\sigma F, \mathbf{A}^{(1)}) \geqslant \Lambda(F, \mathbf{A})$.)

Proof Theorem 4.3.8 applies when $E(\theta_1 | F) \leqslant \lambda$ and Theorem 5.0.2 applies otherwise since $E(\theta_1 | \sigma F) \geqslant E(\theta_1 | F)$ by the Cauchy–Schwarz inequality.   □

This result generalizes Lemma 4.6 of Bradt, Johnson, and Karlin (1956) who considered finite horizon uniform discount sequences; Theorem 2 of Bellman (1956) who considered geometric discounting; and Theorem 4.1 of Berry and Fristedt (1979) who showed it for *regular* nonincreasing discount sequences (see Definition 5.2.1).

Theorem 5.1.3 applies when **A** is nonincreasing to show that there are optimal strategies that stay with the unknown arm if it is successful. But we shall see that there are $(F, \lambda; \mathbf{A})$-bandits where all optimal strategies switch from the known arm; and since an observation on the known arm – success or not – has no effect on future payoff, this means that such strategies switch on a success. As is shown in the next section, this cannot happen when **A** is sufficiently 'smooth'.

5.2 Regular discount sequences

In comparing the possible selections at any stage, arm 1 has two potential benefits: it can yield immediate success and it can give information to aid in future selections. On the other hand, arm 2 has no information value since its characteristics are completely known. This means that there can be no advantage in basing future selections on results from arm 2. One might therefore speculate that arm 1 can be set aside for the indefinite future once arm 2 becomes optimal. This is true in a variety of circumstances, but it is not generally true. When it is true, the problem is one of optimal stopping: when should experimentation with arm 1 cease? This section characterizes discount sequences **A** for which the $(F, \lambda; \mathbf{A})$-bandit is an optimal stopping problem for all F and λ. This characterization is of interest, but the fact that a simple characterization is possible may be more interesting.

Example 3.3.3 shows that it may be uniquely optimal to begin with arm 2 and change to arm 1 at a subsequent stage. In that example the discount sequence is a mixture of geometrics. The next example provides a much simpler setting in which this phenomenon occurs. As in Example 5.1.1, to which it is very similar, the critical aspect of the example is the discount sequence, which encourages gain rather than information-gathering at stage 1 and vice versa at stage 2.

Example 5.2.1 Let $\mathbf{A} = (1, 0, 1, 0, 0, \ldots)$, $\lambda > 1/2$, and $F = \frac{1}{2}\delta_0 + \frac{1}{2}\delta_1$ (cf. Example 1.2.1). Since $\alpha_2 = 0$, complete information can be obtained on arm 1 at stage 2 without risk. The first three selections of every optimal strategy are therefore clear. Since its mean is greater than that of arm 1, arm 2 should be selected initially. Then arm 1 should be selected (at stage 2). If successful, arm 1 should be used again at stage 3 (since θ_1 is then known to be 1) and if it is not successful, arm 2 should be used at stage 3 (since θ_1 is then known to be 0). Continuations beyond stage 3 are immaterial. The maximal expected payoff is

$$V(F, \lambda; \mathbf{A}) = \lambda + (1/2 + \lambda/2). \qquad \square$$

The discount sequence in the previous example was chosen to make the issue transparent. The following example makes it clear that $\alpha_2 = 0$ and the lack of monotonicity in the discount sequence play no essential role.

Example 5.2.2 Let $\mathbf{A} = (4, 1, 1, 0, 0, 0, \ldots)$, $\lambda = 0.6$, and F be as in the previous example. It is easy to check that $V(F, 0.6; \mathbf{A}) = 3.7$ and that this expected payoff can be obtained only by the strategies indicated as optimal in the previous example. $\square$

The phenomenon described in these examples cannot arise if the discount sequence has the regularity property that we now define; cf. (2.4.8).

Definition 5.2.1 For any discount sequence

$\mathbf{A} = (\alpha_1, \alpha_2, \ldots)$ let $\gamma_m = \sum\limits_{j=m}^{\infty} \alpha_j$; $\mathbf{A}$ is *regular* if, for $m = 1, 2, \ldots$,

$$\frac{\gamma_{m+2}}{\gamma_{m+1}} \leqslant \frac{\gamma_{m+1}}{\gamma_m} \tag{5.2.1}$$

provided that $\gamma_{m+1} > 0$.

The infinite sum γ_m is the maximum available worth from stage m onward, that is, it is the total worth of successes into the indefinite future.

The condition of regularity is somewhat weaker than the corresponding condition with α's in place of γ's in (5.2.1).

Proposition 5.2.1 The discount sequence $\mathbf{A}$ is regular if, for each $m = 1, 2, \ldots$, either

$$\frac{\alpha_{m+2}}{\alpha_{m+1}} \leqslant \frac{\alpha_{m+1}}{\alpha_m} \tag{5.2.2}$$

or $\alpha_{m-j} = 0$ for $j = 0, \ldots, m-1$ or $\alpha_{m+j} = 0$ for $j = 1, 2, \ldots$.

Remark Such sequences are called *superregular* in Berry and Fristedt (1983). $\square$

Proof The hypothesis implies

$$\alpha_{m+1}\alpha_{m+j} \geqslant \alpha_m \alpha_{m+j+1}$$

for all positive integers m and j. Therefore,

$$\alpha_{m+1} \sum_{j=1}^{\infty} \alpha_{m+j} \geqslant \alpha_m \sum_{j=1}^{\infty} \alpha_{m+j+1}.$$

It follows that, for all m,

$$\alpha_{m+1}\gamma_{m+1} + \alpha_{m+1}\gamma_{m+2} + \gamma_{m+2}^2 \geqslant \alpha_m\gamma_{m+2} + \alpha_{m+1}\gamma_{m+2} + \gamma_{m+2}^2;$$

that is,

$$\gamma_{m+1}^2 \geqslant \gamma_m\gamma_{m+2}$$

and (5.2.1) follows. □

The two most important classes of regular discount sequence are the finite horizon uniform and the geometric; both have been discussed a number of times and will be considered again in this and subsequent chapters. But geometric sequences are barely regular: each inequality in (5.2.1)–and also in (5.2.2)–holds with equality for all m. Furthermore, by an application of Jensen's inequality, any nontrivial linear combination of geometrics is not regular (see $\mathbf{A}_8$ below).

Other regular sequences are:

$\mathbf{A}_1 = (1, \alpha, \alpha^2, \ldots, \alpha^{n-1}, 0, 0, \ldots), \alpha \geqslant 0,$
$\mathbf{A}_2 = (0, 0, \ldots, 0, 1, 0, 0, \ldots),$
$\mathbf{A}_3 = (0, 0, \ldots, 0, 1, 1, \ldots, 1, 0, 0, \ldots),$
$\mathbf{A}_4 = (0, 1, 2, 3, 4, 3, 2, 1, 1, 0, 0, \ldots),$
$\mathbf{A}_5 = (3, 4, 3, 4, 3, 0, 0, \ldots),$
$\mathbf{A}_6 = e^{-\alpha}(1, \alpha, \alpha^2/2, \ldots, \alpha^{m-1}/(m-1)!, \ldots),$
$\mathbf{A}_7 = (1, 1-e^{-\alpha}, 1-(1+\alpha)e^{-\alpha}, \ldots, e^{-\alpha}\sum_m^{\infty}\alpha^{i-1}/(i-1)!, \ldots).$

($\mathbf{A}_6$ is the sequence of Poisson probabilities and $\mathbf{A}_7$ is the cumulative Poisson sequence.)

Some nonregular sequences are:

$$\mathbf{A}_8 = \left(\frac{1}{2}, \frac{5}{16}, \frac{7}{32}, \ldots, \frac{1}{2}\left(\frac{3}{4}\right)^{m-1} + \frac{1}{2}\left(\frac{1}{4}\right)^{m-1}, \ldots\right),$$

$\mathbf{A}_9 = (4, 1, 1, 0, 0, \ldots),$
$\mathbf{A}_{10} = (1, 0, a, 0, 0, \ldots), \quad a > 0.$

($\mathbf{A}_8$ was considered in Example 3.3.3 and $\mathbf{A}_9$ in Example 5.2.2; $\mathbf{A}_{10}$ was considered in Example 5.1.1 and in Example 5.2.1 with $a = 1$.) If $\mathbf{A}$ is regular and $\alpha_m = 0$ then either $\alpha_1 = \ldots = \alpha_{m-1} = 0$ or

$\alpha_{m+1} = \alpha_{m+2} = \ldots = 0$ (compare A_2 and A_{10}). And if $\alpha_{m+1} \geq \alpha_m$ then condition (5.2.1) is always satisfied; that is, regularity cannot be violated on an increasing part of a discount sequence.

The importance of regularity of discount sequences derives from the next result. It is an extension of Theorem 2.1 of Berry and Fristedt (1979). A modification of the proof given there applies in this more general setting in which A need not be monotone, and is given below. (The result holds beyond the present Bernoulli context and so will be discussed again in Sections 5.5 and 5.7).

Theorem 5.2.2 Consider a discount sequence A. The following two statements are equivalent:

(i) For every $(F, \lambda; A)$-bandit there is an optimal strategy under which every selection of the known arm, arm 2, is followed by another selection of arm 2;

(ii) A is regular.

Furthermore, if $\alpha_1 > 0$ and A is regular, then if ever arm 2 becomes optimal, an optimal continuation is to select arm 2 exclusively and indefinitely.

The most useful half of this result is (ii) $\Rightarrow$ (i). Before proving the theorem we give an intuitive demonstration of this implication when A is the n-horizon uniform that is due to Bradt, Johnson, and Karlin (1956, Lemma 4.1); the reader will see a resemblance between this demonstration and the proof that is given below. (It is surprising to us that the existence of such a trivial argument in this case has not been appreciated by a number of authors.)

Suppose at some stage – there is no loss in saying the first stage – that arm 2 is uniquely optimal. It is selected for a number of stages and eventually, but before stage n, arm 1 becomes optimal. Is this possible? Suppose this latter selection of arm 1 is exchanged with the initial selection of arm 2. Nothing is lost since successes are exchangeable when the discount factors are equal. Therefore arm 1 is also optimal initially and so the assumption that arm 2 is uniquely optimal initially is contradicted. And (i) of the theorem follows in this special case.

Actually, selecting arm 1 early (but not committing to arm 2 thereafter) has some benefit over using it late; namely, there is more opportunity to use any information that is gained. If the discount sequence is not uniform something may be lost when two selections are interchanged. It will develop in the following proof that regularity is precisely the condition on the discount sequence that assures that this loss will not outweigh the gain from obtaining early information.

Proof of Theorem 5.2.2

Part I, (ii) ⟹ (i). Let $\mathscr{S}_n$ denote the set of all regular discount sequences with horizon n. For any $\mathbf{A} = (\alpha_1, \alpha_2, \ldots)$ the notational conventions $\Sigma_{m=\infty}^{\infty} \alpha_m = \gamma_\infty = 0$ will be useful. The proof is by induction on n.

Clearly, (i) holds for every member of $\mathscr{S}_0 \cup \mathscr{S}_1$. Suppose $n \geqslant 2$. Assume it holds for every member of $\mathscr{S}_{n-1}$. Let $\mathbf{A} = (\alpha_1, \alpha_2, \ldots) \in \mathscr{S}_n$; then $\mathbf{A}^{(1)} = (\alpha_2, \alpha_3, \ldots) \in \mathscr{S}_{n-1}$. Assume it is optimal to select arm 1 initially. The inductive hypothesis applies to show that there is an optimal continuation which never switches back to arm 1 after a switch to arm 2. If it is optimal to select arm 2 initially, then the inductive hypothesis applies immediately to show (i) unless an optimal strategy has the form τ^*: select arm 2 initially, select arm 1 at stages $2, \ldots, N$, and arm 2 subsequently. The stage N is random with $P(N > 1) = 1$; it may be infinite with positive probability and it may depend on the history of observations. We may assume that τ^* does not depend on the initial observation on arm 2. So for each m, $\{N > m\}$ is measurable with respect to the σ-field generated by the outcomes of the selections of arm 1 at stages 2 through m, that is, the σ-field generated by $(X_{12}, \ldots, X_{1m})$. By Theorem 5.0.2 we may assume with no loss of generality that if s successes and $f = m - s - 1$ failures have been obtained with arm 1 at stages 2 through m when following τ^*, then

$$N = m \Rightarrow E(\theta_1 | \sigma^s \varphi^f F) < \lambda. \tag{5.2.3}$$

We show that there is a strategy τ which starts with arm 1 and is at least as good as τ^*. We choose τ by modifying τ^* as follows: select arm 1 initially and imitate τ^* subsequently by selecting the indicated arm one stage earlier. The worth of τ^* is

$$W(F, \lambda; \mathbf{A}; \tau^*) = E_{\tau^*}(\lambda \alpha_1 + \sum_{m=2}^{N} X_{1m} \alpha_m + \lambda \sum_{m=N+1}^{\infty} \alpha_m | F), \tag{5.2.4}$$

which, since τ^* is optimal, is no smaller than $\lambda \gamma_1$. Hence,

$$\sum_{m=2}^{\infty} E_{\tau^*}[(X_{1m} - \lambda) \mathbf{1}_{\{N \geqslant m\}} | F] \alpha_m = E_{\tau^*}[\sum_{m=2}^{N} (X_{1m} - \lambda) \alpha_m | F] \geqslant 0. \tag{5.2.5}$$

The worth of τ is

$$W(F, \lambda; \mathbf{A}; \tau) = E_{\tau^*}\left[\sum_{m=2}^{N} X_{1m}\alpha_{m-1} + \lambda \sum_{m=N+1}^{\infty} \alpha_{m-1}|F\right], \quad (5.2.6)$$

which we show to be at least as large as the worth of τ^*. Subtracting (5.2.4) from (5.2.6) gives

$$\sum_{m=2}^{\infty} E_{\tau^*}[(X_{1m} - \lambda)\mathbf{1}_{\{N \geq m\}}|F] (\alpha_{m-1} - \alpha_m). \quad (5.2.7)$$

That (5.2.7) is nonnegative follows from (5.2.5) by showing that

$$\sum_{m=2}^{\infty} b_m \alpha_m \geq 0 \Rightarrow \sum_{m=2}^{\infty} b_m(\alpha_{m-1} - \alpha_m) \geq 0, \quad (5.2.8)$$

where

$$b_m = E_{\tau^*}[(X_{1m} - \lambda)\mathbf{1}_{\{N \geq m\}}|F].$$

Write the first sum in (5.2.8) as follows:

$$\sum_{m=2}^{\infty} b_m \alpha_m = \sum_{m=2}^{\infty} b_m(\gamma_m - \gamma_{m+1}) = b_2\gamma_2 + \sum_{m=2}^{\infty} (b_{m+1} - b_m)\gamma_{m+1}.$$

We multiply the first inequality in (5.2.8) by α_1/γ_2 to obtain

$$b_2\alpha_1 + \sum_{m=2}^{\infty} (b_{m+1} - b_m)\gamma_{m+1}\alpha_1/\gamma_2 \geq 0. \quad (5.2.9)$$

The second sum in (5.2.8) can be written

$$\sum_{m=2}^{\infty} b_m(\alpha_{m-1} - \alpha_m) = b_2\alpha_1 + \sum_{m=2}^{\infty} (b_{m+1} - b_m)\alpha_m.$$

Accordingly, the second inequality in (5.2.8) becomes

$$b_2\alpha_1 + \sum_{m=2}^{\infty} (b_{m+1} - b_m)\alpha_m \geq 0. \quad (5.2.10)$$

In view of (5.2.9) and (5.2.10), (5.2.8) follows from two facts. The first is immediate for regular discount sequences: $\gamma_{m+1}\alpha_1/\gamma_2 \leq \alpha_m$, $m = 2, 3, \ldots$. The second is that the sequence $(b_1, b_2, \ldots)$ is nondecreasing. To show the latter, write

$$\begin{aligned} b_{m+1} - b_m &= E_{\tau^*}[(\lambda - X_{1, m+1})\mathbf{1}_{\{N=m\}}|F] \\ &\quad + E_{\tau^*}[(X_{1, m+1} - X_{1m})\mathbf{1}_{\{N \geq m\}}|F] \\ &= E_{\tau^*}[(\lambda - X_{1, m+1})\mathbf{1}_{\{N=m\}}|F] \geq 0. \end{aligned}$$

For the second equality we used the exchangeability of the X_{1j}'s and the fact that $\{N \geqslant m\}$ is measurable with respect to the σ-field generated by $\{X_{12}, \ldots, X_{1,m-1}\}$; for the inequality we used (5.2.3).

We have proved (i) for any regular discount sequence having finite horizon. Now assume $\mathbf{A}$ is a regular discount sequence having infinite horizon. Let τ be the optimal strategy identified in the proof of Theorem 2.5.2; τ indicates a selection of arm 2 initially if and only if the maximum in (2.5.2) is attained for $i = 2$ but not for $i = 1$. A similar statement holds at any stage for the appropriate modification of (2.5.2). Because both sides of (2.5.2) depend continuously on $\mathbf{A}$, we see that if τ indicates a selection of arm 2 at some stage, then arm 2 is also uniquely optimal for finite horizon approximations for a sufficiently large horizon. For each such approximation arm 2 is optimal thereafter and so, again using the fact that both sides of (2.5.2) depend continuously on $\mathbf{A}$, we conclude that arm 2 is optimal thereafter for the infinite horizon bandit.

We have shown that (ii) $\Rightarrow$ (i). Now assume (ii) and $\alpha_1 \neq 0$. We show that whenever arm 2 becomes optimal, an optimal continuation is to select arm 2 exclusively and indefinitely. The proof is by contradiction. Suppose that arm 2 is optimal at stage 1 and arm 1 is uniquely optimal at stage 2. We have proven that there is an optimal strategy that indicates arm 1 through a random stage $N \geqslant 2$ (possibly $+\infty$) after which arm 2 is indicated exclusively and indefinitely; call this strategy τ^*. To contradict the optimality of τ^* we imitate the above development but now we must ensure that the second inequality in (5.2.8) is strict. The unique optimality of arm 1 at stage 2 gives strict inequality at (5.2.5) and therefore in the first inequality in (5.2.8). With the additional assumption $\alpha_1 \neq 0$, the argument used above to prove (5.2.8) applies to give the desired strict inequality.

We have shown that arm 1 cannot be uniquely optimal at stage 2 when arm 2 is optimal and selected initially. We generalize by supposing that, with positive probability, it is optimal to select arm 2 at some stage m ($m > 1$) and uniquely optimal to select arm 1 at stage $m + 1$. Then $\alpha_m \neq 0$ follows from the assumptions that $\alpha_1 \neq 0$ and $\mathbf{A}$ is regular, for were $\alpha_m = 0$, then α_{m+j} would equal 0 for $j = 1, 2, \ldots$, and neither arm would be uniquely optimal at stage $m + 1$. So we can simply drop the first $m - 1$ stages, leaving the mth stage as the new first stage and a discount sequence whose first member is nonzero. The argument of the preceding paragraph now applies to give a contradiction.

Part II, not (ii) $\Rightarrow$ not (i). Suppose that

$$\gamma_M \gamma_{M+2} > \gamma_{M+1}^2 \qquad (5.2.11)$$

for some M. We shall prove that (i) fails by finding a pair (F, λ) for which there is a strategy τ that follows a selection of arm 2 with one of arm 1 (on a history that has positive probability under τ) and which is strictly better than every strategy that does not.

For an $\varepsilon \in (0, 1)$ still to be specified, define F so that after $M - 1$ failures on arm 1, the probabilities that $\theta_1 = 0$ and $\theta_1 = 1 - \varepsilon$ both are $1/2$; that is

$$\varphi^{M-1} F = \tfrac{1}{2}\delta_0 + \tfrac{1}{2}\delta_{1-\varepsilon}.$$

This can be accomplished by setting

$$F = \frac{\varepsilon^{M-1}}{1 + \varepsilon^{M-1}} \delta_0 + \frac{1}{1 + \varepsilon^{M-1}} \delta_{1-\varepsilon}.$$

Since $\sigma F = \delta_{1-\varepsilon}$ we may restrict our attention to strategies under which a success with arm 1 is followed indefinitely by selections of arm 1. Among such strategies, the only ones having the property that no selection of arm 2 is followed by a selection of arm 1 are the strategies $\tau_J, J = 0, 1, \ldots, \infty$: select arm 1 at the first J stages; thereafter select arm 1 or 2 according as a success was or was not observed at one or more of the first J stages. A straightforward calculation gives:

$$W(F, \lambda; \mathbf{A}; \tau_J) = \frac{(1-\varepsilon)\gamma_1}{1 + \varepsilon^{M-1}} + \lambda \frac{(\varepsilon^{M-1} + \varepsilon^J)\gamma_{J+1}}{1 + \varepsilon^{M-1}} - \frac{(1-\varepsilon)\varepsilon^J \gamma_{J+1}}{1 + \varepsilon^{M-1}}.$$
$$(5.2.12)$$

We shall define a strategy τ and show that it is better than each τ_J, though it is not necessarily optimal: select arm 1 at the first $M - 1$ stages; continue indefinitely with arm 1 thereafter if a success was observed at one or more of the first $M - 1$ stages, and, if not, select arm 2 at stage M, arm 1 at stage $M + 1$, and thereafter select arm 1 or 2 indefinitely according as a success was or was not obtained at stage $M + 1$. A straightforward calculation gives:

$$W(F, \lambda; \mathbf{A}; \tau) = \frac{(1-\varepsilon)\gamma_1}{1 + \varepsilon^{M-1}} + \lambda \frac{\varepsilon^{M-1}[2\alpha_M + (1+\varepsilon)\gamma_{M+2}]}{1 + \varepsilon^{M-1}}$$
$$- \frac{(1-\varepsilon)(\varepsilon^{M-1}\alpha_M + \varepsilon^M \gamma_{M+2})}{1 + \varepsilon^{M-1}}. \qquad (5.2.13)$$

For $J \leqslant M - 2$, subtraction of (5.2.12) from (5.2.13) gives the following necessary and sufficient condition for τ to be better than τ_J:

$$\lambda < \frac{(1 - \varepsilon)(\gamma_{J+1} - \varepsilon^{M-1-J}\alpha_M - \varepsilon^{M-J}\gamma_{M+2})}{\gamma_{J+1} - \varepsilon^{M-1-J}[2\alpha_M + (1 + \varepsilon)\gamma_{M+2} - \gamma_{J+1}]} \quad (5.2.14)$$

which is asymptotic to $1 - \varepsilon$ as $\varepsilon \downarrow 0$. For $J = M - 1$. (5.2.14) is again the correct condition, and it can be rewritten in this case as

$$\lambda < \frac{(1 - \varepsilon)(\gamma_{M+1} - \varepsilon\gamma_{M+2})}{2\gamma_{M+1} - \gamma_{M+2} - \varepsilon\gamma_{M+2}}. \quad (5.2.15)$$

For $J \geqslant M$ and ε small, the necessary and sufficient condition for τ to be better than τ_J is

$$\lambda > \frac{(1 - \varepsilon)(\alpha_M + \varepsilon\gamma_{M+2} - \varepsilon^{J-M+1}\gamma_{J+1})}{2\alpha_M - \alpha_{M+1} + (\gamma_{M+1} - \gamma_{J+1}) + \varepsilon(\gamma_{M+2} - \varepsilon^{J-M}\gamma_{J+1})} \quad (5.2.16)$$

which is asymptotically (as $\varepsilon \downarrow 0$) bounded above by

$$\frac{(1 - \varepsilon)\alpha_M}{2\alpha_M - \alpha_{M+1}}$$

uniformly in $J \geqslant M$. Conditions (5.2.14), (5.2.15), and (5.2.16) can all be satisfied by an appropriately small ε provided

$$\frac{\alpha_M}{2\alpha_M - \alpha_{M+1}} < \frac{\gamma_{M+1}}{2\gamma_{M+1} - \gamma_{M+2}}.$$

This is an easy consequence of (5.2.11). $\qquad\square$

The fact that (ii) implies (i) in Theorem 5.2.2 means that whenever the discount sequence is regular the decision maker need only decide when, if ever, to stop selecting arm 1. The worth of using arm 2 exclusively from stage m onward is $\lambda\gamma_m$. Thus $\lambda\gamma_m$ could be offered at stage m as a lump-sum alternative to selecting arm 1 and the decision problem would not be changed. So the name 'one-armed bandit' is appropriate for a two-armed bandit with one known arm in the case of regular discounting.

Example 3.3.3, in which **A** is a mixture of two geometric sequences (**A**$_8$ above) shows how complicated the solution of a two-armed bandit with one arm known can be when the discount sequence is not regular. In that example, F is the simplest nondegenerate distribution possible in a bandit problem, and yet arm 2 can be selected an arbitrary number of times (depending on λ) before testing arm 1.

Theorem 5.2.2 does not address the question of unique optimality. It is possible for both arms to be optimal at stage 1. Suppose both arms are optimal and that the discount sequence is geometric: $\mathbf{A} = (1, \alpha, \alpha^2, \ldots)$. If arm 2 is selected then no information is gained; the new discount sequence is $(\alpha, \alpha^2, \ldots) = \alpha\mathbf{A}$ and the problem is essentially unchanged. So again both arms are optimal – and so on. There are many optimal strategies and, in particular, there is one which specifies an arbitrary fixed number of selections of arm 2 and then arm 1, regardless of the results obtained with arm 2. But, of course, the theorem is not violated: there are optimal strategies that stay with arm 2 once it is used.

There are obviously a variety of technically different optimal strategies when the horizon is finite ($\alpha_{n+1} = \ldots = 0$ for some n) since it makes no difference what decisions are made at stages later than n. On the other hand, during an initial segment of 0's in the discount sequence, arm 1 is always optimal; but even here, if there is no information on arm 1 to be obtained that will help in later decisions, or if information gathering can safely be delayed (for instance, if θ_1 is either 0 or 1 and $\alpha_2 = 0$ as well as $\alpha_1 = 0$), then arm 2 will also be optimal. In the latter instance arm 2 would not be used indefinitely.

In Section 1.2 we indicated that patients arriving late in a clinical trial were likely to be treated better than those arriving early when an optimal assignment procedure is followed. The next result verifies this characteristic when but one arm is unknown and the discount sequence is regular. It says that the expected value of an observation increases with time when an optimal strategy is followed. It should be compared with Example 5.2.2, which provides a counterexample in the nonregular case. It should also be compared with Example 2.4.5, which provides a counterexample in a case which is regular but which involves two unknown arms.

Corollary 5.2.3 If $\mathbf{A}$ is regular there exists an optimal strategy τ for the $(F, \lambda; \mathbf{A})$-bandit such that

$$E_\tau(Z_{m+1}|F, \lambda) \geqslant E_\tau(Z_m|F, \lambda) \tag{5.2.17}$$

for $m = 1, 2, \ldots$.

Proof If arm 2 is uniquely optimal initially then take $\tau(\varnothing) = 2$; otherwise take $\tau(\varnothing) = 1$. Define τ inductively to be an optimal strategy that indicates arm 1 whenever possible without requiring a selection of arm 1 following a selection of arm 2. (The fuss in the preceding sentences is required to take care of the possibility that some discount factors may be zero.) We prove (5.2.17) only for $m = 1$; the general case is similar, but requires conditioning on the first $m - 1$ stages.

If $\tau(\varnothing) = 2$, then $\tau(z_1) = 2$ for all z_1, so $E_\tau(Z_2|F, \lambda) = \lambda$ as does $E_\tau(Z_1|F, \lambda)$. If $\tau(\varnothing) = 1$, then $\tau(1) = 1$ and there are two cases. First suppose that $\tau(0) = 1$. Then $E_\tau(Z_2|F, \lambda) = E(\theta_1|F)$ as does $E_\tau(Z_1|F, \lambda)$. Next suppose $\tau(0) = 2$. Then

$$E(Z_2|F, \lambda) = E(\theta_1^2|F) + E(1 - \theta_1|F)\lambda$$

which, by Theorem 5.0.2, is at least as large as

$$
\begin{aligned}
E(\theta_1^2|F) &+ E(1 - \theta_1|F)E(\theta_1|\varphi F) \\
&= E(\theta_1^2|F) + E(1 - \theta_1|F)E(\theta_1(1 - \theta_1)|F)/E(1 - \theta_1|F). \\
&= E(\theta_1|F) = E_\tau(Z_1|F, \lambda). \qquad \square
\end{aligned}
$$

5.3 Optimal strategies – regular discounting

When **A** is regular, the major result of the previous section allows us to obtain a number of results concerning optimal strategies. Though we assume regularity throughout this section, we apply it mainly through its consequence that a bandit with a regular discount sequence is a stopping problem. Results thus obtained generalize to any $(F, \lambda; \mathbf{A})$-bandit that is a stopping problem. When **A** is not regular, deciding whether it is a stopping problem seems as difficult as actually finding an optimal strategy. So we do not incorporate this extra generality explicitly.

The present context is that arm 1 is Bernoulli. But the first two results in this section are true more generally and will be discussed again in Section 5.5.

Corollary 5.1.2 shows the existence of a unique Λ when **A** is nonincreasing. Example 5.1.1 shows that the result does not hold when **A** is not monotone. But it does hold for nonmonotone sequences that are regular. The following theorem is the same as Corollary 5.1.2 with 'regular' in place of 'nonincreasing'. The proof is essentially that of Berry and Fristedt (1979, Theorem 2.2) applied to this more general setting.

Theorem 5.3.1 For each regular discount sequence $\mathbf{A}$ with $\alpha_1 > 0$ and each distribution F on $[0, 1]$, there exists a unique $\Lambda(F, \mathbf{A}) \in [0, 1]$ such that arm 1 is optimal initially in the $(F, \lambda; \mathbf{A})$-bandit if and only if $\lambda \leqslant \Lambda(F, \mathbf{A})$ and arm 2 is optimal if and only if $\lambda \geqslant \Lambda(F, \mathbf{A})$.

Proof Let $\Lambda(F, \mathbf{A}) = \inf\{\lambda \in [0, 1]: \text{arm 2 is optimal for the } (F, \lambda; \mathbf{A})\text{-bandit}\}$. So arm 1 is uniquely optimal if $\lambda < \Lambda(F, \mathbf{A})$. An easy modification of the appropriate portion of the proof of Corollary 5.1.2 shows that both arms are optimal initially for the $(F, \Lambda(F, \mathbf{A}); \mathbf{A})$-bandit. Therefore,

$$V(F, \Lambda(F, \mathbf{A}); \mathbf{A}) = \gamma_1 \Lambda(F, \mathbf{A}), \tag{5.3.1}$$

where, as previously defined, $\gamma_1 = |\mathbf{A}|_1 = \sum_{m=1}^{\infty} \alpha_m$.

It remains to show that arm 1 is not optimal for the $(F, \lambda; \mathbf{A})$-bandit when $\lambda > \Lambda(F, \mathbf{A})$. Consider the $(F, \Lambda(F, \mathbf{A}); \mathbf{A})$-bandit and a strategy τ^* that begins with a selection of arm 1. Then

$$V(F, \Lambda(F, \mathbf{A}); \mathbf{A}) \geqslant W(F, \mathbf{A}); \mathbf{A}; \tau^*). \tag{5.3.2}$$

Selections of arm 2 determine the difference in payoffs for the $(F, \Lambda(F, \mathbf{A}); \mathbf{A})$- and $(F, \lambda; \mathbf{A})$-bandits when strategy τ^* is used in both. Thus

$$W(F, \lambda; \mathbf{A}; \tau^*) - W(F, \Lambda(F, \mathbf{A}); \mathbf{A}; \tau^*) \leqslant [\lambda - \Lambda(F, \mathbf{A})]\gamma_2$$
$$< [\lambda - \Lambda(F, \mathbf{A})]\gamma_1. \tag{5.3.3}$$

From (5.3.1), (5.3.2), and (5.3.3) we see that

$$W(F, \lambda; \mathbf{A}; \tau^*) < \lambda\gamma_1 \leqslant V(F, \lambda; \mathbf{A}),$$

the second inequality following because selecting arm 2 exclusively has worth $\lambda\gamma_1$. So τ^* is not optimal for the $(F, \lambda; \mathbf{A})$-bandit. $\square$

The next result gives a method for calculating Λ, and therefore, for finding optimal strategies. It has been shown by a number of authors for particular discount sequences; notably, Bradt, Johnson, and Karlin (1956, Theorem 4.1) and Gittins and Jones (1979).

Corollary 5.3.2 For **A** regular with $\alpha_1 > 0$, the function Λ is given by

$$\Lambda(F, A) = \max_{\tau(\varnothing) = 1} \frac{E_\tau\left(\sum_{m=1}^{M} \alpha_m X_{1m} | F \right)}{E_\tau\left(\sum_{m=1}^{M} \alpha_m | F \right)} \tag{5.3.4}$$

where M is the (random) stage (possibly $+ \infty$) at which arm 1 is used for the last time when following strategy τ. Among those τ's that begin with arm 1 and never switch back to arm 1 after a selection of arm 2, those that are optimal for the $(F, \Lambda(F, A); A)$-bandit are those for which the maximum in (5.3.4) is attained.

Remarks Various example calculations illustrating (5.3.4) will be carried out in the next section. The right-hand side of (2.4.9) is a special case of (5.3.4). $\quad\square$

Proof of Corollary 5.3.2. In view of Theorem 5.2.2, the only strategies that need be considered are those that begin with arm 1 and never return to arm 1 after switching to arm 2. For such a strategy τ,

$$E_\tau\left(\sum_{m=1}^{M} \alpha_m Z_m | F \right) + \Lambda(F, A) E_\tau\left(\sum_{m > M} \alpha_m | F \right)$$

$$= W(F, \Lambda(F, A); A; \tau)$$

$$\leqslant V(F, \Lambda(F, A); A) = \Lambda(F, A) \sum_{m=1}^{\infty} \alpha_m,$$

where the last equality follows because, by Theorems 5.2.2 and 5.3.1, arm 2 is optimal indefinitely for the $(F, \Lambda(F, A); A)$-bandit. Hence,

$$E_\tau\left(\sum_{m=1}^{M} \alpha_m Z_m \,\middle|\, F \right) \leqslant \Lambda(F, A) E_\tau\left(\sum_{m=1}^{M} \alpha_m \,\middle|\, F \right). \tag{5.3.5}$$

Equality holds in (5.3.5) if and only if τ is an optimal strategy. $\quad\square$

The statement of the next result would be meaningful in the general setting and not just the Bernoulli, but an example similar to Example 2.5.1 can be constructed to show it would not hold.

Corollary 5.3.3 Assume **A** is regular and $\alpha_1 > 0$. Then $\Lambda(F, A)$ is a continuous function of F.

Proof We use superscripts on F to identify distributions that are terms of a sequence. Suppose that $F^n \to F$ as $n \to \infty$ and $\Lambda(F^n, \mathbf{A}) \to \lambda$. We will show that both arms are optimal initially for the $(F, \lambda; \mathbf{A})$-bandit and, therefore, $\lambda = \Lambda(F, \mathbf{A})$. By Theorem 4.1.1,

$$V^{(i)}(F^n, \Lambda(F^n, \mathbf{A}); \mathbf{A}) \to V^{(i)}(F, \lambda; \mathbf{A})$$

for $i = 1, 2$. Since

$$V^{(1)}(F^n, \Lambda(F^n, \mathbf{A}); \mathbf{A}) = V^{(2)}(F^n, \Lambda(F^n, \mathbf{A}); \mathbf{A}),$$

we obtain $V^{(1)}(F, \lambda; \mathbf{A}) = V^{(2)}(F, \lambda; \mathbf{A})$ as desired. $\square$

We turn to results whose statements or proofs are set in the Bernoulli context. They may be true more generally, perhaps in modified form. The weak inequality in the first result follows from Theorem 4.1.6; some additional work is required to obtain the strict inequality.

Corollary 5.3.4 Assume $\mathbf{A}$ is regular, $\alpha_1 > 0$, and F^* is strongly to the right of F. Then

$$\Lambda(F^*, \mathbf{A}) \geqslant \Lambda(F, \mathbf{A}) \tag{5.3.6}$$

with equality if and only if $F^* = F$.

Proof For a proof by contradiction suppose that F^* is strongly to the right of F, $F^* \neq F$, and $\Lambda(F^*, \mathbf{A}) \leqslant \Lambda(F, \mathbf{A})$. Arm 1 is optimal initially for both the $(F^*, \Lambda(F^*, \mathbf{A}); \mathbf{A})$-bandit and the $(F, \Lambda(F^*, \mathbf{A}); \mathbf{A})$-bandit. By Corollary 2.5.3, Lemmas 4.1.3 and 4.1.4, and Theorem 4.1.6,

$$\begin{aligned}
&V(F^*, \Lambda(F^*, \mathbf{A}); \mathbf{A}) - V(F, \Lambda(F^*, \mathbf{A}); \mathbf{A}) \\
&= \alpha_1 E(X_{11} | F^*) + P(X_{11} = 1 | F^*) V(\sigma F^*, \Lambda(F^*, \mathbf{A}); \mathbf{A}^{(1)}) \\
&\quad + P(X_{11} = 0 | F^*) V(\varphi F^*, \Lambda(F^*, \mathbf{A}); \mathbf{A}^{(1)}) \\
&\quad - \alpha_1 E(X_{11} | F) - P(X_{11} = 1 | F) V(\sigma F, \Lambda(F^*, \mathbf{A}); \mathbf{A}^{(1)}) \\
&\quad - P(X_{11} = 0 | F) V(\varphi F, \Lambda(F^*, \mathbf{A}); \mathbf{A}^{(1)}) \\
&> [P(X_{11} = 1 | F^*) - P(X_{11} = 1 | F)] V(\sigma F, \Lambda(F^*, \mathbf{A}); \mathbf{A}^{(1)}) \\
&\quad - [P(X_{11} = 0 | F) - P(X_{11} = 0 | F^*)] V(\varphi F, \Lambda(F^*, \mathbf{A}); \mathbf{A}^{(1)}),
\end{aligned}$$

which is nonnegative since

$$V(\sigma F, \Lambda(F^*, \mathbf{A}); \mathbf{A}^{(1)}) \geqslant V(\varphi F, \Lambda(F^*, \mathbf{A}); \mathbf{A}^{(1)}),$$

a consequence of Lemma 4.1.5 and Theorem 4.1.6. On the other hand, since the strategy that indicates arm 2 at every stage is optimal for the $(F^*, \Lambda(F^*, \mathbf{A}); \mathbf{A})$-bandit,

$$V(F^*, \Lambda(F^*, \mathbf{A}); \mathbf{A}) - V(F, \Lambda(F^*, \mathbf{A}); \mathbf{A})$$
$$= \gamma_1 \Lambda(F^*, \mathbf{A}) - V(F, \Lambda(F^*, \mathbf{A}); \mathbf{A}) \leqslant 0,$$

a contradiction. □

The following characterization of optimal strategies is the analogue of Theorem 5.1.3. It says that when $\mathbf{A}$ is regular, arm 1 should be selected again if it was optimal and gave a success. Since Theorem 5.1.3 applies when $\mathbf{A}$ is nonincreasing, we now have two separate proofs of the result when $\mathbf{A}$ is nonincreasing and also regular.

Theorem 5.3.5 Let $\mathbf{A} = (\alpha_1, \alpha_2, \alpha_3, \ldots)$ be a regular discount sequence with $\alpha_1 > 0$ and $\alpha_2 > 0$. Then for all F,

$$\Lambda(F, \mathbf{A}) \leqslant \Lambda(\sigma F, \mathbf{A}^{(1)}),$$

with equality if and only if F is a one-point distribution.

Remarks While it is picturesque, this stay-with-a-winner characterization offers little help in finding an optimal strategy via dynamic programming, though some calculations can be avoided. On the other hand, as we shall see in Theorem 5.3.7, it does aid in the use of Corollary 5.3.2 since it substantially reduces the number of strategies to be considered in (5.3.4). □

Proof of Theorem 5.3.5 For a proof by contradiction we suppose that $\Lambda(F, \mathbf{A}) \geqslant \Lambda(\sigma F, \mathbf{A}^{(1)})$ and F is not a one-point distribution. By Corollary 5.3.4 $\Lambda(\sigma F, \mathbf{A}^{(1)}) \geqslant \Lambda(\varphi F, \mathbf{A}^{(1)})$. Therefore, an optimal strategy for the $(F, \Lambda(F, \mathbf{A}); \mathbf{A})$-bandit indicates arm 1 at stage 1 and arm 2 thereafter. So,

$$V(F, \Lambda(F, \mathbf{A}); \mathbf{A}) = \alpha_1 E(\theta_1 | F) + \gamma_2 \Lambda(F, \mathbf{A}),$$

which must be at least as large as $\alpha_1 \Lambda(F, \mathbf{A}) + \gamma_2 \Lambda(F, \mathbf{A})$, the worth of the strategy that indicates arm 2 at every stage. Hence,

$$\Lambda(F, \mathbf{A}) \leqslant E(\theta_1 | F) < E(\theta_1 | \sigma F),$$

where strict inequality applies since F is not a one-point distribution.

This contradicts

$$E(\theta_1 | \sigma F) \leqslant \Lambda(\sigma F, \mathbf{A}^{(1)}) \leqslant \Lambda(F, \mathbf{A}),$$

which is a consequence of Theorem 5.0.2 and our supposition.   □

The next result is the analogue of Theorem 5.3.5 in case of failure. It indicates that the inclination to select arm 1 is not increased when a failure is observed. It is, however, not a 'switch-with-a-loser' rule; for, though a failure on arm 1 decreases its relative utility, it may still be preferable to arm 2. The theorem generalizes the corresponding result of Bellman (1956, Theorem 2) in the geometric case.

Theorem 5.3.6 Let **A** be a regular discount sequence with $\alpha_1 > 0$ and $\alpha_2 > 0$. Then for all F,

$$\Lambda(\varphi F, \mathbf{A}^{(1)}) \leqslant \Lambda(F, \mathbf{A}),$$

with equality if and only if F is a one-point distribution.

Proof By Corollary 5.3.4 and Lemma 4.1.5, $\Lambda(\varphi F, \mathbf{A}^{(1)}) \leqslant \Lambda(F, \mathbf{A}^{(1)})$ with equality if and only if F is a one-point distribution. By Theorem 5.2.2, $\Lambda(F, \mathbf{A}^{(1)}) \leqslant \Lambda(F, \mathbf{A})$.   □

The next theorem is a corollary of the previous two theorems. It says that the search for optimal strategies can be restricted in the regular case to the one using arm 2 exclusively and those of the following form. Use arm 1 as long as it is successful. When and if it fails, continue with arm 1 only if there were a sufficient number of successes on arm 1, say r_1, before the first failure and otherwise switch permanently to arm 2. If the number of successes prior to the first failure is at least r_1 and so arm 1 is continued, then it is again used as long as it is successful. When and if arm 1 fails for the second time, continue with arm 1 if and only if the total number of successes on arm 1 is at least $r_1 + r_2$. Proceeding in this fashion, arm 1 is continued after the jth failure if and only if it has produced at least $r_1 + \ldots + r_j$ successes. Let $\tau^{(\mathbf{R})}$ denote the strategy just described that corresponds to the sequence $\mathbf{R} = (r_1, r_2, \ldots), r_l \in \{0, 1, \ldots, \infty\}$.

Theorem 5.3.7 Suppose **A** is regular and $\alpha_1 > 0$. If $\lambda \leqslant \Lambda(F, \mathbf{A})$, then some $\tau^{(\mathbf{R})}$ is an optimal strategy. If $\lambda = \Lambda(F, \mathbf{A})$, r_1 can be chosen to be positive.

Remark As a consequence of this theorem, the maximum in (5.3.4) can be taken over strategies $\tau^{(\mathbf{R})}$ with $r_1 > 0$.   □

Proof of Theorem 5.3.7 Since $\lambda \leqslant \Lambda(F, \mathbf{A})$, there is an optimal strategy for the $(F, \lambda; \mathbf{A})$-bandit which begins with arm 1 and, by Theorem 5.3.5, switches to arm 2, if ever, only after a failure on arm 1. Corollary 5.3.4 and Lemma 4.1.5 apply to show the existence of $j_i \in \{0, 1, \ldots, \infty\}$ such that it is optimal to switch after the ith failure if the number of successes is less than j_i and not to switch otherwise. That some $\tau^{(\mathbf{R})}$ is an optimal strategy will follow by setting $r_1 = j_1$ and $r_i = j_i - j_{i-1}$ for $i > 1$ (with $\infty - \infty$ defined arbitrarily) if we can show $j_i \geqslant j_{i-1}$.

Suppose to the contrary that $j_i < j_{i-1}$ for some $i > 1$. Then, with $\mathbf{A}^{(m)}$ denoting the discount sequence $(\alpha_{m+1}, \alpha_{m+2}, \ldots)$, arm 2 is optimal initially for the $(\sigma^{j_i} \varphi^{i-1} F, \lambda; \mathbf{A}^{(j_i + i - 1)})$-bandit and arm 1 is optimal initially for the $(\sigma^{j_i} \varphi^i F, \lambda; \mathbf{A}^{(j_i + i)})$-bandit. Hence,

$$\Lambda(\varphi \sigma^{j_i} \varphi^{i-1} F, \mathbf{A}^{(j_i + i)}) \geqslant \Lambda(\sigma^{j_i} \varphi^{i-1} F, \mathbf{A}^{(j_i + i - 1)}) \qquad (5.3.8)$$

(assuming both sides are meaningful), which contradicts Theorem 5.3.6 provided $\sigma^{j_i} \varphi^{i-1} F$ is not a one-point distribution. If it is a one-point distribution, then without loss we can redefine $j_i = \infty$ if $j_{i-1} > 0$, and $j_i = 0$ if $j_{i-1} = 0$. Suppose $\alpha_{j_i + i} = 0$, so the left-hand side of (5.3.8) is not defined. Then, since $\mathbf{A}$ is regular and $\alpha_1 \neq 0$, it follows that $\alpha_m = 0$ for $m > j_i + i$. Hence, we can redefine $j_i = \infty$ without losing optimality. All distributions $\sigma^{j_i} \varphi^i F$ in (5.3.8) are meaningful unless $F(\{1, 0\}) = 1$. But in this case $\tau^{(\mathbf{R})}$ with $\mathbf{R} = (\infty, \infty, \ldots)$ is optimal.

For the second part of the theorem suppose $\lambda = \Lambda(F, \mathbf{A})$. Suppose arm 1 is selected initially and yields a failure. Then by Theorem 5.3.6, arm 2 must be selected at stage 2 provided F is not a one-point distribution. Thus, r_1 must be positive for $\tau^{(\mathbf{R})}$ to be optimal in this case. If F is a one-point distribution, then $F = \delta_\lambda$ since $\lambda = \Lambda(F, \mathbf{A})$ and so all strategies are optimal; in particular, any $\tau^{(\mathbf{R})}$ with $r_1 > 0$ is optimal.   □

By way of illustration, suppose the discount sequence $\mathbf{A}$ is regular with $\alpha_1 > 0$ and $\alpha_m = 0$ for $m \geqslant 6$. To find Λ via Corollary 5.3.2 we can apply Theorem 5.3.7 to see that only eight essentially different sequences **R** need be considered:

$$(4, \ldots), \qquad\qquad (1, 2, \ldots),$$
$$(3, \ldots) \qquad\qquad (1, 1, \ldots),$$
$$(2, 0, \ldots), \qquad\qquad (1, 0, 1, \ldots),$$
$$(2, 1, \ldots), \qquad\qquad (1, 0, 0, \ldots).$$

The parts of these sequences indicated with dots are of no consequence in any calculation. The list is complete in the sense that no other sequence corresponds to a strategy essentially different from those that correspond to listed sequences. For instance, $(5, \ldots)$ is essentially the same as the listed sequence $(4, \ldots)$: both require a switch to arm 2 if a failure occurs at any of the first four stages, and subsequent selections are immaterial if the first failure occurs after stage 4.

We now have two ways of finding an optimal strategy in a particular problem. One is to use dynamic programming as discussed in Chapter 2, where an approximation is required if the horizon is infinite. Another is to use Corollary 5.3.2, with Theorem 5.3.7 eliminating the need to consider many strategies.

An advantage of the latter approach is that, for the pair $(F, \mathbf{A})$, it finds Λ and therefore indicates an optimal selection for any $\lambda \in (0, 1)$. A disadvantage is that it gives only the optimal initial selections. When arm 1 is optimal initially, deciding on a second selection requires finding $\Lambda(\sigma F, \mathbf{A}^{(1)})$ or $\Lambda(\varphi F, \mathbf{A}^{(1)})$, which entails additional calculation.

On the other hand, if $\Lambda(F, \mathbf{A})$ is desired, and not just an optimal strategy, then dynamic programming is handicapped; for successive approximations are necessary and a separate dynamic program must be carried out at each step. One way to proceed is to guess a value for Λ, decide which arm is optimal, adjusting the guess accordingly (upwards if arm 1 is optimal and downwards if it is not). Repeating this process until both arms are optimal gives Λ. Still, such a computer program is easy to write, and we found ours to be more efficient than our program that finds Λ using Corollary 5.3.2 and Theorem 5.3.7.

Calculations of Λ are given in the next section for various examples. These are compared with easy-to-compute lower and upper bounds.

5.4 Bernoulli examples and bounds – regular discounting

In view of Theorem 5.3.1, when $\mathbf{A}$ is regular the optimal initial selection is specified by $\Lambda(F, \mathbf{A})$; namely, arm 1 or arm 2 according as $\lambda \leqslant$ or $\geqslant \Lambda(F, \mathbf{A})$. In view of Theorem 5.2.2, if $\lambda \geqslant \Lambda(F, \mathbf{A})$ then, assuming arm 2 is selected initially, at least one optimal continuation is clear: select arm 2 indefinitely. When $\lambda < \Lambda(F, \mathbf{A})$ then optimal continuations can be found from $\Lambda(\sigma^s \varphi^f F, \mathbf{A}^{(s+f)})$ for nonnegative integers s and f. In any case, optimal strategies are specified by the various Λ's.

As indicated in the previous section, calculating $\Lambda(F, \mathbf{A})$ can require extensive numerical calculations. However, for certain combinations of F and $\mathbf{A}$ it is possible to give an explicit formula for $\Lambda(F, \mathbf{A})$. Our main objective in this section is to give bounds for Λ. These can be used to approximate Λ and will be most useful when Λ is not available. But we begin with an example in which explicit calculation of Λ is possible. The discount sequence in this example is geometric. So in view of the forthcoming Theorem 6.1.1, the calculation of Λ has special significance. This theorem says that when there are k independent arms, so that $G = F_1 \times \ldots \times F_k$, arm i is optimal initially if and only if

$$\Lambda(F_i, \mathbf{A}) = \max_{1 \leqslant j \leqslant k} \Lambda(F_j, \mathbf{A}).$$

Examples 6.1.1 and 6.1.2 apply Theorem 6.1.1 by exploiting the calculations given in the next example.

Example 5.4.1 Suppose $\mathbf{A} = (1, \alpha, \alpha^2, \ldots)$ and

$$F = p\delta_a + (1-p)\delta_b$$

where $0 < b < a < 1$. Berry and Fristedt (1979, Example 4.3) show that

$$\Lambda(F, \mathbf{A}) = \frac{b(1-p)g(\alpha, b, c) + apg(\alpha, a, c)}{(1-p)g(\alpha, b, c) + pg(\alpha, a, c)},$$

where

$$c = \frac{\log[(1-b)/(1-a)]}{\log[(1-b)/(1-a)] + \log[a/b]},$$

$$g(\alpha, u, c) = \exp\left[-\sum_{m=1}^{\infty} \frac{\alpha^m}{m} P(S_m < 0)\right],$$

and $(S_1, S_2, \ldots)$ is a random walk starting at 0 with individual steps of

$1 - c$ with probability u

$- c$ with probability $1 - u$.

As a special case, when $a + b = 1$ then $c = 1/2$ and

$$g(\alpha, u, 1/2) = 1 - \frac{1 - [1 - 4\alpha^2 u(1 - u)]^{1/2}}{2\alpha u}$$

After considerable algebra,

$$\Lambda(F, \mathbf{A}) = \frac{[ap + b(1 - p)][1 + (1 - 4\alpha^2 ab)^{1/2}] - 2\alpha ab}{1 + (1 - 4\alpha^2 ab)^{1/2} - 2\alpha[ap + b(1 - p)]}.$$

Specializing further by taking $p = \frac{1}{2}$, we obtain

$$\Lambda(F, \mathbf{A}) = \frac{[1 + (1 - 4\alpha^2 ab)^{1/2}] - 4\alpha ab}{2[1 + (1 - 4\alpha^2 ab)^{1/2}] - 2\alpha}.$$

Corresponding to the intuitive notion that experimenting with arm 1 has more potential benefit for larger α, this is an increasing function of α that equals $1/2 = E(\theta_1 | F)$ at $\alpha = 0$ and approaches a as $\alpha \uparrow 1$. $\square$

The lower and upper bounds we derive in this section will be compared with each other and with the exact Λ in special cases. Two such cases are provided by the next two examples. Explicit formulas for Λ are not possible in these examples; of the two methods for finding Λ that were compared at the end of the previous section, we used dynamic programming since it uses computer time more efficiently.

Example 5.4.2 Suppose F has a beta density: for $a, b > 0$,

$$dF(u) \propto u^{a-1}(1 - u)^{b-1} du. \tag{5.4.1}$$

Take $\mathbf{A}$ to be the n-horizon uniform: $\alpha_1 = \ldots = \alpha_n = 1, \alpha_{n+1} = \ldots = 0$. It is clear that $\Lambda(F, \mathbf{A}) = E(\theta_1 | F) = a/(a + b)$ if $n = 1$. From Theorem 5.4.1 it will follow that $\Lambda(F, \mathbf{A}) \to 1$ as $n \to \infty$. Figure 5.1 shows $\Lambda(F, \mathbf{A})$ up to $n = 500$ for $(a, b) = (1, 5)$ and $(1, 1)$; the latter corresponds to the uniform distribution on $(0, 1)$. $\square$

Example 5.4.3 We return to the discounting of Example 5.4.1: $\mathbf{A} = (1, \alpha, \alpha^2, \ldots)$. As in Example 5.4.2, θ_1 has a distribution in the

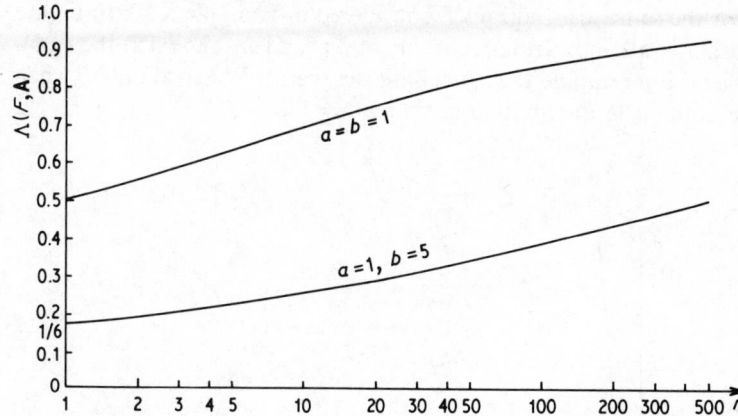

Fig. 5.1 *The break-even value of the known arm as a function of the horizon n for two beta distributions.*

beta family (5.4.1). For reasons similar to those given in that example for the finite-horizon case, $\Lambda(F, \mathbf{A}) = E(\theta_1 | F) = a/(a + b)$ when $\alpha = 0$ and $\Lambda(F, \mathbf{A}) \to 1$ as $\alpha \to 1$. Figure 5.2 shows $\Lambda(F, \mathbf{A})$ for $\alpha \in (0, 1)$

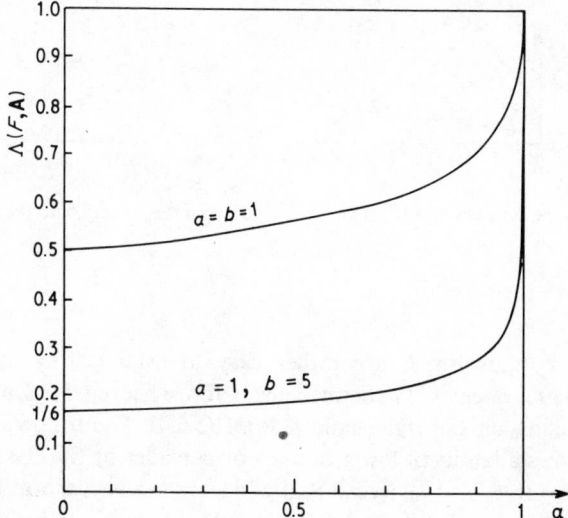

Fig. 5.2 *The break-even value of the known arm as a function of the discount factor α for two beta distributions.*

for the same distributions F considered in Example 5.4.2. In Figure 5.3, $\alpha = 0.9$ and various contours of $\Lambda(F, \mathbf{A})$ are shown in the (a, b)-plane. For example, the figure indicates that $\Lambda(F, \mathbf{A})$ is about 0.7 when F is the beta distribution given by $a = 5$, $b = 3$.   □

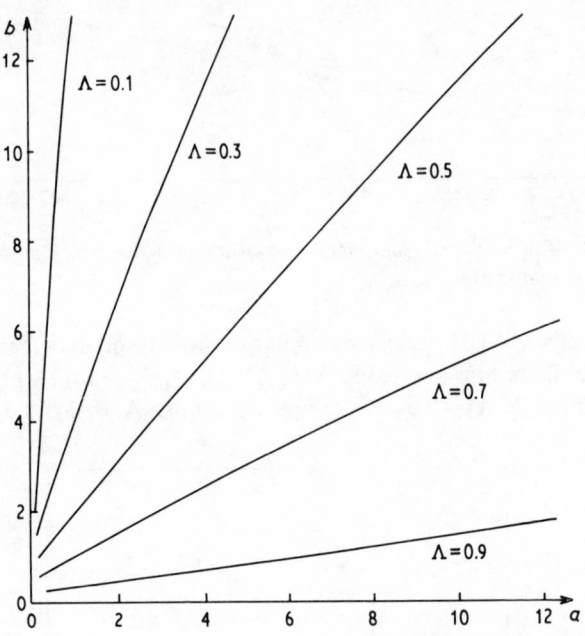

Fig. 5.3   *Contours of $\Lambda(F, \mathbf{A})$ when $\alpha = 0.9$ as a function of beta parameters a and b.*

Lower bounds for Λ are rather easy to establish by using any particular strategy $\tau^{(\mathbf{R})}$, as defined just before Theorem 5.3.8, instead of maximizing on the right-hand side of (5.3.4). The following result establishes a family of lower bounds by considering the class of $\tau^{(\mathbf{R})}$'s with $\mathbf{R} = (r, 0, 0, \ldots)$. Recall that $\tau^{(r, 0, 0, \ldots)}$ indicates arm 1 until it fails; if this occurs before the rth success (i.e., by the rth stage) then arm 2 is used indefinitely thereafter, and if it occurs later then arm 1 is continued indefinitely.

Theorem 5.4.1 Suppose that A is a regular discount sequence with $\alpha_1 > 0$. Then for each $r \in \{1, 2, \ldots, \infty\}$, $\Lambda(F, A) \geq \Lambda_r(F, A)$, where

$$\Lambda_r(F, A) = \frac{\sum\limits_{m=1}^{\infty} \alpha_m E(\theta_1^{m \wedge (r+1)} | F)}{\sum\limits_{m=1}^{\infty} \alpha_m E(\theta_1^{(m-1) \wedge r} | F)}. \tag{5.4.2}$$

Proof That $\Lambda \geq \Lambda_r$ follows by evaluating the ratio in (5.3.4) at $\tau^{(\mathbf{R})}$ where $\mathbf{R} = (r, 0, 0, \ldots)$, instead of maximizing over $\tau(\varnothing) = 1$. $\quad\square$

The bounds Λ_r given in Theorem 5.4.1 do not usually equal Λ. The next result lists the circumstances under which they are exact in the infinite horizon case. The result is given without proof.

Theorem 5.4.2 In addition to the assumptions of Theorem 5.4.1, assume that the horizon of A is infinite. Then $\Lambda(F, A) = \Lambda_r(F, A)$ if and only if either (i) F is supported by $\{0, \Lambda_r(F, A), 1\}$, or (ii) $r = 1$ and F is supported by $\{0\} \cup [\Lambda_1(F, A), 1]$, or (iii) $r = \infty$ and F is supported by $[0, \Lambda(F, A)] \cup \{1\}$.

Let

$$\Lambda_*(F, A) = \sup \{\Lambda_r(F, A): r = 1, 2, \ldots, \infty\}.$$

The next result says that this supremum is attained and the sequence $(\Lambda_1, \Lambda_2, \ldots, \Lambda_\infty)$ is quite well-behaved.

Proposition 5.4.3 For any F and regular A with $\alpha_1 > 0$,

$$\{\Lambda_r(F, A): r = 1, 2, \ldots, \infty\}$$

is unimodal and $\Lambda_\infty(F, A) = \lim\limits_{r \to \infty} \Lambda_r(F, A)$.

Proof Unimodality will follow from

$$\Lambda_{r-1} - \Lambda_r \geq 0 \Rightarrow \Lambda_r - \Lambda_{r+1} \geq 0,$$

or equivalently,

$$E\left((\theta_1 - \Lambda_r) \sum_{m=1}^{\infty} \alpha_m \theta_1^{(m-1) \wedge (r-1)} \middle| F\right) \geq 0$$

$$\Rightarrow E\left((\Lambda_r - \theta_1) \sum_{m=1}^{\infty} \alpha_m \theta_1^{(m-1) \wedge (r+1)} \middle| F\right) \geq 0. \tag{5.4.3}$$

In view of (5.4.2),

$$E\left((\theta_1 - \Lambda_r) \sum_{m=1}^{\infty} \alpha_m \theta_1^{(m-1)\wedge r} | F\right) = 0. \tag{5.4.4}$$

Subtracting (5.4.4) from the first expression in (5.4.3) and adding it to the second reduces the problem to showing

$$\left(\sum_{m=r+1}^{\infty} \alpha_m\right) E\left((\theta_1 - \Lambda_r)\theta_1^{r-1}(1-\theta_1)\bigg| F\right) \geqslant 0$$

$$\Rightarrow \left(\sum_{m=r+2}^{\infty} \alpha_m\right) E\left((\theta_1 - \Lambda_r)\theta_1^{r}(1-\theta_1)\bigg| F\right) \geqslant 0.$$

If $\sum_{m=r+2}^{\infty} \alpha_m = 0$ then the implication holds trivially. So we need only show that

$$E\left((\theta_1 - \Lambda_r)\theta_1^{r-1}(1-\theta_1)\bigg| F\right) \geqslant 0 \tag{5.4.5}$$

implies

$$E\left((\theta_1 - \Lambda_r)\theta_1^{r}(1-\theta_1)\bigg| F\right) \geqslant 0. \tag{5.4.6}$$

Subtracting Λ_r times the left-hand side of (5.4.5) from (5.4.6) gives

$$E\left((\theta_1 - \Lambda_r)^2 \theta_1^{r-1}(1-\theta_1)\bigg| F\right),$$

which is obviously nonnegative. Therefore $(\Lambda_1, \Lambda_2, \ldots, \Lambda_\infty)$ is unimodal for all (F, A). The continuity of $\Lambda_r(F, A)$ at $r = \infty$ is immediate. ☐

In case the horizon n is finite, Λ_* can be found with a finite number of calculations since $\Lambda_r = \Lambda_n$ for $r \geqslant n$. The next result indicates that Λ_* can also be found with a finite number of calculations when n is infinite. The result gives an easy-to-check condition for the sequence $(\Lambda_1, \ldots, \Lambda_\infty)$ to be nondecreasing. The proof is not trivial but is omitted since the result is not important in the sequel.

Proposition 5.4.4 Assume A is regular with $\alpha_1 > 0$. When the horizon of A is infinite, $\Lambda_*(F, A) = \Lambda_\infty(F, A)$ if and only if F is supported by $[0, \Lambda_\infty(F, A)] \cup \{1\}$.

Let r_* denote the smallest r for which (5.4.2) gives the best lower bound:

$$r_* = \min\{r: \Lambda_r(F, \mathbf{A}) = \Lambda_*(F, \mathbf{A})\}.$$

So if a decision maker were required to select arm 1 until it fails once and then commit permanently to one or the other arm, it would be optimal to commit to arm 1 if at least r_* successes had been observed and to arm 2 otherwise.

We now use Theorem 5.4.2 in case F is supported by $\{1, 0\}$. This easy example (cf. Example 1.2.1) is given as a setting in which a simple expression for $\Lambda(F, \mathbf{A})$ is possible, and is given by the bounds in Theorem 5.4.1. Theorem 5.4.2 provides generalizations of this example in two directions when the horizon is infinite.

Example 5.4.4 Suppose $\mathbf{A}$ is an arbitrary regular discount sequence with $\alpha_1 > 0$ and

$$F = p\delta_1 + (1 - p)\delta_0.$$

Then every bound in (5.4.2) is exact. So

$$\Lambda(F, \mathbf{A}) = \frac{p \sum_{m=1}^{\infty} \alpha_m}{p \sum_{m=1}^{\infty} \alpha_m + (1 - p)\alpha_1} = \frac{p\gamma_1}{p\gamma_2 + \alpha_1},$$

where, as defined earlier, $\gamma_2 = \sum_{m=2}^{\infty} \alpha_m$. $\qquad\square$

The next two examples illustrate Theorem 5.4.1 with the discount sequences considered in Examples 5.4.2 and 5.4.3 and with θ_1 uniform on $(0, 1)$.

Example 5.4.5 Consider n-horizon uniform discounting: $\mathbf{A} = (1, 1, \ldots, 1, 0, \ldots)$. Then $\Lambda(F, \mathbf{A})$ is at least

$$\Lambda_r(F, \mathbf{A}) = \frac{\sum_{m=1}^{n} E(\theta_1^{m \wedge (r+1)} | F)}{\sum_{m=1}^{n} E(\theta_1^{(m-1) \wedge r} | F)}$$

for $r = 1, 2, \ldots, \infty$; these bounds were obtained by Bradt, Johnson, and Karlin (1956).

For $F = U(0, 1), \Lambda(F, \mathbf{A})$ is shown in Figure 5.1 (where it is given by the curve labelled $a = b = 1$). For $r \leqslant n$,

$$\Lambda_r(F, \mathbf{A}) = \frac{\sum_{m=1}^{r} \frac{1}{m+1} + \frac{n-r}{r+2}}{\sum_{m=1}^{r} \frac{1}{m} + \frac{n-r}{r+1}},$$

and $\Lambda_{n-1} = \Lambda_n = \ldots = \Lambda_\infty$. It can be shown that $r_* \sim \sqrt{n}$ as $n \to \infty$.

Table 5.1 compares Λ with Λ_* for various values of n, and gives r_*. The values of Λ were calculated to four-decimal accuracy using the BASIC program listed below. It involves dynamic programming and iterates on λ until both arms are optimal. The example output for $n = 100$ determines Λ to be 0.8685. The nine iterations required a total of six minutes on a Commodore 64 personal computer and only 15 memory locations were used for V. Because $\Lambda(F, \mathbf{A})$ increases with n, the locations and time required increase substantially less rapidly than n and n^2, respectively. Still, computation time can be prohibitive

Table 5.1   *Lower and upper bounds for $\Lambda(F, \mathbf{A})$ where $F = U(0, 1)$ and $\mathbf{A}$ is the n-horizon uniform.*

n	r_*	$\Lambda_*(F, \mathbf{A})$	$\Lambda(F, \mathbf{A})$	$\Lambda^*(F, \mathbf{A})$
1	1	0.5000	0.5000	0.5000
2	1	0.5556	0.5556	0.5556
5	3	0.6357	0.6357	0.6490
10	4	0.6954	0.6981	0.7189
20	6	0.7512	0.7576	0.7806
50	8	0.8164	0.8262	0.8466
100	12	0.8575	0.8685	0.8850
200	16	0.8914	0.9024	0.9148
500	24	0.9258	0.9351	0.9436
1 000	34	0.9452	0.9536	0.9590
10 000	103	0.9811	0.9850	0.9863
100 000	321	0.9938	0.9953	0.9953
$\to \infty$	$\sim \sqrt{n}$	1.0000	1.0000	1.0000

for large n. Such limitations can be substantially alleviated using a sequence of truncations.

```
10  DIM V(15): L = .87: N = 100: H = .0064
20  M = N: M1 = INT ((1 − L)*N) + 1
30  FOR I = 0 TO M1: V(I) = 0: NEXT I
40  IF M = 0 THEN GOTO 110
50  M = M − 1
60  FOR I = 0 TO M1: R = M − I: P = (R + 1)/(M + 2)
70  V1 = P*(1 + V(I)) + (1 − P)*V(I + 1): V2 = (N − M)*L
80  IF V1 > V2 THEN V(I) = V1: GOTO 100
90  V(I) = V2: GOTO 40
100 NEXT I
110 D = V(0) − N*L
120 PRINT "L = "L, "D = "D
130 IF H < .00004 THEN END
140 IF D = 0 THEN L = L − H: H = H/2: GOTO 20
150 L = L + H: H = H/2: GOTO 20
```

EXAMPLE OUTPUT

L = .87	D = 0
L = .8636	D = .0587253608
L = .8668	D = .0198577987
L = .8684	D = 6.55137934E − 04
L = .8692	D = 0
L = .8688	D = 0
L = .8686	D = 0
L = .8685	D = 0
L = .86845	D = 7.6983124E − 05

Table 5.1 also gives an upper bound $\Lambda^*(F, \mathbf{A})$ to be discussed shortly.

$\square$

Example 5.4.6 Suppose $\mathbf{A}$ is geometric: $\mathbf{A} = (1, \alpha, \alpha^2, \ldots)$. Then $\Lambda(F, \mathbf{A})$ is at least

$$\Lambda_r(F, \mathbf{A}) = \frac{E[\theta_1(1 - \alpha^r\theta_1^r)(1 - \alpha\theta_1)^{-1} + \alpha^r\theta_1^{r+1}(1 - \alpha)^{-1}|F)}{E[(1 - \alpha^r\theta_1^r)(1 - \alpha\theta_1)^{-1} + \alpha^r\theta_1^r(1 - \alpha)^{-1}|F]}$$

for $r = 1, 2, \ldots, \infty$. While not generally the best lower bound, that given by $r = \infty$ simplifies:

$$\Lambda_\infty(F, \mathbf{A}) = \frac{\psi(\alpha) - 1}{\alpha\psi(\alpha)},$$

where the generating function ψ is given by

$$\psi(\alpha) = E[(1 - \alpha\theta_1)^{-1} | F].$$

Suppose $F = U(0, 1)$, then $\Lambda(F, \mathbf{A})$ is shown in Figure 5.2 (by the curve labelled $a = b = 1$). For $r = 1, 2, \ldots, \infty$,

$$\Lambda_r(F, \mathbf{A}) = \frac{\displaystyle\sum_{m=1}^{r} \frac{\alpha^{m-1}}{m+1} + \frac{\alpha^r}{(r+2)(1-\alpha)}}{\displaystyle\sum_{m=1}^{r} \frac{\alpha^{m-1}}{m} + \frac{\alpha^r}{(r+1)(1-\alpha)}}.$$

Using

$$\psi(\alpha) = -\log(1-\alpha)/\alpha,$$

the limit of this sequence is

$$\Lambda_\infty(F, \mathbf{A}) = \frac{1}{\alpha} + \frac{1}{\log(1-\alpha)}.$$

As a corollary of the corresponding fact with $n \to \infty$ in Example 5.4.5, $r_* \sim 1/\sqrt{(1-\alpha)}$ as $\alpha \to 1$. Table 5.2 compares Λ with Λ_* and Λ_∞ for various values of α. To calculate Λ we used a modification of the program given in the previous example and the truncations discussed in Section 2.6. Table 5.2 also gives an upper bound to be discussed shortly. Note the similarity between Tables 5.1 and 5.2 with n corresponding to $1/(1-\alpha)$.   □

To obtain a lower bound, one need only specify a strategy and evaluate the ratio in (5.3.4) for that strategy. An upper bound requires another method. One that applies to decision problems generally is to assume that any unknowns will be revealed at some future time, and that the decision maker behaves optimally with this new source of information taken into consideration. Such a program is carried out next. The result is not as readily applied as the lower bound given by Theorem 5.4.1 since the solution of the forthcoming equation (5.4.7) will require iteration.

Table 5.2 *Lower and upper bounds for* $\Lambda(F, \mathbf{A})$ *where* $F = U(0, 1)$ *and* $\mathbf{A}$ *is geometric with parameter* α.

α	r_*	$\Lambda_\infty(F, \mathbf{A})$	$\Lambda_*(F, \mathbf{A})$	$\Lambda(F, \mathbf{A})$	$\Lambda^*(F, \mathbf{A})$
0	1	0.5000	0.5000	0.5000	0.5000
0.1	2	0.5088	0.5088	0.5088	0.5110
0.2	2	0.5186	0.5187	0.5187	0.5234
0.3	2	0.5297	0.5299	0.5300	0.5374
0.4	2	0.5424	0.5431	0.5432	0.5536
0.5	2	0.5573	0.5588	0.5590	0.5728
0.6	2	0.5753	0.5781	0.5788	0.5962
0.7	3	0.5980	0.6028	0.6046	0.6260
0.8	3	0.6286	0.6385	0.6413	0.6667
0.9	4	0.6768	0.6974	0.7029	0.7317
0.95	6	0.7188	0.7525	0.7614	0.7888
0.97	7	0.7457	0.7900	0.8004	0.8252
0.99	12	0.7930	0.8579	0.8699	0.8874
0.999	34	0.8562	0.9452	0.9538	0.9593
0.9999	103	0.8915	0.9811	0.9851	0.9864
0.99999	321	0.9132	0.9938	0.9953	0.9953
1.0	$\sim(1-\alpha)^{-1/2}$	1.0000	1.0000	1.0000	1.0000

Theorem 5.4.5 Suppose that $\mathbf{A}$ is a regular discount sequence with $\alpha_1 > 0$. Then $\Lambda(F, \mathbf{A})$ is not greater than the unique solution in $[0, 1]$, say $\lambda = \Lambda^*(F, \mathbf{A})$, of the equation:

$$\lambda[\alpha_1 + \gamma_2 E(\theta_1 | F)] - [\alpha_1 E(\theta_1 | F) + \gamma_2 E(\theta_1^2 | F)]$$

$$- \sum_{m=3}^{\infty} \alpha_m E[(\lambda - \theta_1)^+ \theta_1 (1 - \theta_1^{m-2}) | F] = 0,$$
(5.4.7)

where $\gamma_2 = \sum_{m=2}^{\infty} \alpha_m$.

Proof The left-hand side of (5.4.7) is nonpositive when $\lambda = 0$, nonnegative when $\lambda = 1$, and strictly increasing for $\lambda \in [0, 1]$. So we need only show that it is nonpositive when $\lambda = \Lambda(F, \mathbf{A})$.

For calculational convenience we temporarily assume that $F(\{1\}) = 0$.

Both arms are optimal initially in the $(F, \Lambda(F, \mathbf{A}); \mathbf{A})$-bandit in view of Theorem 5.3.1. First consider an optimal strategy, say τ, that starts with arm 1. In view of Theorem 5.3.6 we can let τ indicate arm 2 indefinitely if arm 1 fails on the initial selection: $\tau(0) = 2$ and $\tau(0, z_2, \ldots, z_m) = 2$ for all $m \geq 2$ and all $z_2, \ldots, z_m$. And in view of Theorem 5.3.5 we can take $\tau(1) = 1$ and, more generally, $\tau(1, 1, \ldots, 1) = 1$.

We have no general results which indicate $\tau(1, 1, \ldots, 1, 0)$, the arm to select after a failure (on arm 1) that occurs later than the first stage. But we know that the value of the $(F, \lambda; \mathbf{A})$-bandit is no greater than the corresponding 'value' should θ_1 become known in such a circumstance. So, where $\lambda = \Lambda(F, \mathbf{A})$,

$$
\begin{aligned}
V(F, \lambda; \mathbf{A}) = E_\tau \Bigg[& (1 - \theta_1)\lambda\gamma_2 \\
& + \sum_{n=1}^{\infty} \theta_1^n (1 - \theta_1) \Bigg(\sum_{m=1}^{n} \alpha_m + \sum_{m=n+2}^{\infty} \alpha_m Z_m \Bigg) \Bigg| F \Bigg] \\
\leq E \Bigg[& (1 - \theta_1)\lambda\gamma_2 \\
& + \sum_{n=1}^{\infty} \theta_1^n (1 - \theta_1) \Bigg(\sum_{m=1}^{n} \alpha_m + \sum_{m=n+2}^{\infty} \alpha_m (\theta_1 \vee \lambda) \Bigg) \Bigg| F \Bigg]. \quad (5.4.8)
\end{aligned}
$$

Simple calculations give

$$
\sum_{n=1}^{\infty} \theta_1^n (1 - \theta_1) \sum_{m=1}^{n} \alpha_m = \sum_{m=1}^{\infty} \alpha_m \sum_{n=m}^{\infty} \theta_1^n (1 - \theta_1) = \sum_{m=1}^{\infty} \alpha_m \theta_1^m
$$

and

$$
\begin{aligned}
\sum_{n=1}^{\infty} \theta_1^n (1 - \theta_1)(\theta_1 \vee \lambda) \sum_{m=n+2}^{\infty} \alpha_m &= \sum_{m=3}^{\infty} \alpha_m \sum_{n=1}^{m-2} \theta_1^n (1 - \theta_1)(\theta_1 \vee \lambda) \\
&= \sum_{m=3}^{\infty} \alpha_m (\theta_1 - \theta_1^{m-1})(\theta_1 \vee \lambda) \\
&= \sum_{m=3}^{\infty} \alpha_m (\lambda - \theta_1)^+ \theta_1 (1 - \theta_1^{m-2}) + \sum_{m=3}^{\infty} \alpha_m \theta_1^2 (1 - \theta_1^{m-2}) \\
&= \sum_{m=3}^{\infty} \alpha_m (\lambda - \theta_1)^+ \theta_1 (1 - \theta_1^{m-2}) + \gamma_3 \theta_1^2 - \sum_{m=3}^{\infty} \alpha_m \theta_1^m.
\end{aligned}
$$

So rewriting (5.4.8),

$$V(F, \lambda; \mathbf{A}) \leqslant E\left[(1-\theta_1)\lambda\gamma_2 + \alpha_1\theta_1 + \gamma_2\theta_1^2 \right.$$

$$\left. + \sum_{m=3}^{\infty} \alpha_m(\lambda-\theta_1)^+\theta_1(1-\theta_1^{m-2})|F \right]. \quad (5.4.9)$$

Since the expression in square brackets in (5.4.9) is appropriate as well when $\theta_1 = 1$, we can lift the restriction $F(\{1\}) = 0$.

Now suppose arm 2 is selected in the $(F, \Lambda(F, \mathbf{A}); \mathbf{A})$-bandit. Theorem 5.2.2 applies to show that arm 2 is optimal indefinitely and

$$V(F, \lambda; \mathbf{A}) = \lambda\gamma_1, \quad (5.4.10)$$

where $\lambda = \Lambda(F, \mathbf{A})$. Subtracting the right-hand side of (5.4.9) from the right-hand side of (5.4.10) gives the left-hand side of (5.4.7) and shows that it is nonpositive when $\lambda = \Lambda(F, \mathbf{A})$, as desired. $\qquad\square$

The bound given implicitly in (5.4.7) can be difficult to evaluate. A cruder but more easily computable bound is provided next. Its proof is immediate upon replacing $(\lambda-\theta_1)^+$ with the larger quantity $\lambda(1-\theta_1)$ in (5.4.7) and solving.

Corollary 5.4.6 Under the assumptions of Theorem 5.4.5,

$$\Lambda(F, \mathbf{A}) \leqslant \frac{\alpha_1 E(\theta_1|F) + \gamma_2 E(\theta_1^2|F)}{\alpha_1 + \gamma_2 E(\theta_1|F) - \sum_{m=3}^{\infty} \alpha_m E[\theta_1(1-\theta_1)(1-\theta_1^{m-2})|F]}.$$

Example 5.4.7 Suppose $\mathbf{A}$ is regular with $\alpha_1 > 0$ and the support of F is contained in $[0, \Lambda(F, \mathbf{A})] \cup \{1\}$. Theorem 5.4.2 applies to show that $\Lambda(F, \mathbf{A}) = \Lambda_\infty(F, \mathbf{A})$ which is defined at (5.4.2); cf. Example 5.4.4. Since a single failure on arm 1 reveals the better arm, the proof of Theorem 5.4.5 makes it clear that $\Lambda(F, \mathbf{A})$ also coincides with upper bound $\Lambda^*(F, \mathbf{A})$. $\qquad\square$

Example 5.4.8 As in Example 5.4.5, assume the discount sequence is the n-horizon uniform. Then (5.4.7) becomes

$$\lambda[1 + (n-1)E(\theta_1|F)] - [E(\theta_1|F) + (n-1)E(\theta_1^2|F)]$$
$$- E[(\lambda-\theta_1)^+\theta_1[n-2-(n-1)\theta_1+\theta_1^{n-1}](1-\theta_1)^{-1}|F] = 0.$$

Specializing to $F = U(0, 1)$:

$$\lambda(n+1)/2 - (2n+1)/6$$

$$- \int_0^\lambda (\lambda - u)u[n - 2 - (n-1)u + u^{n-1}](1-u)^{-1} \, du = 0.$$

The solution $\lambda = \Lambda^*(F, \mathbf{A})$ of this equation is the upper bound given in Table 5.1 for various values of n. $\square$

Example 5.4.9 As in Example 5.4.6, assume the discount sequence is geometric. Then (5.4.7) becomes (after multiplying by $(1-\alpha)$),

$$\lambda[1 - \alpha + \alpha E(\theta_1 | F)] - [(1-\alpha)E(\theta_1 | F) + \alpha E(\theta_1^2 | F)]$$
$$- \alpha^2 E[(\lambda - \theta_1)^+ \theta_1 (1 - \theta_1)(1 - \alpha \theta_1)^{-1} | F] = 0.$$

And when $F = U(0, 1)$ it becomes

$$\lambda[1 - \alpha/2] - [1/2 - \alpha/6] - \alpha^2 \int_0^\lambda (\lambda - u) \, u(1-u)(1 - \alpha u)^{-1} \, du = 0.$$

The solution $\lambda = \Lambda^*(F, \mathbf{A})$ of this equation is the upper bound given in Table 5.2. $\square$

In the next section we leave the Bernoulli setting that has occupied most of our attention so far in this chapter and throughout Chapter 4.

5.5   The non-Bernoulli case with regular discounting

In Sections 5.1 to 5.4, arm 1 has been assumed to produce only 0's and 1's. In this section we take the point of view that Q_1 has arbitrary form; the distribution measure F reflects the information available initially concerning Q_1. In the next section we give an example in which the support of F can be quite large.

We assume regular discounting and continue to assume that arm 2 is known to produce λ at every stage, now with $\lambda \in \mathbb{R}$. By subtracting the known constant produced by arm 2 from every distribution in the support of F, we could always assume $\lambda = 0$. This is especially convenient when discussing stopping problems where the alternative to making an observation from Q_1 is no observation. But we allow λ to be arbitrary since, as in the Bernoulli case, we want to discuss the value of λ, as a function of F and discount sequence $\mathbf{A}$, such that arm 1 and arm 2 are both optimal.

The advantage of arm 1 over arm 2 is easily adaptable to general distributions: for all F, λ, and $\mathbf{A}$,

$$\Delta(F, \lambda; \mathbf{A}) = V^{(1)}(F, \lambda; \mathbf{A}) - V^{(2)}(F, \lambda; \mathbf{A}), \qquad (5.5.1)$$

where, as defined in Theorem 2.5.1,

$$V^{(1)}(F, \lambda; \mathbf{A}) = E[\alpha_1 X_{11} + V((X_{11})F, \lambda; \mathbf{A}^{(1)})|F], \qquad (5.5.2)$$

$$V^{(2)}(F, \lambda; \mathbf{A}) = \alpha_1 \lambda + V(F, \lambda; \mathbf{A}^{(1)}), \qquad (5.5.3)$$

and $V = V^{(1)} \vee V^{(2)}$.

In the Bernoulli case the decision maker has two reasons to like a success at a particular stage: it increases immediate income and, because of exchangeability, it enhances future expectations. We have seen (Theorem 5.3.5) that this preference for success can translate into a preference for the arm that was successful. But when more than two outcomes are possible it is easy to construct examples in which a large observation can be quite distasteful both in terms of value and desirability of the arm that produced it.

Example 5.5.1 Let $\mathbf{A}$ be the 2-horizon uniform and take $\lambda = 0$. Suppose

$$F = \tfrac{1}{2}\delta_Q + \tfrac{1}{2}\delta_{Q^*}.$$

where

$$Q = p\delta_6 + (1-p)\delta_{-4},$$
$$Q^* = \delta_4.$$

Using (5.5.2) gives

$$V^{(1)}(F, 0; \mathbf{A}) = 5p + \tfrac{1}{2}V((4)F, 0; \mathbf{A}^{(1)}) + \tfrac{1}{2}V((6)F, 0; \mathbf{A}^{(1)}),$$

since

$$V((-4)F, 0; \mathbf{A}^{(1)}) = V((6)F, 0; \mathbf{A}^{(1)}).$$

Continuing, using $\mathbf{A}^{(1)} = (1, 0, 0, \ldots)$,

$$V^{(1)}(F, 0; \mathbf{A}) = 5p + \tfrac{1}{2}[4 \vee 0] + \tfrac{1}{2}[(10p - 4) \vee 0] = 5p + (5p \vee 2).$$

Similarly, (5.5.3) gives

$$V^{(2)}(F, 0; \mathbf{A}) = 0 + [5p \vee 0] = 5p.$$

So

$$\Delta(F, 0; \mathbf{A}) = 5p \vee 2 > 0$$

and

$$\Delta((6)F, 0; \mathbf{A}^{(1)}) = 10p - 4,$$

which is negative for $p < 2/5$. For such a value of p, arm 1 is uniquely optimal initially and switching is optimal if '6' is observed but not if '4' is observed.

In addition, $6 + V((6)F, 0; \mathbf{A}^{(1)}) = (10p + 2) \vee 6$ is less than $4 + V((4)F, 0; \mathbf{A}^{(1)}) = 8$ for $p < 3/5$; for such a value of p, the decision maker would rather observe $X_{11} = 4$ than $X_{11} = 6$. If the horizon were greater than 2 the difference between the V's would be even greater.   □

While a bandit problem with general F is less wieldy than one for which F concentrates its mass on Bernoulli distributions, many of the results given in the early sections of this chapter hold more generally. Some proofs are unchanged and others can be modified easily. We will repeat those results formally in this section.

The first is Theorem 5.2.2, the proof of which applies without change.

Theorem 5.5.1 Consider a discount sequence $\mathbf{A}$. The following two statements are equivalent:

(i) For every $(F, \lambda; \mathbf{A})$-bandit there is an optimal strategy under which every selection of the known arm, arm 2, is followed by another selection of arm 2;

(ii) $\mathbf{A}$ is regular.

Furthermore, if $\alpha_1 > 0$ and $\mathbf{A}$ is regular, then if ever arm 2 becomes optimal, an optimal continuation is to select arm 2 exclusively and indefinitely.

The modification of the proof of Corollary 5.2.3 required for the present context is sufficiently involved that we give it here. First we state the result.

Corollary 5.5.2 For each regular discount sequence $\mathbf{A}$ and each distribution F on $\mathcal{D}$, there exists an optimal strategy τ for the $(F, \lambda; \mathbf{A})$-bandit such that

$$E_\tau(Z_{m+1} | F, \lambda) \geq E_\tau(Z_m | F, \lambda) \tag{5.5.4}$$

for $m = 1, 2, \ldots$.

Proof If arm 2 is uniquely optimal initially then take $\tau(\varnothing) = 2$; otherwise take $\tau(\varnothing) = 1$. Define τ inductively to be an optimal strategy that indicates arm 1 whenever possible without requiring a selection of arm 1 following a selection of arm 2. (The fuss in the preceding sentences is required to take care of the possibility that some discount factors may be zero.) We prove (5.5.4) only for $m = 1$; the general case is similar, but requires conditioning on the first $m - 1$ stages.

If $\tau(\varnothing) = 2$, then $\tau(z_1) = 2$ for all z_1 so $E_\tau(Z_2|F, \lambda) = \lambda$ as does $E_\tau(Z_1|F, \lambda)$. Suppose $\tau(\varnothing) = 1$. Then

$$E_\tau(Z_2|F, \lambda) = E(X_{12}\mathbf{1}_{\{\tau(X_{11})=1\}}|F) + E(\lambda\mathbf{1}_{\{\tau(X_{11})=2\}}|F).$$

If $\tau(X_{11}) = 2$, then Theorem 5.0.2 applies to show that $\lambda > E(X_{12}|X_{11}, F)$. Hence,

$$
\begin{aligned}
E(\lambda\mathbf{1}_{\tau(X_{11})=2}|F) &\geqslant E(E(X_{12}|X_{11}, F)\mathbf{1}_{\{\tau(X_{11})=2\}}|F) \\
&= E(E(X_{12}\mathbf{1}_{\{\tau(X_{11})=2\}}|X_{11}, F)|F) \\
&= E(X_{12}\mathbf{1}_{\{\tau(X_{11})=2\}}|F).
\end{aligned}
$$

So

$$
\begin{aligned}
E(Z_2|F, \lambda) &\geqslant E(X_{12}\mathbf{1}_{\{\tau(X_{11})=1\}}|F) + E(X_{12}\mathbf{1}_{\{\tau(X_{11})=2\}}|F) \\
&= E(X_{12}|F) = E(X_{11}|F) = E_\tau(Z_1|F, \lambda),
\end{aligned}
$$

as desired. $\square$

Example 5.5.1 shows that there is no 'break-even' observation, an X_{11} such that arm 1 continues to be optimal for larger observations and not for smaller. But the next theorem says there is a break-even value of λ. The theorem is adapted from Theorem 5.3.1, whose proof applies in this setting as well.

Theorem 5.5.3 For each regular discount sequence **A** with $\alpha_1 > 0$ and each distribution F on $\mathscr{D}$, there exists a unique $\Lambda(F, \mathbf{A}) \in \mathbb{R}$ such that arm 1 is optimal initially in the $(F, \lambda; \mathbf{A})$-bandit if and only if $\lambda \leqslant \Lambda(F, \mathbf{A})$ and arm 2 is optimal if and only if $\lambda \geqslant \Lambda(F, \mathbf{A})$.

Example 5.5.2 Consider the $(F, \lambda; \mathbf{A})$-bandit from Example 5.5.1, but with λ not restricted to equal 0. Then

$$\Delta(F, \lambda; \mathbf{A}) = 5p + \tfrac{1}{2}[4 \vee \lambda] + \tfrac{1}{2}[(10p - 4) \vee \lambda] - \lambda - [5p \vee \lambda].$$

When this equals 0, and so $\lambda = \Lambda(F, \mathbf{A})$, Theorem 5.5.1 applies to give $\Lambda(F, \mathbf{A}) \geqslant 5p = E(X_{11}|F)$. Solving

$$5p + \tfrac{1}{2}[4 \vee \lambda] + \tfrac{1}{2}[(10p - 4) \vee \lambda] - 2\lambda = 0$$

for λ in terms of p gives

$$\Lambda(F, \mathbf{A}) = \begin{cases} \dfrac{10p + 4}{3}, & p \in [0, \tfrac{4}{5}] \\[2mm] \dfrac{20p - 4}{3}, & p \in [\tfrac{4}{5}, 1]. \end{cases}$$

In view of Theorem 5.5.1, $V(F, \Lambda(F, \mathbf{A}); \mathbf{A}) = 2\Lambda(F, \mathbf{A})$. □

An example in which the unknown arm has a normal distribution was considered in Section 2.1. We return to that example to calculate Λ.

Example 5.5.3 Consider the $((\mu, \rho), \lambda; \mathbf{A})$-bandit discussed in Section 2.1. Observations X_{1m} on arm 1 are normally distributed with unknown mean θ_1 and variance 1. Distribution F can be regarded as a distribution on θ_1; it is itself normal with mean μ and variance $\rho^2 > 0$. The discount sequence is $\mathbf{A} = (1, 1, 0, 0, 0, \ldots)$.

In Section 2.1 we showed that arm 1 is optimal initially if and only if

$$(\lambda - \mu)\rho^{-2} \sqrt{(1 + \rho^2)} \leqslant t_0,$$

where $t_0 \approx 0.2760$. Therefore,

$$\Lambda((\mu, \rho), \lambda; \mathbf{A}) = \mu + t_0 \rho^2 / \sqrt{(1 + \rho^2)}. \tag{5.5.5}$$

This is another instance of the phenomenon indicated by Theorem 5.0.2: arm 2 is optimal only if it offers higher immediate payoff than does arm 1, and it must offer substantially higher payoff if arm 1 involves much uncertainty (ρ^2 large). (Cf. discussion in Section 2.1.) □

The next result gives a general method for calculating $\Lambda(F, \mathbf{A})$. Its statement and proof are identical to those of Corollary 5.3.2.

Corollary 5.5.4 For $\mathbf{A}$ regular with $\alpha_1 > 0$, the function Λ is given by

$$\Lambda(F, \mathbf{A}) = \max_{\tau(\varnothing) = 1} \frac{E_\tau\left(\sum_{m=1}^{M} \alpha_m X_{1m} \,\middle|\, F \right)}{E_\tau\left(\sum_{m=1}^{M} \alpha_m \,\middle|\, F \right)} \tag{5.5.6}$$

where M is the (random) stage (possibly $+\infty$) at which arm 1 is used for the last time when following strategy τ. Among those τ's that begin with arm 1 and never switch back to arm 1 after a selection of arm 2, those that are optimal for the $(F, \Lambda(F, \mathbf{A}); \mathbf{A})$-bandit are those for which the maximum in (5.5.6) is attained.

In applying this result it is advisable – perhaps essential – to first pare down the set of strategies to be considered in (5.5.6). For example, the formula for Λ given by (5.5.5) can be obtained using Corollary 5.5.4, but the strategies to be considered are too numerous to be manageable. However, it is possible to argue that one need only consider those strategies in (5.5.6) that indicate arm 1 at stage 2 if the initial observation (on arm 1) is sufficiently large, and arm 2 otherwise.

In the next section we turn to a rather comprehensive example that uses the results of the present section.

5.6 Arms with Dirichlet measures

In this section we give an application of the considerations in Section 5.5. The form of distribution Q_1 is not known, perhaps not even up to a parameter with a countable number of dimensions. Data can be gathered on its form by observing arm 1, but the decision maker is not willing in advance to specify that Q_1 is normal, say, as in Example 5.5.3, with only the mean of Q_1 unknown. In this section we consider a bandit problem incorporating the nonparametric approach to statistical inference of Ferguson (1973). In doing so we exploit more fully the general notation developed in Section 2.2. The section is based on Clayton and Berry (1984). Following Sethuraman and Tiwari (1982) we use the term 'Dirichlet measure' rather than 'Dirichlet process prior' used by Ferguson.

The distribution F of Q_1 is a Dirichlet measure with parameter v, which is itself a measure. The parameter v is a finite non-null measure on $\mathbb{R}$ (though we will consider the limit as its total measure approaches 0 or ∞). For any measurable partition $(B_1, \ldots, B_m)$ of $\mathbb{R}$ and any m, the random vector $(Q_1(B_1), Q_1(B_2), \ldots, Q_1(B_m))$ has a Dirichlet distribution with parameter $(v(B_1), v(B_2), \ldots, v(B_m))$. So, for example,

$$E(Q_1(B_1)|F) = v(B_1)/v(\mathbb{R}). \tag{5.6.1}$$

Let $I = v(\mathbb{R})$ and let $\pi(\mathrm{d}x) = v(\mathrm{d}x)/I$ be the normalized form of v;

we sometimes write πI in place of v. In view of (5.6.1), π is the prior mean of Q_1:

$$E(Q_1(dx)|F) = \pi(dx).$$

The total measure I (which stands for amount of 'information') can be interpreted as the weight of the prior in terms of sample number (Ferguson, 1973). The prior mean of an observation on arm 1 is the mean of π:

$$E(X_{11}|F) = \int_R x\pi(dx),$$

which we assume to be finite.

The parameter v summarizes prior information concerning Q_1. Given $X_{11}, \ldots, X_{1m}$, the posterior distribution of Q_1 is a Dirichlet measure with parameter $v + \sum_{j=1}^{m} \delta_{X_{1j}}$ (Ferguson, 1973, Theorem 1).

The case $I = 0$ gives rise to an improper Dirichlet measure which we define as follows. When $v = \pi \cdot 0$, observations $X_{11}, \ldots, X_{1n}$ are almost surely equal and each has unconditional distribution π. As shown by Sethuraman and Tiwari (1982), as $I \to 0$ the Dirichlet measure with parameter πI tends to the improper measure with parameter $\pi \cdot 0$ defined here.

The limiting case $I \to \infty$ corresponds to knowing in advance that $Q_1 = \pi$.

Ferguson (1973, Proposition 3) shows that, with respect to the topology of convergence in distribution, the support of F is the set of all distributions whose supports are contained in the support of v. So the Dirichlet measure allows for modelling settings in which responses can take on values in any specified set. It actually encompasses the Bernoulli model when the parameter of the Bernoulli has a beta distribution. To see this, let $v = a\delta_1 + b\delta_0$ (in which case $I = a + b$); then $X_{11}, X_{12}, \ldots$ are distributed as conditionally independent Bernoulli variables whose parameter has the density given in (5.4.1). Also, the improper Dirichlet measure with $v = (a\delta_1 + b\delta_0) \cdot 0$ corresponds to the two-point prior on the Bernoulli parameter that is assumed in Example 5.4.4 (by setting $p = a/(a+b)$).

Assume $\mathbf{A}$ is the n-horizon uniform. Since it depends only on the horizon n we shall adopt the convention (to be used again in Chapter 7) of using n where $\mathbf{A}$ normally appears. In a similar vein, we write v in place of F. So $(F, \lambda; \mathbf{A})$ is now written $(v, \lambda; n)$.

For $n \geqslant 1$ we have

$$V^{(1)}(v, \lambda; n) = E(X_{11}|v) + E[V(v + \delta_{X_{11}}, \lambda; n - 1)|v], \quad (5.6.2)$$

$$V^{(2)}(v, \lambda; n) = \lambda + V(v, \lambda; n - 1)$$

and, of course, $V(v, \lambda; 0) = 0$. Theorem 5.5.1 applies to show that the $(v, \lambda; n)$-bandit is a stopping problem. So

$$V(v, \lambda; n) = V^{(1)}(v, \lambda; n) \vee n\lambda.$$

Since **A** is regular, Theorem 5.5.2 indicates that there exists a $\Lambda(v, n)$ such that arm 1 is optimal if and only if $\lambda \leqslant \Lambda(v, n)$.

We shall find $\Lambda(v, n)$ in a simple example.

Example 5.6.1 Suppose $n = 2$ and v is the following discrete measure:

$$v(\{x\}) = \left(\frac{1}{2}\right)^{x+1} I, \qquad x = 0, 1, 2, \ldots$$

To obtain $\Lambda(v, 2)$ we will solve the equation

$$V^{(1)}(v, \lambda; 2) = 2\lambda. \quad (5.6.3)$$

To find $V^{(1)}(v, \lambda; 2)$ we require

$$\begin{aligned}
V(v + \delta_{X_{11}}, \lambda; 1) &= E(X_{12}|v + \delta_{X_{11}}) \vee \lambda \\
&= E\left[\left(\frac{I}{I+1}E(X_{12}|v) + \frac{1}{I+1}X_{11}\right) \vee \lambda \;\middle|\; v\right] \\
&= E\left[\frac{I + X_{11}}{I + 1} \vee \lambda \;\middle|\; v\right],
\end{aligned}$$

since $E(X_{12}|v) = 1$. Therefore $\Lambda(v, 2)$ is the unique solution of (5.6.3) in view of Theorems 5.5.1 and 5.5.3.

Consider equation (5.6.3) for $\lambda \in [1, (I+2)/(I+1)]$. We obtain

$$\frac{I + X_{11}}{I + 1} \vee \lambda = \begin{cases} \lambda & \text{if } X_{11} \in \{1, 0\} \\ \dfrac{I + X_{11}}{I + 1} & \text{if } X_{11} \in \{2, 3, \ldots\}. \end{cases}$$

Hence, for such λ, a straightforward calculation gives

$$E[V(v + \delta_{X_{11}}, \lambda; 1)|v] = \frac{3\lambda}{4} + \frac{I + 3}{4(I + 1)}.$$

From (5.6.2),

$$V^{(1)}(v, \lambda; 2) = \frac{3\lambda}{4} + \frac{5I + 7}{4(I + 1)}.$$

Setting this equal to 2λ we obtain $\lambda = (5I + 7)/(5I + 5)$, which does satisfy $1 \leqslant \lambda \leqslant (I + 2)/(I + 1)$. So we have found the unique solution of (5.6.3):

$$\Lambda(v, 2) = \frac{5I + 7}{5I + 5},$$

valid for $0 \leqslant I \leqslant \infty$.

This example provides another instance in which arm 1 is more attractive when less is known about it, which in this case occurs when I is smaller.   □

In the above example there is a number such that arm 1 continues to be optimal if X_{11} is greater than that number and arm 2 is optimal otherwise. Example 5.5.1 shows that there may not exist such a number when X_{1m} can take on more than two values. But the Dirichlet model exhibits sufficient smoothness that such an example is not possible. Clayton and Berry (1984) give the following result; see their paper for the proof.

Theorem 5.6.1 Fix $\lambda, n \geqslant 2$, and $I < \infty$. For any $(v, \lambda; n)$-bandit there exists a unique $b^* = b^*(v, \lambda; n)$ such that if arm 1 is selected (whether or not it is optimal) then arm 1 is optimal at stage 2 if $X_{11} \geqslant b^*$ and arm 2 is optimal if $X_{11} \leqslant b^*$.

Remarks When λ is greater than the support of v, so is $b^*(v, \lambda; n)$. While the probability is 0 that X_{11} is greater than the support of v, the Dirichlet measure with parameter $v + \delta_{X_{11}}$ is well-defined nonetheless.

In view of the theorem, $b^*(v, \lambda; n)$ is the unique solution of the equation

$$\lambda = \Lambda(v + \delta_{b^*}, n - 1).$$ □

The 'break-even' observation $b^*(v, \lambda; n)$ obviously depends on λ. The special case $\lambda = \Lambda(v, n)$ gives the next result. The result says there is a quantity such that the attractiveness of arm 1 (as measured by Λ) increases if X_{11} is greater than that quantity. Such an X_{11} might be called a 'winner', so the coming result generalizes the stay-with-a-winner property of Bernoulli bandits.

Corollary 5.6.2 For $n \geqslant 2$ and all v with $I < \infty$, there exists a unique $b(v, n)$ such that

$$\Lambda(v + \delta_{X_{11}}, n - 1) \geqslant \Lambda(v, n) \text{ if } X_{11} \geqslant b(v, n)$$
$$\Lambda(v + \delta_{X_{11}}, n - 1) \leqslant \Lambda(v, n) \text{ if } X_{11} \leqslant b(v, n).$$

Remark Clearly, $b(v, n) = b^*(v, \Lambda(v, n); n)$. $\qquad\qquad\square$

The function b^* determines all optimal strategies but the function b does not. On the other hand, calculating b is easier than calculating b^* since b does not depend on λ. Still, the task of finding $b(v, n)$ is formidable if not impossible when n is moderate and the support of v is at all complicated. Not only is dynamic programming required, but there are substantial computational difficulties when dealing with Dirichlet measures (see Berry and Christensen, 1979, for example).

The next two examples give some calculations of $\Lambda(v, n)$ and $b(v, n)$ for small n.

Example 5.6.2 Suppose π is the uniform distribution on $(0, 1)$. Table 5.3 gives $\Lambda(\pi I, n)$ and $b(\pi I, n)$ for $I = 0, 0.1, 1, 5, \infty$, and $n = 2, 3, 4$. $\qquad\qquad\square$

Example 5.6.3 Suppose π is the standard normal distribution. Table 5.4 gives $\Lambda(\pi I, n)$ and $b(\pi I, n)$ for the same combinations of I and n considered in the previous example. $\qquad\qquad\square$

The specification of an optimal strategy is quite complicated, depending on v and n, when $\lambda < \Lambda(v, n)$. Arm 1 is optimal initially. After observing X_{11}, one calculates $\Lambda(v + \delta_{X_{11}}, n - 1)$ and compares it with λ. If arm 1 is optimal again, one calculates $\Lambda(v + \delta_{X_{11}} + \delta_{X_{12}}, n - 2)$ and compares it with λ. And so on.

Example 5.6.4 As in Example 5.6.2 suppose $\pi = U(0, 1)$ and now assume $n = 4$. Table 5.3 indicates which arm is optimal initially. Suppose arm 1 is selected; the new bandit is $(\pi I + \delta_{X_{11}}, \lambda; 3)$. Table 5.5 gives $\Lambda(\pi I + \delta_{X_{11}}, 3)$ for various X_{11} and $I = 0.1, 1, 5$; of course the 'I' for the new bandit is $I + 1$. Table 5.5 also gives $b(\pi I + \delta_{X_{11}}, 3)$.

Table 5.3 $\Lambda(\pi I, n)$ and $b(\pi I, n)$ where $\pi = U(0, 1)$

I	$\Lambda(\pi I, 2)$	$b(\pi I, 2)$	$\Lambda(\pi I, 3)$	$b(\pi I, 3)$	$\Lambda(\pi I, 4)$	$b(\pi I, 4)$
0	0.586	0.586	0.634	0.634	0.667	0.667
0.1	0.578	0.586	0.623	0.630	0.654	0.661
1	0.543	0.586	0.570	0.610	0.590	0.630
5	0.514	0.586	0.524	0.584	0.532	0.589
∞	0.500	–	0.500	–	0.500	–

Table 5.4 $\Lambda(\pi I, n)$ and $b(\pi I, n)$ where π is standard normal

I	$\Lambda(\pi I, 2)$	$b(\pi I, 2)$	$\Lambda(\pi I, 3)$	$b(\pi I, 3)$	$\Lambda(\pi I, 4)$	$b(\pi I, 4)$
0	0.276	0.276	0.436	0.436	0.549	0.549
0.1	0.251	0.276	0.400	0.424	0.505	0.529
1	0.138	0.276	0.228	0.359	0.295	0.421
5	0.046	0.276	0.079	0.276	0.105	0.284
∞	0.000	–	0.000	–	0.000	–

Table 5.5 $\Lambda(\pi I + \delta_{X_{11}}, 3)$ [and $b(\pi I + \delta_{X_{11}}, 3)$] where $\pi = U(0, 1)$

X_{11}	0.5	0.6	0.7	0.8	0.9	1.0
$I = 0.1$	0.510	0.598	0.690	0.783	0.877	0.971
	[0.516]	[0.602]	[0.695]	[0.790]	[0.885]	[0.981]
$I = 1$	0.529	0.575	0.630	0.685	0.741	0.798
	[0.562]	[0.602]	[0.662]	[0.723]	[0.786]	[0.851]
$I = 5$	0.518	0.535	0.552	0.570	0.588	0.607
	[0.572]	[0.588]	[0.608]	[0.630]	[0.652]	[0.674]

Suppose $I = 1$ and $\lambda = 0.58$. Then arm 1 is optimal initially since $\lambda < \Lambda(\pi I, 4) = 0.590$ from Table 5.3. According to Table 5.5, if $X_{11} = 0.5$ or 0.6, then arm 2 should be observed next (and for the remaining two selections as well). On the other hand, arm 1 remains optimal if $X_{11} \geqslant 0.7$. So $0.6 < b^*(\pi \cdot 1, 0.58; 4) < 0.7$. Suppose $X_{11} = 0.7$. Then the tabulated value $b(\pi \cdot 1 + \delta_{0.7}, 3) = 0.662$ indicates that arm 1 continues to be optimal if $X_{12} \geqslant 0.662$ (arm 1 is also optimal for X_{12} as small as $b^*(\pi \cdot 1 + \delta_{0.7}, 0.58; 3) < 0.662$). □

Examples 5.6.2 to 5.6.4 suggest that $\Lambda(v, n) \leqslant b(v, n)$; Clayton and Berry (1984) conjecture that this holds generally. This conjecture has the following motivation. Suppose the decision maker is indifferent regarding the two arms, arm 2 producing the known constant $\lambda = \Lambda(v, n)$. If arm 1 is selected, it is selected with the hope that it is better than the alternative, which is arm 2. So if it yields less than arm 2 is known to deliver, the hope for it fades. And since it was barely optimal to start with, arm 1 must no longer be optimal if $X_{11} < \Lambda(v, n)$.

Consider the $(\pi \cdot 0, \lambda; n)$-bandit. Since arm 1 becomes known with just a single observation, any $X_{11} \geqslant \lambda$ will make arm 1 optimal again (and thereafter) and any $X_{11} \leqslant \lambda$ will make arm 2 optimal. Therefore, $b^*(\pi \cdot 0, \lambda; n) = \lambda$. On the other hand, when $I > 0$ one should not have to observe as large an X_{11} to want to stay with arm 1. So the following result of Clayton and Berry (1984) seems reasonable and is supported by Tables 5.3 and 5.4; see their paper for its proof.

Theorem 5.6.3 For all $v = \pi I$ and $n \geqslant 2$, $b(v, n) \leqslant \Lambda(\pi \cdot 0, n)$.

This theorem gives an easily computable bound for $b(v, n)$.

In the next section we turn to the setting in which the stages occur at random times.

5.7 Real-time discounting

So far in this chapter we have considered the observation at any stage to be discounted independent of the time t at which the stage occurs. In this section we suppose, as described in Section 3.5, that the observation is weighted by α_t which is unrelated to the stage number. We allow more general discounting than that of the exponential function emphasized in Section 3.5. We require the times between stages to be independent and exponentially distributed with known mean κ. In this section we shall indicate which results in the previous sections of this chapter carry over to the current setting.

There are at least two ways of thinking about a strategy as defined in Section 3.5. Selections can depend on t. We can imagine that the decision maker waits until a stage occurs (a patient arrives, a part fails, etc.), selects an arm and instantaneously observes the result. Or, we can suppose that the decision maker keeps an arm operative at all times, changing arms depending on t as well as on any observations on the arms; when a stage occurs the operative arm is observed. This latter view is particularly handy for describing various aspects of optimal strategies.

Suppose the first stage occurs at time $s > 0$. The appropriate bandit is not $(F, \lambda; \alpha_t, \kappa)$, but rather $(F, \lambda; \alpha_t^{(s)}, \kappa)$, where $\alpha_t^{(s)}$ is α_t shifted by s units; that is, $\alpha_t^{(s)} = \alpha_{t+s}$ for $t \geq 0$.

Theorems 5.0.1 and 5.0.2 apply in the present setting. We do not know whether the '(i) $\Rightarrow$ (ii)' portion of Theorem 5.5.1 applies; the randomness in the present setting may enable 'regular portions' of a discount function to compensate for 'irregular portions'. The remaining results of Section 5.5 do apply, provided appropriate definitions are used. In analogy with $\gamma_m = \sum\limits_{i=m}^{\infty} \alpha_i$ in discrete time, let

$$\gamma_t = \int_t^{\infty} \alpha_s \, \mathrm{d}s.$$

Definition 5.7.1 A discount function α_t is *regular* if α_t/γ_t is nondecreasing at all t for which $\gamma_t > 0$.

The following proposition makes it clear that this definition is a natural continuous-time version of Definition 5.2.1.

Proposition 5.7.1 If α_t is regular, then $\gamma_{t+2s}\gamma_t \leq \gamma_{t+s}^2$ for every nonnegative t and s.

Proof Assume that α_t is regular and, with no loss, that $\gamma_{t+2s} > 0$. Then

$$\log(\gamma_t/\gamma_{t+s}) = \int_t^{t+s} (\alpha_u/\gamma_u) \, \mathrm{d}u \leq \int_{t+s}^{t+2s} (\alpha_u/\gamma_u) \, \mathrm{d}u$$
$$= \log(\gamma_{t+s}/\gamma_{t+2s}). \qquad \Box$$

Remark The converse of this proposition is not true since, for instance, α_s can be changed at any particular s, leaving γ_t unchanged for all $t \geq 0$. $\qquad \Box$

With this definition of regularity the '(ii) $\Rightarrow$ (i)' and 'furthermore' portions of Theorem 5.5.1 hold. Therefore we can view a bandit with a regular discount function as a stopping problem. Corollary 5.5.2 also holds in the setting of real-time discounting. This means that when α_t is regular there is an optimal strategy for which later observations have no smaller expectations than earlier ones.

The analogue of Theorem 5.5.3 holds here as well. This means that there is a break-even value $\Lambda(F, \alpha_t, \kappa)$ that can be calculated for arm 1 and that indicates the optimal arm by comparing it with the constant λ yielded by arm 2. Suppose we imagine the decision maker as keeping one arm operative at all times anticipating the occurrence of a stage. When α_t is regular we have as an immediate corollary of the stopping nature of the bandit that the decision maker need never consider switching control from arm 2 to arm 1, whether or not stages occur. But it is easy to find F, λ, regular α_t, and κ such that an optimal decision maker keeps arm 1 operative and, should no stage occur within some period of time, switch and put arm 2 into operation. Summarizing, we have the following result.

Corollary 5.7.2 Suppose α_t is regular with $\alpha_0 > 0$. Then $\Lambda(F, \alpha_t^{(s)}, \kappa)$ is a nonincreasing function of s when $\alpha_s > 0$.

For the present setting the appropriate modification of the formula (5.5.6) for Λ is

$$\Lambda(F, \alpha_t, \kappa) = \max_{\tau(\varnothing; 0) = 1} \frac{E_\tau\left(\sum_{m=1}^{M} \alpha_{T_m} X_{1m} \middle| F, T_1 = 0\right)}{E_\tau\left(\sum_{m=1}^{M} \alpha_{T_m} \middle| F, T_1 = 0\right)}, \quad (5.7.1)$$

where T_m denotes the random time at which stage m occurs.

Example 5.7.1 Suppose that arm 1 is Bernoulli with $F = p\delta_1 + (1-p)\delta_0$. To use (5.7.1) we can assume without loss that a stage occurs at $t = 0$ and arm 1 is selected. The only strategy that need be considered in (5.7.1) is staying with arm 1 indefinitely if a success is obtained and switching permanently to arm 2 if failure is obtained. Accordingly,

$$\Lambda(F, \alpha_t, \kappa) = \frac{p \sum_{m=1}^{\infty} E(\alpha_{T_m} | T_0 = 1)}{p \sum_{m=1}^{\infty} E(\alpha_{T_m} | T_0 = 1) + (1-p)\alpha_0}. \qquad \square$$

The construction leading to Theorem 3.5.1 does not give Bernoulli discrete-time approximations to a Bernoulli real-time bandit. Nevertheless, the inductive proof of Theorem 4.1.6 can be mimicked by considering real-time bandits that terminate after a fixed number of stages, no matter when they occur. Hence, $V(F^*, \lambda; \alpha_t, \kappa)$ $\geqslant V(F, \lambda; \alpha_t, \kappa)$ if F^* is strongly to the right of F. If, in addition, α_t is regular and $\alpha_0 \neq 0$, then $\Lambda(F^*, \alpha_t, \kappa) \geqslant \Lambda(F, \alpha_t, \kappa)$ (cf. Corollary 5.3.4).

Many of the results obtained in the Bernoulli case considered in Chapter 4 and earlier in this chapter do not carry over easily to this setting. Suppose, for example, $\alpha_t = \mathbf{1}_{[0,1]}(t)$. It is easy to choose F, λ, and κ so that the following hold. Arm 1 is optimal for the $(F, \lambda; \alpha_t, \kappa)$-bandit at $t = 0$ if stage 1 occurs then. It is not optimal at stage 2 if stage 2 occurs at time t close to 1, even if a success is observed at stage 1. As an analogue of Theorem 5.3.5, when α_t is regular it is possible to show that $\Lambda(F, \alpha_t, \kappa) \leqslant \Lambda(\sigma F, \alpha_t^{(s)}, \kappa)$ for all sufficiently small s, but the result is of little consequence since it is not possible to predict the timing of the stages.

However, most of the results in the Bernoulli case with geometric discounting have natural and useful analogues in the current setting if we assume an exponential discount function. For, if $\alpha_t = \exp(-\beta t)$ (cf. Section 3.5) then $(F, \lambda; \alpha_t^{(s)}, \kappa)$ is effectively the same bandit for all $s \geqslant 0$. So it is easy to show that the stay-with-a-winner rule (Theorem 5.3.5) holds in this case.

References

Bellman, R. (1956) A problem in the sequential design of experiments. *Sankhyā A* **16**: 221–229.

Berry, D. A. and Christensen, R. (1979) Empirical Bayes estimation of a binomial parameter via mixtures of Dirichlet processes. *Ann. Statist.* **7**: 558–568.

Berry, D. A. and Fristedt, B. (1979) Bernoulli one-armed bandits – arbitrary discount sequences. *Ann. Statist.* **7**: 1086–1105.

Berry, D. A. and Fristedt, B. (1983) Maximizing the length of a success run for many-armed bandits. *Stochastic Process. Appl.* **15**: 317–325.

Bradt, R. N., Johnson, S. M. and Karlin, S. (1956) On sequential designs for maximizing the sum of n observations. *Ann. Math. Statist.* **27**: 1060–1074.

Clayton, M. K. and Berry, D. A. (1984) Bayesian nonparametric bandits. Statistics Tech. Rep. No. 427, Univ. of Minnesota, USA.

Ferguson, T. S. (1973) A Bayesian analysis of some nonparametric problems. *Ann. Statist.* **1**: 209–230.

Gittins, J. C. and Jones, D. M. (1974) A dynamic allocation index for the sequential design of experiments. In *Progress in Statistics* (eds J. Gani *et al.*), pp. 241–266, North-Holland, Amsterdam.

Gittins, J. C. and Jones, D. M. (1979) A dynamic allocation index for the discounted multiarmed bandit problem. *Biometrika* **66**: 561–565.

Sethuraman, J. and Tiwari, R. C. (1982) Convergence of Dirichlet measures and the interpretation of their parameter. In *Statistical Decision Theory and Related Topics III*, Vol. 2 (eds S. Gupta and J. O. Berger), pp. 305–315, Academic Press, New York.

CHAPTER 6

Many independent arms; geometric discounting

We have indicated several times that the two most widely studied discount sequences are the n-horizon uniform and the geometric. The former applies when the problem is to maximize the sum of n observations. When n is unknown the corresponding random discount sequence can be taken to be nonrandom (see Section 3.1); it can be any nonincreasing sequence depending on the uncertainty in n. As a special case suppose n has a geometric distribution; so the opportunity for gain ceases at each stage with constant probability α. Then, and in many other circumstances as well, the appropriate discount sequence is geometric: $\mathbf{A} = (1, \alpha, \alpha^2, \alpha^3, \ldots)$.

This fact gives geometric discounting special significance and, in our view, makes the geometric sequence the single most important discount sequence. Luckily, it is also the most mathematically tractable (except, of course, for sequences with very small horizons). The fact that, alone among discount sequences, the geometric does not change from one stage to the next (after a suitable normalization at each stage) gives a decision problem with geometric discounting a special quality and makes possible the elegant result of Gittins and Jones (1974) presented in Section 6.1. This result says that for k independent arms and geometric discounting, the desirability of an arm can be completely specified by a number that depends only on that arm (and on α) and not on the other arms. Gittins and Jones (1974) call this number the arm's *dynamic allocation index*. Any arm with the largest index is optimal. In view of independence, an arm's index can change only when the arm is observed. So optimal strategies are conceptually easy to describe: always select an arm with the largest index, switching (to the arm that was second-best) only when its index drops from first place.

Suppose we define, as seems reasonable, the dynamic allocation index of an arm with known mean to be that mean. Then the index of arm i can only be $\Lambda(F_i, \mathbf{A})$ as defined in Theorem 5.5.3. So in order to decide how good an arm is, one need only compare it with known arms, defining an arm's index to be the mean of a known arm such that both arms would be optimal in a two-armed bandit problem. And the optimal initial selections in a k-armed bandit with independent arms can be found by solving k families of two-armed bandits, comparing each arm in turn with various known arms.

We show in Section 6.2 (Theorem 6.2.1) that when discounting is regular and strategies specified by dynamic allocation indices are optimal, the discount sequence can only be geometric. So the Gittins–Jones characterization is only possible in the geometric case. An equivalent and rather picturesque characterization is as follows, given in the context of $k = 3$. Consider three independent arms and suppose they are compared in pairs in three two-armed bandits. Is it possible that arm 1 is uniquely optimal in the $(F_1, F_2; \mathbf{A})$-bandit, arm 2 is uniquely optimal in the $(F_2, F_3; \mathbf{A})$-bandit, and arm 3 is uniquely optimal in the $(F_3, F_1; \mathbf{A})$-bandit? Such lack of transitivity is impossible if optimal strategies are specified by dynamic allocation indices. For $\Delta(F_1, F_2; \mathbf{A}) > 0$ would imply $\Lambda(F_1, \mathbf{A}) > \Lambda(F_2, \mathbf{A})$, and $\Delta(F_2, F_3; \mathbf{A}) > 0$ would imply $\Lambda(F_2, \mathbf{A}) > \Lambda(F_3, \mathbf{A})$; so $\Delta(F_3, F_1; \mathbf{A}) > 0$ would be impossible. On the other hand, lack of transitivity is possible when optimal strategies are not specified by dynamic allocation indices. The following example shows that such strategies are not optimal when $\mathbf{A}$ is the 2-horizon uniform. The example is actually a special case of the family of examples used to prove Theorem 6.2.1.

Example 6.0.1. Suppose $\mathbf{A} = (1, 1, 0, 0, 0, \ldots)$. Consider two Bernoulli arms with, as has been our custom in this setting, F_i taken to be the distribution of the Bernoulli parameter θ_i. Suppose

$$F_1 = \tfrac{1}{2}\delta_0 + \tfrac{1}{2}\delta_1,$$
$$F_2 = \tfrac{5}{7}\delta_{1/2} + \tfrac{2}{7}\delta_1.$$

From Example 5.4.7 or Theorem 5.4.2.

$$\Lambda(F_1, \mathbf{A}) = 2/3 < \Lambda(F_2, \mathbf{A}) = 31/46.$$

Nevertheless, as we show below, arm 1 is optimal in the $(F_1, F_2; \mathbf{A})$-bandit.

To show lack of transitivity, consider a third arm with $F_3 = \delta_c$. Any $c \in (2/3, 31/46)$ will do; we take $c = 185/276$ to be specific. Since the maximum worth when selecting arm i in the $(F_i, F_j; \mathbf{A})$-bandit is

$$V^{(i)}(F_i, F_j; \mathbf{A}) = [2E(\theta_i | F_i)]$$
$$\vee [E(\theta_i | F_i) + E(\theta_i^2 | F_i) + (1 - E(\theta_i | F_i))E(\theta_j | F_j)]$$

and similarly for selecting arm j (cf. (4.2.1) and (4.2.2)), it follows that

$$\Delta(F_1, F_2; \mathbf{A}) = [\tfrac{1}{2} + \tfrac{1}{2} + \tfrac{1}{2} \cdot \tfrac{9}{14}] - [2 \cdot \tfrac{9}{14}] + \tfrac{1}{28},$$
$$\Delta(F_2, F_3; \mathbf{A}) = [\tfrac{9}{14} + \tfrac{13}{28} + \tfrac{5}{14}c] - [2c] = \tfrac{1}{168},$$
$$\Delta(F_3, F_1; \mathbf{A}) = [2c] - [\tfrac{1}{2} + \tfrac{1}{2} + \tfrac{1}{2}c] = \tfrac{1}{184}.$$

(Alternatively, the latter two Δ's are known to be positive in view of the choice of c as compared with $\Lambda(F_1, \mathbf{A})$ and $\Lambda(F_2, \mathbf{A})$.) Therefore arm 1 is 'better than' arm 2, as advertised above, and arm 2 is 'better than' 3, which is 'better than' arm 1!

Incidentally, arm 2 is optimal in the $(F_1, F_2, F_3; \mathbf{A})$-bandit and

$$V(F_1, F_2, F_3; \mathbf{A}) = V(F_2, F_3; \mathbf{A}) = \tfrac{9}{14} + \tfrac{13}{28} + \tfrac{5}{14}c \approx 1.347. \qquad \square$$

In Section 6.1 we prove the Gittins–Jones result and in Section 6.2 we show that geometric discounting is necessary for the result. Some readers may find it useful to read Sections 6.1 and 6.2 in reverse order since the proof of Theorem 6.2.1 can help the intuition by showing why nongeometric sequences are ruled out.

6.1 Theorem of Gittins and Jones

When $\mathbf{A} = (1, \alpha, \alpha^2, \ldots)$ we will substitute α for $\mathbf{A}$ and write, for example, $(F_1, \ldots, F_k; \alpha)$ in place of $(F_1, \ldots, F_k; \mathbf{A})$. We note that the results in this section apply with occasional evident modification if $\mathbf{A} = (b, b\alpha, b\alpha^2, \ldots)$ with $b > 0$.

As indicated previously, the dynamic allocation indices $\Lambda(F_i, \alpha)$ play a central role in the $(F_1, \ldots, F_k; \alpha)$-bandit. This role is shown in the following fundamental result of Gittins and Jones (1974), stated as it applies with our definition of bandit problems. Their definition is broader than ours, so their result is more general than the version we state.

Theorem 6.1.1 The optimal initial selections in the $(F_1, \ldots, F_k; \alpha)$-bandit are those i for which

$$\Lambda(F_i, \alpha) = \bigvee_{j=1}^{k} \Lambda(F_j, \alpha).$$

Moreover,

$$V(F_1, \ldots, F_k; \alpha)$$

$$= \frac{1}{1-\alpha} \lim_{\rho \to \infty} \left\{ \rho - (1-\alpha)^k \int_{-\infty}^{\rho} \prod_{j=1}^{k} \frac{\partial}{\partial \lambda} V(F_j, \lambda; \alpha) \mathrm{d}\lambda \right\} \quad (6.1.1)$$

Remark In view of Theorems 5.2.2 and 5.3.1, when $\lambda \geqslant \Lambda(F_i, \alpha)$,

$$V(F_i, \lambda; \alpha) = \lambda/(1-\alpha). \tag{6.1.2}$$

So

$$(1-\alpha)^k \prod_{j=1}^{k} \frac{\partial}{\partial \lambda} V(F_j, \lambda; \alpha) = 1$$

for sufficiently large λ and the limit in (6.1.1) can be replaced with the expression evaluated at any ρ at least as large as $\bigvee_{j=1}^{k} \Lambda(F_j, \alpha)$. $\square$

We will essentially use the argument of Whittle (1982, Section 14.4) to prove Theorem 6.1.1. We need a modification of $(F_1, \ldots F_k; \alpha)$; this modification is not a bandit problem because of a restriction on the available strategies. To get this quasi-bandit, adjoin a $(k+1)$st arm with $X_{(k+1),m} = \lambda^*$ for all m and consider only those strategies under which arm $k+1$ is continued indefinitely once it has been selected. An asterisk indicates that this quasi-bandit is being considered; for instance, $V^*(F_1, \ldots, F_k, \lambda^*; \alpha)$ is its value. Slight modifications of the results in Sections 2.3 and 2.5 apply; so, for instance,

$$V^*(F_1, \ldots F_k, \lambda^*; \alpha) = [\lambda^*/(1-\alpha)]$$

$$\vee \bigvee_{i=1}^{k} E(X_{i1} + \alpha V^*(F_1, \ldots, (X_{i1})F_i, \ldots, F_k, \lambda^*; \alpha)|F_i). \quad (6.1.3)$$

We need the following result.

Lemma 6.1.2 For each F and each $F_1, \ldots, F_k$, the functions $\lambda \mapsto V(F, \lambda; \alpha)$ and $\lambda^* \mapsto V^*(F_1, \ldots, F_k, \lambda^*; \alpha)$ are nondecreasing and concave upwards.

Proof For every τ that stays with arm 2 once arm 2 is selected, the function $\lambda \mapsto W(F, \lambda; \alpha; \tau)$ is obviously linear and nondecreasing. Similarly for $\lambda^* \mapsto W^*(F_1, \ldots, F_k, \lambda^*; \alpha; \tau^*)$, where τ^* is any strategy for the quasi-bandit $(F_1, \ldots, F_k, \lambda^*; \alpha)$. The supremum of a collection of nondecreasing linear functions is nondecreasing and concave upwards.   □

Remark From Lemma 6.1.2 we conclude that $V(F, \cdot; \alpha)$ is absolutely continuous, and a nondecreasing right-continuous version of $\partial V(F, \lambda; \alpha)/\partial \lambda$ can be chosen. This derivative is integrable at $-\infty$ since $V(F, \lambda; \alpha) \to E(X_{11}|F)/(1 - \alpha)$ as $\lambda \to -\infty$. Therefore,

$$(1 - \alpha)^k \int_{-\infty}^{\rho} \prod_{j=1}^{k} \frac{\partial}{\partial \lambda} V(F_j, \lambda; \alpha) \, d\lambda < \infty$$

for each finite ρ.   □

Proof of Theorem 6.1.1 For the quasi-bandit $(F_1, \ldots, F_k, \lambda^*; \alpha)$ we will prove that arm $k + 1$ is optimal initially if and only if

$$\lambda^* \geqslant \bigvee_{j=1}^{k} \Lambda(F_j, \alpha), \tag{6.1.4}$$

that for $i = 1, \ldots, k$, arm i is optimal if and only if

$$\Lambda(F_i, \alpha) = \lambda^* \vee \bigvee_{j=1}^{k} \Lambda(F_j, \alpha), \tag{6.1.5}$$

and that $V^*(F_1, \ldots, F_k, \lambda^*; \alpha)$ equals

$$\xi(F_1, \ldots, F_k, \lambda^*; \alpha)$$
$$= [1/(1 - \alpha)] \lim_{\rho \to \infty} \left\{ \rho - (1 - \alpha)^k \int_{\lambda^*}^{\rho} \prod_{j=1}^{k} \frac{\partial}{\partial \lambda} V(F_j, \lambda; \alpha) \, d\lambda \right\}. \tag{6.1.6}$$

The theorem will then follow by letting $\lambda^* \to -\infty$.

Define

$$\zeta_i(F_1, \ldots, F_k, \lambda; \alpha) = (1 - \alpha)^{k-1} \prod_{j \neq i} \frac{\partial}{\partial \lambda} V(F_j, \lambda; \alpha).$$

By (6.1.2) and Lemma 6.1.2, ζ_i is nonnegative, nondecreasing in λ, and it equals 1 when

$$\lambda \geqslant \bigvee_{j \neq i} \Lambda(F_j, \alpha).$$

In view of (6.1.2), integrating by parts in (6.1.6) gives

$$\xi(F_1, \ldots, F_k, \lambda^*; \alpha) = V(F_i, \lambda^*; \alpha)\zeta_i(F_1, \ldots, F_k, \lambda^*; \alpha)$$

$$+ \int_{\lambda^*}^{\infty} V(F_i, \lambda; \alpha) \, d_\lambda\zeta_i(F_1, \ldots, F_k, \lambda; \alpha).$$

$$(6.1.7)$$

That ξ satisfies (6.1.3) will follow by showing

$$\xi(F_1, \ldots, F_k, \lambda^*; \alpha) \geqslant \lambda^*/(1-\alpha) \qquad (6.1.8)$$

with equality if and only if (6.1.4) holds, and that

$$\xi(F_1, \ldots, F_k, \lambda^*; \alpha)$$
$$\geqslant E(X_{i1} + \alpha\xi(F_1, \ldots, (X_{i1})F_i, \ldots, F_k, \lambda^*; \alpha)|F_i) \qquad (6.1.9)$$

with equality if and only if (6.1.5) holds.

Suppose (6.1.4) holds. Then $V(F_i, \lambda^*; \alpha) = \lambda^*/(1-\alpha)$ and $\zeta_i(F_1, \ldots, F_k, \lambda; \alpha) = 1$ for $\lambda \geqslant \lambda^*$. Hence, (6.1.7) gives equality at (6.1.8).

Now suppose (6.1.4) fails and choose i satisfying (6.1.5). For $\lambda^* \leqslant \lambda < \Lambda(F_i, \alpha)$, $V(F_i, \lambda; \alpha) > \lambda^*/(1-\alpha)$. So from (6.1.7) we obtain

$$\xi(F_1, \ldots, F_k, \lambda^*; \alpha) > [\lambda^*/(1-\alpha)] [\zeta_i(F_1, \ldots, F_k, \lambda^*; \alpha)$$

$$+ \int_{\lambda^*}^{\Lambda(F_i, \alpha)} d_\lambda\zeta_i(F_1, \ldots, F_k, \lambda; \alpha)]$$

$$= [\lambda^*/(1-\alpha)]\zeta_i(F_1, \ldots, F_k, \Lambda(F_i, \alpha); \alpha)$$

$$= \lambda^*/(1-\alpha),$$

and strict inequality at (6.1.8) follows.

Using (6.1.7), we can write the difference between the two sides of (6.1.9) as follows:

$$\{ V(F_i, \lambda^*; \alpha) - E(X_{i1} + \alpha V((X_{i1})F_i, \lambda^*; \alpha)|F_i)\}$$

$$\cdot\zeta_i(F_1, \ldots, F_k, \lambda^*; \alpha) + \int_{\lambda^*}^{\infty} \{ V(F_i, \lambda; \alpha)$$

$$- E(X_{i1} + \alpha V((X_{i1})F_i, \lambda; \alpha)|F_i)\} d_\lambda\zeta_i(F_1, \ldots, F_k, \lambda; \alpha).$$

$$(6.1.10)$$

The factors within braces are nonnegative. The first term equals 0 if and only if $\lambda^* \leqslant \Lambda(F_i, \alpha)$. And when $\lambda^* \leqslant \Lambda(F_i, \alpha)$, the second term equals 0 if and only if

$$\zeta_i(F_1, \ldots, F_k, \Lambda(F_i, \alpha); \alpha) = \zeta_i(F_1, \ldots, F_k, +\infty; \alpha) = 1.$$

Thus, (6.1.10) equals 0 when (6.1.5) holds, and is positive when (6.1.5) fails; correspondingly, (6.1.9) holds with equality and strict inequality.

It remains to prove that $\zeta = V^*$. They both satisfy (6.1.3); we address the question of uniqueness for solutions of (6.1.3). Consider λ^* and α to be fixed and temporarily restrict consideration to those F_j's for which

$$P(|X_{j1}| > c | F_j) = 0 \qquad (6.1.11)$$

for some c. Using the supremum norm on bounded measurable functions of $F_1, \ldots, F_k$ and the fact that the operator on the right-hand side of (6.1.3) is a contraction on the space of such functions we conclude that (6.1.3) has a unique solution.

We will use a limiting argument to remove restriction (6.1.11), even though V^* is not continuous (Example 2.5.1). We leave it to the reader to check that

$$\lim_{c \to \infty} V^*(F_{1c}, F_{2c}, \ldots, F_{kc}, \lambda^*; \alpha) = V^*(F_1, F_2, \ldots, F_k, \lambda^*; \alpha)$$

where F_{jc} is the measure on $\mathcal{D}$ induced by F_j and the mapping $Q_j \mapsto Q_{jc}$ and distribution function $Q_{jc}(x)$ is

$$Q_{jc}(x) = \begin{cases} 0 & \text{if} \quad x < -c \\ Q_j(x) & \text{if} \quad |x| < c \\ 1 & \text{if} \quad x > c. \end{cases}$$

The functions $F_j \mapsto V(F_j, \lambda; \alpha)$ and $F_j \mapsto \dfrac{\partial}{\partial \lambda} V(F_j, \lambda; \alpha)$ also have this 'restricted continuity' property. Hence, so does ξ and, therefore, $V^* = \xi$. $\qquad \square$

Remarks The restriction on switching from arm $k + 1$ in the quasi-bandit $(F_1, \ldots, F_k, \lambda^*; \alpha)$ was necessary in the proof of Theorem 6.1.1. But in view of the theorem it is clear that this restriction is of no consequence in determining an optimal strategy in the quasi-bandit and so, effectively, the quasi-bandit is a bandit. $\qquad \square$

The following two results are immediate consequences of Theorem 6.1.1, Corollary 5.3.4, and Lemma 4.1.5. The first says that staying with a winner is optimal; the second says that switching from a loser is optimal when the arm selected was barely optimal. No attempt to discuss unique optimality is made in stating these results.

Corollary 6.1.3 Suppose arm i is Bernoulli and is optimal in the $(F_1, \ldots, F_k; \alpha)$-bandit. If arm i yields a success then it is optimal at stage 2 as well.

Corollary 6.1.4 Suppose arms i and j in the $(F_1, \ldots, F_k; \alpha)$-bandit are both optimal. If arm i is Bernoulli, is selected, and yields a failure, then arm j is optimal at the second stage.

At several places, particularly in Section 5.4, we have considered examples with geometric discounting and one unknown arm. Theorem 6.1.1 gives those examples added significance: namely, it applies to show how to use the $\Lambda(F, \alpha)$ calculated for the unknown arm in solving a bandit for which one of the k arms has distribution F.

We shall give two related examples that show how Theorem 6.1.1 can be applied. These examples are straightforward and easy, but are not meant to suggest any restrictions on the applicability of the theorem. The only real difficulty in applying the theorem to solve a bandit problem with independent arms and geometric discounting arises from difficulties in calculating the various Λ's. The examples avoid such difficulties by choosing distributions F_i for which $\Lambda(F_i, \alpha)$ can be evaluated explicitly.

Example 6.1.1. Consider the $(F_1, F_2; 1/2)$-bandit where the arms are Bernoulli and, given as distributions on the Bernoulli parameters,

$$F_1 = (1 - p_1)\delta_{1/3} + p_1\delta_1,$$
$$F_2 = (1 - p_2)\delta_{2/3} + p_2\delta_1.$$

From Theorem 5.4.2 or Example 5.4.7 we find

$$\Lambda(F_1, 1/2) = \frac{1 + 4p_1}{3 + 2p_1},$$

$$\Lambda(F_2, 1/2) = \frac{2 + 2p_2}{3 + p_2}.$$

Theorem 6.1.1 applies to show that the choice between arms 1 and 2 can be made by deciding which of these two quantities is greater.

Suppose the means of the two arms are equal: $p_1 + (1 - p_1)/3 = p_2 + 2(1 - p_2)/3$, or $p_1 = (1 + p_2)/2$. Then

$$\Lambda(F_1, 1/2) = \frac{1 + 4p_1}{3 + 2p_1} = \frac{3 + 2p_2}{4 + p_2} > \frac{2 + 2p_2}{3 + p_2} = \Lambda(F_2, 1/2)$$

and so arm 1 is uniquely optimal. This is consistent with the notion, considered in Theorem 5.0.2, for example, that the best arm to select is the one that is 'riskier' – has larger variance – when the means are the same.   □

Example 6.1.2 Consider the $(F_1, \ldots, F_k; 1/2)$-bandit where the k Bernoulli arms are identical: for $i = 1, \ldots, k$,

$$F_i = (1-p)\delta_{1/2} + p\delta_1.$$

As in the previous example we use Theorem 5.4.2 or Example 5.4.7 to conclude that

$$\Lambda(F_i, 1/2) = \frac{1+2p}{2+p}.$$

Optimal strategies in each $(F_i, \lambda; 1/2)$-bandit are clear. Always select the known arm if $\lambda \geqslant (1+2p)/(2+p)$. Always select the unknown arm if $\lambda \leqslant 1/2$. For other λ's select the unknown arm until (if ever) a failure is observed and then switch to the known arm. Easy calculations give, for $i = 1, \ldots, k$.

$$V(F_i, \lambda; 1/2) = \begin{cases} 2\lambda & \text{if } \lambda \geqslant \dfrac{1+2p}{2+p} \\ 2p + 2(1-p)(1+\lambda)/3 & \text{if } \tfrac{1}{2} < \lambda < \dfrac{1+2p}{2+p} \\ p+1 & \text{if } \lambda \leqslant \tfrac{1}{2} \end{cases}$$

Clearly (even without using Theorem 6.1.1), it is optimal to select any arm and use it until it fails, switching to a second arm if and when it does; this pattern is repeated until and if each arm fails once. If that happens then it is clear that $\theta_1 = \ldots = \theta_k = 1/2$ and so all continuations are optimal.

We can use Theorem 6.1.1 to calculate the value of this bandit. Applying (6.1.1) and letting

$$\rho^* = \bigvee_{i=1}^{k} \Lambda(F_i, \alpha) = \frac{1+2p}{2+p},$$

we obtain

$$V(F_1, \ldots, F_k; \alpha) = 2\left[\rho^* - \left(\frac{1}{2}\right)^k \int_0^{\rho^*} \left(\frac{\partial}{\partial \lambda} V(F_1, \lambda; 1/2)\right)^k d\lambda\right]$$

$$= 2\left[\rho^* - \int_{1/2}^{\rho^*} \left(\frac{1-p}{3}\right)^k d\lambda\right]$$

$$= \frac{2(1+2p)}{2+p} - \frac{3p}{2+p}\left(\frac{1-p}{3}\right)^k.$$

In particular, the value approaches $2(1+2p)/(2+p)$ as $k \to \infty$. This compares with $\gamma_1 = 2$, a payoff that would be achieved using an arm that gives all successes. While, in the limit as $k \to \infty$, an arm that gives all successes will be found with probability 1 (if $p > 0$), there are losses incurred while looking for it. $\square$

As advertised in the introduction to this chapter, we turn in the next section to a converse of Theorem 6.1.1.

6.2 Necessity of geometric discounting for the Gittins–Jones result

In the introduction to this chapter we claimed that the Gittins–Jones result (Theorem 6.1.1) requires geometric discounting. We now demonstrate this fact assuming **A** is regular (Definition 5.2.1) with $\alpha_1 > 0$. In view of Theorem 5.5.3, these assumptions guarantee the existence of the various $\Lambda(F_i, \mathbf{A})$.

Theorem 6.2.1 Assume **A** is regular with $\alpha_1 > 0$. If, for all $(F_1, \ldots, F_k)$, the optimal initial selections in the $(F_1, \ldots, F_k; \mathbf{A})$-bandit are those i for which

$$\Lambda(F_i, \mathbf{A}) = \bigvee_{j=1}^k \Lambda(F_j, \mathbf{A}),$$

then $\mathbf{A} \propto (1, \alpha, \alpha^2, \ldots)$ for some $\alpha \geq 0$.

Proof Fix **A** to be regular with $\alpha_1 > 0$. We shall consider certain distributions (F_1, F_2) on Bernoulli parameters θ_1, θ_2. We restrict consideration to pairs (F_1, F_2) for which

$$\Lambda(F_1, \mathbf{A}) = \Lambda(F_2, \mathbf{A}).$$

The hypothesis of the theorem then implies

$$V^{(1)}(F_1, F_2; \mathbf{A}) = V^{(2)}(F_1, F_2; \mathbf{A})$$

for all such (F_1, F_2). So the theorem will follow when we show that this implies $\mathbf{A}$ is geometric.

The parts of the proof are labelled from (i) to (vii). In (i) we define a set of (F_1, F_2) which is indexed by parameters $t \in (0, 1)$ and $p \in [0, 1)$. The functions $\Lambda(F_1, \mathbf{A})$ and $\Lambda(F_2, \mathbf{A})$ are calculated in (ii) and (iii). In (iv) these are set equal by fixing p to be a certain function of t (for the remainder of the proof). The value of selecting arm i and proceeding optimally thereafter, $V^{(i)}(F_1, F_2; \mathbf{A})$, is calculated for $i = 1$ in (v) and for $i = 2$ and all sufficiently small t in (vi). Finally, in (vii) we set $V^{(1)} = V^{(2)}$ which gives the generating function of $\mathbf{A}$, determining it (up to a constant multiple) as being geometric.

(i) For $t \in (0, 1)$ and $p \in [0, 1)$ let

$$F_1 = (1 - t)\delta_0 + t\delta_1,$$
$$F_2 = (1 - p)\delta_t + p\delta_1.$$

(ii) From Example 5.4.4,

$$\Lambda(F_1, \mathbf{A}) = \frac{t\gamma_1}{t\gamma_1 + (1 - t)\alpha_1},$$

where, as previously defined, $\gamma_1 = \sum\limits_{m=1}^{\infty} \alpha_m$.

(iii) From Theorem 5.4.2 or Example 5.4.7,

$$\Lambda(F_2, \mathbf{A}) = \Lambda_\infty(F_2, \mathbf{A}) = \frac{tp + t(1 - p)\eta(t)}{tp + (1 - p)\eta(t)},$$

where Λ_∞ is defined at (5.4.2) and

$$\eta(t) = \sum\limits_{m=1}^{\infty} (\alpha_m/\gamma_1)t^m,$$

the generating function of $\mathbf{A}/\gamma_1$.

(iv) Setting $\Lambda(F_1, \mathbf{A}) = \Lambda(F_2, \mathbf{A})$ gives

$$\frac{p}{1 - p} = \gamma_2 \eta(t)/\alpha_1. \tag{6.2.1}$$

Hereafter, for each t we fix p to satisfy (6.2.1).

(v) If arm 1 is selected initially then an optimal continuation is clear: use arm 1 exclusively if it was successful and use arm 2 exclusively thereafter if it was not. It is simple to calculate

$$V^{(1)}(F_1, F_2; \mathbf{A}) = t\gamma_1 + (1-t)[p + (1-p)t]\gamma_2. \qquad (6.2.2)$$

(vi) Calculating $V^{(2)}(F_1, F_2; \mathbf{A})$ is substantially more difficult than calculating $V^{(1)}(F_1, F_2; \mathbf{A})$. Should arm 2 fail then it is easy to see, assuming (6.2.1), that arm 1 is then optimal; but an optimal continuation is not clear when it succeeds. We will show that for all sufficiently small t arm 2 continues to be optimal as long as it is successful. Equivalently, we show that staying indefinitely with a successful arm 2 is at least as good as switching after any given number of successes when $\mathbf{A}$ is regular. (We note that staying with a winner is not generally a property of optimal strategies when there are two unknown Bernoulli arms even with regular discounting (cf. Example 4.3.4).)

Define a sequence of strategies τ_n where $\tau_n(\varnothing) = 2$ and arm 2 is continued when following τ_n until it fails or until stage n is reached; in either case arm 1 is then used once and the arm then known to be better used thereafter. Clearly,

$$V^{(2)}(F_1, F_2; \mathbf{A}) = W(F_1, F_2; \mathbf{A}; \tau_\infty) \vee \bigvee_{n=1}^{\infty} W(F_1, F_2; \mathbf{A}; \tau_n). \qquad (6.2.3)$$

Straightforward calculations give

$$E_{\tau_n}(Z_m | \theta_1 = 1, \theta_2 = 1) = 1;$$

$$E_{\tau_n}(Z_m | \theta_1 = 1, \theta_2 = t) = \begin{cases} 1 - t^{m-1}(1-t) & \text{if } m \leqslant n \\ 1 & \text{if } m > n; \end{cases}$$

$$E_{\tau_n}(Z_m | \theta_1 = 0, \theta_2 = 1) = \begin{cases} 0 & \text{if } m = n+1 \\ 1 & \text{if } m \neq n+1; \end{cases}$$

$$E_{\tau_n}(Z_m | \theta_1 = 0, \theta_2 = t) = \begin{cases} t & \text{if } m = 1 \text{ or } m > n+1 \\ t - t^{m-1}(1-t) & \text{if } 1 < m < n+1 \\ t - t^n & \text{if } m = n+1. \end{cases}$$

The sum of these quantities weighted by tp, $t(1-p)$, $(1-t)p$, and

$(1-t)(1-p)$, respectively, is $E_{\tau_n}(Z_m|F_1, F_2)$. An easy calculation gives

$$W(F_1, F_2; \mathbf{A}; \tau_n) = \sum_{m=1}^{\infty} \alpha_m E_{\tau_n}(Z_m|F_1, F_2)$$

$$= \gamma_1 - (1-t)^2(1-p)\gamma_2 - (1-t)(1-p) \sum_{m=1}^{n+1} \alpha_m t^{m-1} - (1-t)p\alpha_{n+1}$$

where $\alpha_{\infty} = 0$. For $n < \infty$,

$$W(F_1, F_2; \mathbf{A}; \tau_{\infty}) - W(F_1, F_2; \mathbf{A}; \tau_n)$$

$$= -(1-t)(1-p) \sum_{m=n+2}^{\infty} \alpha_m t^{m-1} + (1-t)p\alpha_{n+1}.$$

Dividing this by $(1-t)(1-p)$ and substituting for $p/(1-p)$ using (6.2.1) gives

$$- \sum_{m=n+2}^{\infty} \alpha_m t^{m-1} + \gamma_2 \alpha_{n+1} \eta(t)/\alpha_1,$$

which is no smaller than

$$-t^2 \gamma_{n+1} + t\gamma_2 \alpha_{n+1}/\gamma_1.$$

That this quantity is nonnegative for all n provided t is sufficiently small follows since, when $\gamma_{n+1} \neq 0$,

$$\frac{\gamma_{n+1}}{\gamma_{n+1}} = 1 - \frac{\gamma_{n+2}}{\gamma_{n+1}}$$

is nondecreasing for regular $\mathbf{A}$.

We have shown that the supremum in (6.2.3) is achieved by $W(F_1, F_2; \mathbf{A})$ for $n = \infty$. Hence

$$V^{(2)}(F_1, F_2; \mathbf{A}) = \gamma_1 - (1-t)^2(1-p)\gamma_2 - (1-t)(1-p)t^{-1}\eta(t)\gamma_1.$$

$$(6.2.4)$$

(vii) We set $V^{(1)}(F_1, F_2; \mathbf{A}) = V^{(2)}(F_1, F_2; \mathbf{A})$ while continuing to assume $\Lambda(F_1, \mathbf{A}) = \Lambda(F_2, \mathbf{A})$ which is effected by guaranteeing that p and t satisfy (6.2.1). Equating (6.2.2) and (6.2.4) we find that

$$(1-p)\eta(t) = t(1 - [(1-p)(1-t) + p + (1-p)t]\gamma_2/\gamma_1)$$

$$= t\alpha_1/\gamma_1.$$

$$(6.2.5)$$

From (6.2.1) we find

$$1 - p = \frac{\alpha_1}{\alpha_1 + \gamma_2 \eta(t)};$$

substituting into (6.2.5) we obtain

$$\gamma_1 \eta(t) = \frac{t\alpha_1}{1 - t(\gamma_2/\gamma_1)},$$

the generating function of the geometric sequence. We conclude from the uniqueness of generating functions that $\mathbf{A} = \alpha_1(1, \alpha, \alpha^2, \ldots)$ where $\alpha = \gamma_2/\gamma_1$. □

Since $\Lambda(F_i, \mathbf{A})$ is not defined for all $\mathbf{A}$, Theorem 6.2.1 will not generalize to include all nonregular discount sequences and sequences for which $\alpha_1 = 0$. We conjecture, however, that an analogous result holds in complete generality. Such a result can be stated along the lines of transitivity of arms as discussed in the introduction to this chapter. Fix $\mathbf{A}$ and for all k-armed bandits $(F_1, \ldots, F_k; \mathbf{A})$ suppose that $\Delta(F_1, F_2; \mathbf{A}) > 0$ and $\Delta(F_2, F_3; \mathbf{A}) > 0$ imply $\Delta(F_1, F_3; \mathbf{A}) > 0$. Then, we conjecture, $\mathbf{A}$ is geometric.

References

Gittins, J. C. and Jones, D. M. (1974) A dynamic allocation index for the sequential design of experiments. In *Progress in Statistics* (eds J. Gani *et al.*), pp. 241–266, North-Holland, Amsterdam.

Whittle, P. (1982) *Optimization Over Time: Dynamic Programming and Stochastic Control*, Vol. I, Wiley, New York.

CHAPTER 7

Two independent Bernoulli arms; uniform discounting

In this chapter we assume that $k = 2$ and the arms are Bernoulli. The arms are independent initially, and therefore also henceforth. The discount sequence **A** has horizon n and is uniform: $\alpha_1 = \ldots = \alpha_n = 1$ and $\alpha_{n+1} = \alpha_{n+2} = \ldots = 0$. Such uniform discounting has been considered extensively through examples in the first five chapters of this book, and in the literature generally. The objective implicit in uniform discounting is to maximize the expected sum of the first n observations.

When one of the arms has known probability of success, the arms are obviously independent. So this chapter generalizes those parts of Chapter 5 that deal with Bernoulli arms and a uniform discount sequence.

Some of the results in this chapter are in Berry (1972). Section 7.1 specializes aspects of Chapter 4 to the uniform-discounting setting. The use of the recursion formula developed in Section 4.2 to determine optimal strategies is discussed in Section 7.2; in particular, this formula is applied to the easy case $n = 2$. In Section 7.3 the horizon is general but the relationship between F_1 and F_2 is special; each is taken to be some posterior distribution resulting from a common underlying distribution F. A central feature of the discussion in Section 7.3 is a necessary and sufficient condition for $F_1 \overset{n-1}{\succsim} F_2$.

7.1 Preliminaries

In this chapter the discount sequence **A** is completely described by the single parameter n, the horizon. We therefore use n in place of **A** throughout. And, consistent with this convention, $\mathbf{A}^{(j)}$ is replaced by

$n - j$ since the horizon is reduced by j at stage $j + 1$. For example, we write $W(F_1, F_2; n; \tau)$ for the worth of strategy τ.

The results of Chapter 4 hold in the special setting of this chapter. For example, Lemma 4.2.1 gives the following recursion: for $n = 1$,

$$\Delta(F_1, F_2; 1) = E(\theta_1 | F_1) - E(\theta_2 | F_2);$$

for $n \geqslant 2$,

$$\begin{aligned}
\Delta(F_1, F_2; n) = {} & E(\theta_1 | F_1)\Delta^+(\sigma F_1, F_2; n - 1) \\
& + E(1 - \theta_1 | F_1)\Delta^+(\varphi F_1, F_2; n - 1) \\
& - E(\theta_2 | F_2)\Delta^-(F_1, \sigma F_2; n - 1) \\
& - E(1 - \theta_2 | F_2)\Delta^-(F_1, \varphi F_2; n - 1). \quad (7.1.1)
\end{aligned}$$

Thus, for $n \geqslant 2$,

$$\begin{aligned}
\Delta(F_1, F_2; n) \leqslant {} & E(\theta_1 | F_1)\Delta^+(\sigma F_1, F_2; n - 1) \\
& + E(1 - \theta | F_1)\Delta^+(\varphi F_1, F_2; n - 1).
\end{aligned}$$

The stay-with-a-winner rule, Theorem 4.3.8, is now easily expressed:

Corollary 7.1.1 Suppose $n \geqslant 2$ and the support of either F_1 or F_2 consists of more than one point. If $\Delta(F_1, F_2; n) \geqslant 0$ then $\Delta(\sigma F_1, F_2; n - 1) > 0$.

An easy consequence of this result is given next. It says that if the mean of arm 1 is sufficiently large (depending on arm 2) then arm 1 is optimal. In particular, it says that if arm 1 would have a larger probability of success than arm 2 even if arm 2 produces $n - 1$ successes in $n - 1$ trials, then arm 1 is optimal initially. While the bound is very crude, it can be instructive.

Corollary 7.1.2 If

$$E(\theta_1 | F_1) \geqslant E(\theta_2 | \sigma^{n-1} F_2) \quad (7.1.2)$$

then arm 1 is optimal initially.

Proof If F_2 is supported by one point then $E(\theta_2 | \sigma^{n-1} F_2)$ equals $E(\theta_2 | F_2)$ and the result follows from Theorem 5.0.2. The result is also immediate if $n = 1$.

Suppose F_2 is not supported by one point and $n > 1$. Then if arm 2 were optimal initially it would, by Corollary 7.1.1, be uniquely optimal from stage 2 onwards as long as successes were obtained. In particular, it would be uniquely optimal at stage n if $n - 1$ successes were obtained. This contradicts (7.1.2). (Moreover, this argument shows arm 1 to be uniquely optimal initially in case F_1 is not supported by one point and $n > 1$.) $\square$

In the next section we give a complete characterization of optimal strategies when $n = 2$, carrying out the detailed calculations of $\Delta(F_1, F_2; n)$. The case $n = 2$ is relatively easy of course, but calculating Δ is still not a trivial matter.

7.2 Optimal selections when the horizon is two

The recursion given in (7.1.1) allows for explicit calculation of Δ. We will describe this procedure, specializing to the case $n = 2$ to obtain simple explicit formulas. The complexity of the calculations for large n will be apparent.

The calculation of $\Delta(F_1, F_2; n)$ requires $\Delta^+(\sigma F_1, F_2; n-1)$, $\Delta^+(\varphi F_1, F_2; n-1)$, $\Delta^-(F_1, \sigma F_2; n-1)$, and $\Delta^-(F_1, \varphi F_2; n-1)$. Since there are two possibilities – positive and zero – for each of these four quantities, there might be as many as sixteen different formulas for $\Delta(F_1, F_2; n)$ in terms of $\Delta(\cdot, \cdot; n-1)$. But eight of these cases are ruled out by Theorem 4.3.6 and Lemma 4.3.5; the remaining eight will be given below. The cases will be identified by quadruples; for example, $(+0++)$ indicates that $\Delta^+(\sigma F_1, F_2; n-1)$, $\Delta^-(F_1, \sigma F_2; n-1)$, and $\Delta^-(F_1, \varphi F_2; n-1)$ are positive and $\Delta^+(\varphi F_1, F_2; n-1) = 0$.

Since $\Delta(\cdot, \cdot; 1)$ is just the difference of expectations, the eight possible formulas can be written simply and explicitly as follows (the designations F_1 and F_2 have been suppressed when expectations involving θ_1 and θ_2 are with respect to these distributions):

$$\Delta(F_1, F_2; 2) =$$

$(+000)$ $E(\theta_1^2) - E(\theta_1)E(\theta_2)$ if $E(\theta_1|\sigma F_1) > E(\theta_2) \geqslant E(\theta_1|\varphi F_1)$
                                          & $E(\theta_1) \geqslant E(\theta_2|\sigma F_2)$

$(++00)$ $E(\theta_1) - E(\theta_2)$ if $E(\theta_1|\varphi F_1) > E(\theta_2)$
                                          & $E(\theta_1) \geqslant E(\theta_2|\sigma F_2)$

$(+++0)$ $\quad E(\theta_1) - E(\theta_2) - E(\theta_2^2)$ $\quad$ if $\quad E(\theta_1|\varphi F_1) > E(\theta_2)$

$\qquad\qquad\qquad\quad + E(\theta_1)E(\theta_2)$ $\quad$ & $\quad E(\theta_2|\sigma F_2) > E(\theta_1)$

$(+0+0)$ $\quad E(\theta_1^2) - E(\theta_2^2)$ $\quad$ if $\quad E(\theta_1|\sigma F_1) > E(\theta_2) \geqslant E(\theta_1|\varphi F_1)$

$\qquad\qquad\qquad\qquad\qquad$ & $\quad E(\theta_2|\sigma F_2) > E(\theta_1) \geqslant E(\theta_2|\varphi F_2)$

$(+0++)$ $\quad E(\theta_1) - E(\theta_2) + E(\theta_1^2)$ $\quad$ if $\quad E(\theta_1|\sigma F_1) > E(\theta_2)$

$\qquad\qquad\qquad\quad - E(\theta_1)E(\theta_2)$ $\quad$ & $\quad E(\theta_2|\varphi F_2) > E(\theta_1)$

$(00++)$ $\quad E(\theta_1) - E(\theta_2)$ $\quad$ if $\quad E(\theta_2) \geqslant E(\theta_1|\sigma F_1)$

$\qquad\qquad\qquad\qquad\qquad$ & $\quad E(\theta_2|\varphi F_2) > E(\theta_1)$

$(00+0)$ $\quad E(\theta_1)E(\theta_2) - E(\theta_2^2)$ $\quad$ if $\quad E(\theta_2) \geqslant E(\theta_1|\sigma F_1)$

$\qquad\qquad\qquad\qquad\qquad$ & $\quad E(\theta_2|\sigma F_2) > E(\theta_1) \geqslant E(\theta_2|\varphi F_2)$

(0000) $\quad 0$ $\quad$ if $\quad E(\theta_2) \geqslant E(\theta_1|\sigma F_1)$

$\qquad\qquad\qquad\qquad\qquad$ & $\quad E(\theta_1) \geqslant E(\theta_2|\sigma F_2)$.

$$(7.2.1)$$

The expectations involving σF_i and φF_i are simply related to those involving F_i:

$$E(\theta_i|\sigma F_i) = E(\theta_i^2)/E(\theta_i),$$
$$E(\theta_i|\varphi F_i) = E(\theta_i(1-\theta_i))/E(1-\theta_i).$$

From Lemma 4.3.5 and Theorem 4.3.6, we see that the case $(0\,0\,0\,0)$ occurs only when F_1 and F_2 are the same one-point distribution.

The case $(+0++)$ in (7.2.1) was used for the calculation of $\Delta(F_1, F_2; \mathbf{A})$ in Example 6.0.1. We now give a more comprehensive example.

Example 7.2.1 We consider $n = 2$ and the special case where F_1 and F_2 are beta distributions:

$$F_i(\mathrm{d}u) \propto u^{a_i - 1}(1 - u)^{b_i - 1}\,\mathrm{d}u, \qquad (7.2.2)$$

where $a_i > 0$ and $b_i > 0, i = 1,2$. We denote $a_i + b_i$ by I_i (which stands for information for the same reason that I was used to stand for information in Section 5.6). We write $\Delta(a_1, b_1, a_2, b_2; 2)$ in place of $\Delta(F_1, F_2; 2)$. To determine the sign of Δ (and thus the optimal initial selection) we shall find the set of (a_1, b_1, a_2, b_2) for which Δ is 0. When it equals 0, and so both arms are optimal initially, $\Delta^+(\sigma F_1, F_2; 2)$ and

$\Delta^-(F_1, \sigma F_2; 2)$ are positive by Corollary 7.1.1. Therefore, $(+ + + 0)$, $(+0+0)$, and $(+0++)$ are the only cases possible in (7.2.1). Straightforward calculations show that in case $(+ + + 0)$, Δ cannot be 0 when $I_1 \leqslant I_2$. Similarly, in case $(+0++)$, Δ cannot be 0 when $I_1 \geqslant I_2$; we assume $I_1 \geqslant I_2$.

To use (7.2.1) we need the following:

$$E(\theta_i) = a_i/I_i,$$
$$E(\theta_i|\sigma F_i) = (a_i + 1)/(I_i + 1),$$
$$E(\theta_i|\varphi F_i) = a_i/(I_i + 1).$$

Evaluating cases $(+ + + 0)$ and $(+0+0)$, we find that the locus of points for which

$$\Delta(a_1, b_1, a_2, b_2; 2) = 0 \qquad (7.2.3)$$

is given as follows:

$$\frac{a_1}{I_1} - \frac{a_2}{I_2}\left[1 + \frac{a_2 + 1}{I_2 + 1} - \frac{a_1}{I_1}\right] = 0 \qquad \text{for } \frac{a_2}{I_2} \leqslant \frac{a_1}{I_1 + 1}$$

$$\frac{a_1}{I_1}\frac{a_1 + 1}{I_1 + 1} - \frac{a_2}{I_2}\frac{a_2 + 1}{I_2 + 1} = 0 \qquad \text{for } \frac{a_2}{I_2} \geqslant \frac{a_1}{I_1 + 1}. \qquad (7.2.4)$$

The solutions of the two equations in (7.2.4) intersect at the point

$$a_1^* = \frac{(I_1 + 1)(I_1 - I_2 - 1)}{I_1 + I_2 + 1}, \qquad a_2^* = \frac{I_2(I_1 - I_2 - 1)}{I_1 + I_2 + 1}, \quad (7.2.5)$$

with the first equation in (7.2.4) applying for values of (a_1, a_2) $\leqslant (a_1^*, a_2^*)$. We are only interested in $0 < a_i < I_i$, $i = 1,2$. The conditions $a_i < I_i$, $i = 1, 2$, are satisfied by (a_1^*, a_2^*). The conditions $a_1^* > 0$ are not satisfied if $I_1 \leqslant I_2 + 1$; in this case the second equation at (7.2.4) gives the points (a_1, a_2) for which $\Delta = 0$. This curve has a corner at (a_1^*, a_2^*) given by (7.2.5) if $I_1 > I_2 + 1$.

To show graphically the dependence of (7.2.4) on the quantities of information I_1 and I_2, we fix the ratio $I_1/I_2 = 3$. The solution is shown in Figure 7.1 for $I_2 = 1$ (and therefore, $I_1 = 3$). The corner point is $(a_1^*, a_2^*) = (4/5, 1/5)$, though it's not discernibly a corner in the figure. The dashed line in Figure 7.1 is the set of points for which the two means are equal, that is, $a_2 = a_1/3$.

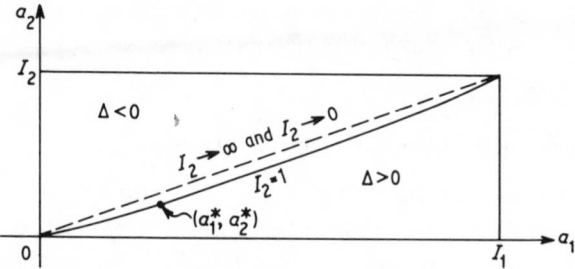

Fig. 7.1 *The sign of* $\Delta(a_1, I_1 - a_1, a_2, I_2 - a_2; 2)$ *for* $I_1 = 3I_2$. *The curve indicates the set for which* $\Delta = 0$ *when* $I_2 = 1$.

This dashed line has additional significance for the issue at hand: it is the set of (a_1, a_2) for which Δ is 0 for $I_i \to \infty$ and also for $I_i \to 0$ (with $I_1/I_2 = 3$). That is, it is the limit of (7.2.4) for both $I_2 \to \infty$ and $I_2 \to 0$ with $I_1/I_2 = 3$. For both I_1 and I_2 large, there is little information to be gained from an observation relative to the information already available; so only the means are relevant. On the other hand, for very small I_i, essentially complete information is gained by the first observation on *either* arm – and so, again, only the means matter.

The curve for which $\Delta = 0$ that is drawn in Figure 7.1 for $I_1 = 3$ and $I_2 = 1$ is approximately the lower envelope of such curves with $I_1/I_2 = 3$ and $0 < I_2 < \infty$. It approximates (7.2.4) for moderate values of I_2 (say between 1/10 and 10).

The region above the $\Delta = 0$ curve in Figure 7.1 is labelled '$\Delta < 0$' and indicates that arm 2 is optimal for these points. Arm 1 is optimal for points in the '$\Delta > 0$' region in the figure. □

Conditions for $\Delta \geqslant 0$ show that arm 1 is optimal initially if and only if

$$[E(\theta_1^2) - E(\theta_1)E(\theta_2)] \vee [E(\theta_1) - E(\theta_2)]$$
$$\geqslant [E(\theta_2^2) - E(\theta_2)E(\theta_1)] \vee [E(\theta_2) - E(\theta_1)],$$

a result given by Bradt, Johnson, and Karlin (1956, Lemma 3.1).

The relatively easy case $n = 2$ treated in this section is actually somewhat cumbersome. As indicated in Chapter 2, complete solutions are increasingly complicated for larger horizons. In the next section we partially determine the optimal initial selection by finding the sign of Δ for certain pairs (F_1, F_2).

7.3   Arms with identical underlying priors

Arm 1 is optimal initially in the $(F_1, F_2; n)$-bandit if F_1 is strongly to the right of F_2 or if the weaker condition $F_1 \overset{n-1}{\succsim} F_2$ holds (Corollary 4.3.7). In this section we find conditions for these relations when (F_1, F_2) belongs to the class of distributions in which the arms are comparable in the sense that information concerning them can be viewed as having arisen from a common, but arbitrary, distribution. That is, some time in the 'past' they had the same distribution, say F. We develop this notion next.

Let F be a distribution measure on $[0, 1]$ such that $F(\{1, 0\}) = 0$ and F is not supported by one point. Suppose that F is the prior distribution of a Bernoulli parameter θ and that a successes and b failures are observed. Then the posterior distribution of θ is

$$\frac{x^a(1-x)^b F(dx)}{\displaystyle\int_{[0,1]} u^a(1-u)^b F(du)} ;$$

that is,

$$(\sigma^a \varphi^b F)(dx) \propto x^a (1-x)^b F(dx). \tag{7.3.1}$$

We assume that both F_1 and F_2 can be written as in (7.3.1) for some F which is common to both arms; so

$$F_i = \sigma^{a_i} \varphi^{b_i} F, \qquad i = 1, 2, \tag{7.3.2}$$

where a_i and b_i are regarded as known but arbitrary.

Two natural generalizations can be made. We permit a_i and b_i to be positive real numbers, not necessarily integers. We require only proportionality in (7.3.1). So a positive measure F that is not a probability measure may be used provided that

$$\int_{[0,1]} u^a (1-u)^b F(du) < \infty$$

for $a, b > 0$.

The most important example in this setting is given by

$F(du) = u^{-1}(1-u)^{-1} du$. Then F_i is a beta distribution with parameters a_i and b_i.

Theorem 7.3.2 below gives sufficient conditions for the relationship of strongly to the right to hold. For its proof we first show the corresponding result with 'to the right' in place of 'strongly to the right'.

Lemma 7.3.1 Let F_1 and F_2 be given by (7.3.2). Then F_1 is to the right of F_2 if $a_1 \geq a_2$ and $b_1 \leq b_2$. The relationship is strict if $a_1 > a_2$ or $b_1 < b_2$.

Proof Suppose that $a_1 \geq a_2$, $b_1 \leq b_2$, and at least one of these inequalities is strict. We want to show that $F_1[x, 1] \geq F_2[x, 1]$ for all $x \in (0, 1)$ with strict inequality for at least one x. The weak inequality is obvious if $F(0, x) = 0$, so we assume $F(0, x) > 0$ in which case $F_i(0, x) > 0$ for $i = 1, 2$. The inequality of interest can be rewritten as

$$\frac{F_1[x, 1]}{F_1[0, x)} \geq \frac{F_2[x, 1]}{F_2[0, x)}, \tag{7.3.3}$$

which follows from

$$\frac{\displaystyle\int_{[x, 1]} u^{a_1}(1-u)^{b_1} F(du)}{\displaystyle\int_{[0, x)} u^{a_1}(1-u)^{b_1} F(du)}$$

$$\geq \frac{x^{a_1-a_2}(1-x)^{b_1-b_2}\displaystyle\int_{[x, 1]} u^{a_2}(1-u)^{b_2} F(du)}{x^{a_1-a_2}(1-x)^{b_1-b_2}\displaystyle\int_{[0, x)} u^{a_2}(1-u)^{b_2} F(du)} \tag{7.3.4}$$

Moreover, the inequality in (7.3.4), and therefore in (7.3.3), is strict if x is chosen so that the support of F contains a number larger than x and a number smaller than x.   □

Theorem 7.3.2 Let F_1 and F_2 be given by (7.3.2). Then F_1 is strongly to the right of F_2 if $a_1 \geq a_2$ and $b_1 \leq b_2$. The relationship is strict if $a_1 > a_2$ or $b_1 < b_2$.

Proof Assume $a_1 \geq a_2$ and $b_1 \leq b_2$. For any nonnegative integers s and f, $a_1 + s \geq a_2 + s$ and $b_1 + f \leq b_2 + f$; so, by Lemma 7.3.1, $\sigma^s \varphi^f F_1$

is to the right $\sigma^s \varphi^f F_2$. Hence F_1 is strongly to the right of F_2. The strictness of this relationship in case $a_1 > a_2$ or $b_1 < b_2$ follows from Lemma 7.3.1. $\square$

Remark In case $F(du) = u^{-1}(1-u)^{-1} du$, the converse of Theorem 7.3.2 holds. However, the converse does not hold in general; to see this let F be a symmetric two-point distribution. $\square$

On the basis of Theorem 7.3.2, arm 1 is seen to be optimal for certain (a_1, b_1, a_2, b_2), arm 2 is optimal for others, and the theorem does not indicate an optimal selection for the rest.

Example 7.3.1 Let $F(du) = u^{-1}(1-u)^{-1} du$ and fix $a_1 + b_1 = I_1$ and $a_2 + b_2 = I_2$ with $I_1 = 3I_2$. So F_1 and F_2 are the same distributions considered in Figure 7.1. The regions in the (a_1, a_2)-plane where Theorem 7.3.2 indicates an optimal initial selection are shown in Figure 7.2.

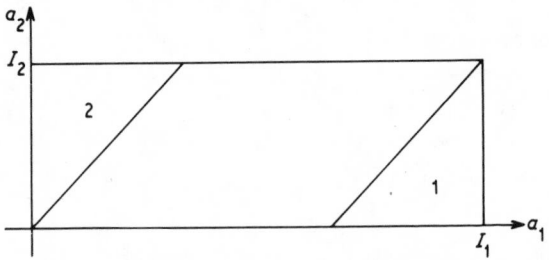

Fig. 7.2   *Regions where the indicated arms are known to be optimal on the basis of Theorem 7.3.2.*

If I_2/I_1 were closer to 1, the region of uncertainty would be smaller. If $I_2 = I_1$, then one (or both) of F_1 or F_2 is strongly to the right of the other and there is no uncertainty region. $\square$

Theorem 7.3.2 gives conditions for optimality that do not depend on n. The condition given by the next theorem depends on n, and the result is stronger. For fixed I_1, I_2, and n the result will allow us to make the region of uncertainty in Figure 7.2 smaller by focusing on the relation $\overset{n-1}{\gtrsim}$ rather than on 'strongly to the right.'

Theorem 7.3.3 Suppose $I_1 > I_2$. For $n > 1$, a necessary and sufficient condition for $F_1 \overset{n-1}{\succeq} F_2$ is

$$E(\theta_1 | \sigma^{n-1} F_1) \geqslant E(\theta_2 | \sigma^{n-1} F_2). \qquad (7.3.5)$$

Remark Another way to write (7.3.5) is

$$\frac{E(\theta_1^n | F_1)}{E(\theta_1^{n-1} | F_1)} \geqslant \frac{E(\theta_2^n | F_2)}{E(\theta_2^{n-1} | F_2)}. \qquad \square$$

For the proof we need the following lemma.

Lemma 7.3.4 Suppose $I_1 > I_2$ and $E(\theta_1 | F_1) = E(\theta_2 | F_2)$. Then $E(\theta_1 | \sigma F_1) < E(\theta_2 | \sigma F_2)$ and $E(\theta_1 | \varphi F_1) > E(\theta_2 | \varphi F_2)$.

Proof The second implication will follow from the first with the roles of success and failure exchanged; we will show the first.

Fix a_2 and b_2 and write

$$E(\theta_1 | F_1) = \frac{f_1(a_1, b_1)}{f_0(a_1, b_1)}$$

and

$$E(\theta_1 | \sigma F_1) = \frac{f_2(a_1, b_1)}{f_1(a_1, b_1)}, \qquad (7.3.6)$$

where

$$f_j(a_1, b_1) = \int_{[0, 1]} u^{a_1 + j} (1 - u)^{b_1} F(du).$$

For $b_1 \geqslant b_2$, let a_1 be the function of b_1 for which the means of the two arms are equal, that is,

$$\frac{f_1(a_1, b_1)}{f_0(a_1, b_1)} = E(\theta_2 | F_2) \qquad (7.3.7)$$

Clearly a_1 is a well defined strictly increasing function of b_1 that equals a_2 when $b_1 = b_2$ and approaches ∞ as $b_1 \to \infty$. Implicit differentiation in (7.3.7) gives

$$\frac{da_1}{db_1} = \frac{f_1 \mathbf{D}_2 f_0 - f_0 \mathbf{D}_2 f_1}{f_0 \mathbf{D}_1 f_1 - f_1 \mathbf{D}_1 f_0},$$

where $\mathbf{D}_i$ indicates differentiation with respect to the ith variable.

Differentiation of (7.3.6) assuming (7.3.7) gives

$$f_1[f_0 \mathbf{D}_1 f_1 - f_1 \mathbf{D}_1 f_0]\frac{\mathrm{d}}{\mathrm{d}b_1}E(\theta_1 | \sigma F_1)$$

$$= f_2[\mathbf{D}_1 f_0 \mathbf{D}_2 f_1 - \mathbf{D}_2 f_0 \mathbf{D}_1 f_1]$$

$$+ \mathbf{D}_1 f_2[f_1 \mathbf{D}_2 f_0 - f_0 \mathbf{D}_2 f_1] + \mathbf{D}_2 f_2[f_0 \mathbf{D}_1 f_1 - f_1 \mathbf{D}_1 f_0]. \quad (7.3.8)$$

Obviously $f_1 > 0$, $f_2 > 0$, and easy calculations shown that $\mathbf{D}_1 f_2 < 0$ and $\mathbf{D}_2 f_2 < 0$. Writing

$$\pi(\mathrm{d}u) = u^{a_1}(1 - u)^{b_1}F(\mathrm{d}u),$$

we have

$$f_0 \mathbf{D}_1 f_1 = \int_{[0,1]} \pi(\mathrm{d}u) \int_{[0,1]} v \log v \; \pi(\mathrm{d}v)$$

$$= \int_{[0,1] \times [0,1]} v \log v \, (\pi \times \pi)(\mathrm{d}(u,v))$$

and

$$f_1 \mathbf{D}_1 f_0 = \int_{[0,1] \times [0,1]} v \log u \, (\pi \times \pi)(\mathrm{d}(u,v)).$$

Hence,

$$f_0 \mathbf{D}_1 f_1 - f_1 \mathbf{D}_1 f_0 = \int_{[0,1] \times [0,1]} v \log(v/u)(\pi \times \pi)(\mathrm{d}(u,v))$$

$$= \int_{u<v} v \log(v/u)(\pi \times \pi))(\mathrm{d}(u,v))$$

$$+ \int_{u>v} v \log(v/u)(\pi \times \pi)(\mathrm{d}(u,v))$$

$$= \int_{u<v} v \log(v/u)(\pi \times \pi)(\mathrm{d}(u,v))$$

$$+ \int_{v>u} u \log(u/v)(\pi \times \pi)(\mathrm{d}(u,v))$$

$$= \int_{u<v} (v-u) \log(v/u)(\pi \times \pi)(\mathrm{d}(u,v)) > 0.$$

Similar calculations show

$$\mathbf{D}_1 f_0 \mathbf{D}_2 f_1 - \mathbf{D}_2 f_0 \mathbf{D}_1 f_1 < 0,$$
$$f_1 \mathbf{D}_2 f_0 - f_0 \mathbf{D}_2 f_1 > 0,$$
$$f_0 \mathbf{D}_1 f_1 - f_1 \mathbf{D}_1 f_0 > 0.$$

From (7.3.8) we conclude that $\dfrac{\mathrm{d}}{\mathrm{d}b} E(\theta_1 | \sigma F_1) < 0$.   □

Proof of Theorem 7.3.3. Necessity follows by definition. We prove sufficiency by induction starting at $n = 2$. Suppose that $E(\theta_1 | \sigma F_1) \geqslant E(\theta_2 | \sigma F_2)$, but that $E(\theta_1 | F_1) \leqslant E(\theta_2 | F_2)$. By increasing a_1 (or perhaps keeping it the same) we can achieve both $E(\theta_1 | \sigma F_1) \geqslant E(\theta_2 | \sigma F_2)$ and $E(\theta_1 | F_1) = E(\theta_2 | F_2)$, contradicting Lemma 7.3.4. Therefore, $E(\theta_1 | \sigma F_1) \geqslant E(\theta_2 | \sigma F_2)$ implies $E(\theta_1 | F_1) > E(\theta_2 | F_2)$. This in turn implies $E(\theta_1 | \varphi F_1) \geqslant E(\theta_2 | \varphi F_2)$. For suppose that $E(\theta_1 | \varphi F_1) < E(\theta_2 | \varphi F_2)$. By increasing b_1 we can obtain both $E(\theta_1 | F_1) = E(\theta_2 | F_2)$ and $E(\theta_1 | \varphi F_1) < E(\theta_2 | \varphi F_2)$. This also contradicts Lemma 7.3.4.

We have shown that $E(\theta_1 | \sigma F_1) \geqslant E(\theta_2 | \sigma F_2)$ implies both $E(\theta_1 | F_1) > E(\theta_2 | F_2)$ and $E(\theta_1 | \varphi F_1) \geqslant E(\theta_2 | \varphi F_2)$ and, hence $F_1 \overset{1}{\succ} F_2$.

Let $n > 2$ and assume $E(\theta_1 | \sigma^{m-1} F_1) \geqslant E(\theta_2 | \sigma^{m-1} F_2)$ implies $F_1 \overset{m-1}{\succ} F_2$ for $m < n$. Assume $E(\theta_1 | \sigma^{n-1} F_1) \geqslant E(\theta_2 | \sigma^{n-1} F_2)$. To show that $E(\theta_1 | \sigma^s \varphi^f F_1) \geqslant E(\theta_2 | \sigma^s \varphi^f F_2)$ for all (s, f) with $s + f \leqslant n - 1$, we first prove it when $s \geqslant 1$. Then we prove it for $s = 0$, obtaining, in particular, the desired strict inequality when $s = 0 = f$.

Since $E(\theta_1 | \sigma^{n-1} F_1) \geqslant E(\theta_2 | \sigma^{n-1} F_2)$ can be written as $E(\theta_1 | \sigma^{n-2} \sigma F_1) \geqslant E(\theta_2 | \sigma^{n-2} \sigma F_2)$, application of the induction hypothesis to σF_1 and σF_2 with $m = n - 1$ gives $\sigma F_1 \overset{n-2}{\succ} \sigma F_2$. Hence, $E(\theta_1 | \sigma^s \varphi^f \sigma F_1) \geqslant E(\theta_2 | \sigma^s \varphi^f \sigma F_2)$ for $s + f \leqslant n - 2$; that is, $E(\theta_1 | \sigma^s \varphi^f F_1) \geqslant E(\theta_2 | \sigma^s \varphi^f F_2)$ when $s \geqslant 1$ and $s + f \leqslant n - 1$.

In particular, $E(\theta_1 | \sigma \varphi^f F_1) \geqslant E(\theta_2 | \sigma \varphi^f F_2)$ for $f \leqslant n - 2$. The induction hypothesis applied to $\varphi^f F_1$ and $\varphi^f F_2$ for $m = 2$ gives $E(\theta_1 | \varphi^f F_1) > E(\theta_2 | \varphi^f F_2)$ and $E(\theta_1 | \varphi^{f+1} F_1) \geqslant E(\theta_2 | \varphi^{f+1} F_2)$, for $f \leqslant n - 2$. Hence, $E(\theta_1 | \varphi^f F_1) \geqslant E(\theta_2 | \varphi^f F_2)$ for $f \leqslant n - 1$ with strict inequality if $f = 0$.   □

Remark A little more care in the preceding proof gives $E(\theta_1 | \sigma^s \varphi^f F_1) > E(\theta_2 | \sigma^s \varphi^f F_2)$ whenever $s + f \leqslant n - 1$ and $s < n - 1$.   □

The following complementary result follows immediately by interchanging the roles of successes and failures.

Corollary 7.3.5 Suppose $I_1 > I_2$. For $n > 1$, a necessary and sufficient condition for $F_2 \overset{n-1}{>} F_1$ is

$$E(\theta_2 | \varphi^{n-1} F_2) \geqslant E(\theta_1 | \varphi^{n-1} F_1).$$

We will now use this corollary and Theorem 7.3.3 to obtain a more complete picture than that provided by Figure 7.2.

Example 7.3.2 As in Example 7.3.1, suppose $I_1 = 3I_2$ and $F(du) = u^{-1}(1-u)^{-1} du$. To obtain formulas with simple appearance, we specialize to $I_1 = 3, I_2 = 1$.

From Theorem 7.3.3 we see that the set of (a_1, b_1, a_2, b_2) for which

$$\frac{a_1 + (n-1)}{a_1 + (n-1) + b_1} = \frac{a_2 + (n-1)}{a_2 + (n-1) + b_2}$$

is of interest. Using $b_1 = 3 - a_1$ and $b_2 = 1 - a_2$, we obtain

$$a_1 = \frac{n+2}{n} a_2 + \frac{2n-2}{n}, \qquad 0 \leqslant a_2 \leqslant 1.$$

For various values of n, these line segments are shown in the lower right half of Figure 7.3. For a particular value of n, arm 1 is optimal initially if (a_1, a_2) is either on or below the corresponding line segment.

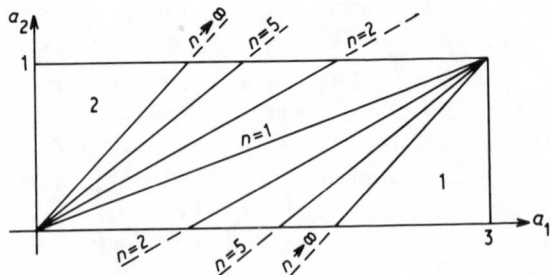

Fig. 7.3 $I_1 = 3, I_2 = 1$. *Regions where the indicated arms are known to be optimal on the basis of Theorem 7.3.3 and Corollary 7.3.5.*

Similarly, a region where arm 2 is optimal initially can be identified from Corollary 7.3.5. The boundaries are given by

$$a_1 = \frac{n+2}{n} a_2, \qquad 0 \leqslant a_2 \leqslant 1$$

and, for various values of n, these are shown in the upper left half of Figure 7.3. As expected, the picture becomes that of Figure 7.2 as $n \to \infty$.

It is somewhat surprising that the elementary Corollary 7.1.2 can be used to enlarge the region where arm 2 is known to be optimal. Setting $E(\theta_2 | F_2) = E(\theta_1 | \sigma^{n-1} F_1)$, we obtain

$$a_2 = \frac{a_1 + (n-1)}{n+2},$$

or

$$a_1 = (n+2)a_2 - (n-1), \qquad \frac{n-1}{n+2} \leqslant a_2 \leqslant 1.$$

Using this in combination with Figure 7.3, we obtain Figure 7.4. For each $n > 1$ there are two curves (one of which is a line segment and the other of which is the union of two line segments). For (a_1, a_2) on or below the lower of the two, arm 1 is optimal. For (a_1, a_2) on or above the higher of the two, arm 2 is optimal.

On the other hand, Corollary 7.1.2 cannot be used to increase the region where arm 1 is known to be optimal. Figures 7.5 and 7.6 are the analogues of Figure 7.4 for other values of I_1 and I_2 satisfying $I_2 = 3I_2$. $\qquad\qquad \square$

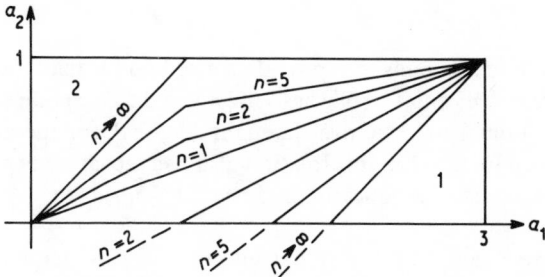

Fig. 7.4 $I_1 = 3, I_2 = 1$. *Regions where the indicated arms are known to be optimal on the basis of Theorem 7.3.3, Corollary 7.3.5, and Corollary 7.1.2.*

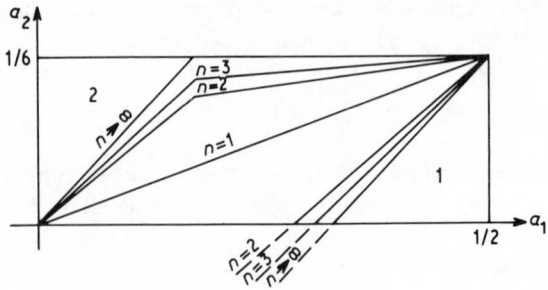

Fig. 7.5 $I_1 = 1/2, I_2 = 1/6$. *Regions where the indicated arms are known to be optimal on the basis of Theorem 7.3.3, Corollary 7.3.5, and Corollary 7.1.2.*

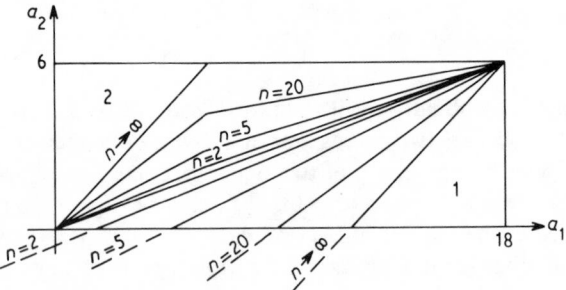

Fig. 7.6 $I_1 = 18, I_2 = 6$. *Regions where the indicated arms are known to be optimal on the basis of Theorem 7.3.3, Corollary 7.3.5, and Corollary 7.1.2.*

We would like theorems that provide better bounds than those given by Theorem 7.3.3 and Corollaries 7.1.2 and 7.3.5. We conjecture that arm 2 is optimal on or above the main diagonal, for there the arm about which there is less information also has the larger expectation. (Cf. our discussion concerning Joshi (1975) in the Annotated Bibliography.)

We have focused on the initial selection in this chapter. This is entirely appropriate since optimal strategies are composed of initial selections, but for different bandits. While the figures shown in this section may be appropriate in the original bandit, different figures will be appropriate in the other bandits. At the second stage of the original bandit, for example, the rectangle in the appropriate figure will be lengthened by one unit in the dimension corresponding to the arm chosen, the point (a_1, a_2) will move or not one unit in the same direction according as the result obtained is success or failure, and n will be decreased by 1.

References

Berry, D. A. (1972) A Bernoulli two-armed bandit. *Ann. Math. Statist.* **43**: 871–897.

Bradt, R. N., Johnson, S. M. and Karlin, S. (1956) On sequential designs for maximizing the sum of n observations. *Ann. Math. Statist.* **27**: 1060–1074.

Joshi, V. M. (1975) A conjecture of Berry regarding a Bernoulli two-armed bandit. *Ann. Statist.* **3**: 189–202.

CHAPTER 8

Continuous-time bandits

For processes that are evolving and can be observed continuously, we would like to formulate a bandit problem in a manner analogous to our formulation for discrete time. First, we would like a strategy to indicate a process to select and observe at time t, for all t, depending on the history of observations prior to time t. Then we would define the worth of a strategy to be a weighted sum of the increments of the processes during the time they are being observed. Unfortunately, there are serious technical difficulties with various aspects of this plan. These difficulties are frequently ignored in the literature. While this may not be unreasonable when there are good discrete-time approximations, the technical problems must be understood and overcome before progress can be made on the general problem.

In Section 8.3 we present several examples that indicate the need for care in formulating a definition of strategy. We formulate a definition of strategy using these examples as guides. But we are not sure that there do not exist other examples that would show our definition to be inappropriate.

For the examples of Sections 8.1 and 8.2 we adopt a narrower definition of strategy. This definition seems adequate for those examples and it allows us to make arguments involving discrete-time approximations. The question of how the definition given in Section 8.3 would apply to these examples remains open.

One reason that the definition of strategy used in Sections 8.1 and 8.2 seems appropriate in the examples of those sections is that the discount functions assumed are regular (cf. Definition 5.7.1) and there is only one unknown arm. This is also the case for the examples of Section 8.4, although the definition of strategy used in Section 8.1 is modified slightly in Example 8.4.1. We regard finding the most appropriate formulation of a continuous-time bandit to be an open problem.

The unknown arm in each of Sections 8.1 and 8.2 involves Brownian motion added to an unknown linear drift. In Section 8.1 the drift is known to be one of two values. This special case plays a central role in the minimax approach of Section 9.2. In Section 8.2 we discuss the work of Chernoff and Ray (1965) in which the drift is normally distributed. We hope these examples motivate the reader to consider the foundational questions that arise in Section 8.3.

The examples in Section 8.4 involve Lévy processes having jumps.

Continuous-time bandits for which there is a cost for switching arms may, from a foundational viewpoint, be more tractable than those for which there is no such cost. The cost makes it unattractive to switch back and forth instantaneously between arms. When there is no cost such switching may be attractive, say, when there are two arms that are Brownian motions added to unknown drifts. We have not considered the possibility of a cost for switching arms.

8.1 Brownian motion with unknown drift: two-point prior

In this and the next section we present two continuous-time two-armed bandit problems. The characteristics of one arm, arm 2 say, are completely known. In particular, arm 2 generates the deterministic process with constant rate λ:

$$Y_2(t) = \lambda t.$$

Arm 1, on the other hand, has unknown drift which is obscured by noise. We assume that arm 1 generates a Brownian motion

$$Y_1(t) = \theta_1 t + B(t), \tag{8.1.1}$$

where $B(t)$ is the standard Wiener process (i.e., standard Brownian motion) having mean 0 and variance 1 at $t = 1$. The drift coefficient θ_1 is itself random and, in this section, has a two-point distribution:

$$F = p\delta_a + (1-p)\delta_b$$

where $b < a$ and F is the distribution of θ_1. To avoid annoying trivialities we assume $b < \lambda < a$.

In this section the discounting is the continuous-time analogue of the geometric—namely, the discount function is $e^{-\beta t}$ for some constant β, $0 < \beta < \infty$. The worth of a strategy τ is

$$W(F, \lambda; e^{-\beta t}; \tau) = E_\tau \int_0^\infty e^{-\beta t} dY_{\tau(t)}(t), \tag{8.1.2}$$

where $\tau(t)$ indicates the arm being observed at time t. In continuous time some care is needed to give a precise meaning to the phrase 'the arm being observed at time t'.

Before giving a precise definition of *strategy*, we make some comparisons with the discrete-time setting. In discrete time, arm 1 is governed by a probability distribution Q_1 which is itself random. Given Q_1, the outcomes X_{1m} on arm 1 are independent and identically distributed. Were we to replace m by a real parameter t while keeping the independence requirement, we would obtain a nonmeasurable process X_{1t}. This would be quite unsatisfactory since we also need to introduce integrals as analogues of discrete-time sums. A different view is needed. For the discrete-time setting let

$$Y_{1m} = \sum_{l=1}^{m} X_{1l}, \qquad m = 1, 2, 3, \ldots; \qquad Y_{10} = 0.$$

Following the approach of previous chapters, the value is incremented by the quantity $\alpha_m X_{1m} = \alpha_m (Y_{1m} - Y_{1,m-1})$ when arm 1 is observed at stage m and where α_m is the mth term of the discount sequence. The process Y_{1m}, consisting of the partial sums of independent identically distributed random variables, does have a convenient continuous-time analogue – namely, a stochastic process with stationary independent increments (i.e., a *Lévy process*), the best known example of which is Brownian motion. So (8.1.2) is natural as a continuous-time analogue of the discrete-time definition of worth given in Chapter 2.

Since $d(\lambda t) = \lambda\, dt$, arm 2 in this example plays a role analogous to that of a discrete-time known arm λ. As in Example 2.4.7, nothing in this example would be changed were the deterministic process $Y_2(t) = \lambda t$ replaced by a Lévy process (or any other process) with known mean λt at time t. (The mean of a Lévy process is either a multiple of t or else does not exist for $t > 0$.)

We can rewrite (8.1.2) as follows:

$$W(F, \lambda; e^{-\beta t}; \tau) = E_\tau \int_0^\infty e^{-\beta t} \mathbf{1}_{\{\tau(t)=1\}} dY_1(t)$$

$$+ E_\tau \int_0^\infty e^{-\beta t} \mathbf{1}_{\{\tau(t)=2\}} \lambda dt. \qquad (8.1.3)$$

The latter term is meaningful provided the random set $\{t: \tau(t) = 2\}$ is almost surely a measurable subset of $[0, \infty]$. In order for τ to qualify

as a strategy we impose this measurability requirement. The first term
in (8.1.3) is a stochastic integral. We require $\mathbf{1}_{\{\tau(t)=1\}}$, and therefore
$\mathbf{1}_{\{\tau(t)=2\}}$, to satisfy the assumption of progressive measurability that is
standard in the definition of a stochastic integral with respect to
continuous stochastic processes; the process $f(t, \omega)$ is *progressively
measurable* with respect to an increasing family of σ-fields $\mathscr{F}_t$ if, for
each t_0, $f(t, \omega)$, $0 \leqslant t \leqslant t_0$, is measurable with respect to the product
of two σ-fields: the σ-field of Borel subsets of $[0, t_0]$ and $\mathscr{F}_{t_0}$. The
relevant $\mathscr{F}_{t_0}$ is the σ-field generated by $\{Y_1(t): 0 \leqslant t \leqslant t_0\}$.

In the discrete-time case we permit the arm indicated by a strategy τ
at time m to depend only on the increments X_{1l} of Y_{1l} for those
$l < m$ for which arm 1 was actually observed. We proceed to develop
an analogous restriction when time is continuous. We require
$\{t: \tau(t) = 1\} \cup [0, t_0)$ to be the union of a finite (possibly random)
number of left-closed, right-open intervals for each finite t_0. Hence,

$$\{t: \tau(t) = 1\} = \bigcup_i [r_i, s_i),$$

where the union may be empty, finite, or infinite, r_i and s_i are random,
and, when defined, $s_i < r_{i+1}$. We require that each r_{i+1} be measurable
with respect to the σ-field generated by the family of increments

$$\{Y_1(t) - Y_1(r_j): r_j \leqslant t \leqslant s_j, j \leqslant i\}$$

and that each event $\{s_{i+1} \leqslant t_0\}$ be measurable with respect to the
σ-field generated by

$$\{Y_1(t) - Y_1(r_{i+1}): r_{i+1} \leqslant t \leqslant t_0\} \cup \{Y_1(t) - Y_1(r_j): r_j \leqslant t \leqslant s_j, j \leqslant i\}.$$

(In general, we would permit the arm indicated by a strategy τ at time t
to depend on increments observed prior to time t on the other arms,
but this would make no difference in this example since the only other
arm is known.)

Now that we have a precise definition of strategy, we can define V in
terms of W as in the discrete-time setting:

$$V(F, \lambda; e^{-\beta t}) = \sup_\tau W(F, \lambda; e^{-\beta t}; \tau),$$

where the supremum is taken over the strategies τ just described.

Exponential discounting has the property possessed by the geometric discount sequence, its discrete-time analogue: except for normalization, the discount function does not change with time. Hence, once

arm 2 becomes optimal and is selected, it remains optimal. Strongly-to-the-right arguments carry over from the discrete case by approximation. Therefore, the larger the probability that $\theta_1 = a$, the stronger is the inclination to use arm 1. Accordingly, there is a constant $C \in (0, 1)$ such that arm 1 is optimal at time t if the current probability that $\theta_1 = a$ is larger than C, and arm 2 is optimal if the current probability is less than C. If this current probability equals C, only arm 2 is optimal; for if arm 1 were selected initially the wild oscillations of Brownian motion would push the new probability that $\theta_1 = a$ below C in an infinitesimal time, indicating that arm 2 should already have been in use.

Summarizing, if $p = F(\{a\}) \leqslant C$, the decision maker should select arm 2 indefinitely into the future. If $p > C$, the decision maker should select arm 1 initially and stay with arm 1 until $p(t, Y_1(t)) = C$, where $p(t, y)$ denotes the conditional probability that $\theta_1 = a$ given that $Y_1(t) = y$. At such time the decision maker should switch permanently to arm 2. Since $Y_1(t)$ and $p(t, y)$ are continuous functions, the switching time can be identified on the basis of the observations of Y occurring strictly before that time, as required in the definition of a strategy.

The current probability that $\theta_1 = a$ is

$$p(t, y) =$$

$$\frac{p \dfrac{1}{\sqrt{(2\pi t)}} \exp\left[-(y - at)^2/(2t)\right]}{p \dfrac{1}{\sqrt{(2\pi t)}} \exp\left[-(y - at)^2/(2t)\right] + (1 - p)\dfrac{1}{\sqrt{(2\pi t)}} \exp\left[-(y - bt)^2/(2t)\right]}$$

$$= \frac{p \exp(ay - a^2 t/2)}{p \exp(ay - a^2 t/2) + (1 - p) \exp(by - b^2 t/2)}. \qquad (8.1.4)$$

Accordingly, the condition $p(t, y) > C$ becomes

$$\frac{p \exp(ay - a^2 t/2)}{(1 - p) \exp(by - b^2 t/2)} > \frac{C}{1 - C},$$

which is equivalent to

$$y > \frac{(a + b)}{2} t - \frac{1}{a - b} \log \frac{(1 - C)p}{C(1 - p)}.$$

(For the special case $t = 0 = y$ this reduces to $p > C$, as previously indicated.)

We proceed to find C. Let τ_x denote the strategy: switch to arm 2 permanently at the random time T (possibly 0 or ∞), the smallest t for which

$$Y_1(t) \leqslant \frac{(a+b)}{2} t - \frac{1}{a-b} \log \frac{(1-x)p}{x(1-p)} \qquad (8.1.5)$$

or, equivalently,

$$B(t) + \frac{2\theta_1 - a - b}{2} t \leqslant -\frac{1}{a-b} \log \frac{(1-x)p}{x(1-p)}. \qquad (8.1.6)$$

Then τ_C is an optimal strategy.

Let us consider an F for which $p > x$ so that the inequalities in (8.1.5) and (8.1.6) hold with equality when $t = T$. Conditioned on θ_1, the Laplace–Stieltjes transform of the distribution of T is known (for instance, Fristedt, 1974, Corollary 9.9 and the formula at top of page 351):

$$E(e^{-\beta T} | \theta_1 = b) = \left[\frac{x(1-p)}{(1-x)p} \right]^{v-1/2} \qquad (8.1.7)$$

and

$$E(e^{-\beta T} | \theta_1 = a) = \left[\frac{x(1-p)}{(1-x)p} \right]^{v+1/2}, \qquad (8.1.8)$$

where

$$v = \frac{\sqrt{[8\beta + (a-b)^2]}}{2(a-b)} \qquad (8.1.9)$$

To calculate $W(F, \lambda; e^{-\beta t}; \tau_x)$ we will use

$$E\left(\int_0^T e^{-\beta t} \, dY_1(t) \,\bigg|\, \theta_1 = a \right)$$

$$= \frac{a}{\beta} E(1 - e^{-\beta t} | \theta_1 = a) + E\left(\int_0^\infty e^{-\beta t} \mathbf{1}_{\{T > t\}} \, dB(t) \,\bigg|\, \theta_1 = a \right)$$

and the similar formula for conditioning on $\theta_1 = b$. The latter term is zero as it is the expected value of a stochastic integral with respect to standard Brownian motion. (To draw this conclusion one should observe that B is a Brownian motion adapted to the σ-fields generated by it and θ_1.) Hence,

$$W(F, \lambda; e^{-\beta t}; \tau_x)$$

$$= \left\{ \frac{a}{\beta} E_{\tau_x}(1 - e^{-\beta T} | \theta_1 = a) + E_{\tau_x}\left(\int_T^\infty e^{-\beta t} \lambda dt \,\bigg|\, \theta_1 = a \right) \right\} p$$

$$+ \left\{ \frac{b}{\beta} E_{\tau_x}(1 - e^{-\beta T} | \theta_1 = b) + E_{\tau_x}\left(\int_T^\infty e^{-\beta t} \lambda dt \,\bigg|\, \theta_1 = b \right) \right\} (1 - p)$$

$$= \beta^{-1}[(E(\theta_1 | F) + (\lambda - a)E_{\tau_x}(e^{-\beta T} | \theta_1 = a)p$$

$$+ (\lambda - b)E_{\tau_x}(e^{-\beta T} | \theta_1 = b)(1 - p)]. \qquad (8.1.10)$$

By using (8.1.7) and (8.1.8) in (8.1.10) and then differentiating with respect to x, setting equal to 0, and solving for $x = C$, we find

$$C = \frac{(\lambda - b)(b - a + \sqrt{[8\beta + (a - b)^2]})}{(a - b)(a + b - 2\lambda + \sqrt{[8\beta + (a - b)^2]})}. \qquad (8.1.11)$$

If $p \leqslant C$ then arm 2 is optimal throughout:

$$V(F, \lambda; e^{-\beta t}) = \beta^{-1}\lambda.$$

If $p > C$ then (8.1.7), (8.1.8), and (8.1.9) imply

$$V(F, \lambda; e^{-\beta t}) = \beta^{-1}[E(\theta_1 | F) + (\lambda - (1 - C)b - Ca)]$$

$$\times \left[\frac{C(1 - p)}{(1 - C)p} \right]^\nu \left[\frac{p(1 - p)}{C(1 - C)} \right]^{1/2} \qquad (8.1.12)$$

where ν and C are defined in (8.1.9) and (8.1.11).

In the spirit of Chapter 5, we insert p for C in (8.1.11) and solve for λ. In notation consistent with Chapter 5, we obtain the solution

$$\Lambda(F, e^{-\beta t}) = \frac{E(\theta_1^2 | F) - ab + E(\theta_1 | F)\sqrt{[8\beta + (a - b)^2]}}{(a - b)(2p - 1) + \sqrt{[8\beta + (a - b)^2]}}. \qquad (8.1.13)$$

If $\lambda < \Lambda(F, e^{-\beta t})$ then arm 1 is optimal initially. If $\lambda \geqslant \Lambda(F, e^{-\beta t})$ then arm 2 is optimal initially and henceforth. Figure 8.1 shows the

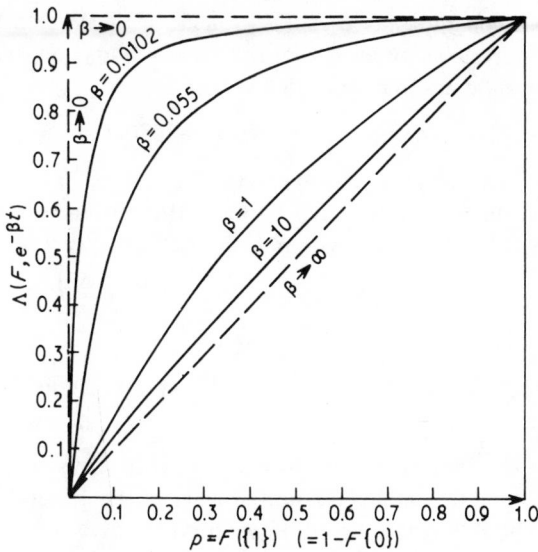

Fig. 8.1 Λ as a function of F which is supported by $\{1, 0\}$.

graph of Λ versus p for $a = 1, b = 0$, and several values of β. When β is small, arm 1 should be tested unless λ is very close to 1, or p is very close to 0. This is highlighted on the figure by two dashed line segments indicating the limits as $\beta \to 0$. When β is large, arm 1 should not be tried unless $E(\theta_1 \,|\, F)$ is rather close to λ. This is indicated on the figure by the dashed line $E(\theta_1 | F)$ which is the limit of the graph as $\beta \to \infty$.

Another approach to arriving at (8.1.13) or (8.1.11) is to regard $p(t, Y_1(t))$ as a diffusion process on $[0, 1]$. Letting w denote the initial value, we have, from (8.1.4),

$$p(t, Y_1(t)) = \frac{w \exp(aY_1(t) - a^2 t/2)}{w \exp(aY_1(t) - a^2 t/2) + (1 - w) \exp(bY_1(t) - b^2 t/2)}.$$

$$(8.1.14)$$

We would like a formula for $dp(t, y)$ in terms of a standard Brownian motion. It is tempting to try (8.1.14), (8.1.1), and Itô's formula, but this scheme fails. The process $p(t, Y_1(t))$ is not adapted to the σ-fields generated by $B(t)$ and it does not have the Markov property with respect to the σ-fields obtained by adjoining the σ-field generated by

θ_1 to those generated by $B(t)$. The σ-fields $\mathscr{F}_t^Y$ generated by Y_1 are those of interest. Theorem 7.12 of Lipster and Shiryayev (1977) gives the appropriate alternative to (8.1.1):

$$dY_1(t) = E(\theta_1 \mid \mathscr{F}_t^Y)dt + d\bar{B}(t)$$

where $\bar{B}(t)$ is a Brownian motion adapted to $(\mathscr{F}_t^Y: t \geqslant 0)$. A straightforward calculation using Itô's formula (for instance, Lipster and Shiryayev, 1977, Theorem 4.4) and

$$E(\theta_1 \mid \mathscr{F}_t^Y) = b + (a - b)p(t, Y_1(t))$$

gives

$$dp(t, Y_1(t)) = (a - b)p(t, Y_1(t))[1 - p(t, Y_1(t))]d\bar{B}(t).$$

The $(F, \lambda; e^{-\beta t})$-bandit is a stopping problem for the diffusion $p(t, Y_1(t))$. Karatzas (1984) solves this problem in terms of solutions of an ordinary differential equation involving the generator of the process. The differential equation we need to solve is

$$\frac{1}{2}[(a - b)w(1 - w)]^2 u''(w) - \beta u(w) = 0.$$

Two linearly independent solutions of the form $w^q(1 - w)^r$ can be obtained. According to Karatzas (1984), it is a decreasing solution that is of interest:

$$w^{1/2 - v}(1 - w)^{1/2 + v}, \tag{8.1.15}$$

where v is defined in (8.1.9).

The expected income from arm 1 between times t and $t + dt$ is

$$e^{-\beta t}[wa + (1 - w)b]dt, \tag{8.1.16}$$

where w is the initial value of $p(t, Y_1(t))$ and, thus, is its expected value for all t. According to Karatzas (1984), another function of interest is the integral from 0 to ∞ of (8.1.16):

$$\beta^{-1}[wa + (1 - w)b]. \tag{8.1.17}$$

Karatzas's (1984) theorem involves a quotient; the denominator is the derivative of (8.1.15) and the numerator is the product of (8.1.17) and the derivative of (8.1.15) minus the product of (8.1.15) and the derivative of (8.1.17). His theorem asserts that $\Lambda(F, e^{-\beta t})$ is obtained by replacing w by p in the quotient. The solution thus obtained agrees with (8.1.12).

8.2 Brownian motion with unknown drift: normal prior

As in Section 8.1 we consider two arms. Arm 1 is Brownian motion with unknown drift: $Y_1(t) = \theta_1 t + B(t)$, and arm 2 is deterministic: $Y_2(t) = \lambda t$. Without loss of generality we take $\lambda = 0$. So the problem is to decide when to gather information and payoff from arm 1 and when to sit idle.

This section differs from Section 8.1 in two ways. First, the prior distribution on θ_1 is normal. Second, discounting is uniform: $\alpha_t = 1_{[0,S]}(t)$, where S is a constant.

According to Definition 5.7.1, α_t is regular. So Proposition 5.7.1 applies to show that there are discrete-time approximations with regular discount sequences. This implies that there is an optimal strategy that indicates arm 2 permanently once it is selected. So, as in the problem considered in Section 8.1, this bandit is a stopping problem. (This conclusion depends on the fact that we are using the same concept of strategy as used in Section 8.1. As indicated in Section 8.3, it is not clear that this is the most appropriate concept.)

Since we are assuming $\lambda = 0$ and uniform discounting, the optimization problem is to choose, for $Y_1(t)$, a stopping time $T \in [0, S]$ that maximizes $E(Y_1(T)|F)$, where F denotes the distribution of θ_1. We assume F is normal with mean μ and variance ρ^2, as in Section 2.1. This problem was studied by Chernoff and Ray (1965) using some results from Chernoff (1961) and Breakwell and Chernoff (1964). We follow their method rather closely.

Consistent with our earlier notation, we use $V((\mu, \rho), 0; S)$ to denote the value of the bandit. A change of variables will prove useful:

$$V_c^*(w, s) = V((ws^{-1}, s^{-1/2}), 0; c - s), \qquad 0 < s \leqslant c,$$

where we have used

$$\mu = ws^{-1}, \quad \rho = s^{-1/2}, \quad S = c - s,$$
$$s = \rho^{-2}, \quad w = \mu\rho^{-2}, \quad c = S + \rho^{-2}.$$

A discrete-time approximation involving a modification of the proof of Theorem 4.3.6 shows that there exists a function $f(\rho, S)$ such that it is optimal, beginning at time 0, to select arm 1 if $\mu > f(\rho, S)$ and to choose arm 2 immediately if $\mu \leqslant f(\rho, S)$. The condition $\mu > f(\rho, S)$ can be written as $w > f_c^*(s)$, where $f_c^*(s) = sf(s^{-1/2}, c - s)$.

It is clear that V and the boundary function f can be recovered from V_c^* and f_c^*, $c > 0$. More is true: for an initial (μ, ρ) and S, there exists a single fixed c – namely, $S + \rho^{-2}$ – such that f_c^* completely determines an optimal strategy for the $((\mu, \rho), 0; S)$-bandit, and not just an optimal initial selection. The reason is that after time t has elapsed the new horizon is $S - t$ and, if arm 1 is selected throughout the interval $[0, t)$, the new variance is $(\rho^{-2} + t)^{-1}$. Hence, the updated value of c is

$$(S - t) + (\rho^{-2} + t) = S + \rho^{-2},$$

and so c has not changed. Accordingly, we turn to the study of the functions V_c^* and f_c^* for fixed but arbitrary $c > 0$.

Clearly, $V_c^*(w, c) = 0$ for all w, and $V_c^*(w, s) = 0$ for $w \leqslant f_c^*(s)$. Also, $f_c^*(s) \leqslant 0$ for all $s \in (0, c]$ and $V_c^*(w, s) \geqslant 0$ for all $w \in (-\infty, \infty)$ and $s \in (0, c]$. When w is very large the probability is close to 1 that arm 1 will be selected exclusively when following an optimal strategy. The worth of using arm 1 exclusively is $\mu S = ws^{-1}(c - s)$. Therefore,

$$\lim_{w \to \infty} V^*(w, s)/w = (c - s)/s$$

for each $s \in (0, c]$. Also, $V^*(w, s)$ is an increasing function of w for each s.

We now argue that $V_c^*(w, s)$ satisfies

$$\frac{1}{2} \frac{\partial^2}{\partial w^2} V_c^*(w, s) + \frac{w}{s} \frac{\partial}{\partial w} V_c^*(w, s) + \frac{\partial}{\partial s} V_c^*(w, s) = -\frac{w}{s} \quad (8.2.1)$$

for $w > f_c^*(s)$, $0 < s \leqslant c$. Consider a point (w_0, s_0) for which $w_0 > f_c^*(s_0)$. It is optimal to begin by selecting arm 1. A switch to arm 2 should be made when and if a (w, s) for which $w = f_c^*(s)$ is reached. (Because of continuity, hitting such a point can be anticipated, so arm 2 can be selected at the time of impact as well as thereafter.) After Y_1 has been observed up to time t, the new variance of θ_1 equals $(\rho_0^{-2} + t)^{-1}$, as described earlier. Here the subscript 0 on ρ indicates that ρ_0 is determined by (μ_0, s_0) – in fact, $\rho_0 = s_0^{-1/2}$. The posterior mean of θ_1, the mean conditioned by the observed values of Y_1, equals

$$\frac{\mu_0 + \rho_0^2 Y_1(t)}{1 + \rho_0^2 t},$$

where, of course, $\mu_0 = w_0 s_0^{-1}$. Thus, the point (w_t, s_t), reached at time t, is given by

$$s_t = [(\rho_0^{-2} + t)^{-1}]^{-1} = \rho_0^{-2} + t = s_0 + t$$

and

$$w_t = s_t \frac{\mu_0 + \rho_0^2 Y_1(t)}{1 + \rho_0^2 t}$$

$$= \rho_0^{-2}(\mu_0 + \rho_0^2 Y_1(t)) = w_0 + Y_1(t).$$

The stochastic process (w_t, s_t) is a continuous Markov process – that is, a diffusion. For ε less than the distance from (w_0, s_0) to the graph of f_c^*, let T_ε denote the first time that this process reaches a point a distance ε from (w_0, s_0). Since $s_t = s_0 + t$, $T_\varepsilon \leqslant \varepsilon$. Writing w, s, μ, ρ, and S with corresponding subscripts throughout, we calculate

$$V_c^*(w_0, s_0) = V((\mu_0, \rho_0), 0; S_0)$$

$$= \mu_0 E(T_\varepsilon) + E\left(V((\mu_{T_\varepsilon}, \rho_{T_\varepsilon}), 0; S_{T_\varepsilon}) \right)$$

$$= \mu_0 E(T_\varepsilon) + E(V_c^*(w_{T_\varepsilon}, s_{T_\varepsilon})).$$

Therefore,

$$-\mu_0 E(T) = E(V_c^*(w_{T_\varepsilon}, s_{T_\varepsilon}) - V_c^*(w_0, s_0))$$

$$= E\left((w_{T_\varepsilon} - w_0)\frac{\partial V_c^*}{\partial w}(w_0, s_0) + \frac{1}{2}(w_{T_\varepsilon} - w_0)^2 \frac{\partial^2 V_c^*}{\partial w^2}(w_0, s_0) \right.$$

$$\left. + (s_{T_\varepsilon} - s_0)\frac{\partial V_c^*}{\partial s}(w_0, s_0) + \dots \right)$$

$$= E\left(Y_1(T_\varepsilon)\frac{\partial V_c^*}{\partial w}(w_0, s_0) + \frac{1}{2}[Y_1(T_\varepsilon)]^2 \frac{\partial^2 V_c^*}{\partial w^2}(w_0, s_0) \right.$$

$$\left. + T_\varepsilon \frac{\partial V_c^*}{\partial s}(w_0, s_0) + \dots \right).$$

So we need to calculate

$$E(Y_1(T_\varepsilon)) = \mu_0 E(T_\varepsilon)$$

and

$$E[Y_1(T_\varepsilon)]^2 = \text{var}[Y_1(T_\varepsilon)] + [E(Y_1(T_\varepsilon))]^2$$

$$= \text{var}[\theta_1 T_\varepsilon + B(T_\varepsilon)] + \mu_0^2[E(T_\varepsilon)]^2$$

$$= E(T_\varepsilon) + \text{var}(\theta_1 T_\varepsilon) + \mu_0^2[E(T_\varepsilon)]^2,$$

which is asymptotic to $E(T_\varepsilon)$ as $\varepsilon \downarrow 0$ since $T_\varepsilon \leqslant \varepsilon$. Keeping only terms of order $E(T_\varepsilon)$ we obtain, as $\varepsilon \downarrow 0$,

$$-\mu_0 E(T_\varepsilon) \sim \mu_0 E(T_\varepsilon) \frac{\partial V_c^*}{\partial w}(w_0, s_0) + \frac{1}{2} E(T_\varepsilon) \frac{\partial^2 V_c^*}{\partial w^2}(w_0, s_0)$$

$$+ E(T_\varepsilon) \frac{\partial V_c^*}{\partial s}(w_0, s_0).$$

The desired partial differential equality (8.2.1) follows by dividing by $E(T_\varepsilon)$, using $\mu_0 = w_0 s_0^{-1}$, and then dropping the subscript 0 throughout.

It is straightforward to prove that V_c^* is a continuous function by an argument similar to that used to prove Theorem 4.1.1. Since V_c^* is identically 0 in the stopping region, the directional derivative of V_c^* along the boundary of the stopping region equals 0. (The question of establishing boundary conditions for a more general reward structure is considered by Bather (1970). He studies the problem for the heat equation but his results transfer to general diffusions by scaling and using a speed measure.)

The problem of finding V_c^* and f_c^* satisfying the various conditions obtained above is called a 'free boundary problem' in the partial differential equations literature. This name derives from the fact that the boundary of the region where the partial differential equation is to be solved is not given, but rather obtaining the boundary is part of the problem. In fact, obtaining the boundary is for us the most important aspect of the problem, for the boundary determines optimal strategies. Chernoff and Ray (1965) indicate that the argument of Breakwell and Chernoff (1964) applies to show that the above conditions determine f_c^* uniquely.

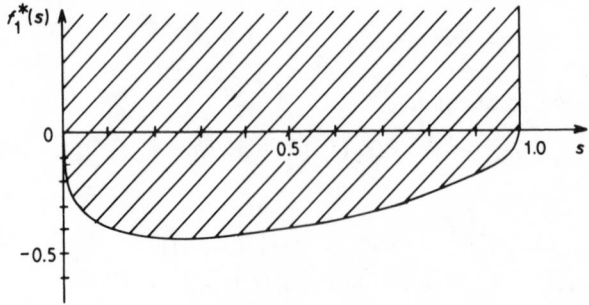

Fig. 8.2 *The continuation region in terms of f_1^*.*

The graph of f_1^* is shown in Figure 8.2, on which the continuation region is shaded. Since $w/\sqrt{s} = \mu\sqrt{s}$ is standard normal when $\theta_1 = 0$, it is helpful to view the stopping boundary in terms of $z = f_1^*(s)/\sqrt{s}$; this is shown in Figure 8.3. Both figures were constructed using Table 21 of Chernoff and Petkau (1983).

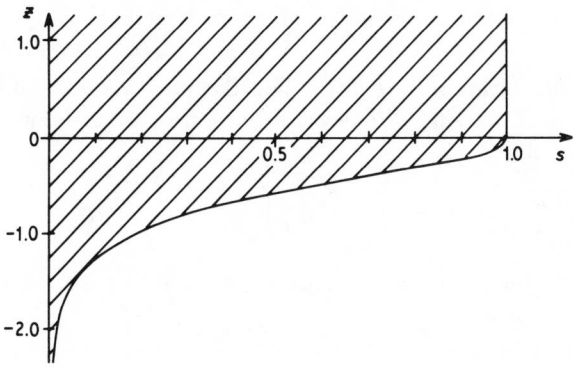

Fig. 8.3   *The continuation region in terms of* $z = f_1^*/\sqrt{s}$.

8.3   General setting

The parameters of a continuous-time bandit are similar to those of a discrete-time bandit. A distribution G describes the probabilistic character of the k arms, and a discount function α_t indicates the utility of the observation at time t. We also require definitions for strategy and for the worth of a strategy.

As in Chapter 2, G represents prior information concerning the k arms. We continue to assume $G \in \mathscr{D}^*(\mathscr{D}^k)$; that is, that each component Q_i of $(Q_1, \dots, Q_k) \in \mathscr{D}^k$ has a finite first absolute moment with G-probability one and, moreover, that this moment has finite G-expectation. Since time is continuous we now require that G be supported by

$$\{(Q_1, \dots, Q_k) \in \mathscr{D}^k : Q_i \text{ is infinitely divisible for each } i\}.$$

Given $(Q_1, \dots, Q_k)$, the k arms produce k independent Lévy processes Y_i, $i = 1, \dots, k$, with the distribution of Y_i at time 1 being Q_i. Lévy processes are stochastic processes with stationary independent increments and are the continuous-time analogues of sums of independent identically distributed random variables. Conditioned on Q_i, the distribution of $Y_i(t)$ is Q_i^{*t}, the tth convolution power of Q_i.

The *discount function* α_t is a nonnegative measurable function on $[0, \infty)$ that has a finite L_1 norm:

$$|\alpha_t|_1 = \int_0^\infty \alpha_t \, dt < \infty. \tag{8.3.1}$$

It is not clear how to formulate the most natural and useful definition of *strategy*. Whatever definition is used, the *worth* of a strategy τ is

$$W(G; \alpha_t : \tau) = \sum_{i=1}^k E_\tau \int_0^\infty \alpha_t \, \mathbf{1}_{\{\tau(t)=i\}} \, dY_i(t), \tag{8.3.2}$$

where $\tau(t)$ denotes the arm indicated by τ at time t. The *value* of the $(G; \alpha_t)$-bandit is

$$V(G; \alpha_t) = \sup_\tau W(G; \alpha_t; \tau).$$

A strategy τ is *optimal* if

$$W(G; \alpha_t; \tau) = V(G; \alpha_t).$$

Compound Poisson processes are among the simplest Lévy processes. They are constant between jumps. The times elapsed between successive jumps have a common exponential distribution and are independent of each other. The sizes of the jumps are identically distributed. They also are independent of each other as well as of the times at which they occur. Parts of Sections 3.5 and 5.7 involve stages occurring at random times determined by exponential waiting times. Thus the random processes of those sections can be regarded as compound Poisson processes. The random structure has the additional property that there exists a $\kappa > 0$ such that, with G-probability one, the random process associated with any arm has time lapses of mean κ between jumps. The magnitudes of the jumps for the known arm 2 in Section 5.7 all equal λ. According to (8.3.2), the intervals of constancy do not contribute to the worth of a strategy; this is consistent with Sections 3.5 and 5.7.

The meanings of *value*, *worth*, and *optimal strategy* obviously depend on the notion of strategy that is used. In particular, this notion must be sufficiently restrictive for the stochastic integrals in (8.3.2) to be meaningful. We also want $\tau(t)$ to depend only on increments of the Lévy processes observed before time t. Whatever notion is used, the analogue of (2.5.1) should hold:

$$- \infty < \bigvee_{i=1}^{k} E(Y_t(1)|G)|\alpha_t|_1 \leqslant V(G; \alpha_t)$$

$$\leqslant E(\bigvee_{i=1}^{k} E(Y_i(1)|Q_1)|G)|\alpha_t|_1 \leqslant E(\bigvee_{i=1}^{k} Y_i(1)|G)|\alpha_t|_1 < \infty. \quad (8.3.3)$$

The strict inequalities involving $-\infty$ and $+\infty$ arise from (8.3.1) and the assumption that $G \in \mathscr{D}^*(\mathscr{D}^k)$.

Conditional on $(Q_1, \ldots, Q_k)$, $(Y_1, \ldots, Y_k)$ is a k-dimensional Lévy process having independent components and finite first moments. The observed process is

$$Z(t) = \sum_{i=1}^{k} \int_{[0,t]} \mathbf{1}_{\{\tau(s)=i\}} \, dY_i(s). \quad (8.3.4)$$

Let $\mathscr{F}_t^Z$ and $\mathscr{F}_t^\tau$ denote the σ-fields generated by $\{Z(s): s \leqslant t\}$ and $\{\tau(s): s \leqslant t\}$. Our vaguely stated requirement that $\tau(t)$ depend only on the increments of the Lévy processes observed before time t can be given a precise meaning as follows: for each t, $\tau(t)$ is $\mathscr{F}_{t-}^Z$-measurable, or, equivalently, $\mathscr{F}_t^\tau \subset \mathscr{F}_{t-}^Z$. This condition also implies that τ is predictable (cf. Dellacherie and Meyer, 1978, page 121) with respect to the σ-fields generated by $(Q_1, \ldots, Q_k)$ and

$$\{(Y_1(s), \ldots, Y_k(s)): \quad 0 \leqslant s \leqslant t\}.$$

(Predictability, rather than just progressive measurability, is the standard assumption in the theory of stochastic integrals when the stochastic processes may have jumps.) Since $Y_i(t)$ is a semimartingale with respect to these σ-fields, the stochastic integrals in (8.3.4) are defined (cf. Dellacherie and Meyer, 1980, Chapter VIII) – or, more precisely, they would be defined if (8.3.4) were merely a sum of stochastic integrals and not a stochastic differential equation. We do not know if additional conditions need be imposed on τ in order that (8.3.4) have a solution and that it be unique.

The following two examples indicate other reasons for imposing additional conditions on τ for it to be called a strategy.

Example 8.3.1 Modify the example of Section 8.1 as follows. Let

$$Y_1(t) = \begin{cases} at + \sigma B(t) \\ bt + \eta B(t) \end{cases}$$

with probability $1/2$ each, where $\sigma \neq \eta$ and $b < \lambda < a$. We continue to assume exponential discounting. If arm 1 is observed initially for some (possibly random) positive amount of time, then the standard deviation can be identified with probability one according to the law of the iterated logarithm. Hence, the mean can be identified with probability one. If arm 1 is observed for a positive length of time, with probability $1/2$ it will be clear that $Y_1(t) = bt + \eta B(t)$ and therefore that arm 1 has already been observed too long. On the other hand, a brief observation of arm 1 seems necessary to determine whether $Y_1(t) = at + \sigma B(t)$. The net loss in observing arm 1 when $Y_1(t) = bt + \eta B(t)$ can be made arbitrarily small by observing arm 1 a sufficiently short time. This makes it seem that there is no optimal strategy among those that stay with either particular arm for an interval of time. Sudderth (1984) has observed, however, that, for strategies as characterized above, there is an optimal strategy – namely, observe arm 1 when $t = 0$ and for $t > 0$ observe arm 1 or arm 2 according as $Y_1(t) = at + \sigma B(t)$ or $Y_1(t) = bt + \eta B(t)$. This peculiar strategy satisfies the condition that, for each t, the events $\{\tau(t) = 1\}$ and $\{\tau(t) = 2\}$ depend only on past observations since by time t, the value of the standard deviation is known from the law of the iterated logarithm.

□

The following example is similar and is due to Heath (1984).

Example 8.3.2 Consider two arms:

$$Y_1(t) = \theta_1 t + \sigma B(t),$$
$$Y_2(t) = t,$$

where $B(t)$ is standard Brownian motion, σ is a large (known) constant, and θ_1 is uniformly distributed on $[0, 2]$. The discount function is $e^{-\beta t}$.

Intuitively, one expects that if β is sufficiently small (depending on how large σ is), then arm 1 should be observed initially for a considerable duration of time so that the decision maker can learn about θ_1. However, there is a better strategy: observe arm 1 at

$t = 0$, and for $t > 0$ observe arm 1 or arm 2 according as $\theta_1 > 1$ or $\theta_1 \leqslant 1$. □

The preceding two examples indicate that the condition $F_t^\tau \subset F_{t-}^Z$ is not sufficiently strong to reflect our intuitive notion of strategy. We suggest adjoining the reasonable condition $\mathscr{F}_{t+}^\tau \subset \mathscr{F}_t^Z$, which is strong enough to rule out the 'undesirable' strategies of the preceding two examples. The condition $\mathscr{F}_{t+}^\tau \subset \mathscr{F}_t^Z$ is implied by $\mathscr{F}_u^\tau \subset \mathscr{F}_{u-}^Z$ for every u in case the family $(\mathscr{F}_u^Z : u \geqslant 0)$ is right-continuous. We believe this will be the case if there is a set of Lévy processes satisfying the following two conditions: (i) the measures induced on the function space $D[0, 1)$ by any two of the processes in the set are absolutely continuous with respect to each other; (ii) with G-probability one, each Q_i determines a Lévy process belonging to this set. ($D[0, 1)$ is the space of right-continuous functions on $[0, 1)$ having left limits.) This absolute continuity property holds in the examples of Sections 8.1 and 8.2. Skorokhod (1965, Section 4.3), for example, gives conditions for mutual absolute continuity of Lévy processes.

It should be emphasized that we do not know if the additional condition $\mathscr{F}_{t+}^\tau \subset \mathscr{F}_t^Z$ is either necessary or sufficient for the stochastic differential equation (8.2.4) to have a solution. Our purpose in inserting this condition is only to rule out 'strategies' that do not correspond to our intuitive notion of strategy.

The problem of defining $\tau(t, \omega)$ as a function of ω deserves some consideration. Since we want $\mathscr{F}_t^\tau \subset \mathscr{F}_{t-}^Z$, we take

$$\tau(t, \omega) = \hat{\tau}(t, Z_{[0, t)}(\omega))$$

where $\hat{\tau}(t, \cdot)$ is a measurable function on $D[0, t)$ and $Z_{[0, t)}$ denotes Z restricted to $[0, t)$. For each t regard $\hat{\tau}(t, \cdot)$ as being defined on $D[0, \infty)$ as follows:

$$\hat{\tau}(t, h) = \hat{\tau}(t, h_{[0, t)})$$

for $h \in D[0, \infty)$ with $h_{[0, t)}$ denoting the restriction of h to $[0, t)$. Then $\mathscr{F}_t^{\hat{\tau}} \subset \mathscr{D}(D[0, \infty))$ for each t, so it is meaningful to speak of $\mathscr{F}_{t+}^{\hat{\tau}}$. In order that $\mathscr{F}_{t+}^\tau \subset \mathscr{F}_t^Z$ we require $\mathscr{F}_{t+}^{\hat{\tau}} \subset \mathscr{D}(D[0, t])$.

If the program indicated above can be completed, it would be desirable to have a theorem saying that the value is the limit of the sequence of values of discrete approximations. One might think that a counterexample could be built along the lines of Example 8.3.1 since a discrete approximation will not have the information implied by the law of the iterated logarithm. However, the variance of Brownian motion at some time is the square variation of Brownian motion up to that time, and with high probability this square variation can be estimated with a discrete approximation. A variational argument can also be used to show that the value of the continuous-time bandit in the next example is the limit of the sequence of values of discrete-time approximations.

Example 8.3.3 Suppose that, with probability $\frac{1}{2}$, Y_1 is standard Brownian motion and that, with probability $\frac{1}{2}$, it is a stable Lévy process of index 3/2 having mean 2 at time 1. If Y_1 is stable of index 3/2, jumps will occur with probability one at a random sequence of arbitrarily small positive times. Suppose that arm 2 is known: $Y_2(t) = t$. For this bandit, as in Example 8.3.1, the value equals the upper bound given in (8.3.3). The decision maker can come arbitrarily close to achieving the value $(3/2)|\alpha_t|_1$ by observing arm 1 for a short interval of time and then using arm 1 or arm 2 according as jumps have or have not been observed.   □

The practical reader might feel that the detailed local structure of the random processes is irrelevant because it cannot be observed in practice. Thus, for instance, one would not be able to draw the conclusions indicated in Examples 8.3.1 and 8.3.3 after observing arm 1 for a very short time. Such a reader might assign probability one to a set of Lévy processes having common local structure – for example, the set of processes each of which is the sum of a compound Poisson process, a linear function of time, and standard Brownian motion. Locally, all such processes look like standard Brownian motion.

We expect there to be continuous-time bandits without optimal strategies. However, if the definition of strategy is too narrow, optimal strategies may fail to exist in cases where it is natural to expect that such strategies should exist. For instance, a right-continuity requirement would rule out the natural optimal strategy in the following example.

Example 8.3.4 For an arbitrary discount function, suppose that $k = 2$, $Y_2(t) \equiv 0$, and that $Y_1(t)$ is, with probability $\frac{1}{2}$ each, either the standard (right-continuous) Poisson process with mean t or its negative. An optimal strategy is to observe arm 1 up to and including the first jump time and thereafter use arm 1 or arm 2 according as that jump is or is not positive. □

The next example also indicates the importance of the class of strategies being considered.

Example 8.3.5 Suppose discounting is exponential and there are two independent arms having the characteristics of arm 1 in Section 8.1; the two possible drifts are the same for both arms but the probabilities may be different. Consistent with the very intuitive result of Theorem 4.3.9, we would expect that it is optimal to observe arm i at time t if the conditional probability that the drift on arm i equals at is larger than or equal to the corresponding conditional probability for the other arm; the conditioning is on the observed process $Z(s)$ for $s < t$. (Because of continuity this conditioning is equivalent to conditioning on $\{Z(s), s \leqslant t\}$.)

We have arranged for $\mathscr{F}_t^\tau \subset \mathscr{F}_{t-}^Z$. The conjecture mentioned earlier that $\{F_u^Z : u \geqslant 0\}$ is right-continuous can be verified in this case; hence $\mathscr{F}_{t+}^\tau \subset \mathscr{F}_t^Z$. This leaves open the question of whether the stochastic differential equation (8.3.4) has a unique solution (assuming τ is specified uniquely by, say, requiring $\tau(t) = 1$ whenever the conditional probabilities for the drifts are identical for the two arms). Extrapolation from Karatzas (1984) suggests that it does have a unique solution – Karatzas works in a slightly different context in which processes are stopped during periods when they are unobserved. □

Regularity, as defined in Section 5.7, is likely to play an important role in any general theory. It is important for the examples of the next section.

8.4 Examples

In this section we present two examples in a rather informal manner. Both examples compare a known arm and an unknown arm. In the second example the unknown arm is either deterministic drift or is better than the known arm. In the first example the unknown arm is either deterministic drift or is worse than the known arm.

In the first example there would be no optimal strategy were we to restrict ourselves, as we did in Sections 8.1 and 8.2, to strategies that select a particular arm at times in the union of intervals of the form $[r, s)$; so the optimal strategy obtained will not have this property. The strategy involves only three possibilities: select arm 2 exclusively, select arm 1 exclusively, or select arm 1 up to and *including* a certain random time and then switch to arm 2.

Example 8.4.1 Suppose there are two arms and the discount function α_t is regular. Let $Y_2(t) = \lambda t$ for some known $\lambda \geqslant 0$. Let Y_1 be the Lévy process

$$Y_1(t) = t + \sum_{s \leqslant t} [Y_1(s) - Y_1(s-)],$$

where $Y_1(s) - Y_1(s-)$ is the jump of Y_1 at time s (Lévy processes are right-continuous and so $Y_1(s) = Y_1(s+)$).

Temporarily conditional on the distribution of $Y_1(1)$. Since Y_1 is a Lévy process, for a Borel subset $C \subset \mathbb{R} - \{0\}$ the cardinality of the set

$$\{t_1 < s \leqslant t_2 : Y_1(s) - Y_1(s-) \in C\} \qquad (8.4.1)$$

is a Poisson random variable with mean $(t_2 - t_1)v_1(C)$ for some $v_1(C)$; $v_1(C) = \infty$ is possible, in which case the random set at (8.4.1) has countably infinitely many members with probability one. Clearly, v_1 is a measure on $\mathbb{R} - \{0\}$; it is called *Lévy measure*.

Since we regard the distribution of Y_1 as random, v_1 is also random. So the distribution F on the space of distributions for $Y_1(1)$ can instead be regarded as a distribution on the space of measures v_1. Assume that F is supported by

$$\{v_1 : v_1(0, \infty) = 0, \; v_1(-\infty, 0) < \infty,$$

$$\int y \, v_1(dy) < -1 \text{ or } v_1(-\infty, 0) = 0\}.$$

This assumption means that Y_1 has no positive jumps and is either deterministic drift t (if $v_1(-\infty, 0) = 0$) or is inferior to arm 2 (if $v_1(-\infty, 0) > 0$ and hence $E(Y_1(1)) = 1 + \int y v_1(dy) < 0 \leqslant \lambda$).

Clearly, any jump in Y_1 calls for an immediate switch to arm 2. The results of Sections 4.3 and 5.2 suggest that our search for optimal strategies can be restricted to two strategies: τ_2, always use arm 2; and τ_1, use arm 1 until (if ever) a jump occurs and then switch permanently to arm 2. This assertion can be demonstrated using discrete approximations and applying the results of Chapters 4 and 5.

The worth of τ_2 is $\lambda\gamma_0$, where γ_t is defined in Section 5.7. The worth of τ_1 is $W_1 + W_2$, where W_1 is the contribution from arm 1 until the time T of the first jump and W_2 is the contribution from arm 2 thereafter; T may be $+\infty$. Clearly, $W_2 = \lambda E(\gamma_T | F)$ where γ_∞ is defined to be 0. Also,

$$W_1 = \int E\left(\int_0^T \alpha_t \, dY_1(t) \,\middle|\, v_1\right) F(dv_1)$$

$$= \int E(\gamma_0 - \gamma_T | v_1) \, E(Y_1(1) | v_1) \, F(dv_1).$$

Hence, τ_2 is optimal if

$$\lambda \geqslant \frac{\int E(\gamma_0 - \gamma_T | v_1) \, E(Y_1(1) | v_1) \, F(dv_1)}{E(\gamma_0 - \gamma_T | F)}; \tag{8.4.2}$$

τ_1 is optimal if the inequality is reversed. These conclusions depend on our assumption that $\lambda \geqslant 0$. But if the right-hand side of (8.4.2) is positive then τ_1 is also optimal if $\lambda < 0$.

To be specific, suppose F assigns probability $\frac{1}{2}$ to each of two Lévy processes. One process corresponds to $v_1 = 0$ (and is thus the function t since jumps occur with probability 0) and the Lévy measure v_1 of the other has a single atom of size q at $-2/q$. Whatever the value of q, $E(Y_1(1) | F) = -1$. Nevertheless, the optimal strategy depends on q. The right-hand side of (8.4.2) is

$$\frac{\int_0^\infty q e^{-qt} \gamma_t \, dt}{2\gamma_0 - \int_0^\infty q e^{-qt} \gamma_t \, dt},$$

which is a nondecreasing function of q that approaches 1 as $q \to \infty$ and 0 as $q \downarrow 0$. This is intuitively appealing: when q is large, the cost of beginning with arm 1 when it is not the better arm is small since $E(T) = 1/q$. $\qquad\square$

The unknown arm in the next example is either deterministic drift or is superior to a known arm.

Example 8.4.2 Suppose there are two arms and the discount function α_t is regular. Let $Y_2(t) = \lambda t$ for some known $\lambda \leqslant 0$. Let

$$Y_1(t) = -t + \sum_{s \leqslant t} [Y_1(s) - Y_1(s-)].$$

As in the previous example, we regard F as a measure on the space of Lévy measures v_1, the jumps of Y_1 being governed by v_1. We suppose that F is supported by

$$\{v_1 : v_1(-\infty, 0) = 0, \ v_1(0, \infty) < \infty, \ \int yv_1 \, dy > 1 \text{ or } v_1(0, \infty) = 0\}.$$

Assuming the results of Chapter 5 apply in this continuous-time setting, we restrict consideration to strategies that stay with arm 2 indefinitely once it is selected. It is clear that arm 2 should never be used if a jump has been observed on arm 1. Accordingly, the only strategies that need be considered are of the form τ_u: arm 1 is used until time u and arm 2 is used indefinitely thereafter unless a jump has been observed on arm 1; $u \in [0, \infty]$ and u is fixed.

Let T (possibly $+\infty$) denote the time of the first jump of Y_1. The worth of strategy τ_u is given by

$$W(F, \lambda; \mathrm{e}^{-\beta t}; \tau_u) = \int E\left(\mathbf{1}_{\{T \leq u\}} \int_0^\infty \alpha_t \, dY_1(t) \right.$$

$$\left. + \mathbf{1}_{\{T > u\}} \Big(\int_0^u \alpha_t dY_1(t) + \gamma_u \lambda \Big) \Big| v_1 \right) F(dv_1)$$

$$= \int \Bigg((1 - \mathrm{e}^{-uv_1(0, \infty)}) \gamma_0 E(Y_1(1)|v_1) $$

$$+ \mathrm{e}^{-uv_1(0, \infty)} \big([\gamma_0 - \gamma_u] E(Y_1(1)|v_1) + \gamma_u \lambda \big) \Bigg) F(dv_1).$$

We will rely on the local behaviour of $W(F, \lambda; \mathrm{e}^{-\beta t}; \tau_u)$ at $u = 0$ to decide which arm is optimal initially. Its derivative evaluated at $u = 0$ equals

$$\int \Bigg([\gamma_0 v_1(0, \infty) + \alpha_0] E(Y_1(1)|v_1) - [\gamma_0 v_1(0, \infty) + \alpha_0] \lambda \Bigg) F(dv_1).$$

It follows that exclusive use of arm 2 is optimal if

$$\lambda \geq \frac{\int [\alpha_0 + \gamma_0 v_1(0, \infty)] E(Y_1(1)|v_1) F(dv_1)}{E(\alpha_0 + \gamma_0 v_1(0, \infty)|F)}; \qquad (8.4.3)$$

otherwise arm 1 is optimal initially and for some (possibly infinite) period of time. Since we have assumed $\lambda \leq 0$, arm 1 is optimal initially when the right-hand side of (8.4.3) is positive. However, the above argument also applies for $\lambda > 0$ provided λ is less than the right-hand side of (8.4.3). If the right-hand side of (8.4.3) is positive and the inequality holds, then no conclusion can be easily drawn. If the right-hand side of (8.4.3) is nonpositive, then arm 2 is optimal in case $\lambda = 0$ and thus, also, in case $\lambda > 0$.

Returning to the original assumption that $\lambda \leqslant 0$, we observe that if (8.4.3) does not hold, arm 1 should be used until (8.4.3) holds with equality, with the distribution of v_1 conditioned by the observation of arm 1 playing the role of F.

To be specific, suppose F assigns probability $1 - p$ to the deterministic Lévy process $-t$ and probability p to the Lévy process with measure v_1 consisting of a single atom of size q at $2/q$. Whatever the value of q, $E(Y_1(1)|F_1) = 2p - 1$. From (8.4.3), arm 1 is optimal initially when $\lambda \leqslant 0$ if

$$\lambda < \frac{\alpha_0(2p - 1) + \gamma_0 pq}{\alpha_0 + \gamma_0 pq}.$$

Suppose, in addition, that the discounting is exponential. Then in case $\lambda > -1$ and no jump has occurred on arm 1, a switch to arm 2 is optimal at time

$$q^{-1} \log \frac{p(\alpha_0 + \gamma_0 q)(1 - \lambda)}{(1 - p)\alpha_0(1 + \lambda)}.$$

If $\lambda < -1$, it is optimal to observe arm 1 exclusively and indefinitely.

□

References

Bather, J. A. (1970) Optimal stopping problems for Brownian motion. *Adv. in Appl. Probab.* **2**: 259–286.

Breakwell, J. and Chernoff, H. (1964) Sequential tests for the mean of a normal distribution II. *Ann. Math. Statist.* **35**: 162–173.

Chernoff, H. (1961) Sequential tests for the mean of a normal distribution, *Fourth Berkeley Symp. of Math. Statist. and Prob.* **1**: 79–91.

Chernoff, H. and Petkau, A. J. (1983) Numerical methods for Bayes sequential decision problems. Univ. of British Columbia Applied Mathematics and Statistics Tech. Rep. No. 83–126.

Chernoff, H. and Ray, S. N. (1965) A Bayes sequential sampling inspection plan. *Ann. Math. Statist.* **36**: 1387–1407.

Dellacherie, C. and Meyer, P.-A. (1978) *Probabilities and Potential*, Part 1, North-Holland, Amsterdam.

Dellacherie, C. and Meyer, P.-A. (1980) *Probabilities and Potential*, Part 2, North-Holland, Amsterdam.

Fristedt, B. (1974) Sample functions of stochastic processes with stationary, independent increments. *Advances in Probability*, Vol. 3 (eds P. Ney and S. Port), pp. 241–396, Marcel-Dekker, New York.

Heath, D. C. (1984) Personal communication.

Karatzas, I. (1984) Gittins indices in the dynamic allocation problem for diffusion processes. *Ann. Prob.* **12**: 173–192.

Lipster, R. S. and Shiryayev, A. N. (1977) *Statistics of Random Processes I*, Springer-Verlag, New York.

Skorokhod, A. V. (1965) *Studies in the Theory of Random Processes*, Addison-Wesley, Reading, Massachusetts.

Sudderth, W. D. (1984) Personal communication.

CHAPTER 9

Minimax Approach

A bandit problem is interesting only if there are arms with unknown characteristics. To choose among the available arms a decision maker must first decide how to handle this uncertainty. In the first eight chapters of this monograph the approach used is to average the payoff over the unknown characteristics with respect to a specified prior distribution – a Bayesian approach, in statistical parlance.

Some people prefer an approach that does not require specifying a prior distribution. One such is the minimax or game-theoretic approach in which the decision maker uses a strategy designed to maximize the minimum payoff – a 'worst case' approach.

If the means are assumed to be as small as possible then the decision maker should simply choose the arm for which the smallest possible mean is largest, and the decision problem is trivial. In this chapter we modify the objective in a way which makes the problem nontrivial. Namely, the objective is to minimize *regret*, the expected difference between the payoff that could be obtained if the characteristics of the arms were known and the payoff actually achieved.

One may view the minimax approach as one in which a protagonist – nature – chooses the characteristics of the arms. In trying to make life difficult for the decision maker, there is no particular reason for nature to make the means small, for it may then be very easy for the decision maker to do as well as possible and thus hold the regret to zero. Indeed, for any particular choice of characteristics of the arms the decision maker always has a strategy that holds the regret to zero. But this is not possible if nature uses a random strategy. The decision maker's objective is to do as well as possible against the best random strategy of nature (this provides a connection with the Bayesian approach since a prior distribution for the decision maker is a random strategy for nature).

In Section 9.1 we discuss two Bernoulli arms in a discrete-time

191

setting. We study the asymptotic behaviour of the value for geometric discounting with the discount factor approaching 1 and uniform discounting with a growing horizon. In Section 9.2 a continuous-time example is treated in some detail.

9.1 Discrete time, two Bernoulli arms

In this section we consider two Bernoulli arms with success probabilities θ_1 and θ_2. The discount sequence $\mathbf{A}$ is arbitrary. Mixed (or randomized) strategies will be introduced shortly. We temporarily consider strategies as defined in Chapter 2 but we now call them 'pure strategies' to distinguish them from mixed strategies. For known θ_1 and θ_2, the worth of a pure strategy τ is

$$W(\theta_1, \theta_2; \mathbf{A}; \tau) = E_\tau \sum_{m=1}^{\infty} \alpha_m Z_m$$

$$= \sum_{m=1}^{\infty} \alpha_m (\theta_1 P_\tau\{\tau^m = 1\} + \theta_2 P_\tau\{\tau^m = 2\}),$$

where τ^m denotes the arm indicated by τ at stage m and the dependence of τ^m on the history of successes and failures preceding stage m has been suppressed in the notation.

The *regret* (sometimes called 'opportunity loss') is defined to be the expected loss (conditioned on (θ_1, θ_2)) resulting from using τ rather than the best arm at every stage:

$$R(\theta_1, \theta_2; \mathbf{A}; \tau) = \sum_{m=1}^{\infty} \alpha_m (\theta_1 \vee \theta_2) - W(\theta_1, \theta_2; \mathbf{A}; \tau)$$

$$= \sum_{m=1}^{\infty} \alpha_m (\theta_1 \vee \theta_2 - \theta_1 \wedge \theta_2) P_\tau\{\tau^m = i, \theta_i = \theta_1 \wedge \theta_2\}.$$

$$(9.1.1)$$

The topology on the space of pure strategies is defined as follows. The distance between two pure strategies is $1/m$, where m is the first stage for which there exists a history for which the two strategies indicate different arms. A *mixed strategy* is defined to be a probability measure S on the Borel field of subsets of this space. The *regret* for such a strategy is defined as

$$R(\theta_1, \theta_2; \mathbf{A}; S) = \int R(\theta_1, \theta_2; \mathbf{A}; \tau) S(d\tau) \qquad (9.1.2)$$

where the integrand is defined in (9.1.1).

For any subset Θ of $[0, 1] \times [0, 1]$, define the $(\Theta, \mathbf{A})$-*upper regret* to be

$$R_U(\Theta; \mathbf{A}) = \inf_S \sup_{(\theta_1, \theta_2) \in \Theta} R(\theta_1, \theta_2; \mathbf{A}; S). \qquad (9.1.3)$$

The mixed strategy S_0 is $(\Theta; \mathbf{A})$-*minimax* if the infimum in (9.1.3) is attained for $S = S_0$.

For each fixed $\mathbf{A} \in \mathscr{A}$ the regret $R(\theta_1, \theta_2; \mathbf{A}; \tau)$ is a uniformly continuous function of $(\theta_1, \theta_2, \tau)$. For mixed strategies, $R(\theta_1, \theta_2; \mathbf{A}; S)$ inherits this property, with the convergence-in-distribution topology being used for the mixed strategies. The space of mixed strategies S also inherits compactness from the space of pure strategies τ. From these facts we obtain the following theorem.

Theorem 9.1.1 For each discount sequence $\mathbf{A}$ and each subset Θ of $[0, 1] \times [0, 1]$, there exists a $(\Theta, \mathbf{A})$-minimax mixed strategy.

A more symmetrical view of this minimax setting is sometimes useful. Suppose nature is a competitor of the decision maker and uses a mixed strategy G.

For Θ measurable, the $(\Theta, \mathbf{A})$-*lower regret* is defined by

$$R_L(\Theta; \mathbf{A}) = \sup_G \left\{ \inf_\tau \int_\Theta R(\theta_1, \theta_2; \mathbf{A}; \tau) G(\mathrm{d}(\theta_1, \theta_2)) : G(\Theta) = 1 \right\},$$
$$(9.1.4)$$

and G_0, supported by Θ, is $(\Theta; \mathbf{A})$-*maximin* if the supremum in (9.1.4) is attained for $G = G_0$. In view of the compactness of $[0, 1] \times [0, 1]$ and the continuity of R, this 'two-person zero-sum game' can be approximated by a game in which each of the two players has only finitely many pure strategies. Such approximations also apply with Θ replaced by its closure $\mathrm{cl}(\Theta)$. Accordingly, from the classical theorems on finite matrix zero-sum two-person games, we obtain the following result.

Theorem 9.1.2 For each measurable $\Theta \subset [0, 1] \times [0, 1]$ and each discount sequence $\mathbf{A}$,

$$R_U(\Theta; \mathbf{A}) = R_L(\Theta; \mathbf{A}) = R_U(\mathrm{cl}(\Theta); \mathbf{A}) = R_L(\mathrm{cl}(\Theta); \mathbf{A}).$$

If Θ is closed, there exists a $(\Theta; \mathbf{A})$-maximin strategy G.

The common value of R_U and R_L is usually called the 'value of the game'. We will use *regret value* to avoid confusion with *value* as introduced in Chapter 2.

The arguments leading to Theorem 9.1.2 are standard. Indeed, rather than rely on finite matrix theorems we could have quoted theorems, for instance, Parthasarathy and Raghavan (1971, Chapter 5), about games with continuous payoff functions.

Generalizations of Theorems 9.1.1 and 9.1.2 may be valid for general bandits, not just those with Bernoulli arms. However, such extensions will have to overcome the lack of continuity mentioned in Example 2.5.1.

We now calculate the regret value and optimal strategies for a particular $(\Theta; \mathbf{A})$.

Example 9.1.1 Let $\mathbf{A} = (1, 1, 1, 0, 0, \ldots)$

and

$$\Theta = \{(\theta_1, \theta_2): 0 \leqslant \theta_1 \leqslant 1, \theta_2 = 0 \text{ or } \theta_2 = 1\};$$

so arm 1 is arbitrary but arm 2 either yields all failures or all successes. If arm 2 is selected at any stage m, then, of course, arm 1 or arm 2 should be selected subsequently according as failure or success is observed on arm 2. So there is no loss in limiting the pure strategies available to the decision maker to those having this property. We temporarily assume (and we will show that the assumption is correct) that the decision maker should use a mixture of the following three pure strategies:

   τ_1: select arm 1 and switch to arm 2 at stage 2 or stay with arm 1
        at stages 2 and 3 according as failure or success is observed at
        stage 1;
   τ_2: select arm 1 until a failure is observed and then switch to
        arm 2;
   τ_3: select arm 2 initially.

The regrets for these strategies calculated from (9.1.1) are given in Table 9.1.

Let us assume (correctly as it will develop) that τ_1, τ_2, and τ_3 need be nature's only concern. There exists a maximum G_0 in view of Theorem 9.1.2. Using the downward concavity of the six relevant functions of θ_1, that there exists an optimal G_0 for nature that is supported by just two points: one of the form $(\theta_1, 0)$ and the other of the form $(\theta_1, 1)$.

Table 9.1 *Values of* $R(\theta_1, \theta_2; (1, 1, 1, 0, 0, \ldots); \tau_i)$

(θ_1, θ_2)	τ_1	τ_2	τ_3
$(\theta_1, 0)$	$\theta_1 - \theta_1^2$	$\theta_1 - \theta_1^3$	θ_1
$(\theta_1, 1)$	$1 + \theta_1 - 2\theta_1^2$	$1 - \theta_1^3$	0

Consider a G which assigns probability p to $(a, 0)$ and probability $1 - p$ to $(b, 1)$. The minimum of the three losses equals

$$ap \wedge [1 - b^3 - (1 - b^3 - a + a^3)p]$$
$$\wedge [1 + b - 2b^2 - (1 + b - 2b^2 - a + a^2)p]. \tag{9.1.5}$$

In view of (9.1.4) and Theorem 9.1.2, G_0 maximizes (9.1.5).

Both $1 - b^3$ and $1 + b - 2b^2$ are decreasing for $b > 1/4$; so restrict consideration to $b \in [0, 1/4]$. As a function of p, (9.1.5) is the minimum of three linear functions whose values at 0 are, respectively, $0 \leqslant 1 - b^3 \leqslant 1 + b - 2b^2$ and whose values at 1 are, respectively, $a \geqslant a - a^3 > a - a^2$. The first of these linear functions has a nonnegative slope, whereas the other two slopes are nonpositive. Accordingly, for fixed a and b, the p that maximizes (9.1.5) is the smaller of two values of p – one makes the first two linear functions equal and the other makes the first and third linear functions equal. When p is so chosen, (9.1.5) equals

$$\frac{a(1 - b^3)}{1 - b^3 + a^3} \wedge \frac{a(1 + b - 2b^2)}{1 + b - 2b^2 + a^2}. \tag{9.1.6}$$

For fixed b, the two quantities in (9.1.6) are equal for two values of a: 0, which need not be considered, and

$$a = \frac{1 - b^3}{1 + b - 2b^2} = \frac{1 + b + b^2}{1 + 2b} \geqslant 7/8 \tag{9.1.7}$$

(the inequality holding since $b \leqslant 1/4$).

As functions of a, the first quantity in (9.1.6) is decreasing for $a \geqslant 7/8$ and the second quantity is increasing on $[0, 1]$. Hence, for fixed b, the maximum of (9.1.6) is attained for a given in (9.1.7). The maximum equals $f(b)/g(b)$, where

$$f(b) = 1 + 4b + 4b^2 - b^3 - 4b^4 - 4b^5 \tag{9.1.8}$$

and

$$g(b) = 2 + 7b + 9b^2 - 2b^3 - 7b^4. \tag{9.1.9}$$

By differentiating $f(b)/g(b)$ we see that it attains its maximum value at the unique $b \in [0, 1/4]$ for which

$$1 - 2b - 8b^2 - 2b^3 - 41b^4 - 128b^5 - 107b^6 + 16b^7 + 28b^8 = 0.$$
$$(9.1.10)$$

We now fix b to be this solution and use the previously derived conditions on a and p to fix these quantities. When so fixed, the three quantities in (9.1.5) are equal. Numerical calculations, using (9.1.10), (9.1.7), and the equality of the (first two) quantities in (9.1.5), give $b \approx 0.218, a \approx 0.881, p \approx 0.591$, and $f(b)/g(b) \approx 0.521$, where $f(b)$ and $g(b)$ are given in (9.1.8) and (9.1.9).

It is now straightforward to calculate

$$\int_\Theta R(\theta_1, \theta_2; \mathbf{A}; \tau) G_0(\mathrm{d}(\theta_1, \theta_2))$$

for each of the other 15 pure strategies available to the decision maker. The result in each case is larger than 0.6, thereby affirming our earlier conjectures that the decision maker should use a mixed strategy involving only τ_1, τ_2, and τ_3 and that nature need only be concerned with these three pure strategies of the decision maker. We also conclude that the regret value equals

$$\frac{f(b)}{g(b)} = \frac{(1-b)(1+b+b^2)(1+2b)^2}{2 + 7b + 9b^2 - 2b^3 - 7b^4} \approx 0.521. \qquad (9.1.11)$$

A minimax strategy is specified by ε_1, ε_2, and ε_3, the optimal probabilities of τ_1, τ_2, and τ_3, respectively. We will calculate these probabilities by finding various linear equations that they must satisfy. The first of these is

$$\varepsilon_1 + \varepsilon_2 + \varepsilon_3 = 1. \qquad (9.1.12)$$

From game theory we know that the regret value $f(b)/g(b)$ must equal the regret when the decision maker uses a minimax strategy and nature uses one of the two pure strategies having positive probability in the maximin strategy. From Table 9.1 we obtain

$$(a - a^2)\varepsilon_1 + (a - a^3)\varepsilon_2 + a\varepsilon_3 = R_U, \qquad (9.1.13)$$
$$(1 + b - 2b^2)\varepsilon_1 + (1 - b^3)\varepsilon_2 = R_U. \qquad (9.1.14)$$

The system consisting of (9.1.12), (9.1.13), and (9.1.14) has infinitely many solutions, but we can find another condition. Temporarily regarding b to be a variable, the derivative of the left-hand side of (9.1.14) at the maximin b must equal 0; for otherwise $(\varepsilon_1, \varepsilon_2, \varepsilon_3)$ when inserted for S in (9.1.3) would yield something greater than R_U. Hence,

$$(1 - 4b)\varepsilon_1 - 3b^2\varepsilon_2 = 0. \qquad (9.1.15)$$

The system consisting of (9.1.12), (9.1.13), (9.1.14), and (9.1.15) has a unique solution, which we express directly in terms of b by using the expression at (9.1.11) for the risk value R_U:

$$\varepsilon_1 = \frac{3b^2(1 + b + b^2)(1 + 2b)^2}{(1 - b)(1 - 2b - 2b^2)(2 + 7b + 9b^2 - 2b^3 - 7b^4)} \approx 0.257,$$

$$\varepsilon_2 = \frac{(1 - 4b)(1 + b + b^2)(1 + 2b)^2}{(1 - b)(1 - 2b - 2b^2)(2 + 7b + 9b^2 - 2b^3 - 7b^4)} \approx 0.235,$$

$$\varepsilon_3 = \frac{1 + b - 3b^2 - 15b^3 - b^4 + 30b^5 + 14b^6}{(1 - 2b - 2b^2)(2 + 7b + 9b^2 - 2b^3 - 7b^4)} \approx 0.508. \qquad \square$$

The next two results give bounds for the regret value in case $\Theta = [0, 1] \times [0, 1]$. In the first, the discount sequence is geometric with a discount factor close to 1, and in the second it is the n-horizon uniform with large n. Such results were obtained by Vogel (1960b) for uniform discount sequences. We use ideas from there as well as from Vogel (1960a) and Bather and Simons (1985), although our treatment will be brief since we will not be concerned with obtaining 'good' values for certain constants.

Theorem 9.1.3 There exist positive finite constants c_1 and c_2 such that for every $\alpha \in (0, 1)$,

$$c_1(1 - \alpha)^{-1/2} \leqslant R_U([0, 1)] \times [0, 1]; (1, \alpha, \alpha^2, \alpha^3, \ldots))$$
$$\leqslant c_2(1 - \alpha)^{-1/2}. \qquad (9.1.16)$$

Proof For the proof of the first equality in (9.1.16), Example 5.4.1 is relevant. We require that θ_1 be one of two values:

$$F_1 = \tfrac{1}{2}\delta_{[1 - \sqrt{(1 - \alpha)}]/2} + \tfrac{1}{2}\delta_{[1 + \sqrt{(1 - \alpha)}]/2}$$

and take arm 2 to have the known success probability $\Lambda(F_1, (1, \alpha, \alpha^2, \ldots))$, a function which is defined in Corollary 5.1.2. By Corollary 5.1.2 and Theorem 5.2.2, the value (not 'regret value') equals $(1 - \alpha)^{-1}\Lambda$ which can be realized by always selecting arm 2. From

(9.1.1) and (9.1.4) the regret value is greater than or equal to

$$\left(\frac{1 + \sqrt{(1-\alpha)}}{2} - \Lambda\right)\Big/[2(1-\alpha)]$$

which, by Example 5.4.1, is asymptotic to $(3 - \sqrt{3})/[12\sqrt{(1-\alpha)}]$ as $\alpha \uparrow 1$. The first inequality in (9.1.16) follows.

For the proof of the second inequality in (9.1.16) there is no loss in considering only those α for which $(1-\alpha)^{-1}$ is the square of an integer K. Consider this strategy for the decision maker: select arms 1 and 2 alternatively until, after some even-numbered stage $2M$, the number of successes observed on one arm is at least K greater than that on the other arm; after that stage select the arm on which the greater number of successes has been observed. By pairing the stages we can interpret the phenomenon through the random stage $2M$ as a random walk with steps

$$-1 \text{ with probability } \theta_1(1-\theta_2)$$
$$0 \text{ with probability } \theta_1\theta_2 + (1-\theta_1)(1-\theta_2)$$
$$1 \text{ with probability } \theta_2(1-\theta_1)$$

and which is stopped at time M, the first hitting time of the set $\{K, -K\}$. The regret (cf. (9.1.1)) for this strategy is no larger than

$$|\theta_2 - \theta_1|E(M) + |\theta_2 - \theta_1|(1-\alpha)^{-1}\sum_{m=K}^{\infty} p_m\alpha^m, \qquad (9.1.17)$$

where p_m is the probability that $M = m$ and the random walk at time m has the sign opposite to that of $\theta_2 - \theta_1$. We find that $E(M) \leqslant K/|\theta_2 - \theta_1|$ (cf. Feller, 1968, answer to problem 5, Chapter XIV), so the first term in (9.1.17) is bounded by $K = (1-\alpha)^{-1/2}$, as desired.

We assume $\theta_2 > \theta_1$ with no loss of generality. The generating function $\sum p_m\alpha^m$ can be obtained via a difference equation (Feller, 1968, Section XIV.4). The generating function multiplied by $(\theta_2 - \theta_1)K^2 = (\theta_2 - \theta_1)(1-\alpha)^{-1}$ (cf. (9.1.17)) equals

$$K^2(\theta_2 - \theta_1)[2\theta_1(1-\theta_2)\alpha]^K\bigg(\{(\theta_1 + \theta_2 - 2\theta_1\theta_2) +$$
$$+ [\theta_1\theta_2 + (1-\theta_1)(1-\theta_2)](1-\alpha) + [(\theta_2 - \theta_1)^2$$
$$+ 2(\theta_1 - \theta_1^2 + \theta_2 - \theta_2^2)(1-\alpha)$$
$$+ (\theta_1 + \theta_2 - 1)^2(1-\alpha)^2]^{1/2}\}^K$$

$$+ \{(\theta_1 + \theta_2 - 2\theta_1\theta_2) + [\theta_1\theta_2 + (1 - \theta_1)(1 - \theta_2)](1 - \alpha)$$

$$- [(\theta_2 - \theta_1)^2 + 2(\theta_1 - \theta_1^2 + \theta_2 - \theta_2^2)(1 - \alpha)$$

$$+ (\theta_1 + \theta_2 - 1)^2 (1 - \alpha)^2]^{1/2}\}^K \Big)^{-1}$$

$$\leqslant K^2(\theta_2 - \theta_1)[2\theta_1(1 - \theta_2)]^K \{(\theta_1 + \theta_2 - 2\theta_1\theta_2) + (\theta_2 - \theta_1)\}^{-K}$$

$$= K^2(\theta_2 - \theta_1)[\theta_1(1 - \theta_2)/\theta_2(1 - \theta_1)]^K$$

$$= K^2(\theta_2 - \theta_1)\left[\left(1 - \frac{\theta_2 - \theta_1}{\theta_2}\right)\left(1 - \frac{\theta_2 - \theta_1}{1 - \theta_1}\right)\right]^K$$

$$\leqslant K^2(\theta_2 - \theta_1)[1 - (\theta_2 - \theta_1]^{2K} = K^2(\theta_2 - \theta_1)e^{2K\log[1 - (\theta_2 - \theta_1)]}$$

$$\leqslant K^2(\theta_2 - \theta_1)e^{-2K(\theta_2 - \theta_1)}$$

which is no larger than K multiplied by the maximum of xe^{-2x}, $0 \leqslant x \leqslant \infty$. $\qquad\qquad\Box$

Corollary 9.1.4 Suppose $\mathbf{A} = (1, \ldots, 1, 0, \ldots)$ with horizon n. There exist positive finite constants c_3 and c_4 such that

$$c_3 n^{1/2} \leqslant R_U([0, 1] \times [0, 1]; (1, \ldots, 1, 0, 0, \ldots)) \leqslant c_4 n^{1/2}$$

for every n.

Proof Fix n and choose α to satisfy $1 - \alpha = n^{-1}$. Consider the strategy τ (depending on n via α) constructed in the above proof to give an upper bound of (9.1.1):

$$\sum_{m=1}^{\infty} \alpha^{m-1}(\theta_1 \vee \theta_2 - \theta_1 \wedge \theta_2)P_\tau(\tau^m = i, \theta_i = \theta_1 \wedge \theta_2)$$

$$\leqslant c_2 n^{-1/2}.$$

Hence

$$\sum_{m=1}^{n} (\theta_1 \vee \theta_2 - \theta_1 \wedge \theta_2)P_\tau(\tau^m = i, \theta_i = \theta_1 \wedge \theta_2) \leqslant c_2 \alpha^{-(n-1)} n^{-1/2}.$$

The upper bound of the corollary follows from

$$\alpha^{-n-1} = (1 - n^{-1})^{-n-1} \to e < \infty \text{ as } n \to \infty.$$

We now turn to the lower bound. Let G_0 denote a maximin strategy for $\Theta = [0, 1] \times [0, 1]$ and the geometric discount sequence

$\mathbf{A} = (1, (1 - n^{-1}), (1 - n^{-1})^2, \ldots)$. By Theorem 9.1.3,

$$c_1 n^{1/2} \leqslant \sum_{m=1}^{\infty} (1 - n^{-1})^m [E(\theta_1 \vee \theta_2 | G_0) - E_\tau(Z_m | G_0)] \qquad (9.1.18)$$

for every strategy τ and, in particular, if τ is optimal for the $(G_0; (1, \ldots, 1, 0, 0, \ldots))$-bandit with horizon n. Fix such a τ that proceeds after stage n as well as possible for the $(G_0; ((1 - n^{-1})^n, (1 - n^{-1})^{n+1}, (1 - n^{-1})^{n+2}, \ldots))$-bandit without relying on information from the first n stages. Then, suppressing the dependence of the expectations on G_0,

$$\sum_{m=n+1}^{\infty} (1 - n^{-1})^m [E(\theta_1 \vee \theta_2 | G_0) - E_\tau(Z_m | G_0)]$$

$$= (1 - n^{-1})^n \sum_{m=1}^{\infty} (1 - n^{-1})^m [E(\theta_1 \vee \theta_2 | G_0) - E_\tau(Z_{m+n} | G_0)] \quad (9.1.19)$$

$$\leq (1 - n^{-1})^n \sum_{m=1}^{\infty} (1 - n^{-1})^m [E(\theta_1 \vee \theta_2 | G_0) - E_\tau(Z_m | G_0)].$$

From (9.1.18) and (9.1.19) we conclude that

$$c_1 [1 - (1 - n^{-1})^n] n^{1/2}$$

$$\leqslant [1 - (1 - n^{-1})^n] \sum_{m=1}^{\infty} (1 - n^{-1})^m [E(\theta_1 \vee \theta_2 | G_0) - E_\tau(Z_m | G_0)]$$

$$\leqslant \sum_{m=1}^{n} (1 - n^{-1})^m [E(\theta_1 \vee \theta_2 | G_0) - E_\tau(Z_m | G_0)]$$

$$\leqslant \sum_{m=1}^{n} [E(\theta_1 \vee \theta_2 | G_0) - E_\tau(Z_m | G_0)].$$

This completes the proof since $c_1 [1 - (1 - n^{-1})^n]$ is bounded below by a positive constant. $\square$

In the proof of Theorem 9.1.3 the strategies constructed for nature were not symmetric even though Θ was symmetric. Theorem 9.1.5 below says that only symmetric strategies need to be considered when Θ is symmetric.

Definition 9.1.1 The set Θ of possible pairs (θ_1, θ_2) of Bernoulli parameters is *symmetric* if

$$(\theta_1, \theta_2) \in \Theta \Leftrightarrow (\theta_2, \theta_1) \in \Theta.$$

Definition 9.1.2 A probability measure G on a symmetric $\Theta \subset [0, 1] \times [0, 1]$ is *symmetric* if $G(\Psi) = G(\{(\theta_1, \theta_2) : (\theta_2, \theta_1) \in \Psi\})$ for every measurable $\Psi \subset \Theta$.

Definition 9.1.3 A probability measure S on the space of pure strategies for the decision maker is *symmetric* if

$$S(T) = S(\{\tau : \tau_c \in T\})$$

for every measurable set T of pure strategies, where τ_c denotes the strategy obtained from τ by interchanging the roles of the two arms.

Theorem 9.1.5 Suppose that the set Θ of possible pairs (θ_1, θ_2) of Bernoulli parameters is symmetric. Then there exists a symmetric minimax strategy, and the supremum in (9.1.4) may be taken over symmetric probability measures on Θ. If, in addition, Θ is closed, there exists a symmetric maximin strategy.

Proof Let S_0 denote a $(\Theta, \mathbf{A})$-minimax strategy (cf. Theorem 9.1.1). Define S_c by $S_c(T) = S_0(\{\tau : \tau_c \in T\})$, with τ_c as in Definition 9.1.3. From (9.1.1), (9.1.2), and the symmetry of Θ it is clear that S_c is also minimax. Let $S = (S_0 + S_c)/2$. From (9.1.1) and (9.1.2), we obtain

$$R(\theta_1, \theta_2; \mathbf{A}; S) = [R(\theta_1, \theta_2; \mathbf{A}; S_0) + R(\theta_1, \theta_2; \mathbf{A}; S_c)]/2.$$

So,

$$\sup_{(\theta_1, \theta_2) \in \Theta} R(\theta_1, \theta_2; \mathbf{A}; S)$$

$$\leq \left[\sup_{(\theta_1, \theta_2) \in \Theta} R(\theta_1, \theta_2; \mathbf{A}; S_0) + \sup_{(\theta_1, \theta_2) \in \Theta} R(\theta_1, \theta_2; \mathbf{A}; S_c) \right] \bigg/ 2$$

which, by (9.1.3) and the fact that S_0 and S_c are minimax, equals $R_U(\Theta; \mathbf{A})$. Hence, S is minimax – and it is obviously symmetric. The proof for symmetric strategies for nature is very similar and is omitted. $\square$

9.2 A continuous-time example

The continuous-time minimax example presented here is due largely to Bather (1983). There are two arms. Arm 2 generates a process which is identically zero. The process Y_1 generated by arm 1 is given by

$$Y_1(t) = \theta_1 t + B(t),$$

where B denotes standard Brownian motion and θ_1 is an unknown real number. The discount function is $e^{-\beta t}$ for some fixed $\beta > 0$.

We will not develop a general continuous-time theory and so no general terminology or notation will be introduced. Obvious analogues of concepts and notation introduced in Section 9.1 will be used.

We first hypothesize a particular form of the solution and then verify the correctness of that form while simultaneously obtaining conditions that the parameters in that form must satisfy. We conjecture that there exist three constants $b < 0$, $a > 0$, and $p \in (0, 1)$ such that a maximin strategy assigns probability p to $at + B(t)$ and probability $1 - p$ to $bt + B(t)$. In view of Section 8.1 we also conjecture that a minimax strategy is the optimal pure strategy τ for the decision maker obtained in Section 8.1 for a two-point prior distribution on θ_1; τ indicates arm 1 as long as $Y_1(t) > qt + r$ and arm 2 after any t for which $Y_1(t) \leqslant qt + r$, where

$$q = (a+b)/2 \qquad (9.2.1)$$

$$r = -\frac{1}{a-b}\log\frac{(1-C)p}{C(1-p)}, \qquad (9.2.2)$$

with

$$C = \frac{-b(b-a+\sqrt{[8\beta+(a-b)^2]})}{(a-b)(b+a+\sqrt{[8\beta+(a-b)^2]})}. \qquad (9.2.3)$$

The regret associated with τ is

$$R(\theta_1; e^{-\beta t}; \tau) = \beta^{-1}\bigg(\theta_1 \vee 0$$
$$-\theta_1 E(1-\exp[-\beta\inf\{t: \theta_1 t + B(t) \leqslant qt+r\}])\bigg).$$

To substantiate the conjectures we show that b and a can be chosen so that $R(\theta_1; e^{-\beta t}; \tau)$ takes on its maximum value at both b and a. Take $p > C$; then $r < 0$ and

$$R(\theta_1; e^{-\beta t}; \tau) = \beta^{-1}\bigg(\theta_1 \vee 0$$
$$-\theta_1 E(1-\exp[-\beta\inf\{t: B(t) = (q-\theta_1)t+r\}])\bigg). \qquad (9.2.4)$$

The Laplace transform in (9.2.4) is known (for instance, Fristedt, 1974, page 351); (9.2.4) reduces to

$$R(\theta_1; e^{-\beta t}; \tau) = \beta^{-1}\Bigg(\theta_1 \vee 0$$

$$- \theta_1(1 - \exp(r\{\theta_1 - q + \sqrt{[2\beta + (\theta_1 - q)^2]}\}))\Bigg). \quad (9.2.5)$$

For $\theta_1 \geqslant 0$,

$$\frac{dR}{d\theta_1} = \beta^{-1}\left(1 + r\theta_1\left\{1 + \frac{\theta_1 - q}{\sqrt{[2\beta + (\theta_1 - q)^2]}}\right\}\right)$$

$$\exp(r\{\theta_1 - q + \sqrt{[2\beta + (\theta_1 - q)^2]}\}).$$

Differentiation shows that

$$1 + r\theta_1\left\{1 + \frac{\theta_1 - q}{\sqrt{[2\beta + (\theta_1 - q)^2]}}\right\}$$

is a decreasing function of θ_1 since $r < 0$. This function is positive at 0 and negative for large θ_1. Hence, on $[0, \infty)$, $R(\cdot; e^{-\beta t}; \tau)$ increases to a maximum and then decreases. This maximum will occur at a if and only if

$$1 + ra\left\{1 + \frac{a - q}{\sqrt{[2\beta + (a - q)^2]}}\right\} = 0. \quad (9.2.6)$$

For $\theta_1 < 0$, let

$$v_1 = -r\{[2\beta + (\theta_1 - q)^2]^{1/2} + \theta_1 - q\}.$$

When $\theta_1 = 0$, $v_1 = -r\{[2\beta + q^2]^{1/2} - q\} > 0$. As $\theta_1 \to -\infty$, $v_1 \to 0$. Moreover, v_1 is an increasing function of θ_1 and so that relationship can be inverted:

$$\theta_1 = \frac{2\beta r^2 + 2qrv_1 - v_1^2}{2rv_1}. \quad (9.2.7)$$

Thus, the problem of maximizing $R(\theta_1; e^{-\beta t}; \tau)$ for $-\infty < \theta_1 \leqslant 0$ is equivalent to the problem of maximizing

$$f(v_1) = \frac{-2\beta r^2 - 2qrv_1 + v_1^2}{2\beta rv_1}(1 - e^{-v_1}), \quad 0 < v_1 \leqslant -r\{[2\beta + q^2]^{1/2} - q\}.$$

Differentiating f we find

$$2\beta|r|v_1^2 e^{v_1} f'(v_1) = -(2\beta r^2 + v_1^2)(e^{v_1} - 1) + 2\beta r^2 v_1 + 2qrv_1^2 - v_1^3 = g(v_1), \tag{9.2.8}$$

say, and differentiating again,

$$g'(v_1) = -(2\beta r^2 + 2v_1 + v_1^2)(e^{v_1} - 1) + 4qrv_1 - 4v_1^2. \tag{9.2.9}$$

Suppose $g(v) = 0$. Then from (9.2.8),

$$4qrv_1 = (4\beta r^2 v_1^{-1} + 2v_1)(e^{v_1} - 1) - 4\beta r^2 + 2v_1^2,$$

which we substitute into (9.2.9) for $4qrv_1$ to obtain

$$v_1 g'(v_1) = 4\beta r^2(e^{v_1} - 1) - (2\beta r^2 v_1 + v_1^3)(e^{v_1} + 1)$$
$$\leqslant 2\beta r^2 [2(e^{v_1} - 1) - v(e^v + 1)]$$

which is negative for $v_1 > 0$ since all the coefficients in the Taylor's series (in powers of v_1) are nonpositive. Hence, $g(v_1) = 0 \Rightarrow g'(v_1) < 0$, so g is zero for at most one value of v_1 and, in case such a v_1 exists, g is positive for smaller values of v_1, and negative for larger values. In view of (9.2.7) and (9.2.8), the maximum of R for $\theta_1 < 0$ will occur at b if and only if

$$b = \frac{2\beta r^2 + 2qrv_1 - v_1^2}{2rv_1} \tag{9.2.10}$$

and

$$2\beta r^2 v_1 + 2qrv_1^2 - v_1^3 = (2\beta r^2 + v_1^2)(e^{v_1} - 1). \tag{9.2.11}$$

In view of (9.2.5), the requirement that $R(a; e^{-\beta t}; \tau) = R(b; e^{-\beta t}; \tau)$ becomes

$$a \exp(r\{a - q + \sqrt{[2\beta + (a-q)^2]}\})$$
$$= -b(1 - \exp(r\{b - q + \sqrt{[2\beta + (b-q)^2]}\})). \tag{9.2.12}$$

According to the preceding discussion we wish to find a, b, p, q, r, v_1, and C satisfying (9.2.1), (9.2.2), (9.2.3), (9.2.6), (9.2.10), (9.2.11), and (9.2.12). In addition, these inequalities must hold: $r < 0$, $0 < C < p < 1$, $b < 0 < a$, $0 < v_1 < -r\{[2\beta + q^2]^{1/2} - q\}$. Verifying that these conditions can be satisfied simultaneously will show that a minimax strategy is to use arm 1 until $Y_1(t) = qt + r$ and thereafter use arm 2, and that a 'least favorable distribution', a maximin strategy for nature, is given by $\theta_1 = a$ with probability p and $\theta_1 = b$ with probability $1 - p$.

For any $b < 0 < a$, C as given by (9.2.3) lies in $(0, 1)$. For $C \in (0, 1)$, $r < 0$ and $b < 0 < a$, (9.2.2) determines a unique $p \in (C, 1)$. Accordingly, we may focus our attention on finding a, b, q, r, and v_1 satisfying (9.2.1), (9.2.6), (9.2.10), (9.2.11), and (9.2.12) together with the inequalities $b < 0 < a$ and $0 < v_1 < -r\{[2\beta + q^2]^{1/2} - q\}$.

In view of (9.2.1), equation (9.2.6) can be rewritten:

$$r\{[2\beta + (b-q)^2]^{1/2} + b - q\}$$
$$+ r^2 a\{[2\beta + (b-q)^2]^{1/2} - b + q\} = r(b-q). \qquad (9.2.13)$$

Setting $b = \theta_1$ in (9.2.2) we obtain

$$v_1 = -r\{[2\beta + (b-q)^2]^{1/2} + b - q\}.$$

Since

$$[2\beta + (b-q)^2]^{1/2} - b + q = 2\beta\{[2\beta + (b-q)^2]^{1/2} + b - q\}^{-1}$$
$$= -2\beta r v_1^{-1},$$

(9.2.13) can be replaced by

$$-v_1 - 2\beta r^3 v_1^{-1} a = r(b-q). \qquad (9.2.14)$$

Equations (9.2.1), (9.2.10), and (9.2.14) may be regarded as three linear equations in the unknowns a, b, and q. With k denoting $2\beta v_1^{-2} r^2$, the solution of this system of linear equations is

$$b = \sqrt{2\beta}\,(1 + k + 2kv_1 - 2k^2 v_1)/(2k^{3/2}v_1), \qquad (9.2.15)$$

$$a = \sqrt{2\beta}\,(1 + k)/(2k^{3/2}v_1), \qquad (9.2.16)$$

$$q = \sqrt{2\beta}\,(1 + k + kv_1 - k^2 v_1)/(2k^{3/2}v_1). \qquad (9.2.17)$$

Substitution for q in (9.2.11) and replacing r by $-v_1(k/2\beta)^{1/2}$ gives a quadratic equation for k. It has one positive solution, namely

$$k = \frac{2v_1 + e^{v_1} + [(2v_1 + e^{v_1})^2 + 4(2v_1 + 1 - e^{v_1})]^{1/2}}{2(2v_1 + 1 - e^{v_1})}, \qquad (9.2.18)$$

if $2v_1 + 1 - e^{v_1} > 0$, and no positive solutions otherwise. Accordingly, we want a positive v_1 small enough for $2v_1 + 1 - e^{v_1}$ to be positive.

Substitution for a, b, q, and $r = -v_1(k/2\beta)^{1/2}$ in (9.2.12) gives an equation in the single variable v_1. The intermediate value theorem applied to that equation gives the existence of a solution v_1. That solution determines k, $r < 0$, q, a, and b via (9.2.18), the relation $r = -v_1(k/2\beta)^{1/2}$, (9.2.17), (9.2.16), and (9.2.15). Moreover, numerical answers are easy to obtain: $v_1 \approx 0.1292$, $k \approx 12.27$, $r \approx -0.3199\beta^{-1/2}$, $q \approx -0.5845\beta^{1/2}$, $a \approx 1.690\beta^{1/2}$, $b \approx -2.859\beta^{1/2}$. The reader can easily check that the desired inequalities are satisfied. From (9.2.3) and (9.2.2) we obtain $C \approx 0.1212$ and $p \approx 0.3715$.

The minimax strategy is to observe arm 1 until the process it generates meets the line given by

$$-0.3199\beta^{1/2} - 0.5845\beta^{1/2}t,$$

and then switch to arm 2. The maximin strategy has θ_1 equal to $1.690\beta^{1/2}$ with probability 0.3715, and θ_1 equal to $-2.859\beta^{1/2}$ otherwise.

If $\theta_1 > -0.5845\beta^{1/2}$, then there is positive probability that the decision maker using the minimax strategy will never use arm 2, even if $\theta_1 < 0$.

References

Bather, J. A. (1983) Personal communication.

Bather, J. A. and Simons, G. (1985) The minimax risk for clinical trials. *J.R. Statist. Soc.* B (to appear).

Feller, W. (1968) *An Introduction to Probability Theory and Its Applications*, Vol. I, 3rd edn, Wiley, New York.

Fristedt, B. (1974) Sample functions of stochastic processes with stationary, independent increments. *Advances in Probability*, Vol. 3 (eds P. Ney and S. Port), pp. 241–396, Marcel-Dekker, New York.

Parthasarathy, T. and Raghavan, T. E. S. (1971) *Some Topics in Two-Person Games*, American Elsevier, New York.

Vogel, W. (1960a) A sequential design for the two-armed bandit. *Ann. Math. Statist.* 31: 430–443.

Vogel, W. (1960b) An asymptotic minimax theorem for the two-armed bandit problem. *Ann. Math. Statist.* 31: 444–451.

Annotated bibliography

The papers included here deal with 'bandit problems' as described in this monograph. The essential criterion for inclusion is that the problem addressed, directly or implicitly, is one of maximizing a sum (or an integral in the continuous case), perhaps with discounting. The discount factors can be random but they cannot depend on the observations on the arms. This means that we have excluded the very large literature of 'ranking and selection' problems in which various allocation rules are proposed and data-dependent stopping rules are used.

On the other hand, we include papers that consider particular allocation rules and allow a shift in allocation to a single arm at a stage that is determined by the data, as long as the results of those stages occurring later than the shift-point are considered in the objective function. The papers in this category are listed under Colton (1963); 'See Colton (1963)' is indicated under the 'related references' for such a paper.

We have included another set of papers that does not fit in perfectly with our formulation. These are the so-called 'finite-memory' and 'finite-state' bandit problems. All are listed under Robbins (1956); and 'See Robbins (1956)' is indicated under the 'related references' for such a paper.

We make no claim of thoroughness either in the papers included or in our descriptions of the contents of the papers. Entries without annotations were not available to us. For consistency we use our own terminology and notation throughout. We point out what we perceive to be errors in the original papers, and occasionally say why we disagree with an approach or a viewpoint of the authors. Many of our criticisms apply to a number of papers even though they may be given only once.

Not all of the papers listed make contributions. A surprising number of papers duplicate results proven earlier; ofttimes the earlier proof is more elegant. Certain early papers – notably Thompson (1933), Robbins (1952), and Bradt, Johnson, and Karlin (1956) – continue to be important for their creativity and elegance. Many later papers have not made full use of the ideas in these early papers.

Abdel Hamid, A. R. (1981) Randomized sequential decision rules. D. Phil. thesis, Univ. of Sussex, England.

Anscombe, F. J. (1963) Sequential medical trials. *J. Amer. Statist. Assoc.* **58**: 365–383.

The setting is essentially the same as that of Colton (1963). Anscombe considers the sequential stopping problem during the experimental phase from a Bayesian point of view. He considers both fixed n and a fixed length for the terminal phase; only the first is consistent with our bandit approach. Stopping boundaries are provided in terms of the difference between sample means on the two arms. These have been considered by a number of authors since (e.g., Lai, Robbins, and Siegmund, 1983) and have been shown to be asymptotically optimal.

Anscombe notes that assessing the horizon n is critical and that his results are quite sensitive to n (see also Upton and Lee, 1976). While we agree, we have three reasons for being somewhat less concerned than Anscombe. Firstly, information concerning the horizon can be updated during the course of the trial. So the boundary can be changed as the first phase develops, or a second 'first phase' can be started during the 'terminal phase'! A related approach allows multiple phases in a 'two-phase look-ahead' fashion as suggested by Cornfield, Halperin, and Greenhouse (1969). Secondly, when, as is likely to be dictated in practice by logistical limitations, the length of the first phase is fixed in advance, then this length varies only as $\sqrt{n}$ (Colton, 1963; Canner 1970; and others). Finally, an unknown horizon can be reflected in a prior distribution on n in the same way that a distribution is placed on the other unknown parameters in the problem (cf. our Chapter 3). (Witmer (1983) considers a special case in which n can be replaced by its expectation.) Indeed, n may depend on these parameters – at least to an extent.

The importance of Anscombe's paper greatly transcends the eminently reasonable study described in the first paragraph above. Anscombe criticizes a book by P. Armitage [1961, *Sequential Medical Trials*, Blackwell Scientific Publications, Oxford, England] in a most elegant fashion. Anscombe's paper should be studied carefully by all biostatisticians. The sequential procedures proposed by Armitage pay heed to the possibility of obtaining data that were not obtained and to decisions that might have been made but were not. The result is that a sequential trial can actually involve a greater expected number of patients than a trial with a fixed sample size! Anscombe argues that decisions (to stop sampling, for example) should be made on the basis of the data at hand and considering the costs due to ineffective treatment. Unfortunately, it is Armitage's work and not Anscombe's that has influenced current biostatistical practice. The designs of modern clinical trials are dictated by considerations of statistical power which allow 'proper' inferences. However, classical statistical tests do not attempt to balance the welfare of the present patient against the value of eventually obtaining a 'sound' statistical conclusion. While explicit consideration of the patient horizon may be objectionable on ethical grounds, at least it makes clear the extent to which current patients are being sacrificed for future patients. Ethical dilemmas are not solved by avoiding them.

Related references: See Colton (1963).

Bather, J. A. (1977) A simple bandit problem. *Markov Decision Theory* (eds H. C. Tijms and J. Wessels), pp. 213–220, Mathematisch Centrum, Amsterdam.

There are two Bernoulli arms: $\theta_2 = 1/2$ and θ_1 is known to be either a or $1 - a$ where $a \in (1/2, 1)$. Discounting is uniform. Bather focuses on fixed strategies as the horizon approaches ∞. He shows that there is no nonrandomized stationary strategy for which the proportion of successes approaches $\max\{\theta_1, 1/2\}$ as the horizon approaches ∞ unless the prior probability of $\{\theta_1 = a\}$ is 0 or 1. He gives a class of stationary randomized strategies that do have this characteristic.

Related references: Our Example 5.4.1; Bather (1980).

Bather, J. A. (1980) Randomized allocation of treatments in sequential trials. *Adv. in Appl. Probab.* **12**: 174–182.

There are k independent Bernoulli arms with uniform discounting. Bather considers a family of strategies, one of which is the following: arm i is selected at stage m if i is such that $[s_i + Y_i(m)]/[s_i + f_i]$ is maximized, where s_i and f_i are the numbers of successes and failures on arm i in the first $m - 1$ stages, and $\{Y_i(m): 1 \leqslant i \leqslant k, m \geqslant 1\}$ is a family of independent exponentially distributed random variables.

Suppose the horizon approaches ∞. Bather shows that the proportion of successes approaches the maximum of the k Bernoulli parameters. Percus and Percus (1984) consider a similar strategy in which a large constant plays the role of $Y_i(m)$; in the latter circumstance there is a positive probability of selecting an inferior arm at every stage (except for some trivial cases).

Related references: Bather (1977), Fox (1974), Percus and Percus (1984), Robbins (1952), Thompson (1933, 1935).

Bather, J. A. (1981) Randomized allocation of treatments in sequential experiments (with discussion). *J. R. Statist. Soc. B* **43**: 265–292.

This is a comprehensive article that discusses various contributions of the papers listed below as 'related references', presents some new ideas, and evaluates a large number of procedures. Also notable is the lengthy and varied discussion included with this article.

Bather's main objective is to investigate strategies for k-armed Bernoulli bandits that have good asymptotic characteristics but that also perform well in the finite-horizon case. He recommends a randomized strategy (investigated further in Bather, 1984) that selects each arm sufficiently often to be asymptotically optimal; moreover, as calculations for $k = 2$ show, it performs reasonably well for moderate n over a variety of (θ_1, θ_2) values. (We prefer a Bayesian approach which averages over the uncertainty in (θ_1, θ_2) as well as n, and gives rise to a nonrandomized strategy. However, we admit that randomized strategies have a certain appeal to practitioners.)

Bather uses diffusion approximations as an aid to understanding the asymptotic behaviour of randomized strategies (cf. Bather, 1983a).

Related references: Bather (1980), Bellman (1956), Berry (1978), Berry and Fristedt (1979), Bradt, Johnson, and Karlin (1956), Fabius and van Zwet (1970), Feldman (1962), Fox (1974), Gittins (1979), Glazebrook (1980), Kelly (1981), Poloniecki (1978), Robbins (1952), Rodman (1978), Vogel (1960a, 1960b), Wahrenberger, Antle, and Klimko (1977), Whittle (1980), Zelen (1969).

Bather, J. A. (1983a) The minimax risk for the two-armed bandit problem. *Mathematical Learning Models – Theory and Algorithms* (eds U. Herkenrath, D. Kalin and W. Vogel), pp. 1–11, Springer-Verlag, New York.

It is shown that c_3 in our Corollary 9.1.4 may be chosen to be 0.305, provided that n is sufficiently large. To show this, Bather introduces a modified bandit in which two arms must be selected at each stage. One of the two arms selected (which one is designated by the decision maker) does *not* contribute immediately to the worth of the strategy, but does give information that may be useful in the future. The other arm selected contributes to the worth but is *not* observed. When the decision maker is constrained to select the same arm for observation and for information, then of course the modified bandit is equivalent to the original bandit. So the *regret* (see Section 9.1) is no larger in the modified problem than in the original problem.

Another technique that Bather uses is that of approximating Bernoulli processes with diffusion processes.

Related references: Our Chapter 9; Bather (1981), Bather and Simons (1985), Fabius and van Zwet (1970), Vogel (1960a, 1960b).

Bather J. A. (1983b) Optimal stopping of Brownian motion: a comparison technique. *Recent Advances in Statistics; Papers in Honor of Herman Chernoff on his Sixtieth Birthday* (eds M. H. Rizvi, J. S. Rustagi and D. Siegmund), pp. 19–49, Academic Press, New York.

Time is continuous and the discount function is $e^{-\beta t}$ (see our Chapter 8). Bather considers an arm that is the sum of Brownian motion and a linear drift whose value at $t = 1$ is normally distributed with mean μ and variance ρ^2. Bather shows that the break-even value of such an arm is $\Lambda((\mu, \rho), \beta) = \mu + \beta^{1/2}\psi(\rho/\beta)$, where ψ is a function satisfying: $\psi(v)/v \to 2^{-1/2}$ as $v \to 0$; $\psi(v)/v \leqslant 2^{-1/2}$ for $v > 0$; $\psi(v)/v^{1/2} \uparrow \infty$ as $v \uparrow \infty$. Bather gives some evidence which suggests that $\psi(v)/(2v \log v)^{1/2}$ approaches 1 as $v \to \infty$.

Bather addresses this only as a stopping problem, but one may be tempted to apply his result in k-armed bandits as in Gittins and Jones (1974). However, as far as we know, a general continuous-time version of the Gittins–Jones theorem has not been developed. So we do not know whether the value of Λ for each of the available arms determines an optimal strategy for a multi-armed bandit.

Related references: Our Chapter 8; Breakwell and Chernoff (1964), Chernoff and Ray (1965), Karatzas (1984).

Bather, J. A. (1984) Towards a more rational allocation of treatments in medical trials. Unpublished.

A strategy for two-armed Bernoulli bandits proposed in Bather (1981) is investigated and compared with two other strategies outside a bandit setting. Namely, a conclusion concerning the sign of $\theta_1 - \theta_2$ is desired without exposing an excessive number of patients to inferior treatment.

This paper is included here for its similarity with various listed papers.
Related references: See Colton (1963).

Bather, J. A. and Simons, G. (1985) The minimax risk for clinical trials. *J. R. Statist. Soc. B* (to appear).

Numerical evidence is given supporting the conjecture that c_3 and c_4 in our Corollary 9.1.4 may be chosen to be 0.371 and 0.372, respectively, provided that n is sufficiently large. Evidence is also given that the following strategy is very close to minimax for a Bernoulli two-armed bandit with n-horizon uniform discounting: calculate the nearest integer K_n to $0.292\,n^{1/2}$; use the arms equally until the number of successes on one arm exceeds the number on the other arm by K_n; thereafter use the arm on which the greater number of successes has been observed.

The methods of Vogel (1960a, 1960b) are explained, used, and extended. In particular, the process of maximizing the risk over (θ_1, θ_2) in the unit square and then minimizing over K_n is described in detail.

Related references: Our Section 9.1; Bather (1983a), Fabius and van Zwet (1970); and see Colton (1963).

Beckmann, M. J. (1973) Der diskontierte Bandit. *OR-Verfahren* **XVIII**: 9–18.

Begg, C. B. and Mehta, C. R. (1979) Sequential analysis of comparative clinical trials. *Biometrika* **66**: 97–103.

The problem as posed by Anscombe (1963) is reconsidered. The authors consider the rule which terminates the experimental phase when no fixed size continuation of the phase is an improvement on stopping. Chernoff and Petkau (1981) consider continuous-time versions of this rule and the rule of Anscombe (1963) and conclude that the former tends to stop much too soon.

Related references: See Colton (1963).

Bellman, R. (1956) A problem in the sequential design of experiments. *Sankhyā A* **16**: 221–229.

There are two Bernoulli arms: θ_1 has distribution F and θ_2 is known (cf. our Chapter 5). Discounting is geometric: $\mathbf{A} = (1, \alpha, \alpha^2, \ldots)$. Bellman shows that the dynamic programming equations (our (2.5.2)) have a unique solution that is the limit of a recursion (see our Chapter 2 for a more general treatment). He shows the existence of a break-even value of θ_2, $\Lambda(F, \mathbf{A})$, such that arm 1 is optimal if and only if $\Lambda(F, \mathbf{A}) \geqslant \theta_2$ (our Corollary 5.1.2 and Theorem 5.3.1 give two different generalizations). He also shows that $\Lambda(\sigma F, \mathbf{A}) > \Lambda(F, \mathbf{A}) > \Lambda(\varphi F, \mathbf{A})$; cf. our Corollary 5.3.3 (we note that the strict inequalities mean that Bellman is implicitly assuming that F is not a one-point distribution).

Related references: Our Chapters 2, 5, and 6; Berry and Fristedt (1979), Gittins and Jones (1974), Gittins and Jones (1979), Kakigi (1983), Yakowitz (1969).

Bellman, R. (1961) *Adaptive Control Processes: A Guided Tour*, Princeton University Press, Princeton, New Jersey.

The dynamic programming equations of Bellman (1956) are given in Section 16.15; they represent part of a more general discussion.

Related reference: Bellman (1956).

Benzing, H., Hinderer, K. and Kolonko, M. (1984) On the k-armed Bernoulli bandit: Monotonicity of the total reward under an arbitrary prior distribution. *Math. Operationsforsch. Statist. Ser. Optim.*, **15**: 583–595.

There are k Bernoulli arms with n-horizon uniform discounting. Using the result indicated in our discussion of Hengartner, Kalin, and Theodorescu (1981), the authors allow θ_1 and θ_2 to be dependent and determine conditions on G which satisfy the hypothesis of the result of that paper.

Related references: Hengartner, Kalin, and Theodorescu (1981), Kolonko and Benzing (1985).

Benzing, H. and Kolonko, M. (1984) Structured policies for a sequential design problem with general distributions. Unpublished.

There are two arms with one arm known. The unknown arm has an arbitrary distribution. Discounting is uniform. The authors prove special cases of results in Berry and Fristedt (1979); see also our Chapter 5.

Related references: Our Chapter 5; Berry and Fristedt (1979), Bradt, Johnson, and Karlin (1956), Kolonko and Benzing (1983, 1985).

Berry, D. A. (1972) A Bernoulli two-armed bandit. *Ann. Math. Statist.* **43**: 871–897.

There are two independent Bernoulli arms. The discount sequence is the

n-horizon uniform. The stay-with-a-winner rule (our Theorem 4.3.8) is proved in this setting. This paper contains some of the results given in our Chapter 7, but the methods are somewhat different.

Related references: Our Sections 4.2 and 4.3 and Chapter 7; Berry (1978), Bradt, Johnson and Karlin (1956).

Berry, D. A. (1978) Modified two-armed bandit strategies for certain clinical trials. *J. Amer. Statist. Assoc.* **73**: 339–345.

There are two Bernoulli arms and discounting is finite-horizon uniform. Easy-to-use strategies are suggested for clinical trials. These strategies are optimal when there are two Bernoulli arms with two-point prior distributions, as in our Corollary 4.3.10. These strategies are adapted to beta distributions and their worths are compared with optimal worths for various distributions and horizons.

Related references: Our Chapter 7; Bather (1981), Berry (1972), Percus and Percus (1984), Thompson (1933, 1935).

Berry, D. A. (1983) Bandit problems with random discounting. *Mathematical Learning Models – Theory and Algorithms* (eds U. Herkenrath, D. Kalin and W. Vogel), pp. 12–25, Springer-Verlag, New York.

Many-armed bandits with random discount sequences are considered. The main results in this paper are given in our Chapter 3.

Related references: Our Chapter 3; Berry and Fristedt (1979).

Berry, D. A. (1985). One- and two-armed bandit problems. *Encyclopedia of Statistical Sciences*, Vol. VI (eds by S. Kotz and N. L. Johnson), Wiley, New York (to appear).

This is a survey article. Its perspective is similar to that of this monograph.

Berry, D. A. and Fristedt, B. (1979). Bernoulli one-armed bandits – Arbitrary discount sequences. *Ann. Statist.* **7**: 1086–1105.

Bernoulli bandits with one unknown arm are considered when discounting is arbitrary. This paper provides the basis for parts of Chapter 5 of this monograph. In particular, it gives a necessary and sufficient condition for the bandit to be a stopping problem for all θ_2 and F_1.

Related references: Our Chapter 5; Bellman (1956), Berry (1983), Bradt, Johnson and Karlin (1956), Gittins (1979), Gittins and Jones (1974).

Berry, D. A. and Fristedt, B. (1980a) Two-armed bandits with a goal, I. One arm known. *Adv. in Appl. Probab.* **12**: 775–798.

The problem considered is an adaptation of a classical bandit problem. One of two arms is selected at each of a possibly infinite number of stages and learning takes place as the results are observed. But now the decision maker strives to turn a current fortune into a given larger fortune (the goal) before it becomes a smaller fortune (ruin). Observing an arm either increases or decreases the fortune by one unit. One of the arms has a known probability of increase and that of the other is unknown. Optimal strategies are shown to exist in great generality and are characterized for some special prior distributions on the arms. In some settings it is shown that the 'risky' arm – the unknown arm – should be selected when the decision maker is near ruin, but not otherwise.

Related references: Berry and Fristedt (1980b), Gittins (1983), Vogel (1961b).

Berry, D. A. and Fristedt, B. (1980b) Two-armed bandits with a goal, II. Dependent arms. *Adv. in Appl. Probab.* **12**: 958–971.

The setting is similar to that of Berry and Fristedt (1980a). Now both probabilities of increase are known, but not which goes with arm 1. Optimal strategies are shown to exist. Myopic strategies (cf. our Section 1.2 and Corollary 4.3.10) are shown to be optimal for various pairs of probabilities but not optimal in general.

Related references: Berry and Fristedt (1980a), Gittins (1983), Vogel (1961b).

Berry, D. A. and Fristedt, B. (1983) Maximizing the length of a success run for many-armed bandits. *Stochastic Process. Appl.* **15**: 317–325.

The same problem is considered as in Berry and Viscusi (1981). Now the discount sequence is arbitrary. It is shown that there is always an optimal strategy that uses a single arm indefinitely whenever the arms are independent and the discount sequence is *superregular* (see our Proposition 5.2.1). Optimal strategies may or may not have this characteristic when the discount sequence is not superregular.

Related references: Berry and Viscusi (1981), Viscusi (1979a, 1979b).

Berry, D. A. and Pearson, L. (1984) Optimal designs for two-stage clinical trials with dichotomous responses. Univ. of Minnesota Tech. Rep. (To be in *Statist. in Med.*).

There are two Bernoulli arms with parameters θ_1 and θ_2. Discounting is n-horizon uniform. Two-phase designs are considered in which the first is an experimental or learning phase and a single arm is selected in the second. Allocations in the first phase must be set in advance. The length of the first phase is either given or to be optimized as a function of n and the prior information (cf. Pearson, 1980; Witmer, 1983). Two forms of prior distributions are considered: (i) θ_2 is known while the distribution of θ_1 is arbitrary, and (ii) the values of θ_1 and θ_2 are known but not which is θ_1 (cf. Feldman, 1962). Graphs are provided to show optimal first-phase lengths and maximal success proportions for various n.

When θ_2 is known and θ_1 has a uniform distribution on $(0, 1)$, the authors show that the optimal first-phase length is approximately $[(n+1)(1/\theta_2 - 1)]^{1/2} - 2$. This is similar to Canner's (1970) result which applies for both θ_1 and θ_2 uniform and equal first-phase allocations.

Related references: See Colton (1963).

Berry, D. A. and Viscusi, W. K. (1981) Bernoulli two-armed bandits with geometric termination. *Stochastic Process. Appl.* **11**: 35–45.

The standard Bernoulli bandit problem is modified so that all payoff ceases with the first failure. The discount sequence is either $\mathbf{A} = (0, \ldots, 0, 1, 0, \ldots)$ or geometric (as in our Chapter 6). It is shown that the search for an optimal strategy can be restricted without loss to 'single-arm' strategies – those which stay indefinitely with the arm selected initially. An example shows that the optimal strategies do not have this form for all discount sequences.

Related references: Berry and Fristedt (1983), Viscusi (1979a, 1979b).

Bradt, R. N., Johnson, S. M. and Karlin, S. (1956) On sequential designs for maximizing the sum of n observations. *Ann. Math. Statist.* **27**: 1060–1074.

There are two Bernoulli arms. Discounting is n-horizon uniform. The most important contributions of this paper are for the case in which one arm is known. In particular, a number of the results in our Chapter 5 are generalizations of results in this paper. One such result is worth mentioning here only because it has not been appreciated by a number of later authors: Lemma 4.1 says that if the known arm is optimal at any stage then it is also optimal thereafter (see the discussion preceding the proof of our Theorem 5.2.2).

The authors touch on a number of issues when both arms are unknown. They completely characterize optimal strategies when $n = 2$ (cf. our Section 7.2). They give an example (our Example 2.4.1) in which 'stay-with-a-loser/switch-from-a-winner' is optimal. Also, our Example 2.4.5 is similar to one of their examples. They indicate that it is possible for a one-step look-ahead strategy (i.e., a myopic strategy) to be better than a two-step look-ahead strategy. And they discuss the problem (they call it the 'classical' problem) that was eventually solved by Feldman (1962).

This paper is both innovative and comprehensive. Almost everyone writing about bandit problems refers to this paper, but not all have read it! Results proven by Bradt, Johnson, and Karlin in most elegant ways are sometimes credited to other authors who were both very late and much less elegant.

Related references: Our Chapter 5 and Section 7.2; Berry and Fristedt (1979), Feldman (1962).

Brand, H., Sakoda, J. M. and Woods, P. J. (1957) Effects of a random versus pattern instructional set in a contingent partial reinforcement situation. *Psychol. Rep.* **3**: 473–479.

The hypothesis being tested is that subjects who select the more successful arm exclusively are those convinced of randomness while those who mix selections feel that there is a pattern in the responses. We note that if the subjects were actually using optimal strategies (assuming randomness) they would not select the more successful arm exclusively but would attempt to balance information – unless available information is sufficiently one-sided. So subjects trying to anticipate patterns may be confused with subjects who use good strategies.

Some subjects were told that there was a pattern concerning which of the two arms would result in success; others were told that successes were random. (The latter were given instructions similar to those described in Brand, Woods, and Sakoda (1956) even though $\theta_1 + \theta_2$ was not always equal to 1.) The experiment was actually conducted as independent Bernoulli processes using electronic devices. The authors note a difference in behaviour between the two groups only when θ_1 or θ_2 is 1.

As a general matter, it seems to us that the problem of how the subject perceives the producing mechanism is greatly alleviated by using physical rather than electronic devices. Physical devices would seem to be beneficial to any study since conclusions would not be muddled by such questions. Thus, for example, if the subject is given a choice between two urns, each containing balls of two colours one of which is preferable, then little explanation is required to make it clear that the two arms (urns) are independent. On the other hand, if one urn is used and the subject is asked to predict the colour of the ball drawn then it is transparent that $\theta_1 = 1 - \theta_2$ even though both θ_1 and θ_2 may be unknown. Of course, sampling with replacement is necessary to preserve conditional independence (given the θ_i), but sampling without replacement would also give rise to interesting experiments.

Related references: Brand, Woods, and Sakoda (1956), Bush and Mosteller (1955), Estes (1950), Horowitz (1973), Murray (1971), Woods (1959).

Brand, H., Woods, P. J. and Sakoda, J. M. (1956) Anticipation of reward as a function of partial reinforcement. *J. Exp. Psychol.* **52**: 18–22.

Results of an experiment are reported. Subjects were allowed to choose between two arms (electronic devices) in a series of trials and were told to try to get as many successes as possible. Various pairs (θ_1, θ_2) were assigned by the experimenter. While $\theta_1 + \theta_2$ was not always equal to 1, the subjects were told, 'On a particular trial, then, one level will be correct and one incorrect.' We do not see what is to be gained from giving the subjects this misinformation.

Related references: Brand, Sakoda, and Woods (1957), Bush and Mosteller (1955), Estes (1950), Horowitz (1973), Murray (1971), Woods (1959).

Breakwell, J. and Chernoff, H. (1964) Sequential tests for the mean of a normal distribution II (large t). *Ann. Math. Statist.* **35**: 162–173.

This paper is not directly related to bandit problems. The authors consider free boundary problems for partial differential equations. In particular, they address the question: Does a solution of the free boundary problem necessarily solve the original probabilistic problem? This question is important for Chernoff and Ray (1965) (cf. our Section 8.2).

Related references: Our Section 8.2; Chernoff and Ray (1965).

Bush, R. R. and Mosteller, F. (1955) *Stochastic Models for Learning*, Wiley, New York.

Chapter 13 deals with experiments involving sequential choices by subjects in various Bernoulli two-armed bandit settings; three-armed bandits are also discussed. An *ad hoc* linear model is proposed in which the probability that a subject selects an arm increases or decreases according as the arm was successful or not. The parameters in the model are estimated for various data and the model is found to fit quite well on an overall basis; cf. Cane (1962).

Since averaging results of all players combines 'good' and 'bad' strategies, both are obscured. Of greater interest to us is the possibility that *individual* players select optimally for some prior distribution. (Horowitz (1973) addresses this issue.) For example, if a player switches on a success then there is no prior distribution for which the player is behaving optimally (unless the arms are dependent or observations on the same arm are conditionally dependent).

Related references: Brand, Sakoda, and Woods (1957), Brand, Woods, and Sakoda (1956), Cane (1962), Estes (1950), Horowitz (1973), Murray (1971), Schmalansee (1975).

Cane, V. R. (1962) Learning and inference (with discussion). *J. R. Statist. Soc. A* **125**: 183–209.

Two-armed bandit experiments involving humans and rats are described. Cane discusses models that hypothesize recursions for the probability that a subject selects arm 1, say. She cites studies which she claims invalidate linear models as discussed by Bush and Mosteller (1955).

Related references: Bush and Mosteller (1955), Schmalansee (1975).

Canner, P. L. (1970) Selecting one of two treatments when the responses are dichotomous. *J. Amer. Statist. Assoc.* **65**: 293–306.

There are two independent Bernoulli arms with parameters θ_1 and θ_2 having beta distributions (though a minimax approach is also considered). Discounting is n-horizon uniform. The two arms must be used equally in the first $2r$ stages (cf. Berry and Pearson (1984) who allow unequal allocation). Subsequent to this experimental phase, the arm with greater current mean (probability of success) is used exclusively. The only unspecified part of the strategy is the choice of r. Canner gives tables of optimal values of r. When θ_1

and θ_2 are uniformly distributed he shows that r is approximately $\sqrt{(n/2 + 1)}$ $- 1$ (cf. Berry and Pearson, 1984; Kelley 1976).

Related references: See Colton (1963).

Chernoff, H. (1967) Sequential models for clinical trials. *Fifth Berkeley Symp. of Math. Statist. and Probab.* **4**: 805–812.

There are two arms with arm 2 known. Discounting is n-horizon uniform. Chernoff approximates the Bernoulli case using a continuous-time approximation from Chernoff and Ray (1965); see also our Section 8.2. As an interesting sidelight, Chernoff calculates that the number of immediate failures that the decision maker should tolerate with arm 1 has order of magnitude $\log n$ when θ_1 is uniform on (0, 1) and θ_2 is fixed and not 0 or 1.

Chernoff touches briefly on various other issues: the possibility that both arms are unknown, geometric discounting, and the two-phase design of Anscombe (1963) and Colton (1963).

Related references: Our Chapter 5 and Section 8.2; Chernoff (1968, 1972), Chernoff and Ray (1965).

Chernoff, H. (1968) Optimal stochastic control. *Sankhyā A* **30**: 221–252.

A number of sequential problems are considered, including bandits. Time is continuous and discounting is uniform. A 'one-armed bandit' of Chernoff and Ray (1965) is discussed. Chernoff considers a two-armed bandit in which the arms are independent and are Brownian motions with unknown drifts and the same (known) variance. He conjectures that the solution of a one-armed bandit provides a bound for the solution of a two-armed bandit in the following sense. Consider modifying the two-armed bandit so that whenever arm 2 is selected it must be selected exclusively and indefinitely. If arm 2 is optimal with this restriction then it must be optimal without it. But the modified problem is equivalent to a one-armed bandit. The conjecture is still open although some variations of it have been resolved by Gittins (1975).

Related references: Chernoff (1972), Chernoff and Ray (1965), Gittins (1975).

Chernoff, H. (1972) *Sequential Analysis and Optimal Design*, SIAM, J. W. Arrowsmith, Bristol, England.

This monograph surveys a broad range of sequential statistical problems, including bandit problems (in Section 18). The discussion of bandits is similar to that in Chernoff (1968).

Related references: Chernoff (1968), Chernoff and Ray (1965), Gittins (1975).

Chernoff, H. (1975) Approaches in sequential design of experiments. In *A Survey of Statistical Design and Linear Models* (ed. J. N. Srivastava), pp. 67–90, North-Holland, Amsterdam.

This is a review article which discusses a variety of problems, including bandits.

Related reference: Feldman (1962).

Chernoff, H. and Petkau, A. J. (1976) An optimal stopping problem for sums of dichotomous random variables. *Ann. Probab.* **4**: 875–889.

An optimal stopping problem is discussed and shown to be related to the problem addressed by Petkau (1978), among others.

Related references: Chernoff and Petkau (1983, 1985), Petkau (1978).

Chernoff, H. and Petkau, A. J. (1981) Sequential medical trials involving paired data. *Biometrika* **68**: 119–132.

The problem posed by Anscombe (1963) is considered for continuous time. Anscombe's procedure is compared with various other procedures (including that of Begg and Mehta (1979)) and is shown to be well-approximated by the optimal continuous-time procedure.

Related references: See Colton (1963).

Chernoff, H. and Petkau, A. J. (1983) Numerical methods for Bayes sequential decision problems. Applied Mathematics and Statistics Tech. Rep. No. 83–26, Univ. of British Columbia, Canada.

The authors discuss computational issues for a number of statistical decision problems involving sums of random variables, including bandit problems. They use continuous-time approximations for discrete-time problems. The computational techniques go in the other direction: backwards induction is used for discrete-time processes to obtain approximations for continuous-time processes.

Information for constructing our Figures 8.2 and 8.3 was taken from Table 21 of this report.

Related references: Our Section 8.2; Chernoff (1967, 1968, 1972), Chernoff and Petkau (1976, 1981, 1985a, 1985b), Chernoff and Ray (1965), Petkau (1978).

Chernoff, H. and Petkau, A. J. (1985a) Sequential medical trials with ethical cost. *Proceedings of Kiefer-Neyman Conference* (to appear).

The setting is that of Anscombe (1963) and Colton (1963) except that an 'ethical cost' is now assessed for each pair of observations made in the experimental phase. This cost is proportional to the absolute difference in the current means of the two arms. The authors give optimal stopping boundaries for various proportionality factors.

Obviously, it is better to stop experimenting sooner with such a cost than without. A similar effect can be obtained by discounting future observations (or, approximately, by specifying a smaller horizon), a modification we prefer to the one made in this paper.

Related references: Chernoff and Petkau (1976, 1981, 1983a); and see Colton (1963).

Chernoff, H. and Petkau, A. J. (1985b) Numerical solutions for Bayes sequential decision problems. *SIAM J. Sci. Statist. Comput.* (to appear).

This is an abbreviated version of Chernoff and Petkau (1983).

Chernoff, H. and Ray, S. N. (1965). A Bayes sequential sampling inspection plan. *Ann. Math. Statist.* **36**: 1387–1407.

There are two arms with one arm known; time is continuous and discounting is uniform. The decision problem is to decide when to stop observing the unknown arm. The unknown arm is the sum of Brownian motion and a normally distributed linear drift. The boundary of the stopping region is the free boundary for a partial differential equation with certain boundary conditions. (Our Section 8.2 is based on this development.)

The authors use asymptotic methods to calculate the boundary. They discuss applying these solutions as approximations to certain discrete-time bandits.

Related references: Our Section 8.2; Bather (1983b), Breakwell and Chernoff (1964), Chernoff (1968, 1972), Chernoff and Petkau (1983, 1985b).

Chung, F. (1984) Contributions to the multiarmed bandit problem. Ph.D. thesis, Columbia Univ., USA.

Clayton, M. K. (1983) Bayes sequential sampling for choosing the better of two populations. Ph.D. thesis, Univ. of Minnesota, USA.

A number of problems are considered; these include the following bandit problem. There are two independent arms whose prior distributions are Dirichlet measures, as described in our Section 5.6. The discount sequence is $A = (0, 0, \ldots, 0, 1, 0, 0, \ldots)$, which means that one of the arms is to be selected at the termination of a number of sequential 'learning observations'.

Related references: Our Sections 3.6 and 5.6.

Clayton, M. K. and Berry, D. A. (1984) Bayesian nonparametric bandits. Statistics Tech. Rep. No. 427, Univ. of Minnesota, USA. (To be published.)

The substance of this paper is treated in our Section 5.6.

Related references: Our Chapter 5; Berry and Fristedt (1979), Bradt, Johnson, and Karlin (1956).

Colton, T. (1963) A model for selecting one of two medical treatments. *J. Amer. Statist. Assoc.* **58**: 388–400.

There are two independent normal arms and discounting is n-horizon uniform. An initial experimental phase is followed by a terminal phase in which the arm with greater mean after the initial phase is used exclusively. Allocations in the initial phase are equally divided between the two arms. Two settings are considered: (i) the length r of the first phase is fixed in advance, and (ii) the first phase is stopped as a function of the accumulating data which are assumed to be available immediately.

One might view the second as the more realistic approach and the first as an approximation; however, responses are delayed in most clinical trials and so sequential assignments may not be possible. An important set of problems and a fertile area of research is provided by the case in which information becomes available gradually after an arm is selected, or the response occurs only after a number of other selections must be made.

In the first problem Colton uses a 'local maximin' argument to show that the best value of r is $n/3$. We do not feel that this approach has merit here and, for reasons indicated in our discussion of Zelen (1969), we think that r/n should tend to 0 as $n \to \infty$. Using a Bayesian approach, Colton finds that r has order of magnitude $\sqrt{n}$. In the second problem, Colton studies procedures in which the first phase is stopped when the absolute difference in the sums on the two arms is greater than a constant times $\sqrt{n}$; this is similar to boundaries calculated by Anscombe (1963) and Vogel (1960a), but see Lai, Robbins, and Siegmund (1983).

The papers listed below all consider clinical trials in stages. However, unlike Colton, those listed with an asterisk restrict consideration to rules for which the ability to make classical statistical inferences is a primary consideration. Hence, the problem (though perhaps not the solution!) is on the fringe of what we call bandit problems. Those listed with two asterisks do not consider observations at stages beyond the experimental phase as part of the objective; these are listed in this bibliography only because there are a number of ways in which they are similar to other papers that are listed.

Related references: Anscombe (1963), Bather (1984)**, Bather and Simons (1985), Begg and Mehta (1979), Berry and Pearson (1984), Canner (1970), Chernoff and Petkau (1981), Colton (1965), Cornfield, Halperin, and Greenhouse (1969), Day (1969a, 1969b), Flehinger and Louis (1971)**, Flehinger and Louis (1972)**, Flehinger, Louis, Robbins, and Singer (1972)**, Fox (1974), Goto, Sugimura, and Asano (1971)*, Kelley (1976), Lai, Levin, Robbins, and Siegmund (1980)*, Lai, Robbins, and Siegmund (1983), Langenberg and Srinivasan (1981), Louis (1975)**, Meeter (1973), Oudin and Lellouch (1972), Pearson (1980), Petkau (1978), Robbins and Siegmund (1974)**, Upton and Lee (1976), Vogel (1960a, 1960b), Witmer (1983), Zelen (1969).

Colton, T. (1965) A two-stage model for selecting one of two treatments. *Biometrics* **21**: 169–180.

The setting is that of Colton (1963) except that there is a transitional phase between the experimental and terminal phases; really a three-phase design. Two simple procedures are compared.

Related references: See Colton (1963).

Cornfield, J., Halperin, M. and Greenhouse, S. W. (1969) An adaptive procedure for sequential clinical trials. *J. Amer. Statist. Assoc.* **64**: 759–770.

The setting is that of Anscombe (1963) and Colton (1963). The approach of these latter papers is generalized by allowing arbitrary prior mean and arbitrary imbalance in the first phase. They also consider applying their two-phase design repeatedly in an *r*-phase trial (a kind of 'two-phase look-ahead').

The authors derive an asymptotic expression for the optimal length of the first phase that generalizes that of Colton (1963); again it is of order $\sqrt{n}$.

Related references: See Colton (1963).

Cover, T. M. (1968) A note on the two-armed bandit problem with finite memory. *Inform. and Control* **12**: 371–377.

There are two Bernoulli arms. For a memory of size 2 and allowing the choice of arms to depend on the stage at which it is to be used, Cover argues that it is possible to achieve $\max\{\theta_1, \theta_2\}$ as the limiting proportion of success.

Related references: See Robbins (1956).

Cover, T. M. and Hellman, M. E. (1970) The two-armed bandit problem with time-invariant finite memory. *IEEE Trans. Inform. Theory* **16**: 185–195.

There are two arms and two distributions (arbitrary but known) – which distribution goes with which arm is not known. One of the two distributions is regarded as better than the other. The decision maker is allowed to remember only a finite-valued statistic. The authors find the supremum of the long-run proportion of observations from the better distribution; this supremum depends on the two distributions and on the size of the memory. They argue that there are no strategies which attain the supremum but exhibit a family whose members come arbitrarily close.

B. Chandrasekaran [1970, *IEEE Trans. Inform. Theory* **16**: 494–496; and 1971, *IEEE Trans. Inform. Theory* **17**: 104–105] argues that the strategies used actually require infinite memory. The authors respond in [1970, *IEEE Trans. Inform. Theory* **16**: 496–497].

Related references: Feldman (1962); and see Robbins (1956).

Cover, T. M. and Wagner, T. J. (1976) Topics in statistical pattern recognition. *Communication and Cybernetics* 10: *Digital Pattern Recognition* (ed. K. S. Fu), pp. 15–46, Springer-Verlag, Berlin.

This is a survey article of a variety of statistical areas of which finite-memory bandit problems constitute a small part.

Related references: See Robbins (1956).

Day, N. E. (1969a) Two-stage designs for clinical trials. *Biometrics* **25**: 111–118.

Extensions of Colton's (1965) three-phase designs are considered in the normal case. Various designs are compared numerically. Day proves in an appendix that equal allocation is optimal in the first phase when the arms are exchangeable *a priori*. While we believe the result, we note that his proof seems to apply as well when the arms are Bernoulli, a case in which Pearson (1980) shows that the result is false.

Related references: See Colton (1963).

Day, N. E. (1969b) A comparison of some sequential designs. *Biometrika* **56**: 301–311.

There are two independent normal arms with *n*-horizon discounting. Lucid descriptions and comparisons of three designs are given: (i) optimal (unrestricted) bandit strategies, (ii) pairwise allocation in the first of two phases as in Anscombe (1963) and Colton (1963), and (iii) myopic strategies.

Related references: See Colton (1963).

DeGroot, M. H. (1970) *Optimal Statistical Decisions*, McGraw-Hill, New York.

In Sections 14.5 to 14.7 DeGroot discusses two Bernoulli bandit problems with *n*-horizon uniform discounting. He solves an easy example in which one arm is known and gives the problem and development of Feldman (1962).

Related references: Our Chapter 5; Berry and Fristedt (1979), Bradt, Johnson, and Karlin (1956), Feldman (1962), Kelley (1974).

Dubins, L. E. and Savage, L. J. (1976) *Inequalities for Stochastic Processes: How to Gamble If You Must*, Dover, New York.

This book provides the fundamental theory for determining optimal decisions in a wide variety of problems. Chapters 2 and 3 constitute especially relevant background for bandit problems. In particular, the theorems concerning excessivity and thriftiness are useful for finding optimal strategies. Bandit problems in discrete time are considered in Sections 12.5 and 12.6 to show that they fall within the scope of the book. In a discussion similar to that which opens our Chapter 6, the authors argue that bandit problems with geometric discounting are more realistic than those with finite horizon.

These authors credit F. Mosteller with coining the term 'two-armed bandits'.

Related reference: Our Chapter 2.

Emrich, L. J. (1983) Optimal decision making using non-expert opinions. Ph.D. thesis, State Univ. of New York at Buffalo, USA.

The standard dynamic programming equations are developed and applied to a number of problems, including bandits. Finite horizon and geometric discounting are both considered. One of the problems Emrich considers is essentially the arm-acquiring bandit of Whittle (1981); the arms are independent and responses are normal.

Related references: Our Chapter 2; Fahrenholtz (1982), Gittins (1979), Whittle (1981, 1982).

Estes, W. K. (1950) Towards a statistical theory of learning. *Psychol. Rev.* **57**: 94–107.

This is an early paper in the literature of probabilistic learning models. However, it does not deal explicitly with bandit situations.

Fabius, J. and van Zwet, W. R. (1970) Some remarks on the two-armed bandit. *Ann. Math. Statist.* **41**: 1906–1916.

There are two Bernoulli arms which may be dependent. The discount sequence is n-horizon uniform. Randomized strategies are allowed. The authors give thorough demonstrations of the relationships among Bayes, admissible, and minimax strategies. Their results concerning Bayes strategies generalize the main result of Feldman (1962).

Related references: Our Corollary 4.3.10 and Section 9.1; Feldman (1962), Vogel (1960b, 1960c, 1964).

Fahrenholtz, S. K. (1982) Normal Bayesian two-armed bandits. Ph.D. thesis, Iowa State Univ., USA.

There are two normal arms and time is discrete. Discounting is n-horizon uniform. Two families of prior distributions are considered; variances are known in both. In one, the arms are independent with normally distributed means. In the second, the arms are dependent with means that sum to zero; the difference in means is normally distributed. Various properties are shown and bounds are calculated; some numerical calculations are given. Fahrenholtz shows that myopic strategies are optimal in the dependent case (cf. Rodman, 1978).

Related references: Our Section 2.1 and Chapter 7; Berry (1972), Feldman (1962), Rodman (1978).

Feldman, D. (1962) Contributions to the 'two-armed bandit' problem. *Ann. Math. Statist.* **33**: 847–856.

This important paper solved a problem that had eluded a number of statisticians: there are two Bernoulli arms with (θ_1, θ_2) equal to either (a, b) or

(b, a) and discounting is uniform. Feldman showed that myopic strategies are optimal: always select the arm with greater mean. This result has been generalized in a number of ways (see related references listed below).

In our view there is an unfortunate aspect of the notoriety associated with Feldman's problem. Feldman poses the problem in terms of maximizing the expected number of successes. He notes correctly that this is equivalent to minimizing the expected number of selections of the inferior arm. But while they are equivalent when the support of the prior distribution contains two points, this is not true generally. Many authors have focused on minimizing the number of selections of the inferior arm. For reasons indicated in our discussion of Percus and Percus (1984), we feel that this emphasis is ill-placed.

Related references: Our Corollary 4.3.10; Berry (1972), Bradt, Johnson, and Karlin (1956), DeGroot (1970), Fabius and van Zwet (1970), Kelley (1974), Rodman (1978), Vogel (1960c, 1964), Zaborskis (1976).

Fischer, J. (1979) Der diskontierte Einarmige Bandit. *Metrika* **26**: 195–204.

There are two Bernoulli arms; $\theta_2 = 1/2$ and θ_1 is uniform on $(0, 1)$. Discounting is geometric. Fischer finds that when the discount factor α is no larger than $3 - \sqrt{5} \approx 0.7639$, then an optimal strategy is as follows: select arm 1 until (if ever) the number of failures exceeds the number of successes, then switch permanently to arm 2.

We can improve this result by showing that $3 - \sqrt{5}$ may be replaced by the solution in $(1/2, 1)$ of $4\alpha - 4\alpha^4 = 1$, which is approximately 0.8968 and that this is the maximum possible α for which the indicated strategy is optimal. The proof can be accomplished using the fundamental theorem of gambling (Dubins and Savage, 1976, Theorem 2.12.1).

More generally, Fischer shows that for each discount factor α it is optimal to select arm 2 if the number of failures on arm 1 exceeds the number of successes by $1 + \text{int}[\alpha(2 - \alpha)/(4(1 - \alpha))]$. (This formula does not agree with Fischer's formula (30) which we believe contains a misprint; however, Fischer's subsequent table is correct.)

Results are also given in case $\theta_2 = 1/r$ where r is a known integer.

Related references: Our Chapter 5; Beckmann (1973), Glazebrook and Jones (1983).

Flehinger, B. J. and Louis, T. A. (1971) Sequential treatment allocation in clinical trials. *Biometrika* **58**: 419–426.

There are two arms whose responses are exponential variables with parameters θ_1 and θ_2. These are observed in real time and arm selections are separated by equal intervals of time. So the event in question (a patient's death, say) may be observed after other selections have been made, however information (of a positive nature) concerning an arm accrues even though the event has not yet been observed.

There are three possibilities: $\theta_1 = \theta_2$, $\theta_1 = k\theta_2$, $\theta_1 = \theta_2/k$. A family of procedures is proposed in which allocations are made according to death rates and numbers of deaths on the two arms; generally, an arm is selected if it is performing better, except that there is a tendency toward balance. The length of the trial is determined sequentially using a likelihood ratio test. (Presumably, the arm indicated to be better in the trial is used subsequent to the trial.) For various of the allocation schemes proposed, the numbers of selections of the inferior arm are compared with that of equal allocation using simulation. While the latter tends to result in shorter trials, it results in a greater number of selections of the inferior arm.

This paper differs from most of the papers in this bibliography in that it does not expressly consider the results of selections beyond the experimental phase. Put another way, there is no discount sequence that is assumed or implicit. The paper is included because of its similarity with other listed papers that do treat sum-maximization problems.

Related references: See Colton (1963).

Flehinger, B. J. and Louis, T. A. (1972) Sequential medical trials with data dependent treatment allocation. *Sixth Berkeley Symp. of Math. Statist. and Probab.* **4**: 43–51.

There are two arms with unknown means θ_1 and θ_2. The hypotheses to be tested are: $\theta_1 - \theta_2 = 0$, $\theta_1 - \theta_2 \leqq -\delta$, $\theta_1 - \theta_2 \geqq \delta$. A sequential generalized likelihood ratio test is used to decide when to stop sampling, and the allocation schemes that are suggested are designed to yield a small number of observations on the inferior arm.

This is not a bandit problem in our sense; the paper is included for the same reason that Flehinger and Louis (1971) is included.

Related references: See Colton (1963).

Flehinger, B. J., Louis, T. A., Robbins, H. and Singer, B. H. (1972) Reducing the number of inferior treatments in clinical trials. *Proc. Nat. Acad. Sci. USA* **69**: 2993–2994.

This note provides some mathematical explanation for the results shown by simulation in Flehinger and Louis (1971, 1972).

Related references: See Colton (1963).

Fox, B. L. (1974) Finite horizon behavior of policies for two-arm bandits. *J. Amer. Statist. Assoc.* **69**: 963–965.

There are two Bernoulli arms with parameters θ_1 and θ_2. Discounting is uniform and there is an experimental phase followed by a phase in which the evidently inferior arm is used, but only occasionally.

Monte Carlo methods are used to compare various first-phase strategies for various horizons. One class of strategies is similar to that proposed by

Zelen (1969). Another is similar to that considered by Vogel (1960a). Fox exhibits worths for various lengths of the first phase. The long-run success proportion is $\max\{\theta_1, \theta_2\}$ for strategies in both classes. Fox concludes from his simulations for various pairs (θ_1, θ_2) that 'play-the-winner/switch-from-a-loser' (Zelen, 1969) is not a very good strategy. (See our discussion of Percus and Percus (1984).) Wahrenberger, Antle, and Klimko (1977) conclude that none of the strategies considered by Fox are as good as Bayes strategies, even with an incorrect prior.

Related references: See Robbins (1956); see Colton (1963).

Furukawa, N. (1964) On some properties of an optimal strategy in the 'two-armed bandit' problem. *Mem. Fac. Sci. Kyushu Univ. A* **18**: 74–88.

Gait, P. A. (1972) Optimal allocation and control under uncertainty. Ph.D. thesis, Cambridge Univ., England.

Gittins, J. C. (1975) The two-armed bandit problem: variations on a conjecture by H. Chernoff. *Sankhyā A* **37**: 287–291.

Chernoff's (1968) conjecture is resolved in the negative when time is discrete and discounting is uniform. Two related two-armed bandits are compared for the discount sequence $(1, 1, 1, 0, 0, 0, \ldots)$. The first bandit has an unknown arm with prior distribution F_1 and a known arm 2 with mean $\Lambda(F_1, (1, 1, 1, 0, 0, \ldots))$. Thus either arm – in particular the known arm – is optimal. The second bandit has independent arms. The prior distribution for arm 1 is the same F_1 mentioned above. The prior for arm 2 is F_2 whose mean is $\Lambda(F_1, (1, 1, 1, 0, 0, \ldots))$. Arm 1 is uniquely optimal in the second example. (Cf. Theorem 6.2.1.)

Gittins observes that such an example is not possible in case the discount sequence is geometric. An easy proof is given which uses the Gittins and Jones (1974) result (our Theorem 6.1.1) and a result equivalent to our Theorem 5.0.2.

Chernoff's conjecture remains open for the setting in which it was proposed: continuous time and uniform discounting.

Related references: Our Chapter 6; Chernoff (1968), Gittins and Jones (1974).

Gittins, J. C. (1979) Bandit processes and dynamic allocation indices (with discussion). *J. R. Statist. Soc. B* **41**: 148–177.

The setting is that of Gittins and Jones (1974). Methods for calculating dynamic allocation indices are explored. Applications are given.

Related references: Our Sections 5.3 and 6.1; Bellman (1956), Gittins (1982), Gittins and Jones (1974).

Gittins, J. C. (1982) Forwards induction and dynamic allocation indices. In *Deterministic and Stochastic Scheduling* (eds M. A. H. Dempster, J. K. Lenstra and A. H. G. Rinnooy Kan), pp. 125–156, Reidel, Hingham, Massachusetts, USA.

The discounting is geometric. The formula for Λ given in our Corollary 5.3.2 is rewritten to be both meaningful and true in a setting more general than that of this monograph.

Related references: Our Sections 5.3 and 6.1; Gittins and Jones (1974), Gittins (1979).

Gittins, J. C. (1983). Dynamic allocation indices for Bayesian bandits. In *Mathematical Learning Models – Theory and Algorithms* (eds U. Herkenrath, D. Kalin and W. Vogel), pp. 50–67, Springer-Verlag, New York.

The extent to which Bather's (1983b) results give information for discrete-time bandits is studied both theoretically and numerically. Gittins also considers a bandit in which the goal is to observe a value in a certain range as soon as possible.

Related references: Bather (1983b), Berry and Fristedt (1980a, 1980b), Vogel (1961b).

Gittins, J. C. and Jones, D. M. (1974) A dynamic allocation index for the sequential design of experiments. In *Progress in Statistics* (eds J. Gani *et al.*), pp. 241–266, North-Holland, Amsterdam.

Discounting is geometric. The setting involves k independent arms each of which is a Markov process. In the authors' setting, our posterior distributions are 'states'. They prove that the desirability of selecting an arm can be found by finding a known arm such that both the arm under consideration and the known arm are optimal in a two-armed bandit (with the same geometric discount sequence); the arm's 'dynamic allocation index' is the mean of the known arm. This theorem is the cornerstone for most of the current work on bandit problems having geometric discounting.

(*Note:* The inequalities in Lemma 2 of this paper should be reversed.)

Related references: Our Chapter 6; Bellman (1956), Gittins (1979), Glazebrook (1983a), Roberts and Weitzman (1980), Varaiya, Walrand, and Buyukkoc (1983), Whittle (1980, 1982).

Gittins, J. C. and Jones, D. M. (1979) A dynamic allocation index for the discounted multiarmed bandit problem. *Biometrika* **66**: 561–565.

Dynamic allocation indices (Gittins and Jones, 1974) are calculated for beta distributions of Bernoulli parameters and the discount sequence $(1, 0.75, (0.75)^2, \ldots)$.

Related references: Our Sections 5.3 and 6.1; Bellman (1956), Gittins and Jones (1974).

Gittins, J. C. and Nash, P. (1977) Scheduling, queues, and dynamic allocation indices. *Proc.* 1974 *EMS, Prague A* 191–202, Czechoslovak Academy of Sciences, Prague.

Various sequential decision problems, including bandits, are surveyed. The authors indicate for which problems 'dynamic allocation index' strategies are optimal; such a strategy in a bandit context assigns to each arm a number that depends only on that arm and the discount sequence, and not on the other arms (cf. Gittins and Jones, 1974, or our Section 6.1). (In Section 6.2 we show that they are optimal *only* for geometric discounting, assuming regularity.)

Related references: Our Chapter 6; Gittins (1975), Gittins and Jones (1974).

Glazebrook, K. D. (1978) On the optimal allocation of two or more treatments in a controlled clinical trial. *Biometrika* **65**: 335–340.

Assuming geometric discounting, Glazebrook studies the problem of calculating the break-even value Λ for an unknown arm. He uses the approach we employ in Section 2.6 when we approximate the maximal worth of a bandit problem by beginning the backwards induction from the left side of (2.6.1). Letting Λ_n denote this approximation (n is defined in our (2.6.1)), Glazebrook observes that it is easy to show that $\Lambda_n \uparrow \Lambda$ as $n \to \infty$. The geometric discounting plays no essential role although it does simplify some formulas.

For $\quad \mathbf{A} = (1, 0.5, (0.5)^2, (0.5)^3, \ldots) \quad$ and $\quad 1 \leq a \leq 15, \quad 1 \leq b \leq 15$, Glazebrook takes $n = 30 - a - b$ and calculates an approximation of Λ in case the unknown arm is Bernoulli and its probability of success has a beta distribution with parameters a and b. The results are presented in tabular form.

Related references: Our Sections 2.6 and 6.1; Gittins (1979), Gittins and Jones (1974).

Glazebrook, K. D. (1980) On randomized dynamic allocation indices for the sequential design of experiments. *J. R. Statist. Soc. B* **42**: 342–346.

There are k independent Bernoulli arms and discounting is geometric. The randomization scheme used by Bather (1980) is applied to dynamic allocation index strategies (Gittins and Jones, 1974). Such strategies are asymptotically optimal in the sense that the long-run success proportion is $\max\{\theta_i: i = 1, \ldots, k\}$. Glazebrook shows for beta priors that there are strategies in this class that are ε-optimal (i.e., ε-Bayes).

Related references: Bather (1980, 1981), Gittins (1979), Gittins and Jones (1974, 1979), Robbins (1952).

Glazebrook, K. D. (1982) On the evaluation of suboptimal policies for families of alternative bandit processes. *J. Appl. Probab.* **19**: 716–722.

There are k independent arms. Discounting is geometric with discount factor α. Glazebrook proves that

$$V(F_1, \ldots, F_k; \alpha) - W(F_1, \ldots, F_k; \alpha; \tau)$$
$$\leqslant E_\tau \sum_{m=1}^{\infty} \alpha^{m-1} [\max\{\Lambda(F_1^{(m-1)}, \alpha), \ldots, \Lambda(F_k^{(m-1)}, \alpha)\} - \Lambda(F_{\tau_m}^{(m-1)}, \alpha)]$$

where τ_m is the arm indicated by τ at stage m and $F_i^{(m-1)}$ is the distribution of arm i at stage m. He applies this result to two Bernoulli arms; θ_1 is uniformly distributed on $(0, 1)$ and θ_2 is known to be $1/2$.

Related references: Our Sections 5.3 and 6.1; Beckmann (1973), Fischer (1979), Gittins (1979), Gittins and Jones (1974), Glazebrook and Jones (1983).

Glazebrook, K. D. (1983a) Optimal strategies for families of alternative bandit processes. *IEEE Trans. Autom. Control* **28**: 858–861.

This is an expository paper on dynamic allocation indices. It contains a proof of the fundamental theorem of Gittins and Jones (1974).

Related references: Our Section 6.1; Gittins (1979), Gittins and Jones (1974), Roberts and Weitzman (1980), Varaiya, Walrand, and Buyukkoc (1983), Whittle (1980, 1982).

Glazebrook, K. D. (1983b) The role of dynamic allocation indices in the evaluation of suboptimal strategies for families of bandit processes. *Mathematical Learning Models–Theory and Algorithms* (eds U. Herkenrath, D. Kalin and W. Vogel), pp. 68–77, Springer-Verlag, New York.

Discounting is geometric. The bound for the discrepancy between the worth of any given strategy and the optimal worth given in Glazebrook (1982) is applied to a job scheduling problem in continuous time and to a job scheduling problem in discrete time for which there are constraints on the order in which jobs must be carried out.

A number of references pertinent to job scheduling and related problems are provided. Most of these are not included in this bibliography.

Related references: Our Section 6.1; Bather (1981), Gittins and Jones (1974), Glazebrook (1982).

Glazebrook, K. D. and Cox, T. F. (1980) On the design of efficient experiments for choosing between two Bernoulli populations. *Comm. Statist. A* **9**: 255–264.

Glazebrook, K. D. and Jones, D. M. (1983) Some best possible results for a discounted one armed bandit. *Metrika* **30**: 109–115.

There are two Bernoulli arms; θ_1 is uniformly distributed on $(0, 1)$ and θ_2 is known to be $1/2$. Discounting is geometric with discount factor α. The authors address the following question: is it optimal to select arm 1 until (if ever) more failures than successes have been observed and then to switch to arm 2? They use computer calculations to find that the answer is affirmative if and only if $\alpha \leqslant \alpha^* = 0.801$ (to 3 decimal accuracy). As indicated in our discussion of Fischer (1979), we disagree: we use analytical methods to find $\alpha^* \approx 0.8968$.

Related references: Beckmann (1973), Fischer (1979).

Goto, M., Sugimura, M. and Asano, C. (1971) Numerical tables of optimum sequential designs based on Markov chains for selecting one of two medical treatments. *Bull. Math. Statist.* **14**: 27–56.

This paper contains a large bibliography of articles by these and other authors that concern problems marginally related to bandits. We have chosen to include only this paper in our bibliography as a source for the interested reader. Generally, all the papers listed by the authors deal with two-phase trials and there is an implicit objective of selecting good treatments over the course of the trial. So they have similarities with papers listed under Colton (1963). However, the papers listed by the authors are not easily categorized. Allocations in the first phase are usually *ad hoc* and are investigated numerically. Many of the papers are closely related to the 'ranking and selection' literature that we have chosen not to review here, though a few resemble papers dealing with finite-memory bandits as listed under Robbins (1956).

Related references: See Colton (1963).

Gray, K. B., Jr (1968) Sequential selection of experiments. *Ann. Math. Statist.* **39**: 1953–1977.

As an example of a general theory, Gray shows that selections in a Bernoulli two-armed bandit can be based on the sufficient statistics.

Related references: Our Chapter 2; Fabius and van Zwet (1970), Rieder (1975).

Hamada, T. (1978) A uniform two-armed bandit problem: The parameter of one distribution is known. *J. Jap. Statist.* **8**: 29–36.

Hengartner, W., Kalin, D. and Theodorescu, R. (1981) On the Bernoulli two-armed bandit problem. *Math. Operationsforsch. Statist. Ser. Optim.* **12**: 307–316.

[This paper was not available to us. The following account is culled from Benzing, Hinderer, and Kolonko (1984).]

There are two Bernoulli arms with parameters θ_1 and θ_2; discounting is n-

horizon uniform. Consistent with notation introduced in our Chapter 2, write the prior distribution of (θ_1, θ_2) in terms of a joint distribution measure G as follows: $(\sigma_1^a \varphi_1^b G)(du, dv) \propto u^a (1-u)^b G(du, dv)$. The authors show that the value is increasing in a and decreasing in b whenever the same is true for $E(\theta_1 | \sigma_1^a \varphi_1^b G)$, which holds when θ_1 and θ_2 are independent under G. This latter implication is a special case of our Theorem 4.1.6.

Related references: Our Section 4.1, Chapters 5 and 7; Benzing, Hinderer, and Kolonko (1984).

Herkenrath, U. (1983) The N-armed bandit with unimodal structure. *Metrika* **30**: 195–210.

There are k arms which have a special kind of dependence: the arms are labelled so that the sequence of k means is unimodal, reaching a (unique) maximum at θ, which is unknown. Observations are made sequentially. There are two objectives: (i) estimate θ, and (ii) maximize the limiting proportion of selections of arm θ. Herkenrath discusses various randomized strategies with regard to the rate of convergence of error in estimating θ and the average rate of payoff.

Related reference: Robbins (1952).

Herkenrath, U. and Theodorescu, R. (1978a) On certain aspects of the two-armed-bandit problem. *Elektron. Informationsverarb. Kybernet.*

There are two Bernoulli arms. The authors consider two families of strategies. One is a class of randomized strategies in which the probability of selecting arm 1 is modified as in Bush and Mosteller (1955). The authors show that these strategies achieve a long-run success proportion of $\max\{\theta_1, \theta_2\}$. The other class is that considered by Vogel (1960a).

Related references: Bush and Mosteller (1955), Herkenrath and Theodorescu (1978b), Witten (1973); and see Colton (1963).

Herkenrath, U. and Theodorescu, R. (1978b) On a stochastic approximation procedure applied to the bandit problem. Preprint 230, Sonderforschungsbereich 72, Univ. Bonn, FRG.

In a two-armed bandit setting with arbitrary unknown distributions, the authors give a strategy which approaches an average payoff per stage that is the maximum of the means of the two distributions. They obtain similar results in a particular many armed bandit that is motivated by a problem in market pricing suggested by Rothschild (1974).

Related references: Herkenrath and Theodorescu (1978a), Rothschild (1974).

Hill, C. and Sancho-Garnier, H. (1978) The two-armed-bandit problem: a decision theory approach to clinical trials. *Biomedicine* **28 Special Issue**: 42–43.

This is an expository article which compares the use of bandit strategies with the use of completely randomized designs in clinical trials. Though the authors give little evidence for their position, they claim that a balanced randomized allocation is a 'good procedure' in a bandit setting. We disagree; randomized clinical trials can sacrifice many patients in order to gain information concerning the treatments involved. The size of a randomized clinical trial is usually determined according to considerations of statistical power. We think that statistical power is irrelevant in many cases. It would usually be better to choose trial size on the basis of a cost/benefit analysis: What is the value of the information that is to be gained? Are there many patients with the condition in question, or few? How much of a sacrifice are patients who receive treatment randomly being asked to make? Costs and benefits should be weighed in specifying the discount sequence in a bandit problem: a large horizon, or a large discount factor when the sequence is geometric, corresponds to a setting in which it is likely that many patients will be treated with one of treatments involved in the trial. Our attitude in this regard is similar to that of Anscombe (1963).

Horowitz, A. D. (1973) Experimental study of the two-armed bandit problem. Ph.D. thesis, Univ. of North Carolina at Chapel Hill, USA.

There are two independent Bernoulli arms and discounting is n-horizon uniform. Horowitz compares optimal strategies (obtained by dynamic programming – see our Chapter 2) with three suboptimal strategies: alternating choice (or an arbitrary data-independent strategy), myopic, and 'play-the-winner/switch-from-a-loser'.

Horowitz's main objective is to analyse the way in which subjects actually behave in a bandit setting. He gives a number of helpful summaries of the results. The instructions to the subjects are well conceived and well written. But we do have one complaint: the subjects should not have been offered a bonus for the best performance among those participating. The presence of a bonus (or simply a competitive atmosphere) could change a subject's behaviour – a subject trying to achieve the largest number of successes would act differently from a subject who is trying to maximize the expected number of successes.

Related references: Our Example 2.4.3 and Chapter 7; Berry (1972), Bush and Mosteller (1955), Estes (1950).

Iosifescu, M. and Theodorescu, R. (1969) *Random Processes and Learning*, Springer-Verlag, New York.

The authors devote a portion of this book to linear (cf. Bush and Mosteller, 1955) and nonlinear learning models.

Related references: Bush and Mosteller (1955).

Isbell, J. R. (1959) On a problem of Robbins. *Ann. Math. Statist.* **30**: 606–610.

There are two unknown Bernoulli arms. The objective is to maximize the long-run success proportion. The current selection can depend only on the previous r selections and observations. Isbell improves on strategies suggested by Robbins (1956).

Related references: See Robbins (1956).

Jones, D. M. (1970) A sequential method for industrial chemical research. M.Sc. thesis, Univ. of Wales, Aberystwyth.

Jones, D. M. (1974) Search procedures for industrial chemical research. Ph.D. thesis, Cambridge Univ., England.

Jones, P. W. (1975) The two-armed bandit. *Biometrika* **62**: 523–524.

There are two independent Bernoulli arms (with beta priors) and discounting is n-horizon uniform. The values (maximal expected payoffs) of two families of bandits ($n = 2$ to 15) are compared in tabular form with the worths of the myopic strategy and the 'play-the-winner/switch-from-a-loser' strategy. Not surprisingly (see our discussion of Percus and Percus (1984)), the latter strategy fares badly.

Related references: Our Example 2.4.3 and Chapter 7; Berry (1972), Jones and Kandeel (1983), Percus and Percus (1984).

Jones, P. W. (1976) Some results for the two-armed bandit problem. *Math. Operationsforsch. Statist. Ser. Optim.* **7**: 471–475.

There are two independent Bernoulli arms with beta priors; discounting is n-horizon uniform. In terms of the notation used in our Example 7.2.1, Jones shows that the value of the bandit is increasing in a_i for fixed $a_i + b_i$, $i = 1, 2$; this result is an instance of our Theorem 4.1.6. He conjectures that if arm i is optimal for beta parameters a_i and b_i, it is also optimal if a_i is increased with $a_i + b_i$ fixed; that this is true is immediate from Theorem 5.2 of Berry (1972).

Related references: Our Example 2.4.3 and Chapters 4 and 7; Berry (1972), Jones and Kandeel (1983).

Jones, P. W. (1977) Some designs for the two-armed bandit with one probability known. *Biom. J.* **19**: 693–695.

There are two Bernoulli arms with one arm known; the parameter of the unknown arm has a beta distribution. Discounting is n-horizon uniform. Jones gives tables comparing the worths of the following three strategies with that of an optimal strategy: the better single-arm strategy, the myopic strategy, and a variant of the myopic strategy in which the unknown arm is also selected if its mean is slightly less than that of the known arm.

Related references: Our Chapter 5; Berry and Fristedt (1979), Jones (1978), Jones and Kandeel (1983).

Jones, P. W. (1978) On the two-armed bandit with one probability known.
  Metrika **25**: 235–239.

There are two Bernoulli arms with one arm known. Discounting is n-horizon
uniform. Jones gives two theorems which have hypotheses concerning
posterior means; we note that these hypotheses follow easily from the
Cauchy–Schwarz inequality. The second theorem is incorrect. If this theorem
were correct then it would imply, in conjunction with our Theorem 5.0.2, that
the break-even value of the unknown arm is its mean; so myopic strategies
would be optimal.

  Related references: Our Chapter 5; Berry and Fristedt (1979), Bradt,
Johnson, and Karlin (1956), Jones (1977), Jones and Kandeel (1983).

Jones, P. W. and Kandeel, H. A. (1983) Numerical investigation of the two
  armed bandit. *Mathematical Learning Models – Theory and Algorithms* (eds
  U. Herkenrath, D. Kalin and W. Vogel), pp. 101–107, Springer-Verlag,
  New York.

The discount sequence considered is $\mathbf{A} = (1, \alpha, \ldots, \alpha^{n-1}, 0, 0, \ldots)$ for
$0 < \alpha \leqslant 1$; special emphasis is given to $\alpha = 1$, in which case $\mathbf{A}$ is the n-horizon
uniform. In the first of two settings considered, there are two independent
Bernoulli arms whose parameters have beta distributions. When $\alpha = 1$,
optimal worths are obtained by dynamic programming and are compared in
tabular form with the worths of myopic strategies. The analytical results the
authors give are proved in Berry (1972) and also in our Chapter 7. The
authors conjecture a special case of the conjecture we discuss in the
penultimate paragraph of Chapter 7; cf. Berry (1972), Joshi (1975).

  In the other setting considered by the authors, one arm is known. They
briefly discuss the effect of taking $\alpha < 1$ and compare myopic and optimal
strategies. They argue that the number of stages for which the unknown arm
is selected is never greater for $\alpha < 1$ than it is for $\alpha = 1$. It is assumed that once
the known arm is selected it should be selected thereafter. That this is correct
for any α (even $\alpha > 1$!) follows from our Theorem 5.2.2, or from Berry and
Fristedt (1979). However, when $\alpha = 1$ the proof is easy and is given by Bradt,
Johnson, and Karlin (1956, Lemma 4.1) – see our discussion immediately
preceding the proof of Theorem 5.2.2.

  Related references: Our Example 2.4.3 and Chapters 5, 6, and 7; Berry
(1972), Berry and Fristedt (1979), Bradt, Johnson, and Karlin (1956), Gittins
and Jones (1979), Jones (1977), Jones (1978), Jones and Kandeel (1985), Joshi
(1975), Kalin and Theodorescu (1982), Percus and Percus (1984).

Jones, P. W. and Kandeel, H. A. (1985) A comparison of sampling rules for a
  Bernoulli two armed bandit. *Comm. Statist. C* **3** (to appear).

The setting is that of Jones and Kandeel (1983). Various properties of
strategies considered in that paper are examined.

  Related reference: Jones and Kandeel (1983).

Joshi, V. M. (1975) A conjecture of Berry regarding a Bernoulli two-armed bandit. *Ann. Statist.* **3**: 189–202.

The setting is that of our Section 7.3; there are two independent Bernoulli arms having common underlying prior distributions and the discount sequence is the n-horizon uniform. In our notation, it is proved in Corollary 2.1 of Joshi that if $b_1 \leqslant b_2$ and $E(\theta_1 | \sigma^s \varphi^f F_1) \geqslant E(\theta_1 | F_1) \geqslant E(\theta_2 | F_2)$, then $E(\theta_1 | \sigma^s \varphi^f F_1) \geqslant E(\theta_2 | \sigma^s \varphi^f F_2)$.

Joshi claims to prove the conjecture of Berry (1972) that is discussed in the penultimate paragraph of our Section 7.3. However, there is an error in his proof that cannot be easily repaired. The proof relies on his Corollary 3.1, which is not correct. This is easily seen by taking, in Joshi's notation, μ equal to Lebesgue measure on $(0, 1)$, $r_0 = r_0' = l_0 = l_0' = m_2 = n_2 = 0$, $m_1 = n_1 > 0$, and $n = 2$, and then applying our Section 7.2. (In response to our query regarding this matter, Professor Joshi agrees with our assessment.) Therefore, we regard the conjecture as unresolved.

Related references: Our Chapter 7; Berry (1972).

Kakigi, R. (1983) A note on discounted future two-armed bandits. *Ann. Statist.* **11**: 707–711.

There are two possibly dependent Bernoulli arms: (θ_1, θ_2) has a two-point distribution on the unit square. Discounting is geometric. Kakigi applies [Blackwell, D. (1965). *Ann. Math. Statist.* **36**: 226–235] to find the optimal strategies for some special cases.

Related references: Our Corollary 4.3.10 and Chapter 5; Berry and Fristedt (1979), Feldman (1962), Kelley (1974), Rodman (1978), Zaborskis (1976).

Kalaba, R. E. and Tesfatsion, L. (1978) Two solution techniques for adaptive reinvestment: a small sample comparison. *J. Cybernet.* **8**: 101–111.

There are two investment opportunities; arm 2 always gives 0 rate of return and arm 1 gives $+r$ and $-r$ with unknown probabilities θ_1 and $1 - \theta_1$. The decision maker has an initial capital and at each of n stages is to allocate the current capital between the two arms. The objective is to maximize the expected value of the logarithm of total capital at stage n. So there are a number of ways that this differs from the problems we consider. Foremost among these is that both arms are observed at all stages, and so payoff and information-gathering are separate considerations. The authors show that the decision problem can be decomposed into n simple maximization problems.

Related reference: Tesfatsion (1978).

Kalin, D. (1979) Über Markoffsche Entscheidungsmodelle mit halbgeordnetem Zustandsraum. *Methods of Operations Research* **33**: 233–245.

Kalin, D. (1981) Beiträge zu strukturierten Markoffschen Entscheidung-smodellen. *Habilitationsschrift*, Univ. Bonn, FRG.

Kalin, D. (1982) Zum Problem des zweiarmigen Bernoulli-Banditen mit einer bekannten Erfolgswahrscheinlichkeit und unendlich vielen Spielen. *Metrika* **29**: 261–270.

There are two Bernoulli arms, one of which is known. Discounting is geometric and truncated geometric. Kalin gives special cases of our Theorems 4.1.6 and 5.0.2 (cf. Berry and Fristedt, 1979). The material presented is similar to that of Kalin and Theodorescu (1980).

Related references: Our Chapters 4 and 5; Berry and Fristedt (1979), Jones (1978), Kalin and Theodorescu (1980).

Kalin, D. and Theodorescu, R. (1980) Sur le probleme du bandit á deux bras quand une probabilité est connue. *Publ. Inst. Statist., Univ., Paris* **XXV**: 49–60.

There are two Bernoulli arms with one arm known. Discounting is n-horizon uniform; a truncated geometric is also considered. The authors give special cases of our Theorems 4.1.6 and 5.0.2 (cf. Berry and Fristedt, 1979). The material presented is similar to that of Kalin (1982).

Related references: Our Chapters 4 and 5; Berry and Fristedt (1979), Jones (1978), Kalin (1982).

Kalin, D. and Theodorescu, R. (1982) A note on structural properties of the Bernoulli two-armed bandit problem. *Math. Operationsforsch. Statist. Ser. Optim.* **13**: 469–472.

[This paper was not available to us. We understand from other sources that it contains an independent proof of Berry's (1972) result that optimal strategies stay with a winner when there are two independent Bernoulli arms and discounting is uniform.]

Kalin, D. and Theodorescu, R. (1983) On a stopping rule for a class of sequential decision problems. *Metrika* **30**: 117–123.

There are two Bernoulli arms with one arm known. Discounting is n-horizon uniform. The authors show that once the known arm becomes optimal it remains optimal. This was proved by Bradt, Johnson, and Karlin (1956, Lemma 4.1). It is a simple instance of Theorem 2.1 of Berry and Fristedt (1979) – see our Theorem 5.2.2. The proof in this easy special case is given in the discussion immediately preceding our proof of Theorem 5.2.2.

Related references: Our Chapter 5; Berry and Fristedt (1979), Bradt, Johnson, and Karlin (1956), Jones and Kandeel (1983).

Karatzas, I. (1984) Gittins indices in the dynamic allocation problem for diffusion processes. *Ann. Probab.* **12**: 173–192.

Time is continuous and discounting is exponential. The arms are one-dimensional diffusion processes. The theorem of Gittins and Jones (1974) is extended to this setting. Explicit calculations of dynamic allocation indices are made. Karatzas's setting includes instances of our setting (for instance, our Section 8.1) in which posterior distributions can be represented by a real parameter undergoing a diffusion.

Related references: Our Section 8.1; Gittins (1979), Gittins and Jones (1974).

Keener, R. W. (1984) Further contributions to the 'two-armed bandit' problem. Statistics Tech. Rep. No. 124, Univ. of Michigan, USA.

There are two dependent arms and two states of nature: G is supported by two ordered pairs of distributions on **R**. Discounting is geometric (with $\alpha = 1$ being allowed). Keener obtains optimal strategies in terms of expected values of ladder epochs of random walks.

Related references: Our Corollary 4.3.10; Feldman (1962), Kelley (1974), Quisel (1965), Rodman (1978).

Kelley, T. A. (1974) A note on the Bernoulli two-armed bandit problem. *Ann. Statist.* **2**: 1056–1062.

There are two dependent Bernoulli arms with the prior for (θ_1, θ_2) concentrated on two points: (a, b) and (c, d). The discount sequence is the n-horizon uniform. For $n \geqslant 3$ Kelley shows that, except for some simple special cases, Feldman's (1962) assumption that $a = d$ and $b = c$ is necessary for the conclusion that myopic strategies are optimal. Kelley shows that myopic strategies are optimal for $n \geqslant 3$ and two-point distributions if and only if either (i) $a \leqslant b$ and $c \leqslant d$, (ii) $a \geqslant b$ and $c \geqslant d$, (iii) $a + b = c + d = 1$, or (iv) $(a, b) = (d, c)$.

Related references: Our Corollary 4.3.10; Berry (1972), DeGroot (1970), Fabius and van Zwet (1969), Feldman (1962), Keener (1984), Rodman (1978), Zaborskis (1976).

Kelley, T. A. (1976) Two-stage procedures for the Bernoulli two-armed bandit. Statistics Tech. Rep. No. 103, Univ. of Florida, USA.

The problem studied by Canner (1970) is treated (independently). Kelley finds $(\sqrt{(2n + 5)} - 1)/2$ as the (approximate) optimal first-phase length when the prior is uniform (which compares with Canner's $(\sqrt{(2n + 4)} - 2)/2)$.

Related references: See Colton (1963).

Kelly, F. P. (1981) Multi-armed bandits with discount factor near one: the Bernoulli case. *Ann. Statist.* **9**: 987–1001.

There are k independent Bernoulli arms with common underlying prior F (cf. our Section 7.3). Thus, $F_i = \sigma^{a_i} \phi^{b_i} F$, $1 \leqslant i \leqslant k$. Discounting is geometric. Kelly assumes that $F[x, 1]$ is regularly varying as $x \to 1$. For the discount factor α sufficiently close to 1, Kelly proves that an optimal initial arm i must satisfy $b_i = \min\{b_j\}$, and that among such i the only optimal selections are those for which a_i is maximum. (How close α must be to 1 depends on the numbers a_i and b_i, $1 \leqslant i \leqslant k$.)

Whether $F[x, 1]$ regularly varying can be replaced by $F[x, 1] > 0$ for $x < 1$ is an open problem.

Berry (1972) conjectures that this 'least failures rule' is optimal for n-horizon uniform discounting as $n \to \infty$.

Related references: Our Sections 6.1 and 7.3; Berry (1972), Gittins and Jones (1974), Woodroofe (1976).

Kolonko, M. and Benzing, H. (1983) The sequential design of Bernoulli experiments including switching costs. Unpublished.

There are two Bernoulli arms with one arm known. Discounting is n-horizon uniform (though truncated geometric is also considered). Various results shown by Bradt, Johnson, and Karlin (1956) – and incorrectly attributed by Kolonko and Benzing to other authors – are proven in the presence of fixed costs for switching arms. In particular, optimal strategies 'stay with a winner'. (This statement includes staying with the known arm once it is optimal because the state of information is the same whether the known arm is a 'winner' or 'loser'.)

When one arm is known and the discount sequence is *regular* (our Definition 5.2.1), the two-armed bandit with no switching costs is a stopping problem. It is not surprising that a decision maker would be less likely to switch when there is a fixed cost for so doing! We would like to see conceptual proofs using this notion. The following conceptual argument (cf. our discussion just before the proof of Theorem 5.2.2) shows that a bandit with the same cost for switching in either direction is a stopping problem when discounting is uniform. Suppose every optimal strategy indicates arm 2 (the known arm) initially; so arm 2 is uniquely optimal. If the problem is not a stopping problem then there is a nonrandom stage, say $r + 1$, at which some optimal strategy τ first indicates arm 1. Now consider strategy τ^* which, for stages 1 through $n - r$, indicates the arm indicated by τ at respective stages $r + 1$ through n; for stages $n - r + 1$ to n, τ^* indicates the arm indicated by τ at stages 1 through r (arm 2 in all cases). Defined thus, the expected number of successes is the same for both τ^* and τ. Moreover, the number of switches is either the same or one fewer on τ^* than on τ. Therefore, τ^* is at least as good as τ and, since τ^* indicates arm 1 initially, arm 2 cannot be uniquely optimal. It follows that arm 2 can be used exclusively once it becomes uniquely optimal.

Costs for switching are not as interesting when one arm is known as otherwise. For in the former case there is at most one 'switch'; so the cost is really one imposed for stopping.

Related references: Our Chapter 5; Berry and Fristedt (1979), Bradt, Johnson, and Karlin (1956), Kalin and Theodorescu (1982, 1983).

Kolonko, M. and Benzing, H. (1985) On monotone optimal decision rules and the stay-on-a-winner rule for the two-armed bandit. *Metrika* (to appear).

The stay-with-a-winner rule of Berry (1972) is generalized to certain instances of dependence between the arms.

Related references: Our Examples 2.4.1 and Theorem 4.3.8; Benzing, Hinderer, and Kolonko (1984), Berry (1972), Bradt, Johnson, and Karlin (1956), Hengartner, Kalin, and Theodorescu (1981), Kalin and Theodorescu (1982).

Kôno, K. (1966) How to deal with the a priori probability on the 'two-armed' bandit problem. *Math. Rep.* **4**: 27–34.

A special case of Feldman (1962) is considered. The author treats the possibility that the prior distribution is not known; we fail to appreciate the motives or results of this discussion.

Related references: Bradt, Johnson, and Karlin (1956), Feldman (1962).

Kumar, P. R. and Seidman, T. I. (1981) On the optimal solution of the one-armed bandit adaptive control problem. *IEEE Trans. Automat. Control* **26**: 1176–1184.

There are two Bernoulli arms; θ_2 is known and θ_1 has a beta distribution with parameters a and b. Discounting is geometric with discount factor α. The authors give results that are special cases of our Theorems 5.0.2 and Corollaries 5.3.3 and 5.3.4 (cf. Berry and Fristedt, 1979). They give upper and lower bounds for $\Lambda(F, \mathbf{A})$ that are weaker than those given in our Section 5.4 and in Berry and Fristedt (1979).

The authors also consider the question studied by Fischer (1979). They obtain the weaker bound $1/2$ in lieu of Fischer's $3 - \sqrt{5}$.

Related references: Our Chapter 5; Bellman (1956), Berry and Fristedt (1979), Fischer (1979), Glazebrook and Jones (1983).

Lai, T. L., Levin, B., Robbins, H. and Siegmund, D. (1980) Sequential medical trials. *Proc. Natl. Acad. Sci. USA* **77**: 3135–3138.

There are n patients (n-horizon uniform discounting) in a clinical trial. The first $2r$ patients will be allocated in pairs to arms 1 and 2; each difference (arm 1 response minus that of arm 2) is distributed with unknown mean δ and variance σ^2 (known and unknown σ^2 are dealt with). The remaining $n - 2r$ patients are assigned arm 1 if the sum of the pairwise differences is positive

and arm 2 otherwise. The objective is equivalent to maximizing the sum of the n observations by choosing r sequentially. The authors consider three stopping rules (including that of Anscombe (1963)) and state theorems which indicate that the three are asymptotically optimal ($n \to \infty$). These theorems are proven by Lai, Robbins, and Siegmund (1983).
Related references: See Colton (1963).

Lai, T. L. Robbins, H. and Siegmund, D. (1983) Sequential design of comparative clinical trials. *Recent Advances in Statistics; Papers in Honor of Herman Chernoff on his Sixtieth Birthday* (eds M. H. Rizvi, J. S. Rustagi and D. Siegmund), pp. 51–68, Academic Press, New York.

The setting is that of Anscombe (1963) and Colton (1963). The authors give properties of strategies (stopping rules) discussed by Lai, Levin, Robbins, and Siegmund (1980), including that suggested by Anscombe (1963). They prove results stated in Lai, Levin, Robbins, and Siegmund (1980) and give asymptotic properties of suboptimal rules suggested by Begg and Mehta (1979) and Colton (1963).
Related references: See Colton (1963).

Lakshmanan, K. B. and Chandrasekaran, B. (1978) On finite memory solutions to two-armed bandit problem. *IEEE Trans. Inform. Theory* **24**: 244–248.

There are two unknown Bernoulli arms. The objective is to maximize the long-run proportion of selections of the better arm subject to having but m memory states available. The authors show that at most 1 bit of memory is saved (for all m) by knowing the success probability of one of the arms. They also show that optimal deterministic strategies require no more than two bits over that required for randomized strategies.
Related references: See Robbins (1956).

Langenberg, P. and Srinivasan, R. (1981) On the Colton model for clinical trials with delayed observations – normally-distributed responses. *Biometrics* **37**: 143–148.

The setting is that of Colton (1963) and Colton (1965). The authors consider a transitional phase between the experimental and terminal phase. Two simple nonsequential procedures are compared for assigning arms in this middle phase.
Related references: See Colton (1963).

Langholz, G. (1977) Interaction between stochastic automata and random environment. *Internat. J. Man-Mach. Stud.* **9**: 223–231.

Two-armed bandits with finite memory are related to stochastic automata learning models with finite memory. Simulations are used to compare three strategies.

Related references: See Robbins (1956).

Louis, T. A. (1975) Optimal allocation in sequential tests comparing two Gaussian populations. *Biometrika* **62**: 359–370.

The problem considered is similar to that of Flehinger and Louis (1972); it is assumed that the arms are normally distributed with the same known variance. The allocation rule that is optimal (in Louis's setting) in the continuous-time analogue is shown to be asymptotically optimal when time is discrete; simulations show that it performs well in the discrete case.

Related references: See Colton (1963).

Mallows, C. L. and Robbins, H. (1964) Some problems of optimal sampling strategy. *J. Math. Anal. Appl.* **8**: 90—103.

An arbitrary number (∞ is allowed) of arms with unknown distributions (subject to mild regularity conditions – boundedness is sufficient) is considered. The authors show that there are strategies that are asymptotically optimal in the sense that, with probability one, the limiting average of the observations is the supremum of the means of the distributions. Some related problems are considered.

Related references: Bather (1981); and see Robbins (1956).

Meeter, D. A. (1975) A two-armed bandit with terminal decision (Bayes rule). *A Survey of Statistical Design and Linear Models* (eds J. N. Srivastava), pp. 419–426, North-Holland, Amsterdam.

Results from Fabius and van Zwet (1970) are generalized to the case in which the discount sequence is $\mathbf{A} = (1, 1, \ldots, 1, T, 0, 0, \ldots)$; cf. Wahrenberger, Antle, and Klimko (1977). Fabius and van Zwet (1970) treat the n-horizon uniform: $T = 1$ (or 0). An equivalent way of viewing the problem is that the decision maker's strategy is restricted to selecting the same arm for the last T stages. (We note that $\mathbf{A}$ is regular and so the appropriate results in our Chapter 5 would apply if one arm were known.)

The distinction between Meeter's problem and most of the references listed under Colton (1963) is that Meeter allows optimal sequential selections in the first phase, while most of the others place restrictions on first-phase allocations.

Related references: Fabius and van Zwet (1970); and see Colton (1963).

Meybodi, M. R. and Lakshmivarahan, S. (1983) On a class of learning algorithms with symmetric behaviour under success and failure. *Mathematical Learning Models – Theory and Algorithms.* (eds U.

Herkenrath, D. Kalin and W. Vogel), pp. 145–155, Springer-Verlag, New York.

There are k Bernoulli arms. A class of randomized strategies is proposed such that, loosely speaking, the probability of selecting an arm increases should the arm yield a success and it decreases should the arm yield a failure. Conditions are given under which such a strategy results in a long-run success proportion within ε of that of the best arm.

Related references: Bather (1981); and see Robbins (1956).

Morrison, D. F. (1967) On the two-armed bandit problem. *Psychology of Management Decision* (ed. G. Fisk), pp. 186–195, C. W. K. Gleerup, Lund, Sweden.

There are two arms and two densities f and f^*; the case of normal densities with equal variances is given special consideration. Which density goes with which arm is not known. Discounting is n-horizon uniform. (Feldman (1962) proved that myopic strategies are optimal in the Bernoulli case. Rodman (1978) removed the Bernoulli hypothesis. Morrison incorrectly attributes the general result to Feldman.) Morrison considers the n stages divided into r phases of fixed length. Allocation is balanced in the first phase (cf. Anscombe, 1963; Colton, 1963, etc.), and thereafter the arm with greater probability of having density f^* (say f^* is preferred) at the beginning of a phase is used exclusively during that phase.

An *ad hoc* strategy based on sample means is discussed.

Related references: Feldman (1962), Rodman (1978), Zaborskis (1976); and see Colton (1963).

Murray, F. S. (1971) Multiple probable situation: A study of a five one-armed bandit problem. *Psychon. Sci.* **22**: 247–249.

This reports an experimental study involving 14 subjects. There are five independent Bernoulli arms (electronic devices) with n-horizon uniform discounting, $n = 250$. The subjects were given an M&M candy for each success. (We are not convinced that this is sufficient reward to induce a subject to try to maximize the expected number of successes.) The probabilities of success on the arms were not known to the subjects, but were in fact $\theta_1 = 0.5$, $\theta_2 = 0.25$, $\theta_3 = 0.25$, $\theta_4 = 0.125$, $\theta_5 = 0$. Murray reports that the overall proportion of the stages numbered 211 to 250 in which the subjects used the arm with greatest current frequency of success (not necessarily arm 1 – he does not indicate whether it was ever a different arm) was 52%. This compares with 24% for the arm with the second highest success frequency and 7% for the lowest (presumably the latter was always arm 5).

We note that the subjects were given only 3 seconds between selections. Results from a similar experiment with unlimited time between selections would be more interesting to us.

Related references: Brand, Sakoda, and Woods (1957), Brand, Woods, and Sakoda (1956), Bush and Mosteller (1955), Estes (1950), Horowitz (1973).

Nakajima, N. and Noshi, T. (1978) On the behaviour of the shrewd automaton. Technol. Rep. No. 25, Seikei Univ., Japan.

Nash, P. (1973) Optimal allocation of resources between research projects. Ph.D. thesis, Cambridge Univ., England.

Nash, P. (1980) A generalized bandit problem. *J. R. Statist. Soc. B* **42**: 165–169.

There are k independent arms with geometric discounting. The setting is that of Gittins (1979). The payoff at any stage depends on the arm selected at that stage but also (in a multiplicative way) on the other arms. Nash shows that the theorem of Gittins and Jones (1974) applies to his problem.

This is not a bandit problem in our setting since the payoff at any stage cannot be viewed as an observation from the distribution of the arm selected.

Related references: Gittins (1979), Gittins and Jones (1974)

Obregon, I. (1968) The N-armed bandit problem and other topics in sequential decision processes. Operations Research Tech. Rep., Massachusetts Institute of Technology, USA.

Oudin, C. and Lellouch, J. (1972) La comparison de quelques stratégies dans la conduite des essais thérapeutiques *Rev. Statist. Appl.* **XX**: 5–21.

This is mainly an expository article. There are two independent Bernoulli arms with beta priors on θ_1 and θ_2. Discounting is n-horizon uniform. Maximizing the expected number of successes among the n ('ethique collective') is contrasted with maximizing the probability of immediate success ('ethique individuelle'). Two-phase and unrestricted bandit problems are put in the first category (the authors give special attention to the case in which one arm is known) and myopic strategies in the second. Geometric discounting truncated after stage n is recommended as a compromise.

For the two-phase design the approach is similar to that of Pearson (1980). When the distribution of (θ_1, θ_2) is uniform the authors conclude 'by symmetry' that the two arms should be allocated equally in the first phase. Pearson (1980) shows that equal allocation is never optimal in this case! With the condition of equal allocation the authors rederive the approximation to the optimal first-phase length of Canner (1970).

Related references: Our Chapters 5 and 7; and see Colton (1963).

Pearson, L. M. (1980) Treatment allocation for clinical trials in stages. Ph.D. thesis, Univ. of Minnesota, USA.

The setting is that of Berry and Pearson (1984). A number of extensions of the problems considered in that paper are given and numerous tables are provided.

Pearson shows that optimal first-phase allocations may not be symmetric even though the prior is symmetric in θ_1 and θ_2.

Related references: See Colton (1963).

Percus, O. E. and Percus, J. K. (1984) Modified Bayes technique in sequential clinical trials. *Comput. Biol. Med.* **14**: 127–134.

This paper deals with myopic strategies (called 'ethical' by the authors) for assigning patients to one of two medical treatments. The arms are Bernoulli. Both θ_1 and θ_2 have a beta distribution with parameters a and 1. The authors consider the myopic strategy which always uses the arm with greater $[s_1 + a]/[s_i + f_i + a + 1]$ where s_i and f_i are the current numbers of successes and failures on arm i; cf. Bather (1980). The authors use simulation to find the long-run proportion of stages for which the inferior treatment is used for various combinations of Bernoulli parameters. (The prior distribution serves only as a means of finding a strategy which pays heed to the accumulating data rather than as a reflection of initial information. In our view the resulting strategy is not 'ethical' unless this distribution corresponds to that of the clinician.) Not surprisingly, they find that this proportion is smaller when a is large.

The authors compare these 'ethical' strategies with 'play-the-winner/switch-from-a-loser'. Again not surprisingly, the former perform better according to the authors' criterion; playing a winner is reasonable, but switching on a loser can be arbitrarily bad and as a general policy has no merit whatever.

This paper is an example of a large literature (most of which is not mentioned here) which compares strategies with respect to the proportion of patients who are treated with the inferior treatment. Our approach (using $A = (1, 1, \ldots, 1, 0, 0, \ldots)$) of minimizing the expected number of *failures* is quite different: a success on a 'bad' arm is as good as a success on a 'good' arm. And to us, using an arm whose success probability is substantially smaller than an alternative should be penalized more heavily than using an arm whose success probability is only slightly smaller than the alternative.

Related references: Bather (1980, 1981), Berry (1978), Feldman (1962), Fox (1974), Jones and Kandeel (1983), Oudin and Lellough (1972), Robbins (1952), Thompson (1933, 1935).

Petkau, A. J. (1978) Sequential medical trials for comparing an experimental with a standard treatment. *J. Amer. Statist. Assoc.* **73**: 328–338.

There are two Bernoulli arms; θ_1 has a beta prior distribution and θ_2 is known. Normal distributions are also considered. Discounting is n-horizon uniform. There is a constant cost of observation for each stage until a decision is made to select a single arm for the duration, for which there are no costs for observation. The objective is to maximize the expected number of successes overall minus cost of observation in the experimental phase; the problem is to decide when to stop paying for observing arm 1 at which time either arm 1 or arm 2 is selected exclusively.

Assuming n is large, Petkau adopts the continuous-time approximation suggested by Chernoff and Ray (1965). Optimal stopping boundaries are shown for various costs of observations. When there is no such cost the problem is the same as that treated in our Section 8.2; our Figure 8.2 is essentially the same as the boundary shown in Figure A of this paper corresponding to $\gamma_1 = 0.1$, the smallest cost of observation considered (the upper boundary having disappeared), and Figure 8.3 corresponds to Figure B of this paper.

Petkau considers the sensitivity of optimal strategies to misspecification of n. He also evaluates three suboptimal strategies: (i) stop experimenting when the number of successes minus failures is sufficiently large or small – horizontal boundaries in Figure 8.2; (ii) stop experimenting when the best fixed-sample-size continuation strategy is to stop (cf. Begg and Mehta, 1979, and Chernoff and Petkau, 1981); and (iii) the best fixed-sample-size strategy – a vertical boundary in Figure 8.2.

Related references: Our Chapter 5 and Section 8.2; Chernoff (1967, 1968, 1972), Chernoff and Petkau (1976, 1983a, 1984), Chernoff and Ray (1965); and see Colton (1963).

Poloniecki, J. D. (1978) The two armed bandit and the controlled clinical trial. *Statistician* **27**: 97–102.

Presman, E. L. and Sonin, I. M. (1982) *Posledovatel'noe Upravlenie po Nepolnym Dannym*, Izdatelstvo Nauka, Moscow.

The authors devote a considerable portion at their book to the study of bandits in both discrete and continuous time. Unfortunately, we do not read Russian well enough to give a detailed review.

The title is *Sequential Control with Partial Information*. The chapter headings are: (1) Fundamental scheme for discrete and continuous time, (2) Formulation of the problem and methods of solution for discrete time, (3) The solution of certain problems in the fundamental scheme for discrete time, (4) Formulation of the problem and methods of solution for continuous time, (5) The solution of problems in the fundamental scheme for continuous time, (6) Application of the maximum principle of Pontryagin to control problems with random jumps, (7) Some other problems.

Related references: This monograph; Presman and Sonin (1983).

Presman, E. L. and Sonin, I. M. (1983) 'Two and many-armed bandit' problems with infinite horizon. *Proceedings of the Fourth USSR-Japan Symp. in Probab.Theory and Math. Statist.* (eds K. Itô and J. V. Prokhorov), pp. 526–540, Springer-Verlag, Berlin.

There are k not necessarily independent Bernoulli arms; discounting is n-horizon uniform. Randomized strategies are permitted; this is relevant since the authors are interested in limit theorems for $n \to \infty$. They assume that the distribution G of the vector $(\theta_1, \ldots, \theta_k)$ of Bernoulli parameters is supported by a finite number of atoms. Necessary and sufficient conditions are given on G for $n[E(\max(\theta_1, \ldots, \theta_k)|G] - V(G; n)$ to approach a finite limit as $n \to \infty$. In case the limit is finite, a strategy τ is indicated for which $V(G; n) - W(G; n; \tau) \to 0$ as $n \to \infty$. The authors refer the reader to Presman and Sonin (1982) for the proof.

The analogous result is obtained in a continuous-time setting. Each arm is a Poisson process with a random intensity. At each instant the decision maker is permitted to mix the intensities and observe the mixture. For example, the decision maker is permitted to observe arm 1 with weight 2/3 and arm 2 with weight 1/3.

Related references: Our Section 8.3; Bather (1981), Kelley (1974), Presman and Sonin (1982).

Quisel, K. (1965) Extensions of the two-armed bandit and related processes with on-line experimentation. Tech. Rep. No. 137, Institute for Mathematical Studies in the Social Sciences Stanford Univ., USA.

This report gives an extensive description of bandit and related problems. The point of view is the same as ours.

Quisel considers the case of two independent arms; both distributions are bounded and are in the same family, for which there is a one-dimensional sufficient statistic (for fixed sample size). The discount sequence $A = (\alpha_1, \alpha_2, \ldots)$ has finite horizon and is nonincreasing, but is arbitrary otherwise. Quisel assumes that the prior distribution is in a conjugate family – an example is the beta distribution when the observations are Bernoulli. He claims to prove the 'stay-with-a-winner' rule in this context; this says that when an arm is optimal and yields the maximum possible observation ('success' or '1' in the Bernoulli setting) then it is optimal again (see our Theorem 4.3.8 and Corollary 5.6.2). But his Theorem 4.2 is not a stay-with-a-winner rule as he claims since the result applies for fixed discount sequence and does not allow for it to change from one stage to the next. It is easy to see by adapting our Examples 3.3.3 or 5.2.2 that such a result cannot hold in the generality indicated. Our Example 5.2.2 assumes $\theta_2 = 0.6$ and $\theta_1 \in \{1, 0\}$; to adapt it to Quisel's setting with conjugate priors take the beta parameters for θ_2 very large and in the ratio $6:4$ and take those for θ_1 very small and equal.

Another setting considered by Quisel has k independent arms, each of which is governed by one of two known distributions. He shows that it is optimal at any stage to select an arm having the largest probability of having the distribution with larger mean (cf. Rodman 1978).

Related references: Our Chapters 4, 5, and 7; Berry (1972), Bradt, Johnson, and Karlin (1956), Feldman (1962), Keener (1984), Rodman (1978).

Reimnitz, P. (1978) A study of 'two-armed bandits'. Ph.D. thesis, Univ. of Rochester, Rochester, New York, USA.

The extension by Vogel (1964) of Vogel (1960c) is repeated and carried further. The approach uses an *oscillating random walk*, used also by Vogel (1960c) and named by Kemperman [(1974) The oscillating random walk. *Stochastic Process. Appl.* **2**: 1–29]. The tool is studied in detail.

Reimnitz considers the myopic strategy for two Bernoulli arms where (θ_1, θ_2) is either (a, b) or (b, a). Following such a strategy, the probability that the inferior arm is used at stage m decreases exponentially. Reimnitz considers the possibility that (θ_1, θ_2) is different from (a, b) or (b, a). He shows that the exponentially decreasing property still holds when $|\log(a/b)/\log[(1-a)/(1-b)]|$ lies between $(1-\theta_1)/\theta_1$ and $(1-\theta_2)/\theta_2$. A weaker result about the boundedness of the regret is discussed in Vogel (1964).

Related references: Our Section 9.1; Feldman (1962), Rodman (1978), Vogel (1960c, 1961a, 1964).

Rieder, U. (1975) Bayesian dynamic programming. *Adv. in Appl. Probab.* **7**: 330–348.

Rieder discusses Markov decision problems. For a given problem having random parameters he gives a method for finding an equivalent problem that has no random parameters. His approach provides an alternative to some of the technical issues we address in Chapter 2.

Related references: Our Chapter 2; Gray (1968).

Robbins, H. (1952) Some aspects of the sequential design of experiments. *Bull. Amer. Math. Soc.* **58**: 527–535.

The problem of choosing sequentially from two populations to maximize the sum of n observations is posed. Robbins considers two Bernoulli arms with parameters θ_1 and θ_2 and shows that the 'play-the-winner/switch-from-a-loser' strategy results in a greater long-run success proportion than do rules that do not depend on accumulating data – uniformly in (θ_1, θ_2). He exhibits a family of strategies, each of which achieves the long-run success proportion of $\max\{\theta_1, \theta_2\}$.

This innovative paper spawned research in a number of directions; the references listed under Robbins (1956) deal with finite memory-bandit problems that generalize Robbins' 'play-the-winner/switch-from-a-loser' strategy.

Related references: See Robbins (1956).

Robbins, H. (1956) A sequential decision problem with a finite memory. *Proc. Nat. Acad. Sci. USA* **42**: 920–923.

There are two Bernoulli arms. Memory of the history of arm selections and results is limited to the previous r stages. Robbins calculates the long-run success proportion for the following rule: select arm 1 and switch to arm 2 if it fails; if it is successful then use it only until it gives r failures in a row; repeat this sequence alternating from one arm to the other.

Related references: Cover (1968), Cover and Hellman (1970), Cover and Wagner (1976), Fox (1974), Isbell (1959), Lakshmanan and Chandrasekaran (1978), Langholz (1977), Meybodi and Lakshmivarahan (1983), Robbins (1952), Samuels (1968), Smith and Pyke (1965), Witten (1973, 1974, 1976, 1977a, 1977b, 1983).

Robbins, H. and Siegmund, D. (1974) Sequential tests involving two populations. *J. Amer. Statist. Assoc.* **69**: 132–139.

There are two normal arms with means θ_1 and θ_2 and unit variance. The hypotheses $\theta_1 = \theta_2 \pm \delta$, where δ is known, are to be tested. A sequential likelihood ratio test is used to decide when to stop sampling, and the allocation scheme is designed to yield a small number of observations of the inferior arm.

This is not a bandit problem in our sense; the paper is included for the same reason that Flehinger and Louis (1971) is included.

Related references: See Colton (1963).

Roberts, K. W. S. and Weitzman, M. L. (1980) On a general approach to search and information gathering. Economics Working Paper 263, Massachusetts Inst. of Technology, USA.

The authors prove a somewhat more general version of the theorem of Gittins and Jones (1974). Applications to a number of problems are discussed. These include optimal search and job scheduling as well as bandits.

Related references: Our Section 6.1; Gittins (1979), Gittins and Jones (1974), Glazebrook (1983a), Varaiya, Walrand, and Buyukkoc (1983), Whittle (1980, 1982).

Robinson, D. (1983) A comparison of sequential treatment allocation rules. *Biometrika* **70**: 492–495.

There are two Bernoulli arms and discounting is n-horizon uniform. Several strategies are compared using simulation for various values of θ_1 and θ_2 and

$n = 50$ and 100. Among those considered are three index strategies of Gittins and Jones (1974) designed for geometric discounting and assuming beta prior distributions on θ_1 and θ_2 (the author does not say which betas – perhaps both are uniform on $(0, 1)$). These three strategies are optimal for discount factors of 0.99, 0.995, and 0.9999. (If these strategies are to be applied in this setting for which they are not designed, then the appropriate discount factor would seem to be $1 - 1/n$.)

This is a curious exercise since finding optimal strategies for these values of n and any prior distribution is quite easy – easier than finding optimal strategies for geometric discounting with factor 0.9999, for example.

Robinson also considers the possibility that half of the responses are delayed for 20 stages.

Related references: Our Chapters 3 and 6; Bather (1981, 1983), Fox (1974), Gittins (1979), Gittins and Jones (1974), Jones (1975), Percus and Percus (1984).

Rodman, L. (1978) On the many-armed bandit problem. *Ann. Probab.* **6**: 491–498.

Consider two known distributions Q^* and Q_* with different means. There are k arms with one of the arms having distribution Q^* (which one is not known) and the other $k-1$ having distribution Q_*. The discount sequence is $\mathbf{A} = (1, \alpha, \ldots, \alpha^{n-1}, 0, 0, \ldots)$ where $\alpha \in (0, 1]$ and $n \in \{1, 2, \ldots, \infty\}$. (The case $\alpha = 1$ and $n = \infty$ is not allowed in our context but makes sense in Rodman's; cf. Zaborskis (1976).) Rodman shows that myopic strategies are optimal: the decision maker can select any arm that has the highest current probability of having the distribution with larger mean. So this generalizes the result of Feldman (1962) in a number of ways. Feldman assumed $k = 2$, Q^*, and Q_* to be Bernoulli, and $\alpha = 1$ with $n < \infty$. (Our Corollary 4.3.10 generalizes Feldman's result to arbitrary nonincreasing discount sequences.)

The uniform discounting version of this result for Bernoulli arms with Q^* having a larger mean than Q_* was proven first by Zaborskis (1976), which was not known to Rodman.

Related references: Our Corollary 4.3.10; DeGroot (1970), Fabius and van Zwet (1970), Feldman (1962), Keener (1984), Kelley (1974), Quisel (1965), Vogel (1960c, 1964), Zaborskis (1976).

Ross, S. M. (1983) *Introduction to Stochastic Dynamic Programming*, Academic Press, New York.

Chapter VII treats bandit problems with geometric discounting. The Gittins–Jones (1974) result is proved using the approach of Whittle (1980, 1982), in a vein similar to our Section 6.1. Ross also considers the extension of Whittle (1981) in which new arms are generated according to a Poisson process.

Related references: Our Section 6.1; Bellman (1956), Gittins (1979), Gittins and Jones (1974), Whittle (1980, 1981, 1982).

Rothschild, M. (1974) A two-armed bandit theory of market pricing. *J. Econ. Theory* **9**: 185–202.

A theory to explain the way in which stores should set prices is constructed using bandit problems. Rothschild also gives some analytic results assuming geometric discounting and two Bernoulli arms. The arms may not be independent; the distribution on parameters θ_1 and θ_2 is arbitrary and has support $(0, 1) \times (0, 1)$. Rothschild shows that optimal strategies will, with positive probability, lead to an infinite number of selections of arm 1 and a finite number of selections of arm 2 when $0 < \theta_2 < \theta_1 < 1$. He also shows that optimal strategies will, with probability one, lead to an infinite number of selections of only one arm when $\theta_1 \neq \theta_2$ (of course, it may not be the better arm that is selected).

Related references: Bellman (1956), Percus and Percus (1984), Schmalansee (1975).

Samuels, S. M. (1968) Randomized rules for the two-armed-bandit with finite memory. *Ann. Math. Statist.* **39**: 2103–2107.

Robbins (1956), Isbell (1959), and Smith and Pyke (1965) consider strategies that stay with an arm until it gives r consecutive failures; a switch is made to the other arm if the other arm passes some predetermined test. The objective is to maximize the long-run proportion of heads. Samuels shows that all these strategies can be improved by randomizing to decide whether to perform the test or stay with the current arm; strategies which give a higher probability of staying with the current arm are better (except that this probability cannot be one).

Related references: See Robbins (1956).

Schmalansee, R. (1975) Alternative models of bandit selection. *J. Econom. Theory* **10**: 333–342.

Some 'learning models' of Bush and Mosteller (1955) are applied to the pricing problem considered by Rothschild (1974). These are *ad hoc* randomized strategies that are not optimal but are easy to apply and may reasonably fit actual behaviour (but see Cane, 1962).

Related references: Bush and Mosteller (1955), Cane (1962), Rothschild (1974), Thompson (1933, 1935).

Smith, C. V. and Pyke, R. (1965) The Robbins–Isbell two-armed bandit problem with finite memory. *Ann. Math. Statist.* **36**: 1375–1386.

There are two unknown Bernoulli arms. The objective is to maximize the long-run success proportion. The current selection can depend only on the previous r selections and observations. The authors improve on strategies considered by Isbell (1959).

Related references: See Robbins (1956).

Sudderth, W. D. (1982) Dynamic programming. *Encyclopedia of Statistical Sciences*, Vol. II (eds S. Kotz and N. L. Johnson) Wiley, New York.

Bandits are discussed as examples of dynamic programming problems. Sudderth discusses the scope of dynamic programming problems and gives a guide to the literature. We have not discussed general theories of dynamic programming either in the body of this monograph or in this bibliography; in particular, articles that treat such general theories are cited here only if they give bandit problems as examples.

Tesfatsion, L. (1978) A new approach to filtering and adaptive control. *J. Optim. Theory Appl.* **25**: 247–261.

A two-armed bandit is posed as an example of a more general set of control problems. The author wants to bypass the use of prior probabilities and Bayes's theorem. This does not seem possible to us. We do not understand whether the author is considering only suboptimal strategies (one strategy discussed is myopic and so backwards induction is avoided) or if she is making a broader claim.

Related reference: Kalaba and Tesfatsion (1978).

Thompson, W. R. (1933) On the likelihood that one unknown probability exceeds another in view of the evidence of two samples. *Biometrika* **25**: 275–294.

This is the paper that started it all! There are two independent Bernoulli arms (Thompson did not use the 'arm' and 'bandit' terminology) with θ_1 and θ_2 having uniform prior distributions on $(0, 1)$. While recognizing that his randomized strategy is 'not the best possible', Thompson proposes selecting arm 1 with probability equal to the current probability that $\theta_1 > \theta_2$. He devotes most of the paper to calculational aspects of this probability.

Even setting aside this paper's historical importance, it is our view that it should be read by all researchers in this area (though the calculational issues are not as important now as they were in 1933); many have not and have failed to improve on the original!

Related references: Our Chapter 7; Bather (1980, 1981), Berry (1978), Thompson (1935).

Thompson, W. R. (1935) On the theory of apportionment. *Amer. J. Math.* **57**: 450–456.

Further calculations are made along the line of Thompson (1933). A crude but effective simulation is carried out to evaluate the randomized strategy suggested in that earlier paper. Even for the small horizons considered, the strategy performs quite well.

It is surprising to us that modern-day authors have not compared their strategies with Thompson's. For example, we think the strategies discussed by Fox (1974) are markedly inferior to Thompson's.

Related references: Bather (1980), Fox (1974), Percus and Percus (1984), Thompson (1933).

Upton, G. J. G. and Lee, R. D. (1976) The importance of the patient horizon in sequential analysis of binomial clinical trials. *Biometrika* **63**: 335–342.

Most of the papers in this bibliography regard the discount sequence as known. This paper makes the point that ignorance of the discount sequence, the horizon in particular, can make discussions concerning differences in allocation strategies largely irrelevant. The discussion is couched in terms of two quite special strategies but we think it applies generally. It is for this reason that we allow the horizon to be random; see our Chapter 3, especially Section 3.1, and also Witmer (1983).

Related references: Our Chapter 3; and see Colton (1963).

Varaiya, P., Walrand, J. and Buyukkoc, C. (1983) Extensions of the multi-armed bandit problem. Electrical Engineering Tech. Rep., Univ. of California, Berkeley, USA.

A generalization of the result of Gittins and Jones (1974) is proved. Various applications are discussed, including discrete- and continuous-time bandit problems.

Related references: Our Section 6.1 and Chapter 8; Gittins (1979), Gittins and Jones (1974), Roberts and Weitzman (1980), Whittle (1980, 1981, 1982).

Viscusi, W. K. (1979a) *Employment Hazards: An Investigation of Market Performance*, Harvard Univ. Press, Cambridge, Massachusetts, USA.

Viscusi, W. K. (1979b) Job hazards and worker quit rates: An analysis of adaptive worker behaviour. *Internat. Econom. Rev.* **20**: 29–58.

An interesting application of two-armed bandits to job selections is discussed. There are two arms. One arm is known and is Bernoulli. The other is unknown (a potentially hazardous job) and results in 0 or w. Discounting is truncated geometric. Various examples and calculations of optimal strategies are carried out.

Related references: Our Chapter 5; Berry and Fristedt (1979, 1983), Berry and Viscusi (1981), Viscusi (1979a).

Vogel, W. (1960a) A sequential design for the two-armed bandit. *Ann. Math. Statist.* **31**: 430–443.

There are two Bernoulli arms. The discount sequence is the n-horizon uniform. The author studies the strategy that indicates the arms alternatively until, at some even-numbered stage $2r$, the discrepancy between the numbers of successes on the two arms is some constant, and thereafter indicates a single arm: the one yielding more successes. Variants of this two-phase design have been considered by many authors – see Colton (1963). Methods of this paper are used in the proof of our Theorem 9.1.3.

Related references: Our Section 9.1; Bather and Simons (1985), Vogel (1960b); and see Colton (1963).

Vogel, W. (1960b) An asymptotic minimax theorem for the two-armed bandit problem. *Ann. Math. Statist.* **31**: 444–451.

Using the results of Vogel (1960a), it is shown that the constants c_3 and c_4 in our Corollary 9.1.4 may be chosen to be 0.187 and 0.377, respectively, provided that n is sufficiently large.

Related references: Our Section 9.1; Bather (1983a), Bather and Simons (1985), Fabius and van Zwet (1970), Vogel (1960a).

Vogel, W. (1960c) Ein Irrfahrten-Problem und seine Anwendung auf die Theorie der sequentiellen Versuchs-Plane. *Arch. Math.* **XI**: 310–320.

There are two Bernoulli arms and the discount sequence is n-horizon uniform. Vogel considers the strategy that indicates the arm for which the current number of successes minus failures is larger. The regret (cf. Section 9.1) when the Bernoulli parameters are θ_1 and θ_2 is obtained in terms of probabilities associated with a particular Markov chain that moves one step to the right or to the left with probabilities depending on the sign of the current state. Limit theorems, as $n \to \infty$, are obtained. In particular, the regret approaches a finite limit if and only if $\theta_1 = \theta_2$ or $1/2$ is between θ_1 and θ_2.

Related references: Feldman (1962), Vogel (1961a, 1964).

Vogel, W. (1961a) Bemerkungen zu einem sequentiellen Versuchsplan. *Z. Angew. Math. Mech.* **41**: 179–181.

The portion of Vogel (1960c) concerned with a bound on the risk independent of n is generalized to a non-Bernoulli setting.

Related reference: Vogel (1960c).

Vogel, W. (1961b) Sequentielle Versuchs-Pläne. *Metrika* **4**: 140–157.

There are two unknown Bernoulli arms. The problem is to minimize the number of stages required to achieve a given number of successes.

Related references: Berry and Fristedt (1980a, 1980b), Gittins (1983).

Vogel, W. (1964) Sequentielle Versuchspläne. *Unternehmensforshung, Operations Research – Recherche Opérationnelle* **8**: 65–74.

The myopic strategy is considered for two Bernoulli arms where (θ_1, θ_2) is either (a, b) or (b, a). Discounting is n-horizon uniform. As a function of $(\theta_1, \theta_2) \in [0, 1] \times [0, 1]$, the regret approaches a finite limit as $n \to \infty$ if $|\log(a/b)/\log[(1-a)/(1-b)]|$ lies between $(1-\theta_1)/\theta_1$ and $(1-\theta_2)/\theta_2$; it does, for example, when $(\theta_1, \theta_2) = (a, b)$ or (b, a).

Related references: Our Section 9.1; Feldman (1962), Reimnitz (1978), Vogel (1960c).

Wahrenberger, D. L., Antle, C. E. and Klimko, L. A. (1977) Bayesian rules for the two-armed bandit problem. *Biometrika* **64**: 172–174.

There are two independent Bernoulli arms and n stages. Allocations are sequential in an experimental phase consisting of r stages; a single arm is used thereafter. However, we cannot tell whether $\mathbf{A} = (1, 1, \ldots 1, T, 0, 0, \ldots)$ as in Meeter (1973), or if $\mathbf{A}$ is the r-horizon uniform with the better arm being used in a follow-on second stage not expressly considered in the objective. (Also, we cannot tell which value of r was used in constructing their tables but we think it was 50 throughout, which means $r = n$ in half the cases considered and $r = n/2$ in the other half.) Simulations are performed assuming that θ_1 and θ_2 have the same beta distribution. The effect of having the wrong prior distribution is considered by generating θ_1 and θ_2 using different beta distributions. The authors conclude that the strategies are robust. We feel that the priors considered are not sufficiently disparate to warrant this conclusion.

The strategy that was best among those strategies considered by Fox (1974) does not fare well compared with the strategies discussed above.

Related references: Our Chapter 7; Berry (1972), Fox (1974); and see Colton (1963).

Whittle, P. (1980) Multi-armed bandits and the Gittins index. *J. R. Statist. Soc. B* **42**: 143–149.

Whittle gives an alternative proof of the Gittins–Jones (1974) theorem. This result is given as Theorem 6.1.1 in our Chapter 6 and the proof we use is essentially that of Whittle.

Related references: Our Section 6.1; Gittins (1979), Gittins and Jones (1974), Glazebrook (1983a), Roberts and Weitzman (1980), Varaiya, Walrand, and Buyukkoc (1983), Whittle (1982).

Whittle, P. (1981) Arm-acquiring bandits. *Ann. Probab.* **9**: 284—292.

Each of a number of independent arms can be in one of a finite number of 'states', which are distributions in our description of the problem. (This finiteness restriction limits the paper's applicability, but Whittle claims it 'could almost certainly be relaxed'.) Time is discrete and the number of arms in each state can increase with time. Such an increase is random and independent of the number of arms in each state and of the strategy used.

Discounting is geometric. Using an approach similar to that of Whittle (1980), the author extends the Gittins–Jones (1974) result to this case: namely, he shows that 'Gittins index policies' are optimal (see Section 6.1).

Related references: Our Section 6.1; Gittins and Jones (1974), Gittins and Nash (1977), Whittle (1980, 1982).

Whittle, P. (1982) *Optimization Over Time: Dynamic Programming and Stochastic Control*, Vol I, Wiley, New York.

Chapter 14 treats bandit problems with geometric discounting. Whittle proves the Gittins–Jones (1974) result along the lines of Whittle (1980), see our Section 6.1. A number of applications are discussed.

Related references: Our Section 6.1; Gittins (1979), Gittins and Jones (1974), Glazebrook (1983a), Roberts and Weitzman (1980), Varaiya, Walrand, and Buyukkoc (1983), Whittle (1980, 1981).

Whittle, P. (1983). *Optimization Over Time: Dynamic Programming and Stochastic Control*, Vol. II, Wiley, New York.

This is included here since it is a continuation of Whittle (1982); it does not treat bandit problems.

Witmer, J. A. (1983) Bayesian multistage decision problems. Ph.D. thesis, Univ. of Minnesota, USA.

The problem is the same as the two-phase trial considered by Berry and Pearson (1984) with arm 2 known, except that the horizon n is unknown. A strategy specifies a first-phase length r, which arm to use at each stage up to r, and then, depending on results from the first phase, which arm to use for the duration of the trial (if n turns out to be larger than r). Optimal strategies depend on the distributions of n and θ_1, which are assumed to be independent. For reasons discussed in our Chapter 3, the problem is equivalent to discounting observations assuming $A = (\alpha_1, \alpha_2, \ldots)$ where α_m is the probability that $n \geqslant m$. Witmer characterizes optimal strategies when A is *regular*, (see our Definition 5.2.1). He gives special consideration to the case in which n has a geometric distribution; he concludes in that case that the problem is then equivalent to one in which n is assumed known, and equal to the mean of the geometric.

Related references: Our Chapters 3 and 5; and see Colton (1963).

Witten, I. H. (1973) Finite-time performance of some two-armed bandit controllers. *IEEE Trans. Systems Man Cybernet*, **3**: 194–197.

A finite-memory strategy is suggested and compared with two other strategies, one of which was put forth by Cover and Hellman (1970). All three strategies achieve the maximum possible long-run success proportion (among

finite-memory strategies), but have quite different finite behaviour.
Related references: See Robbins (1956).

Witten, I. H. (1974) On the asymptotic performances of finite-state two-armed bandit controllers. *IEEE Trans. Systems Man Cybernet*, **4**: 465–467.

Witten notes that many strategies have the same long-run success proportion. As an alternative, he suggests a two-parameter asymptotic measure of the expected number of selections of the inferior arm.
Related references: See Robbins (1956).

Witten, I. H. (1976) The apparent conflict between estimation and control – a survey of the two-armed bandit problem. *J. Franklin Inst.* **301**: 161–189.

This is a survey article which gives an excellent review of the then existing literature concerning finite-memory and finite-state bandit problems, and some discussion of the Bayesian approach.
Related references: Bellman (1956), Bradt, Johnson, and Karlin (1956); and see Robbins (1956).

Witten, I. H. (1977a) An adaptive optimal controller for discrete-time Markov environments. *Informat. and Control* **34**: 286–295.

A general Markov control setting is considered; discounting is geometric. The arm selected next is allowed to depend on the previous and also the penultimate selection.
Related references: See Robbins (1956).

Witten, I. H. (1977b) Exploring, modelling, and controlling discrete sequential environments. *Internat. J. Man-Mach. Stud.* **9**: 715–735.

Witten considers a number of control problems, including bandit problems, in a manner similar to Witten (1977a).
Related references: See Robbins (1956).

Witten, I. H. (1983) Non-deterministic modelling and its application in adaptive optimal control. *Mathematical Learning Models – Theory and Algorithms* (eds U. Herkenrath, D. Kalin and W. Vogel), pp. 213–226, Springer-Verlag, New York.

This is a survey paper in which the author discusses bandit problems along the lines of Witten (1977a).
Related references: See Robbins (1956).

Woodroofe, M. (1976) On the one arm bandit problem. *Sankhyā A* **38**: 79–91.

There are two arms. Arm 2 is known to always give 0. Discounting is geometric with factor α. Woodroofe is interested in results for $\alpha \to 1$.

In one setting arm 1 is uniformly distributed on $(-1 + 2\theta_1, 1)$, where the conditional (conditional on Q_1 in our notation) mean θ_1 is distributed according to the distribution measure $F_1(\mathrm{d}u) = ab^a(1-u)^{-(a+1)}\mathrm{d}u$, $u < 1-b$, where $a > 1$ and $b > 0$. If $b \leqslant 1$ it is clearly optimal to always select arm 2 and $V((a, b), 0; \alpha) = 0$. For $0 < b < 1$, Woodroofe shows that $E[\max\{0, \theta_1\}|F_1]/(1-\alpha) - V((a, b), 0; \alpha) \to ab^a/(a-1)$ as $\alpha \uparrow 1$. He shows in addition that the same expression holds if the maximal worth V is replaced by the worth of the strategy that indicates arm 1 until (if ever) it gives a number $\leqslant -1$, and arm 2 thereafter.

In a second setting, the unknown distribution Q_1 on arm 1 is normal with variance 1. The mean θ_1 is unknown and its distribution F_1 is normal with mean μ and variance ρ^2 (cf. our Section 2.1). It is shown that $E[\max\{0, \theta_1\}|F_1] - V((\mu, \rho), 0; \alpha) \sim [2\rho\sqrt{(2\pi)}]^{-1} \cdot \exp[-\mu^2/(2\rho^2)]\log^2[1/(1-\alpha)]$ as $\alpha \uparrow 1$. A family of strategies is discussed in this setting.

Woodroofe makes significant use of the methods employed in [Chow, Y. S., Robbins, H., and Siegmund, D. (1970). *Great Expectations*. Houghton Mifflin, Boston].

Related references: Our Chapter 5; Kelly (1981).

Woodroofe, M. (1979) A one-armed bandit problem with a concomitant variable. *J. Amer. Statist. Assoc.* **74**: 799–806.

This paper is unique to our knowledge in that it considers the following modification of a bandit problem that has much practical relevance. The response of an arm depends on a concomitant variable (or 'covariate') which can be measured and that has a known distribution. For example, this covariate may be severity of a disease, age of a machine, difficulty of a task, etc. The objective is the same: maximize the expected discounted sum of the observations.

Discounting is geometric. There are two arms. Arm 2 is known to always give 0. The outcome X_{1m} on arm 1 at stage m can be written as $X_{1m} = R_m + Y_m + \theta_1$, where $E(Y_m) = 0$, $\{R_m : m = 1, 2, \ldots\} \cup \{Y_m : m = 1, 2, \ldots\} \cup \{\theta_1\}$ is an (unconditionally) independent family of random variables, and the vectors (R_1, Y_1), (R_2, Y_2), ... are identically distributed. At stage m, R_m is the concomitant variable – it is observed (but is not added to the payoff) before an arm is selected for stage m.

Woodroofe makes certain assumptions concerning the various distributions; in particular, he assumes that the support of the distribution of R_m is unbounded above. He proves that the discrepancy between $(1-\alpha)^{-1} E(\max\{X_{1m}, 0\})$ and the maximal worth is asymptotic to $c \log(1/(1-\alpha))$ as $\alpha \uparrow 1$, where c is a constant that depends on the distribution of R_m and Y_m. He shows that the same is true for the worth of a myopic strategy. The reason such a result is possible is that the presence of an unbounded R_m assures that a myopic strategy will indicate arm 1 infinitely often.

We note that the result of Gittins and Jones (1974) applies to this concomitant-variable setting.

Related references: Bather (1980, 1981), Gittins (1979), Gittins and Jones (1974), Woodroofe (1976).

Woods, P. J. (1959) The effects of motivation and probability of reward on two-choice learning. *J. Exp. Psychol.* **57**: 380–385.

The experimental procedure is similar to that of Brand, Sakoda, and Woods (1957) and Brand, Woods, and Sakoda (1956). However, an additional instruction makes that current setting different from what we term a 'bandit'. Namely, the subject is told that she is to try to outguess the experimenter. This makes the problem more like a sequential two-person game than a problem of decision under uncertainty.

Related references: Brand, Sakoda, and Woods (1957), Brand, Woods, and Sakoda (1956).

Yakowitz, S. J. (1969) *Mathematics of Adaptive Control Processes*, Prentice-Hall, Englewood Cliffs, N.J.

Chapter 4 contains an elementary account of two-armed bandits with one arm known. Examples are carried out for n-horizon uniform discounting. The development of Bellman (1956) for geometric discounting is given. Yakowitz also discusses the history of the two-armed bandit problem.

Related references: Our Chapters 1 and 5; Bellman (1956), Berry and Fristedt (1979), Bradt, Johnson, and Karlin (1956).

Zaborskis, A. A. (1976) Sequential Bayesian plan for choosing the best method of medical treatment. *Avtomatika i Telemekhanika* **II**: 144–153.

There are k Bernoulli arms. One of the arms has known success probability a (but which one is not known), and the other $k-1$ have success probability $b < a$. (Zaborskis also allows these latter success probabilities to be arbitrary but none can be greater than b.) The discount sequence is n-horizon uniform, and $n = \infty$ is allowed by considering regrets, which are shown to be finite. Zaborskis shows that myopic strategies are optimal for any n: the decision maker should always select an arm which has the largest probability of being the a-arm. This result generalizes Feldman (1962) and is itself generalized by Rodman (1978), who allows arbitrary distributions and also considers geometric discounting.

An interesting feature of this paper is that it cites an actual clinical trial that was conducted as a bandit problem (this trial is unique in this respect as far as we know); namely, the myopic strategy described above was followed.

Unfortunately, while we applaud such an application, we do not believe that the prior distribution was appropriate. The three arms involved were three different doses of the same drug (isoptin, which is used to treat cardiac arrhythmias). It does not seem reasonable to suppose that one dose is $100a$ percent effective – equally likely to be low, medium, and high – while the other two are only $100b$ percent effective. (It was assumed that $b = 0.6$ and $a = 0.8$. Predictably, after the first few patients were given the lower doses, the rest received the high dose.) Rather, this seems to be a problem in estimating a dose–response relationship: higher doses are bound to be more effective but the minimum effective dose should be used to reduce side effects.

Related references: Our Corollary 4.3.10; DeGroot (1970), Fabius and van Zwet (1970), Feldman (1962), Kelley (1974), Quisel (1965), Rodman (1978), Vogel (1960c, 1964).

Zelen, M. (1969) Play the winner rule and the controlled clinical trial. *J. Amer. Statist. Assoc.* **64**: 131–146.

There are two Bernoulli arms with n-horizon discounting. The trial is in two phases. Two strategies are considered for the first or experimental phase: (i) 'play-the-winner/switch-from-a-loser', and (ii) equal allocation. In both cases the arm yielding more successes is selected exclusively in the terminal phase.

Zelen concludes that the optimal length of the first phase should be $n/3$ for both types of allocation. For the equal allocation case he cites Colton (1963) who uses an unusual 'local maximin' argument. For the play-the-winner/switch-from-a-loser case Zelen uses a Taylor series argument on the top of page 143 that we think is incorrect. In either case we fail to see how any fixed fraction of n has merit. Suppose $r(n)$ is the first-phase length. Both allocations discussed require that a fraction of the $r(n)$ be used on the inferior arm (unless, in the case of play-the-winner/switch-from-a-loser, $\max\{\theta_1, \theta_2\}$ $= 1$). To achieve a long-run success proportion of $\max\{\theta_1, \theta_2\}$, the design must satisfy $r(n) \to \infty$ and $r(n)/n \to 0$ as $n \to \infty$ (unless $\theta_1 = \theta_2$ or $\max\{\theta_1, \theta_2\} = 1$). Colton (1963) and Cornfield, Halperin, and Greenhouse (1969) find Bayesian solutions for which r has order of magnitude $\sqrt{n}$, as do Canner (1970) and Berry and Pearson (1984), among others.

Using tables and assuming particular values of θ_1 and θ_2, Zelen gives the overall proportion of successes achieved by using play-the-winner/switch-from-a-loser allocation in the first phase. Such a design is better than equal allocation, but, as we have indicated elsewhere (see Percus and Percus, 1984, for example), play-the-winner/switch-from-a-loser allocation has little to recommend itself in any absolute sense. In the case at hand, suppose the values of θ_1 and θ_2 are known but not which is θ_1. Then an extension of the Feldman (1962) result to the current two-phase setting applies to show that myopic strategies are optimal. Such strategies stay with a winner but there are a great many instances in which they stay with a loser as well. One way of viewing Zelen's approach (he is far from being unique in this regard) is that he uses the two-phase design to have a way of getting out from using a bad first-phase allocation. If the decision maker's hands are not tied then there is no reason at all to want the first phase to end (so $r(n)$ would equal n). In practice, of course, there are good reasons which force one to stop experimenting eventually. But putting unnecessary restrictions on strategies to bring this about artificially gives mathematical exercises devoid of relevance beyond the mathematics.

Related references: See Colton (1963).

Name index

Abdel Hamid, A. R., 207
Anscombe, F. J., 208, 211, 219, 221, 222, 223, 233, 241, 243
Antle, C. E., 3, 8, 210, 227, 242, 255
Armitage, P., 208
Asano, C., 221, 231

Barnett, B. N., 3, 7, 63, 64
Bather, J. A., viii, 178, 189, 197, 201, 206, 209, 210, 211, 213, 220, 221, 228, 229, 230, 242, 243, 245, 247, 250, 252, 254, 259
Beckmann, M. J., 211, 225, 230, 231
Begg, C. B., 211, 219, 221, 241, 246
Bellman, R., 2, 7, 86, 88, 134, 210, 212, 213, 227, 228, 229, 240, 251, 257, 259
Benzig, H., 212, 231, 232, 239, 240
Berger, J. O., 135
Berry, D. A., 50, 64, 73, 82, 88, 90, 92, 99, 107, 125, 128, 129, 131, 134, 150, 165, 210, 212, 213, 214, 215, 217, 218, 220, 221, 223, 224, 225, 228, 233, 234, 235, 236, 238, 239, 240, 245, 248, 252, 253, 254, 255, 256, 259, 260
Blackwell, D., 236
Bradt, R. N., 2, 3, 7, 28, 49, 86, 88, 92, 114, 134, 155, 165, 207, 210, 212, 213, 215, 220, 223, 225, 235, 237, 239, 240, 248, 257, 259

Brand, H., 215, 216, 217, 244, 259
Breakwell, J., 175, 178, 189, 211, 216, 220
Bush, R. R., 216, 217, 232, 233, 244, 251
Buyukkoc, C., 228, 230, 249, 253, 255, 256

Cane, V. R., 217, 251
Canner, P. L., 208, 217, 221, 238, 244, 260
Chandrasekaran, B., 222, 241, 249
Chernoff, H., 167, 175, 178, 179, 189, 211, 216, 217, 218, 219, 220, 221, 227, 246
Chow, Y. S., 20, 49, 258
Christensen, R., 134
Chung, F., 220
Clayton, M. K., viii, 63, 125, 128, 131, 134, 220
Colton, T., 207, 208, 209, 211, 215, 218, 219, 220, 221, 222, 223, 226, 227, 231, 232, 238, 241, 242, 243, 244, 245, 246, 249, 253, 254, 255, 260, 261
Cornfield, J., 208, 221, 222, 260
Cover, T. M., 222, 249
Cox, T. F., 230

Day, N. E., 221, 223
DeGroot, M. H., 11, 12, 49, 80, 82, 223, 225, 238, 250, 260
Dellacherie, C., 181, 190
Dempster, M. A. H., 228
Dubins, L. E., 19, 45, 49, 223, 225

Eick, S. G., viii, 20, 36, 49

263

Kelley, T. A., 81, 82, 218, 221,
   223, 225, 238, 239, 247, 250,
   260
Kelly, F. P., 210, 238, 258
Kemperman, J. H. B., 248
Klimko, L. A., 3, 8, 210, 227, 242,
   255
Kolonko, M., 212, 231, 232, 239,
   240
Kôno, K., 240
Kotz, S., 213, 252
Kumar, P. R., 240

Lai, T. L., 208, 221, 240, 241
Lakshmanan, K. B., 241, 249
Lakshmivarahan, S., 242, 249
Langenberg, P., 221, 249
Lee, R. D., 208, 221, 253
Lellouch, J., 221, 244, 245
Lenstra, J. K., 228
Levin, B., 221, 240, 241
Lindley, D. V., 63, 64
Lipster, A. S., 174, 190
Louis, T. A., 221, 225, 226, 242,
   249

Mallows, C. I., 242
McCulloch, R. E., viii
Meeter, D. A., 221, 242, 255
Mehta, C. R., 211, 219, 221, 241,
   246
Meybodi, M. R., 242, 249
Meyer, P.-A., 181, 190
Morrison, D. F., 243
Mosteller, F., 216, 217, 223, 232,
   233, 244, 251
Murray, F. S., 216, 217, 243

Nakajima, N., 244
Nash, P., 229, 244, 256
Ney, P., 190, 206
Nordbrock, E., 79, 82
Noshi, T., 244

Obregon, I., 244
Orey, S., viii
Oudin, C., 221, 244, 245

Parthasarathy, K. R., 13, 14, 15,
   21, 49
Parthasarathy, T., 194, 206
Pearson, L. M., 35, 49, 214, 217,
   218, 221, 244, 245, 256, 260
Percus, J. K., 209, 213, 225, 227,
   234, 235, 245, 250, 251, 253,
   260
Percus, O. E., 209, 213, 225, 227,
   234, 235, 245, 250, 251, 253,
   260
Petkau, A. J., viii, 179, 189, 211,
   219, 220, 221, 245, 246
Poloniecki, J. D., 210, 246
Port, S., 190, 206
Pratt, J. W., 63, 64
Presman, E. L., 246, 247
Prokhorov, J. V., 247
Pyke, R., 3, 7, 249, 251

Quisel, K., 70, 82, 238, 247, 248,
   250, 260

Raghavan, T. E. S., 194, 206
Ray, S. N., 63, 64, 167, 175, 178,
   189, 211, 217, 218, 219, 220,
   246
Reimnitz, P., 248, 255
Rhenius, D., 19, 49
Rieder, U., 231, 248
Rinnooy Kan, A. H. G., 228
Rizvi, M. H., 210
Robbins, H., 3, 7, 79, 82, 207, 208,
   209, 210, 221, 222, 223, 226,
   227, 229, 231, 232, 234, 240,
   241, 242, 243, 245, 248, 249,
   251, 252, 257, 258
Roberts, K. W. S., 228, 230, 249,
   253, 255, 256
Robinson, D., 249
Rodman, L., 81, 82, 210, 224, 225,
   238, 243, 248, 250, 259, 260
Ross, S. M., 250
Rothschild, M., 232, 251
Rustagi, J. S., 210

Sakoda, J. M., 215, 216, 217, 244,
   259

Subject index

Symbol index

General

Specific to real (continuous) time

BRITISH AND AMERICAN

1750–1920

General editors: Martin Banham an

**George Colman the Younger
and Thomas Morton**

Plays by
George Colman the Younger
and Thomas Morton

INKLE AND YARICO
THE SURRENDER OF CALAIS
THE CHILDREN IN THE WOOD
BLUE BEARD, or FEMALE CURIOSITY!
SPEED THE PLOUGH

Edited with an introduction and notes by
Barry Sutcliffe

CAMBRIDGE UNIVERSITY PRESS

Cambridge
London   New York   New Rochelle
Melbourne   Sydney

Published by the Press Syndicate of the University of Cambridge
The Pitt Building, Trumpington Street, Cambridge CB2 1RP
32 East 57th Street, New York, NY 10022, USA
296 Beaconsfield Parade, Middle Park, Melbourne 3206, Australia

First published 1983

Printed in Great Britain at the University Press, Cambridge

Library of Congress catalogue card number: 83–5156

British Library Cataloguing in Publication Data
Colman, George
Plays by George Colman the Younger and
Thomas Morton — (British and American playwrights
1750—1920)
1. English drama — 18th century
I. Title   II. Morton, Thomas   III. Sutcliffe,
Barry   IV. Series
822′.6′08   PR1269

ISBN 0 521 24019 0 hard covers
ISBN 0 521 28400 7 paperback

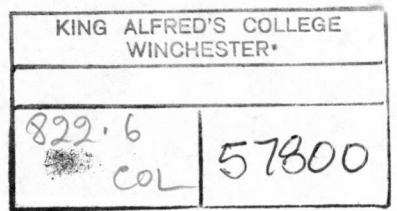
WD

GENERAL EDITORS' PREFACE

It is the primary aim of this series to make available to the British and American theatre plays which were effective in their own time, and which are good enough to be effective still.

Each volume assembles a number of plays, normally by a single author, scrupulously edited but sparingly annotated. Textual variations are recorded where individual editors have found them either essential or interesting. Introductions give an account of the theatrical context, and locate playwrights and plays within it. Biographical and chronological tables, brief bibliographies, and the complete listing of known plays provide information useful in itself, and which also offers guidance and incentive to further exploration.

Many of the plays published in this series have appeared in modern anthologies. Such representation is scarcely distinguishable from anonymity. We have relished the tendency of individual editors to make claims for the dramatists of whom they write. These are not plays best forgotten. They are plays best remembered. If the series is a contribution to theatre history, that is well and good. If it is a contribution to the continuing life of the theatre, that is well and better.

We have been lucky. The Cambridge University Press has supported the venture beyond our legitimate expectations. Acknowledgement is not, in this case, perfunctory. Sarah Stanton's contribution to the series has been substantial, and it has enhanced our work.

Martin Banham
Peter Thomson

CONTENTS

ILLUSTRATIONS

INTRODUCTION

During the final decade of the eighteenth century and the early decades of the nineteenth century, a period which gave witness to the playwriting pre-eminence of George Colman the Younger and Thomas Morton, there arose an acute problem in the critical perception of theatre which still persists today. John Genest, a meticulous scholar of theatre history, writing in 1830, saw the difficulty in the following terms: 'Good plays,' he observes, 'generally succeed — bad plays are generally unsuccessful — there are however so many exceptions to the general rule that success cannot with propriety be considered as the criterion of merit.'[1] If that, indeed, is the case, then by what yardstick should merit be measured? Genest approaches this question by quoting from an earlier authority:

> Before a play can be concluded to be good because it pleases, we ought to consider who are pleased by it: they who understand, or they who do not? They who understand? Alas, they are but few . . . He who writes to the many at present, writes only to them, and his works are sure never to survive his admirers; but he who writes to the knowing few at present, writes to the race of mankind in all succeeding ages.[2]

Genest's source here is John Morris, who was writing over 130 years earlier, at the close of the seventeenth century. Morris's remarks are striking for the accuracy with which they anticipate the critical sympathies of succeeding generations of dramatic commentators. Genest, for example, in spite of the fact that *The Surrender of Calais* was undoubtedly the play most popular with audiences in the London of 1791, tartly insists that 'it met with much greater success than it deserved'.[3] Similarly, summing up the career of Colman the Younger, whose playwriting achievement dominated the thirty years between 1790 and 1820, Allardyce Nicoll is prepared to concede only that 'he was among the worthiest of the unworthy dramatists who followed Sheridan'.[4] In spite of the century which separates Nicoll and Genest, the criterion of merit implicit in their judgments is remarkably similar and, in both cases, fails even to acknowledge the existence of a theatrical or a social context which might account for the particular success of Colman's work. Indeed, if popular acclaim and palpable artistic integrity appear, to the critic, to be parting company, is he justified in dismissing the play, or should he readjust his criteria to take into account the new set of prevailing standards?

Frederick Reynolds, himself one of the unworthy dramatists who followed Sheridan yet whose comedies were extraordinarily popular with audiences, puts the case from the playwright's point of view. In the preface to his play *Begone Dull Care* (1808), he protests against what he sees as the blinkered vision and inappropriate terms of reference of the majority of theatre critics of the period, whose

1

'constant cry,' writes Reynolds, 'is, why don't you give us the good old legitimate drama, such as flourished in the days of Shakespeare, Jonson, Vanbrugh and Congreve?'[5] And the reason why this type of dramatic fare was not presented at this time is quite simply explained, by Genest, ironically enough, in his closing remarks on the Covent Garden season for 1819—20: 'The plays most frequently acted this season were either bad or indifferent — the fault however was more in the audience . . . when the taste of the public is become so vitiated that they can applaud one of Shakespeare's plays turned into an opera and hiss Congreve and Vanbrugh, what can a manager do?'[6]

Reynolds's phrase 'the good old legitimate drama' is of particular significance here and provides the first clue towards understanding the ethos of public taste and theatrical performance by which the critical keepers of drama's conscience were at first distressed and then alienated during the late eighteenth and early nineteenth centuries. Made to look to its laurels by moralists like Morris, perpetuated, if only in spirit, by scholars like Genest, 'the good old legitimate drama' is specifically located by Reynolds as the force against which he is fully prepared to define his identity as a playwright. Central to this definition is the particular stress Reynolds places on the word *legitimate* to indicate the point of comparison between the work of previous generations of playwrights and that of his own. To Morris, the use of *legitimate* as a critical expression would have seemed distinctly peculiar, yet, by the end of the eighteenth century, it had become one of the most contentious terms in the whole vocabulary of theatrical life, capable not only of polarising opinion instantly, but of triggering a complicated series of value judgments.

Legitimate, in fact, had a twofold significance at this time. To begin with, it could simply mean lawful, the law in question being the 1737 Licensing Act, passed ostensibly as a public order measure although in effect an important tool of government in enforcing censorship. Strict control over theatrical performance was achieved by means of careful licensing both of theatres and of individual plays, responsibility for this resting with the Lord Chamberlain. However, as far as London was concerned, the situation was complicated by the existing rights of two Letters Patent granted in perpetuity by Charles II to Thomas Killigrew and Sir William Davenant, enabling them to share a monopoly over the performance of 'spoken drama' in the City of Westminster. By the year the Licensing Act came into operation, these Letters Patent had passed into the hands of the managers of the Covent Garden and Drury Lane theatres, a section of the Act appearing to endorse their theatrical monopoly by prohibiting the performance of plays in any London theatre outside the immediate Westminster area. Indeed, the Covent Garden/Drury Lane monopoly continued to be protected in law until 1843, the only substantial alteration during this extensive time-span occurring in 1766, when the Haymarket Theatre (later to be managed by Colman the Younger) was granted the privilege of receiving an annual licence to perform plays through the summer months, the two other theatres traditionally being closed over this period.

However, the controversy surrounding *legitimate* as a theatrical term might never have become so fierce had it not been for an additional Act of 1752 which allowed a degree of ambiguity to creep into the definition of performance. This Act permitted local magistrates in and within twenty miles of London and Westminster to license 'places of public entertainment', its purpose being to facilitate the policing of areas where crowds were likely to gather. Nevertheless, by the end of the century, several auditoria had been licensed in this way, amongst them Sadler's Wells, Astley's Amphitheatre and the Royal Circus. On the assumption that the two patent monopolies and the Licensing Act removed their right to perform plays, the managers of these auditoria exhibited shows of singing, dancing, horsemanship and scenic spectacle which, as the century advanced, became increasingly quasi-dramatic in form, following strong narrative lines derived either from stirring topical events (*The Destruction of the Bastille*, Royal Circus, 1789) or pirated from popular plays produced at the patent theatres. In spite of this deliberately low profile kept by the minor theatres, John Palmer, a Drury Lane actor, opened a playhouse in the east of London in 1787 on the dubious authority of a magistrate's 'entertainment' licence and proceeded to fly in the face of existing laws and rights by producing the two spoken dramas of *As You Like It* and *Miss In Her Teens*. Retribution was swift and effective. The combined might of Covent Garden, Drury Lane and Haymarket vested interests, together with the invocation of the Licensing Act and veiled threats about Palmer's future employability, crushed this encroachment upon the monopoly outright. In a vigorous pamphlet and newspaper campaign, the message was spelled out to other would-be trespassers that it was *legitimate* for the patent theatres alone to present regular tragedy, comedy and farce to a London public.

The second application of *legitimate* as a theatrical term is more intriguing. Not only could it describe a lawful type of performance, but it also came to be used to indicate the quality — almost the moral worth — of that performance, even if wholly legitimate in the legal sense. Richard Cumberland was a writer of comedies in the generation immediately preceding that of Morton and Colman who lived to see his plays and his style slip out of fashion. The following passage from his *Memoirs*, although heavily tainted with resentment, gives a clear outline of the sort of questions being raised in some quarters at the turn of the eighteenth century about the very legitimacy of the legitimate theatre itself.

> I have stood firm for the corps into which I enrolled myself, and never disgraced my colours by abandoning the cause of *legitimate* comedy, to whose service I am sworn, and in whose defence I have kept the field for nearly half a century, till at last I have survived all true national taste, and lived to see buffoonery, spectacle and puerility so effectually triumph, that now to be repulsed from the stage is to be recommended to the closet, and to be applauded by the theatre is little else than a passport to the puppet-show.[7]

Playwriting, he adds, is a process 'referable only to the unwritten law of the heart, and that is nature'.[8]

There are two aspects to Cumberland's attack on his successors in the legitimate drama. In the first place, by identifying their work with the 'buffoonery, spectacle and puerility' in terms of which the entertainment ethic of the minor theatres operated, Cumberland is laying a charge of illegitimacy which goes beyond a mere allusion to categories of performance. Indeed, implicit in his criticism is the condemnation of an act of moral subversion and artistic irresponsibility. To Cumberland, the new playwrights represent a serious threat to the notion of drama as rational entertainment, in which the spoken word is hallowed because it conveys an idea, by appearing to be hastening in a new dark age of puppet-show hedonism. This charge is reinforced by the second element in Cumberland's attack, which carries the positive suggestion that his successors are not simply breaking a few artistic rules but are active anarchists, dedicated to challenging the very laws of nature on which legitimate drama, in his view, should be founded. What is so interesting about this latter accusation is that Cumberland is clearly in no two minds about the fact that this form of anarchy largely represents the spirit of the times. 'It is a gaudy, thoughtless age,' he observes, 'and they, who live up to the fashion of it, live in a continual display of scenery; their pleasures are all pantomimical.'[9]

Some of the commentators who identified contemporary playwrights as being implicated in an anarchic process saw their actions in stronger terms than just the breaking of accepted theatrical traditions. In 1794, William Gifford produced *The Baviad*, a satirical poem which lampoons several dramatic authors of the period for reducing their art to 'the drivellings of idiotism',[10] and singles out Colman and Morton for particular mention. Colman's work is damned as 'flippant trash',[11] whilst Morton is said to 'have happily found "In the *lowest* deep, a *lower* still".'[12] In a revealing footnote, however, Gifford extends these observations to his own experience of their plays in performance, all of which met, he says, 'with peals of applause. I cannot believe it myself, though I have witnessed it.'[13] If Gifford's satire was an intellectual riposte aimed at correcting 'the growing depravity of the public taste',[14] his poem certainly supplied powerful ammunition to others with a more overtly political message to deliver. When the American edition of *The Baviad* appeared in 1799, it was prefixed with a poetical epistle to the author written by William Cobbett, whose famous *Rural Rides* was published in 1830. Cobbett interprets the trends pinpointed by Gifford as part of a far more sinister force then beginning to infect society:

> Touch'd with the mania, now, what millions rage
> To shine the laureat blockheads of the age.
> The dire contagion creeps thro' every grade,
> Girls, coxcombs, peers and patriots drive the trade:
> And e'en the hind, his fruitful fields forgot,
> For rhyme and misery leaves his wife and cot . . .
> Thus swarming wits, of all materials made,

Their gothic hands on social quiet laid,
And, as they rave, unmindful of the storm,
Call lust refinement, anarchy reform.[15]

What Cobbett detects in this phenomenon of 'dire contagion', represented in the artistic world by the popular appeal of a new taste which applauds the work of 'laureat blockheads', amongst them Colman and Morton, is a circumstance analogous, in the political world, to the growing movement for reform, or anarchy, as Cobbett prefers to see it (although it is worth bearing in mind that Cobbett had reversed his political position by 1804, itself interesting evidence of the volatility of society at the turn of the eighteenth century). Reform, during this period, meant the struggle to assert a totally representative form of parliamentary government, large sections of the population effectively being disenfranchised by the electoral system, which equated votes with property ownership. At a time when Britain was governed by a consensus restricted to certain social classes, any movement, whether political or artistic, which asserted a popular identity was by its very nature radical and a threat to the *status quo*. Writers of plays in the populist tradition were as guilty as the reformers in one crucial respect: both sought in some degree to level the differences in late eighteenth-century society.

Heartened by the Revolution in France, radical opinion became more organised and more outspoken in England during the 1790s than at any previous time. Tom Paine's *Rights of Man*, published in 1792, achieved widespread circulation within a single year. In London, the Corresponding Society was formed and grew into a popular forum of radical debate, its cardinal principle being that 'every adult person, in possession of his reason, and not incapacitated by crimes, should have a vote for a Member of Parliament'.[16] Amongst others, some playwrights of the period are known to have had strong radical sympathies. Elizabeth Inchbald, for instance, was a keen supporter of the French armies in the Revolutionary Wars in Europe, a political stance which put her in direct opposition to English foreign policy of the period. Thomas Holcroft, a fellow dramatist of Colman and Morton, was one of the leading members of the Constitutional Society, which was also pledged to electoral reform, arrested in the summer of 1794 on an indictment of high treason. Their subsequent trial became a contemporary *cause célèbre* and terminated, for Holcroft, with his case being dismissed. However, by the end of the decade, all lawful outlets for the expression of radical sentiment had been closed. The *Rights of Man* was a proscribed book, meetings of the London Corresponding Society and all similar institutions were prohibited, and the struggle for a more equitable society was no longer regarded by the lawmakers of England as a legitimate activity.

Rousseau, in his *Social Contract* (1762), one of the most famous of the political statements to emerge from the eighteenth century and certainly the central philosophical impulse behind the concerted pursuit of popular liberty in America and France during this period, anticipates the exact nature of the crisis that was to occur in England in the 1790s. 'Laws,' he wrote, 'are really nothing other than the conditions on which civil society exists. A people, since it is subject to laws, ought to be the author of them. The right of laying down the rules of society belongs only to those who form the society; but how can they exercise it?'[17] In England, the existing political balance of power was challenged by the growing body of opinion, eventually and inevitably suppressed, which recognised that the law as it then stood was, in general, nothing more than a Magna Carta granted to vested interests. Nevertheless, although such interests were successful in depriving the reformers of their dream for all men to participate in the democratic process, as far as London theatre life was concerned they failed, even after invoking the combined weight of 'legitimacy' both civil and moral, to preserve the elitist cachet of drama handed down by the aristocratic playhouse of the Restoration period and the intellectual snobbery of the Augustan age. In the words of a contemporary observer: 'The present was the age of revolutions. The most surprising events had occurred on the stage of real life, and the mimic world followed the course which seemed to strike down all reasonable expectations.'[18] As will be seen, public opinion, when translated into audience reaction, was an uncompromising and unstoppable force in the face of which a term like *legitimate* became, at best, little more than the rhetorical, if insistent, bark of those who realised that their bite was virtually lost.

Clearly, the relationship between the desire for change in society at large and the substantial shift of emphasis in theatrical taste, both of which took place simultaneously, should be seen as one of analogy only. The rigid screening of all scripts prior to their first performance eliminated any possibility of overt political sentiment — apart, that is, from patriotic effusions — receiving systematic exposition in the plays of the period. Nor were all playwrights, by any means, friendly to the idea of social change or even conscious that they were capable, simply through their choice of modes of presentation, of contributing to a process which some observers were interpreting as a subversive act. Yet, there can be little doubt that the shock given to the social *status quo* by the American Rebellion followed by the French Revolution set up a sympathetic reaction against traditional expectations amongst theatre audiences. In a letter addressed to James Winston, one of his partners in the Haymarket Theatre, in 1810, Colman acknowledges the increasing inconsistency of audiences, and adds: 'I feel it a duty to our interests to neglect nothing that is proposed to us in these times of dramatic revolution.'[19] He is also aware of how dangerously charged the London theatre atmosphere had become. In another letter to Winston, dated earlier in the same year, Colman is extremely nervous about the possible effects of a managerial *faux pas*:

> My dear Winston, I have just been inform'd, to my utter astonishment, that *Over the Water to Charley* is intended to be sung tonight in the last act of *The Royal Oak* [then in performance at the Haymarket]. This is putting a lighted match to a barrel of gunpowder, in the midst of the theatre. Surely you must be aware, with all the world, that this is a *rebel* song.[20]

One of the roots of Colman's anxiety that night must have been in the knowledge that his audiences harnessed all the social ingredients necessary for a colossal human explosion. Bearing in mind what has already been indicated of the gradual fragmentation of English society into opposing interest groups during the last quarter of the eighteenth century, the existence of monopolist theatres whose audiences comprised a whole social cross-section, elements of which were potentially at odds with each other, was certainly a bizarre and artificial feature of London life at this time. Indeed, by the middle years of the nineteenth century, theatres existed in the metropolis in sufficient numbers for audiences to exercise the sort of entertainment choice that could express their social identity and cultural preference. But in 1810, and for the preceding century and a half, as a direct legacy of the patent system and the restrictions imposed by the Licensing Act, the provision of acted drama in London was dominated by the three theatres-royal. By 1810, in spite of the alternative attractions of the Italian Opera for the aristocracy and the increasing numbers of illegitimate theatres catering for a less fashionable clientèle, newspaper reports from the period confirm that theatre-royal audience cross-sections remained substantially intact whilst their audience sizes actually increased. And, to these audiences, the opportunity of free choice in entertainment was still as remote as was universal male suffrage to the reformers and radicals.

However, in the theatre, one particular course open to those sections of the public wishing to establish their identity under such restricted circumstances, a course which, in the world of politics, only the most committed of reformers would have been prepared to take, was that of direct action. Throughout the eighteenth and early nineteenth centuries, rioting was the crudest but most effective form of self-assertion practised by audiences. In fact, what probably made Colman so sensitive to the risk of inflaming audience feeling in 1810 was the all-too-vivid recollection of the riots which had erupted when the rebuilt Covent Garden Theatre had opened the previous autumn. Taking the opportunity of a fresh start after Covent Garden had been gutted by fire, the managers imposed several architectural and administrative innovations upon the new building which offended the social sensibilities of its returning audience. The most prominent innovation, which became a rallying-point for the rioters, was the raising of the price of admission to the pit that had remained unaltered for over half a century — hence the name generally given to the protest of 'Old Price' riots, or 'O.P.', from the ticket bearing these initials worn by the rioters in their hatbands as a sign of allegiance. Additional provocation was provided by the rioters' discovery that the entire third circle of the theatre, which comprised some of the best viewing in the house, had been given over to private boxes. In order to bring an end to the riots, which took place nightly over a period of almost ten weeks in protest at this attempt to discriminate against the interests of the less privileged theatre-goer, the managers had no choice but to restore the old price of admission and substantially reduce the number of the controversial private boxes. Elizabeth Inchbald, writing to a friend in 1809, puts a strident interpretation on this development: 'If the public force the managers to reduce their prices,' she prophesied, 'a revolution in England is effected.'[21]

But, to a large extent, another revolution, the 'dramatic revolution' observed by Colman in English theatrical taste, which overthrew all the criteria cherished by Genest, Cumberland and Gifford, had already taken place by this time, achieved, once again, by direct audience action. Although fettered in terms of exercising choice, audiences in the patent theatres were left with one powerful weapon with which they could bludgeon their way towards a statement of preference, and that was the right of veto. Not only did they 'hiss Vanbrugh and Congreve' in 1819, but they frequently demonstrated their unequivocal vote of no confidence in any new play or performer failing to please. Theatrical accounts of the period teem with descriptions of opening nights brought to a standstill by the sheer ferocity of disapproval, and of new plays, and sometimes revivals, subsequently 'laid aside on the prompter's shelf'. However, the pattern of eighteenth-century audience reaction is not simply a catalogue of negative assertion. From time to time, a development occurred with which audiences could identify with vigorous enthusiasm, the common purpose shared in these cases providing a reliable index of the state of mind which underpinned popular taste.

One of the most phenomenal manifestations of audience response in the immediate period leading up to 1787, the year in which Colman wrote *Inkle and Yarico*, the earliest of the plays in the present selection, can be seen in the reception given to the acting style of Sarah Siddons, who first took audiences by storm at Drury Lane in the season of 1782–3. James Boaden, a contemporary playwright and theatrical biographer, gives the following eyewitness account of the impact of her early performances on the public:

> I well remember (how is it possible I should ever forget?), the *sobs*, the *shrieks* among the tenderer part of her audiences; or those *tears*, which manhood, at first, struggled to suppress, but at length grew proud of indulging . . . the nerves of many a gentle being gave way . . . and fainting fits long alarmed the decorum of the house . . . We hear much of the moral effect of the stage, and from our youth onward we all repeat, after Aristotle, the important truth, that the mind is purified by terror and pity. But I have never met with any clear demonstration of the process; and, for the most part, the stage is imagined to improve us by its DOCTRINE alone.[22]

Sarah Siddons was, of course, a tragic actress, and Boaden seems to go to considerable lengths to remind himself of the fact. Later in the same account he can be found returning once again to this theme in search of a *legitimate* explanation of the power of Sarah Siddons's acting. 'In a crowded theatre,' he tries to reassure himself, 'with beauty before you, and the most affecting thing in the world, a woman's voice, thrilling to your soul, the *nerve* is gained, and the judgment dethroned . . . However delightful the charming agonies may be . . . we should yield only to *true* emotion; and, even in ecstasy itself, be found CUM RATIONE INSANIRE.'[23] But there is something desperate in the reason—madness paradox

Boaden has used here in order to articulate his response. Indeed, it strongly suggests that he was having to stretch his perceptions to the limit to maintain his intellectual involvement with the tragic idea during Sarah Siddons's performances. If this crisis could take place in the mind of a man whose critical and artistic awareness was highly developed, the extent to which the larger part of Sarah Siddons's audience was prepared simply to let itself yield to the dethroning of its judgment and the thrilling of its soul, as documented by Boaden, is hardly surprising. In the words of John Galt, another eyewitness of the Siddons phenomenon: 'In the performances of Mrs Siddons, the spectator sat astonished . . . No-one ever felt that he beheld in reality Mrs Siddons, but something more sublime, – the poetry of human nature.'[24]

The 'terror and pity' formula, in terms of which Boaden saw the effect inspired by Sarah Siddons in her audiences, is rather fascinating. According to strict Aristotelian definition, terror and pity should properly be produced by tragic action, whereas the pity evoked by Sarah Siddons seems to have been achieved by means of a portrayal of the human predicament in such a way as to have induced powerful and spontaneous sympathetic emotions in her audiences. Far from illustrating Aristotelian principles of tragedy, this performance technique would appear to have a good deal more in common with the spirit of Romanticism which was beginning to emerge at this time. In this context, the terror excited by Sarah Siddons's acting of emotional naturalism can be seen as being psychologically distinct from the process which, in tragedy, leads to catharsis. Just as Aristotle subordinates character to action, implicit in his idea of catharsis is a purging of the emotions of pity and terror in order to restore them to their correct proportions, a theorem which places audience response firmly within the rational perspective of tragic plot structure. Nevertheless, Boaden's own experience would suggest that the response he identifies as terror was in fact a sensation over which no intellectual considerations could prevail and which his own imagination was responsible for generating whilst in a state of heightened consciousness. What is particularly striking about this sensation is that it seems to have been largely self-gratifying, a significant indication that the atmosphere predominating over Sarah Siddons's rendering of tragedy was virtually indistinguishable from the thrill orientation of gothic horror. As William Hazlitt observed, her performance of a tragic character such as Lady Macbeth was 'little less appalling in its effects than the apparition of a preternatural being'.[25]

The unavoidable inference to be drawn from eyewitness accounts of the Siddons phenomenon is that the overwhelming acceptance shown by audiences towards the new performance mode of which Sarah Siddons was the chief exponent represents a radical reshaping of the needs and preferences expressed by public taste. During the 1770s, audience identity had been catalysed by the prodigious interest shown in the plays of Sheridan. According to time-honoured tradition, Sheridan's comedy held the mirror up to nature, allowing the spectator to see an image of himself and his society depicted in a searching satire of contemporary manners. The dramatisation of a social critique places, by definition, a considerable premium on the exercise of intellectual discernment, and Sheridan's particularly literate and witty style made this an essential feature of audience involvement. Yet, within a matter of years, the consensus of audience sympathy was beginning to indicate an opposite pattern of expectations, which the acting of Sarah Siddons went a long way towards satisfying. The possibility that the intellectual response might not be infallible, that the idea might not be sacrosanct and that the drama might not merely be the means of representing these things, found its expression in the fascination audiences discovered in the instinctive response, in the feeling of emotions and sensations, and in stage settings and acting styles which were impressive and inspiring. Moreover, theatre was being challenged, for the first time, when holding up its mirror to nature, to reflect more than just the familiar face of external reality. Sarah Siddons, by allowing audiences to experience an alternative response to tragedy, was a prime mover in the 'dramatic revolution' which was to displace the sovereignty of the legitimate drama, and 'Siddonsmania' can certainly be viewed as a turning-point in patent theatre taste in the last quarter of the eighteenth century.

Of course, Sarah Siddons cannot be credited with having invented the gothic mood, whose effects she exploited. The eighteenth century had already seen a significant gothic revival, both in architecture and in literature. Horace Walpole, for example, had written one of the first novels to capitalise upon the popular idea of the gothic (*The Castle of Otranto*, 1764), which had come, after speculations originally made about the Dark Ages following the general fascination with ruined, medieval castles and monuments, to mean the spirit of gloom, mystery and terror. With hindsight, it is even possible to locate a handful of plays which might be said to display strong gothic characteristics written before Sarah Siddons's astonishing acting début. However, without doubt, she was the first to demonstrate not only that gothic could be theatrical but that theatre was the ideal environment into which to release the potent psychology of the paranormal that gothic had come to express. In subsequent years, playwrights were to increase markedly the effort they invested in exploring the gothic perception, its influence being the most persistent, if not always the major, quality discernible in their work during the last decade and a half of the eighteenth century, the period when the plays included in the present selection were written. Indeed, gothic quickly became an accomplished artistic infiltrator, a perceptual common denominator which undermined the established foundations of British dramatic tradition by making the emotional excitement of audiences a first consideration. Just as the subversive power of the 'gothic hand' identified as a social menace by Cobbett turned Sarah Siddons from the classic, rational presentation of tragedy towards a performance which was significantly more accessible to theatre-goers but the actual effects of which might not properly be described as tragic at all, so the same perception began to insinuate itself across, and then demolish, the barriers which had previously separated areas of differing theatrical experience. The plays in this volume, for instance, include a comic opera, two musical dramas, a pantomime and a five-act comedy, and yet all contain gothic elements of such strength that their formal distinctions become largely peripheral.

As will be seen later, writers like Colman and Morton were now prepared to create within a single play a psychologically volatile blend of tragedy and comedy structured not to focus attention on a moral truth but to create a theatrical switchback ride of thrilling dips and peaks, whisking its audience through extremes of laughter, tears and terror to emerge emotionally exhausted and eager for more. Whilst its novelty lasted and disbelief could still be suspended, entire audiences were kept spellbound by gothic-orientated drama.

The crucial lever into the gothic frame of mind was the object of terror. In performance terms, Sarah Siddons, in her various tragic personifications, was herself that object, although the written drama, which could have no specific control over the style in which it might be acted, included instead a gothic motif in its setting as a concrete means of mood inducement. Some plays, in order to heighten that mood and weave it into the fabric of their dramatic composition, also made use of an agent of terror, a character whose past was shrouded in mystery and whose physical presence was a constant focus of menace and latent horror. And so, of the plays in the present selection, gothic influence makes itself apparent in the wilds of America and the cave which houses the beautiful Yarico in *Inkle and Yarico*; the medieval setting and the sub-Shakespearean dialogue used throughout *The Surrender of Calais* are evidence of an absorption in the mists of history which is particularly gothic; the wicked Sir Rowland and his castle in *The Children in the Wood* and the fiendish Abomelique and his Blue Chamber in *Blue Beard* are objects and agents of gothic terror of a type still found to be cinematographically valid today; and in *Speed the Plough*, the mystery which surrounds Sir Philip Blandford is traced, ultimately, to the secret of a fatal room in the east wing of his gothic mansion. In each case, the play derives its theatrical impact not from its ability to demonstrate a rationale but from the immense and irresistible power of suggestion which emanates from the gothic components it contains.

Gothic is the most easily distinguishable of the aspects of Romanticism which surfaced in late eighteenth-century playwriting in response to the growing anti-intellectual taste of theatre audiences. However, the ethos of Romanticism, itself more difficult to define, also seems to pervade the drama at this time. There are, for instance, grounds for supposing that the decline of the comedy of manners, in particular, over this period was given impetus by the same readjustment of values identifiable with Romanticism that also led to sympathy for the oppressed becoming the major theme not only of the five plays included in this volume, but of many others of the era. The comedy of manners, which had tended to direct its satire towards the follies of the fashionable, simply became isolated when the dramatic attention of late eighteenth-century plays shifted, in general, towards the predicament of the ordinary man, towards characters like Walter the carpenter (*The Children in the Wood*), Farmer Ashfield (*Speed the Plough*) and the humble French townsman, Eustache de St Pierre (*The Surrender of Calais*). In fact, even though, in some cases, the skeleton of a comedy of manners form remained, as it does in *Speed the Plough* in the treatment of Sir Abel Handy, and his wife and son, the perceptions it presents are of whimsical behavioural eccentricities rather than of grotesque moral flaws. The function of comedy became noticeably more humane, no longer being used as the vehicle of acerbic and biting social comment or of the verbal gymnastics with which the playwright would formerly have sought to display his intellectual superiority both over his material and over the bulk of his audience. And these modified notions of the correlation of drama and society reveal just to what extent the late eighteenth-century dramatic revolution ran in parallel with the spirit of its political counterpart. Fundamentally, each shared the same point of departure, articulated by René-Charles Pixérécourt in the following question and answer taken from his play *Victor, ou l'enfant de la forêt* (1798), itself the product of the revolutionary theatre in France and the first play to appear under the title of melodrama:

> VICTOR: Qui t'en donne le droit?
> ROGER: Mon amour pour l'humanité.[26]

And so, this outline of some of the conditions prevailing at the time when George Colman the Younger and Thomas Morton came to write their plays should be beginning to indicate that the crisis of critical perception, which led to a hue and cry in some quarters against the achievements of writers of their generation, although a genuine response, was one which has a substantial part of its significance in the historical context of the late eighteenth century. In the theatre, as in other areas of artistic activity which came under the influence of Romanticism, this period saw a positive movement away from the intellect as a guiding principle. The critic, who found his own terms of reference no longer recognised by playwrights, and himself no longer capable of endorsing the opinion reached by audiences, was effectively trapped: to review his criteria would constitute the nearest equivalent to a political act, running contrary to his own sympathies and interests, and that left him with very little choice but to complain against the subversion of public taste and the treasonable indifference of authors towards the play of ideas, the legitimate drama. Yet, to accuse the public taste of being 'vitiated' was a ludicrous impertinence. By the end of the eighteenth century, John Morris's intellectually motivated dissection of a theatre audience into the worthy ('they who understand') and the unworthy ('they who do not'), although still the standard acknowledged by Genest as late as 1830, had become a hopeless anachronism. Audiences had already taken the criterion of merit out of the hands of critics in every meaningful respect, their judgments being formed and their applause awarded on the wholly subjective basis of whether or not a play or a performer had the power to excite and to satisfy the needs of their collective consciousness. Ironically, the most revolutionary act committed by late eighteenth-century audiences involved the perception of what has since become an accepted critical nuance: they discovered the point at which the theatrical experience can pass beyond the frontiers of textual representation and yet still retain significance. For the modern reader of plays of this period, an awareness of the fact that the audiences for which they were written had begun to pay less attention to the words they were hearing and more to the theatrical image as a whole is not only deeply fascinating but vitally important to a proper understanding. But to the highly rational mind of the late eighteenth-century critic – and to others who have since followed in his tradition – this was a phenomenon of marvellous perversity, impenetrable to the intellect and fit only for the severest condemnation.

If George Colman the Younger and Thomas Morton were instrumental in subverting the legitimate drama in England during the 1790s, they were almost certainly innocent of the charge, frequently implied by their contemporaries, that their work represented an act of deliberate social sabotage. Nothing is less indicative of the revolutionary, even of the tired revolutionary, than the entrenched and unpopular positions in which both men found themselves at the twilight of their careers. Colman, for instance, became a bastion of the theatrical establishment reinforced by the 1737 Licensing Act when he took up the office, on 19 January 1824, of Examiner of Plays. That Colman, whose first poetic effusion, scribbled in a Laurencekirk inn's visitors' book in 1782, alludes to the freshness of the air in the following stunning simile:

> You would think, for the scent is so pow'rful become,
> Your nose was stuff'd up in another man's bum . . . [27]

should eventually have become the nation's dramatic censor (let alone a professional playwright) is a curious caprice of fate. However, Colman took full advantage of the scope of his responsibilities to be censorious of the work of fellow dramatists. In 1832, before the Select Committee on Dramatic Literature, W.T. Moncrieff testified that Colman, as Examiner, 'would not let one mention the word "thighs", in the *Bashful Man*; he said those were indecent; and he would not let me insert "goblin damned", for he said it was blasphemy . . . '[28] More significantly, Colman also used his position to exert a rigorous control over freedom of political speech. Explaining his reasons for refusing a licence to one play in particular, Colman exposes, although probably in zealous overstatement, his own natural sympathies in the following lines from a letter to Sir William Knighton, George IV's private secretary: 'Although the ferment of the times has greatly subsided, still plays which are built upon conspiracies, and attempts to revolutionize a state, stand upon ticklish ground; and the proposed performance of such plays is to be contemplated with more jealousy when they portray the disaffected as gallant heroes and hapless lovers.'[29] Colman's loyalty had always been one of his more conspicuous affectations. On George IV's accession to the throne, he wrote three additional stanzas to 'God Save the King' and rushed them to Covent Garden Theatre for immediate performance. In reward for his fervour, he was swiftly appointed Lieutenant of the Yeoman Guard; the previous holder of this post, as it later transpired, had been committed for forgery.

In a similar way, Thomas Morton can be seen applying his elderly efforts to propping up the restrictive practices of English theatre. In 1828, he accepted the post at Covent Garden of Reader of Plays, a situation which involved him in sifting through the mountain of unsolicited scripts, received every year by the theatre management, in search of suitable material for presentation. To many contemporary observers, especially those critical of the monopoly system, Morton and the office he performed were evidence of the extent to which the patent theatres were seeking to impose their own definition over the style and content of new playwriting through the process of careful selection based on their own self-interested criteria. Morton transferred his services as Reader of Plays in 1831 to Drury Lane, and it was here that he received the most outspokenly critical of the public attacks upon his personal and professional integrity as a dramatic arbiter. 'You, Sir,' declared F.A. Wilson, in his *Epistolary Remonstrance to Thomas Morton, Esquire, Dramatic Writer and Professed Critic and Reader to Captain Polhill and His Majesty's Servants of Drury Lane Theatre*, 'with an admirable confidence in your own infallible taste and judgment . . . and in virtue of some mysterious authority . . . have engaged yourself . . . to sit in judgment on your contemporaries, repelling, or promoting, at your own good will and pleasure, any whose ambition hungers for a small modicum of that public favour, once so indulgently accorded yourself.'[30] But this pointed assault, which in particularly unminced words portrays the man whose plays had once been considered a threat to the national well-being as an incorrigible die-hard, reveals more than the familiar notion of the whirligig of time bringing in his revenges. To Morton as much as to Colman, any overt challenge to legitimacy, either in the theatre or in any other aspect of life, would simply have been inconceivable. Indeed, not once in their entire careers did either playwright knowingly support the cause of any other but the legitimate theatre. And this was the dedication recognised by the legitimate theatre itself when it bestowed on two of its most esteemed elders these positions of dramatic trusteeship. That both men used their respective offices as vehicles for expressing their essential conservatism is proof that the trust was not misplaced.

As an extension of their allegiance to the legitimate theatre, there can be little doubt that Colman and Morton would have described their playwriting contributions to the patent houses as legitimate dramas. Indeed, Wilson, continuing his diatribe against Morton, possibly touches very near the truth when he offers the ironical suggestion that, as far as the ageing Reader of Plays was concerned, the broadest definition of legitimate drama was to be preferred, such as 'what in virtue of an exclusive patent, Drury Lane regales the people withal'.[31] Yet, as we have already seen, the concepts of legitimate theatre and legitimate drama were not closely enough related to be linked so conveniently by mere association. The paradox remained that a legitimate play performed in a legitimate theatre might be condemned, especially by the more sensitive of contemporary critics, as a piece of illegitimate drama and that a playwright of moderate political sympathies, by tampering with new methods of presentation, could find himself being arraigned as a subversive. Some of the reasons for this intriguing state of affairs have been discussed. However, for the playwright of the period, the dilemma posed by the widening gulf between popular taste and critical respectability was acute. As the anonymous author of *The Children of Apollo* (1794), a satirical poem of London theatre life, laments of the way in which Colman was to resolve this important choice of identity:

> Oh Colman! Colman! Wherefore stoop so low,
> When you can reach the head, to touch the toe?[32]

What, in fact, was to limit the creative autonomy of Colman and Morton to a servile crouch to mass appeal was their almost total dependence upon the income generated by their playwriting. This condition in itself might not seem to the modern reader to be of overpowering novelty. However, during the second half of the eighteenth century, an important change of status was to follow the substantial increase in real terms which took place in the earning power of a successful dramatist. In the years up to 1790, although authors such as Arthur Murphy or John O'Keeffe could maintain a tolerable standard of living on their dramatic earnings alone, provided their output was prolific and of a reliable enough quality, many others required the security of an additional, non-theatrical income, either private or earned, in order to subsidise their literary activities. Richard Jodrell, for instance, dedicates his play *A Widow and No Widow* (1780) to nobody, because 'NOBODY respects an Author: NOBODY gives Authors any thing.'[33] Nevertheless, in 1803, Colman received the sum of £1,200 from Covent Garden Theatre for his comedy of *John Bull*, and for *Town and Country* (1807) Morton was paid £1,000. Nor were these payments unusually high for this later period. Mrs Inchbald received £800 for *Wives as They Were and Maids as They Are* (1797) and, in his autobiography, Frederick Reynolds estimated with some pride that, over his forty-year career as a playwright, his total income had exceeded £19,000, 'a sum hitherto unequalled in the history of dramatic writing'.[34] By comparison, during the 1770s and 1780s, the average amount obtained per play by O'Keeffe was a mere £150. But not only were the payments made to dramatists significantly more generous after 1790, the terms on which they were made distinctly enhanced author prestige within the theatre. Established writers were now able to negotiate a guaranteed fixed sum for their work, in some cases payable in advance. Indeed, in 1808, Colman could virtually extort £1,100 from the trustees of the Haymarket Theatre for his play *The Africans* simply by threatening, after the work had been advertised for performance there, to sell it instead to Covent Garden. A few years earlier, this would have been considered, even by playwrights, as scandalous conduct, whereas Colman seemed to view it as the legitimate use of a strong bargaining position. Nothing could have been further from the tradition which, until the 1790s, had obliged reputable playwrights to submit to the lottery of a series of author's benefit nights, held successively on the third, sixth, ninth and twentieth nights of their plays, the proceeds being surrendered only after the costs of the house had been deducted. And so, in these various ways, the pastime of scribbling dramas began to give place to the profession of playwriting. Certainly, one of the unique claims that can be made for Colman and Morton is that they were members of one of the first generations of authors to have become professional playwrights in the modern sense of the term.

Sadly, perhaps, money was without doubt the most potent motivating force behind the careers of both Colman and Morton. Many of his contemporaries agree that Colman, for instance, led the life of a careless spendthrift. 'He paid at least twice over for every hundred pounds he obtained,' wrote Boaden,[35] to which Cyril Maude adds the suggestive comment that Colman 'loved to pose as being an extremely smart person, and tried to live in the style of people who were worth double his income'.[36] But these observations give a misleadingly fickle impression of the reason why Colman's adult life was devoted to the pursuit of lucre. Jeremy Bagster-Collins, Colman's twentieth-century biographer, is factually inaccurate when he suggests that the difficulties which beset the Haymarket Theatre, inherited by the dramatist from his father, 'unquestionably lay in Colman's notorious improvidence'.[37] In fact, Bagster-Collins rather glosses over the counter-claim, made by Anne Mathews, wife of the famous nineteenth-century actor who was a close personal friend of Colman, that 'the witty man and dramatist . . . inherited nothing from his father but his talents and his debts'.[38] However, new evidence supports Mrs Mathew's view of Colman's predicament in a somewhat spectacular way. It would now appear that, shortly after Colman took over the management of the Haymarket Theatre following his father's lapse into senility, unpaid bills and other debts began to come to light, amounting, in total, to a staggering £16,260. This was in addition to the discovery that, by a codicil to his will, Colman the Elder had made over the family home in Gower Street, together with its contents, to a favourite prostitute. From 1789 to 1818, his years as manager and proprietor of the Haymarket Theatre, Colman himself received little more than a token payment from its annual revenue in recognition of his services to its administration and scarcely ever participated in its profits, since his holding in the enterprise had been committed in trust to its creditors. Indeed, his famous arrest for debt and confinement in the Rules of the King's Bench Prison, which lasted for eleven years, was the result not of his prodigality but of a desperate attempt to raise money for the theatre by selling an annuity in it. His detention followed the trustees' failure to meet the annual payments on his behalf when they fell due. Yet it is some measure of Colman's business acumen as a playwright that his work continually commanded a high enough premium to keep him from serious distress under such circumstances.

If Colman's preoccupation with money originated from need, Morton's originated from greed. Asked by a member of the Select Committee on Dramatic Literature, 'What description of theatre do you prefer writing for, the large stages or the minor; I mean for your own reputation?', Morton had replied, 'I prefer, certainly, for remunderation, the large theatres.'[39] Indicative of Morton's financial self-confidence is the bid he made at the beginning of the nineteenth century for a stake in theatrical proprietorship. 'He wished to purchase a share in Covent-garden

I   The Haymarket Theatre in the 1790s

theatre,' wrote Thomas Gilliland in the *Dramatic Mirror* (1808), 'and was promised the first vacant one, by Mr Harris: the preference, however, was given to Mr Kemble in 1803, which was naturally resented by Mr Morton, conscious that the theatre had derived more benefit from his comic productions, than ever they could from the other's tragic talents.'[40] The share in question, which constituted one-sixth of the property, was sold to John Kemble at a value of £23,000. Such a sum would simply have been outside Morton's means in any case, despite the small legacy left to him by his uncle, John Maddison, in 1801. Either Morton lacked judgment when he had initially expressed interest in a transaction of this sort, or else the manoeuvre was conceived as a deliberate gesture of professional arrogance to bring to the notice of the theatrical world the new-found wealth and commercial significance of top-ranking playwrights. Interestingly, after 1800, as the size of Morton's dramatic income began to increase, both the quality and the quantity of the plays he produced noticeably diminished. Boaden, not unaware of the acquisitive shortcomings of his playwriting colleague, concludes that Morton 'never would admit himself to be excited by anything but the hope of gain, and always pleasantly ridiculed the ambition that looked to fame as its pre-eminent reward'.[41]

The question, 'Oh Colman! Colman! Wherefore stoop so low?' has been partially answered by locating a greater awareness of the marketability of their talents amongst the new professional playwrights. A brief examination of the market-place where they hoped to sell their wares should explain why metaphysical considerations were not uppermost amongst the criteria that fixed their worth. One tends not to imagine that the relatively genteel ambience suggested for the late eighteenth century by the various prints of gracious theatre façades, of enchantingly pretty actresses and of handsome actors poised in the midst of heroic exploits could have had, as its underside, the uncompromising pursuit of the commercial principle on a scale which was ultimately to ruin the patent theatres. Boaden indicates something of the destruction done by the hard pace of playhouse economy during this period when he expresses his regret that there could be no subsidised national theatre to release theatre managers from the vicious circle of needing to make more and more money, 'the government itself supplying funds to raise, renew, and petpetuate the literary glories of the stage'.[42] To a large extent, this was a pace that had been set by the audiences of the period, who, as has already been demonstrated, had now grown restless of the old and familiar styles of dramatic entertainment. Frederick Reynolds describes exactly what can happen when two such powerful forces as vigorous commercial management and volatile audience taste begin to interact. Referring to the vogue for animal performers on the patent theatre stages in the early years of the nineteenth century, a development which produced critical reactions verging upon apoplexy, he forcibly puts the case that

> if beings of human form and talents, will not satisfy the cormorant
> appetite of the public, and allow the manager to pay his necessarily heavily
> nightly expenditure, he must indeed be a simpleton, who will not empty

every menagerie, foreign and native, rather than incur ruin, by allowing an emptiness in his own treasury. Certain pedants, however, demand the classical, legitimate drama . . . these pedantic gentlemen ought to pay handsomely to see [it], for it appears that nobody else will.[43]

When Samuel Johnson indulged in the following piece of gentle audience flattery in his *Prologue at the Opening of Drury Lane Theatre* in 1747:

> The stage but echoes back the public voice.
> The drama's laws the drama's patrons give,
> For we that live to please, must please to live.

he could scarcely have imagined he was laying the basis of a creed which was to place the legitimate theatres on so furious a commercial treadmill that they had to abdicate any sense of artistic responsibility simply in order to survive. By 1790, the consensus behind that 'public voice' had begun a radical reorientation; its tone was now more strident and less resistible, its demands increasingly difficult to predict. Novelty became, of necessity, an essential component of managerial strategy. In exploiting it, the patent houses became committed to a form of competition which ultimately tested not their ability to produce theatre but their power to invest capital. Money was poured in colossal amounts into constructing new and larger playhouses. During the 1790s, both Covent Garden and Drury Lane were rebuilt, the former having a new capacity of 3,013 and the latter of 3,611, the figure in each case representing an increase of upwards of 40 per cent over their mid-century auditorium sizes. And when Covent Garden burnt down in 1808, reconstruction was to cost £150,000. However, larger theatres, especially at a time of uncertain audience tastes, required greater investment in production if they were to have sufficient drawing power to maintain profitability. The raising of payments to authors was one aspect of this trend. Another was the extent to which managers were prepared to sink speculative capital into extravagant and spectacular performances. When, for instance, horses were used to add to the sensationalism of a revival of Colman's *Blue Beard* at Covent Garden in 1811, the managers were put to the considerable expense of reinforcing the stage and providing stable quarters in the wings. William Dunlap, writing in 1813, blames, for 'the depreciation of the drama' during this period, 'that monopolizing and money-making spirit, which excludes from the stage the works of those who are not the hirelings or connections of managers, or renders access to it so irksome, that genius will not bend to obtain it'.[44] Dunlap is correct in his diagnosis up to a point. But what he fails to perceive is that their own financial dependence upon their theatres' revenues had effectively robbed the managers of their ability to make artistic choices. Indeed, there were really only two alternatives facing the patent theatre managers at this time, and these were as simple as they were stark. In Boaden's words, either 'a commercial speculation must be profitable, or it must close'.[45]

George Colman and Thomas Morton were certainly in no position to initiate any major offensive upon late eighteenth-century dramatic taste by openly confronting their audiences with a new style of playwriting. Nor were they capable, conversely, of single-handedly championing the cause of the legitimate drama, even had they been so inclined. Both courses of action would have brought them headlong against the harsh economic realities of contemporary theatrical life. Yet, to have emerged from this period as successful professional playwrights should not automatically suggest that their careers were a catalogue of creative compromise spiralling ever downwards into mediocrity. What characterises their work most significantly, especially during the period up to 1800 from which the plays in this volume have been chosen, is the obvious resourcefulness and enterprise with which they made use of the only infallible dramatic material available to them in their precarious state of financial dependency − the national mood. Colman and Morton, perhaps because of their need to understand as accurately as possible the workings of the sensitive commercial barometer of audience response, demonstrate an awareness of the spirit of their age, for which their plays find a series of stylistic correlatives, that is startlingly vivid to the modern reader and was clearly of immense importance to their contemporaries. Indeed, since the plays of Colman and Morton undoubtedly bred so much initial controversy as a result of the reflection they supplied to the late eighteenth-century public of its own image during a period of intense identity crisis, their modern value as theatrical statement and social document more than compensates for their failure to rank as masterpieces of English literature.

George Colman the Younger was twenty-four when his fourth play, *Inkle and Yarico*, received its première, on Saturday, 4 August 1787, at the Haymarket Theatre. *Inkle and Yarico* was the product of a theatrical career which had formally commenced in 1782 with the anonymous production of a maiden effort, a two-act musical farce written whilst Colman was still in his teens, entitled *The Female Dramatist*. But the seeds of the wilful social deviance in terms of which the eighteenth century tended to view theatrical ambition had already been sown. Colman had been toiling assiduously at the glittering prizes of non-conformity: he had succeeded in being sent down from Oxford, achieved the unusual prestige of banishment to King's College, Aberdeen, married so conspicuously badly that he was forced to keep his wife hidden for over five years until 'our author's father's paralysis removed his dread of parental displeasure',[46] and then finally let all higher aspirations be dashed by following the well-worn dramatic path of throwing aside his legal studies. Colman the Elder had himself borrowed from the same social pattern-book when considering schemes to cast himself outside the respectable ambit of his successful guardian, William Pulteney, reputedly one of the greatest orators and parliamentarians of his generation. Nevertheless, after a lifetime as an important theatrical manager and dramatist, the elder Colman seems to have cherished the hope of consolidating his own achievements by fashioning his son into a pillar of the establishment. His efforts, needless to say, merely brought him up against the intransigence of his material: Colman the Younger was destined to become a chip off the old block.

According to Colman's own account, *The Female Dramatist* was 'uncommonly hiss'd',[47] although newspaper evidence fails to endorse this proud underestimation of his professional début. The most likely explanation for the truncated stage life of this play of average competence (it received one performance) would seem to lie in Colman the Elder's method of bringing it out at his theatre. His son's authorship was not acknowledged, whilst the piece itself was scheduled for performance at one of the most inert points of the theatrical calendar, a benefit night. The reasoning behind these tactics is quite clear: the managerial father intended to humour his stage-struck son, having taken precautions to minimise the risk of promoting him by mistake. By 1784, Colman the Elder was prepared to admit the collapse of his parental expectations. In fact, the public anticipation he nurtured around the production of his son's second play, the musical comedy of *Two to One* (1784), had created an atmosphere reminiscent of a theatrical coming-out party. A prologue, written especially for the occasion, conjured up the image of the parent bird, 'with beating Heart, and anxious Eye', watching 'his vent'rous Youngling strive to fly'.[48] The flight proved a promising one. *Two to One* ran for nineteen nights during the 1784 Haymarket season, a respectable achievement bearing in mind that the ninth night was usually considered to mark the threshold of tolerable success. But another musical comedy, *Turk and No Turk*, Colman's third play and the predecessor of *Inkle and Yarico*, performed at the Haymarket in the summer of 1785, saw a significant slump in the interest of audiences in his work. Newspaper critics were careful in their judgments, possibly out of deference to the elder Colman, but Colman the Younger was astute enough to recognise the signs of failure. As he admits in his memoirs, after *Turk and No Turk*, 'I could not be so blinded by youthful coxcombry as not to suspect that I had been a *little* mistaken in the *measure I had taken of the Town*.'[49] Beneath the irony of the older sage is the crucial recognition that here can be pinpointed a moment in his career when he knew he had lost the sympathies of his audiences. Indeed, in a very revealing comment, the reviewer for the *Public Advertiser* locates a fundamental weakness in Colman's plays to date, suggesting that they wanted 'something more of the heart than the head . . . a something that calls forth the affections'.[50] In his next play, *Inkle and Yarico*, Colman went to some lengths to redress that balance.

'The story,' observed the *General Magazine* in its comments on *Inkle and Yarico*, 'as related in the *Spectator*, is universally known, and is not greatly promising of dramatic incident. The genius of the author has happily supplied this deficiency.'[51] The story in question, from the pen of Sir Richard Steele, appeared as part of a narrative discussing the relative merits of the sexes in the *Spectator* of 10 March 1711, having originated in Richard Ligon's *True and Exact History of the Island of Barbadoes* (1673). Ligon's version outlines the circumstances which led to an American Indian called Yarico becoming a Barbadian slave-girl.

This Indian dwelling near the Seacoast, upon the Main, an English ship put in to a bay, and sent some of her men a shoar [*sic*], to try what victuals or waters they could find, for in some distress they were: But the Indians perceiving them to go up so far into the Country, as they were sure they could not make a safe retreat, intercepted them in their return, and fell upon them, chasing them into a Wood, and being dispersed there, some were taken, and some kill'd: but a young man amongst them stragling [*sic*] from the rest, was met by this Indian Maid, who upon first sight fell in love with him, and hid him close from her country-men . . . in a Cave, and there fed him, till they could safely go down to the shoar, where the ship lay at anchor, expecting the return of their friends. But at last, seeing them upon the shoar, sent the long-boat for them, took them aboard, and brought them away. But the youth, when he came ashoar in the Barbadoes, forgot the kindness of the poor maid, that had ventured her life for his safety, and sold her for a slave, who was as free born as he.[52]

By way of a coda, Steele adjusts his version to stress the poignancy of Yarico's predicament by depicting her as pregnant with the Englishman's child at the time she comes to be sold, a development which the Englishman exploits to inflate his asking price. Additionally, Steele sharpens the tale into a weapon of particular social critique when he provides for the amorphous Englishman the identity of Inkle, 'the third son of an eminent citizen, who had taken particular care to instil into his mind an early love of gain, by making him perfect master of numbers, and consequently giving him a quick view of loss and advantage, and preventing the natural impulses of his passions by prepossession towards his interests'.[53] After Steele, Colman's treatment of this highly charged narrative was without doubt the most significant, especially since the themes of slavery in particular and human rights in general were of such urgent topicality at the time when it appeared. In order to make his dramatisation, Colman extended his material freely, adding, for instance, the Trudge—Wowski subplot, Inkle's engagement to Narcissa, and Campley's complicating love interest. Trudge and Wowski, Sir Christopher Curry and Medium, Patty and Campley are all Colman inventions. However, in one essential respect Colman abandoned his source and thereby asserted a substantially different moral statement: he changed the ending.

As a result of Colman's decision to supply a happy resolution to the main plot of *Inkle and Yarico*, considerable tension is set up between the central idea of the play and the logic implied by its form of presentation. On the basic level of the creative process, it would almost seem that Colman had sold out his subject matter to the prevailing conventions of sentimental drama. Evidence, however, can be found which suggests that his intention, never actually carried into execution, was considerably less compromising. In his *Random Records*, Colman is at some pains to give the impression that his writing of plays was a haphazard operation which he illustrates with reference to the final act of *Inkle and Yarico*:

Critics have been pleased to observe that it was a good hit when I made Inkle offer Yarico for sale to the person whom he afterwards discovers to be his intended father-in-law; — The hit, good or bad, only occur'd to me when I came to that part of the Piece in which it is introduced, and arose from the accidental turn which I had given to previous scenes.[54]

Possibly it was at this point that the following development involving Jack Bannister, one of the Haymarket company's leading performers who was assigned to play the part of Inkle, might have taken place. John Adolphus, Bannister's biographer, has stated that he often heard Bannister say 'that the thought of Inkle's repentance, which brings the piece to a satisfactory, if an awkward conclusion, was suggested by him. "But, after all," said Colman, "what are we to do with Inkle?" "Oh!" said Bannister, "let him repent"; and so it was settled.'[55] Astonishingly enough, a documentary basis for this anecdote exists and certainly indicates that Colman rewrote the entire final section of *Inkle and Yarico* very close to its opening night, perhaps after representations made by Bannister, who is known to have found the character of Inkle oppressive. In the manuscript copy, submitted to the Examiner of Plays just ten days before the first performance, the final pages have been cut away and others, written in a different hand and bearing Inkle's reformation according to subsequently printed versions, inserted in their place. One half-page of the original text has survived, and this fragment offers the following tantalising glimpse of what might have been a much more harrowing ending to the play:

NARCISSA: I have this moment heard a story of a transaction in the forest which, I own, would have rendered compliance with your former commands very disagreeable.

PATTY: Yes, sir, I told my mistress he had brought over a Hottypot gentlewoman.

SIR CHRISTOPHER: (*to* NARCISSA) Yes, but he would have left her for you and you for his interest, and sold you, perhaps, as he has this poor girl, as a requital for preserving his life.

ALL: How?

INKLE: (*going*) This is too much! I can bear this place no longer!

YARICO: Stay, oh stay, or take me with you! Remember your oaths of eternal constancy, your earnest promises that if ever you returned to your own country, Yarico should be your companion.

INKLE: No more! I cannot answer you. This is no time or place for remonstrance or expostulation.

YARICO: No time, or place, can ever change your Yarico. Did she not, before she left her home, risk every danger to preserve your life? And she will follow you and still watch over you, go where you will, till she gives up her own.

INKLE: Oh heaven! Where shall I turn myself? Such affection and fidelity strike deeper than the bitterest reproaches. But, oh, remorse and

penitence may cut me to the heart yet never can restore my own
peace, or retrieve the good opinion of others!

SIR CHRISTOPHER: Why, you'll find it a pretty hard taskmaster, Inkle,
that's the truth on't. And yet, the worse you speak or think of yourself, the better we and the rest of the world shall speak and think of
you. What have you to say?[56]

Sir Christopher's unanswered question raises a host of others. Has the way been
opened for Inkle's repentance, for example? And if so, would Colman, who was not
usually inclined to anguish over detail, have bothered to substitute one concluding
scene for another written in the same vein, especially after the play had been to the
copyist? What would seem to be one of the most significant qualities of this variant
ending is its emotional temperature, which is clearly in the process of rising to a
level higher than that of its published counterpart. If Colman, in this first experiment with sentimentality in his own work, had allowed the climax of *Inkle and
Yarico*, clearly conceived as a sentimental comedy, to reach too high an emotional
temperature, there is every likelihood that it might have been brought to the brink
of boiling over into what was soon to be called melodrama. For the time being,
however, this was an area of theatrical potential which Colman lacked the confidence to share with his audiences.

The theatre-goers of London, judging Colman's new play simply on the basis of
what they saw, were certainly quite satisfied. For the young playwright, *Inkle and
Yarico* was an important break, a dazzling success which 'at once established his
character as an author'.[57] Boaden declares that he cannot mention the production
'without testifying the delight with which it was received by all ranks'.[58] It was
given twenty performances during the five weeks that remained of the 1787 Haymarket season and, over the period up to 1800, was acted a total of 164 times at
the London patent theatres, a figure which places it second in the league table of
the most frequently produced plays of the last quarter of the eighteenth century,
first place being held by Sheridan's *School for Scandal*. Since *Inkle and Yarico* was
so regularly presented to the public, many of the great performers of the day can
be found to have played in it at some time or another. Elliston, for instance,
brought his enormous passionate talent to bear on the character of Inkle at Drury
Lane in 1806, the idiosyncratic comic actor, Joe Munden, became a popular Sir
Christopher Curry at the Haymarket in 1812, and others, amongst them Farren,
Blanchard, Fawcett and the Listons, chose the piece for their benefit nights. However, the massive extent to which *Inkle and Yarico* caught the public imagination
and left its imprint both in the consciousness of playgoers and the policies of
theatres is almost impossible to indicate. Perhaps the play's impact is best summed
up by the enthusiasm expressed in the following lines, written in 1795:

Hark! the loud roar, and now the sable train
Pursue with dreadful yell — the man of trade,
The buyer of his race, to chaines condemn'd,

And all the direful ills, which await the slave!
He seeks the cavern, and, yet trembling views
The ebon maid. – The faithful Yarico,
Asleep with sacred innocence and peace!
She wakes! She looks! She lives! and to her heart
Clasps th'ingrate who meditates her ruin;
E'en in the hour in which she saves his life;
And saves it in the hazard of her own! –
The tale is known to all: Simplicity
Has plac'd its stamp on *Yarico*! the stage
Receives her – Colman's classic pen
Has raised the interesting scene, to last
Till time and nature close, and ALL IS STILL.[59]

One unique advantage possessed by Colman as a playwright with managerial connections (he took over the management of the Haymarket Theatre from his father in 1789) was the detailed knowledge he had at his disposal of the company who were to perform his plays, and this he would appear to have used to excellent effect when planning *Inkle and Yarico*. Central to the moral statement and theatrical power of the play is the character of Yarico herself, whose romantic plight probably roused Thomas Gilliland's comment that 'from pieces of this description, an auditor retires impressed with those lessons of morality that animate his feelings to the best duties of his nature, and even soften the obdurate into a sense of Charity, and a love of the species'.[60] Nothing was destined to soften the obdurate more effectively than the acting of Mrs Stephen Kemble, the original performer of Yarico, whose presentational style is described in this typical eyewitness account: 'Those sweet and pathetic tones and that exquisite plantiveness by which Mrs Kemble, in Yarico, brought tears into the eyes of the audience, defy the powers of panegyric.'[61] Shock waves of sympathetic emotion seemed to have dispossessed audiences of their self-control wherever Mrs Kemble performed this character. Robert Burns, for example, having seen her in in Yarico seven years later, at Dumfries, wrote the following stanza in tribute:

Kemble, thou cur'st my unbelief
Of Moses and his rod:
At Yarico's sweet notes of grief,
The rock with tears had flowed.[62]

Clearly, Colman owed a great deal to this actress's rendering of a key role and especially to her ability to convert literary pathos into an irresistibly potent performance phenomenon. But to what extent might his knowledge of her potential to achieve this result have been a pre-condition of his writing the play in the first place? A particularly intriguing description of Mrs Kemble's personal and professional qualities comes from Boaden, who was obviously on the brink of total infatuation:

The stage never in my time exhibited so pure, so interesting a candidate
... her modest timidity — her innocence — the tenderness of her tones,
and the unaffected alarm that sat upon her countenance — all together
won for her at once a high place in the public regard ... I have often
listened to the miserable counterfeit of what she was, and would preserve,
if language could but do it, her lovely impersonation of artless truth ...
The FANCY may restore her, or be contented with its own creation. That
of Steele, in one of its softest inspirations, first saw her about the year
1674, on the continent of America, fondly bending over a young
European, whom she had preserved from her barbarous countrymen; she
was banquetting him with delicious fruits, and playing with his hair. He
called the vision Yarico.[63]

And so, in Boaden's opinion, Mrs Kemble was Yarico personified. The possibility
that Colman, watching this actress as she went about her business at the Haymarket
Theatre, where she had been engaged since 1780, had not only come to the same
conclusion but allowed it to determine the outline of his next playwriting idea
should not be discounted. After all, Colman was in an ideal position to observe the
delicate balance of theatrical power tilt, as it had done in the wake of 'Siddonsmania', in the direction of the performer, and to harness for his own purposes the
capability of the new romantic performance ethic to reach 'the heart by the plain
path'.[64]

Colman's careful consideration of the human resources available in his theatre to
carry his plays into performance was to become a recognised hallmark of his
approach to authorship and an essential aspect of his success. Boaden alludes to it
for the first time in his remarks on *The Surrender of Calais*, Colman's eighth play,
which was brought out at the Haymarket Theatre on Saturday, 30 July 1791. 'The
summer manager,' he comments, 'appeared to write for his green-room, and to
value himself upon exactly fitting his actors with proper vehicles for their talents.
What he did in this way for Bannister, was really an achievement.'[65] But what
exactly were the talents of Bannister that Colman went so far to accommodate?
'The main feature of his acting,' suggests Thomas Munden, reconstructing the
opinions he had often heard his father express, 'was what the French term *bon-
homie*, which carried the auditor's feelings with him.'[66] Reynolds, more particularly, draws attention to what he calls Bannister's 'strong serio-comic power',[67] an
unusual quality during a period when the comedian—tragedian demarcation was still
actively practised, and one which could scarcely have failed to catch the notice of
a discerning manager. Hardly surprising, then, that Bannister, having been given the
character of La Gloire, 'from the first scene to the last always delighted the audience; sometimes touching the heart with exquisite sensibility, at others provoking
irresistible merriment'.[68] The opening scene with Ribaumont he would seem to
have made his own, and it 'stamped the character of Bannister as a first-rate performer. His good-humoured descant upon their danger; the honourable attachment

he professes to the Count, mixed with a ludicrous description of his own early service; his soldier-like resentment of the opprobious term [of coward], and his reluctant resignation to the word of command, – all were admirable',[69] and all had probably been anticipated in Colman's imagination as he sat down to write the scene. Equally well calculated in advance was the effect that the older and rather more mannered actor, Robert Bensley, might have had in the role of Eustache de St Pierre. As Colman himself has pointed out, Bensley's acting was flawed, and although he 'always maintained an upper rank upon the stage' and 'was respectable in *all* the characters he undertook', the presence, as he performed, 'of a stalk, and a stare, – a stiffness of manner, and a nasal twang of utterance . . . prevented his being very popular in *most* of them'.[70] Boaden, however, has perhaps put his finger on the particular weakness in Bensley out of which Colman saw he could extract enormous theatrical impact. Bensley's 'very voice,' he says, 'rendered all *deception* . . . impossible. He acted, as he always did, with terrible energy'[71] – adding that the old actor 'implored pity in the noisy shout of defiance'.[72] From these natural tendencies in the performer, Colman formed the essential characteristics of Eustache, predicting, in all likelihood, their pathetic potential at the hands of Bensley, who subsequently 'soared above the reach of adverse criticism'[73] in this role and 'in the grand tirade before Edward . . . became absolutely sublime, from the virtuous energy, that seemed to dilate his person, and thunder from his sonorous organ. It really was a display not to be forgotten.'[74]

The need to supply two popular actors with cameo parts must also have had its effect on Colman's preparation of *The Surrender of Calais*. The first of these was written especially for Jack Johnstone, 'the arch, witty and humorous' representative of stage Irishmen, who 'possessed talents that rendered him a prime favourite of the public'[75] throughout this period. Johnstone particularly excelled at singing Irish songs, and the theatrically redundant character of O'Carrol was no doubt introduced by Colman simply to contrive an opportunity for this popular performer to regale the audience with the two plaintive airs of 'Savourna Deligh Shighan Ogh!' and 'Corporal Casey'. Not for the first time did Johnstone crop up in one of Colman's plays in this fashion. The other cameo was designed for William Parsons, a much-loved comedian who had loyally supported the Haymarket Theatre every summer since 1776, but was now in poor health, being 'very thin and much afflicted with an asthma',[76] as a result of which his ability to perform was seriously limited. For this reason, and because audiences still demanded a glimpse of their old favourite, Parsons's name being 'of great effect in the playbill',[77] Colman probably hit upon the expedient of the scaffold-makers' scene, which was sufficiently self-contained to show Parsons to good effect without running the risk of tiring him. On one particular occasion, however, Parsons found enough of his old energy to burst out of his theatrical safe-keeping. George III was in the audience. 'It was in Parsons's part to say "So, the King is coming: an the King like not my scaffold, I am no true man." The humorous player gave the passage quite a different turn: advancing very near the royal box, he exclaimed, "An the King were here, and did

not admire my scaffold, I would say 'Damn him! he has no taste.' " This innovation,' concludes the anecdote, rather ambiguously, 'produced an irresistible
effect.'[78] When Parsons died in 1795, it was for his contribution to *The Surrender
of Calais* that Colman chose to honour him publicly, in the dialogue of the prelude
written for the opening of that summer's Haymarket season:

> CARPENTER: We want a new scaffold for *The Surrender of Calais*.
> PROMPTER: Ah, but where shall we get another such hangman? – Poor
> fellow! Poor Parsons! The old cause of our mirth is now the cause of
> our melancholy. He who so often made us forget our cares may well
> claim a sigh to his memory.
> CARPENTER: He was one of the comicalest fellows I ever see!
> PROMPTER: Ay, and one of the honestest, Master Carpenter.[79]

For Colman, *The Surrender of Calais* represented the refining of a new development in his writing technique which had been initiated in a previous drama, *The
Battle of Hexham* (1789). After *Inkle and Yarico*, Colman's reputation had hit the
doldrums with the production, in quick succession, of a mediocre three-act
comedy, erected mistakenly on the crumbling foundations of the Restoration
intrigue plot structure, and a totally forgettable farce. *The Battle of Hexham*, which
had revived Colman's fortunes, was an attempt at serious drama and derived much
of its force from its manner of blending historical narrative, romantic legend,
knockabout humour and atmospheric music, a feat previously untried on such a
scale in a London theatre. Genest viewed with some grief the birth of this bastard
child of the history play, this 'jumble of Tragedy, Comedy and Opera' which
honoured its material not with moral truths but with theatrical sensationalism. For
Genest, and those who shared his notions of drama, this was indeed a black day.
'The success which the Battle of Hexham met with,' he declared, 'encouraged
Colman and others to persist in this despicable species of the Drama, in defiance of
nature and common sense.'[80] Underpinning Colman's experiment was a crucial
stylistic departure which he describes as follows:

> plays which exhibit incidents of former ages should have the language of
> the characters conform to their dress. To copy Shakespeare, in the general
> *tournure* of his phraseology, is a mechanical task, which may be accom-
> plished with a common share of industry and observation: – and this I
> have attempted (for the reason assigned); endeavouring, at the same time,
> to avoid a servile quaintness, which would disgust.[81]

But the result was uneven, and several critics refused to be impressed by Colman's
dramatic half-timbering, one going so far as to publish his reactions in recipe form:
'Take one edition of SHAKESPEARE . . . the more black letter the better – select
all the obsolete terms rejected by later editors – be sure to pick out a large
quantity, such as "he comes me o'er the sconce" . . . and let the serious scenes be
distinguished from the comic by the thundering drum and ear-piercing fife, and the

shrill trumpet.'[82] Nevertheless, if Colman was guilty of writing to a recipe, there were plenty who were willing to consume its product. As the *Diary* reviewer, noting the popularity of *The Battle of Hexham* with audiences, was drawn to conclude: 'the play possesses the three best qualities of a mixed drama in an eminent degree; it attracts, entertains, and affects very forcibly'.[83] The experiment clearly warranted repetition.

Having transferred his canvas from the Wars of the Roses to the opening phase of the Hundred Years War, rich in chivalric exploits and knife-edge excitement of the kind encountered when Edward III laid down his grisly terms for the surrender of Calais in 1347, Colman wrote his second historical play and, as the following account indicates, scored an important victory over his sceptics:

> In point of language we have met with few plays so well written in every part; the style is Shakespearean throughout, but appears to be a successful effort at Shakespeare's general manner . . . The sentiments are dignified and moral, and the images well suited . . . The conduct of the plot and the management of the incidents, remind one equally of the plays of Shakespeare . . . It evinces improvement in the author's pen, and is superior to *The Battle of Hexham*.[84]

Such was the opinion of the *London Chronicle*. However, many of Colman's Shakespearean derivatives can be traced back to specific sources. The Sergeant of Act I Scene 2 would seem to share a number of behavioural similarities with Falstaff in *Henry IV Part II* whilst Eustache's railing brings him very close at times to Shakespeare's portrayal of Timon of Athens. Certain phrases, too, are clearly echoes from Colman's Shakespearean reading: 'peevish town', 'wind his sluggish courage to the pitch', and 'shakes my manhood to the centre' might have had a ring of familiarity about them to the more erudite in his audiences. But nothing gives away Colman's attitude towards these borrowings more effectively than his bare-faced plagiarism from *Hamlet* in the scaffold-makers' scene. According to Boaden, Colman would often apply the term 'twaddler' to anyone who tried to thwart his theatrical schemes with pedantic objections. Since the scheme at present was quite simply to entertain London theatre-goers with an acceptable, romantic chronicle play with music, to suggest that other considerations had any validity whatsoever would have seemed, from the manager-playwright's point of view, to have been twaddle of the most incomprehensible sort. And this extended to all questions of provenance.

Audiences, however, were clearly less perturbed by Colman's sources than they were by the play itself. 'We seldom ever witnessed more tears shed in a theatre,' exclaimed one delighted reviewer, substantiating his claim with a description of what was rapidly becoming a typical event during a Colman drama: 'a young Lady in the Pit was carried out in hysterics: and the *English Character* was well shown, in stopping the performance while humanity was doing its office'.[85] And the currency of *The Surrender of Calais* was still strong in 1812, when the authors of *Biographia Dramatica* noted that the play, 'originally acted with great success . . . still continues to be a favourite performance',[86] indeed, 'the most popular acting drama of the day',[87] if an account in the *European Magazine* of 1817 is to be believed. According to Mrs Inchbald, in her 'Remarks' accompanying its first printed edition in 1808 (Colman was often coy about immediately disposing of the copyright of his plays), *The Surrender of Calais* also had the distinction of being 'considered, by every critic, as the very best of the author's numerous and successful productions'.[88] Twenty-eight performances were given in its first season at the Haymarket, eighteen during its second, and thereafter it became a stock piece both in London and at the provincial theatres-royal. Mrs Jordan played Julia in 1792; John Palmer took the role of Eustache in 1796. Significantly, the play also found favour with the rising generation of what might crudely be called the romantic-melodramatic performers: Charles Kemble appeared as King Edward in 1795, Henry Johnston as Ribaumont in 1801 and Elliston as Eustache de St Pierre in 1803. But possibly the most stirring of the later performances of Eustache was given by Edmund Kean, leading one critic to assert that the actor had never 'been seen to greater advantage in any character he has undertaken'.[89] And twenty-three years on, Jack Bannister, when selecting a suitable piece for his benefit night at Drury Lane, chose *The Surrender of Calais* as an effective audience catcher. On this occasion he repeated his early triumph as La Gloire, supported by Elliston's equally popular Eustache, to an ecstatic house.

Colman's determination in his two historical plays, which raised the curtain on the final decade of the eighteenth century, to squeeze current scriptwriting practice into a new mould, would seem to have derived its urgency from the need to marshal in a more lucrative direction the obsession of audiences with the gothic perception of tragedy. A handful of reviewers shrewdly guessed at Colman's purpose, which, as one of them indicated, was 'to blend the grave and the gay, with suitable music to each, so as in one piece to please many minds'.[90] But the theatrical implications raised by this eclectic pilfering from traditionally distinct dramatic forms did not pass unnoticed. Some commentators were willing to be impressed. 'In this drama,' observes Mrs Inchbald, discussing *The Surrender of Calais*, 'are comprised tragedy, comedy, opera, and some degree of farce — yet so happily is the variety blended, that one scene never diminishes the interest of another.'[91] Others, like Genest, were altogether less sanguine about such a marriage of convenience. The great majority of critics, however, were more preoccupied with finding a generic name for this new species, so that it could be compartmentalised, along with other branches of the eighteenth-century *phylum dramaticum*. The name they eventually settled for was 'mixed drama', a term already applied to plays of dubious parentage and one which was to grow rapidly in prominence over the next ten years. Indeed, the idea of mixed drama became part of an even more significant definition when, in 1802, Henry Harris, son of the manager of Covent Garden, wrote from Paris attempting to explain the etymology of the title given there to a new dramatic phenomenon. Harris, who had actually witnessed what he was attempting to describe, suggested that it derived its name from the French verb *mêler* (to mix), a deduction which was later to prove his poor instinct for language although revealing, at the same time, a fascinating line of theatrical reasoning. The phenomenon in question was melodrama.

Of the ingredients in Colman's particular mixture, perhaps one of the most conspicuous was his use of music. This in itself was no innovation. Ballads and songs had appeared in English drama throughout the eighteenth century, from John Gay's *The Beggar's Opera* (1728) to Sheridan's *The Duenna* (1775). But Colman was the first playwright of the period systematically to explore their use in a non-comic context, and the introduction of songs in the serious dramas of *The Battle of Hexham* and *The Surrender of Calais* signalled a substantial break with tradition. They were also well placed to indicate the two other major ingredients in the mixture whose juxtaposition and balance, fundamental to the nature of the plays, they largely maintained: comedy and tragedy. In his two historical plays, Colman established the pattern of supplying songs where he needed to create atmospheric counterpoint or to reinforce the sharp contrasts of mood he had set out to achieve between moments of high pathos and broad farce. Superficially, this might seem to be a carbon copy of the structural and stylistic approach already used in *Inkle and Yarico*. Indeed, a comparison between that play and *The Surrender of Calais* would suggest a number of basic similarities: broad farce, for example, provided by the various verbal and behavioural absurdities of Trudge and La Gloire, offers a continual foil to the serious dramatic exposition of the uncertainties of human existence, resolved into the romantic notion of virtue struggling against oppression and presented through the predicaments of the innocent Yarico and the honest Eustache de St Pierre, while the whole is kept from a fatal mid-air collision by the timely arrival of songs. Ultimately, however, Colman's choice of the comic opera mode for *Inkle and Yarico* deprived him, as perhaps he realised when he came to rewrite the play's final scene, of the opportunity of developing the tragic potential implicit in his source beyond the level of sentimentality. By doing little more than mixing his ingredients in slightly different proportions and taking a starting-point suggestive of a greater degree of gloom, Colman found himself capable, in his historical dramas, not only of keeping the potential for tragedy continually alive, but of maximising its theatrical capital by letting it be seen to hang over his play by a thread. In this way, he first threw the fatal switch that passed the life-giving force of horror and suspense into the assembled components of his gothic musicalised tragicomedy.

There can be little doubt that with *The Battle of Hexham* and *The Surrender of Calais* Colman had set an important fashion in contemporary playwriting, particularly through his style of balancing comic and gothic within a single play. Some years later the following rather graphic description appeared in a daily newspaper: 'This Gentleman's muse reminds us of those hocus pocus prints, in which a face that seems in a broad grin one way, shall appear in a most doleful whine another. Thus if you ever see the first scene of one of . . . [his] plays in a horse laugh, you have nothing to do but to turn to the second, and you will find every body in tears.'[92]

One might be forgiven for assuming that it is Colman who is being depicted. In fact, the subject of the sketch is Thomas Morton.

Morton, born in 1764, was just two years Colman's junior, although his active playwriting career did not begin until he was twenty-nine when, a year after the first performance of *The Surrender of Calais*, his début piece, the historical drama of *Columbus*, was brought out at Covent Garden. The son of a country gentleman living in County Durham, Morton might never have been in a position even remotely to consider writing for the stage had his father not suddenly died, leaving him a four-year-old orphan and throwing 'the care of his education and fortunes . . . on his uncle, Mr Maddison, an eminent stockbroker',[93] and an acknowledged progressive, then resident in London. Maddison sent his young charge to Soho Square Academy, an establishment whose chief attraction was in its promotion of drama as an educational tool, involving 'the elaborate and expensive performance, at least as often as once a year, of an English tragedy or comedy'.[94] The enlightened principal of this seminary was one Mr Barwis.

> Barwis's view [contends Boaden] in not merely permitting, but urging and correcting such performances, was confessedly to give the pupils a free and unembarrassed manner and an accurate and powerful elocution, which he concluded to be essential to the display of the sound erudition which occupied their studies. I am not able to state whether the church, or the bar, or the senate, have derived any accession of graceful oratory from the plan; it, I confess, seemed to me, if I may parody the poet, –
> Stage-born, and destin'd to the *stage* again.
> Two of the school, have certainly *trod* the stage with great distinction, Holman and Fawcett; and Morton is likely to occupy it as an author, at least as long as any one of his contemporaries.[95]

Holman, in fact, was Morton's particular schoolfellow, and the friendship proved to be lifelong. But perhaps the greatest professional service Holman did for Morton was to introduce him, in 1786, to the young and aspiring, soon to be the highly successful, dramatic writer, Frederick Reynolds.

When Morton and Reynolds first met, in a room above Exeter 'Change, they had a violent quarrel over a point at billiards: 'our first and last . . . during an intimacy of forty years,' remarks Reynolds, 'and notwithstanding, that we have passed thirty of this period, in constant contact and competition, as rival dramatists'.[96] Reynolds's influence, which massively increased in theatrical circles after the popularity of his play *The Dramatist* at Covent Garden in 1789, together with his encouragement were certainly crucial elements in Morton's early playwriting development. Two events in 1791, for both of which Reynolds was responsible, were to boost Morton's sense of theatrical identity to the point where he gained the self-confidence to seek his own involvement. The first of these was the forming of the voguish literary and theatrical 'Keep the Line' club, which thrust him into the midst of a coterie of the most powerful and the most talented in the dramatic and critical

worlds. The second was Reynolds's invitation to him to write a song for Jack
Johnstone in his forthcoming comedy of *Notoriety*. To Morton's surprise, his song
became something of a hit, 'and its success inducing him to proceed,' adds
Reynolds, 'he has often reproached me, as the cause of his entanglement, in the
theatrical labyrinth'.[97]

Columbus, Morton's first solo effort, would seem to owe significantly more to
the creative leadership of George Colman than to the whimsical satire which was
beginning to appear under Reynolds's signature. Indeed, its sombre tale of dark
deeds in the New World, of Europeans oppressing the natives, of a doomed love
between an invading Spaniard and a girl from the Inca Temple, mirrored by the
comic attachment of Harry Herbert (an expatriate Englishman) and Nelti, appear to
be not totally without precedent. As the *Thespian Magazine* observed, perhaps a
little unfairly, 'Harry Herbert and Nelti reminded us of Trudge and Wowski.'[98] But,
if Morton had trespassed into areas where others had staked their claim, it was by
negligence rather than by design. *Columbus* showed that Morton had as much to
offer as the most original of his contemporaries: the ability to avert almost inevit-
able tragedy at the eleventh hour and transform it into a happy ending, and an
acute sense of the appropriate use of stage spectacle, this particular piece including
'the most horrible *earthquake* and *storm* the stage has ever felt'.[99] The whole venture was considered to be highly promising, one critic ranking Morton's *Columbus*
as 'the best composition . . . that has appeared on the Covent Garden boards for a
length of time',[100] and the young author's overnight success was clearly an important watershed in his life: 'often has the author been heard to regret,' states one
report, 'that it was not hissed from the boards: it . . . drew him from a profession
in which he might have acquired fortune and distinction'.[101] That neglected profession, inevitably enough, was the law. At various periods during the late 1780s
and 1790s Morton occupied chambers at Lincoln's Inn and the Temple, but was
never called to the Bar, perhaps never even received legal instruction. Nonetheless,
for the time being, the Inns of Court were a congenial enough place for the writing
of his second play, *The Children in the Wood*.

First performed on Tuesday, 1 October 1793, the two-act musical entertainment
that comprised Morton's *Children in the Wood* was one of the first novelties of an
irregular season at the Haymarket Theatre. Under normal conditions, Colman's
annual licence to perform plays would have expired on 15 September. However, in
the autumn of 1792, the demolition and rebuilding of Drury Lane Theatre had
commenced, the intention having been to open a new and larger edifice on the site
in time for the season starting in September 1793. During the intervening winter of
1792–3, the Drury Lane Company had operated a makeshift season at the Italian
Opera House, but when it was realised that the target date for the completion of
the new theatre could not be met, steps were taken by the managers, as an
expedient for keeping the company together, to approach Colman with an offer
of loaning him the protection of the Drury Lane patent for a winter season at the
Haymarket if he, in return, would provide employment for as many Drury Lane

performers as possible. To this Colman assented, and although it soon became clear that he was prepared to honour the agreement only to the extent of replacing those actors from his summer company who were under contract to Covent Garden for the winter, the resulting ensemble was unusually strong. And so Morton was extremely fortunate that, 'having read his piece to Reynolds at Chambers, who admired it' and agreed to be 'the bearer of it to Colman', an old friend of Reynolds since their days together at Westminster School, *The Children in the Wood* was selected for performance by Colman's special winter company. But even under such propitious circumstances, Morton's new play 'succeeded beyond the most sanguine hopes of the parties'.[102]

Beneath the criticism which had been levelled at *Columbus* is the suggestion that reviewers had not been convinced by one of the basic features of Morton's style, his method of opposing the comic and pseudo-tragic; 'the conjunction of the two species of Drama,' the *Thespian Magazine* had declared, 'banishes all effect from either'.[103] However, in dramatising the traditional English ballad of *The Children in the Wood*, Morton seems to have taken considerably greater care to achieve a more sophisticated fusion of the extremes of theatrical experience he has chosen to arouse. A comparison, for instance, between the play and its source amply demonstrates the sensitivity for dramatic effect which guided him through the various structural modifications he found it necessary to make in order to convert the ballad into theatre. By juxtaposing serious and comic episodes, Morton establishes a tension which builds progressively through the play, replacing and redistributing that which is generated in the original by the final catastrophe, far too shocking for him to have depicted on stage, of the children's death through exposure and their burial under leaves by the robins. Morton's final scene, inevitably, is one of jubilant reunion of children and parents. The critics fought with each other for eulogiums to express their delight with the new playwright's second attempt. 'If not the how MUCH, but the how WELL, be to be considered, this is the greater performance of the two. Here is nothing that Nature does not own – nothing that Sensibility should be ashamed to enjoy,'[104] wrote one. 'Though the Author has necessarily departed from the anecdote in the ballad,' added another, 'he has imparted such tenderness and simplicity to the fable, and in the gayer scenes thrown so much cheerfulness into his characters, that we are confident it will prove the most attractive performance that has appeared for some time in the Theatre.'[105] The *Times* reviewer was particularly struck by the way that 'the sentiments arise from the subject naturally, making a powerful appeal to the heart'.[106] Morton's adaptation clearly showed his skill and judgment. As one commentator concluded: 'to dramatise a familiar, and a popular tale, like an attempt to modify a beautiful ruin, is difficult, hazardous, and generally ineffectual. The Author has fully succeeded . . .'[107]

Almost every theatre document in which Jack Bannister is mentioned makes at least some reference to his phenomenal acting achievement as Walter in *The Children in the Wood*. The role became his professional *tour de force* and an outstanding example of romantic acting at its best. 'Your whole conscience stirred with Bannister's performance of Walter,'[108] recalled Charles Lamb in his *Essays of Elia*. And the authors of the *Biographia Dramatica* explain the basis of the rare blend of playwriting and performing talent out of which developed one of the most power-ful theatrical statements of the last quarter of the eighteenth century: 'the charac-ter of Walter possesses those traits of excellence, which furnish an opportunity to the actor to seize and expand them to a perfection of which the author perhaps never dreamt. Read the character of Walter, and then see Mr Bannister perform it! You are astonished at the effect . . . it perhaps exceeds every other effort of the modern drama.'[109] Adolphus explains in more detail the nature of the impression Bannister created. In the forest, for instance, Bannister's Walter 'returns, and not finding the little ones, as he expected, endeavours to persuade himself that they are hiding to frighten him; he calls; his agitation increases; he tries the playful art of pretending that he sees them, and they may as well come out; and, driven by despair almost to insanity, he rushes out, screaming through the torture of his sensations . . . '[110] The eyewitness account provided by Adolphus is probably at its most valuable when it comes to describe Bannister's portrayal of Walter at the climax of the play. Inside his cottage, his mother and Josephine having just been expressing their various apprehensions

> in a manner most calculated to increase the uneasiness of his mind . . . *mal à-propos*, Josephine, to cheer him, begins a song about a murder and a ghost. His agony is at the height, and displayed by incoherent expressions and stifled groans, when, suited exactly to the words of Josephine's song, a violent knocking shakes the cottage. The door opens, and the parents enter, bearing with them the children, – the cause of all Walter's distress. His wild screams of joy, – his affectionate caressing of the little ones, – his revulsion from extreme despair to an agony of delight, formed a most exquisite dramatic spectacle. His recall to the plain feelings of common sense was not less admirable, when the boy, having declared that he was very hungry, Walter seized the chicken on the table, and, with clumsy eagerness, mauled rather than carved it into pieces for the supper of his famished darlings.[111]

With so much concentrated energy needed to keep the character of Walter buoyant throughout such rapid transitions of mood, the two-act play would certainly not have seemed too short from Bannister's point of view. Nor did its brevity appear to limit the acclaim awarded by his audiences. In less than an hour he wrung out of them the whole gamut of responses, 'tears and sobs in some passages, heart-easing laughter in others, a hearty applause in all'.[112] Few, indeed, would have disputed the verdict that, in Walter, Bannister was 'displayed to the highest advantage, and in the utmost perfection'.[113]

The Children in the Wood was performed sixty-three times in its first year alone, becoming one of the most popular afterpieces of the next decade and continuing to appear in the playbills of the London theatres with regularity until the early 1820s. The original production benefited from a highly competent supporting cast which included the popular comic actor and singer, Dicky Suett, whose own tendency to be 'too much of a *bon vivant*',[114] although sadly condemning him to a career of acting stage drunks, gave a great deal of force to the character of Apathy. Miss De Camp, soon to make a mighty melodramatic impact in Colman's *Blue Beard*, played Lady Helen with hysterical charm, her screams being described as 'truly Siddonian',[115] whilst Mrs Bland, as Josephine, contributed effectively to heightening the atmosphere of the scene in Walter's cottage by singing her tragic and ghostly ballad 'with a captivation that subdues every heart'.[116] But the two children, the focal point of pathos in the piece, demonstrate Colman's skilful handling of crucial casting decisions, as the admiration of their performance by the *Times* reviewer would seem to indicate: 'the acting of the two MENAGES must be seen — we cannot describe its excellence'.[117] One of the most original interpretations of Walter to emerge in later years was that of Elliston at the Haymarket in 1797. This was during the season in which Bannister, who since 1778 had been a regular member, latterly a leader, of Colman's summer company, had refused his usual engagement there on the grounds that he could earn significantly more from provincial touring. Unwittingly, Bannister opened the way for one of the most serious challenges he was to face during his acting career, the performer chosen by Colman to replace him in Walter being none other than the young and ebullient Robert Elliston. Elliston's contemporary biographer, George Raymond, suspected that, 'although the pathos of the part was a material which he was expected successfully to deal with', there were indications that Elliston 'was somewhat too stilted for the impersonation of lowly and familiar scenes — *Difficile est communia dicere* — it was the pathetic of tragedy, not comedy'.[118] However, this consideration did not deter audiences from applauding him in the role nor prevent Elliston from adopting Walter as one of his stock characters. As late as 1820, he was praised in this character by Charles Lamb, who described his performance as 'an admirable representation of rough honesty, and manly sorrow'. In Lamb's opinion, 'nothing could be happier' than Elliston's 'valorious resolution, springing naturally; or the mingled humour and feeling of his triumph over the assassin — or his efforts to appear composed, when the fate of the children was doubtful — or the broken accents of joy when he folded them in his arms'.[119] All of which would suggest a more disciplined rendering than that supplied by Bannister. But Bannister persistently claimed Walter as his own. At his farewell performance to mark his formal retirement from the stage at Drury Lane on 1 June 1815, it was as Walter that he stood to receive the final tribute of London theatre-goers.

In the spring of 1794, the new Drury Lane Theatre, with its comparatively massive stage and auditorium, was at last opened and, together with Covent Garden, where enlargements had been completed in 1792, ushered in the age of the macroplayhouse in London. Architecturally, these new theatre environments, which proportionally boosted the physical dimensions of the standard playhouse design to an unprecedented size, severely distorted the spatial relationship of the actor not only with his audience but also with his immediate context within the scenic image. Describing the new Drury Lane, Boaden has taken particular note of the dimensions of the stage, which 'required scenery, certainly, thirty-four feet in height, and about forty-two feet in width, so that an entire suite of new scenes was essential on great occasions, though where *display* was not material, the old pieced flats [from the former theatre] might be run on still, and the huge gaps beneath them and the wings, filled up by any other scenes drawn forward, merely "to keep out the wind" '.[120] However, many of the contemporary published reactions to the growth in size of the London theatres run with deceptive undercurrents of personal and professional bias. Cumberland expresses the most typical reservation, written, of course from a playwright's point of view:

> Since the stages of Drury Lane and Covent Garden have been so enlarged in their dimensions as to be henceforward theatres for spectators rather than playhouses for hearers, it is hardly to be wondered at if their managers and directors encourage those representations, to which their structure is best adapted. The splendour of the scenes, the ingenuity of the machinist and the rich display of dresses, aided by the captivating charms of music, now in a great degree supercede the labours of the poet. There can be nothing very gratifying in watching the movements of an actor's lips, when we cannot hear the words that proceed from them: but when the animating march strikes up, and the stage lays open its recesses to the depth of a hundred feet for the procession to advance, even the most distant spectator can enjoy his shilling's-worth of show.[121]

And Colman's continual irony at the expense of the giant stages, displayed at its most acerbic in his prelude *Poor Old Haymarket* (1792), was undoubtedly rooted in his frustration as a manager who suddenly found himself having to pay the inflated wages generated by the new large theatre economy in order to retain his own summer ensemble. The Haymarket, unlike the other London theatres-royal, resisted aggrandisement, with the result that it remained, in every respect, a smaller concern, its size equivalent to just over half that of the two winter houses. Evidence would also suggest that the performers themselves disliked the dimensions of the enlarged theatres. For comedians, especially, the Haymarket, subsequently nick-named 'Georgey Colman's little snuggery in the Hay-mow',[122] became a welcome refuge to which even the younger performers, brought up to the large stages, returned with relief at the end of their winter engagements. Indeed, Boaden goes as far as to suggest that the career of actor James Dodd, a specialist in genteel comedy roles, was actually extinguished by the advent of the macro-playhouse, which simply swallowed up the nuances of his performance. But great care is needed here. Although it might well have been the case that 'never did the really good actors appear to such advantage' as on the Haymarket Theatre's modestly proportioned stage,[123] and that 'large theatres were of detriment to fine acting' by demanding 'extravagance in the three articles of *action, expression* and *utterance*',[124] the increase in theatre size seems, nevertheless, to have provoked significantly little alarm from audiences of the period. In fact, Raymond makes the following fascinating observation on the fortunes of the Covent Garden Company who, after being driven from their own theatre by the fire of 1808, spent part of the ensuing winter at the Haymarket:

> Scarcely will it be believed that, with this extraordinary assemblage of talent . . . the receipts were sometimes under the *nightly expenses* . . . This identical company, a few weeks before, acting in the Italian Opera House [of comparable size to the winter theatres], rarely failed attracting full audiences – a fact which, undoubtedly, proves that, whether or not large theatres be more beneficial to dramatic representations, the public, at least, like them and prefer them.[125]

Why should this have been? The answer would seem to be suggested by the prominent place, discernible since the days of 'Siddonsmania', which was now reserved in the taste of London audiences for plays and performances of conspicuous theatrical size. The evidence, for instance, of the popular approval given to Jack Bannister's Walter and Colman's *The Surrender of Calais* indicates the extent to which, even before the opening of the larger theatres, audiences were drawn towards productions offering a theatrical experience of considerably inflated dimensions, being sensational, impressive and awe-inspiring enough to induce in their spectators extreme states of imaginative and emotional involvement. Not only did the new macro-playhouses possess fewer physical limitations upon the extension to still greater heights of the psychologically potent effects audiences craved in the theatre, the increased actor-audience distance alone substantially enhancing their amplification, but the sheer expanse of their stages focussed attention, as never before, on the theatrical power of size in scenic presentation. And so, when Colman, writing especially to exercise the technical resources of the new Drury Lane, saw their fittest use in supporting the gothic extravaganza he called *Blue Beard*, which received its first performance on Tuesday, 16 January 1798, new chapters opened simultaneously in the history of patent theatre stagecraft and the demise of the spoken drama.

Colman has described *Blue Beard* as a 'Grand Dramatick Romance' although, as he explains himself, the piece was designed with a rather more specific aim in view:

> I am far from endeavouring to vitiate the taste of the Town, and to overrun the Stage with Romance, and Legends, but English Children, both old and young, are disappointed without a Pantomime, at Christmas; – and, a Pantomime not being forth-coming in Drury-Lane, I was prevailed upon to make out the subsequent Sketch, expressly for that season, to supply the place of Harlequinade.[126]

Pantomime, an invention of the early eighteenth-century English theatre, had by this period established its own detailed set of conventions, setting it apart from standard theatrical fare as a stylised vehicle for the display of the acrobatic and clowning skills of a Harlequin actor, and Colman was wise to leave it well alone. But, as a pantomime surrogate, *Blue Beard* was raised, from the outset, to a unique and crucial position in the theatrical calendar which Boaden explains as follows: 'the manager of a theatre ... looks at Pantomime as the grand source of emolument; and so entirely is this taken for granted, that the forecast of a season divides itself into but two parts – "how to get on till the Pantomime", and, after it has done its duty, "till the Benefits".'[127] *Blue Beard* was also required to fulfil other expectations normally satisfied by pantomime. A bizarre anecdote from Reynolds illustrates one of the main elements of the eighteenth-century pantomime's popular appeal. In 1787, whilst touring in Switzerland, he came across an expatriate English waiter who proceeded to deride the local geography in the following terms:

> as to Swiss scenery, pooh! why the Alps, Mont Blanc, and the Glacières, are always the same; but when I saw them at Drury Lane, in the last new pantomime, they not only looked better at first sight, but when Harlequin in a twinkling changed them into the Adam and Eve tea-gardens, with playing grounds for *skittles* and *bumble puppy*! – oh, capital! Sir, there is nothing like a playhouse for fine prospects; and when, without fatigue, and trouble, one can see all Europe, *well lighted for a shilling*.[128]

Such scenic spectacles, and particularly scenes of rapid transition brought about by Harlequin's magic wand, were part of the essential stock-in-trade of pantomimic entertainment at this time, the investment required to bring such elaborate entertainments before the public justifying Boaden's claim that, for the winter managers, 'the Pantomime is usually a trial of strength'.[129] This compares significantly with the scant attention still being paid during this period to the rest of the year's productions, 'scenery, dress, decoration of every kind', according to a retrospective assessment, being 'reserved for Christmas prodigalities; and the legitimate drama in those days, it was thought, might be kept alive by the *pathos* or the *humour* of the performer'.[130] Tradition, therefore, entitled Colman to consider stage impact a priority when writing *Blue Beard* for the Christmas audience, and he is quite prepared to acknowledge that his 'Dialogue and Songs [were] subservient' to the efforts of Greenwood the scene-painter and Johnston the machinist.[131] However, the Christmas entertainment privileged by tradition was pantomime; its scenic effects were to contribute to innocent wonderment. Colman had written something totally alien, a gothic fantasy, using scenery to heighten its blood-chilling horror. The influence of this simple innovation upon *fin-de-siècle* ideas of theatre was to prove both irresistible and fatal.

Michael Kelly, who composed the music of *Blue Beard*, has claimed part of the initiative for Colman's theatrical venture. In August 1790, when in Paris with his *chère amie* Mrs Crouch, who played the role of Fatima in the original production of the piece, Kelly had seen Grétry's highly successful opera of *Barbe Bleue*. Shortly afterwards, Mrs Crouch 'wrote down the programme of the drama'[132] and, several years later, this document was produced in front of Colman by Kelly, who was anxious that the author dramatise it for him. 'I told him, moreover,' continues the loquacious songsmith,

> that my object was to endeavour to establish my name as a composer, by furnishing the music for it; that I was perfectly sure a week's work would accomplish the literary part of the two acts, for which I would give him a couple of hundred pounds. After having discussed the subject, and two bottles of wine, the witty dramatist agreed . . . and before the week was ended, the piece was complete.[133]

Kelly's music, in fact, attracted more than its full measure of publicity, the *London Chronicle* praising it as 'at once grand, simple and scientific',[134] the *Authentic Memoirs of the Green Room* later lambasting it as plagiarism so extensive that it branded its composer as nothing better than a 'musical quack'.[135] Nor could Colman run the critical gauntlet unscathed. The *Morning Post* grimly warned the author of *Blue Beard* that he 'will add by it nothing to his reputation',[136] the *London Chronicle* describing the dialogue as 'weak and incoherent',[137] whilst the *Monthly Mirror* included amongst its barrage of invective such statements as 'beneath contempt . . . in the very worst of Colman's very worst productions . . . a patchwork of buffoonery and bombast'.[138] The critical voice rose in unison to rue the day that Colman had hired himself out 'in the service of the Smithfield Muses'.[139] But at least one critic was prepared to restore a sense of perspective to the furore, albeit from a distance of twenty-five years.

> The author, who had little in view beyond manufacturing a convenient vehicle for the display of gorgeous scenery and shewy processions, has effected his intention with a cleverness, which many who think meanly of the performance, would find some difficulty in equalling; and, what probably to him was but the pastime of an evening, has on numerous succeeding evenings imparted gratification to thousands.[140]

Blue Beard's success with audiences was absolutely prodigious. In just two years it stood unopposed as the most frequently performed afterpiece of the eighteenth century. Surprisingly, perhaps, one of the main ingredients of its original popularity was the 'exquisite and well-employed talent'[141] of members of its prestigious cast, Boaden declaring that 'it was performed so well, and was so truly splendid, that it has never been surpassed in my remembrance'.[142] Adolphus once again, has documented the most accurate account of the play in performance:

> Not one part could have been better filled. Palmer condescended to perform the savage bashaw; he gave to his love the proper haughtiness, and roared out his impatience to fill up the number of his murders with characteristic force. Dicky Suett, as Ibrahim . . . punned, exulted, shivered,

and ran away, with his usual drollery . . . even the part of a little negro
[Hassan], brought on merely to waste a few minutes, was made of value
by Hollingsworth. Bannister, in Shacabac . . . highly increased his reputation. The workings of his mind when indignation at the past murders of
his patron, and a desire to prevent that which was then in contemplation,
were ill restrained by fear . . . gave scope to his fine display of blended
tragic and comic power . . . It is not possible to praise too highly the
exquisite feeling with which Mrs Crouch played and sung the oppressed
and unfortunate Fatima. Mrs Bland . . . embellished Beda with her sweetest notes. Her song, 'His sparkling eyes were dark as jet', and the duet,
'Tink-a-tink' . . . were among her most popular exertions.[143]

But by far the most significant performance offered in *Blue Beard* was that of Miss
De Camp in the character of Irene. As some indication of the imprint made upon
the consciousness of the theatrical world by this event, six years later the *Authentic
Memoirs of the Green Room* published as its only illustration a frontispiece engrav-
ing of De Camp's Irene. Boaden stresses her contribution to *Blue Beard* as 'the most
prominent merit it had', observing in her performance an important stylistic departure: she 'looked, and acted, and sang in such a way as to prove herself the first
melodramatic actress that had been seen among us'.[144] Adolphus elaborates:

> The high achievement of the character was her interesting grief at the
> menaced woeful catastrophe; and in the quartette, where . . . the author
> places her at the top of a tower, to 'look out if she can see anybody
> coming', her advance from infant hope to a full-grown assurance of aid,
> her progressive animation from the moment when she sees 'a cloud of dust
> arise', to that when she sees 'them galloping', her scream of joy, and the
> agitation of her whole frame when she 'waves her handkerchief', – all
> these constituted the high perfection of the dramatic art; and there was
> not in the house an eye nor a hand which did not give signs of sensibility,
> and pay a tribute of applause.[145]

Indeed, so great was the following she achieved in this role that Boaden wonders
whether it might not in fact have been 'injurious to the fascinating actress, and
stopped her progress in the profession'.[146]

Of course, the scenic spectacle of *Blue Beard*, the detail of which has so fortunately been preserved in the text, was equally mesmeric. According to the *London
Chronicle*, £2,000 was lavished on its preparation, a sum almost equivalent to the
entire profits of the Haymarket Theatre in 1791. Most contemporary commentators are agreed that the play's most impressive effect was the grand cavalcade
across the mountains in Act I Scene 1 which was considered to rank as 'one of the
finest ever produced on our stage'.[147] A gigantic piece of animated scenery presented a perspective panorama of the march wending its way between rocky out-
crops and through craggy passes, its model figures and animals growing larger at

each successive reappearance. However, 'in the course of the representation, many blunders in working the scenery, which are unavoidable in a first representation of this nature, occurred, and the delays which took place were frequently very great . . . It was twelve o'clock before the curtain dropped.'[148] On this occasion, the cavalcade scene alone took half an hour, 'the small elephants, needing the Gulliver-like aid of the scene-shifter, to get them through the defiles', whilst the band stalwartly tried to cover the confusion by playing 'the same march' throughout.[149] Kelly, who appeared himself in the character of Selim, also became embroiled in a scenic fiasco.

> At the end of the piece, when Blue Beard is slain by Selim, a most ludicrous scene took place. Where Blue Beard sinks under the stage, a skeleton rises, which, when seen by the audience, was to sink down again; but not one inch would the said skeleton move. I, who had just been killing Blue Beard, totally forgetting where I was, ran up with my drawn sabre, and pummelled the poor skeleton's head with all my might, vociferating, until he disappeared, loud enough to be heard by the whole house. 'Damn you! Damn you! Why don't you go down?' The audience were in roars of laughter at this ridiculous scene, but good-naturedly appeared to enter into the feelings of an infuriated composer.

By the following day, 'the scenery and machinery were quite perfect', and at its next performance *Blue Beard* 'was received with the most unqualified approbation, by overflowing houses, and has kept its standing for six-and-twenty-years'.[150]

Johnston, the Drury Lane machinist, whose designs for the scenic apparatus of *Blue Beard* should have raised him to co-author status, remains the least public of the figures associated with the play. Colman has described him as a man 'celebrated for his superior taste and skill in the construction of flying chariots, triumphal cars, palanquins, banners, wooden children to be tossed over battlements, and straw heroes and heroines to be hurl'd down a precipice; — he was, further, famous for wickerwork lions, paste-board swans, and all the sham birds and beasts appertaining to a theatrical menagerie'.[151] Gratitude is owing to Kelly for providing the rare glimpse of this proud craftsman of theatrical ephemera that follows:

> The second act of Blue Beard opened with a view of the Spahis' horses, at a distance; these horses were admirably made of pasteboard, and answered every purpose for which they were wanted. One morning, Mr Sheridan [the playwright, then proprietor of Drury Lane], John Kemble [the great tragic actor, then a member of the Drury Lane Company], and myself, went to the property-room of Drury Lane Theatre, and there found Johnston . . . at work upon the horses, and on the point of beginning the elephant, which was to carry Blue Beard. Mr Sheridan said to Johnston, — 'Don't you think, Johnston, you had better go to Pidcock's, at Exeter 'Change, and hire an elephant for a couple of nights?' — 'Not I, Sir,'

replied the enthusiastic machinist; 'if I cannot make a better elephant than that at Exeter 'Change, I deserve to be hanged.'[152]

Sheridan's suggestion that Johnston use a live animal on stage, no doubt intended in jest, nevertheless contained an element of prophecy which, on 18 February 1811, came to its fulfilment when a glittering revival of *Blue Beard*, featuring a troupe of sixteen white performing horses, took place at Covent Garden. Boaden seems to have been totally enthralled by the antics of the beasts which were ridden by the Spahis and made their first appearance early in the second act:

> Their various and incessant action produced a delightful effect upon the eye, and when they were afterwards seen ascending the heights with inconceivable velocity, the audience were in raptures, as at the achievement of a wonder. Subsequently, however, they seemed still more astonished at the sagacity, or recollection, of the noble animals before them; — in the charge, some of the horses appeared to be wounded, and with admirable imitation fainted away. One of them, who in the anguish of his wounds had thrown off his rider, and was dying on the field, on hearing the report of a pistol sprung suddenly upon his feet, as if again to join, or enjoy the battle; but his ardour not being seconded by strength, he fell again as if totally exhausted.[153]

Genest was stunned. As far as he was concerned, the *raison d'être* of human existence perished that night. 'One might as well have sitten in a stable' as a playhouse, he thundered,[154] offering as a wicked afterthought the speculation that perhaps, if the animals 'had "drowned the stage with tears" . . . we should have had the acme of rational amusement'.[155] But audiences of the day did not share his misgivings. 'The Piece was announced for repetition with thunders of applause,' recorded the *European Magazine*. 'Louder acclamations were never heard. In its progress it was twice greeted with three regular cheers. The House was crowded almost to suffocation — a great number of persons, who could not procure seats, filled the coffee-rooms and corridors, and hundreds went away.'[156] Greater success could not have been hoped for. 'The first forty-one nights of *Blue Beard*, revived with the horses, produced above *twenty-one thousand pounds*,' declared Reynolds,[157] always sensitive to the financial issues of theatrical life. But no amount of money could provide adequate compensation for the long-term damage ultimately inflicted on the public image of the legitimate theatre by its own promoters when they took the decision to introduce horses on a patent house stage in *Blue Beard*. Indeed, their compromise of their own legal privilege was self-evident: the sixteen elegant horses belonged not to Covent Garden but to Mr Parker of Astley's Amphitheatre, where they were exhibited only because that establishment was prevented by the monopoly system from acting the 'spoken drama'. The vogue for animal entertainments at the patent houses which the revival of *Blue Beard* created was to strain to its breaking-point the once inviolable notion of theatrical legitimacy.

The counsels which prevailed at Covent Garden in 1800, when on 8 February Morton's seventh play, *Speed the Plough*, received its first performance, were altogether more moderate. *Speed the Plough* was Morton's fourth successive production to have appeared at Covent Garden since *The Children in the Wood* and subsequently *Zorinski* were brought out by Colman at the Haymarket, a fact significant enough to have attracted the comment that 'the friendly trio of dramatists, Reynolds, Holman, and Morton, were now in possession of the Covent-Garden stage'.[158] An abrupt and rather fascinating stylistic modification in Morton's playwriting accompanied his break from Haymarket patronage. *Zorinski* (1795) was the last in a series of three early Morton plays deriving their themes and mood from legend or history. *The Way to Get Married* (1796), which marked his return to Covent Garden, displayed, outwardly at least, the distinguishing features of quite a different species of drama: it was a five-act comedy with terms of reference firmly rooted in the modern idiom. This new departure led to instant success, demonstrating, in the opinion of the *True Briton*, 'a degree of dramatic excellence that far exceeds the merit of his former productions . . . it unveils the artifices of the town, raises considerable merriment, exercises the noblest affections of the heart, and leaves a strong moral impression'.[159] At the same time, the basic preoccupation of Morton's earlier plays remained the central impulse of his new style, as the *Monthly Mirror* reviewer was shrewd enough to notice when he ventured the claim that 'no modern comedy can be produced where the *serious* and *comic* departments are blended with so much ingenuity and success . . . The particular *forte* of this writer is evidently PATHOS . . . All the serious business is finely and beautifully wrought.'[160] Morton could scarcely have ignored the impact of his new playwriting formula upon contemporary playhouse fashion. 'No theatre in these kingdoms,' remarks Boaden, 'was long without *The Way to Get Married*. It is no very great improbability, that at one hour, in some one evening of the week, the whole playgoing part of the community of Great Britain, through all her cities, was applauding the work of Morton.'[161] Since popular appeal was the primary concern of the late eighteenth-century professional playwright, it is understandable that Morton should have used the broad outlines of *The Way to Get Married* to establish the mould which was to typify many more of his plays, including *Speed the Plough*. However, unlike Colman, Morton had continually to bear in mind a crucial secondary factor when devising his theatrical pieces: the preferences and policies of those who controlled the outlets for their performance. A revealing comment, for instance, which appeared at the time when *Zorinski* was produced at the Haymarket, describes the play as 'of that kind, to which Mr COLMAN has given popularity'.[162] Moreover, reports indicate that, as a result of *Zorinski*'s appeal, 'COLMAN's Theatre never contained larger audiences'.[163] If that, indeed, was the case, then what induced Morton to take the apparently superfluous step of altering his style? A brief examination of the complicated politics of patent theatre management during this period should throw some light on this intriguing question.

Of the three London theatres protected in law, the Haymarket stood to suffer the least from direct competition with the others. In fact, Colman's particular enemy was the summer climate. During the tropical summer of 1793, for instance, when 'a Beehive shaded from the sun . . . became so heated as to cause the honey to run from the hive',[164] the receipts at the Haymarket fell by 75 per cent. Even moderate rises in temperature drove the fashionable out of London, whilst those who remained in town tended to frequent the pleasure gardens unless a convincing reason for them to sweat it out under Colman's roof was produced. Indeed, it is to Colman's credit that his resourceful management enticed sizeable audiences into the Haymarket Theatre, in spite of such conditions, throughout his regime. But the circumstances which gradually shaped the particular policy identities of the winter houses during the 1790s evolved less distinctly out of the personalities of their proprietors. Thomas Harris was the chief proprietor of Covent Garden and a man who received, in his epitaph, an especial tribute for the way in which he 'mingled so much benevolence with justice towards the numerous individuals under his control, that, whilst he commanded their respect as a manager, he gained their attachment as a friend'.[165] In bitter contrast to this is the oppressive atmosphere at Drury Lane that Sheridan, through his behaviour as its chief proprietor, had created. 'He was seldom agreeable in the presence of actors,' recalls Anne Mathews. 'I perfectly well remember one particular evening, when Miss De Camp, after a somewhat *animated* colloquy with him, closed it by telling him, "that the performers were all very happy before he entered the room, and that he never came but to make everybody uncomfortable".'[166] Another curiosity she relates is of Sheridan's manner of entering 'his own theatre as if stealthily and unwillingly'.[167] There was a perfectly simple reason for this: he did not wish to be recognised. If Sheridan had been known to be on the premises he ran the risk of being beset by the theatre's many creditors who, at times, might also have included members of the acting company. So irresponsible was Sheridan's financial stewardship at Drury Lane that two of his stage managers, Thomas King and John Philip Kemble, at different periods resigned their positions in protest. Harris, on the other hand, had been trained in business and, as a result, 'did not consider himself entitled to delay, much less to alienate, the stipulated payment for which he had received the valuable labours of his performers'.[168] Boaden found Harris an impressive figure. 'I never knew a gentleman better calculated to be at the head of a theatrical concern,' he declared.[169] And playwrights were attracted by his reliability. Munden has remarked upon 'the large prices which he cheerfully paid for the productions of dramatists . . . who preferred the ready money of Covent-Garden to the promissory notes of the rival house'.[170] However, one of Sheridan's most significant acts was to appoint, in 1788, the tragedian and classical enthusiast, John Philip Kemble, as Drury Lane's stage manager. Although little evidence has survived to suggest that Sheridan was greatly esteemed amongst contemporary commentators of this period, Mrs Mathews, for instance, dismissing him as 'burnt out',[171] there is probably more than just malice in the accusation that, with age, the former dramatist grew jealous of his past achievements, becoming 'ill-inclined . . . to see his theatre in possession of any other comic writer'.[172] If such an embargo was his intention, Kemble was just the man to guarantee its success. 'Kemble cared little for any *comedy*,' explains Boaden, 'and for modern comedy not at all. Sheridan was satisfied if he saw the *Rivals* and the *School for Scandal* sparkling among the gloom of "dark December" . . . But for a comedy, that was to take its rank with one of the modern Congreve [an epithet used to describe Sheridan in his heyday] , *that* never occurred to the cabinet of this political and bankrupt theatre.'[173] Against this background, Harris developed the strategy which was eventually to lead to his creative monopoly of the talents of writers such as Morton. Guided by the basic principle that '*new* plays, with the established company, were less expensive, and more productive than *old* plays, with stars',[174] the latter being Drury Lane policy, Harris proceeded to the 'judicious adoption of contemporary light comedy',[175] otherwise known during this period as the blue-coat-and-white-waistcoat play. Reynolds elaborates on the financial implications of this managerial preference, making a comparison with the net profits of *Pizarro*, Sheridan's exotic (and belated: his last full-length play had appeared in 1779) adaptation of a German historical drama which was performed at Drury Lane in 1799:

> *Pizarro*, as I have previously stated, brought more money into the theatre [estimated £30,000] ; but, at the same time it should be remembered that it also took more money out of it. When the treasurer strikes the balance, he will necessarily find a vast difference between the expenses attached to the production of what is technically termed a 'blue coat and white waist-coat play', and those of a 'spangled and processional play'.[176]

Indeed, by the mid 1790s Covent Garden's 'succession of simple coat-and-waistcoat plays, running their one-and-twenty nights to full houses, rendered almost ridiculous the sturdy persistence of the other theatre in more regular, costly and classical exhibitions'.[177] In this way, Harris astutely engineered for Covent Garden, out of the retrenchment of his competitors, a desirable and recognisable market, and there can be little doubt that, in return for the opportunity and security he offered to competent playwrights, he would have expected their fullest co-operation in supplying it with appropriate material. That *Speed the Plough* was described as one of Morton's 'inexpensive comedies'[178] might suggest the author had, for some time, appreciated the managerial expectations of him. The play rendered Harris 'safe for the rest of the season';[179] Covent Garden protected Morton for the bulk of his career.

Generally, *Speed the Plough* received the critical accolade which it deserved. However, enthusiasm got the better of the *Morning Chronicle* reviewer, who marvelled that the play generated 'a popularity which has rarely been attained. The claims which it possesses to the patronage of the Public are genuine, and derived from a just knowledge of the national character. With no inconsiderable portion of the wit of CONGREVE, and the sprightliness of FARQUHAR, it happily combines the elegant expression and affecting sentiment of STEELE, and the broad humour of O'KEEFFE.'[180] At this, the *True Briton*, raising a sardonic eyebrow, was tempted to ask: 'Is it *game* you are making?'[181] Morton's adept handling and presentation of his materials was admired by the *Morning Post*: 'The Plot, in complexity and variety of incident and embellishment, far exceeds the scope of most modern dramas, even the most diversified. This complexity is however free from intricacy, and the business, though greatly varied, is conducted with such masterly skill as not to perplex, but' — and here the idea becomes unfortunately expressed — 'merely require the lowest degree of attention which it is capable of exciting, to follow the story.'[182] Nevertheless, one particular aspect of *Speed the Plough* gave rise to widespread critical unease which the previous reviewer articulates as follows: 'A piece like this, composed of one half broad farce, the other half tragedy, we know not how to refer to any distinct class of dramatic composition.' This was certainly no new response to Morton's work. However, the reasoning which accompanied it was. 'We can only place it,' continues the critic, 'under the head of those German Dramas so highly favoured.'[183] After a long and cantankerous discussion of *Speed the Plough*, Thomas Dutton, editor of the *Dramatic Censor*, reaches a very similar, if more acrimoniously worded, verdict. 'And you may have, further,' he summarises, addressing the author in the critical dock,

> the additional advantages of being able to rank your Play in which ever of these two classes of the drama [comedy or tragedy] you think proper. Or you may refer them to both, and have your Play acted alternately, as a *Comedy* one night, and as a *Tragedy* the other, by closing it alternately with a *Dance* or the exhibition of a *Statuary's Shop*! after the example of our most approved modern masters, and completely *à la Kotzebue*![184]

Dutton's final reference is to the German dramatist, August Friedrich Ferdinand von Kotzebue.

During the final years of the eighteenth century, but especially after the performance at Drury Lane in March 1798 of *The Stranger*, Benjamin Thompson's adaptation of *Menschenhass und Reue* which had been written by Kotzebue in 1789, a mania developed amongst audiences for the so-called 'German drama'. This was satisfied chiefly by further renderings from Kotzebue, twenty of which appeared before 1801, and reached its fashionable zenith in 1799 with Sheridan's *Pizarro*, a production which was by far the most conspicuous in the canon of Kotzebue derivatives. Reynolds was of the opinion that Kotzebue 'had no small knowledge of human nature . . . [and a] full command over the human passions',[185] credentials which could do nothing less than assure him an immediate appeal amongst audiences who now put a high value on any performance offering them the means of their own emotional laceration. 'There is no reserve at all in *German passion*,' observed Boaden, derisively,[186] perhaps conveniently forgetting that at Drury Lane in 1798 he had found 'the *Stranger*, as he looked and moved in Kemble . . . of all exhibitions that I have ever seen, the most affecting'.[187] From the evidence of his plays in their adapted forms, Kotzebue would seem to have pushed forward to a considerable extent dramatic treatment of psychological motivation. However, in doing so, he broadened the humane element then appreciated in the English theatre beyond what some critical observers regarded as its proper limits. Boaden was not alone in seeing in Kotzebue a dangerous liberal who

> exhibited the *adultress*, and the *seducer*, and the *robber*, and even the *murderer*, as the most generous of the species. The sort of thing became popular, from the *passion* it set in motion . . . Thus sympathy usurped the place of censure, and a door was opened to that fatal fallacy, of making a *compromise* with morals, and setting the vices to which we were *not* inclined, as a sort of balance to those in which we were determined to indulge.[188]

But audiences saw in Kotzebue only the romantic; whether or not he was also a subversive, as Hannah More maintained,[189] was a question offering entertainment of a different and more exclusive sort.

To suggest, as some contemporary critics have suggested, that the entanglement of Morton's style with the German drama in *Speed the Plough* is indicated simply by the play's use of contrasting comic and serious scenes would seem to show little awareness of the author's stylistic hallmark, which is visible in exactly those terms from the very outset of his career. However, whether by instinct or by forensic skill, several reviewers narrowed their particular concern down to Morton's treatment of his serious scenes. One recommends 'curtailment in the first three acts. These are chiefly broad farce . . . What follows is of a very different kind, and, perhaps, not the less interesting from being unexpected.'[190] Another, maybe recognising the German influence at work, takes a more severe line, condemning *Speed the Plough* as 'faulty, and even monstrous, in the greater part of the serious incidents'.[191] In both cases, the reviewers were right to draw attention to the non-comic content of the play; eight years later Elizabeth Inchbald was to explain why: 'the plot, and serious characters . . . are said to be taken from a play of Kotzebue's, called "The Duke of Burgundy" . . . condemned or withdrawn at Covent Garden Theatre, not very long before "Speed the Plough" was received with the highest marks of admiration'.[192] Her account is substantially accurate. In 1798, Anne Plumptre published an English translation of Kotzebue's *Der Graf von Burgund*, and this text, under the title of *The Count of Burgundy*, received an unsuccessful single performance at Covent Garden on 12 April 1799. Here lie, in embryo, the basic features of Morton's *Speed the Plough*: a murdering brother (although, in this case, he makes a more conscientious job of it) and an estranged son (also called Henry) brought up in the country in ignorance of the true identity of his parents. However, appearing in Kotzebue's original in a totally dissimilar context, these ingredients provided only the broadest outlines to Morton's own fertile creativity and were probably adopted as a means of lacing his forthcoming comedy with a heavy compound of well-proven strength. Above all, *Speed the Plough* demonstrates its author's 'passion for a country life',[193] critics agreeing, for instance, that 'the whole character of farmer Ashfield is delightfully drawn'.[194] Those who quibbled over his flirtation with elements they suspected of being of German origin, Dutton managing snideness even over the most innocent of Morton's borrowings from Kotzebue – the tableau with which *Speed the Plough* closes – could have taken no notice of the careful way in which Morton has resolved his plot: he deftly contrives, 'without the sacrifice of any principle',[195] in the play's final moments to retrieve Sir Philip Blandford from the charge of murder. Nevertheless, the notion of German drama was a sensitive trigger of critical prejudice; its themes were thought by some observers to be a threat to the moral base of the political *status quo*, and once again attention was focussed on the anarchic implications of changes in playhouse taste.

The popularity of *Speed the Plough* shows evidence of few political or moral scruples amongst its audiences. 'No piece was ever received with greater applause, or concluded with stronger manifestations of desire from a crowded audience to see it again,' announced the *Morning Post*.[196] Indeed, it went on to become the 'most successful production of the season,' according to Covent Garden playwright, Thomas Dibdin, which 'added some thousands to the treasury' of the theatre.[197] Performed forty-one times in its first season, *Speed the Plough* soon had an established place as 'a favourite stock piece'[198] in the repertoire of the British theatre, appearing regularly in the schedules of the London patent houses for the next thirty years. In Boaden's opinion, the comedy was 'admirably acted in all its parts',[199] although with perhaps an exception being made for Mrs Thomas Dibdin in Lady Handy, whose 'performance induces us to conclude,' grumbled the *Authentic Memoirs of the Green Room for 1801*, 'that she owes her engagement more to her husband's interest, than to any intrinsic merit of her own'.[200] The casting reflects Morton's theatrical mode: Alexander Pope, the tragedian, took the role of Sir Philip Blandford whilst Sir Abel Handy received the exuberant attention of broad comedian Joe Munden. A rather different quality of performance was injected into the play through the selection of Mr and Mrs Harry Johnston to represent the characters of Henry and Miss Blandford. This was a decision of particular significance since 'Harry Johnston . . . and his wife . . . were, as melodramatists, of much consequence.'[201] Indeed, Johnston himself had already displayed the ability 'to convey sentiments, and describe the passions, unaccompanied by that useful assistant, speech',[202] and this would have considerably heightened his effect as the gloomy and Kotzebue-inspired fatherless swain. Also contributing to the picturesqueness of the play in performance would have been the acting of the pretty, sixteen-year-old Miss Murray as Susan Ashfield, which was guaranteed to take her into the hearts of audiences by the single virtue of its acknowledged power 'in scenes which require the felicitous union of *pathos* with simplicity, and the genuine delineation of *unsophisticated* nature'.[203] However, Farmer Ashfield was clearly the character most attractive to audiences, and he was played 'in a masterly style'[204] by Thomas Knight. 'In this piece,' recollects Anne Mathews, 'Mr Knight introduced the Somersetshire dialect with great effect, which was, I believe, banished from the stage by the more humorous dialect of Yorkshire, which Emery made so popular, and left as a sort of legacy to all succeeding actors.'[205] Genest traced Knight's dialectal prowess back to its source after reminiscing with an old actress at Bath, where Knight's career began, who described how 'the Bath Company, on their return to Bristol [where they also played], used to stop and sup at an inn on the road — the ostler of this inn was a country fellow with a good deal of drollery about him — Knight used to get into conversation with him, for the sake of improving himself in country parts'.[206] Emery was also a skilful imitator of accent; in Yorkshire 'he was perfect to an aspirate, or the want of one'.[207] And in this respect, both comedians brought a measure of discipline to the hitherto mediocre presentation of countrymen on the stage. Nevertheless, a curious hybrid must have been produced by Emery's appearance as Farmer Ashfield of Hampshire at the Haymarket in 1801. Fortunately, the play was infinitely accommodating and adaptable, containing, as Boaden observed, a core of material 'so powerful as to support every company of comedians who have performed it'.[208]

After 1800, the playwriting careers of George Colman the Younger and Thomas Morton were to continue for another twenty-five and thirty years respectively. Indeed, by the end of the eighteenth century, both men had completed less than a half of their lifetime's total of dramatic output. However, the social, political and theatrical conditions which were to prevail over the ensuing decades established terms of reference against which the plays of Colman and Morton, when measured, produced a markedly different result from that which has been documented for the 1790s. In 1802, for instance, Holcroft's translation of the first play to appear in England under the formal description of melodrama, *A Tale of Mystery*, opened a new era in theatrical endeavour after its production at Covent Garden. But when plays of this type began to be written by native playwrights, no contribution was forthcoming either from Colman or from Morton. Additionally, Reynolds has pointed out a significant directional change in audience taste during the early part of the nineteenth century which led to 'the difficulty of writing a successful five-act comedy. I will state,' he continues, 'with accuracy I believe, that only two have turned the twentieth nights, during the last eighteen years [i.e. after 1809].'[209] In spite of this, comedy remained the mode in which both men persisted in writing the majority of their remaining plays, Colman having virtually abandoned serious drama in favour of comedy and farce after the humiliating failure of *The Iron Chest* at Drury Lane in 1796. If the two authors were successful in the latter part of their careers, it was in preserving their professional standing intact until their retirement. In the years after 1800, the playwriting of Colman and Morton began to slide slowly away from the cutting edge of the process which carved out the creative identity of the London theatre.

NOTE ON THE TEXT

Few late eighteenth-century plays, those in the present volume being no exception, would have survived for many nights of their first season without some form of curtailment. Cutting was commonplace, vigorous and often arbitrary, its rationale being to expunge lines, sometimes whole scenes, which attracted any measure of public disapproval, regardless of whether or not that disapproval was generated by faults in the production. Such deletions were made in the first instance in the prompt copies of the plays held at the theatres of their performance, these modified texts subsequently forming the basis of the authorised published versions offered for sale by booksellers. The present editions represent an attempt to reconcile the full texts as originally written by Colman and Morton (preserved in the Larpent Collection of Licensing MSS in the Huntington Library, California) with the curtailed forms in which they were published. In each case, the first authorised publication (*Inkle and Yarico* by G.G.J. and J. Robinson, London, 1787; *The Surrender of Calais* by Longman, Hurst, Rees and Orme, 1808; *The Children in the Wood* printed privately for the author, London, 1794; *Blue Beard* by Cadell and Davies, London, 1798; *Speed the Plough* by Longman and Rees, London, 1800) has provided the textual basis used here. Omissions have been supplied from the appropriate Larpent MS where earlier mutilations required repair.

The Children in the Wood posed additional problems of occasional divergence between its first published version and its Larpent MS over details of phrasing. Since Morton himself had the play printed at his own expense, it is highly likely that he yielded to the temptation to put a more literary gloss on his earlier attempt before offering it to a reading public. At such points, lines are rendered as originally written.

NOTE ON THE MUSIC

All five plays in this volume contain music to a greater or a lesser degree and the original scores written for each have survived in the following published forms:

Dr Samuel Arnold, *Inkle and Yarico. A Comick Opera . . . adapted for the Voice, Harpsichord, Piano-Forte &c. Op. 30.*, London 1787

Dr Samuel Arnold, *The Overture, Songs, Chorusses &c in the Surrender of Calais. Op. 33.*, London 1791

Dr Samuel Arnold, *The Children in the Wood. A Comic Opera in two Acts for the Piano-Forte, Harpsichord and Violin &c. Op. 35.*, London 1793

Michael Kelly, *The Grand Dramatic Romance of Blue Beard . . . the Music Composed and Selected by M. Kelly.*, London 1798

John Moorehead, *The favourite Dance introduced in the New Comedy called 'Speed the Plough' . . . Arranged as a Rondo for the Piano-Forte.*, London 1800

Dr Samuel Arnold began his musical career in the theatre in 1764 when he became harpsichordist to Covent Garden. His first major success as a composer followed in 1765 when he compiled the pastiche opera with Bickerstaffe of *The Maid of the Mill*. In 1769 he left Covent Garden to enter what was to prove a disastrous financial speculation involving Marylebone Gardens, a summer resort specialising in fireworks and concerts. Seven years later Arnold abandoned that venture and joined Colman the Elder's Haymarket Theatre as resident composer when it opened under his direction in 1777. This position he continued to hold until his death in October 1802, latterly becoming an important associate and close friend of Colman the Younger. In 1797 he invested £1,100 in the Haymarket Theatre in an attempt to ease the financial burden of the young manager. Ironically, this act of generosity was to lead directly to Colman's eleven years of confinement for debt.

Michael Kelly was an ebullient figure in the theatrical world of this period who had made his début as a tenor in Bickerstaffe's comic opera of *Lionel and Clarissa* at Drury Lane in 1787. Whilst maintaining his connections with the music at Drury Lane, he became stage manager of the King's Theatre (which performed the Italian Opera in London) in 1793 and, following the death of Storace, its composer. *Blue Beard* was the first professional collaboration of Kelly and Colman and the prototype of two further joint productions. After the death of Arnold, he composed a large proportion of the music used at the Haymarket Theatre over the next ten years.

John Moorehead took up the position of principal violinist in the Covent Garden orchestra in 1798 and soon began composing that theatre's incidental music. In 1802 he was committed to Tothill Fields Prison on the grounds of insanity.

REFERENCES

1   *Some Account of the English Stage*, Bath 1830, IX, 564.
2   Ibid.
3   Ibid., IX, 568.
4   'George Colman the Younger', *Review of English Studies*, XXIII (1947), 370.
5   P. iii.
6   *Some Account of the English Stage*, IX, 59.
7   *Memoirs*, London 1807, I, 270.
8   Ibid., I, 303.
9   Ibid., II, 362.
10 *The Baviad and Maeviad*, revised edition, London 1799, p. 21.
11 Ibid., p. 24.
12 Ibid., p. 27.
13 Ibid., p. 28.
14 Ibid., p. xx.
15 *The Baviad and Maeviad*, New York 1799, p. x.
16 Quoted in E.P. Thompson, *The Making of the English Working Class*, Harmondsworth 1968, p. 19.

17   *Social Contract*, translated by Maurice Cranston, Harmondsworth 1968, p. 83.
18   James Boaden, *Memoirs of Mrs Siddons*, London 1827, II, 345.
19   Autograph letter, Broadley Collection (Little Haymarket Theatre, fo. 93), Westminster Public Libraries.
20   Ibid.
21   Quoted in James Boaden, *Memoirs of Mrs Inchbald*, London 1833, II, 143.
22   *Memoirs of Mrs Siddons*, I, 327–8.
23   Ibid., II, 71.
24   *The Lives of the Players*, London 1831, II, 242.
25   *Hazlitt on Theatre*, edited by William Archer and Robert Lowe, New York 1957, p. 122.
26   Quoted in Paul Ginisty, *Le Mélodrame*, Paris 1910, p. 54.
27   *Public Characters of 1800–1801*, edited by Henry Woodfall, London 1801, p. 144.
28   *Parliamentary Papers*, 1832, VII, 178.
29   Quoted in Richard Brinsley Peake, *Memoirs of the Colman Family*, London 1841, II, 400.
30   London 1832, p. 7.
31   Ibid., p. 17.
32   P. 16.
33   P. iii.
34   *The Life and Times of Frederick Reynolds*, London 1826, II, 421.
35   *Memoirs of the Life of John Philip Kemble*, London 1825, II, 411.
36   *The Haymarket Theatre*, London 1903, p. 60.
37   *George Colman the Younger*, New York 1946, p. 178.
38   Anne Mathews, *Tea Table Talk*, London 1857, I, 123.
39   *Parliamentary Papers*, 1832, VII, 142.
40   I, 471.
41   *Memoirs of Kemble*, II, 74.
42   *Memoirs of Mrs Siddons*, II, 176.
43   Reynolds, I, 264–5.
44   *The Life of George Frederick Cooke*, second edition, London 1815, II, 104–5.
45   *Memoirs of Mrs Siddons*, II, 176.
46   *Public Characters of 1800–1801*, p. 151.
47   *Random Records*, London 1830, II, 112.
48   George Colman the Elder, *Prose on Several Occasions*, London 1787, III, 252.
49   *Random Records*, II, 275.
50   11 July 1785.
51   I (1787), 161.
52   London 1673, pp. 54–5.
53   *Spectator*, no. 11 (10 March 1711), 53.
54   II, 180.
55   *Memoirs of John Bannister*, London 1839, I, 167–8.
56   *Inkle and Yarico* MS, Larpent Collection, Henry E. Huntington Library, California.
57   *Public Characters of 1801–1802*, p. 149.
58   *Memoirs of Kemble*, I, 371.
59   Thomas Bellamy, *The London Theatres*, London 1795, pp. 8–9.

60 *A Dramatic Synopsis*, London 1804, p. 81.
61 *General Magazine and Impartial Review*, August 1787.
62 *The Mirror of the Stage*, II (1823), 6.
63 *Memoirs of Mrs Siddons*, I, 214–15.
64 *Public Advertiser*, 7 July 1788.
65 *Memoirs of Kemble*, II, 44.
66 *Memoirs of Joseph Shepherd Munden*, London 1844, p. 244.
67 Reynolds, II, 318.
68 Adolphus, I, 265.
69 Ibid., I, 265–6.
70 *Random Records*, II, 9.
71 *Memoirs of Mrs Siddons*, I, 320.
72 Ibid., II, 65.
73 Adolphus, I, 264.
74 Boaden, *Memoirs of Kemble*, II, 44–5.
75 John Taylor, *Records of My Life*, London 1832, II, 110–11.
76 Genest, VII, 196.
77 John O'Keeffe, *Recollections*, London 1826, II, 300.
78 Adolphus, I, 267–8.
79 George Colman, *New Hay at the Old Market*, London 1795, pp. 20–1.
80 Genest, VI, 569.
81 George Colman, *The Battle of Hexham*, London 1808, p. iv.
82 *The Times*, 19 August 1789.
83 12 August 1789.
84 *London Chronicle*, 30 July 1791.
85 *World*, 1 August 1791.
86 David Baker, Isaac Reed and Stephen Jones, *Biographia Dramatica*, London 1812, III, 309.
87 *European Magazine*, LXXI (1817), 438.
88 *The Surrender of Calais*, p. 5 in *Inchbald's British Theatre*, XX, London 1808.
89 *British Stage*, I (1817), 130.
90 *General Magazine and Impartial Review*, August 1789.
91 *Inchbald's British Theatre* (Surrender of Calais), p. 3.
92 *The Times*, 11 March 1807.
93 *Monthly Mirror*, II (1796), 67.
94 J.H. Cardwell, *Men and Women of Soho*, London 1903, p. 243.
95 *Memoirs of Kemble*, I, 218–19.
96 Reynolds, I, 316.
97 Ibid., II, 128.
98 I (1793), 209.
99 *Oracle*, 3 December 1792.
100 *Thespian Magazine*, I (1793), 211.
101 *Morning Post*, 3 April 1838.
102 Boaden, *Memoirs of Kemble*, II, 105.
103 I (1793), 206.
104 *Oracle*, 2 October 1793.
105 *Morning Chronicle*, 2 October 1793.
106 1 October 1793.
107 Unidentified newspaper cutting dated 3 October 1793, Gabrielle Enthoven Collection, Victoria and Albert Museum.
108 *Essays of Elia*, London 1823, p. 318.
109 II, 97.

110 Adolphus, I, 318–19.
111 Ibid., I, 319–20.
112 Ibid., I, 320–1.
113 Ibid., I, 320.
114 O'Keeffe, *Recollections*, II, 308.
115 *The Times*, 2 October 1793.
116 *Morning Chronicle*, 2 October 1793.
117 2 October 1793.
118 George Raymond, *Memoirs of Robert William Elliston*, second edition, London 1846, I, 102–3.
119 Quoted ibid., II, 330.
120 *The Life of Mrs Jordan*, London 1831, I, 254.
121 *Memoirs*, II, 384.
122 Adolphus, I, 113.
123 Munden, p. 145.
124 Boaden, *Life of Mrs Jordan*, I, 306.
125 *Memoirs of Elliston*, I, 379.
126 *Blue Beard*, London 1798, p. iii.
127 *Memoirs of Mrs Inchbald*, I, 157.
128 Reynolds, I, 364.
129 *Life of Mrs Jordan*, II, 201.
130 Boaden, *Memoirs of Mrs Siddons*, II, 174.
131 *Blue Beard*, p. vi.
132 Michael Kelly, *Reminiscences*, London 1826, I, 348.
133 Ibid., II, 130–1.
134 16 January 1798.
135 London 1801, p. 33.
136 20 January 1798.
137 16 January 1798.
138 V (1798), 109.
139 *Monthly Review*, NS XXXVI (1798), 95.
140 Preface, *Blue Beard*, London 1823, p. iv.
141 Adolphus, II, 13.
142 *Life of Mrs Jordan*, I, 352.
143 Adolphus, II, 13–15.
144 *Life of Mrs Jordan*, I, 352.
145 Adolphus, II, 15.
146 *Memoirs of Kemble*, II, 208.
147 'Life of George Colman' in *The Dramatic Works of George Colman the Younger*, edited by J.W. Lake, Paris 1827, I, p. xix.
148 *London Chronicle*, 18 January 1798.
149 Boaden, *Life of Mrs Jordan*, I, 351.
150 Kelly, *Reminiscences*, II, 132.
151 *Random Records*, I, 228.
152 *Reminiscences*, II, 134.
153 *Memoirs of Kemble*, II, 542.
154 Genest, VIII, 232.
155 Ibid., VIII, 287.
156 LIX (1811), 131–2.
157 Reynolds, II, 404.
158 Boaden, *Memoirs of Mrs Inchbald*, II, 4.
159 25 January 1795.
160 I (1796), 243.

161   *Memoirs of Kemble*, II, 152–3.
162   *Morning Chronicle*, 30 June 1795.
163   Ibid.
164   *The Times*, 17 July 1793.
165   Quoted in Reynolds, II, 412.
166   *Memoirs of Charles Mathews*, London 1838, II, 59–60.
167   Ibid., II, 59.
168   Boaden, *Memoirs of Mrs Siddons*, II, 342.
169   Ibid., II, 343.
170   Munden, pp. 146–7.
171   *Memoirs of Charles Mathews*, II, 61.
172   Boaden, *Memoirs of Mrs Siddons*, I, 258.
173   *Life of Mrs Jordan*, II, 67.
174   Reynolds, II, 402–3.
175   Boaden, *Memoirs of Mrs Jordan*, I, 270.
176   Reynolds, II, 341–2.
177   Boaden, *Memoirs of Mrs Inchbald*, II, 5.
178   Boaden, *Life of Mrs Jordan*, II, 45.
179   Ibid.
180   12 February 1800.
181   18 February 1800.
182   10 February 1800.
183   Ibid.
184   *Dramatic Censor*, II (1800), 96.
185   Reynolds, II, 259.
186   *Memoirs of Mrs Inchbald*, II, 21.
187   *Memoirs of Kemble*, II, 210.
188   Ibid., II, 253.
189   *Vide* Hannah More, *Strictures on the Modern System of Female Education*,
       London 1799.
190   *Morning Post*, 10 February 1800.
191   *Monthly Magazine*, IX (1800), 179.
192   *Speed the Plough*, p. 2 in *Inchbald's British Theatre*, XXV, 1809.
193   Boaden, *Memoirs of Kemble*, I, 219.
194   *Monthly Magazine*, IX (1800), 179.
195   Boaden, *Memoirs of Kemble*, II, 252.
196   10 February 1800.
197   Thomas Dibdin, *Reminiscences of the Theatre Royal*, London 1827, I,
       275.
198   Walley Oulton, *History of the Theatres of London*, London 1818, II, 62.
199   *Memoirs of Kemble*, II, 254.
200   London 1801, p. 70.
201   Boaden, *Life of Mrs Jordan*, II, 143.
202   *Authentic Memoirs of the Green Room for 1803*, London 1803, p. 19.
203   *Authentic Memoirs of the Green Room for 1801*, London 1801, p. 77.
204   Munden, p. 81.
205   *Memoirs of Charles Mathews*, I, 421–2.
206   Genest, VII, 265.
207   Boaden, *Life of Mrs Jordan*, II, 57.
208   *Memoirs of Kemble*, II, 254.
209   Reynolds, II, 387.

IIa George Colman the Younger, from the portrait by John Jackson

IIb Thomas Morton, from the portrait by Sir M.A. Shee

BIOGRAPHICAL RECORD

In spite of his popularity as a playwright, very little detail has survived about the life of Thomas Morton. From an exhaustive check of contemporary published sources, it would seem that he managed the rare feat of maintaining an anonymity impenetrable by the curiosity-ridden theatrical press of the period.

21 October 1762	Birth of George Colman the Younger, in London. The circumstances of his parentage — he was born outside wedlock to an unidentified woman and was referred to ambiguously in official documents as Colman the Elder's 'adopted son' — were to cause particular embarrassment when investigated by the Court of Chancery in 1789 in connection with the transfer of the control of the Haymarket Theatre from his lunatic father.
1764	Birth of Thomas Morton at Whickham, in County Durham.
February 1766	First performance of *The Clandestine Marriage*, a highly successful comedy written at the height of his career by Colman the Elder (in conjunction with David Garrick).
c. 1768	Death of Morton's father. The young Morton was taken to live with his uncle, John Maddison, a London stockbroker.
1770	Beginning of Colman's education, at Marylebone Seminary. In 1772, he went to Westminster, his father's old school.
c. 1774	Morton entered at the idiosyncratic Soho Square Academy where he received a theatrically orientated education and commenced what was to be a lifelong friendship with actor/writer Joseph Holman.
1776	The United States' Declaration of Independence.
1777	Colman the Elder purchased the Haymarket Theatre, together with its right of receiving an annual licence for performing the spoken drama in London during the summer months, from Samuel Foote. Sheridan's *School for Scandal* appeared in this year.
1778	Jack Bannister, soon to become a close personal friend and a crucial instrument in the early playwriting success of Colman, joined the Haymarket acting company.
1780	Colman at Christ Church, Oxford. Since Oxford proved to offer Colman easy access to the lure of London theatrical life, the recalcitrant student was moved, at his father's request, to a more remote cradle of learning: King's College, Aberdeen.

16 August 1782	Anonymous production at the Haymarket Theatre of Colman's first play, *The Female Dramatist*, a two-act musical farce written whilst he was still at Aberdeen.
19 June 1784	Production at the Haymarket Theatre of Colman's second play, the musical comedy of *Two to One*, written on his return from Scottish banishment earlier that year. Colman the Elder formally presented his son to the public as a theatrical *débutant* in a specially written prologue.
July 1784	Morton enrolled at the Temple and took up chambers at Lincoln's Inn. The following month, Colman was enrolled at Lincoln's Inn and took up chambers at the Temple. Their distaste for legal studies kept them from social contact.
3 October 1784	Colman the Younger's clandestine marriage. Performed at Gretna Green, Colman's marriage to Catherine Morris, a minor Haymarket actress, was a personal and professional disaster. Professionally, it led to the involvement in Haymarket affairs of Colman's power-seeking brother-in-law, David Morris, whose commercial war of attrition finally gave him unmerited outright control of the concern.
September 1785	Colman the Elder suffered the first of a series of paralytic strokes which were to rob him of his reason.
1786	Beginning of Morton's crucial association with the young and aspiring dramatist, Frederick Reynolds.
4 August 1787	*First performance of Colman's INKLE AND YARICO.*
July 1789	In the same month as the storming of the Bastille in Paris, Colman took over the management of the Haymarket Theatre after a Commission of Lunacy had pronounced his father insane. His managerial connection with the theatre was to last almost thirty years.
11 August 1789	Haymarket première of Colman's first historical play, *The Battle of Hexham*.
February 1791	Publication of the first part of Thomas Paine's *Rights of Man*.
30 July 1791	*First performance of Colman's SURRENDER OF CALAIS.*
October 1791	Formation of the literary and theatrical 'Keep the Line' club. Morton and Reynolds were founder members.
5 November 1791	Covent Garden première of Reynolds's *Notoriety* which included Morton's first dramatic effort, a song for the Irish comedian, Jack Johnstone. The song was a hit and Morton was persuaded to take his writing seriously.
1 December 1792	Production at Covent Garden of Morton's first play, *Columbus*, a historical drama written in the style popularised by Colman. Covent Garden had recently reopened after extensive enlargements.

3 August 1793	Enthusiastic opening night of *The Mountaineers*, another of Colman's historical romances, at the Haymarket Theatre. John Philip Kemble played Octavian to the hilt, and beyond it.
September 1793	Colman opened the Haymarket Theatre for a special winter season during the rebuilding of Drury Lane and under the protection of the Drury Lane patent.
1 October 1793	*First performance of Morton's CHILDREN IN THE WOOD.* In France, thousands were now dying in the Reign of Terror.
March 1794	Official opening of the new and enlarged Drury Lane Theatre.
3 August 1794	Death of George Colman the Elder. Colman inherited his father's estate and creditors.
c. 1795	Morton joined the M.C.C. and for many subsequent years played cricket at Lord's, eventually becoming the club's senior member.
23 January 1796	Covent Garden première of Morton's *The Way to Get Married*, the first in a series of highly successful 'blue-coat-and-white-waistcoat' contemporary comedies written especially for that theatre. With one exception, Morton was to write his next seventeen plays for Covent Garden.
12 March 1796	Spectacular failure of Colman's *The Iron Chest*, a quasi-melodramatic play based on William Godwin's novel *Caleb Williams* and brought out at Drury Lane Theatre. The piece was performed only four times and, in his notorious Preface to its first edition, Colman lashed the production, and particularly John Philip Kemble, who played Sir Edward Mortimer, for incompetence.
25 June 1796	Colman introduced his protégé, Robert William Elliston, to the London public as Octavian in *The Mountaineers*. Later, he brought him out as Sir Edward Mortimer in his own production of a revised *Iron Chest* which received sufficient acclaim to heal the author's wounded pride. Irving revived the play at the Lyceum in 1879.
10 January 1797	First performance of Morton's *A Cure for the Heart Ache*. As with *The Way to Get Married*, which preceded it, and its successor, *Secrets Worth Knowing*, the play contained material specifically written for Covent Garden comedian, William Lewis. Young Rapid, whom Lewis played here, was to become one of the most popular characters in his extensive acting repertoire.
	Later in this year, the English naval mutinies at Spithead and the Nore took place. Colman was discovering serious

	financial difficulties in keeping the 1797 Haymarket season solvent.
16 January 1798	*First performance of Colman's BLUE BEARD*. The same season at Drury Lane also saw the production of Benjamin Thompson's *The Stranger*, adapted from Kotzebue's *Menschenhass und Reue*, which set the vogue for the 'German drama' at this time.
May 1799	Production at Drury Lane of one of the most lavish and popular of Kotzebue adaptations, Sheridan's *Pizarro*.
8 February 1800	*First performance of Morton's SPEED THE PLOUGH*.
July 1800	Negotiations in progress between Colman and popular actress Mrs Jordan for the sale of a 50-per-cent shareholding in the Haymarket Theatre, the problem of solvency having reached a critical level. A month later all transactions inexplicably ceased.
1802	In this year Morton made two unsuccessful bids for a stake in theatrical proprietorship. The first was undertaken with Frederick Reynolds and actor John Fawcett for a lease on Sadler's Wells, the second was an independent approach to the manager of Covent Garden for the shareholding which was finally sold to John Philip Kemble.
	Thomas Holcroft's *A Tale of Mystery*, the first translation into English of one of the new French *mélodrames*, was produced at Covent Garden in 1802.
5 March 1803	Covent Garden première of Colman's comedy of *John Bull*, his last important play. Some contemporary commentators regarded it as his best. It certainly generated the highest financial reward he was to receive from his playwriting. Admired by Samuel Phelps, *John Bull* was revived by him at Sadler's Wells in 1859 and Drury Lane in 1867.
May 1803	Beset with financial difficulties, Colman tried a new managerial stratagem to increase the Haymarket Theatre's revenue. This involved the formation of an 'independent' company of performers, mainly drawn from the provinces. Traditionally, the Haymarket company was composed largely of performers from Covent Garden and Drury Lane, but if those theatres stayed open beyond 15 May to chase the profits of a lucrative season, Colman had no option but to postpone his performances, sometimes for up to six weeks (out of a sixteen-week maximum run). He repeated the same stratagem in 1804 before he realised its long-term consequences were even more ruinous than the problems he sought to overcome.

16 May 1803	Colman gave Charles Mathews, destined to become one of the greatest comedians of the early nineteenth century, his London début at the Haymarket Theatre as Jabal in *The Jew* (Richard Cumberland, 1794).
15 January 1805	First performance of Morton's *The School of Reform* at Covent Garden. The play provided John Emery, a popular comedian specialising in countrymen, with challenging serio-comic material. The character of Tyke became the supreme performance of his acting career. *The School of Reform* was later part of a minor mid-century Morton revival which also included productions of *The Way to Get Married* and *A Cure for the Heart Ache*.

The original production of *The School of Reform* coincided with the height of the mania for the thirteen-year-old 'Infant Roscius', Master Betty.

June 1805	Mounting debts at the Haymarket Theatre could now be paid off only through the immediate sale of one half of the property. One quarter share was bought by David Morris, Colman's brother-in-law, and the other by James Winston and a business associate. For some reason they failed to take the elementary precaution of filing a legal deed of partnership, the squabbles which soon broke out amongst them taking the partners in and out of the Court of Chancery for over eight years.
10 June 1805	Colman introduced John Liston, later to become one of the favourite comedians of his generation, to the London theatre-going public at the Haymarket Theatre as Sheepface in *The Village Lawyer* (William Macready, 1787).
October 1805	The Battle of Trafalgar.
February 1806	Colman arrested for debt and confined to the Rules of the King's Bench Prison. This type of punishment merely imposed restrictions upon freedom of movement and association although, for a theatre manager, the penalty was almost as severe as physical imprisonment. Colman remained subject to this control for the next eleven years, the dispute with his theatrical partners originating over its alleged hindrance of his work.
12 June 1806	Colman brought out *Catch Him Who Can!*, the first in a series of successful pieces written by Theodore Hook for the Haymarket Theatre. Colman promoted Hook both as a literary talent and as a society figure.
10 March 1807	Opening night of Morton's *Town and Country*, its success manufactured solely by virtue of Kemble's willingness to

	bolster an unusually weak script with a dazzling performance as Reuben Glenroy.
22 June 1807	Colman prepared the début of Charles Young, soon to become one of the most powerful tragic actors of the next two decades, before a London audience at the Haymarket Theatre as Hamlet.
September 1808	Covent Garden Theatre destroyed by fire. In March of the following year, Drury Lane met with an identical fate.
September 1809	Formal opening of the rebuilt Covent Garden and the start of the 'Old Price' riots.
3 January 1811	Birth of Morton's famous playwriting son, John Maddison.
18 February 1811	*Revival of Colman's BLUE BEARD at Covent Garden, with horses.* Later in the year, parliament enacted the Regency Bill.
October 1812	Formal opening of the rebuilt Drury Lane designed by Wyatt.
Summer 1813	Total breakdown in the relations between Colman and his managerial partners. Performances at the Haymarket were suspended for this season.
June 1815	The Battle of Waterloo.
12 November 1816	First performance of Morton's romantic musical drama of *The Slave*, written expressly to capitalise upon the success and talents of the recent Covent Garden acting *débutant*, William Macready. Morton subsequently lived in France for several years. Of the remainder of his playwriting efforts, the majority were derived from French originals.
23 November 1816	Edmund Kean played Sir Edward Mortimer in a revival of Colman's *The Iron Chest* at Drury Lane. This was one of Kean's greatest triumphs.
September 1817	Colman discharged from the Rules of the King's Bench Prison following a change in English bankruptcy law. He also began looking for a purchaser for his remaining half-share in the Haymarket Theatre. Ironically, he needed to sell in order to settle the enormous account of the solicitors who had presented the defence of his theatrical management in the Court of Chancery. The purchaser was David Morris and the legal transfer of the property took place in the autumn of 1818.
February 1820	Colman flamboyantly marked the occasion of the death of George III and his succession by the former Prince Regent by penning additional stanzas to 'God Save the King'. These he had performed at Covent Garden.
13 May 1820	Colman appointed as Lieutenant of the Yeoman Guard. The following year, several of the major London newspapers

	announced his forthcoming knighthood. This honour never materialised.
4 July 1821	Formal opening of the new Haymarket Theatre designed by Nash. The shell of the old building, derelict beside it, must have been a poignant reminder to Colman of the passing of an era.
19 January 1824	Following the death of John Larpent, Colman took up the position of Examiner of Plays in the Lord Chamberlain's office. His handling of this responsibility provoked much controversy and some ill-will. He was to remain as Examiner until his own death.
c. 1828	Morton took up the post at Covent Garden of Reader of Plays.
1830	Publication of Colman's autobiographical *Random Records*. In general, this two-volume work was not well received by the critics.
June 1830	Death of George IV. First signs of serious physical collapse in Colman.
6 September 1831	Colman sold his commission in the Yeoman Guard.
Autumn 1831	Morton transferred his services as Reader of Plays to Drury Lane. Here his professional competence as an arbiter of new material was publicly called into question in *An Epistolary Remonstrance to Thomas Morton*, published in 1832.
June 1832	Both Colman and Morton were called to give evidence before the Select Committee on Dramatic Literature. In July, Morton was recalled.
July 1833	Morton retired from his post as Reader of Plays at Drury Lane.
26 October 1836	Death of George Colman the Younger. In the following year, the reign of Queen Victoria opened.
8 May 1837	Morton elected to honorary membership of the Garrick Club.
28 March 1838	Death of Thomas Morton.

INKLE AND YARICO

An opera in three acts by George Colman the Younger

First performed at the Theatre Royal, Haymarket, Saturday, 8 August 1787, with the following cast:

INKLE	Mr Bannister junior
SIR CHRISTOPHER CURRY	Mr Parsons
CAMPLEY	Mr Davies
MEDIUM	Mr Baddeley
TRUDGE	Mr Edwin
MIDSHIPMAN	Mr Meadows
RUNNER	Mr Farley
SAILORS	Mr Painter and Mr Ledger
PLANTERS	Mr Usher, Mr Gardner and Mr Johnson
YARICO	Mrs Stephen Kemble
NARCISSA	Mrs Bannister
WOWSKI	Miss George
PATTY	Mrs Forster

Music by Dr Samuel Arnold

INKLE AND YARICO

INKLE. — BY HEAVENS! A WOMAN

ACT. I.   SCENE. III.

PAINTED BY HOWARD. A.   PUBLISHD BY LONGMAN AND CO   ENGRAVD BY J. HEATH A

III   Frontispiece from the Inchbald's British Theatre edition of *Inkle and Yarico* (1808).

ACT I

Scene 1. *An American forest.*

MEDIUM: (*calling, without*) Hilli-ho-ho!

TRUDGE: (*calling, without*) Hip! Hollo-ho-ho! Hip!

         (*Enter* MEDIUM *and* TRUDGE.)

MEDIUM: Pshaw, it's only wasting time and breath! Bawling won't persuade him to budge a bit faster. Things are all altered now. And, whatever weight it may have in *some* places, bawling, it seems, don't go for argument here. Plague on't, we are now in the wilds of America!

TRUDGE: Hip! Hillio-ho-hi!

MEDIUM: Hold your tongue, you blockhead, or —

TRUDGE: Lord, sir, if my master makes no more haste, we shall all be put to sword by the knives of the natives! I'm told they take off heads like hats and hang 'em on pegs in their parlours. Mercy on us! My head aches with the very thoughts of it. Hollo! Mr Inkle! Master, hollo!

MEDIUM: (*Stops his mouth.*) Head aches! Zounds, so does mine with your confounded bawling! It's enough to bring all the natives about us, and we shall be stripped and plundered in a minute.

TRUDGE: Aye, stripping is the first thing that would happen to us, for they seem to be woefully off for a wardrobe. I myself saw three, at a distance, with less clothes than I have when I get out of bed, all dancing about in black buff, just like Adam in mourning.

MEDIUM: This it is to have to do with a schemer, a fellow who risks his life for a chance of advancing his interest! Always advantage in view! Trying, here, to make discoveries that may promote his profit in England. Another Botany Bay scheme, mayhap. Nothing else could induce him to quit our foraging party from the ship when he knows every inhabitant here is not only as black as a pepper-corn, but as hot into the bargain. And *I*, like a fool, to follow him and then to let him loiter behind! (*calling*) Why, nephew! Why, Inkle!

TRUDGE: Why Ink — Well, only to see the difference of men! He'd have thought it very hard, now, if I had let him call so often after me. Ah, I wish he was calling after me now, in the old jog-trot way, again! What a fool was I to leave

---

An American Forest (s.d.): Colman has confused his intention (to write a play condemning the eighteenth-century trade in slaves between Africa and the West Indies) with the specific details of his seventeenth-century source (which relates the story of how a North American Indian girl was sold into slavery). Since Colman's most attractive weakness is his disdain for accuracy, the American setting used here should be allowed to denote an imprecise continent containing strong elements of Africa.

Botany Bay scheme: Captain Arthur Phillip's expedition would have been in the middle of its voyage to Australia as the early performances of this play took place. The scheme was to serve the twofold purpose of setting up a new penal colony — the British government being embarrassed at having found nowhere to send criminals sentenced to deportation since the recent loss of the American colonies — and of opening up the Pacific to trade.

London for foreign parts! That ever I should leave Threadneedle Street to thread an American forest, where a man's as soon lost as a needle in a bottle of hay!

MEDIUM: Patience, Trudge, patience! If we once recover the ship —

TRUDGE: Lord, sir, I shall never recover what I have lost in coming abroad. When my master and I were in London, I had such a mortal snug birth of it! Why, I was factotum.

MEDIUM: Factotum to a young merchant is no such sinecure, neither.

TRUDGE: But then the honour of it, think of that, sir. To be clerk as well as *own man*. Only consider, you find very few city clerks made out of a man, nowadays. To be King of the Counting-House as well as Lord of the Bedchamber! Ah, if I had him but now in the little dressing-room behind the office, tying his hair with a bit of red tape, as usual.

MEDIUM: Yes, or writing an invoice in lampblack and shining his shoes with an ink bottle, *as usual*, you blundering blockhead!

TRUDGE: Oh, if I was but brushing the accounts or casting up the coats. Mercy on us! What's that?

MEDIUM: That! What?

TRUDGE: Didn't you hear a noise?

MEDIUM: Y — es — but — hush! Oh, heavens be praised! Here he is at last.
(*Enter* INKLE.)
Now, nephew!

INKLE: So, Mr Medium.

MEDIUM: Zounds, one would think, by your confounded composure, that you were walking in St James's Park instead of an American forest and that all the beasts were nothing but good company: the hollow trees, here, sentry boxes, and the lions in 'em, soldiers; the jackals, courtiers; the crocodiles, fine women; and the baboons, beaux! What the plague made you loiter so long?

INKLE: Reflection.

MEDIUM: So I should think; reflection generally comes lagging behind. What, scheming, I suppose? Never quiet! At it again, eh? What a happy trader is your father, to have so prudent a son for a partner! Why, you are the carefullest co. in the whole city, never losing sight of the main chance. And that's the reason, perhaps, you lost sight of us, here, on the main of America.

INKLE: Right, Mr Medium. Arithmetic, I own, has been the means of our parting at present.

TRUDGE: (*aside*) Ha! A sum in division, I reckon.

MEDIUM: And pray, if I may be so bold, what mighty scheme has just tempted you to employ your head when you ought to make use of your heels?

INKLE: My heels? Here's a pretty doctrine! Do you think I travel merely for motion? A fine expensive plan for a trader, truly. What, would you have a man of business come abroad, scamper extravagantly here and there and

---

Threadneedle Street: one of the main thoroughfares of the mercantile City of London. Here Inkle would have had his counting-house and chambers, although, if he followed contemporary practice, most of his business would have been transacted in the neighbouring coffee-houses, such as Jonathen's, where the eighteenth-century Stock Exchange operated.

everywhere, then return home and have nothing to tell but that he has *been* here and there and everywhere? 'Sdeath, sir, would you have me travel like a lord?

MEDIUM: No, the Lord forbid!

INKLE: Travelling, uncle, was always intended for improvement, and improvement is an advantage, and advantage is profit, and profit is gain. Which, in the travelling translation of a trader, means that you should gain every advantage of improving your profit.

MEDIUM: How? Gain and advantage and profit? Zounds, I'm quite at a loss!

INKLE: You've hit it, uncle! So am I. I have lost my clue by your conversation. You have knocked all my meditations on the head.

MEDIUM: It's very lucky for you nobody has done it before me.

INKLE: I have been comparing the land, here, with that of our own country.

MEDIUM: And you find it like a good deal of the land of our own country: cursedly encumbered with black legs, I take it.

INKLE: And calculating how much it might be made to produce by the acre.

MEDIUM: You were?

INKLE: Yes, I was proceeding algebraically upon the subject.

MEDIUM: Indeed!

INKLE: And just about extracting the square root.

MEDIUM: Hum!

INKLE: I was thinking, too, if so many natives could be caught, how much they might fetch at the West Indian markets.

MEDIUM: Now let me ask you a question or two, young cannibal catcher, if you please.

INKLE: Well?

MEDIUM: Aren't we bound for Barbadoes, partly to trade, but chiefly to carry home the Governor's, Sir Christopher Curry's, daughter, who has till now been under your father's care in Threadneedle Street for polite English education?

INKLE: Granted.

MEDIUM: And isn't it determined, between the old folks, that you are to marry Narcissa as soon as we get there?

INKLE: A fixed thing.

MEDIUM: Then what the devil do you do here, hunting old hairy negroes, when you ought to be ogling a fine girl in the ship? Algebra, too! You'll have other things to think of when you are married, I promise you! A plodding fellow's head in the hands of a young wife, like a boy's slate after school, soon gets all its arithmetic wiped off, and then it appears in its true simple state: dark, empty and bound in wood, Master Inkle.

INKLE: Not in a match of this kind. Why, it's a table of interest from beginning to end, old Medium.

MEDIUM: Well, well, this is no time to talk. Who knows but instead of sailing to a

---

black legs: sharpers; in the eighteenth century, they were associated particularly with gambling at race meetings.

table of interest: i.e. a ready reckoner for calculating interest from principal; possibly punning on Medium's line, 'bound in wood'.

wedding we may get cut up, here, for a wedding dinner: tossed up for a dingy duke, perhaps, or stewed down for a black baronet, or eat raw by an inky commoner.

INKLE: Why, sure you aren't afraid?

MEDIUM: Who, I afraid? Ha! ha! ha! No, not I! What the deuce should I be afraid of? Thank heaven I have a clear conscience and need not be afraid of anything. A scoundrel might not be quite so easy on such an occasion, but it's the part of an honest man not to behave like a scoundrel. I never behaved like a scoundrel, for which reason I am an honest man, you know. But come, I hate to boast of my good qualities.

INKLE: Slow and sure, my good, virtuous Mr Medium! Our companions can be but half a mile before us and, if we do but double their steps, we shall overtake 'em at one mile's end, by all the powers of arithmetic.

MEDIUM: Oh, curse your arithmetic! How are we to find our way?

INKLE: That, uncle, must be left to the doctrine of chances!

(*Exeunt.*)

Scene 2. *Another part of the forest. A ship at anchor in the bay at a small distance. Mouth of a cave.*

(*Enter* SAILORS *and a* MIDSHIPMAN *as returning from foraging.*)

MID: Come, come, bear a hand, my lads. Tho' the bay is just under our bowsprits, it will take a damned deal of tripping to come at it; there's hardly any steering clear of the rocks here. But do we muster all hands? All right, think ye?

1st SAIL: Ey-eye! All to a man, besides yourself and a monkey! The three land lubbers that edged away in the morning goes for nothing, you know. They're all dead, mayhap, by this.

MID: Dead! You be — Why, they're friends of the captain, and, if not brought safe aboard tonight, you may all chance to have a salt eel for your supper, that's all! Moreover, the young plodding spark — he with the grave, foul weather face, there — is to man the tight little frigate, Miss Narcissa what d'ye call her, that is bound with us for Barbadoes. Rot 'em for not keeping under way, I say!

2nd SAIL: Foolish dogs! Suppose they're met with by the natives?

MID: Why, then the natives would look plaguy black upon 'em. But come, let's see if a song will bring 'em to. Let's have a full chorus to the good merchant ship, the *Achilles.*

Song. SAILORS *and* MIDSHIPMAN.

The *Achilles*, tho' christen'd good ship, 'tis surmis'd,

---

under our bowsprits (etc.): although Colman has made relentless use of nautical language in this section of the scene, the reader should bear in mind that all of it is figurative; as the stage directions indicate, the sailors are in fact on dry land (hence, 'under our bowsprits' = 'under our noses').

plodding spark: a deliberate contradiction in terms, the word 'spark', even if used in a derogatory context to suggest foppishness, usually conveying a sense of the flamboyant or larger-than-life personality, such as that of a lover, its particular application here.

From that old man of war, great Achilles, so priz'd;
Was he, like our vessel, pray, fairly baptiz'd?
    Ti tol lol, &c.

Poets sung *that* Achilles — if, now, they've an itch
To sing *this*, future ages may know which is which,
And that one rode in Greece — and the other in pitch.
    Ti tol lol, &c.

What tho' but a merchant ship, sure our supplies;
Now your man of war's gain in a lottery lies,
And how blank they all look when they can't get a prize!
    Ti tol lol, &c.

What are all their fine names? When no rhino's behind,
The *Intrepid* and *Lion* look sheepish, you'll find,
Whilst — alas! — the poor *Aeolus* can't raise the wind!
    Ti tol lol, &c.

Then the *Thunderer*'s dumb, out of tune the *Orpheus*,
The *Ceres* has nothing at all to produce,
And the *Eagle*, I warrant you, looks like a goose.
    Ti tol lol, &c.

But we merchant lads, tho' the foe we can't maul,
Nor are paid, like fine King's Ships, to fight at a call,
Why, we pay ourselves well without fighting at all!
    Ti tol lol, &c.

1st SAIL: Avast! Look ahead there! Here they come chased by a fleet of black devils.

MID: And the devil a *fire* have I to give 'em. We ha'n't a grain of powder left. What must we do, lads?

2nd SAIL: Do? Sheer off, to be sure.

MID: What, and leave our companions behind? No, damn it, I can't. I can't do that, neither.

---

fairly baptiz'd: in this comparison of launchings, the sailors are suggesting that the unassisted first contact of their ship with water was more auspicious than its mythological namesake's baptism, when a mother's helping hand concealed Achilles' heel from immersion in the invulnerability-giving River Styx.

prize: an enemy ship captured as the spoils of maritime war. Traditionally, the British Navy sold all the ships captured in this way and distributed the money thus realised — the 'prize-money' — amongst the crews responsible for such captures. Hence the lottery/prize quibble in this verse.

rhino: ready cash; ironically, ten years after this song was written, the low pay and low morale in naval ships to which it alludes, together with the appalling conditions in which men were expected to serve, precipitated the two wholesale mutinies of British naval fleets at Spithead and the Nore.

Aeolus: son of Poseidon and said to have invented ships' sails, he was the Roman wind-god.

3rd SAIL: Why then, we'll leave you. Who the plague is to stand here and be peppered by a parcel of savages?

MID: Why, to be sure, as it is so — yet — plague on't!

1st SAIL: Paw, mun! They're as safe as we. Why, we're scarce a cable's length asunder, and they'll keep in our wake now, I warrant 'em.

MID: Why, if you will have it so — It makes a body's heart yearn to leave the poor fellows in distress, too. (*reluctantly*) Well, if I must, I must. (*going to the other side, and holloing to* INKLE *and others*) Yoho, lubbers! Crowd all the sail you can, d'ye mind me!

    (*Exeunt* SAILORS *and* MIDSHIPMAN. *Enter* MEDIUM, *running across the stage, as pursued by the blacks.*)

MEDIUM: Nephew! Trudge! Run! Scamper! Scour! Fly! Zounds, what harm did I ever do to be hunted to death by a pack of black bloodhounds? (*calling*) Why, nephew! Oh, confound your long sums in arithmetic! I'll take care of myself! And, if we must have any arithmetic, dot and carry one, for my money!

    (MEDIUM *runs off. Enter* INKLE *and* TRUDGE *hastily.*)

TRUDGE: Oh, that ever I was born! To leave pen, ink and powder for this!

INKLE: Trudge, how far are the sailors before us?

TRUDGE: I'll run and see, sir, directly.

INKLE: Blockhead! Come here! The savages are close upon us; we shall scarce be able to recover our party. Get behind this tuft of trees with me; they'll pass us, and we may then recover the ship with safety.

TRUDGE: (*going behind*) Oh, Threadneedle Street! Thread —

INKLE: Peace!

TRUDGE: (*hiding*) — Needle Street.

    (*They hide behind trees. Natives cross. After a long pause,* INKLE *looks from the trees.*)

INKLE: Trudge?

TRUDGE: (*in a whisper*) Sir.

INKLE: Are they all gone by?

TRUDGE: Won't you look and see?

INKLE: (*looking round*) So, all's safe at last. (*coming forward*) Nothing like policy in these cases. But you'd have run on like a booby! A tree, I fancy, you'll find in future the best resource in a hot pursuit.

TRUDGE: Oh, charming! It's a retreat for a king, sir. Mr Medium, however, has not got up in it. Your uncle, sir, *has* run on like a booby, and has got up with our party, by this time, I take it, who are now most likely at the shore. But what are we to do next, sir?

INKLE: Reconnoitre a little, and then proceed.

TRUDGE: Then pray, sir, proceed to reconnoitre, for the sooner the better.

INKLE: Then look out, d'ye hear, and tell me if you discover any danger.

TRUDGE: Y — Ye — s — Yes. But (*trembling*) as you understand this business better than I, sir, suppose you stick close to my elbow to give me directions?

INKLE: Cowardly scoundrel! Do as you are ordered, sir! Well, is the coast clear?

TRUDGE: Eh? Oh, lord! — Clear? (*rubbing his eyes*) Oh, dear! Oh, dear! The coast will soon be clear enough now, I promise you. The ship is under sail, sir!

INKLE: Death and damnation!

TRUDGE: Aye, death falls to my lot.

INKLE: My property carried off in the vessel.

TRUDGE: All, all, sir, except me.

INKLE: Treacherous villains! My whole effects lost.

TRUDGE: Lord, sir! Anybody but you would only think of effecting his safety in such a situation.

INKLE: They may report me dead, perhaps, and dispose of my property at the next island.

              (The vessel appears under sail.)

TRUDGE: Ah, there they go! *(a gun fired)* That will be the last report we shall ever hear from them, I'm afraid. That's as much as to say, 'Good-bye to ye.' And here we are left, two fine, full-grown babes in the wood!

INKLE: What an ill-timed accident! Just, too, when my speedy union with Narcissa, at Barbadoes, would so much advance my interests. *(thinking)* Something must be hit upon, and speedily. But what resources?

TRUDGE: The old one: a tree, sir. 'Tis all we have for it, now. What would I give, now, to be perched upon a high stool, with our brown desk squeezed into the pit of my stomach, frizzing away at an old parchment! But all my red ink will be spilt by an old black pin of a negro. Hum, I was thinking —

INKLE: Well, well? What? Something to our purpose, I hope.

TRUDGE: I was thinking, sir, if so many natives could be caught, how much they might fetch at the West Indian markets!

INKLE: Scoundrel! Is this a time to jest?

TRUDGE: No, faith, sir! Hunger is too sharp to be jested with. As for me, I shall starve for want of food. Now you may meet a luckier fate: you are able to extract the square root, sir, and that's the very best provision you can find here to live upon. But I —

Song. TRUDGE

    A voyage over seas had not enter'd my head,
    Had I known but on which side to butter my bread.
    Heigho! Sure I — for hunger must die!
    I've sailed like a booby, come here in a squall,
    Where — alas! — there's no bread to be butter'd at all!
        Oho! I'm a terrible booby!
        Oh, what a sad booby am I!

---

frizzing away: putting into curls; as will have been noticed already, Trudge's character turns, in part, on his continual confusion of clerical and valeting terminology.

black pin: since parchment membranes were of large sheet size and not particularly supple, drawing pins would have been an essential commodity, as well as a frequent hazard, in Trudge's Threadneedle Street office.

In London, what gay chop-house signs in the street!
But the only sign here is of nothing to eat.
Heigho! That I — for hunger should die!
My mutton's all lost, I'm a poor starving elf,
And for all the world like a lost mutton myself!
    Oho! I shall die a lost mutton!
    Oh, what a lost mutton am I!

For a neat slice of beef I could roar like a bull,
And my stomach's so empty, my heart is quite full.
Heigho! That I — for hunger should die!
But, grave without meat, I must here meet my grave,
For my bacon, I fancy, I never shall save.
    Oho! I shall ne'er save my bacon!
    I can't save my bacon, not I!

    (noise without)
Mercy on us! Here they come again!

INKLE: Confusion! Deserted on one side and pressed on the other. Which way shall I turn? Ha! What's this? A cavern? This may prove a safe retreat to us for the present. I'll enter, cost what it will.

TRUDGE: Oh Lord! No! Don't, don't! We shall pay too dear for our lodging, depend on't.

INKLE: This is no time for debating. You are at the mouth of it: lead the way, Trudge!

TRUDGE: What! Go in before your honour? I know my place better, I assure you. *(aside)* I might walk into more mouths than one, perhaps.

INKLE: Coward! Then follow me.
    (noise again)

TRUDGE: I must, sir, I must! Ah, Trudge, Trudge, what a damned hole are you getting into!
    (Exeunt into a cavern.)

    Scene 3. *A cave, decorated with skins of wild beasts, feathers &c. In the middle of the scene, a rude kind of curtain by way of door to an inner apartment.*

    (Enter INKLE *and* TRUDGE *as from the mouth of the cavern.)*

TRUDGE: Why, sir, sir, you must be mad to go any farther!

INKLE: So far, at least, we have proceeded with safety. Ha! No bad specimen of savage elegance! These ornaments would be worth something in England. We have little to fear here, I hope. This cave rather bears the pleasing face of a profitable adventure.

---

chop-house signs: chops made a popular meal in the eighteenth century, so it is hardly surprising that London should have had houses which specialised in catering for this form of eating. Colman, indeed, was addicted to mutton chops; John Byng, fifth Viscount Torrington, having witnessed the playwright devouring twelve at one sitting, entered in his *Diary* the conspicuous note that 'no man possesses an happier appetite'.

TRUDGE: Very likely, sir. But for a pleasing face it has the cursedest ugly mouth I
ever saw in my life! Now do, sir, make off as fast as you can. If we once get clear
of the natives' houses, we have little to fear from the lions and leopards: for by
the appearance of their parlours they seem to have killed all the wild beast in
the country. Now, pray, do, my good master, take my advice and run away.

INKLE: Rascal! Talk again of going out and I'll flay you alive!

TRUDGE: That's just what I expect for coming in! All that enter here appear to
have had their skins stripped over their ears, and ours will be kept for
curiosities. We shall stand here, stuffed, for a couple of white wonders.

INKLE: This curtain seems to lead to another apartment. I'll draw it.

TRUDGE: No, no, no! Don't! Don't! We may be called to account for disturbing
the company. You may get a curtain-lecture, perhaps, sir.

INKLE: Peace, booby, and stand on your guard!

TRUDGE: Oh, what will become of us! Some grim, seven-foot fellow ready to scalp
us!

   (*As the curtain draws,* YARICO *and* WOWSKI *discovered asleep.*
   Bows and arrows.)

INKLE: By heaven! A woman!

TRUDGE: (*aside*) A woman? (*aloud*) But let him come on, I'm ready, damme! I
don't fear facing the devil himself! — Faith, it is a woman. Fast asleep too.

INKLE: And beautiful as an angel!

TRUDGE: And, egad, there seems to be a nice little plump bit in the corner! Only
she's an angel of a rather darker sort.

INKLE: Hush! Keep back! She wakes!

   (YARICO *comes forward.* INKLE *and* TRUDGE *retire to opposite*
   sides of the scene.)

Song. YARICO.

> When the chase of day is done
> And the shaggy lion's skin,
> Which, for us, our warriors win,
> Decks our cells at set of sun;
> Worn with toil, with sleep opprest,
> I press my mossy bed, and sink to rest.
>
> Then, once more, I see our train,
> With all our chase, renew'd again:
>   Once more 'tis day,
>   Once more our prey
> Gnashes his angry teeth, and foams in vain.
>   Again, in sullen haste, he flies,
>   Ta'en in the toil, again he lies,
> Again he roars — and, in my slumbers, dies.

---

curtain-lecture: 'a reproof given by a wife to her husband in bed'. Dr Johnson, *Dictionary*,
1755. Trudge's remark comes at a time when Inkle is poised ready to draw back the curtain
covering the mouth of the cave.

(INKLE *and* TRUDGE *come forward.*)

INKLE: Our language!

TRUDGE: Zounds, she has thrown me into a cold sweat!

YARICO: Hark! I heard a noise! Wowski, awake! Whence can it proceed?

(*She wakes* WOWSKI, *and they both come forward:* YARICO
towards INKLE, WOWSKI *towards* TRUDGE.)

TRUDGE: (*Bows to* WOWSKI.) Ma'am, your very humble servant.

YARICO: Ah! What form is this? Are you a man?

INKLE: True flesh and blood, my charming heathen, I promise you.

YARICO: (*gazing*) What harmony in his voice! What a shape! How fair his skin,
too!

TRUDGE: This must be a lady of quality, by her staring.

YARICO: Say, stranger, whence come you?

INKLE: From a far distant island, driven on this coast by distress and deserted by
my companions.

YARICO: And do you know the danger that surrounds you here? Our woods are
filled with beasts of prey. My countrymen too — yet, I think they couldn't
find the heart — might kill you. It would be a pity if you fell in their way. I
think I should weep if you came to any harm.

TRUDGE: Oho! It's time, I see, to begin making interest with the chambermaid.
(*Takes* WOWSKI *apart.*)

INKLE: How wild and beautiful! Sure, there is magic in her shape and she has
rivetted me to the place. But where shall I look for safety? Let me fly and
avoid my death!

YARICO: Oh, no! But — (*as if puzzled*) well, then die, stranger! — but don't
depart! I will try to preserve you, and, if you are killed, Yarico must die too!
Yet, 'tis I alone can save you. Your death is certain without my assistance,
and indeed, indeed, you shall not want it.

INKLE: My kind Yarico! What means, then, must be used for my safety?

YARICO: My cave must conceal you. None enter it since my father was slain in
battle. I will bring you food by day, then lead you to our unfrequented
groves by moonlight, to listen to the nightingale. If you should sleep, I'll
watch you, and awake you when there's danger.

INKLE: Generous maid! Then to you I will owe my life, and, whilst it lasts,
nothing shall part us.

YARICO: And shan't it, shan't it, indeed?

INKLE: No, my Yarico! For when an opportunity offers to return to my country,
you shall be my companion.

YARICO: What? Cross the seas?

INKLE: Yes, help me to discover a vessel and you shall enjoy wonders. You shall
be decked in silks, my brave maid, and have a house drawn with horses to
carry you.

YARICO: Nay, do not laugh at me. But is it so?

INKLE: It is indeed!

YARICO: Oh wonder! I wish my countrywomen could see me. But won't your
warriors kill us?

INKLE: No, our only danger on land is here.

YARICO: Then let us retire further into the cave. Come, your safety is in my keeping.

INKLE: I follow you. Yet, can you run some risk in following me?

> *Duet.* INKLE *and* YARICO.
>
> O say, simple maid, have you form'd any notion
> Of all the rude dangers in crossing the ocean?
> When winds whistle shrilly — ah! — won't they remind you
> To sigh with regret for the grot left behind you?

YARICO: Ah, no! I could follow and sail the world over,
Nor think of my grot when I look at my lover;
The winds which blow round us, your arms for my pillow,
Will lull us to sleep whilst we're rocked by each billow.

INKLE: Then, say, lovely lass, what if haply espying
A rich gallant vessel with gay colours flying?

YARICO: I'll journey with thee, love, to where the land narrows
And fling all my cares at my back with my arrows.

BOTH: O say, then, my true love, we never will sunder,
Nor shrink from the tempest, nor dread the big thunder;
Whilst constant, we'll laugh at all changes of weather
And journey all over the world together.

> (*Exeunt* INKLE *and* YARICO, *as retiring further into the cave.*
> *Manent* TRUDGE *and* WOWSKI.)

TRUDGE: Why, you speak English as well as I, my little Wowski!

WOWSKI: Iss.

TRUDGE: Iss! And you learnt it from a strange man that tumbled from a big boat many moons ago, you say?

WOWSKI: Iss. Teach me. Teach good many.

TRUDGE: Then what the devil made 'em so surprised at seeing us? Was he like me? (WOWSKI *shakes her head.*) Not so smart a body, mayhap. Was his face, now, round and comely and — eh? (*stroking his chin*) — was it like mine?

WOWSKI: Like dead leaf. Brown and shrivel.

TRUDGE: Oh, ho! An old shipwrecked sailor, I warrant. With white and grey hair, eh, my pretty beauty spot?

WOWSKI: Iss. All white. When night come, he put it in pocket.

TRUDGE: Oh! Wore a wig! But the old boy taught you something more than English, I believe.

WOWSKI: Iss.

TRUDGE: The devil he did! What was it?

WOWSKI: Teach me put dry grass, red hot, in hollow white stick.

TRUDGE: Aye? What was that for?

WOWSKI: Put in my mouth. Go poff-poff!

TRUDGE: Zounds, did he teach you to smoke?

WOWSKI: Iss.

TRUDGE: And what became of him at last? What did your countrymen do for the poor fellow?

WOWSKI: Eat him one day. Our chief kill him.

TRUDGE: Mercy on us! What damned stomachs, to swallow a tough old tar! Ah, poor Trudge, your killing comes next!

WOWSKI: (*running to him anxiously*) No, no. Not you. No.

TRUDGE: No? Why, what shall I do if I get in their paws?

WOWSKI: I fight for you!

TRUDGE: Will you? Ecod, she's a brave, good-natured wench! She'll be worth a hundred of your English wives: whenever they fight on their husband's account, it's *with* him instead of *for* him, I fancy. But how the plague am I to live here?

WOWSKI: I feed you. Bring you kid.

Song. WOWSKI.

> White man, never go away —
>   Tell me, why need you?
> Stay with your Wowski, stay;
>   Wowski will feed you.
> Cold moons are now coming in,
>   Ah, don't go grieve me!
> I'll wrap you in leopard's skin —
>   White man, don't leave me.
>
> And, when all the sky is blue,
>   Sun makes warm weather,
> I'll catch you a cockatoo,
>   Dress you in feather.
> When cold comes, or when 'tis hot,
>   Ah, don't go grieve me!
> Poor Wowski will be forgot —
>   White man, don't leave me!

TRUDGE: Zounds! Leopard's skin for winter wear and feathers for a summer's suit! Ha! ha! I shall look like a walking hammer-cloth at Christmas and an upright shuttlecock in the dog days. And for all this, if my master and I find our way to England, you shall be part of our travelling equipage, and, when I get there, I'll give you a couple of snug rooms on a first floor and visit you every evening, as soon as I come from the counting-house. Do you like it?

WOWSKI: Iss.

TRUDGE: Damme, what a flashy fellow I shall seem in the city! I'll get you a *white* boy to bring up the tea-kettle. Then I'll teach you to write and dress hair.

WOWSKI: You great man in your country?

---

hammer-cloth: 'Hammer-cloths are among the principal ornaments of a Carriage' according to a contemporary manual on the subject (W. Felton, *Carriages*, I, p. 153). They were commonly spread over the driver's seat or 'box', although the origin of their name is not known.

TRUDGE: A very great man. I am head clerk of the dressing-room and first valet-de-chambre of the counting-house. I powder parchments, pounce hair, ink shoes, black paper, mend beards and shave pens. But hold! I had forgot one material point: you aren't married, I hope?

WOWSKI: No. You be my chum-chum?

TRUDGE: So I will. It's best, however, to be sure of her being single, for Indian husbands are not quite so complaisant as English ones, and the vulgar dogs might think of looking a little after their spouses. Well, as my master seems King of this palace, and has taken his Indian Queen already, I'll e'en be Usher of the Black Rod here! But you have had a lover or two in your time, eh, Wowski?

WOWSKI: Oh, iss. Great many. I tell you.

Duet. TRUDGE *and* WOWSKI.

WOWSKI: Wampum, Scampum, Yanko, Lanko, Nanko, Pownatowski
　　　　Black men — plenty — twenty — fight for me;
　　　　　　White man, woo you true?
TRUDGE: Who?
WOWSKI: 　　You.
TRUDGE: 　　　　Yes, pretty little Wowski!
WOWSKI: Then I leave all and follow thee.
TRUDGE: 　　　Oh, then turn about, my little tawny tight one!
　　　　　　Don't you like me?
WOWSKI: Iss, you're like the snow!
　　　　If you slight one —
TRUDGE: 　　　　Never, not for any white one;
　　　　　　You are beautiful as any sloe.
WOWSKI: Wars, jars, scars, can't expose ye,
　　　　In our grot —
TRUDGE: 　　　　So snug and cosy!
WOWSKI: Flowers, neatly
　　　　Picked, shall sweetly
　　　　Make your bed.
TRUDGE: Coying, toying,
　　　　With a rosy
　　　　　Posy,
　　　　　　When I'm dozy;
　　　　Bear-skin nightcaps, too, shall warm my head.
BOTH: 　　Bear-skin nightcaps, too, shall warm my head.

　　　　(*Exeunt.*)

---

pounce: the verb used here derives from the noun, pounce being finely powdered cuttle-bone or sanderach resin applied to a freshly written page in order to absorb surplus ink. Hence, to decipher Trudge's malapropos diction here, a parchment would have been pounced and hair powdered.

Indian Queen: title of a play by John Dryden and Sir Robert Howard (first performed 1664) which was made into a popular opera with music composed by Henry Purcell (first performed 1695).

ACT II

Scene 1. *The quay at Barbadoes, with an inn upon it. People employed in unlading vessels, carrying bales of goods, &c.*

(*Enter several* PLANTERS.)

1st PLANT: I saw her this morning, gentlemen, you may depend on't. My telescope never fails me. I popped upon her as I was taking a peep from my balcony. A brave, tight ship, I tell you, bearing down directly for Barbadoes, here.

2nd PLANT: Od's my life! Rare news! We have not had a vessel arrive in our harbour these six weeks.

3rd PLANT: And the last brought only Madam Narcissa, our Governor's daughter, from England, with such a parcel of lazy, idle white folks about her. Such cargoes will never do for our trade, neighbour.

2nd PLANT: No, no, we want slaves. A terrible dearth of 'em in Barbadoes lately! But your dingy passengers, for my money. Give me a vessel like a collier, where all the lading tumbles out as black as my hat. (*to* 1st PLANT.) But are you sure, now, you aren't mistaken?

1st PLANT: Mistaken! 'Sbud, do you doubt my glass? I can discover a gull by it six leagues off. I could see everything as plain as if I was on board.

2nd PLANT: Indeed! And what were her colours?

1st PLANT: Um — why English! — or Dutch, — or French, — I don't exactly remember.

2nd PLANT: What were the sailors aboard?

1st PLANT: Eh? Why, they were English, too! — or Dutch, — or French, — I can't perfectly recollect.

2nd PLANT: Your glass, neighbour, is a little like a glass too much: it makes you forget everything you ought to remember.

(*cry without, 'A sail! A sail!'*)

1st PLANT: Egad, but I'm right, though! Now, gentlemen —

ALL: Aye! aye! The devil take the hindmost!

(*Exeunt hastily. Enter* NARCISSA *and* PATTY.)

Song. NARCISSA.

> Freshly now the breeze is blowing
>   As yon ship at anchor rides;
> Sullen waves, incessant flowing,
>   Rudely dash against the sides.
> So my heart, its course impeded,
>   Beats in my perturbed breast;
> Doubts, like waves by waves succeeded,
>   Rise, and still deny it rest.

PATTY: Well, ma'am, as I was saying —

NAR: Well, say no more of what you were saying! Sure, Patty, you forget where you are. A little caution will be necessary now, I think.

PATTY: Lord, madam, how is it possible to help talking? We are in Barbadoes here, to be sure. But then, ma'am, one may let out a little in a private morning's walk by ourselves.

NAR: Nay, it's the same thing with you indoors.

PATTY: I never blab, ma'am, never, as I hope for a gown.

NAR: And your never blabbing, as you call it, depends chiefly on that hope, I believe. The unlocking of my chest locks up all your faculties: an old silk gown makes you turn your back on all my secrets, a large bonnet blinds your eyes, and a fashionable high handkerchief covers your ears and stops your mouth at once, Patty.

PATTY: Dear ma'am, how can you think a body so mercenary! Am I not always teasing you about gowns and gew-gaws, and fallals and finery? Or do you take me for a conjuror, that nothing will come out of my mouth but ribbons? I have told the story of our voyage, indeed, to old Guzzle, the butler, who is very inquisitive — and, between ourselves, is the ugliest old quiz I ever saw in my life — but when he asks me what you and I think of the matter, why, I look wise, and cry like other wise people who have nothing to say, 'all's for the best!'

NAR: And thus you lead him to imagine I am but little inclined to the match.

PATTY: Lord, ma'am, how could that be? Why, I never said a word about Captain Campley.

NAR: Hush! Hush! For heaven's sake!

PATTY: Aye, there it is now! There, ma'am, I'm as mute as a mackerel. That name strikes me dumb in a moment. I don't know how it is, but Captain Campley somehow or other has the knack of stopping my mouth oftener than anybody else, ma'am.

NAR: His name again! Consider, never mention it, I desire you!

PATTY: Not I, ma'am, not I. But if our voyage from England was so pleasant, it wasn't owing to Mr Inkle, I'm certain. He didn't play the fiddle in our cabin, and dance on the deck, and come languishing with a glass of warm water in his hand when we were seasick. Ah, ma'am, that water warmed your heart, I'm confident. Mr Inkle — no, no, Captain Cam — ! There, he has stopped my mouth again, ma'am.

NAR: There is no end to this! Remember, Patty, keep your secrecy or you entirely lose my favour.

PATTY: Never fear me, ma'am. But if somebody I know is not acquainted with the Governor, there's such a thing as dancing at balls and squeezing hands when you lead up, and squeezing them again when you cast down — and walking on the quay in a morning.

NAR: No more of this!

PATTY: Oh, I won't utter a syllable. (*archly*) I'll go and take a turn on the quay by

---

old quiz: 'quiz' was a word in high vogue at this time and denoted a person thought to have a particularly odd appearance.

myself, if you think proper. But remember, I'm as close as a patch-box. Mum's the word, ma'am, I promise you.

Song. PATTY.

This maxim let everyone hear
    Proclaimed from the north to the south:
Whatever comes in at your ear
    Should never run out at your mouth.
We servants, like servants of state,
    Should listen to all and be dumb;
Let others harangue and debate:
    We look wise, shake our heads, and are mum.

The judge in dull dignity dressed
    In silence hears barristers preach,
And then, to prove silence is best,
    He'll get up and give 'em a speech.
By saying but little, the maid
    Will keep her swain under her thumb,
And the lover that's true to his trade
    Is certain to kiss, and cry mum.

(*Exit* PATTY.)

NAR: How awkward is my present situation! Promised to one who, perhaps, may never again be heard of and who, I am sure, if he ever appears to claim me, will do it merely on the score of interest. Pressed, too, by another who has already, I fear, too much interest in my heart. What can I do? What plan can I follow?

(*Enter* CAMPLEY.)

CAMPLEY: Follow my advice, Narcissa, by all means. Enlist with me under the best banners in the world. General Hymen, for my money! Little Cupid's his drummer; he has been beating a rub-a-dub on our hearts and we have only to obey the word of command, fall into the ranks of matrimony, and march through life together.

NAR: Halt! Halt, Captain! You march too quick. Besides, you make matrimony a mere parade.

CAMPLEY: Faith, I believe many make it so at present. But we are all volunteers, Narcissa, and I am for actual service, I promise you!

NAR: Then consider our situation.

CAMPLEY: That has been duly considered. In short, the case stands exactly thus: your intended spouse is all for money, I am all for love; he is a rich rogue, I am rather a poor honest fellow; he would pocket your fortune, I will take you without a fortune in your pocket.

---

close as a patch-box: another of Patty's platitudes, like her 'mute as a mackerel' and conveying an identical meaning. A patch-box was the highly ornate container in which a fashionable woman would have secreted her face patches and beauty spots. By this time, however, patching was considered *passé*.

NAR: But where's Mr Inkle's view of interest? Hasn't he run away from me?

CAMPLEY: And I am ready to run away *with* you.

NAR: Oh, I am sensible of the favour, most gallant Captain Campley! And my father, no doubt, will be very much obliged to you.

CAMPLEY: Aye, there's the devil of it! Sir Christopher Curry's confounded good character knocks me up at once. Yet I am not acquainted with him either, not known to him even by sight, being here only as a private gentleman on a visit to my old relation — out of regimentals, and so forth — and not introduced to the Governor, as other officers of the place. But then, the report of his hospitality, his odd, blunt, whimsical friendship, his whole behaviour —

NAR: All stare you in the face, eh, Campley?

CAMPLEY: They do, till they put me out of countenance. But then again, when I stare *you* in the face, I can't think I have any reason to be ashamed of my proceedings. I stick here, between my love and my principle, like a song between a toast and a sentiment.

NAR: And if your love and your principle were put in the scales, you doubt which would weigh most?

CAMPLEY: Oh, no! I should act like a rogue, and let principle kick the beam. For love, Narcissa, is as heavy as lead and, like a bullet from a pistol, could never go through the heart if it wanted weight.

NAR: Or rather, like the pistol itself that often *goes off* without any harm done, your fire must end in smoke, I believe.

CAMPLEY: Never, whilst —

NAR: Nay, a truce to protestations at present. What signifies talking to *me* when you have such opposition from others? Why hover about the city instead of boldly attacking the guard? Wheel about, Captain! Face the enemy! March! Charge! Rout 'em! Drive 'em before you! And then —

CAMPLEY: And then?

NAR: Lud ha' mercy on the poor city!

Song. Rondeau. NARCISSA.

> Mars would oft, his conquests over,
>   To the Cyprian Goddess yield;
> Venus gloried in a lover
>   Who, like him, could brave the field.
>   Mars would oft, &c.

---

a song between a toast and a sentiment: a reference to the complicated etiquette which provided extensive pretexts for eighteenth-century social drinking, a toast being a straightforward honour to the health of an individual and a sentiment being drunk to celebrate an idea.

kick the beam: commonly used during this period to suggest the comparative weight or importance of an idea, this figure of speech derives from the action of a balance-scale disproportionally loaded, the lighter of the two scale-pans being flung upwards with such violence that it strikes the fulcrum-arm or 'kicks the beam'.

Cyprian goddess: Venus; Cyprus, in classical literature, was so famed for its worship of this goddess that the eighteenth-century vernacular adopted the word 'Cyprian', both as adjective and noun, in reference to prostitutes, hence providing a double-meaning here.

In the cause of battles hearty,
    Still the god would strive to prove
He who faced an adverse party
    Fittest was to meet his love.
        Mars would oft, &c.

Hear, then, captains, ye who bluster,
    Hear the god of war declare:
Cowards never can pass muster,
    Courage only wins the fair
        Mars would oft, &c.

(*Enter* PATTY, *hastily.*)

PATTY: Oh lud, ma'am, I'm frightened out of my wits! Sure as I'm alive, ma'am, Mr Inkle is not dead; I saw his man, ma'am, just now, coming ashore in a boat with other passengers from the vessel that's come to the island.

NAR: Then one way or other I must determine.

PATTY: But pray, ma'am, don't tell the Captain! I'm sure he'll stick, he'll stick poor Trudge in his passion, and he's the best-natured, peaceable, kind, loving soul in the world!

(*Exit* PATTY.)

NAR: Look ye, Mr Campley, something has happened which makes me waive ceremonies. If you mean to apply to my father, remember that delays are dangerous.

CAMPLEY: Indeed!

NAR: (*smiling*) I mayn't be always in the same mind, you know.

(*Exit* NARCISSA.)

CAMPLEY: Nay, then — Gad, I'm almost afraid, too — but living in this state of doubt is torment. I'll e'en put a good face on the matter: cock my hat, make my bow, and try to reason the Governor into compliance. Faint heart never won a fair lady!

Song. CAMPLEY.

Why should I vain fears discover,
    Prove a dying, sighing swain?
Why turn shilly-shally lover
    Only to prolong my pain?

When we woo the dear enslaver,
    Boldly ask, and she will grant;
How should we obtain a favour
    But by telling what we want?

Should the nymph be found complying,
    Nearly, then, the battle's won;
Parents think 'tis vain denying
    When half our work is fairly done.

(*Exit* CAMPLEY. *Enter* TRUDGE *and* WOWSKI, *as from the ship, with a dirty* RUNNER *to one of the inns.*)

RUNNER: This way, sir. If you will let me recommend —

TRUDGE: Come along, Wows! Take care of your furs and your feathers, my girl!

WOWSKI: Iss.

TRUDGE: That's right. Somebody might steal 'em, perhaps.

WOWSKI: Steal? What that?

TRUDGE: Oh, Lord! See what one loses by not being born in a Christian country!

RUNNER: If you would, sir, but mention to your master the house that belongs to my master. The best accommodations on the quay —

TRUDGE: What's your sign, my lad?

RUNNER: The Crown, sir. Here it is.

TRUDGE: Well, get us a room for half an hour and we'll come. And harkee, let it be light and airy, d'ye hear? My master has been used to your open apartments lately.

RUNNER: Depend on it. Much obliged to you, sir.

(*Exit* RUNNER.)

WOWSKI: Who be that fine man? He great prince?

TRUDGE: A prince? Ha! ha! No, not quite a prince. But he belongs to the Crown! But how do you like this, Wows? Isn't it fine?

WOWSKI: Wonder!

TRUDGE: Fine men, eh?

WOWSKI: Iss. All white like you.

TRUDGE: Yes, all the fine men are like me: as different from your people as powder and ink, or paper and blacking.

WOWSKI: And fine lady. Face like snow.

TRUDGE: What? The fine ladies' complexions? Oh yes, exactly! For too much heat very often dissolves 'em! Then their dress, too —

WOWSKI: Your countrymen dress so?

TRUDGE: Better. Better a great deal. Why, a young flashy Englishman will sometimes carry a whole fortune on his back. But did you mind the women? (*pointing before and behind*) All here — and there. They have it all from us in England. And then the fine things they carry on their heads, Wowski.

WOWSKI: Iss. One lady carry good fish; so fine she call everybody to look at her.

TRUDGE: Pshaw! An old woman bawling flounders! But the fine girls we meet here, on the quay: so round and so plump!

WOWSKI: You not love me now.

TRUDGE: Not love you? Zounds, have I not given you proofs?

WOWSKI: Iss. Great many. But now you get here, you forget poor Wowski!

TRUDGE: Not I. I'll stick to you like wax.

WOWSKI: Ah! I fear! What make you love me now?

TRUDGE: Gratitude, to be sure.

WOWSKI: What that?

TRUDGE: Ha! This it is, now, to live without education! The poor dull devils of her country are all in the practice of gratitude, without finding out what it means, while we can tell the meaning of it with little or no practice at all. Lord! Lord! What a fine advantage Christian learning is! Harkee, Wows!

WOWSKI: Iss.

TRUDGE: Now we've accomplished our landing, I'll accomplish you. You remember the instructions I gave you on the voyage?

WOWSKI: Iss.

TRUDGE: Let's see now. What are you to do when I introduce you to the nobility, gentry, and others — of my acquaintance?

WOWSKI: Make believe sit down. Then get up.

TRUDGE: Let me see you do it.

(WOWSKI *makes a low curtsey.*)

Very well! And how are you to recommend yourself when you have nothing to say, amongst all our great friends?

WOWSKI: Grin. Show my teeth.

TRUDGE: Right! They'll think you've lived with people of fashion. But suppose you meet an old shabby friend in misfortune that you don't wish to be seen to speak to. What would you do?

WOWSKI: Look blind. Not see him.

TRUDGE: Why would you do that?

WOWSKI: 'Cause I can't see good friend in distress.

TRUDGE: That's a good girl! And I wish everybody could boast of so kind a motive for such cursed cruel behaviour. Lord, how some of your flashy bankers' clerks have *cut* me in Threadneedle Street. But come, though we have got among fine folks here, in an English settlement, I won't be ashamed of my old acquaintance. Yet, for my own part, I should not be sorry now to see my old friend with a new face. — Odsbobs, I see Mr Inkle! Go in, Wows. Call for what you like best.

WOWSKI: Then I call for you. Ah, I fear I not see you often now! But you come soon!

Song. WOWSKI.

> Remember when we walked alone
>   And heard, so gruff, the lion growl,
> And when the moon so bright it shone,
>   We saw the wolf look up and howl;
> I led you well, safe to our cell,
>   While tremblingly
>   You said to me
> — And kiss'd so sweet: dear Wowski, tell,
>   How could I live without ye?
>
> But now you come across the sea
>   And tell me here no monsters roar;
> You'll walk alone, and leave poor me,
>   When wolves, to fright you, howl no more.
> But — ah! — think well on our old cell
>   Where tremblingly
>   You kissed poor me;
> Perhaps you'll say: dear Wowski, tell,
>   How can I live without ye?

(*Exit* WOWSKI.)

TRUDGE: Eh? Oh, my master's talking to somebody on the quay.

(*Enter* 1st PLANTER.)

Who have we here?

PLANT: Harkee, young man! Is that young Indian of yours going to our market?

TRUDGE: Not she. She never went to market in all her life.

PLANT: I mean, is she for our sale of slaves? Our Black Fair?

TRUDGE: A Black Fair? Ha! ha! ha! You hold it on a brown green, I suppose.

PLANT: She's your slave, I take it?

TRUDGE: Yes, and I'm her humble servant, I take it.

PLANT: Aye, aye, natural enough at sea. But at how much do you value her?

TRUDGE: Just as much as she has saved me: my own life.

PLANT: Pshaw! You mean to sell her?

TRUDGE: (*staring*) Zounds, what a devil of a fellow! Sell Wows? My poor, dear, dingy wife?

PLANT: Come, come. I've heard your story from the ship. Don't let's haggle. I'll bid as fair as any trader amongst us. But no tricks upon travellers, young man, to raise your price. Your wife, indeed! Why, she's no Christian!

TRUDGE: No, but I am, so I shall do as I'd be done by, Master Black Market. And, if you were a good one yourself, you'd know that fellow-feeling for a poor body who wants your help is the noblest mark of our religion. (*aside*) I wouldn't be articled clerk to such a fellow for all the world!

PLANT: Hey-day, the booby's in love with her! Why, sure, friend, you wouldn't live here with a black?

TRUDGE: Plague on't, there it is! I shall be laughed out of my honesty, here. But you may be jogging, friend! I may feel a little queer, perhaps, at showing her face, but damme if ever I do anything to make me ashamed of showing my own.

PLANT: Why, I tell you, her very complexion —

TRUDGE: Rot her complexion! I'll tell you what, Mr Fair Trader, if your head and heart were to change places, I've a notion you'd be as black in the face as an ink bottle.

PLANT: Pshaw, the fellow's a fool! A rude rascal! He ought to be sent back to the savages again! He's not fit to live amongst us Christians!
(*Exit* PLANTER.)

TRUDGE: Christians! Ah, tender souls they are, to be sure! — Oh, here comes my master, at last.
(*Enter* INKLE *and the* 2nd PLANTER.)

INKLE: Nay, sir, I understand your customs well. Your Indian markets are not unknown to me.

PLANT: And, as you seem to understand business, I need not tell you that dispatch is the soul of it. Her name, you say, is —

INKLE: Yarico. But urge this no more, I beg you. I must not listen to it, for, to speak freely, her anxious care of me demands that here — though here it may seem strange — I should avow my love for her.

PLANT: Lord help you for a merchant! It's the first time I ever heard a trader talk

---

you may be jogging: i.e. 'on your way!'

of love, except, indeed, the love of trade — and the love of the *Sweet Molly*, my ship.

INKLE: Then, sir, you cannot feel my situation.

PLANT: Oh yes, I can! We have a hundred such cases after a voyage, but they never last long on land. It's amazing how constant a young man is in a ship! But, in two words, will you dispose of her or no?

INKLE: In two words, then: meet me here at noon and we'll speak further on this subject. And lest you think I trifle with your business, hear why I wish this pause. Chance threw me, on my passage to your island, among a savage people, deserted, defenceless, cut off from my companions, my life at stake. To this young creature I owe my preservation; she found me like a dying bough, torn from its kindred branches, which, as it drooped, she moistened with her tears.

PLANT: Nay, nay, talk like a man of this world.

INKLE: Your patience. And yet your interruption goes to my present feelings, for, on our sail to this your island, the thoughts of time misspent, doubt, fears — or call it what you will — have much perplexed me. And as your spires arose, reflections still rose with them. For here, sir, lie my interests, great connections, and other weighty matters which now I need not mention —

PLANT: But which her presence here will mar.

INKLE: Even so. And yet the gratitude I owe her —

PLANT: Pshaw! So, because she preserved your life, your gratitude is to make you give up all you have to live upon.

INKLE: Why, in that light, indeed — This never struck me yet. I'll think on't.

PLANT: Aye, aye, do so. Why, what return can the wench wish more than taking her from a wild, idle, savage people and providing for her, here, with repu-table hard work in a genteel, polished, tender, Christian country?

INKLE: Well, sir, at noon —

PLANT: I'll meet you. But remember, young gentleman, you must get her off your hands. You must, indeed. (*aside*) I shall have her a bargain, I see that. (*to him*) Your servant! Zounds, how late it is! — But never be put out of your way for a woman! — I must run. (*aside*) My wife will play the devil with me for keeping breakfast!

(*Exit* 2nd PLANTER.)

INKLE: Trudge!

TRUDGE: Sir?

INKLE: Have you provided a proper apartment?

TRUDGE: Yes, sir, at the Crown, here. A neat, spruce room, they tell me. You haven't seen such a convenient lodging this good while, I believe.

INKLE: Are there no better inns in the town?

TRUDGE: Um. Why, there's the Lion, I hear, and the Bear and the Boar, but we saw them at the door of all our late lodgings and found but bad accom-modations within, sir.

INKLE: Well, run to the end of the quay and conduct Yarico hither. The road is straight before you. You can't miss it.

TRUDGE: Very well, sir. What a fine thing it is, to turn one's back on a master without running into a wolf's belly! One can follow one's nose on a message, here, and be sure it won't be bit off by the way.

(*Exit* TRUDGE.)

INKLE: Let me reflect a little. This honest planter counsels well. Part with her? What is there in it which cannot easily be justified? My interest, honour, engagements to Narcissa: all demand it. My father's precepts, too. — I can remember when I was a boy what pains he took to mould me, schooled me from morn to night, and still the burden of his song was, 'Prudence! Prudence, Thomas, and you'll rise!' Early he taught me numbers which, he said, — and he said rightly — would give me a quick view of loss and profit and banish from my mind those idle impulses of passion which mark young thoughtless spendthrifts. His maxims rooted in my heart, and as I grew, *they* grew, till I was reckoned among our friends a steady, sober, solid, good young man and all the neighbours called me *the prudent Mr Thomas*. And shall I now, at once, kick down the character which I have raised so warily? — Part with her! Sell her! — The thought once struck me in our cabin as she lay sleeping by me, but in her slumbers she passed her arm around me, murmured a blessing on my name, and broke my meditations.

(*Enter* YARICO *and* TRUDGE.)

YARICO: My love!

TRUDGE: I have been showing her all the wigs and bales of goods we met on the quay, sir.

YARICO: Oh, I have feasted my eyes on wonders!

TRUDGE: And I'll go feast on a slice of beef in the inn, here.

(*Exit* TRUDGE.)

YARICO: My mind has been so busy that I almost forgot even you. I wish you had stayed with me. You would have seen such sights!

INKLE: Those sights have become familiar to *me*, Yarico.

YARICO: And yet, I wish they were not; you might partake my pleasures. But now, again, methinks I will not wish so, for, with too much gazing, you might neglect poor Yarico.

INKLE: Nay, nay, my care is still for you.

YARICO: I am sure it is. And if I thought it was not, I'd tell you tales about our poor grot, bid you remember our palm tree near the brook, where, in the shade, you often stretched yourself, while I would take your head upon my lap and sing my love to sleep. I know you'll love me then.

Song. YARICO.

> Our grotto was the sweetest place!
>   The bending boughs, with fragrance blowing,
> Would check the brook's impetuous pace,
>   Which murmur'd to be stopt from flowing:
> 'Twas there we met, and gazed our fill;
> Ah, think on this, and love me still!
>
> 'Twas then my bosom first knew fear,
>   —Fear to an Indian maid a stranger —

The war song, arrows, hatchet, spear,
 All warn'd me of my lover's danger.
For him did cares my bosom fill;
Ah, think on this, and love me still!

For him, by day, with care conceal'd,
 To search for food I climb'd the mountain,
And when the night no form reveal'd,
 Jocund we sought the bubbling fountain.
Then, then would joy my bosom fill;
Ah, think on this and love me still!

   (*Exeunt* INKLE *and* YARICO.)

  Scene 2. *An apartment in the house of* SIR CHRISTOPHER CURRY.
   (*Enter* SIR CHRISTOPHER *and* MEDIUM.)

SIR CHR: I tell you, old Medium, you are all wrong. Plague on your doubts! Inkle *shall* have my Narcissa. Poor fellow! I dare say he's finely chagrined at this temporary parting; eat up with the blue devils, I warrant.

MEDIUM: Eat up by the black devils, I warrant, for I left him in hellish hungry company.

SIR CHR: Pshaw! He'll arrive with the next vessel, depend on't. Besides, have I not had this in view ever since they were children? I must and will have it so, I tell you. Is it not, as it were, a marriage made above? They *shall* meet, I'm positive.

MEDIUM: Shall they? Then they must meet where the marriage was made, for hang me if I think it will ever happen below.

SIR CHR: Ha! And if that is the case, hang me if I think you'll ever be at the celebration of it.

MEDIUM: Yet, let me tell you, Sir Christopher Curry, my character is as unsullied as a sheet of white paper.

SIR CHR: Well said, old fool's-cap! And it's as mere a blank as a sheet of white paper: it bears the traces of neither a bad nor a good hand upon it!

MEDIUM: Well, it is not for me to boast of virtues. That's a vice I never give in to.

SIR CHR: Your virtues? Zounds, what are they?

MEDIUM: I am not addicted to passion. That, at least, Sir Christopher —

SIR CHR: Is like all your other virtues, a mere negative. You are honest, old Medium, by comparison, just as a fellow sentenced to transportation is happier than his companion condemned to the gallows: very worthy because you are no rogue, tender hearted because you never go to fires and executions and an affectionate father and husband because you never pinch your children or kick your wife out of bed!

MEDIUM: And that, as the world goes, is more than every man can say for himself.

---

blue devils: a modern derivative of this is 'the blues'; blue devils were supposed to have appeared in the hallucinations experienced during the *delirium tremens* which follows bouts of excessive drinking. Colman would certainly have been lucky not to have experienced this condition personally. Here the phrase is used to indicate despondency.

Yet, since you force me to speak my positive qualities — But, no matter. You remember me in London and know there was scarcely a laudable institution in town without my name in the list. Haven't I given more tickets to recommend the lopping of legs than any governor of our Hospital? And didn't I, as a member of the Humane Society, bring a man out of the New River who, it was afterwards found, had done me an injury?

SIR CHR: And damme if I would not kick any man into the New River that had done me an injury! There's the difference of our honesty. Oons, if you want to be an honest fellow, act from the impulse of nature! Why, you've no more gall than a pigeon.

MEDIUM: And you have as much gall as a turkey cock, and are as hot into the bargain. You're always so hasty. Among the hodge-podge of your foibles, passion is always predominant.

SIR CHR: Foibles, quotha? Foibles are foils that give additional lustre to the gems of virtue. You have not so many foils as I, perhaps.

MEDIUM: And, what's more, I don't want 'em, Sir Christopher, I thank you.

SIR CHR: Very true, for the devil a gem you have to set off with 'em.

MEDIUM: Well, well, I never mention errors. That, I flatter myself, is no disagreeable quality. It don't become me to say you are hot.

SIR CHR: 'Sblood, but it does become you! It becomes every man, especially an Englishman, to speak the dictates of his heart.

Song. SIR CHRISTOPHER CURRY.

> Oh, give me your plain-dealing fellows
> > Who never from honesty shrink,
> Not thinking on all they should tell us,
> > But telling us all that they think.
>
> Truth from man flows like wine from a bottle,
> > His free-spoken heart's a full cup;

---

my name in the list: this was the list of benefactors subscribing to a particular charitable institution or organisation. To be in the subscription list required an annual contribution above a specified minimum sum, in return for which certain privileges were obtained, such as a vote in determining the organisation's business and, in some cases, the right of nominating an agreed number of people a year as suitable candidates for relief by the charity.

tickets to recommend the lopping of legs: at the beginning of the eighteenth century, London had only five hospitals prepared to give treatment to the poor, out of which two alone had facilities for performing surgical operations. Following the founding of the Westminster Infirmary in 1719, the century saw the rapid expansion of voluntary hospitals funded by charity which aimed at ameliorating this neglect. However, to gain admission to a voluntary hospital, a prospective patient had first to obtain the written recommendation of a subscriber (see note above) certifying that he was a proper object of the charity.

Humane Society: The Royal Humane Society, founded in 1774, on the model of a life-saving charity which operated in Amsterdam specifically for the purpose of reviving those unfortunate enough to have fallen into the city's canals.

New River: completed in 1613 under the auspices of Sir Hugh Myddleton, this was an open channel constructed from Ware, in Hertfordshire, to Clerkenwell, in London, to carry a supply of fresh water into the metropolis.

But when truth sticks halfway in the throttle,
    Man's worse than a bottle cork'd up.

Complaisance is a gingerbread creature
    Us'd for show, like a watch, by each spark;
But truth is a golden repeater
    That sets a man right in the dark.

MEDIUM: But, suppose his heart dictates to anyone to knock up your friend, Sir Christopher?

SIR CHR: Eh? Why, then it becomes me to knock him down.

MEDIUM: Mercy on us! If that was the consequence of scandal in England, nowadays, all our fine gentlemen would cut each other's throats over a bottle. And if it extended to the card tables, our routs would be fuller of black eyes than black aces!

    (*Enter* SERVANT.)

SERV: An English vessel, sir, is just arrived in the harbour.

SIR CHR: A vessel? Od's my life! Now for the news, if it is but as I hope. Any dispatches?

SERV: This letter, sir, brought by a sailor from the quay.

    (*Exit* SERVANT.)

SIR CHR: Now for it! If Inkle is but amongst 'em – Zounds, I'm all in a flutter! My head shakes like an aspen leaf and you, you old fool, are as stiff and steady as an oak. Why aren't you all tiptoe, all nerves?

MEDIUM: Well, read, Sir Christopher.

SIR CHR: (*opening the letter*) Huzza! Here it is: he's safe – safe and sound – at Barbadoes! (*reading*) 'Sir, My master, Mr Inkle, is just arrived in your harbour . . . ' Here, read! Read, old Medium!

MEDIUM: (*reading*) Um. ' . . . your harbour. We were taken up by an English vessel on the 14th ulto. He only waits till I have puffed his hair to pay his respects to you and Miss Narcissa. In the meantime, he has ordered me to brush up this letter for your honour, from your humble servant to command, TIMOTHY TRUDGE.'

SIR CHR: Hey-day! Here's a style! The voyage has jumbled the fellow's brains out of their places; the water has made his head turn round. But no matter. Mine turns round, too. I'll go and prepare Narcissa directly; they shall be married slap-dash, as soon as he comes from the quay. From Neptune to Hymen: from the hammock to the bridal bed, ha, old boy?

MEDIUM: Well, well, don't flurry yourself. You're so hot!

SIR CHR: Hot! 'Sblood, aren't I in the West Indies? Aren't I Governor of Barbadoes? – He shall have her as soon as he sets his foot on shore. But, plague on't, he's so slow! – She shall rise to him like Venus out of the sea! – His hair puffed? He ought to have been puffing here, out of breath, by this time.

---

like Venus out of the sea: perhaps in the style of the famous Botticelli painting, *The Birth of Venus*. Venus was said to have risen from the sea and to have been blown by the west wind towards the coast of Cyprus.

MEDIUM: Very true. But Venus's husband is always supposed to be lame, you know, Sir Christopher.

SIR CHR: Well now, do, my good fellow, run down to the shore and see what detains him. (*hurrying him off*)

MEDIUM: Well, well, I will, I will.

(*Exit* MEDIUM.)

SIR CHR: In the meantime, I'll get ready Narcissa and all shall be concluded in a second, my heart's set upon it. Poor fellow! After all his rumbles and tumbles and jumbles and fits of despair, I shall be rejoiced to see him! I have not seen him since he was that high. – But, zounds, he's so tardy!

(*Enter* SERVANT.)

SERV: A strange gentleman, sir, come from the quay, desires to see you.

SIR CHR: From the quay? Od's my life, 'tis he! 'Tis Inkle! Show him up directly.

(*Exit* SERVANT.)

The rogue is expeditious after all. I'm so happy!

(*Enter* CAMPLEY.)

(*embracing him*) My dear fellow! (*Shakes hands.*) I'm rejoiced to see you! Welcome! Welcome, here, with all my soul!

CAMPLEY: This reception, Sir Christopher, is beyond my warmest wishes. Unknown to you –

SIR CHR: Aye, aye, we shall be better acquainted by and by. Well, and how, eh? Tell me! – But old Medium and I have talked over your affair a hundred times a day, ever since Narcissa arrived.

CAMPLEY: You surprise me! Are you then really acquainted with the whole affair?

SIR CHR: Every tittle.

CAMPLEY: And can you, sir, pardon what is past?

SIR CHR: Pooh! How could you help it?

CAMPLEY: Very true. Sailing in the same ship and –

SIR CHR: Aye, aye. But we have had a hundred conjectures about you: your despair and distress, and all that. Yours must have been a damned situation, to say the truth.

CAMPLEY: Cruel indeed, Sir Christopher, and, I flatter myself, will move your compassion. I have been almost inclined to despair indeed, as you say. And when you consider the past state of my mind, the black prospect before me –

SIR CHR: Ha! ha! Black enough, I dare say.

CAMPLEY: – The difficulty I have felt in bringing myself face to face with you –

SIR CHR: That I am convinced of. But I knew you would come the first opportunity.

CAMPLEY: Very true. – Yet the distance between the Governor of Barbadoes and myself – (*bowing*)

SIR CHR: Yes, a devilish way asunder!

Venus's husband: this was Hephaestus, son of Zeus. The allusion is not a particularly fortunate one in this context: Hephaestus was not only lame but also incredibly ugly and frequently cuckolded.

CAMPLEY: Granted, sir. — Which has distressed me with the cruellest hopes as to our meeting —

SIR CHR: It was a toss up.

CAMPLEY: (*aside*) The old gentleman seems devilish kind. Now to soften him. Perhaps, sir, in your younger days you may have been in the same situation yourself.

SIR CHR: Who I? 'Sblood, no! Never in my life!

CAMPLEY: I wish you had, with all my soul, Sir Christopher —

SIR CHR: Upon my soul, sir, I am very much obliged to you. (*bowing*)

CAMPLEY: — As what I now mention might have greater weight with you.

SIR CHR: Pooh! Prithee, I tell you I pitied you from the bottom of my heart!

CAMPLEY: Indeed! Had you been kind enough to have sent to me, how happy should I have been in attending your commands!

SIR CHR: I believe you would, egad! Ha! ha! Sent to you! Very well! Ha! ha! ha! A dry rogue! You'd have been ready enough to come, I dare say!

CAMPLEY: But now, sir, if with your leave I may still venture to mention Miss Narcissa —

SIR CHR: An impatient, sensible young dog! Like me to a hair! Set your heart at rest, my boy: she's yours, yours before tomorrow morning.

CAMPLEY: Amazement! I can scarce believe my senses!

SIR CHR: Zounds, you ought to be out of your senses! But dispatch, make short work of it ever while you live, my boy. Ha, here she is!

(*Enter* NARCISSA *and* PATTY.)

(*to* NARCISSA) Here, girl. Here's your swain.

CAMPLEY: I just parted with my Narcissa on the quay, sir.

SIR CHR: Did you? Ah, sly dog! Had a meeting before you came to the old gentleman? But here, take him and make much of him. And, for fear of further separations, you shall e'en be tacked together directly. What say you, girl?

NAR: I always obey my father's commands with pleasure, sir. (*aside, to* PATTY) Steal out, Patty, as soon as you can, and prevent Mr Inkle's appearance. My father has mistaken Campley, I am confident.

PATTY: It is not for his daughter, ma'am, to tell him of his mistakes, you know!

SIR CHR: Od, I'm so happy I hardly know which way to turn! But we'll have the carriage directly, drive down to the quay, trundle old Spintext into church, and hey for matrimony!

CAMPLEY: With all my heart, Sir Christopher. The sooner the better!

Song. SIR CHRISTOPHER, CAMPLEY, NARCISSA *and* PATTY.

SIR CHR: Your Colinettes and Arriettes,
    Your Damons of the grove,
  Who, like fallals and pastorals,
    Waste years in love!

Colinettes . . . Damons: Colinette is a variant of Colin, the name commonly given in English and French pastoral poetry to a shepherd. Rustic swains appearing in pastorals were frequently called Damon, following the tradition set by Virgil in his third *Eclogue*.

> But modern folks know better jokes
>> And, courting once begun,
> To church they hop at once — and pop! —
>> Egad, all's done!

ALL: In life we prance a country dance
           Where every couple stands;
       Their partners set, a while curvet,
           But soon join hands.

NAR: When at our feet, so trim and neat,
           The powder'd lover sues,
       He vows he dies, the lady sighs,
           But can't refuse.
       Ah! how can she unmov'd e'er see
           Her swain his death incur?
       If once the squire is seen expire,
           He lives with her.

ALL: In life we prance, &c.

PATTY: When John and Bet are fairly met,
           John boldly tries his luck;
       He steals a buss: without more fuss
           The bargain's struck.
       Whilst things below are going so,
           Is Betty, pray, to blame,
       Who knows upstairs her mistress fares
           Just, just the same?

ALL: In life we prance, &c.

       (*Exeunt omnes.*)

ACT III

Scene 1. *The quay.*

(*Enter* PATTY.)

PATTY: Mercy on us! What a walk I have had of it! Well, matters go on swimmingly
at the Governor's. The old gentleman has ordered the carriage and the young
couple will be whisked here to church in a quarter of an hour. My business is
to prevent young sobersides, young Inkle, from appearing to interrupt the
ceremony. Ha! Here's the Crown, where I hear he is housed. So now to find
Trudge and trump up a story in the true style of a chambermaid.
      (PATTY *goes into the house.*)
(*within*) I tell you it don't signify, and I will come up.

---

partners set: stand apart, facing each other.
curvet: dance a leaping step.

TRUDGE: (*within*) But it does signify, and you can't come up.
(*Re-enter* PATTY *with* TRUDGE.)
PATTY: You had better say at once I shan't.
TRUDGE: Well then, you shan't.
PATTY: Savage! Pretty behaviour you have picked up amongst the Hottypots!
Your London civility, like London itself, will soon be lost in smoke, Mr
Trudge, and the politeness you have studied so long in Threadneedle Street
blotted out by the blacks you have been living with.
TRUDGE: No such thing. I practised my politeness all the while I was in the
woods. Our very lodging taught me good manners, for I could never bring
myself to go into it without bowing.
PATTY: Don't tell me! A mighty civil reception you give a body, truly, after a six
weeks' parting!
TRUDGE: Gad, you're right. I am a little out here, to be sure. (*Kisses her.*) Well,
how do you do?
PATTY: Pshaw, fellow, I want none of your kisses!
TRUDGE: Oh, very well, I'll take it again! (*Offers to kiss her.*)
PATTY: Be quiet. I want to see Mr Inkle; I have a message to him from Miss
Narcissa. I shall get a sight of him now, I believe.
TRUDGE: Maybe not. He's a little busy at present.
PATTY: Busy, ha? Plodding? What, he's at his multiplication table again?
TRUDGE: Very likely, so it would be a pity to interrupt him, you know.
PATTY: Certainly. And the whole of my business was to prevent his hurrying him-
self. Tell him we shan't be ready to receive him at the Governor's till
tomorrow, d'ye hear?
TRUDGE: No?
PATTY: No. Things are not prepared. The place isn't in order and the servants
haven't had proper notice of the arrival.
TRUDGE: Oh, let me alone to give the servants notice! Rat-tat-tat: it's all the
notice we had in Threadneedle Street of the arrival of a visitor.
PATTY: Threadneedle Street? Threadneedle nonsense! I'd have you know we do
everything with an air. Matters have taken another turn. Style, style, sir, is
required here, I promise you.
TRUDGE: Turn? Style? And, pray, what style will serve your turn now, Madame
Patty?
PATTY: A due dignity and decorum, to be sure. Sir Christopher intends Mr Inkle,
you know, for his son-in-law, and must receive him in public form — which
can't be till tomorrow morning — for the honour of his governorship. Why,
the whole island will ring of it.
TRUDGE: The devil it will!
PATTY: Yes, they've talked of nothing but my mistress's beauty and fortune for
these six weeks. Then he'll be introduced to the bride, you know.
TRUDGE: Oh, my poor master!
PATTY: Then a breakfast, then a procession, then — if nothing happens to prevent
it — he'll get into church and be married in a crack.
TRUDGE: Then he'll get into a damned scrape, in a crack.
PATTY: Hey-day! A scrape? The holy state of matrimony? How?

TRUDGE: Yes, it's plaguy holy, and many of its votaries, as in other holy states, live in repentance and mortification. Ah, poor Madame Yarico! My poor pil-garlic of a master, what will become of him?
PATTY: Why, what's the matter with the booby?
TRUDGE: Nothing, nothing. He'll be hanged for poli-bigamy!
PATTY: Polly who?
TRUDGE: It must out, Patty!
PATTY: Well?
TRUDGE: Can you keep a secret?
PATTY: Try me.
TRUDGE: Then (*whispering*) my master keeps a girl.
PATTY: Oh monstrous! Another woman?
TRUDGE: As sure as one and one make two.
PATTY: (*aside*) Rare news for my mistress! (*aloud*) Why, I can hardly believe it. The grave, sly, steady, sober Mr Inkle do such a thing?
TRUDGE: Pooh, it's always your sly, sober fellows that go the most after the girls!
PATTY: Why, I should sooner suspect *you*.
TRUDGE: Me? Oh Lord! He! he! (*conceitedly*) Do you think any smart, tight little black-eyed wench would be struck with my figure?
PATTY: Pshaw, never mind your figure, tell me how it happened!
TRUDGE: You shall hear. When the ship left us ashore, my master turned as white as a sheet of paper: it isn't everybody that's blessed with courage, Patty.
PATTY: True!
TRUDGE: However, I bid him cheer up, told him to stick to my elbow, took the lead and began our march.
PATTY: Well?
TRUDGE: We hadn't gone far when a damned one-eyed black boar, that grinned like a devil, came down the hill in jog trot! My master melted as fast as a pot of pomatum!
PATTY: Mercy on us!
TRUDGE: But what does I do but whips out my desk knife that I used to cut the quills with at home, met the monster, and slit up his throat like a pen.
PATTY: Lord, Trudge, what a great traveller you are!
TRUDGE: Yes, I remember we fed on the flitch for a week.
PATTY: Well, well! But the lady –
TRUDGE: The lady? Oh, true. By and by we came to a cave, a large hollow room underground, like a warehouse in the Adelphi. Well, there we were half an hour before I could get him to go in. There's no accounting for fear, you

---

pil-garlic: lit. 'peeled garlic'; the phrase was first used in graphic reference to a bald pate but subsequently became an affectionate term of pity.
pomatum: an ointment used to preserve hair colour and retention. Trudge's comparison here derives from his obsession with the minutiae of valeting.
warehouse in the Adelphi: completed in 1774, the Adelphi scheme was a triumph for the Adam brothers, who overcame the problem of redeveloping a steeply inclined site between the Strand and the River Thames for luxury housing by creating an artificial level underpropped by a system of massive brick arches. The subterranean area provided rows of cavernous warehouses for city merchants.

know. At last, in we went to a place hung round with skins, as it might be a furrier's shop, and there was a fine lady snoring on a bow and arrows.

PATTY: What? All alone?

TRUDGE: Eh? No, no, no! – hum – she had a young lion – by way of a lap-dog.

PATTY: Gemini! What did you do?

TRUDGE: Gave her a jog, and she opened her eyes. She struck my master immediately.

PATTY: Mercy on us! With what?

TRUDGE: With her beauty, you ninny, to be sure! And they soon brought matters to bear. The wolves witnessed the contract, I gave her away, the crows croaked amen, and we had board and lodging for nothing.

PATTY: And this is she he has brought to Barbadoes?

TRUDGE: The same.

PATTY: Well! And tell me, Trudge, she's pretty, you say: is she fair or brown or – ?

TRUDGE: Um – she's a good comely copper.

PATTY: How? A tawny?

TRUDGE: Yes, quite dark, but very elegant. Like a Wedgwood teapot.

PATTY: Oh, the monster! The filthy fellow! Live with a black-a-moor?

TRUDGE: Why, there's no great harm in't, I hope?

PATTY: Faugh, I wouldn't let him kiss me for all the world! He'd make my face all smutty.

TRUDGE: Zounds, you are mighty nice all of a sudden! But I'd have you to know, Madame Patty, that black-a-moor ladies, as you call 'em, are some of the very few whose complexions never rub off! 'Sbud, if they did, Wows and I should have changed faces by this time. But mum, not a word, for your life.

PATTY: Not I! (*aside*) except to the Governor and his family. (*aloud*) But I must run. And remember, Trudge, if your master has made a mistake here, he has himself to thank for his pains.

Song. PATTY.

Tho' lovers, like marksmen, all aim at the heart,
    Some hit wide of the mark, as we wenches all know,
But, of all the bad shots, he's the worst in the art
    Who shoots at a pigeon and kills a crow – oho!
        Your master has killed a crow.

When younkers go out the first time in their lives,
    At random they shoot and let fly as they go;
So your master, unskilled how to level at wives,
    Has shot at a pigeon and killed a crow – oho!
        Your master has killed a crow.

Wedgwood teapot: although the name of Wedgwood is largely associated with a style of porcelain typified by blue jasper with white relief, Josiah Wedgwood is known to have produced, at this time, a range of staple wares, such as teapots, in an inexpensive unglazed earthenware of dark red to chocolate colour, which he called 'rosso antico'.

Love and money thus wasted in terrible trim,
   His powder is spent and his shot running low,
Yet the pigeon he missed, I've a notion, with him
   Will never for such a mistake pluck a crow — no, no,
      Your master may keep his crow.

     (*Exit* PATTY.)

TRUDGE: Pshaw, these girls are so plaguy proud of their white and red! But I
won't be shamed out of Wows, that's flat. Master, to be sure, while we were
in the forest, taught Yarico to read with his pencil and pocket-book. What
then? Wows comes on fine and fast in her lessons. A little awkward at first, to
be sure. Ha! ha! She's so used to feed with her hands that I can't get her to
eat her victuals in a genteel, Christian way, for the soul of me! When she has
stuck a morsel on her fork, she don't know how to guide it, but pops up her
knuckles to her mouth and the meat goes up to her ear. But no matter. After
all the fine, flashy London girls, Wowski's the wench, for my money.

Song. TRUDGE.

   A clerk I was in London gay,
      Jemmy linkum feedle,
   And went in boots to see the play,
      Merry fiddlem tweedle.
   I march'd the lobby, twirl'd my stick,
      Diddle, daddle, deedle;
   The girls all cried, 'He's quite the kick.'
      Oh, Jemmy linkum feedle.

   Hey, for America I sail!
      Yankee doodle deedle;
   The sailor-boys cried, 'Smoke his tail!'
      Jemmy linkum feedle.
   On English belles I turn'd my back,
      Diddle, daddle, deedle,
   And got a foreign fair, quite black,
      O twaddle, twaddle, tweedle!

   Your London girls, with roguish trip,
      Wheedle, wheedle, wheedle,
   May boast their pouting under-lip,
      Fiddle, faddle, feedle.

---

so plaguy proud of their white and red: references to colour consciousness are frequently made
in this play in terms of facial make-up, the usual allusion being to black skin as a permanent
(and hence superior) form of cosmetic.

in boots to see the play: a particular source of chagrin to the employers of servants at this time
was to see them in public places, such as the theatres, dressed in extravagant outfits and aping
the affectations of high society. Trudge's appearance at the theatre in top-boots and his twirling
a stick in the lobby (outside the entrance to the boxes) would have been a deliberate attempt to
copy the manner of a young buck of the town.

My Wows would beat a hundred such,
    Diddle, daddle, deedle,
Whose upper lip pouts twice as much,
    O, pretty double wheedle!

Rings I'll buy to deck her toes,
    Jemmy linkum feedle;
A feather fine shall grace her nose,
    Waving siddle seedle.
With jealousy I ne'er shall burst,
    Who'd steal my bone of bone-a?
A white Othello, I can trust
    A dingy Desdemona.

(*Exit* TRUDGE.)

Scene 2. *A room in the Crown.*

(*Enter* INKLE.)

INKLE: I know not what to think. I have given distant hints of parting, but still, so strong her confidence in my affection, she prattles on without regarding me. Poor Yarico! I must not, cannot, quit her. When I would speak, her look, her mere simplicity, disarms me. I dare not wound such innocence. Simplicity is like a smiling babe which, to the ruffian that would murder it, stretching its little naked, helpless arms, pleads speechless its own cause. And yet, Narcissa's family —

(*Enter* TRUDGE.)

TRUDGE: There he is, like a beau bespeaking a coat, doubting which one to choose. Sir —

INKLE: What now?

TRUDGE: Nothing unexpected, sir. I hope you won't be angry, but I am come to give you joy, sir!

INKLE: Joy? Of what?

TRUDGE: A wife, sir! A white one. I know it will vex you, but Miss Narcissa means to make you happy — tomorrow morning.

INKLE: Tomorrow!

TRUDGE: Yes, sir. And as I have been out of employ in both my capacities lately, after I have dressed your hair, may I draw up the marriage articles?

INKLE: Whence comes your intelligence, sir?

TRUDGE: Patty told me all that has passed in the Governor's family on the quay, sir. Women, you know, can never keep a secret. You'll be introduced in form, with the whole island to witness it.

INKLE: So public, too. Unlucky!

TRUDGE: There will be nothing but rejoicings, in compliment to the wedding, she tells me; all noise and uproar! Married people like it, they say.

INKLE: Strange that I should be so blind to my interest as to be the only person this distresses!

TRUDGE: They're talking of nothing else but the match, it seems.

INKLE: Confusion! How can I, in honour, retract?

TRUDGE: And the bride's merits.

INKLE: True, a fund of merits! I would not, but from necessity — a case so nice as this — I would not wish to retract.

TRUDGE: Then they call her so handsome.

INKLE: Very true, so handsome! The whole world would laugh at me; they'd call it folly to retract.

TRUDGE: And then they say so much of her fortune.

INKLE: Oh, death, it would be *madness* to retract! Surely my faculties have slept and this long parting from my Narcissa has blunted my sense of her accomplishments. 'Tis this alone makes me so weak and wavering. I'll see her immediately (*going*).

TRUDGE: Stay, stay, sir. I am desired to tell you that the Governor won't open his gates to us till tomorrow morning and is now making preparations to receive you at breakfast with all the honours of matrimony.

INKLE: Well, be it so. It will give me time, at all events, to put my affairs in train.

TRUDGE: Yes, it's a short respite before execution, and if your honour was to go and comfort poor Madame Yarico —

INKLE: Damnation! Scoundrel, how dare you offer your advice? I dread to think of her!

TRUDGE: I've done, sir, I've done. But I know I should blubber over Wows all night if I thought of parting with her in the morning.

INKLE: Insolence! Begone, sir!

TRUDGE: Lord, sir, I only —

INKLE: Get down stairs, sir, directly!

TRUDGE: Ah, you may well put your hand to your head! (*aside*) And a bad head it must be to forget that Madame Yarico prevented her countrymen from peeling off the upper part of it.

(*Exit* TRUDGE.)

INKLE: 'Sdeath, what am I about? How have I slumbered! Is it I, I who, in London, laughed at the younkers of the town and, when I saw their chariots with some fine, tempting girl perked in the corner, would cry, 'Ah, there sits ruin, there flies the green-horn's money!', then wondered with myself how men could trifle time on women or, indeed, think of any women without fortunes? And now, forsooth, it rests with *me* to turn romantic puppy and give up all for love. Give up? Oh, monstrous folly! Thirty thousand pounds!

(TRUDGE *peeps in at the door.*)

TRUDGE: May I come in, sir?

INKLE: What does the booby want?

TRUDGE: Sir, your uncle wants to see you.

INKLE: Mr Medium? Show him up directly!

(*Exit* TRUDGE.)

He must not know of this. Tomorrow! I wish this marriage were more distant,

---

younkers of the town: young men, especially those testing their independence for the first time and indulging in a conspicuously careless and fashionable lifestyle; *younker* probably derived from the German *Junker*.

that I might break it to her by degrees. She'd take my purpose better were it less suddenly delivered. Women's weak minds bear grief as colts do burdens: load them with their full weight at once and they sink under it, but every day add little imperceptibly to little, 'tis wonderful how much they'll carry.

        (*Enter* MEDIUM.)

MEDIUM: Ah, here he is! Give me your hand, nephew! Welcome, welcome to Barbadoes, with all my heart!

INKLE: I am glad to meet you here, uncle!

MEDIUM: That you are, that you are, I'm sure. Lord, lord, when we parted last, how I wished we were in a room together, if it was but the black hole! Since we sundered I haven't been able to sleep o'nights for thinking of you. I've laid awake and fancied I saw you sleeping your last, with your head in the lion's mouth for a night-cap. And I've never seen a bear brought over to dance about the street but I thought you might be bobbing up and down in its belly.

INKLE: I am very much obliged to you.

MEDIUM: Aye, aye, I am happy enough to find you safe and sound, I promise you. Why, I've been hunting you all over the quay, and been in half the houses upon it, before I could find you. I should have been here sooner else. Whew, I'm so warm! I've run as fast —

INKLE: As you did in the forest, eh, Mr Medium?

MEDIUM: Well, well, thank heaven we are both out of the forest! Hounslow Heath at dusk is a trifle to it. I could not walk in a grove again with comfort, though it were in the middle of Paradise. But you have a fine prospect before you now, young man. I come to take you with me to Sir Christopher, who is impatient to see you.

INKLE: Tomorrow, I hear, he expects me.

MEDIUM: Tomorrow? Directly! This moment! In half a second! I left him standing to tip-toe, as he calls it, to embrace you. And he's standing on tip-toe now, in the great parlour, and there he'll stand till you come to him.

INKLE: Is he so hasty?

MEDIUM: Hasty? He's all pepper! Hasty, indeed! Why, he vows you shall have his daughter this very night.

INKLE: What a situation!

MEDIUM: Why, it's hardly fair just after a voyage. But come, bustle, bustle! He'll think you neglect him. He's rare and touchy, I can tell you. And if he once

---

the black hole: of Calcutta, where 156 English prisoners of the Nawab of Bengal died of suffocation in a single cell measuring 18 feet by 15 feet on the night of 20 June 1756. Some indication of how the mood of British audiences fluctuated at this time is given by the survival here of Colman's facetious allusion which, by the turn of the century, would certainly have provoked patriotic outrage.

Hounslow Heath: close enough to London to be crossed by an exceptionally high number of travellers and open enough for an escape to be made effective, this was an area traditionally notorious for its highwaymen and footpads. Even as late as the 1780s, gibbets with bodies of convicted highway robbers hanging in chains stood at regular intervals on every road across the Heath as a grotesque but ineffectual reminder that law and order could still be made to operate there.

takes it into his head that you show the least slight to his daughter it would knock up all your schemes in a minute.

INKLE: Confusion! (*aside*) If he should hear of Yarico!

MEDIUM: But at present you are all and all with him. He has been telling me his intentions these six weeks. You'll be a fine warm husband, I promise you.

INKLE: (*aside*) This cursed connection!

MEDIUM: It is not for me, though, to tell you how to play your cards. You are a prudent young man and can make calculations in a wood. I need not tell you that the least shadow of affront disobliges a testy old fellow.

INKLE: (*aside*) Fool! Fool! Fool!

MEDIUM: Why, what the devil is the matter with you?

INKLE: (*aside*) It must be done effectually or all is lost. Mere parting would not conceal it.

MEDIUM: Ah, now he's got to his damned square root again, I suppose, and Old Nick would not move him! Why, nephew!

INKLE: (*aside*) The planter that I spoke with cannot be arrived. But time is precious. The first I meet — common prudence now demands it. I'm fixed: I'll sell her.

(*Exit* INKLE.)

MEDIUM: Damn me, but he's mad! The woods have turned the poor boy's brains! He's scalped and gone crazy! Hollo, Inkle! Nephew? Gad, I'll spoil your arithmetic, I warrant me!

(*Exit* MEDIUM.)

Scene 3. *The quay.*

(*Enter* SIR CHRISTOPHER CURRY.)

SIR CHR: Od's my life, I can scarce contain my happiness! I've left 'em safe in church, in the middle of the ceremony. I ought to have given Narcissa away, they told me, but I capered about so much for joy that Old Spintext advised me to go and cool my heels on the quay till it was all over. Od, I'm so happy! And they shall see now what an old fellow can do at a wedding!

(*Enter* INKLE.)

INKLE: Now for dispatch! (*to* SIR CHRISTOPHER) Hark'ee, old gentleman!

SIR CHR: Well, young gentleman?

INKLE: If I mistake not, I know your business here.

SIR CHR: Egad, I believe half the island knows it, by this time.

INKLE: Then to the point: I have a female whom I wish to part with.

SIR CHR: Very likely. It's a common case nowadays with many a man.

INKLE: If you could satisfy me you would use her mildly and treat her with more kindness than is usual — for I can tell you she's of no common stamp — perhaps we might agree.

SIR CHR: Oho, a slave! Faith, now I think on't, my daughter may want an attendant or two extraordinary, and as you say she's a delicate girl, above the common run, and none of your thick-lipped, flat-nosed, squabby, dumpling dowdies, I don't much care if —

INKLE: And for her treatment —

SIR CHR: Look ye, young man, I love to be plain: I shall treat her a good deal better than you would, I fancy, for though I witness this custom every day, I can't help thinking the only excuse for buying our fellow creatures is to rescue 'em from the hands of those who are unfeeling enough to bring them to market.

INKLE: Fair words, old gentleman. An Englishman won't put up with an affront.

SIR CHR: An Englishman? More shame for you! Let Englishmen blush at such practices. Men who so fully feel the blessings of liberty are doubly cruel in depriving the helpless of their freedom.

INKLE: Confusion!

SIR CHR: 'Tis not my place to say so much, but I can't help speaking my mind.

INKLE: (*aside*) I must be cool. (*to him*) Let me assure you, sir, 'tis not my occupation. But, for a private reason, an instant pressing necessity —

SIR CHR: Well, well, I have a pressing necessity, too; I can't stand to talk now. I expect company here presently, but if you'll ask for me tomorrow, at the Castle —

INKLE: The Castle!

SIR CHR: Aye, sir, the Castle, the Governor's Castle, known all over Barbadoes.

INKLE: (*aside*) 'Sdeath, this man must be on the Governor's establishment, his steward, perhaps, and sent after me while Sir Christopher is impatiently waiting for me. I've gone too far; my secret may be known. As 'tis, I'll win this fellow to my interest. (*to him*) One word more, sir. My business must be done immediately and, as you seem acquainted at the Castle, if you should see me there — and there I mean to sleep tonight —

SIR CHR: The devil you do!

INKLE: Your finger on your lips! And never breathe a syllable of this transaction!

SIR CHR: No? Why not?

INKLE: Because, for reasons which, perhaps, you'll know tomorrow, I might be injured with the Governor, whose most particular friend I am.

SIR CHR: (*aside*) So, here's a particular friend of mine coming to sleep at my house that I never saw before in my life! (*to him*) I fancy, young gentleman, as you are such a bosom friend of the Governor's, you can hardly do anything to alter your situation with him; I shouldn't imagine anything could bring him to think a bit worse of you than he does at present.

INKLE: Oh, pardon me, but you'll find that hereafter. Besides, you doubtless know his character.

SIR CHR: Oh, as well as I do my own! But let's understand one another; you may trust me now you've gone so far. You are acquainted with his character, no doubt, to a hair?

INKLE: I am. — I see we shall understand each other. — You know him too, I see, as well as I; a very touchy, testy, hot old fellow.

SIR CHR: (*aside*) Here's a scoundrel! I hot and touchy? Zounds, I can hardly contain my passion! But I won't discover myself. I'll see the bottom of this. (*to him*) Well now, as we seem to have come to a tolerable explanation, and as you may be assured I'm incapable of whispering all this in the Governor's ear, let's proceed to business. Bring me the woman.

INKLE: No, there you must excuse me. I rather would avoid seeing her more and wish it to be settled without my seeming interference. The poor thing's fond, sir — you conceive me? — and my presence might distress her.

SIR CHR: (*aside*) Zounds, what an unfeeling rascal! The poor girl's in love with him, I suppose. (*to him*) No, no, fair and open; my dealing is with you and you only. I see her now or I declare off.

INKLE: Well then, you must be satisfied. Yonder's my servant — Ha, a thought has struck me! Come here, sir.

> (*Enter* TRUDGE.)

I'll write my purpose and send it her by him. It's lucky that I taught her to decypher characters; my labour now is paid.

> (INKLE *takes out his pocket-book and writes.*)

(*to himself*) This is somewhat less abrupt; 'twill soften matters. (*to* TRUDGE) Give this to Yarico, then bring her hither with you.

TRUDGE: I shall, sir. (*going*)

INKLE: Stay! Come back! (*aside*) This soft fool, if uninstructed, may add to her distress; his drivelling sympathy may feed her grief instead of soothing it. (*to* TRUDGE) When she has read this paper, seem to make light of it, tell her it is a thing of course, done purely for her good. I here inform her that I must part with her. D'ye understand your lesson?

TRUDGE: Pa—part with Ma—madame Ya—ric—o!

INKLE: Why does the blockhead stammer? I have my reasons. No muttering! And let me tell you, sir, if your rare bargain were gone too, 'twould be the better. She may babble our story of the forest and spoil my fortune.

TRUDGE: I'm sorry for it, sir; I have lived with you a long while. I've half a year's wages, too, due the 25th ulto. for scribbling your hair and dressing your parchments. But take my scribbling, take my frizzing, take my wages, and I and Wows will take ourselves off together. She saved my life, and rot me, sir, if anything but death shall part us!

INKLE: Impertinent! Go and deliver your message!

TRUDGE: I'm gone, sir. Lord, lord, I never carried a letter with such ill-will in all my born days!

> (*Exit* TRUDGE.)

SIR CHR: Well, shall I see the girl?

INKLE: She'll be here presently. One thing I had forgot: when she is yours, I needn't caution you, after the hints I've given, to keep her from the Castle. If Sir Christopher should see her, 'twould lead, you know, to a discovery of what I wish concealed.

SIR CHR: Depend upon me, Sir Christopher will know no more of our meeting than he does at this moment.

INKLE: Your secrecy shall not be unrewarded; I'll recommend you particularly to his good graces.

SIR CHR: Thank ye, thank ye, but I'm pretty much in his good graces, as it is. I don't know anybody he has a greater respect for.

> (*Re-enter* TRUDGE.)

INKLE: Now, sir, have you performed your message?

TRUDGE: Yes, I gave her the letter.

INKLE: And where is Yarico? Did she say she'd come? Didn't you do as you were ordered, didn't you speak to her?

TRUDGE: I couldn't, sir, I couldn't. I intended to say what you bid me, but I felt such a pain in my throat I couldn't speak a word, for the life of me. And so, sir, I fell a-crying.

INKLE: Blockhead!

SIR CHR: 'Sblood, but he's a very honest blockhead! Tell me, my good fellow, what said the wench?

TRUDGE: Nothing at all, sir. She sat down with her two hands clasped on her knees and looked so pitifully in my face I could not stand it. Oh, here she comes. I'll go and find Wows; if I must be melancholy, she shall keep me company.

    (*Exit* TRUDGE.)

SIR CHR: Od's my life, as comely a wench as I ever saw!

    (*Enter* YARICO, *who looks for some time in* INKLE's *face, bursts into tears, and falls on his neck.*)

INKLE: In tears, my Yarico? Why this?

YARICO: Oh do not, do not leave me!

INKLE: Why, simple girl, I'm labouring for your good! My interest here is nothing. I can do nothing from myself — you are ignorant of our country's customs; I must give way to men more powerful who will not have me with you. But see, my Yarico, ever anxious for your welfare, I've found a kind, good person who will protect you.

YARICO: Ah, why not *you* protect me?

INKLE: I have no means. How can I?

YARICO: Just as I sheltered you. Take me to yonder mountains, where I see no smoke from tall, high houses filled with your cruel countrymen. None of your princes, there, will come to take me from you. And should they stray that way, we'll find a lurking place, just like my own poor cave, where many a day I sat beside you and blessed the chance that brought you to it, that I might save your life.

SIR CHR: His life? Zounds, my blood boils at the scoundrel's ingratitude!

YARICO: Come, come, let's go. I always feared these cities. Let's fly and seek the woods, and there we'll wander hand in hand together. No cares shall vex us then. We'll let the day glide by in idleness, and you shall sit in the shade and watch the sunbeam playing on the brook while I sing the song that pleases you. No cares, love, but for your good. And we'll live cheerily, I warrant. In the fresh, early morning you shall hunt down our game and I will pick you berries, and then, at night, I'll trim our bed of leaves and lie me down in peace. Oh, we shall be so happy!

INKLE: This is mere trifling! The trifling of an unenlightened Indian! Hear me, Yarico. My countrymen and yours differ as much in minds as in complexions. We were not born to live in woods and caves. 'Tis misery to us to be reduced to seek subsistence by pursuing beasts. We Christians, girl, hunt money, a thing unknown to you. Here 'tis money which brings us ease, plenty, command, power, everything; and, of course, happiness. You are the bar to my attaining this. Therefore, 'tis necessary for my good — and which, I think, you value —

YARICO: You know I do, so much that it would break my heart to leave you.

INKLE: But we must part. If you are seen with me, I shall lose all.

YARICO: I gave up all for you: my friends, my country, all that was dear to me and still grown dearer since you sheltered there. All, all was left for you, and were it now to do again, I'd cross the seas and follow you all the world over.

INKLE: We idle time! Sir, she is yours; the stated price for women in Barbadoes I shall expect tomorrow. See you obey this gentleman, perform all duties, too, about his house which he commands you; 'twill be the better for you. (*going*)

YARICO: Oh, barbarous! (*holding him*) Do not, do not abandon me!

INKLE: No more!

YARICO: Stay but a little. I shan't live long to be a burden to you; your cruelty has cut me to the heart. Protect me but a little, and I'll obey this man and undergo all hardships for your good; stay but to witness 'em. I shall soon sink with grief; tarry till then, and hear me bless your name when I am dying and beg you, now and then, when I am gone, to heave a sigh for your poor Yarico.

INKLE: I dare not listen. You, sir, I hope, will take good care of her. (*going*)

SIR CHR: Care of her? That I will; I'll cherish her like my own daughter, and pour balm into the heart of a poor, innocent girl that has been wounded by the artifices of a scoundrel!

INKLE: Hah? 'Sdeath, sir, how dare you!

SIR CHR: 'Sdeath, sir, how dare *you* look an honest man in the face!

INKLE: Sir, you shall feel –

SIR CHR: Feel! It's more than you ever did, I believe! Mean, sordid wretch, dead to all sense of honour, gratitude or humanity! I never heard of such barbarity! I have a son-in-law who has been left in the same situation but, if I thought him capable of such cruelty, damme if I would not return him to sea – with a peck loaf, in a cockleshell! (*taking* YARICO *by the hand*) Come, come, cheer up, my girl. You shan't want a friend to protect you, I warrant you.

INKLE: Insolence! The Governor shall hear of this insult!

SIR CHR: The Governor? Liar! Cheat! Rogue! Impostor! Breaking all ties you ought to keep and pretending to those you have no right to! The Governor never had such a fellow in the whole catalogue of his acquaintance. The Governor disowns you, the Governor disclaims you, the Governor abhors you, and, to your utter confusion, here stands the Governor to tell you so!

INKLE: Sir Christopher! Lost and undone!

MEDIUM: (*without*) Hollo! Young Multiplication! Zounds, I've been peeping in every cranny of the house! Why, young Rule of Three!

(*Enter* MEDIUM *from the inn.*)

Oh, here you are at last. Ah, Sir Christopher! What, are you there? Too impatient, I see, to wait at home. But here's one that will make you easy, I fancy. (*clapping* INKLE *on the shoulder*)

SIR CHR: (*aside*) Old Medium! So here's the virtuous gentleman that never speaks

---

Rule of Three: an algebraic method of finding a fourth number from three given numbers if the first is in the same proportion to the second as the third is to the unknown fourth.

ill of his neighbours. And a pretty acquaintance he has picked up here to prove his forbearance upon. (*to* MEDIUM) How came you to know him?

MEDIUM: Ha! ha! Well, that's curious enough, too. So you have been talking here without finding out each other.

SIR CHR: No, no, I have found him out with a vengeance.

MEDIUM: Not you. Why, this is the dear boy; it's my nephew — that is, your son-in-law that is to be. It's Inkle!

SIR CHR: It's a lie! And you're a purblind old booby! And this dear boy is a damned scoundrel!

(INKLE *retires.*)

MEDIUM: Hey-day, what's the meaning of this? One was mad before, and he has bit the other, I suppose.

SIR CHR: But here comes the dear boy, the true boy, the jolly boy, piping hot from church, with my daughter.

(*Enter* CAMPLEY, NARCISSA *and* PATTY.)

MEDIUM: Campley!

INKLE: Campley? Our passenger in the ship?

SIR CHR: Who? Campley? It's no such thing!

CAMPLEY: That's my name indeed, Sir Christopher.

SIR CHR: The devil it is! And how came you, sir, to impose upon me and assume the name of Inkle, a name which every man of honesty ought to be ashamed of?

CAMPLEY: I never did, sir. Since I sailed from England with your daughter, my affection has daily increased, and, when I came to explain myself to you, by a number of concurring circumstances which I am now partly acquainted with, you mistook me for that gentleman. Yet, had I even then been aware of your mistake, I must confess the regard for my own happiness would have tempted me to let you remain undeceived.

SIR CHR: And did you, Narcissa, join in?

NAR: How could I, my dear sir, disobey you?

PATTY: Lord, your honour, what young lady could refuse a captain?

CAMPLEY: I am a soldier, Sir Christopher; 'love and war' is the soldier's motto. Though my income is trifling to your intended son-in-law's, still, the chance of war has enabled me to support the object of my love above indigence. Her fortune, Sir Christopher, I do not consider myself by any means entitled to.

SIR CHR: 'Sblood, but you must, though! Give me your hand, my young Mars, and bless you both together! Thank you, thank you for cheating an old fellow into giving his daughter to a lad of spirit when he was going to throw her away upon one in whose breast the mean passion of avarice smothers the smallest spark of affection or humanity.

INKLE: Damnation!

NAR: I have this moment heard a story of a transaction in the forest which, I own, would have rendered compliance with your former commands very disagreeable.

PATTY: Yes, sir, I told my mistress he had brought over a Hottypot gentlewoman.

SIR CHR: (*to* NARCISSA) Yes, but he would have left her for you and you for his interest, and sold you, perhaps, as he has this poor girl to me as a requital for preserving his life.

NAR: How?

(*Enter* TRUDGE *and* WOWSKI.)

TRUDGE: Come along, Wows! Take a long, last leave of your poor mistress. Throw your pretty ebony arms about her neck.

WOWSKI: No, no, she not go. You not leave poor Wowski. (*throwing her arms about* YARICO)

SIR CHR: Poor girl! A companion, I take it?

TRUDGE: A thing of my own, sir. I couldn't help following my master's example in the woods. Like master, like man, sir.

SIR CHR: But you would not sell her, and be hanged to you, you dog, would you?

TRUDGE: Hang me like a dog if I would, sir.

SIR CHR: So say I to every fellow that breaks an obligation due to the feelings of a man. But, old Medium, what have you to say for your hopeful nephew?

MEDIUM: I never speak ill of my friends, Sir Christopher.

SIR CHR: Pshaw!

INKLE: Then let me speak. Hear me defend a conduct —

SIR CHR: Defend? Zounds! Plead guilty at once; it's the only hope left of obtaining mercy.

INKLE: Suppose, old gentleman, you had a son —

SIR CHR: 'Sblood, then I'd make him an honest fellow and teach him that the feeling heart never knows greater pride than when it's employed in giving succour to the unfortunate. I'd teach him to be his father's own son to a hair.

INKLE: Even so my father tutored me from my infancy, bending my tender mind, like a young sapling, to his will. Interest was the grand prop round which he twined my pliant green affections, taught me in childhood to repeat old sayings — all tending to his own fixed principles — and the first sentence that I ever lisped was 'Charity begins at home.'

SIR CHR: I shall never like a proverb again, as long as I live.

INKLE: As I grew up, he'd prove — and by example: were I in want, I might e'en starve for what the world cared for their neighbours; why then should I care for the world? — men now lived for themselves. These were his doctrines. Then, sir, what would you say should I, in spite of habit, precept, education, fly in my father's face and spurn his counsels?

SIR CHR: Say? Why, that you were a damned honest, undutiful fellow! Oh, curse such principles, principles which destroy all confidence between man and man, principles which none but a rogue could instil and none but a rogue could imbibe, principles —

INKLE: Which I renounce.

SIR CHR: Eh?

INKLE: Renounce entirely. Ill-founded precept too long has steeled my breast, but still 'tis vulnerable. This trial was too much. Nature, 'gainst habit combating within me, has penetrated to my heart, a heart, I own, long callous to the feelings of sensibility. But now it bleeds, and bleeds for my poor Yarico. Oh, let me clasp her to it whilst 'tis glowing, and mingle tears of love and penitence. (*embracing her*)

TRUDGE: (*capering about*) Wows, give me a kiss!

(WOWSKI *goes to* TRUDGE.)

YARICO: And shall we, shall we be happy?

INKLE: Aye, ever, ever, Yarico.

YARICO: I knew we should. And yet I feared. But shall I still watch over you? Oh, love, you surely gave your Yarico such pain only to make her feel this happiness the greater!

WOWSKI: (*going to* YARICO) Oh, Wowski so happy! And yet I think I not glad, neither.

TRUDGE: Eh, Wows? How? Why not?

WOWSKI: 'Cause I can't help cry.

SIR CHR: Then, if that's the case, curse me if I think I'm very glad either. What the plague's the matter with my eyes? (*to* INKLE) Young man, your hand; I am now proud and happy to shake it.

MEDIUM: Well, Sir Christopher, what do you say to my hopeful nephew now?

SIR CHR: Say? Why, confound the fellow, I say, that is ungenerous enough to remember the bad account of a man who has virtue left in him to repent it. (*to* TRUDGE) As for you, my good fellow, I must, with your master's permission, employ you myself.

TRUDGE: Oh, rare! Bless your honour! Wows, you'll be a lady, you jade, to a Governor's factotum!

WOWSKI: Iss, I Lady Jacktotum!

SIR CHR: And now, my young folks, we'll drive home and celebrate the wedding. Od's my life, I long to be shaking a foot at the fiddles, and I shall dance ten times the lighter for reforming an Inkle while I have it in my power to reward the innocence of a Yarico.

Finale. INKLE, YARICO, TRUDGE, WOWSKI, CAMPLEY, NARCISSA, PATTY, SIR CHRISTOPHER.

SIR CHR: Hey for bells and cannonadoes,
        Triple bobs thro' all Barbadoes.

CAMPLEY: Come, let us dance and sing,
        While all Barbadoes bells shall ring:
        Love scrapes the fiddle string
           And Venus plays the lute;
        Hymen gay, foots away,
        Happy at our wedding day,
        Cocks his chin and figures in,
           To tabor, fife and flute.

CHORUS: Come then dance and sing,
        While all Barbadoes bells shall ring, &c.

NAR: Since thus each anxious care
        Is vanished into empty air,
        Ah, how can I forbear
           To join the jocund dance!
        To and fro couples go,

On the light fantastic toe,
While with glee, merrily,
   The rosy hours advance.

CHORUS: Come then, &c.

WOWSKI: Whilst all around rejoice,
Pipe and tabor raise the voice,
It can't be Wowski's choice,
   Whilst Trudge's, to be dumb.
No, no, I, blithe and gay,
Shall, like master, missy play,
Dance and sing, hey ding ding,
   Strike fiddle and beat drum.

CHORUS: Come then, &c.

PATTY: Let Patty but say a word,
A chambermaid may sure be heard:
Sure men are grown absurd,
   Thus taking black for white!
To hug and kiss a dingy miss
Will hardly suit an age like this,
Unless, here, some friends appear
   Who like this wedding night.

CHORUS: Come then, &c.

TRUDGE: 'Sbobs, now I'm fix'd for life!
My fortune's fair, tho' black's my wife;
Who fears domestic strife?
   Who cares now a souse!
Merry cheer my dingy dear
Shall find with her factotum here;
Night and day I'll frisk and play
   About the house with Wows.

CHORUS: Come then, &c.

YARICO: When first the swelling sea
Hither bore my love and me,
What then my fate would be
   Little did I think.
Doomed to know care and woe,
Happy still is Yarico,

---

who cares now a souse: souse = sou. Colman has simply spelt the word phonetically, according to its contemporary pronunciation. (cf. Churchill, *Rosciad*, 1761, p. 212: 'Next came the treasurer of either house; / One with full purse, t'other with not a sous.')

Since her love will constant prove
And nobly scorns to shrink.

CHORUS: Come then, &c.

INKLE: Love's convert here behold,
Banish'd now my thirst of gold,
Bless'd in these arms to fold
    My gentle Yarico.
Hence all care, doubt and fear,
Love and joy each want shall cheer,
Happy night, pure delight,
    Shall make our bosoms glow.

CHORUS: Come then, &c.

(*Exeunt omnes.*)

THE SURRENDER OF CALAIS

A play in three acts by George Colman the Younger

First performed at the Theatre Royal, Haymarket, Saturday, 30 July 1791, with the following cast:

KING EDWARD III	Mr Williamson
HARCOURT	Mr Bland
SIR WALTER MANNY	Mr Usher
ARUNDEL	Mr Powell
WARWICK	Mr Nigh
JOHN DE VIENNE	Mr Aickin
RIBAUMONT	Mr Farren
EUSTACHE DE ST PIERRE	Mr Bensley
LA GLOIRE	Mr Bannister junior
O'CARROL	Mr Johnstone
JOHN D'AIRE	Mr Evatt
PIERRE WISSANT	Mr Henderson
JACQUE WISSANT	Mr Knights
OFFICER	Mr Iliff
SERGEANT	Mr Wilson
CRIER	Mr Rock
OLD MAN	Mr Chapman
CITIZENS	Mr Wewitzer, Mr Abbott and Mr Barrett
WORKMEN	Mr Parsons and Mr Burton
QUEEN PHILIPPA	Mrs Goodall
JULIA	Mrs Stephen Kemble
MADELON	Mrs Bland
NUNS	Miss Fontenelle, Miss De Camp, Mrs Edwin and Mrs Powell
ATTENDANTS	Mrs Taylor, Miss Fontenelle, Miss De Camp and Mrs Powell

Friars, English soldiers, French soldiers, train-bearers, heralds, etc.

Music by Dr Samuel Arnold

The Surrender of Calais: Calais surrendered, after an eleven-month siege, on the 4 August 1347. In the wake of his crushing defeat at Crécy Philip VI offered no attempt to relieve the town. Froissart's *Chronicles* have provided Colman with all the historical details, including Edward III's terms and Queen Philippa's timely intervention, which form the basis of the present play. Froissart would also seem to have suggested a starting-point for the playwright. In his *Chronicles* he admits that his record is deficient with regard to the names of the six who elected to suffer Edward's penalty on behalf of the rest of the Calais community. Eustache de St Pierre, John d'Aire and the two Wissant brothers are on Froissart's list; the identity and circumstances of the missing two have been invented by Colman himself and supply the central interest of his theatrical rendering of *The Surrender of Calais*.

MESS.ᴿˢ PARSONS & BURTON *in the* SURRENDER *of* CALAIS.

IV   William Parsons in the Scaffold-Makers' Scene from *The Surrender of Calais.*

ACT I

Scene 1. *A view of Calais, the sea and the English camp. The scene becomes gradually lighter towards the conclusion of the dialogue to mark the approach of morning.*

(*Enter* RIBAUMONT *and* LA GLOIRE.)

RIBAUMONT: Thus far in safety. All is hush. Our subtle air of France quickens not the temperament of the enemy. These phlegmatic English snore out the night in as gross heaviness as when their senses stagnate in their own native fogs, where stupor lies like lead upon them which the muddy rogues call sleep. We have nearly passed the entrenchments. The day breaks. La Gloire?

LA GLOIRE: My commander!

RIBAUMONT: Our enterprise will be successful, I cannot see a doubt on't.

LA GLOIRE: That's because we have undertaken it in the dark, commander!

RIBAUMONT: Where did you direct our mariners to meet us with the boat?

LA GLOIRE: Marry, I told 'em to meet us with the boat at the seashore.

RIBAUMONT: Vague booby! At what point?

LA GLOIRE: That's the point I was coming to, my lord. And, if a certain jutting out of land in the shape of a white cliff with brown furze on its top, like a bushy head of hair over a pale face, stand where it did —

RIBAUMONT: East of the town. I have marked it.

LA GLOIRE: Look you there, now: what I have hunted after a whole day to fix upon, hath he noted without labour. Oh, the capacious heads of your great officers! No wonder they are so careful of 'em in battle and thrust forward the pitiful pates of the privates to be mowed off like a parcel of daisies. — But there lies the spot, and there will the mariners come. We are now within earshot, and, when they are there, they will whistle.

RIBAUMONT: And till they give the signal, here, if there be aught of safety to be picked from danger, is the least dangerous spot to tarry for them. We are here full early.

LA GLOIRE: I would we were not here at all! This same scheme of victualling a town blockaded by the enemy is a service for which I have little appetite.

RIBAUMONT: Think, La Gloire, on the distress of our countrymen, the inhabitants perishing with hunger —

LA GLOIRE: Truly, my lord, it doth move the bowels of my compassion! Yet, consider your risk, consider your rank: the gallant Count Ribaumont, flower of chivalry, cream of the French army and commander of his regiment, turned cook to the corporation of Calais, carving his way to glory through stubble-rumped capons, unskinned mutton, raw veal and vegetables! And perhaps, my lord, just before we are able to serve up the meat to the town, in comes a raw-boned Englishman and runs his spit through your body!

RIBAUMONT: Prithee, no more objections.

LA GLOIRE: Nay, I object not, I. But I have served your honour, in and out of the army, babe, boy and man, these five-and-twenty years come the next feast of the Virgin, and Heaven forfend I should be out of service by being out of my master!

RIBAUMONT: Well, well, I know thy zeal.

LA GLOIRE: And yet your English rapier is a marvellous sudden dissolver of attachments; 'twill sever the closest connections. 'Twill even whip you, forever, friend head from his intimate acquaintance, neck and shoulders, before they have time to take leave! Not that I object. Yet men do not always sleep. The fat sentinel, as we passed the outpost, might have waked with his own snoring and –

RIBAUMONT: Peace! Remember your duty to me, to your country!
Yet out, alas, I mock myself to name it!
Did not these rugged battlements of Calais,
This tomb, yet safeguard, of its citizens,
Which shuts the sword out and locks hunger in,
Where many a wretch, pale, gaunt and famine-shrunk,
Smiles ghastly at the slaughterer's threat, and dies,
Did not these walls, like Vulcan's swarthy arms
Clasping sweet beauty's queen, encircle now
Within their cold and ponderous embrace
The fair – yet, ah, I fear, the fickle! – Julia,
My sluggish zeal would lack the spur to rouse it.

LA GLOIRE: And, of all the spurs in the race of mortality, love is the only true tickler to a man's motions. But to reconcile a mistress by victualling a town! Well, dark and puzzling is the road to woman's affection. But this is not the first time I ever heard of sliding into her heart through her palate, or choking her anger by stopping her mouth with a meal. An this pantry fashion of wooing should last, woe to the ill-favoured; beauty will raise the price of provisions and poor ugliness soon be starved out of the country!

RIBAUMONT: This enterprise may yet regain her.
Once she was kind, until her father's policy,
Nourished in courts, stepped in and checked her love.
Yet 'twas not love, for true love knows no check;
There is no skill in Cupid's archery
When duty heals a love-wound. Why, why, Julia,
Why did you blush and own to me your love
When never loving had more proved your love
Than once to give and then withdraw your love?
Alas, she sighs not now for her poor Ribaumont!

LA GLOIRE: No, truly. – (*aside*) Nor is she likely to while he provides her such rich entertainments. True lovers' groans are best nourished by fasting. They never sigh when a good dinner has taken the wind off their stomachs! – But, dear my lord, think on the great danger and little reputation –

RIBAUMONT: No more! Mark me, La Gloire, as your officer I may command you onward, but, in respect to your early attachment, your faithful service ere

---

Vulcan's swarthy arms: Vulcan was the Roman god of metal-working. The expression 'Vulcan's badge' for a cuckold's horns derives from his capacity to be deceived by Venus, his wife, a condition with which Ribaumont seems eager to identify.

An: archaic weakened form of 'and' used to convey the sense of 'if' and included here in imitation of the Shakespearean idiom.

you followed me to the army, if your mind misgive you in this undertaking, you have my leave to retreat.

LA GLOIRE: (*amazed*) My lord!

RIBAUMONT: I say you are free to return.

LA GLOIRE: (*after a pause*) Look ye, my lord, I am son to brave old Eustache de St Pierre, as tough a citizen as any in all Calais. I was carried into your lordship's father's family, your lordship being then but just born, at six days old, a mere whelp as a body may say. According to puppy reckoning, my lord, I was with you three days before I could see. I have followed you through life, frisking and trotting after your lordship, ever since, and, if you think me now mongrel enough to turn tail and leave my master in a scrape, why, 'twere kinder e'en to hang me up at the next tree than cut me through the heart with your suspicions!

RIBAUMONT: No, La Gloire, I –

LA GLOIRE: No, my lord! – 'Tis fear for you that makes me bold to speak. To see you running your head through stone walls for a woman, and a woman who, though she be an angel, has – saving your presence – played you but a scurvy sort of a jade's trick and –

RIBAUMONT: 'Sdeath, villain, how dare your slanderous tongue to – ! But 'tis plain 'tis for thy own wretched sake thou art thus anxious, drivelling coward!

LA GLOIRE: Coward! Cow – ! Diable! A French soldier, who has the honour to carry arms under his Christian majesty, Philip the Sixth, King of France, called coward? Sacre bleu! Have I already served in three campaigns, and been thumped and bobbed about by the English, to be called coward at last? Oh, that any but my commander had said it!

RIBAUMONT: Well, well, La Gloire, I may have been hasty. I –

LA GLOIRE: Oh, my lord, it – is – no matter. But, haply, you'd like to be convinced of the courage of your company. And, if such a thing as raising the enemy's camp can clear a man's character (*raising his voice*), I can do it as soon as –

RIBAUMONT: 'Sdeath, blockhead, we shall be discovered!

LA GLOIRE: (*still louder*) Coward? 'Sblood, I'll run into the English entrenchments! I'll go back and tweak the fat sentinel by the nose! I'll –

RIBAUMONT: Peace, I command you, La Gloire! I command you as your officer!

LA GLOIRE: (*sulkily*) I know my duty to my officer, my lord.

RIBAUMONT: Then move not. (*pointing forward*) Here, sir, on this spot.

LA GLOIRE: (*going to the spot*) Coward!

RIBAUMONT: Speak not, for your life!

LA GLOIRE: Cow – ! Umph.

RIBAUMONT: Obey!

(LA GLOIRE *stands motionless and silent. A low whistle is heard.*)
Ha, the signal! The morning breaks; they arrive in the very nick. Now then, La Gloire, for the enterprise! – Why does not the blockhead stir? – Well, well, my good fellow, I have been harsh, but – Not yet? Pshaw, this military enforcement has acted like a spell upon him! How to dissolve it?
(*Another low whistle is heard.*)

Again? – Come, come, La Gloire, I – Dull dolt! – I have it! (*addressing* LA GLOIRE *in a military bark*) March!

> (LA GLOIRE *faces to the left and marches out after* RIBAUMONT.)

Scene 2. *The Place in the town of Calais.* FRENCH SOLDIERS *drawn out as finishing the morning parades.*

> (*Enter an* OFFICER *and* SERGEANT. CITIZENS *enter severally during the scene.*)

OFFICER: Bravely, good fellows! Courage! Why, there's still life in't! Sergeant.

SERGEANT: Your honour!

OFFICER: How do the men bear up? Have they stout hearts still?

SERGEANT: I know not, sir, for their hearts, but I'll warrant them stout stomachs. Hunger is so powerful in them that I fear me they'll munch their way through the stone walls of the city.

OFFICER: This famine pinches, poor rogues! Cheer them with hopes, good sergeant.

SERGEANT: Hope, your honour, is but a meagre mess for a regiment. Hope has almost shrunk them out of their doublets. Hope has made their legs so weary of the lease they had taken of their hose that all their calves have slunk away from the premises. There isn't a stocking in the whole company that can boast of a tolerable tenant. The privates join in the public complaining, the drummers grow noisy, our poor corporal has no body left, and the trumpeter's blown up with the wind.

OFFICER: Do they grow mutinous? Look to them. Check their muttering.

SERGEANT: Troth, sir, I do my best. When they grumble for meat, I make them eat their own words and give them some solid counsel, well seasoned with the pepper of correction.

OFFICER: Well, well, look to them, keep a strict watch, and march the guards to their several posts.

> (*The parade is dismissed according to military form. Exit* OFFICER.
> *The* SERGEANT *comes forward with the* SOLDIERS.)

SERGEANT: Now must I administer consolation and give the rogues their daily meal of encouragement. – Hem! Countrymen, fellow soldiers and Frenchmen, be of good cheer, for famine is come upon you and you are all in danger of starving. Is there anything dearer to a Frenchman than his honour? Isn't honour the greater, the greater the danger? And has anybody ever had the honour of being in greater danger than you? Rejoice, then, for your peril is extreme! Be merry, for you have a glorious dismal prospect before you and as pleasing a state of desperation as the noble heart of a soldier could wish! Come, one cheer for the glory of France, St Dennis and our grand monarch, King Philip the Sixth!

> (*The* SOLDIERS *huzza very feebly.*)

Oons, it sounds as hollow as a churchyard! The voice comes through their wizen mouths like wind from the crack of an old wainscot. Away, rogues, to

---

Saint Dennis: i.e. St Denys, patron saint of France.

your posts. Bristle up your courage and await the event of time! Remember ye are Frenchmen, and bid defiance to famine! Our mistresses are locked up with us in the town; we have frogs in the wells and snuff at the merchants'. An Englishman, now, would hang himself upon this, which is enough to make a gay Frenchman happy. Allons, comrades!

Song. SERGEANT.

SERGEANT: My comrades, so famished and queer
         Hear the drums, how they jollily beat!
      They fill our French hearts with good cheer,
         Although we have nothing to eat.
           Rub a dub.

ALL: Nothing to eat, rub a dub,
      Rub a dub, we have nothing to eat.

SERGEANT: Then hark to the merry-toned fife!
         To hear it will make a man younger.
      I tell you, my lads, this is life
         For anyone dying of hunger.
           Toot a toot.

ALL: Dying with hunger, toot a toot,
      Toot a toot, we are dying with hunger.

SERGEANT: The foe, to inspire you to beat,
         Only list to the trumpet so shrill!
      Till the enemy's killed we can't eat;
         Do the job, you may eat all you kill!
           Ran ta tan.

ALL: We'll eat all we kill, ran ta tan,
      Ran ta tan, we may eat all we kill.

(*Exeunt* SERGEANT *and* SOLDIERS. CITIZENS *come forward.*)

1st CIT: Bonjour, Monsieur Grenouille! Good-day, neighbour!

2nd CIT: Aha, mon voisin! Here's a goodly morning. The sun shines till our blood dances to it like a frisky wench to a tabor.

1st CIT: Yes, truly. But 'tis a dance without refreshments! We are in a miserable plight, neighbour.

2nd CIT: Ma foi, miserable indeed! Mais le soleil —

1st CIT: How fare your wife and family, neighbour Grenouille?

2nd CIT: Ah, my pauvre wife and family! Little to eat now, mon voisin; nothing by-and-by. Lucky for me 'tis fine weather. Great many mouths in my house, very little to put into 'em. But I am French: the sun shines, I am gay. There is myself, my poor, dear wife, half a loaf, seven children, three sprats, a tom-cat and a pipkin of milk. I am hungry, mais il fait beau temps. I dance, my family starves. I sing, toujours gai, the sun shines — tal lal la! Tal lal la!

3rd CIT: Tut, we won't bear it. 'Tis our Governor is in fault. This way we are certain to perish. Why doesn't he surrender to our besiegers that they may either cut our throats at once, or fill our stomachs?

4th CIT: Peste, we'll not endure it! Shut up near eleven months within the walls —

2nd CIT: In fine weather. No promenade! —

3rd CIT: No provisions! — The citizens sacrificed that the city may be saved. Men falling that houses may stand. We'll to the Governor, force the keys, and surrender the town. Allons. Come along, neighbours, to the Governor!

ALL: Aye, aye, to the Governor! Away!

> (*The* CITIZENS, *going in a posse, are met by* EUSTACHE DE ST PIERRE *as he enters. He is carrying a small wallet.*)

EUSTACHE: Why, how now, ho? Nothing but noise and babble?
Whither away so fast? Stand, rogues, and speak!

3rd CIT: Whither away? Marry, we would wither away — from famine! We are for the Governor's, to force the keys of the town.

EUSTACHE: There roared the wrathful mouse! You squeaking braggart,
Whom hunger has made vent'rous, who would thrust
Your starveling nose out to the cat's fell gripe
That watches round the cranny you lie snug in,
Nibble your scraps! Be thankful and keep quiet!
Thou rail on hunger? Why, 'twas hunger bore thee,
'Twas hunger reared thee, fixing in thy cradle
Her meagre stamp upon thy weasel visage;
And, from a child, that half-starved face of thine
Has given full meals the lie. When thou dost eat,
Thou dost digest consumption. Thou'rt of those kine
That wouldst e'en swallow up thy brethren, here,
And still look lean. What, fellow citizens,
Trust you this thing? Can skin and bones mislead you?
If we must suffer, suffer patiently.
Did I e'er grumble, mongrels? What am I?

3rd CIT: You? Why, Eustache de St Pierre are you, one of the sourest old crabs of all the citizens of Calais. And, if reviling your neighbours be a sign of ill-will to one's country and ill-will to one's country a sign of good-will to strangers, why, a man might go near to think you are a friend to the English.

EUSTACHE: I honour them.
They are our enemy, a gallant enemy,
A biting but a blunt, straightforward foe,
Who, when we weave our subtle webs of state
And spin fine stratagems to entangle them,
Come to our doors and pull the work to pieces,
Dispute it, fist to fist, and score their arguments
Upon our politic pates. Remember Crécy!

Crécy: this was the battle that made a hero of Edward III. Initiated by the French, whose army possessed every advantage, including a belief in its own infallibility, the English defending force, numerically inferior and in low spirits, should have lost the Battle of Crécy. That they emerged the victors was largely the product of Philip VI's ineptitude.

We've reason to remember it: they thumped us,
And soundly, there, 'tis but some few months back.
There, in the bowels of our land, at Crécy,
They so bechopped us with their English logic
That our French heads ached sorely for it. Thence,
Marching through Picardy to Calais here,
They have engirded us, fixed the dull tourniquet
Of war upon our town, constraining thus
The life-blood of our commerce with fair France,
Of whom we are a limb, and all this openly.
And therefore, as an open foe, who think
And strike in the same breath, I do esteem
Their valour and their plainness, and were I,
As once I was — my son now fills my place —
A young French soldier, I would do my best
To send their valour to the devil for it!
I view them with a most respectful hatred.
Much may be learnt from these same Englishmen —

4th CIT: Aye? Prithee, what? Hunger and hard blows seem all we are likely to get
from them.

EUSTACHE: Courage — which you may have, 'twas never tried, tho' —
Patience to bear the buffets of the times.
Ye cannot wait till Fortune turns her wheel,
Ye'll to the Governor's and get the keys!
And what would your wise worships do with 'em?
Eat them, mayhap? For ye have ostrich stomachs!
Ye dare not use them otherwise. — Home, home,
And pray for better luck!

> (*The* CITIZENS *exeunt severally. An* OLD MAN, *alone, remains in
> the back of the scene.*)
>   Fie, I am faint

With railing on these cormorants! Three days
And not break bread? 'Tis somewhat. There's not one
Among these trencher-scraping knaves that yet
Has kept a twenty hours' lent, I know it.
Yet how they crave! I've here, by strong entreaty
And a round sum — entreaty's weak without it —
E'en just enough to make Dame Nature wrestle
Another round with famine. (*searching his wallet*) Out, provision!

OLD MAN: (*coming forward*) Oh, Heaven!

---

ostrich stomachs: cf. 'I'll make thee eat iron, like an ostrich', Shakespeare, *Henry VI Pt 2*, IV,
10. Frequently seen to eat non-digestible material (as an aid to the action of the gizzard),
ostriches were thought to be quite literally omnivorous.
cormorants: the name of this voracious sea-bird was often transferred figuratively, as here, to
denote a glutton.

EUSTACHE: Who bid thee bless the meat?
  How now, old greybeard, what cause hast thou —
OLD MAN: I have a daughter —
EUSTACHE: Hungry, I warrant.
OLD MAN: Dying!
  The blessing of my age! I could bear all,
  But for my child, my dear, dear child, to lose her,
  To lose her thus, to see disease so wear her,
  And when a little nourishment — She's starving!
EUSTACHE: Go on. No tears; I hate them!
OLD MAN: She has had no nourishment these four days.
EUSTACHE: (*affected*) Death and — ! Well?
OLD MAN: I care not for myself, I should soon go
  In Nature's course, but my poor darling child,
  Who fifteen years has been my prop, to see her
  Thus wrested from me, then to hear her bless me
  And see her wasting! —
EUSTACHE: Peace! Peace!
  I have not ate, old man, since — Pshaw, the wind
  Affects my eyes! — but yet, I — 'Sdeath, what ails me?
  I have no appetite. Here, take this trash, and —
      (*The* OLD MAN *takes the wallet, falls upon his knees and attempts*
      . *to speak.*)
  Prithee, away, old soul — nay, nay, no thanks! —
  Get home and do not talk; I cannot.
      (*Exit* OLD MAN.)
                        Out on't,
  I do belie my manhood, and if misery,
  With gentle hand, touches my bosom's key,
  I bellow straight, as if my tough old lungs
  Were made of organ pipes! Well, thank Nature still!
  I envy not that sturdy composition
  — I will not call it man — which has not known
  What 'tis, by listening to another's grief,
  To spoil a breakfast.
      (*A huzza is heard without.*)
                Hey, how sits
  The wind now?
      (*Enter* CITIZENS *crying* 'Huzza!' *and* 'Succour!', LA GLOIRE *in*
      the midst of them, loaded with casks of provision, &c.)
LA GLOIRE: Here, neighbours, here! Here I am dropped in among you, like a lump
  of manna. Here have I, following my master, the noble Count Ribaumont,
  brought wherewithal to check the grumbling in your gizzards. Here's meat,
  neighbours, meat — fine, raw, red meat! — to turn the tide of tears from your
  eyes and make your mouths water.
ALL: Huzza!
2nd CIT: Ah, mon Dieu, que je suis gai! Meat, and sun too! Tal lal lal la!

LA GLOIRE: Silence, or I'll stop your windpipe with a mutton cutlet!

ALL: Huzza!

EUSTACHE: Peace ho, I say! Can ye be men and roar thus?
　　　Blush at this clamour! It proclaims you cowards
　　　And tells what your despair has been. Peace, hen-hearts!
　　　Slink home and eat!

LA GLOIRE: Od's my life! Cry you mercy, father, I saw you not! My honest,
　　　hungry neighbours, here, so pressed about me — Marry, I think they are ready
　　　to eat me! Stand aside, friends, and patience till my father has said grace over
　　　me. Father, your blessing. (*kneeling*)

EUSTACHE: Boy, thou hast acted bravely, and thou follow'st
　　　A noble gentleman. What succour brings he?

LA GLOIRE: A snack, a bare snack, father, no more. We scudded round the point
　　　of land under the coast, unperceived by the enemy's fleet and freighted with
　　　a good three days' provender, but the sea, that seems ruled by the English, —
　　　marry, I think they'll always be masters of it, for my part! — stuck the point
　　　of a rock through the bottom of our vessel, almost filled it with water, and,
　　　after tugging hard for our lives, we found the provision so spoiled and pickled
　　　that our larder is reduced to a luncheon. Every man may have a meal, and
　　　there's an end. Tomorrow comes famine again.

2nd CIT: N'importe. We are happy today, c'est assez pour un français.

LA GLOIRE: (*aside to* EUSTACHE) But, father, cheer up! Mum, if, after the dis-
　　　tribution, an odd sly barrel of mine — you take me? — rammed down with good
　　　powdered beef that will stand the working of half a dozen pairs of jaws for a
　　　month should be found in an odd corner of my father's house, why — hum? —

EUSTACHE: Base cur! Insult me? — But I pardon thee;
　　　Thou dost mean kindly. Know thy father better! —
　　　Though these be sorry knaves, I scorn to wrong them.
　　　I love my country, boy. Ungraced by fortune,
　　　I dare aspire to the proud name of patriot.
　　　If any bear that title to misuse it,
　　　Decking their devilships in angel-seeming
　　　To glut their own particular appetites,
　　　If any, 'midst a people's misery,
　　　Feed fat by filching from the public good,
　　　Which they profess is nearest to their hearts,
　　　The curses of their country or, what's sharper,
　　　The curse of guilty conscience, follow them!
　　　The suffering's general; general be the benefit.
　　　We'll share alike. — You'll find me, boy, at home.
　　　　　　(*Exit* EUSTACHE.)

LA GLOIRE: There he goes, full of sour goodness, like a lemon! He's as trusty a
　　　crusty citizen and as good-natured an ill-tempered old fellow as any in
　　　France! But now, neighbours, for provision.

3rd CIT: Aye, marry, we would fain fall to!

LA GLOIRE: I doubt it not, good, hungry neighbours. You'll all remember me for
　　　this succour, I warrant.

ALL: Toujours! Always!

LA GLOIRE: (*aside*) See, now, what it is to bind one's country to one by doing it a
service. Good souls, they are running over with gratitude. (*walking about,*
CITIZENS *following*) I could cluck them all round the town after my tail like
an old hen with a brood of chickens. — (*to* CITIZENS) Now will I be carried
in triumph to my father's, and ye may e'en set about it now —

>   (*Two stout* CITIZENS *take* LA GLOIRE *on their shoulders.*)

— now, while the provisions are sharing at the Governor's house.

ALL: Sharing provisions?

>   (*The* CITIZENS *let* LA GLOIRE *fall.*)

Allons vite! Away! Away!

>   (*Exeunt* CITIZENS *hastily.*)

LA GLOIRE: Oh diable! This is popularity! Adieu, my grateful neighbours! Thus
does many a foolhardy booby, like me, run his head into danger, and a few
empty huzzas, which leave him at the next turning of a corner, are all he gets
for his pains. Now, while all the town is gone to dinner, will I go to woo. My
poor Madelon must be woefully fallen away since I quitted Calais. Heigho!
I've lost, I warrant me, a good half of my mistress since we parted. All the
while we were preparing the succour did my heart ache for her: every morsel
of meat we packed brought tears to my eyes, every crammed fowl that was
stowed did I wish smoking under her nose. I have secured for her the daintiest
bits of our whole cargo as marks of my affection. A butcher couldn't show
her more tenderness than I shall. If love were now weighed out by the pound,
bating my master, the Count Ribaumont, who is in love with Lady Julia, not
all the men in the city could balance the scales with me.

>   (*Exit* LA GLOIRE.)

Scene 3. *A hall in the house of* JOHN DE VIENNE.

>   (*Enter* JULIA *and* O'CARROL.)

JULIA: Now, O'Carrol, what is the time of day?

O'CARROL: Time? Faith, Lady Julia, we might have called it a little past breakfast
time formerly, but since the fashion of eating has been worn out in Calais, a
man may be content to say it bears hard upon ten. Och, if clocks were jacks,
now, time would stand still and the year would go down for the want of
winding up every now and then!

JULIA: Saw you my father, this morning?

O'CARROL: You may say that.

JULIA: How looked he, O'Carrol?

O'CARROL: By my soul, Lady Julia, that old father of yours and master of mine is
a gallant gentleman, and gallantly he bears himself, for certain. And so he
ought, being a knight of Burgundy and Governor of Calais! But if I was

---

if clocks were jacks: a reference to the roasting-jack, a contrivance which, when wound, could
turn meat on a spit. O'Carrol is saying that it is fortunate for time that the lack of winding now
being given to roasting-jacks has not been reflected in the amount of similar attention given to
clocks.

Governor just now, to be sure, I should not like to take a small trip from Calais, one morning, just to see what sort of a knight I was in Burgundy.

JULIA: Who has he in his company?

O'CARROL: Why, madam, why — (*aside*) Now dare I not tell who, for fear of offending her. — Company? Why, to be sure, I have been in his company. For want of finer acquaintance, madam, he was e'en forced to put up, half an hour, with a humble friend.

JULIA: Poor fool, thy words are shrewder than thy meaning.
How many crowd the narrow space of life
With those gay, gaudy flowers of society,
Those annuals called acquaintance, which do fade
And die away ere we can say they blossom,
Mocking the idle cultivator's care
From year to year, while one poor slip of friendship,
Hardy, tho' modest, stands the winter's frost
And cheers its owner's eye with evergreen!

O'CARROL: Troth, lady, one honest potato in a garden is worth a hundred beds of your good-for-nothing tulips. Oh, 'tis meat and drink to me to see a friend, and, truly, 'tis lucky, in this time of famine, to have one in the house to look at, to keep me from starving! Little did I think, eight years ago, when I came over among fifty thousand brave boys, English, Irish and else, to fight under King Edward, who now lies before Calais here, that I should find such a warm soul towards me in a Frenchman's body, especially when the business that brought me was to help give his countrymen a beating.

JULIA: Thy gratitude, O'Carrol, has well repaid the pains my father took in preserving thee.

O'CARROL: Gratitude? Faith, madam, begging your pardon, 'tis no such thing. 'Tis nothing but showing the sense I have of my obligation. There was I, in the year 1339, in the English camp on the fields of Vianfosse, near Capelle, which never came to an action, excepting a trifling bit of skirmish in which my good cruel friends left me for dead out of our lines, when a kind enemy, your father, — a blessing on his friendly heart for it! — picked me up and set the breath going again that was almost thumped out of my body. He saved my life; it is but a poor commodity but, as long as it lasts, by my soul, he and his family shall have the wear and tear of it!

JULIA: Thou hast been a trusty follower, O'Carrol — nay, more a friend than a follower. Thou art entwined in all the interests of our house and art as attached to me as to my father.

O'CARROL: Aye, troth, Lady Julia, and a good deal more, more shame to me for it, because I am indebted for all to the Governor. I don't know how it may be with wiser nations but, if regard is to go to a whole family, there's a something about the female part of it that an Irishman can't help giving the preference to, for the soul of him!

---

one honest potato: the comparison also conveniently includes an Irish reference.
in the year 1339: which saw the first English acts of armed aggression in the Anglo-French dynastic struggle later to be known as the Hundred Years War.

JULIA: But tell me, who is with my father?

O'CARROL: Indeed, that I will not, for a reason —

JULIA: And what may the reason be?

O'CARROL: Truly, the fear of disobeying your bidding.

JULIA: What, when I bid thee tell me!

O'CARROL: Aye, but long before he arrived you bid me never mention his name! It may be, perhaps, the noble gentleman who has just succoured the town. Well, if I must not say who is with my master, I may say who my master is with: it is the Count Ribaumont.

JULIA: Why should I tremble at that name, when
The harshest voice that uttered it seemed sweet
As, through the still and dew-bespangled wood,
The liquid plainings of the nightingale?
Why should my tongue be now constrained to speak
The language of my heart? Oh father, father!

O'CARROL: (*sighing*) Och, ho!

JULIA: Why dost thou sigh, O'Carrol?

O'CARROL: Truly, madam, I was thinking of a piece of a rich old uncle I had in Ireland, who sent me to the French wars to tear me away from a dear little creature I loved better than my eyes.

JULIA: And wast thou ever in love, O'Carrol?

O'CARROL: That I was, faith. Up to my chin! I never think upon it but it remembers me of the song that was wont to be played by honest Clamoran, poor fellow, our minstrel in the north.

JULIA: I prithee, sing it to me, good O'Carrol,
For there is something in these artless ditties,
Expressive of a simple soul in love,
That fills the mind with pleasing melancholy.

Song. O'CARROL.

Oh, the moment was sad when my love and I parted,
    Savourna delizh shighan ogh!
As I kiss'd off her tears, I was nigh broken-hearted,
    Savourna delizh shighan ogh!
Wan was her cheek which hung on my shoulder,
Damp was her hand, no marble was colder;
I felt that I never again should behold her,
    Savourna delizh shighan ogh!

Long I fought for my country, far, far from my true love,
    Savourna delizh shighan ogh!
All my pay and my booty I hoarded for you, love,

Savourna delizh shighan ogh!: 'The Author had this burden from an Irish friend; — but he cannot vouch for its orthography.' Footnote from the first edition (Longman, Hurst, Rees and Orme, 1808) of *The Surrender of Calais*. The Irish friend is likely to have been Jack Johnstone, the singer and comedian, who was also the original performer of O'Carrol.

Savourna deligh shighan ogh!
Peace was proclaimed; escaped from the slaughter,
Landed at home, my sweet girl, I sought her,
But sorrow, alas, to the cold grave had brought her,
Savourna deligh shighan ogh!

(*Enter* JOHN DE VIENNE *and* RIBAUMONT.)

DE VIENNE: Pish, pish, my lord, sans compliment, you're welcome!
You have brought wherewithal to make you so!
Think my poor house your own! Though, to be plain,
Were I more private here, some prudent qualms,
Which you well wot, I trow, my noble lord,
Might cause me flatly sound that full-toned welcome
Which breathes the mellow note of hospitality.
Yet, being Governor of Calais, here,
My duty and your service both considered —
But take me with you, Count. I can discern
Your noble virtues, aye, and love them, too,
Did not a father's care — But let that pass.
Julia, my girl, the Count of Ribaumont:
Thank the brave champion of our city.

JULIA: Sir,
Though one poor simple drop of gratitude
Amid the boisterous tide of general thanks
Can little swell the glory of your enterprise,
Accept it freely. You are welcome, sir.

RIBAUMONT: (*aside*) Cold does it seem to me. 'Sdeath, this is ice,
Freezing indifference! Down, down, my heart! —
I pray you, lady, do not strain your courtesy.
I am a soldier and have merely done
A soldier's duty; yet, in doing it,
If I have reaped a single grain of favour
From your fair self and noble father here,
I have obtained the harvest of my hope.

DE VIENNE: Hey-day, here's bow and jut and cringe and scrape
And poor tormented language so be-frittered
That a plain man, in good straightforward speaking,
Might journey on to many a meaning's end
Ere he could thread the thicket of one compliment
In this same zigzag phraseology!
Count, I have served in battle; witness for me
Some curious scars, the soldier's coxcombry,
In which he struts, fantastically carved,

---

you well wot, I trow: a rather prodigal helping of archaism which can be deciphered according
to the formula *wot* = know if *trow* = trust.

Upon the tough old doublet nature gave him.
Let us then speak like brothers of the field,
Roundly and blunt. Have I your leave, my lord?
RIBAUMONT: As freely, sir, as you have asked it.
DE VIENNE: Thus, then:
I have a daughter — look you, there she stands —
Right fair and virtuous —
            (RIBAUMONT *attempts to speak*.)
                              Nay, Count, spare your speech;
I know I've your assent to the position.
I have a king, too, to whose will we French
Do make our sovereign purpose act the weathercock,
Still pointing toward that quarter whence the breath
Of sovereignty blows. It blows of late,
And I am honoured in't, towards my poor house.
My daughter must be matched with, so 'tis signified,
A certain lord about the royal person.
Now, though there may be some whose gallant bearing —
And glean from this, Count, what it is I aim at —
I might be proud to be allied to, yet,
Being a veteran French soldier stuffed
With right enthusiastic loyalty,
My house, myself, my child — Heaven knows I love her! —
Should perish piecemeal ere I could infringe
The faintest line or trace of the proceeding
The king, our master, honours me in marking.
RIBAUMONT: I do conceive you, sir.
DE VIENNE: Why then, conceiving,
Once more right welcome, Count! I lodge you here
As my good friend, and Julia's friend, the friend
To all our city. You have done us service;
We'll in and talk upon't, and we'll deliberate
How next to act in good, stout, manly freedom,
Unmixed with froth of — Tut, Count, love is boy's play;
A soldier has not time for't, has he, Julia?
See there, the girl knows well enough what's right,
For, to whatsoe'er I say, she does assent!
Come, Count. — (*calling*) Within there, ho, we need refreshment!
(*to* RIBAUMONT) Which you have furnished. — Love? Pish, love's a gewgaw!
Nay, come, Count, come.
            (*Exit* DE VIENNE.)
JULIA: Sir, will it please you follow?
RIBAUMONT: (*aside*) I fain would speak one word, and, 'sdeath, I cannot! —
Pardon me, madam. I attend. — (*aside*) Oh, Julia!
            (*Exit* RIBAUMONT *leading out* JULIA.)
O'CARROL: Och, ho! Poor, dear creatures, my heart bleeds for them! To be sure,
the ould gentleman means all for the best, and what he talks must be right,

but, if love is a gewgaw, as he says, by my soul, 'tis the prettiest plaything for children from sixteen to five-and-twenty that ever was invented!
(*Exit* O'CARROL.)

Scene 4. *The English camp before Calais.*

(*A flourish is heard. Enter* KING EDWARD THE THIRD, SIR WALTER MANNY, HARCOURT, ARUNDEL, WARWICK *and* ATTENDANTS.)

KING: Fie, lords, it slurs our name! The town is succoured!
  'Twas dull neglect to let them pass, a blot
  Upon our English camp, where vigilance
  Should be the watchword. Which way got they in?
SIR WALTER: By sea, as we do learn, my gracious liege.
KING: Where was our fleet, then? Does it ride the ocean
  In idle mockery? It should float to awe
  These Frenchmen here, but, like a log i' the marsh,
  Thrown in to terrify the croaking reptiles,
  It lies a lump, and each frog, grown familiar,
  Hops over it in sport. How are they stored?
HARCOURT: Barely, as it should seem. Their crazy vessel,
  Driven among the rocks that skirt the shore,
  Let in the waves so fast upon the cargo
  The better half is either sunk or spoilt.
  They scarce can hold another day, my liege.
KING: Thanks to the sea for't, not our admiral. —
  They brave it stubborn to the very last,
  But they shall smart for't shortly, smart severely.
  Meantime prepare we for our Queen, who comes
  From England decked in conquest. The wild Scots,
  Who, in the absence of the Royal Shepherd,
  Would play the wolves upon our territory,
  Has she chased back from our fat pasturage
  Into their naked land, to howl at nature
  And feed on thistles. Say, Lord Harcourt,
  Are all prepared to welcome her arrival?
HARCOURT: All, my dread liege. The beach is thickly lined

---

the wild Scots: the Scottish highlanders had defeated Edward II at Bannockburn in 1314, thereby crushing all Plantagenet claims to nominal control over their kingdom. But Edward III was as determined to subdue the Scots as he was to control France and in 1333 launched a major offensive against the armies of David Bruce, King of Scotland, winning territory as far north as the Firth of Forth. By 1346, a Franco-Scottish alliance had been formed, and when Edward and his troops were occupied with the siege of Calais, King David invaded England in an attempt to reoccupy Scottish land and to force an English withdrawal from France. The Scots were defeated at Nevill's Cross by an army under Queen Philippa; David Bruce was taken prisoner. Late eighteenth-century audiences might also have been reminded of a more recent event in Anglo-Scottish relations, the Jacobite Rising of 1745.

With English soldiery in ardent watch,
Fixing their eyes upon the bark
Which bears our royal mistress. It was hoped ere this
'T had reached the harbour —
> (*A shout is heard.*)
>   Hark! The Queen has landed!
KING: Do you then, good my lord, escort her hither.
> (*Exit* HARCOURT.)
Sir Walter Manny?
SIR WALTER: Aye, my gracious sovereign?
KING: Guard well this packet. When the Governor
Of this same peevish town shall call a parley, —
And soon, I trust, despair will urge him to't —
Break you it up and from it speak our pleasure.
Here are the terms, the only terms, on which
We do allow them to capitulate.
> (*Another shout is heard, followed by a grand flourish, then a triumphal march.* QUEEN PHILIPPA *enters with* GUARDS *and* ATTENDANTS *in procession.*)
Oh, welcome, welcome! We shall give you here
Rude martial fare and soldiers' entertainment!
QUEEN: Royal sir,
Well met, and happily! I learn your labours
Draw to a glorious end. When you return,
Besides the loyal subjects who would greet you,
The Scottish King, my lord, waits your arrival,
Who, somewhat partial to his neighbour's land,
Did come an uninvited guest among us.
I doubt he'll think us over-hospitable,
For, dreading his too-quick departure from us,
I have made bold to guard him in the Tower,
And hither have I sailed, my noble liege,
To glad you with the tidings.
KING: My sweet warrior,
We will dispatch our work here, then for England.
Calais will soon be ours, and we shall prove,
Spite of this obstinate and close defence,
Our English excellence. 'Twere policy,
For future expedition in our wars,
Against this town, which has so long held out,
To show some rigour — but of that hereafter.
Think we today on nought but revelry.
You, madam, shall diffuse your influence
Throughout our camp. — Strike, there, our martial music! —
For want of better, good Philippa, take
A soldier's noisy concert. — Strike, I say!

Grand Chorus.

> War has still its melody,
> When blows come thick and arrows fly!
> When the soldier marches o'er
> The crimson field, knee-deep in gore,
> By carnage and grim death surrounded
> And groans of dying men confounded,
> If the warlike drum he hear,
> And the shrill trumpet strike his ear,
> Roused by the spirit-stirring tones,
> Music's influence he owns;
> His lusty heart beats quick and high:
> War still has its melody!
>
> But when the hard-fought day is done
> And the battle's fairly won,
> Oh, then he trolls the jolly note
> In triumph thro' his rusty throat,
> And all the story of the strife
> He carols to the merry fife!
> His comrades join, their feats to tell,
> The chorus then begins to swell,
> Loud martial music rends the sky:
> This is the soldier's melody!

    (*Exeunt omnes.*)

ACT II

Scene 1. *An apartment, poorly furnished, in Calais.* LA GLOIRE *and* MADELON *discovered,* MADELON *seated at a table covered with eatables, wines, &c.,* LA GLOIRE *standing near the table.*

LA GLOIRE: Blessings on her heart, how cleverly she feeds! The meat goes as naturally into her little mouth as if it had been used to the road all the time of the famine, though Heaven knows 'tis a path that has lately been little frequented.

MADELON: A votre santé, mon ami. Your health, La Gloire. (*drinking*)

LA GLOIRE: Nay, I'll answer thee in that, though bumpers were Englishmen and went against my French stomach! (*taking wine*) Heaven bless thee, my poor Madelon! May a woman never tumble into the mire of distress, and, if she is in, ill befall him that won't help her clean out again. (*drinking*)

MADELON: (*coming from the table*) There, enough.

---

bumpers: goblets or glasses of wine filled to the brim and drunk usually as a toast.

LA GLOIRE: So, (*kissing her*) one kiss for a bonne bouche. Dost love me the better for this feast now, Madelon?

MADELON: No, truly, not a jot. I love you e'en as well before dinner as after.

LA GLOIRE: What a jewel is regular affection, to love equally through the week, maigre-days and all! I cannot but own a full meal makes an improvement in the warmth of my feelings: I can eat and drink myself into a glow of tenderness that fasting can never come up to! And what hast thou done in my absence, Madelon?

MADELON: Little, La Gloire, but grieve with the rest. I have thought on you, gone to confession in the morning, seemed happy in the day, to cheer my poor old father, but my heart was bursting, La Gloire, and at night, by myself, I looked at this little cross you gave me and cried.

LA GLOIRE: (*smothering his tears*) Madelon, I — I — I — want another draught of burgundy. (*drinking*)

MADELON: Once, indeed, — I thought it was hard — Father Anthony enjoined me penance for thinking so much about you.

LA GLOIRE: An old — ! What, by putting peas in your shoes, as usual?

MADELON: Yes. But, as it happened, I escaped.

LA GLOIRE: Aye? Marry, how?

MADELON: Why, as the famine pressed, the holy fathers had boiled all our punishments in puddings for the convent, and there wasn't a penitential pea left in the town.

LA GLOIRE: Oh, gluttony! To deprive the innocent of their hard, dry penances and apply 'em soft to their own offending stomachs! I never could abide these pampered friars. They are the pot-bellied children of the Pope, nursed at the bosom of old Mother Church, and plaguy chubby boys they are. One convent of them in a town breeds a famine sooner than an English blockade! But what says thy father within here, Madelon, to our marriage?

MADELON: Truly, he has no objection but in respect to your being a soldier.

LA GLOIRE: Sacre bleu! Object to my carrying arms? My glory? My pride?

MADELON: Prithee, now, 'tis not for that.

LA GLOIRE: Degrade my profession, my — ? Lookye, Madelon, I love thee with all my heart, with an honest soldier's heart, else I could tell your father that a citizen could never get on in the world without a soldier to do his journey-work. And your soldier, lookye — 'sblood, it makes me fret like a hot day's march! — your soldier, in all nations, when he is rusted down to your quiet citizen and so sets up at home for himself, is in double respect for having served such an honourable apprenticeship.

MADELON: Nay, now, La Gloire, my father meant not —

LA GLOIRE: Marry, I would tell your father this to his teeth, which, were it not for my captain and me — two soldiers, mark you me — might not, haply, have been so soon set agoing!

---

maigre-days: fast-days in the calendar of the Roman Catholic church.
pot-bellied children of the Pope: sentiments such as these strayed dangerously close to fuelling the powerful anti-Catholic feeling which had already shown its viciousness in the Gordon Riots of 1780, the most serious civil disturbance ever to have taken place in London.

MADELON: Ungenerous! I could not have spoken such cutting words to you, La Gloire! My poor father only meant that the wars might separate us. But I had a remedy for that, too, for all your unkindness.

LA GLOIRE: Pish! Remedy? Well, — psha! — what was the remedy, Madelon?

MADELON: Why, I could have followed you to the camp.

LA GLOIRE: And wouldst thou follow me, then?

MADELON: Aye, surely, La Gloire. I could follow him I love all over the world.

LA GLOIRE: And bear the fatigue of a campaign, Madelon?

MADELON: Anything with you, La Gloire. I warrant us, we should be happy enough. Aye, and I could be useful, too. I could pack your rucksack, sing canzonets with you to make us merry on a day's march, mix in the soldiers' dance upon occasion and, at sunset, I would dress up our little tent as neat as any captain's in the field. Then, at supper, La Gloire, we should be as cheerful —

LA GLOIRE: Now could I cut my tongue out for what I have said! Cuff me, slap my face, Madelon! Then kiss me and forgive me. And, if ever I bestride my great war-horse again and let him run away with me and trample over the heart of my best friends, I wish he may kick me off and break my neck in a ditch for my pains! But what — ha, ha! — what should we do with our children, Madelon?

MADELON: Ah, mon dieu, I had forgot that! But, if our endeavours be honest, La Gloire, providence will take care of them. I warrant you.

Duet. LA GLOIRE *and* MADELON.

MADELON: Could you to battle march away
        And leave me here complaining?
     I'm sure 'twould break my heart to stay
        When you are gone campaigning.
           Ah, non, non, non,
           Pauvre Madelon
               Could never quit her rover!
           Ah, non, non, non,
           Pauvre Madelon
               Would go with you all the world over!

LA GLOIRE: No, no, my love! Ah, do not grieve!
        A soldier true you'll find me.
     I could not have the heart to leave
        My little girl behind me.
           Ah, non, non, non,
           Pauvre Madelon
               Should never quit her rover!
           Ah, non, non, non,
           Pauvre Madelon
               Should go with me all the world over!

     And can you to the battle go
        To woman's fear a stranger?

MADELON: No fear my breast will ever know
        But when my love's in danger.
          Ah, non, non, non,
          Pauvre Madelon
             Fears only for her rover!
          Ah, non, non, non,
          Pauvre Madelon
             Will go with you all the world over!

BOTH: Then let the world jog as it will
        Let hollow friends forsake us,
      We both shall be as happy still
        As war and love can make us.
          Ah, non, non, non,
          Pauvre Madelon
             Shall never quit her rover!
          Ah, non, non, non,
          Pauvre Madelon
          Shall go with $\left\{ \begin{array}{l} \text{me} \\ \text{you} \end{array} \right.$ all the world over!

LA GLOIRE: By the mass, Madelon, such a wife as thou wilt be would make a man, after another campaign, — for another I must have, to satisfy the cravings of my appetite — go nigh to forswear the wars!

MADELON: Ah, La Gloire, would it were so! But the sound of a trumpet will ever lead thee after it.

LA GLOIRE: Tut! A trumpet? Thy voice, Madelon, will drown it!

MADELON: (*shaking her head*) Ah, La Gloire!

LA GLOIRE: Nay, then I am the veriest poltroon if I think the sound of a trumpet would move me any more than —
    (*A parley is sounded from the walls.*)
Eh? Gad! Oh, ecod, there's a bustle, a parley from the walls, which may end in a skirmish! — or a battle! — or a — ! I'll be with you again in the chopping off of a head!

MADELON: Nay, now, La Gloire, I thought the sound of a trumpet —

LA GLOIRE: A trumpet? Simpleton! That was a — gad, I — wasn't it a drum? Adieu, Madelon, I'll be back again ere —
    (*The parley is repeated.*)
March! Charge! Huzza!
    (*Exit* LA GLOIRE *drawing his sword.*)

MADELON: Welladay, a soldier's wife must have a fearful time on't! Yet I do love La Gloire. He is so kind, so tender, and he has simply the best leg in the army. Heigho! It must feel very odd to sleep in a tent. A camp must be ever in alarms and soldiers always ready for surprise. Dame Toinette, who married a

---

best leg: an unexpected reference to La Gloire's dancing prowess. The context implies a number of less innocent secondary meanings.

corporal ere I was born, told me that for one whole campaign her husband
went to bed in his boots!

Song. MADELON.

> Little thinks the townsman's wife,
>   While at home she tarries,
> What must be the lass's life
>   Who a soldier marries.
> Now with weary marching spent,
>   Dancing now before the tent,
> Lira lira lira lira lira la,
>   With her jolly soldier.
>
> In the camp at night she lies,
>   Wind and weather scorning,
> Only grieved her love must rise
>   And quit her in the morning.
> But, the doubtful skirmish done,
>   Blithe she sings at set of sun,
> Lira lira lira lira lira la,
>   With her jolly soldier.
>
> Should the captain of her dear
>   Use his vain endeavour,
> Whisp'ring nonsense in her ear
>   Two fond hearts to sever,
> At his passion she will scoff,
>   Laughing thus she'll put him off:
> Lira lira lira lira lira la,
>   For her jolly soldier.

(*Exit* MADELON.)

Scene 2. *The Town Hall of Calais. A council table with a chair at its head,
benches, &c. The* CITIZENS *and* SOLDIERS *are discovered, standing in
confusion.* EUSTACHE DE ST PIERRE *is seated on a low stool towards
the front of the stage. In the midst, a* TOWN CRIER, *ringing his bell.*

ALL: Silence! Silence!
CRIER: Silence! An ye all talk thus, there's an end to conversation. Your 'silence',
    my masters, will breed a disturbance. Open your eyes, good people, to reason,
    and hear me! Mass, 'tis hard that I, who am Crier, should be laughed at and
    held at nought among you.
ALL: Hear! Hear!
CRIER: Listen! It seems that we shall all go near to be famished, and your learned
    politicians promise that dearth of provisions is the main cause on't. The good
    John de Vienne, our Governor, — a blessing on his old merry heart! — grieving
    for your distress, has e'en now called a parley on the walls with the English
    and has chosen me, in his wisdom, to ring you all into the Town Hall here,

where, an you abide his coming, you will hear what he shall seem to signify
unto you. And, by our lady, here the Governor comes! — (*ringing*) Silence!

ALL: Silence!

CRIER: Nay, 'tis ever so. An I were to bid a dumb man hold his tongue, by my
troth, I think a' would cry 'silence' till the drum of my ear were bursten.
Silence!

> (*Enter* JOHN DE VIENNE *who seats himself at the head of the
> council table.*)

DE VIENNE: You partly know why I have here convened you.
I prithee, now, I prithee, honest friends,
Summon up all the fortitude within you
Which you are masters of. Now, Heaven forgive me,
I almost wish I had not been a soldier,
For I have here a matter to deliver
Requires a schoolman's preface. 'Tis a task
Which bears so heavy on my poor old heart
That 'twill go nigh to crack beneath the burden.
You know I love you, fellow citizens,
You know I love you well.

ALL: Aye, aye, we know it.

DE VIENNE: I could be well content, in peace or peril,
To 'bide with you forever.

EUSTACHE: No one doubts it.
I never yet did hear of governor,
Spite of the rubs and watchful toil of office,
Would willingly forgo his place!

DE VIENNE: Why, how now!
Why, how now, friend! Dost thou come o'er me thus?
But I shall find a time, it fits not now,
When I will teach thee! 'Sdeath, old John de Vienne,
A veteran, bluff soldier, bearded thus
And sneered at by a saucy — (*rising*) Mark you me! —
Well, let it pass. The general calamity
Will sour the best of us. (*sitting*) — My honest citizens,
I once more pray you, think that ye are men.
I pray you, too, my friends —

EUSTACHE: I pray *you*, sir,
Be somewhat brief; you'll tire else. These same citizens,
These *honest* citizens, would fain e'en know
The worst at once. When members are impatient
For a plain tale, the orator — you'll pardon me —
Should not be too long-winded.

DE VIENNE: Fellow, peace!
Ere now I've marked thee — thou art he, I take it —
'Tis Eustache de St Pierre I think they call thee,
Whom all the town, our very children, point at

As the most growling knave in Christendom.
Yea, thou art he.
EUSTACHE: The same. The mongrels here
Cannot abide rough honesty. I'm hated.
Smooth talking likes them better. You, good sir,
Are popular among them —
ALL: Silence!
EUSTACHE: Buzz!
DE VIENNE: Thus then, in brief: finding we are reduced
By famine and fatigue unto extremity,
And the late succour, though 'twas nobly meant,
All insufficient, seeing this, my friends,
I sounded for a parley from the walls.
E'en now 't has ended. Edward ordered forth
Sir Walter Manny, and I needs must own
A courteous knight, although an enemy.
I told him our distress. Sir Knight, said I,
'Tis near a year your master hath besieged us.
He knows, I ween, that we have done our duty,
But we are sorely pressed: we have not now
A hope left of relief and, to speak truth, —
And here it makes me almost blush to think
An Englishman should see me drop a tear,
But, spite of me, it stole upon my cheek —
To speak the honest truth, Sir Knight, said I,
My gallant men are perishing with hunger,
Therefore I will surrender.
EUSTACHE: Surrender?
               (*The rest look amazed.*)
DE VIENNE: But, conceive me,
On this condition: that I do secure
The lives and liberties of those brave fellows
Who, in this galling and disastrous siege,
Have shared with me in each fatigue and peril.
ALL: Huzza! Long live our Governor! Huzza!
DE VIENNE: I thank you, friends. It grieves me to repay
Your honest love with tidings sure as heavy
As ever messenger was charged withal.
The King of England steels his heart against us.
He calls us stubborn, and our long defence,
He says, will be in future an example
To others who oppose him, thus to harass
And to weary out his men; which to prevent
He does let loose his vengeance, and he wills,
If we would save our city from the sword,
From wild destruction and the general slaughter

Which, ere they enter in, his fretted soldiers
Are bursting ripe for, that I straight do send him
Six of my first and best-reputed citizens,
Bare-headed, tendering the city keys,
And — 'sdeath, I choke! — with vile and loathsome ropes
Circling their necks, in guise of malefactors,
To suffer instant execution.
        (*The* CITIZENS *appear confused. A pause.*)
                    Friends,
I do perceive you're troubled. 'Tis enough
To pose the stoutest of you. To be plain,
Nothing remains but to throw ope our gates
And let the victors find we still can sell
Our lives full dearly to them. Though we perish,
We go like men; 'tis all that now is left us.
For who will venture forth to death like this —
And yet, 'twere glorious in a cause like this —
When he can take his slender chance of life
Or fall among his fellows? Who among you
Can smother Nature's workings, which do prompt
Each, to the last, to struggle for himself? —
Yet, were *I* not objected to, as Governor,
There might be found — no matter. — Who so bold
That, for the welfare of a wretched multitude
Involved with him in one great common cause,
Would volunteer it on the scaffold?
EUSTACHE: (*rising*)   I,
E'en I, the growling knave whom children point at,
To save those children and their hapless mothers,
To snatch the virgin from the ravisher,
To shield the bent and hoary citizen,
To push the sword back from his aged throat —
Fresh reeking, haply, in his house's blood —
I render up myself for sacrifice.
Ye curs, ye chicken-hearted dunghill slaves,
It is to you I speak, you who of late
Did roar it open-mouthed about our streets
Ye would have this and this. Your ease, forsooth,
Must be considered, which, by this defence, —
'T has been a brave one — ye did call neglected
As members of our city. 'Sdeath, such members!
Ye would have nought but good, and now, to show
How much ye merit when your country calls
To prove if ye be worthy of that care
Ye loudly stickled for, ye sit mum, chance,
And hang your scurvy ears in sheepish silence.
'Twere throwing life away to yield it for you!

But there are many patient sufferers
Pent in our city that lay claim to't. What,
Will no one budge? Then let the English in,
Let in the enemy, to find us wasted
And winking in the socket. Let them clap
Fierce Conquest's vile extinguisher upon you,
And out you go in smother. Rouse, for shame!
Rouse, citizens! Think on your wives, your infants,
And let us not be so far shamed in story
That we should lack six men within our walls
To save them thus from slaughter.

DE VIENNE: Noble soul!
I could, for this, fall down and worship thee.
Thou warm'st my heart. Does no one else appear
To back this gallant veteran?

> (*Three of the* CITIZENS – JOHN D'AIRE, JACQUE WISSANT *and*
> PIERRE WISSANT – *come forward.*)

D'AIRE: Eustache,
Myself and these two brothers, my companions,
All of your house and near of kin to you,
Have pondered on your words. We sure must die
If we or go or stay, but, what weighs most,
We would not see our helpless little ones
Butchered before our eyes. We are resolved:
We'll go with thee.

EUSTACHE: Now, by our good Saint Dennis,
I do feel proud! My lowly house's glory
Shall live on record. What are birth and titles?
Feathers for children! The plain, honest mind
That branches forth in charity and virtue
Shrinks lordly pomp to nought and makes vain pedigree
Blush at his frothy boasting. We are four,
Fellows in death and honour. Two remain
To fill our number.

DE VIENNE: Pause awhile, my friends;
We yet have breathing time, though troth but little.
I must go forth, a hostage to the English,
Till you appear. Break up our sad assembly
And, for the rest, agree among yourselves.
(*to* EUSTACHE) Were the time apt, I could well waste a year
In praising this your valour.

EUSTACHE: Break we up. If any
Can wind his sluggish courage to the pitch,

Fierce Conquest's vile extinguisher: the image here is of the vital spark being doused by a candle-snuffer, smother being the smoke produced by a smouldering wick.

Meet me anon i' th' market-place, and thence
We will march forth. Ye have but this, remember:
Either plunge bravely into death, or wait
Till the full tide of blood flows in upon you
And shame and slaughter overwhelm us. (*to* JOHN D'AIRE *and the*
    WISSANTS) Come,
My noble partners, come! (*to the* CITIZENS) Break up and think on't!
    (*Exeunt omnes.*)

Scene 3. *An apartment in the Governor's House.*

(*Enter* JULIA *and* RIBAUMONT.)

RIBAUMONT: Yet hear me, Julia —
JULIA: Prithee, good my lord,
Press not me thus. My father's strict command —
I must not say 'tis harsh — forbids me listen.
RIBAUMONT: Is then the path of duty so precise
That 'twill not for a little deviate?
Sweet, let it wind and bend to recollection.
Think on our oaths. Yes, lady, they are mutual.
You said you loved; I treasured the confession
As misers hoard their gold. Nay, 'twas my all.
I lived on't, fed on't. For a lover, Julia,
Will feed on air: but let his mistress breathe
A word that's kind and he will banquet on it.
But fickleness from her he dotes upon
Is his soul's baneful prison: it corrodes
And preys upon his heart till he dies grief-worn.
Think not I chatter in the idle school
Of whining coxcombs, where despair and death
Are words of course. I swell not fancied ills
With windy eloquence. No, trust me, Julia,
I speak in honest, simple suffering,
And disappointment in my life's best hope
So feeds upon my life and wears me inward
That I am nearly spirit-broken.
JULIA: Why?
Why this, my lord? Think you I cannot feel
Another's grief? You know I can, good sooth.
You urge me past a maiden's modesty.
What should I say? In Nature's course, my lord,
The parent sits at helm, in grey authority,
And pilots the child's action. For my father,
You know what humour sways him.
RIBAUMONT: Yes, court policy,
Time-serving zeal, tame, passive, blind obedience
To the stern will of power, which doth differ
As wide from true, impulsive loyalty

As puppet-work from Nature. Oh, I would
The time were come! Our enemy, the English,
Bid fairest first to show a bright example.
When, 'twixt the ruler and the ruled, affection
Shall be reciprocal, when majesty,
With placid sway, shall gather strength from mildness,
No show of tyranny shall urge the citizen
To break the ties of blood, but general laws
Shall issue from the throne in wholesome care
To guard his birthright, and the grateful subject
Shall look with duteous love upon his sovereign,
As the child eyes its father. Now, by Heaven,
Old John de Vienne is turned a temporiser,
Making his daughter the poor topmost round
Of his vile ladder to preferment. 'Sdeath,
And you to suffer this, you who professed
A love so pure and lasting! Oh, fie, fie, Julia,
'Twould show more noble in you to lay bare
Your mind's inconstancy than thus to keep
The semblance of a passion, meanly veiling
Your broken faith with the excuse of duty!
Out on't, 'tis shallow! You ne'er loved!

JULIA: My lord,
My cup of sorrow was brimful, and you, —
I looked not for it — have thrown in a drop
Which makes it overflow. No more of that. —
You have reviled my father; me, too, Ribaumont.
Heaven knows I little merit it! My lord,
Upon this theme we must not meet again.
Farewell! And do not, do not think unkindly
On her you once did call your Julia.
If it will soothe your anguish, Ribaumont,
To find a fellowship in grief, why, think
That there is one, while struggling for her duty,
Sheds many a tear in private. Heaven be with you!
            (*Exit* JULIA.)
RIBAUMONT: Stay!
Stay and listen to me! Gone, and thus, too,
Washing her sweet rebuke away in tears.
And have I lost thee, and forever, Julia,
Thee whom I made so main a prop of happiness

No show of tyranny: more immediate as a comment on the role of monarchy in the eighteenth
rather than in the fourteenth century, this speech was clearly intended to make a comparative
reference to the political situation (and its causes) then current in France. However, the 'ties of
blood' were to break completely when, on 21 January 1793, the new French Republic
executed the country's former head-of-state, King Louis XVI.

That, ta'en away, this building totters now?
Now do I look on life as the worn mariner,
Stretching his eyes o'er seas immeasurable,
And all is drear and comfortless. Henceforward
My years will be one void, day roll on day
In sameness infinite, without a hope
To chequer the sad prospect. Oh, if death
Came yoked with honour to me, I could now
Embrace it with as warm and willing rapture
As mothers clasp their infants.
        (*Enter* LA GLOIRE.)
Now, La Gloire, what is the news?

LA GLOIRE: Good faith, my lord, the saddest that ever tongue told! I am as full of it as an over-ripe gooseberry, and, before it comes out, I verily think it will burst me.

RIBAUMONT: What is't?

LA GLOIRE: The town has surrendered.

RIBAUMONT: I guessed as much.

LA GLOIRE: Upon conditions.

RIBAUMONT: What are they?

LA GLOIRE: Very scurvy ones, my lord: hanging conditions. 'Twas or neck or nothing with us and, as the enemy could get little by our nothing, they have resolved to make sure of the necks.

RIBAUMONT: How mean you, sirrah? Speak brief and plainly. What has been carried into execution?

LA GLOIRE: Six honest men will be carried to execution anon, my lord, to save the city from sacking. Six citizens must swing for it in Edward's camp. But four have yet been found, and they are —

RIBAUMONT: Who?

LA GLOIRE: Oh lord, all of my own family! All my living relations will go in ropes like a parcel of onions. There's John d'Aire, Jacque and Pierre Wissant, my three good cousins german, my lord. And the fourth, who was the first that offered, is — is —

RIBAUMONT: Who, La Gloire?

LA GLOIRE: (*wiping his eyes*) I crave your pardon, my lord, for being thus unsoldier-like, but 'tis — 'tis — my own father.

RIBAUMONT: Eustache?

LA GLOIRE: He, my lord, he, old Eustache de St Pierre, the honestest, kindliest soul — I cannot talk upon't; grief plays the hangman with me and has almost choked me already.

RIBAUMONT: (*aside*) Why, I am courted to't. The time, example,
Do woo me to my very wish. — (*to* LA GLOIRE) Come hither.
Two, it should seem, are wanting to complete
The little band of those brave men who die
To save their fellows.

LA GLOIRE: Aye, my lord. There is a meeting upon't, half an hour hence, in the market-place.

RIBAUMONT: Mark me, La Gloire, and see that you obey me,
    Even to the very letter of my orders.
    They are the last, perhaps, my honest fellow,
    I e'er shall give thee. Seek thy father out
    And tell him this from me: his gallant bearing
    Doth school his betters; I have studied o'er
    His noble lesson and have learnt my duty.
    Say he will find me in the market-place,
    Disguised in humble-seeming, — and I fain
    Would pass for one allied to him — and thence —
    Dost mark me well? — I will along with him,
    E'en hand in hand, to death.
LA GLOIRE: My lord, I — I —
        (LA GLOIRE *bursts into tears, falls on his knees, takes hold of*
        RIBAUMONT*'s hand and kisses it.*)
I shall lose my father. When he was gone, I looked you would have been my
father. The thought of still serving you was a comfort to me. My heart's
almost broke, my lord. You are my commander, and I hope I have, hitherto,
never disobeyed orders, but if I now deliver your message, drum me out for
ingratitude as the greatest rascal that ever came into a regiment!
RIBAUMONT: (*aside*) 'Sdeath, I am moved to tears! This honest soul,
    For pure simplicity of love, unmans me. —
    Prithee, no more, La Gloire. I am resolved,
    My purpose fixed. It would be bitter to thee
    To see me die in anger with thee, therefore
    Do thou my bidding, close thy service up
    In duty to my will. Go, find thy father.
    I will prepare within the while. Obey me,
    Or the last look from thy expiring master,
    Darting reproach, shall burst thy heart in twain.
    Mark, and be punctual!
        (*Exit* RIBAUMONT.)
LA GLOIRE: Oh, the Virgin, why was I ever attached to man, woman or child? All
    that I stand affected to seem certain to come to the gallows. Nay, had I been
    put out to dry in the wind and dangle by the roadside like a washed jerkin —
        (*Enter* EUSTACHE DE ST PIERRE.)
EUSTACHE: Where's thy commander, boy, Count Ribaumont?
LA GLOIRE: Oh father!
EUSTACHE: Peace. — I must a word with him.
    I have a few short thanks I would deliver
    Touching his care of thee. It is the last
    Of all my worldly packages. That done,
    I may set forward on my journey.
LA GLOIRE: Beshrew me, father, but 'tis the worst manner of travelling that ever
    was hit upon. Never did rider, till now, venture with good-will upon such a
    breakneck road, nor draw on a night-cap as quietly for it as if he were going
    to bed. Bed? I shall never go to bed again in peace as long as I live. Sorrow

will keep my eyes open half the night, and, when I drop into a doze at daybreak, I shall be hanged with you, father, a score of times every morning.

EUSTACHE: (*aside*) I could have spared this meeting. — Boy, I will not,
　　Nor would I, had I time for't, ring a chime
　　Of drowsy document at this our parting,
　　Nor will I stuff the simple plan of life
　　That I would have thee follow with trim angles
　　And petty intersections of nice conduct
　　Which dotards, rotten in their wisdom, oft
　　Will mark in mathematical precision
　　Upon a stripling's mind, until they blur
　　The modest hand of nature. Thou'rt a soldier, —
　　'Tis said a good one — and I ne'er yet knew
　　A rough, true soldier lack humanity.
　　If, then, thou canst, with one hand, push aside
　　The buffets of the world and, with the other
　　Stretched forth in warm and manly charity,
　　Assist the weak, be thankful for the groundwork,
　　And e'en let impulse build upon't. Thou needst
　　No line nor level formal age can give thee
　　To raise a noble superstructure. — Come,
　　Embrace me. When thy father sleeps in honour,
　　Think that —
　　　　　　(EUSTACHE, *embracing* LA GLOIRE, *bursts into tears.*)
　　　　　　My son! My boy! — Psha! Pish! This nature! —
　　Conduct me to —

LA GLOIRE: (*catching hold of* EUSTACHE) Hold! Hold! We shall leap here from
　　bad to worse. Rope-honour is catching nowadays, and my commander has the
　　distemper at the height. The Count —

EUSTACHE: Well?

LA GLOIRE: I — I — am bidden, father, to deliver a message to you.

EUSTACHE: Be quick, then. The time wears.

LA GLOIRE: No, truly, 'twill not come quick; I must force it out in driblets. My
　　captain bids me say that — that brave men are scarce — find six in the town
　　and you find all — so he will join you at the market-cross and — go with you
　　— to —

EUSTACHE: The scaffold?

LA GLOIRE: Yes, the sca—. That word sticks so in my throat I can't squeeze it
　　out, for the life of me.

EUSTACHE: Why, this shows nobly now. Our honest cause
　　Is graced in the addition. Lead me —
　　　　　　(EUSTACHE *observes* LA GLOIRE *weeping.*)
　　　　　　How now?
　　Out on thee, knave, thou'lt bring disgrace upon me!

---

rope-honour: a verbal compound devised for the sake of a pun with 'catching'.

By Heaven, I feel as proud in this, my death —
And thou, the nearest to my blood, to sully
My house's name with womanhood? Shame! Shame!
Where is the noble Ribaumont? (*going*)

LA GLOIRE: Stay, father, stay! I can hold it no longer. I love Madelon too well to keep her waking o'nights with blubbering over her for the loss of my father and my captain. Another neck is wanting to make up the half-dozen, so I'll e'en along, father, as the sixth.

EUSTACHE: (*after a pause*) I know not what to answer; thou hast shaken
My manhood to the centre. Follow, boy.
Thy aim is honour, but the dreary road to't,
Which thou must tread, does stir the father in me.
'Tis such a nice and tickle point between
The patriot and the parent that, Heaven knows,
I need a counsellor. I'll to thy captain.
With him, anon, you'll find me.
    (*Exit* EUSTACHE.)

LA GLOIRE: So, how many a lad with a fair beginning of life comes to an untimely conclusion? I wonder, now, if we find ropes ourselves or if we are provided gratis, at the expense of the enemy. Hanging must have an awkward feel with it. To kick about for something to put one's foot upon doesn't so much matter, but when a man bears his whole weight upon his throat all the while, it isn't altogether so pleasant. Then my poor Madelon, too! She little thinks that —
    (*Enter* MADELON.)

MADELON: (*peeping in*) Hist! Hist! La Gloire.

LA GLOIRE: Eh?

MADELON: Why, where hast thou been, La Gloire? I have been seeking you all over the town. I feared you would get into danger. Finding the Governor's gate thrown open, and all the city in confusion, I e'en ventured in to look for you. Where hast thou been, La Gloire?

LA GLOIRE: Been? Nowhere. But I am going —

MADELON: Where, La Gloire?

LA GLOIRE: A — a — little way with my father. Hast heard the news, Madelon?

MADELON: Only in part. I hear the town has surrendered and that six poor men are to be executed and march from the town gates. But we shall then be in safety, La Gloire. Poor fellows, I would not see them go forth, for the world!

LA GLOIRE: Poor fellows? — Ahem! — Aye, poor fellows! True, Madelon, I would not have thee shocked with the sight, I confess.

MADELON: But prithee, La Gloire, keep at home now with me. You are ever gadding. You soldiers are so wild and turbulent. How can you, La Gloire? You must be present, now, at this horrid ceremony?

LA GLOIRE: Why, truly, I — I — must be present. But it will be for the last time, Madelon. I take little pleasure in it, believe me.

MADELON: I would thou would'st home with me. I have provided, out of thy bounty, a repast for us this evening. My father, who has ne'er stirred out these three weeks, is filled with joy for thy return. He will sit at our table, La Gloire, he will give us his blessing and wish us happy in marriage. Come, you shall not go away this evening, in sooth, now?

LA GLOIRE: I must, Madelon, I must. The throng will press and — and — I may lose somewhat of value. 'Tis seldom a soldier's pocket is heavy, but I carry all my worldly goods about me. I would fain not lose them, so e'en be mistress of them till my return. Here is a casket with five years' wages from my captain, three quarters' pay from my regiment, and eleven marks plucked from the boot of a dead English corporal, with a flask of eau de vie, quite empty, and part of a cracked looking-glass. 'Tis my whole fortune. Keep it, Madelon, for fear of accidents and, if any cross accident ever should befall me, remember you are heir-apparent to the bulk of my property.

MADELON: But why thus particular? I would you would stay quiet with me.

LA GLOIRE: But for this once, Madelon, and I shall be quiet ever after. Kiss me. So, adieu!

MADELON: Adieu, La Gloire! Remember, now, at night —

LA GLOIRE: Adieu! — (*aside*) At night? Mercy on me, should I stay three minutes longer my heart would rescue my neck, for the breaking of one would save the stretching of the other!

(*Exit* LA GLOIRE.)

MADELON: How rich my La Gloire has got in the wars! My father, too, has something to throw in at our wedding, and, when we meet, we shall be the happiest couple in Picardy!

Song. MADELON.

I tremble to think that my soldier's so bold,
To see with what danger he gets all his gold,
Yet, danger all over, 'twill keep out the cold,
    And we shall be warm when we're married!

For riches, 'tis true that I covet them not,
Unless 'tis to better my dear soldier's lot,
And he shall be master of all I have got
    The very first moment we're married!

My heart, how it beats but to look to the day
In church, when my father will give me away!
But that I shall laugh at, I've heard many say,
    A day or two after we're married!

(*Exit* MADELON.)

Scene 4. *A large gateway leading out of the town, the stage filled with* CITIZENS, WOMEN, &c.

1st CIT: Stand back, they are coming!

3rd CIT: Nay, my masters, they will not forth this quarter of an hour. Men seldom move lightly on such a heavy occasion.

4th CIT: Who are the two others that have filled up the number?

3rd CIT: Marry, two more of old Eustache's family: his own son and the other, as
'tis rumoured, a relation in the town that few of us are acquainted withal.

4th CIT: That's strange.

3rd CIT: Why, aye. But when a man chooses a rope for his preferment, few are found
envious enough to dispute the title with him! — By the rood, here they come!

(*Enter* EUSTACHE DE ST PIERRE, RIBAUMONT, LA GLOIRE,
JOHN D'AIRE, JACQUE *and* PIERRE WISSANT, *with ropes on
their necks, going to execution, a procession of* SOLDIERS,
FRIARS, NUNS, *&c. accompanying them.* RIBAUMONT *is rather
more muffled than the rest. They march solemnly, then halt.*)

EUSTACHE: (*to* RIBAUMONT) My lord, —

RIBAUMONT: I prithee, peace, Eustache! I fain would 'scape
Observance from the rabble. Hurry o'er
This irksome march, and straightway to the camp.

EUSTACHE: Enough. — Set forth! We are engaged, my friends,
Upon a business here which most, I wot,
Do think of moment, and we would not waste
The time in idle ceremony. — On! —
Ere we are ushered to the English camp, —
And most of you, I trust, will follow thither —
We will bestow the little time allowed us
In manly leave-taking. — Strike, and set onward!

CITIZENS: Bless our countrymen! Bless our deliverers! Huzza!

Glee. PERSONS OF THE PROCESSION.

Peace to the heroes! Peace, who yield their blood
And perish nobly for their country's good!
Peace to their noble souls! Their bodies die;
Their fame shall flourish long in memory,
Recorded still in future years,
Green in a nation's gratitude and tears!

Sound, sound in solemn strains and slow!
Dully beat the muffled drum!
Bid the hollow trumpet blow
In deadened tones, clear, firm and low,
For see, the patriot heroes come!

(*Towards the end of the chorus the characters proceed on their
march out of the town, and, when the last persons of the procession
are going through the gates, the curtain drops.*)

ACT III

Scene 1. *An apartment in the house of* JOHN DE VIENNE.

(*Enter* JULIA, *in man's apparel, and* O'CARROL.)

JULIA: Come on, bestir thee, good fellow! Thou must be my guide and conduct me.

O'CARROL: Faith, and I'll conduct you, with all my heart and soul. — (*aside*) And some good creature, I warrant, will be kind enough to show me the way! By my troth, now, the sad news and the bustle and my care for the Lady Julia has made such a sort of confusion in my head that, if I could leave it behind and put my eyes into any other part that belongs to me, it would be the readiest way to prevent my mistaking the road.

JULIA: But art thou well assured, O'Carrol, of what thou hast informed me?

O'CARROL: To be sure, I am well informed, for I informed myself, and I never yet catched myself out in telling a lie! There was six of them, as tall fellows as any in France, with ugly ropes about their good-looking necks, going to the town gates, and Count Ribaumont marched second in the handsome half-dozen. The whole town followed them with their eyes till they were as full of water as if they had been peeping into so many mustard pots. And so, madam, knowing he loves you better than dear life — which, to be sure, he seems to hold cheap enough at present — and thinking you would be glad to hear the terrible news, why, I made all the haste I could to come and tell it to you.

JULIA: And thus, in haste, have I equipped myself. Come, good O'Carrol, dost think I shall 'scape discovery in these accoutrements?

O'CARROL: Escape? By my soul, lady, one would think you had been a young man from the very first day you were born. Och, what a piece of work a little trimming and drapery makes in a good fellow's fancy! A foot is a foot all the world over, but take the foot of the sweetest little creature that ever tripped over greensward and, if it doesn't play at bo-peep under a petticoat, faith, I don't know the reason of it but it gives a contrary turn to a man's imagination. But what is it you would be after now, Lady Julia?

JULIA: I scarcely yet know myself. The horror of the circumstance, doubt, nay despair, have e'en bewildered me.

O'CARROL: I'll be shot, now, if that isn't my old luck, to a hair! I never yet set about laying a scheme for a cause I was warm in, but my heart so hurried my brains that it fairly tumbled them out of their places.

JULIA: Something I will do, and it must be speedy. At all hazards, we will to the English camp, O'Carrol. Opportunity must shape the rest.

O'CARROL: The camp? Faith, that's my element, and Heaven send us success in it! If an Irishman's prayers, lady, could make you happy, your little heart should soon be as light as a feather-bed.

JULIA: I thank thee, my honest fellow. Thy care for me shall not go long unrewarded.

O'CARROL: Now the devil fetch rewarding, say I! If a man does his best friends a piece of service, he must be an unconscionable sort of an honest fellow to look for more reward than the pleasure he gets in assisting them.

JULIA: Well, well, each moment is now precious. Haste thee, O'Carrol; time has wings!

O'CARROL: Och, be easy, madam; we'll take the old fellow by the forelock, I

---

we'll take the old fellow by the forelock: i.e. Time. Allegorical representations show Time with a bald pate out of which a solitary forelock of hair protrudes, allowing him to be tugged forward by anyone willing to seize the opportunity.

warrant him. When honest gentlemen's business calls them on a small walk to the gallows, a man may set out a quarter of an hour behind them and be certain of meeting them upon the road. And, now I bethink me, madam, if we go out at the drawbridge from the citadel hard by the house here, we may be at the camp ere the poor souls have marched their body round the battlements.

JULIA: Thou sayest well, and we will forth that way. 'Twill be most private, too. Thou'lt follow me, O'Carrol?

O'CARROL: Aye, that I would, to the end of the wide world. Indeed, now, but I begin to think my disposition has an ill-natured twist with it, for, when them I should wish happy are breaking their hearts, that's the very time I can't bear to be a moment out of their company.

JULIA: Yet tarry here awhile, till I prepare the means of our going forth. Join me a few minutes hence in the hall, O'Carrol.

(*apart*) Hold, hold, my heart, for thou hast need be stout
To bear me through this trial! What I lack
In natural firmness now, despair must execute.
And, Fortune, frown not on a poor weak woman
Who, if she fail in this, her last, sad struggle,
Is so surrounded by a sea of grief
That she must sink forever!
(*Exit* JULIA.)

O'CARROL: And sink or swim, I'll to the bottom along with you! — Och, what a sad thing it is to see sorrow wet the sweet cheeks of a woman. Faith, now, I can't make out that same crying, for the life of me. My sorrow is always of a dry sort that gives me a sore throat without ever troubling my eyes about the business! — The camp? Well, with all my heart, it won't be the first time I have been present at a bit of a bustle!

Song. O'CARROL.

When I was at home I was merry and frisky,
My dad kept a pig and my mother sold whisky,
My uncle was rich but would never be asy
Till I was enlisted by Corporal Casey.
Oh, rub-a-dub, row-de-row, Corporal Casey!
My dear little Sheelah I thought would run crazy
When I trudged away with tough Corporal Casey.

I marched from Kilkenny and, as I was thinking
On Sheelah, my heart in my bosom was sinking,
But soon I was forced to look fresh as a daisy
For fear of a drubbing from Corporal Casey.
Oh, rub-a-dub, row-de-row, Corporal Casey,
The devil go with him, I ne'er could be lazy!
He stuck in my skirts so, old Corporal Casey.

We went into battle; I took the blows fairly
That fell on my pate, but they bothered me rarely.

And who should the first be that dropped? Why, an plase ye,
It was my good friend, honest Corporal Casey.
Oh, rub-a-dub, row-de-row, Corporal Casey,
Thinks I, you are quiet, and I shall be asy!
So eight years I fought without Corporal Casey.

(*Exit* O'CARROL.)

Scene 2. *The English camp. A throne on one side of the stage and a scaffold towards the back of the scene.* TWO WORKMEN *descend as having just erected it.*

1st WORK: There 'tis, and finished, as pleasing a piece of work as man could wish to turn out of hand. If King Edward — Heaven bless him! — give me not a pension for this, let'n make the next scaffold himself. Mass, I would — with reverence be it spoken — build a scaffold and fix a gallows with any king in Christendom!

2nd WORK: Yea, marry, if he had not served his time to the trade?

1st WORK: Yea, marry, or if he had. I have been prime gallows-maker and principal hangman now nine-and-twenty years. Thank Heaven, neighbour, I have long been notorious.

2nd WORK: Thou sayest true, indeed. Thy enemies cannot deny thee that!

1st WORK: And why, I pray you, why have I been so?

2nd WORK: Mass, I know not. I think 'tis thy good luck.

1st WORK: Tut! I will tell thee. My parents, I thank them, bred me to the gallows. Marry, then, how was it? Why, look you, I took delight in my business. An you would be a good workman, ever, while you live, take a delight in your business. I have been an honest, painstaking man, neighbour. No one is notorious without taking pains for it.

2nd WORK: Truly, then, I fear my character is nought. I never can bring myself to take pains for it.

1st WORK: Thou art the more to be pitied. I never made but one small mistake since I entered on business.

2nd WORK: I prithee, now, tell me that.

1st WORK: 'Twas on execution day. We were much thronged and the signal was given full soon, when — a pize on it! — I whips me, in haste, the halter over the neck of an honest stander-by and I jerks me him up to the top of a twenty-foot gibbet! Marry, the true rogue escaped by't, for 'twas a full hour ere the error was noted. But hast heard who the six be that will be here anon?

2nd WORK: Only that they be citizens. They are e'en now coming hitherward; some of our men have seen them. They march, as 'tis reported, wondrous doleful.

1st WORK: No matter. Tarry till they see my work, that's all. An that do not content them, mark them for sour knaves! An a man be not satisfied when a sets foot on my scaffold, say he is hard to please. Rot them! Your condemned men, nowadays, have no discernment! I would I had the hanging of all my fellow craft: I should then have some judges of my skill, and merit would not go praiseless.

(*A flourish is heard.*)
So, the King is coming! Stand clear, now, neighbour. An the King like not my
scaffold, I am no true man!
> (*The* TWO WORKMEN *go over to the scaffold. Enter* KING
> EDWARD, QUEEN PHILIPPA, HARCOURT, SIR WALTER
> MANNY, ARUNDEL, WARWICK, TRAIN-BEARERS,
> STANDARD-BEARERS, *&c.*)

KING: Yes, good Philippa, 'tis our firm decree,
And a full wise one, too. 'Tis but just recompense.
For near twelve weary months their stubbornness
Has caused us linger out before their city.
Should we not now resent, in future story
Our English would be chronicled as dullards;
These French would mock us for the snails of war
Who bring our houses on our sluggish backs
To winter it before their mould'ring walls,
Nay, every village circled by a ditch
Would think itself a town impregnable,
Check the vigour of our march, and worry
Our armies with resistance.

QUEEN: And yet, my liege, I cannot choose but pity
The wretched men who now must suffer for it.

KING: Rather look forward, madam, and rejoice
That this example of their death, diffusing
Through France a terror of the English name,
Shall teach our enemy a quick submission
Which, henceforth in our wars, will timely save
The lives of thousands and let full many a soldier
On either side escape from carnage. Justice,
Minute in her stern exercise of office,
Is comprehensive in effect, and, when
She points her sword to the particular,
She aims at general good. —
> (*Solemn music is heard, at a distance.*)
>   But hark, they come!
Are they within our lines?

SIR WALTER: They are, my liege.

KING: Deliver up Sir John de Vienne.
> (KING EDWARD *and* QUEEN PHILIPPA *seat themselves on the
> throne. Enter* EUSTACHE DE ST PIERRE *with the keys of the
> town,* RIBAUMONT, LA GLOIRE, JOHN D'AIRE, JACQUE *and*
> PIERRE WISSANT *with halters round their necks, a multitude of*
> FRENCH TOWNSPEOPLE *following. The* ENGLISH SOLDIERY
> *are ranged on one side of the stage, the* FRENCH TOWNSPEOPLE
> *on the other.*)
Are these the six must suffer?

EUSTACHE: Suffer? No,

We do embrace our fate; we glory in't.
They who stand forward, sir, to yield their lives, —
A willing forfeit for their country's safety —
When they meet death, meet honour, and rejoice
In the encounter. Suffer is a term
The upright and undaunted spirit blots
From death's vocabulary.

KING: Now beshrew thee,
Knave, thou dost speak bluntly!

EUSTACHE: Aye, and cheerily.
But to our purpose: I am bidden, sir,
I and my noble comrades, here, of Calais,
Thus lowly, at your feet, to tender to you
Our city's keys.

   (EUSTACHE *kneels and lays the keys at the foot of the throne.*)
     And they do guard a treasure
Well worth a king's acceptance, for they yield
A golden opportunity to mightiness
Of comforting the wretched. Take but these
And turn our ponderous portals on the hinge,
And you will find, in every street, a document,
A lesson at each step, for iron power
To feel for fellow men: our wasted soldiers
Dropping upon their watch, the dying mother
Wailing for her famished child, the meagre son
Grasping his father's hand in agony,
Till their sunk eyes exchange a feeble gleam
Of love and blessing, and they both expire.

KING: Your citizens may thank themselves for't!
Had they, when first we did set down in arms
Before their town, — 'twould have shown wisely in them —
Surrendered to our power, those mighty mischiefs
Which you have taught your tongue to dwell upon,
And somewhat saucily, had been prevented!

EUSTACHE: Sworn liegemen to their master and their monarch,
They have performed their duty, sir! I trust
You, who yourself are king, can scarcely blame
Poor fellows for their loyalty? 'Tis plain
You do not, sir, for now your royal nature
O'erflows in clemency and, setting by
All thought of crushing those beneath your feet,
Which, in the heat and giddiness of conquest,
The victor sometimes is seen guilty of,
Our town finds grace and pity at your hands.
Your noble bounty, sir, is pleased consider
Some certain trifles we have suffered, such

As a bare twelvemonths' siege, a lack of food,
Some foolish greybeards dead by't, some few heaps
Of perished soldiers, and, humanely weighing
These nothings as misfortunes, spare our people,
Simply exacting that six useless citizens,
Mere logs in the community, and prized   .
For nothing but their honesty, come forth,
Like malefactors, and be gibbeted!

KING: Villain and slave! For this, thy daring taunt,
Howe'er before we might incline to listen,
Howe'er an humble, penitent demeanour,
Befitting this your state, might once have moulded, —
Spite of the ample cause of our resentment —
Our mind to lenity, we henceforth shut
The ear to supplication!

EUSTACHE: Mighty sir,
We marched not forth to supplicate, but die!
Had we come beggars to your camp, I fear
Our errand had been bootless. Trust me, King,
We could not covet aught in your disposal
Would swell our future name with half the glory
As this same sentence, which, we thank you for't,
You have bestowed unasked.

KING: Conduct them straight to execution!

LA GLOIRE: (*advancing to the left of* EUSTACHE) Father!

EUSTACHE: How now? Thou shakest!

LA GLOIRE: 'Tisn't for myself, then. For my own part, I am a man. But I cannot
     look on our relations, and my captain, and on you, father, without feeling a
     something that makes a woman of me. But I —

EUSTACHE: Briefly, boy. What is't?

LA GLOIRE: Give me thy hand, father. (*kissing* EUSTACHE*'s hand*) So. And now,
     if I part with it while a puff of breath remains in my body, I shall lose one of
     the most sorrowful comforts that ever poor fellow in jeopardy fixed his heart
     upon. Were I but well assured poor Madelon would recover the news, I could
     go off as tough as the stoutest.

RIBAUMONT: (*advancing to the right of* EUSTACHE) Farewell, old heart! Thy
     body doth incase
The noblest spirit soldier e'er could boast
To face grim death withal. Inform our fellows,
At the last moment given on the scaffold,
We will embrace and —
          (*A muffled drum beats.*)
                         Hark, the signal beats!

EUSTACHE: Lead on!
          (*The six condemned men march up to the scaffold.*)

SOLDIER: (*without*) You cannot pass!

JULIA: (*without*) Nay, give me way!
             (*Enter* JULIA *and* O'CARROL.)
    Stay your hands! Desist, or –
KING: How now, stripling?
    Wherefore this boldness?
JULIA: Great and mighty King,
    Behold a youth much wronged. My injuries
    Call loudly for redress. Men do esteem
    The monarch's throne as the pure fount and spring
    Where justice flows, and here I cry for it.
KING: What is the suit this urges?
JULIA: Please you, sir,
    Suspend awhile this fatal ceremony, –
    For therein lies my grief – and I will on.
KING: (*to the execution party*) Pause ye awhile! – Young man, proceed.
JULIA: (*aside*)   Now, Heaven,
    Make firm my woman's heart! – Most royal sir,
    Although the cause of this my suit doth wound
    My private bosom, yet it doth involve
    And couple with me a right noble sharer.
    'Tis you, great sir! You are yourself abused.
    My countrymen do palter with thee, King.
    Your dread decrees are slighted, your resentment
    Is turned to mockery and, by a juggle, –
    It would disgrace a giddy-pated schoolboy
    To be so put upon! – your royal sentence
    Is scorned and flouted at. You did require
    Six of our citizens, first in repute
    And best-considered of our town, as victims
    Of your high-throned anger. Mark you, now,
    How your vengeance is baffled. (*pointing to* RIBAUMONT) Here is one
    I single out and challenge to the proof.
    Let him stand forth, and here I do avouch
    He is no member of our city nor,
    Till some months back, set foot within our walls.
    He does usurp another's right, defeats
    Your mighty purpose, and your rage, which
    Thirsted for a rich draught of vengeance,
    Must be served with the mere dregs of our community.
RIBAUMONT: (*advancing*) Shame! I shall burst! The dregs?
KING: Thou self-willed fool,
    Who would run headlong into death, what art thou?
RIBAUMONT: A man. Let that content you, sir! 'Tis blood
    You crave, – and with an appetite so keen
    'Tis strange to find you nice about its quality! –
    No matter whence it flows. But for this slave,
    Who thus has dared belie me, did not circumstance

Rein in my wish, — oh grant me patience, Heaven!
The dregs? — now, by my soul, I'd crush the reptile
Beneath my feet, now, while his poisonous tongue
Is darting forth its venomed slander on me!

KING: I will be satisfied in this. Speak, fellow!
    Say, what is thy condition?

RIBAUMONT: Truly, sir,
    'Tis waste of royal breath to make this stir
    For one whom, some few minutes hence, your sentence
    Must sink to nothing. I would die in quiet.
    My story matters not. It is a tale
    Of private grief and bitter disappointment:
    Let that suffice. Henceforward I am dumb
    To all interrogation.

KING: Now, by our diadem — ! (*to* JULIA) But answer you.
    What is his state? Say, of whose wretched place
    Is he the bold usurper?

JULIA: Sir, of mine.
    He does despoil me of my title, filches from me
    My rights and franchises, and comes bedecked
    In my just dues which, as a citizen —
    A young one though I be — I here lay claim to.
    I am your victim, sir. Dismiss this man,
    Who, haply, comes in pity to my youth,
    An ill-placed pity which doth wound mine honour
    And plucks the glory from me which this ceremony
    Would grace my name withal, and let me die.

O'CARROL: (*aside*) Die? Och, the devil! Did I come to the camp for this?
    Madam!
    Dear, dear madam!

KING: The glory? Why, by Heaven, these headstrong French
    Toy with our punishments! The weighty scourge
    Our mightiness lets fall upon their heads,
    And justly, too, the villains glory in!
    For thee, rash stripling, who dost brave our vengeance,
    Prepare to meet it! Yoke thee with this knave,
    Whose insolence hath roused our spleen, and straight
    You both shall suffer for't together.

JULIA: (*kneeling*)   Sir,
    Ere I do meet my fate, upon my knees
    I make one poor request. This man, great sir,
    Though now there's reason why he knows me not,
    I own doth touch me nearly. I do owe him
    A debt of gratitude. 'Twould shock me sore
    To see him in his agony, so please you
    Command that, in the order of our deaths,
    I may precede him.

KING: Well, so be it, then.
    Guards, lead them forth!
JULIA: And might he, oh dread sir,
    Might he but live, I then should be at peace.
KING: No more of this! — (*aside*) Had this same strain been sounded
    When first they did appear, it might have worked,
    And in their favour, too.
JULIA: Can no prayers move you?
KING: Conduct them to their fate!
JULIA: (*rising*)   Then, ere we go,
    A word at parting: this, your sentence, sir,
    Tho' it fall heavy on the lower file
    And order of our people, is no jot
    Less bloody in its aim than if it pointed
    To the rich flower of our nation. Yet
    It is the safest dealing. Subtle policy,
    When it bears hard upon the lowly, dreads
    No powerful redresser in the cause
    Of wretched men on whom it gluts its vengeance.
    But here your prudence, sir, doth break its bound;
    The blood you now would spill is pure and noble,
    Nor will the shedding of it lack avengers.
    (*to* RIBAUMONT) Shame on disguise! Off with't, my lord! — Behold
    Our France's foremost champion and remember,
    In many a hardy fight, the gallant deeds —
    For fame has blown them loudly, King — of Ribaumont.
    Oft has he put you to't. Nay, late at Crécy,
    Ask of your Black Prince Edward, there, how long
    Count Ribaumont and he were point to point.
    He has attacked our foe, relieved our people,
    Succoured our town, till cruel disappointment
    Where he had fixed his gallant heart did turn him
    Wild with despairing love. Old John de Vienne
    Denied his daughter to him, drove him hither
    To meet your cruelty, and now that daughter,
    Grown desperate as he, doth brave it, King,
    And we will die together!
            (JULIA *runs and embraces* RIBAUMONT.)
RIBAUMONT: Heaven! My Julia!
    Art thou then true? Oh, give me utterance! —
          (RIBAUMONT *throws off his disguise.*)
    Now, Fortune, do thy worst! — You cannot, King,
    Nor dare not, for your life, lay savage hands

Black Prince Edward: Edward III's eldest son and Prince of Wales; he never succeeded to the throne owing to the unusual longevity of his father.

On female innocence! And for myself,
E'en use your will.
> (KING EDWARD *descends from the throne;* HARCOURT *kneels and offers his arm.* QUEEN PHILIPPA *descends and goes opposite to the* KING.)

KING: This staggers much
My steady purpose. Lady, you are free.
Our British knights are famed for courtesy,
And it will ne'er, I trust, be said an Englishman
Denied protection to a woman. (*to* RIBAUMONT) You
Must, under guard, my lord, abide our pleasure
Till riper thought doth give it utterance.
For the remainder, they have heard our will
And they must suffer. 'Tis but fit we prove,
Spite of their obstinate and close defence,
Our English excellence.

QUEEN: (*kneeling*)   Oh, then, my liege,
Prove it in mercy. Valour and compassion
Do character the Englishman: oppose him
And he's invincible; sue to him, he's conquered.
His adamantine heart hath waxen fibres,
Which, though in perilous hazard, cool and firm,
Is rock against attack, show him the fallen,
'Twill warm and melt for the unfortunate.
War, noble sir, when too far pushed is butchery.
When manly victory o'erleaps its limits,
The tyrant blasts the laurels of the conqueror.
Let it not dwell within your thoughts, my liege,
Thus to oppress these men. And, royal sir,
Since you were free to promise, in reminiscence
Of the poor service which my weak endeavour
Wrought in your absence for your realm, to grant
Whatever boon I begged, now, on my knee,
I beg it, sir: release these wretched men.
Make me the means of cheering the unhappy
And, though my claim were tenfold what it is
Upon your bounty, 'twould reward me nobly.

KING: Arise, our Queen. Though 'twas our fixed intent,
Not from a wanton thirst of blood — for, madam,
Far be it from our nature — but for reason,
Weighty and good, of which you were possessed,
To awe these French by terrible example,
Our promise still is sacred, good Philippa;
Your suit is won and we relax our rigour.
Let them pass free, while we do here pronounce
A general pardon!

QUEEN: Thanks, my gracious liege.

LA GLOIRE: A pardon? No! Oh diable! My father and my commander, too? Huzza!
(LA GLOIRE *takes the rope from* EUSTACHE*'s neck, then from his own, and runs downstage from the scaffold with* JOHN D'AIRE *and the two* WISSANTS.)
Oh, that I should live to unrope my poor old father — (*running to* RIBAUMONT *and taking the rope off his neck*) and master!
(*Enter* MADELON. *She and* LA GLOIRE *rush into each other's arms.*)

MADELON: Oh, my poor La Gloire! My tears —

LA GLOIRE: That's right, cry, Madelon! Cry for joy, wench! Old Eustache is safe, my captain and relations free! Here's a bundle of honest necks recovered; mine's tossed in, in the lump! And we'll be married, Madelon, tomorrow!

EUSTACHE: (*to the* QUEEN) Madam, to you we owe our thanks, and here
We pay them gratefully. (*to the* KING) For you, dread sir,
The labour of your war is now repaid
With this our city. When you enter it,
Look round upon the conquered; if your heart
Be noble, sir, as I do trust it is,
'Twill swell with rapture when you do reflect
Upon the pangs you have escaped by thus
Reversing your o'er-hasty order.

KING: (*to* RIBAUMONT)   Now, my lord,
For you. Old John de Vienne, it seems, denies
His daughter to you. Tell me on what ground.

RIBAUMONT: His sovereign's pleasure, sir. Alas, my lord,
The royal will designs her for another.

KING: We will have more reversals, then. The right
Which conquest gives us we will put to proof.
Nor will we quit your town until we see
Your marriage solemnised.

O'CARROL: Well, if I didn't know what crying was before, I have found it out at last. Faith, it has a mighty pleasant relieving sort of a feel with it!

KING: Prepare we, then, to enter Calais; straight
Give order for our march! And still may ever
Successes crown an English enterprise!
Breathe forth our instruments of war and, as
We do approach the rugged walls, sound high
The strains of victory!

Grand Chorus.

Rear, rear our English banner high
In token proud of victory!
Where'er our god of battle strides,
    Loud sound the trump of fame!
Where'er the English warrior rides,
    May laurelled conquest grace his name.

(*Exeunt omnes.*)

THE CHILDREN IN THE WOOD

A musical play in two acts by Thomas Morton

First performed at the Theatre Royal, Haymarket, Tuesday, 1 October 1793, with the following cast:

WALTER	Mr Bannister junior
LORD ALFORD	Mr Dignum
SIR ROWLAND	Mr Barrymore
OLIVER	Mr Caulfield
APATHY	Mr Suett
GABRIEL	Mr Benson
TWO RUFFIANS	Mr Waldron junior and Mr Cooke
SERVANT	Mr Lyons
BOY	Master Menage
JOSEPHINE	Mrs Bland
LADY HELEN	Miss De Camp
WINIFRED	Mrs Booth
GIRL	Miss Menage

Music by Dr Samuel Arnold

V   Jack Bannister as Walter listening to Josephine's song, 'The Norfolk
Tragedy', in act II, scene 6 of *The Children in the Wood*. Oil on
canvas by Samuel de Wilde

ACT I

Scene 1. *A room in* SIR ROWLAND's *Castle.* APATHY *discovered at a table with books and a bottle and glasses lying before him.*

APATHY: What a set of fools are philosophers, who advise to study away life for the benefit of posterity, that is, die while you live that you may live after you are dead! These things (*showing books*) may do well enough to garnish the brains of fools, but this (*showing the bottle*), this is the true feast of reason (*drinking*). As tutor to these orphans, I lead a tolerably easy life of it: I teach the children idleness — that's no difficult matter — I pimp for my patron, their uncle — that's no difficult matter — muster Latin enough to puzzle the curate — that's no difficult matter — go into the cellar for an hour or two — that's no difficult matter — come out again — that's no — yes, egad, that *sometimes* is a very difficult matter! (*drinking*)

        (*Enter* JOSEPHINE.)

JOSEPHINE: Oh fie, Mr Apathy, shame on you! What, drinking in a morning?

APATHY: Why, my dear, Sir Rowland bid me plead his passion for you and so I was just taking a drop to inspire me.

JOSEPHINE: I wonder Sir Rowland will continue his importunities. What can he have to say to a poor girl like me?

APATHY: He says he's unhappy, and how a man who has such a cellar as Sir Rowland has can be unhappy is to me amazing! But have you no feeling?

JOSEPHINE: Feeling indeed! Don't you remember when poor Walter the carpenter's house was burnt down?

APATHY: I have a shrewd guess that Walter has drilled a hole through your heart.

JOSEPHINE: Don't you remember, I say, that, instead of enquiring after the poor sufferers by the fire, the first question you asked was whether the young sucking pigs were safe? Was *that* feeling?

APATHY: No, that was philosophy.

JOSEPHINE: Philosophy!

APATHY: Yes, *my* philosophy! (*showing the bottle*) And this is the source from whence it springs. By eating, we arrive at the highest preferments of church and state. How do you arrive at the dignity of Lord Mayor? Why, you eat your way to it! And, by drinking, we approach the gods, who never walked — they slid!

Song. APATHY

    The true cause of all philosophical wrangling,
    Affirmatives, negatives and logical jangling,
    Is that truth is not dry, so no Stoic has found it,

Stoic: a member of the Greek school of philosophy founded by Zeno in the fourth century B.C. The central pivot of Stoicism was the belief that man's highest good was virtue, its antithesis being Epicureanism, a hedonistic system of thought which came into being in Greece only a few decades later. History has tended to transmit only the more vivid principles of Epicureanism, the popular imagination preferring those visions of unbridled sensuality which the term came to conjure up and which are portrayed here by Apathy.

And the doctrine of fluids alone can expound it.
 Then hither come each learned ass,
  Listen to me
  And you'll agree
  In vino veritas.

By philosophy, abstinence, study and care,
You may spurn at misfortune and smile on despair,
But I'm no such fool as employ a life in it
When, by help of my bottle, it's done in a minute!
 Then hither come, &c.

'Tis the grand panacea of vivification,
To this doctrine physicians give corroboration,
For, however their patients they julep or bolus,
This, this is the dose they take when they're *solus.*
 Then hither come, &c.

JOSEPHINE: (*looking out*) Ha? What do I see? Yonder comes my dear Walter! I wonder how he got admission. — Mr Apathy, go to the children. (*pushing him*) Now go!
APATHY: But what shall I say to Sir Rowland?
JOSEPHINE: Oh, say anything — what you please — only go! (*pushing him out*)
 (*Exit* APATHY, *reeling.*)

 Song. JOSEPHINE.

When love gets you fast hold in her clutches
 And you sigh for your sweetheart away,
Old Time cannot move without crutches,
 Alack, how he hobbles! Welladay!

But when *Walter* my trembling hand touches
 And love's colourings o'er my cheeks stray,
Old Time throws aside both his crutches,
 Alack, how he gallops! Welladay!

 (*Enter* WALTER.)
WALTER: My dear Josephine!
JOSEPHINE: Well, Walter, how do you do?
WALTER: Very well, Josephine, but I say it's devilish hard to be so poor, I, that

---

doctrine of fluids: the Archimedes Principle, which states that the loss of weight measurable in a body when wholly or partly immersed in a fluid is equal to the weight of the fluid it displaces. Morton's reference, however, does not pretend to take the definition of the term into account.
In vino veritas: 'In wine is truth.'
julep: a medicinal pill. Here Morton has concocted a transitive verb out of the noun in order to convey, in a condensed but recognisable form, the idea of the medicine administered.
bolus: a medicinal pill, especially one that appears to be uncomfortably large. Once again, Morton is using the noun in the place of a verb.

everybody says am such an industrious, clever fellow. Now a coffin, I'd make a coffin with e'er an undertaker in Norfolk. And at a bed, why the carpenters' wives say that at a bed I'm the very thing.

JOSEPHINE: I should not have thought, indeed, of your making beds for the carpenters' wives!

WALTER: Ah, Josephine, I'm making a bed for us, my girl.

Song. WALTER.

> There was Dorothy Dump would mutter and mump
> And cry, 'My dear Walter, heigho!'
> But no step she could take would my constancy shake,
> For she had a timber toe.
>
> There was Deborah Rose, with her aquiline nose,
> Who cried, 'For you, Walter, I die!'
> But I laughed at each glance she threw me askance,
> For she had a gimlet-eye
>
> There was Tabitha Twist had a mind to be kissed
> And made on my heart an attack,
> But her love I derided, for she was lopsided
> And cursedly warped in the back.
>
> There was Barbara Brian, who always was crying,
> 'Dear youth, put an end to my woes!'
> But, to save in her head all the tears that were shed,
> Nature gave her a bottle-nose.
>
> Josephine came at last to nail my heart fast,
> Firm as oak will I prove to my dear,
> And when Parson Feather has tacked us together
> Some chips of the block may appear.

Oh, curse that master of ours! I tell you what, Josephine, if you don't consent to run away from the Castle, I shall believe you listen to Sir Rowland.

JOSEPHINE: Lord, Walter, don't be a fool, now. I must ever fulfil the parting injunction of my dear Lady Helen when she went to India to meet her husband, Lord Alford. 'Josephine,' says she, 'thy affection for my dear infants is my greatest comfort. Should fate separate us, my faithful girl, protect them.' And while the poor little orphans are at the Castle, I am determined not to leave it.

WALTER: Ah, Heaven rest their souls! We have given up all hope of seeing Lord Alford and his sweet lady!

---

gimlet-eye: squint. A *gimlet* is a wood-boring tool, hence providing the allusion in this phrase to the piercing quality of eyes affected by this disability.
bottle-nose: Morton's lack of coyness about making physical deformity the butt of his humour in this song effectively demonstrates the high tolerance level of eighteenth-century society towards casual brutality, depicted disturbingly by Hogarth in his *Four Stages of Cruelty*.

JOSEPHINE: Hush! Sir Rowland's voice!

> (*Enter* SIR ROWLAND *and* OLIVER *with two* RUFFIANS.)

SIR ROWLAND: Ha? Walter here? Slave, your business?

WALTER: My business? Oh, my business, your honour, was a job.

JOSEPHINE: Yes, a job, sir.

SIR ROWLAND: A job was it? Seize that fellow, there!

> (RUFFIANS *seize* WALTER.)

WALTER: Oh Lord, here's a pretty job!

> (*The* CHILDREN *are heard laughing without.*)

SIR ROWLAND: Silence those brats and prepare them for a visit they must pay their gossips.

> (*The* CHILDREN *are heard laughing again.*)

Silence them, I say!

> (*Exit* JOSEPHINE.)

(*aside*) Soon their silence shall be eternal. My brother being concluded dead, that lustrous orb being set in night, shall these pigmy satellites eclipse me? No. That fellow (*pointing to* OLIVER) I am sure of. From his eye remorse is banished, and unmasked murder lowers upon his brow. He shall dispatch them while on this seeming visit, yet to venture him alone will breed distrust. Were it not well to ply this Walter? Relief from present fears, the hopes of Josephine, with large rewards backed with tenements and beeves, will surely ply the conscience of a hind. I'll about it, and use him as the peaceful scabbard to conceal the murdering blade. – (*to* OLIVER) Oliver, I have found you a companion for our purpose. Be ready. (*to* WALTER) Slave, follow me!

> (*Exeunt.*)

Scene 2. *Another apartment in the Castle.* APATHY *discovered asleep on a chair with books at his feet, the* CHILDREN *playing about the room.*

> (*Enter* JOSEPHINE.)

JOSEPHINE: What, asleep, Mr Apathy? (*awaking him*)

APATHY: Egad, I've had a very comfortable nap! What o'clock is it?

JOSEPHINE: Exactly midday. The children are going to visit their godfather directly.

APATHY: Is dinner ready yet?

JOSEPHINE: No, it isn't ordered.

APATHY: Not ordered? Oh Lord! The dinner not ordered? Talk to me of children and nonsense, and dinner not ordered? Here, cook! Cook!

> (*Exit* APATHY.)

BOY: Who goes with us to our godfather?

JOSEPHINE: Oliver, my dear.

BOY: I won't go with Oliver.

JOSEPHINE: Why, my love?

---

gossips: godparents.
beeves: an archaic plural of beef, i.e. oxen or cattle.

BOY: Because of what I heard Walter say.

JOSEPHINE: What was that?

BOY: Why, that Oliver was a damned black-looking rascal.

JOSEPHINE: Heavens, my dear! Hush! I shall scold Walter for talking so wickedly.

GIRL: I know you won't, though you say so.

JOSEPHINE: Why, my dear?

GIRL: Because of what I heard you say last night.

JOSEPHINE: I don't recollect it. What was it?

GIRL: Why, you cried out in the middle of your sleep, 'Oh, Walter, how I love you!' Oh, you did! And now I'm sure it's true because you blush so.

JOSEPHINE: Oh, you little tell-tale! – (*to the* BOY) Have you learnt my song yet?

BOY: I'll try to sing it if you will help me.

Duet. JOSEPHINE *and* BOY.

JOSEPHINE: Young Simon, in his lovely Sue,
Beheld a darling treasure.

BOY: Young Simon, in his lovely Sue,
Beheld a darling treasure.

JOSEPHINE: The toilsome day before him flew,
For love makes labour pleasure.

BOY: The toilsome day before him flew,
For toil makes love a pleasure.

JOSEPHINE: Oh fie, dear boy, can't you discern
'Tis love makes labour pleasure?

BOY: Oh yes, dear girl, I soon shall learn
That love makes labour pleasure!

JOSEPHINE: Oh fie?

BOY: Oh yes!

JOSEPHINE: Dear boy.

BOY: Dear girl.

JOSEPHINE: Oh fie, you can't discern!

BOY: Oh yes, dear girl, I soon shall learn
That love makes labour pleasure.
But I am loth to sour sweet music's strain.
Shall we begin? –

JOSEPHINE: Yes.

BOY: We will begin again!

(*The song is repeated over again by both together.*)

GIRL: Have you finished your song?

BOY: Yes.

GIRL: I'm glad of it, (*to her doll*) an't you, my darling? If you are good when I am gone, you shall have a Lord Mayor for your husband; so my dear Mamma said to me when she went away. And when I am gone, Josephine, don't you listen to the naughty men; my Mamma said that to you, too.

(SIR ROWLAND *is heard without.*)

Oh, here comes our cross uncle. Let's run away. I wish you were our uncle instead of him, dear Josephine.

JOSEPHINE *and* BOY: Come, let's run away!

> (*Exeunt* JOSEPHINE *and the* CHILDREN. *Enter* SIR ROWLAND
> *followed by* WALTER.)

WALTER: What, your honour?

SIR ROWLAND: Murder the children. That's my resolve. The reward: Josephine.

WALTER: Murder innocents? Tempt me in the form of an angel to do the act of a devil? — (*aside*) Damme, I have a great mind to throttle him! Eh, stop! Suppose I only seemingly consent, and then, if I can but save them — the very thought makes me cry for joy!

SIR ROWLAND: What, whimpering, fool?

WALTER: Consider, your honour, I'm not much used to butcher children. It's rather out of my line.

SIR ROWLAND: What's your determination?

WALTER: (*aside*) I must not consent too soon. — But then, to be scorned —

SIR ROWLAND: Look through the world. Where points scorn his finger? At ermined guilt? No, at houseless merit! It is not levelled at the wealthy cheat but at ragged honesty. Be wise! Be wise!

WALTER: Why, to be sure, as your honour says. But *my* honour —

SIR ROWLAND: Honour? That's a tinsel toy! Wise men plate it o'er with gold; that gives the worthless metal currency and brings wealth to the holder of it, think of that.

WALTER: Why, indeed, that's very true again. Very true. — (*aside*) Oh, the devil damn him! — Well then, your honour, I consent, and, if I don't —

SIR ROWLAND: Hush! Take this sword. But first swear.

WALTER: Oh, your honour, I never swear, never swear.

SIR ROWLAND: No trifling, fool, but swear when next we meet this sword shall be sheathed with blood.

WALTER: Well, for once I will swear. (*taking the sword*) By all my hopes of mercy hereafter, it shall be sheathed in blood.

SIR ROWLAND: Oliver will accompany you.

WALTER: Zounds, that blood-thirsty villain? You had better let me do it myself, your honour.

SIR ROWLAND: Silence! Now follow me.

> (*Exit* SIR ROWLAND *and* WALTER.)

Scene 3. *Another apartment in the Castle.*

> (*Enter* JOSEPHINE *and the two* CHILDREN.)

JOSEPHINE: Come, my dears, which of you will have your Mamma's picture?

BOY: I will.

GIRL: I'm sure I ought to have it; I'm a very funny little girl and ought to be made a pet of.

JOSEPHINE: She was an elegant woman.

GIRL: And everybody says I'm very much like her.

JOSEPHINE: (*looking out*) Ha? Walter in earnest conversation with Sir Rowland?

> (*Enter* WALTER *with a sword on.*)

WALTER: Oh, Josephine, I've such news to tell you as will make your hair stand on end: I am in high favour with Sir Rowland and am to go with the children to their godfather's.

CHILDREN: I'm glad Walter is to go with us.

WALTER: Josephine, you must know —

> (*Enter* SIR ROWLAND *with* OLIVER.)

SIR ROWLAND: Well, my little cherubs — What, delighted with your walk?

CHILDREN: Oh yes, uncle!

> (SIR ROWLAND *retires up the stage with the* CHILDREN.)

JOSEPHINE: (*aside to* WALTER) Why, Walter, you have got on a sword.

WALTER: (*confused*) A sword? Have I? Why, yes, it is a bit of a kind of a sword —

JOSEPHINE: What are you going to do with it?

WALTER: What am I going to do with it? Oh, Josephine, I've such a —

> (SIR ROWLAND *and the* CHILDREN *come downstage.*)

SIR ROWLAND: (*to the* CHILDREN) Well, take your leave of Josephine.

CHILDREN: Come, kiss us, Josephine. Good-bye, dear Josephine. Don't cry; we'll come back again, shan't we, uncle?

SIR ROWLAND: Certainly, sweetings. Farewell, and Heaven take you in its care!

WALTER: (*aside*) Amen, say I! — Come.

> (*Exit* OLIVER, WALTER *following with the* CHILDREN, *one in each hand.*)

JOSEPHINE: Heigho! I shall be quite uneasy till they return, and I can't bear melancholy.

SIR ROWLAND: How cruel, then, to inflict it.

JOSEPHINE: Pray, my lord, cease your importunities. Were your passion such as with honour I could listen to, I could not love you.

SIR ROWLAND: Josephine, hear me. I see persuasion is in vain. Mark, what has hitherto been entreaty shall now be force. Though love and gratitude be dead in you, fear, I perceive, exists. My purpose is determined. (*shouting to his attendants*) Confine her to her chamber! (*to* JOSEPHINE) Choose now my love or hate!

> (*Enter* APATHY *with a bill of fare.*)

Air. JOSEPHINE *and* APATHY.

JOSEPHINE: Great sir, consider, my honour is steady.

APATHY: Great sir, consider, the dinner is ready.

JOSEPHINE: A humble domestic is not worth your care.

APATHY: Dear sir, give me leave to present the bill of fare.

JOSEPHINE: Take a lady —

APATHY: Here's tongue —

JOSEPHINE: With honour —

APATHY: And mutton —

JOSEPHINE: If handsome and young —

APATHY: What a feast for a glutton —

JOSEPHINE: Dressed in bodice so fine and in kirtle so tasty —

APATHY: With bittern and quails and a venison pasty —

JOSEPHINE: But, ah, sir, beware of jealousy —

APATHY: And mustard —
JOSEPHINE: Else you'll prove by your care —
APATHY: A goose and a bustard —
JOSEPHINE: Your love is too hot —
APATHY: The mutton will be cold —
JOSEPHINE: My fame you would blot —
APATHY: And the pig will be spoiled —
JOSEPHINE: Believe me, great sir, to my honour I'm steady.
APATHY: And believe me, great sir, the dinner is ready.

(*Exeunt.*)

Scene 4. *A wood and cut wood.*

(*Enter* OLIVER *through cut wood. He looks around, then beckons
to* WALTER *who comes forward with the* CHILDREN.)

BOY: It's a long way, Walter, to our godfather's.

WALTER: (*aside to* BOY) Yes, dreary it is —

OLIVER: I say, Walter, this place will do delightfully!

WALTER: Nay, I don't much like this place. Let's find some other.

OLIVER: I say *this* shall be the place.

WALTER: (*aside*) Shall it? — (*to* CHILDREN) There, little dears, go and play,
    while I talk to Oliver a bit.

(*The* CHILDREN *go to play at the back.*)

    I tell you what, Oliver, I know you have one failing.

OLIVER: Aye? What's that?

WALTER: Why, you are too tender-hearted.

OLIVER: Am I?

WALTER: You are indeed. Now I am, you know, such a blood-thirsty rascal that I
    could murder for amusement. Therefore, I say, Oliver, suppose you leave this
    job to me.

OLIVER: What, you'll dispatch them, will you?

WALTER: Yes, to be sure on't. So, my dear fellow, you may go back to the Castle,
    get the reward, and leave them to me. (*endeavouring to urge him*) Go! Yes,
    yes, you're too tender-hearted, one may see it with half an eye. So, good-bye!
    I'll do for them! Good-bye, Oliver! (*attempting to push him off*)

OLIVER: Why, you must think me a pretty scoundrel —

WALTER: (*aside*) Why, I do, for that matter.

OLIVER: To receive money for doing a bit of work and not completing it. But
    I'll —

WALTER: (*stopping him*) Stay a little. I see you can't bear the thoughts of it. This
    is all put on. Your heart melts — (*aside*) how savage he looks! — and there's a
    tear standing in the corner of your eye (*wiping one from his own*). Oliver,
    how pity becomes you! I say, Oliver, suppose we were — suppose we were —
    just — just to — to save them.

OLIVER: What?

WALTER: I say, my dear fellow (*leaning on* OLIVER's *shoulder*), suppose we were
    — how pretty and innocent they look! — we were to save them.

OLIVER: To save them? Eh?

WALTER: Me? Save them? Oh, very well, we will. You wish it, and I consent!

OLIVER: (*impatiently*) Why should we save 'em?

WALTER: I don't know. There are two or three *trifling* reasons, to be sure. First, it isn't very manly to murder innocents; next, we shall be damned for it; and —

OLIVER: Why, an't you a pretty rascal!

WALTER: Well, Oliver, you must consent to save 'em! Look at 'em, poor little dears!

OLIVER: No more words! (*drawing his sword*) I am determined, so —

    (OLIVER *runs to the* CHILDREN. WALTER *seizes his arm. The* CHILDREN, *frightened, come down and kneel to* WALTER.)

CHILDREN: Oh, Walter, save us!

WALTER: Stop, Oliver. Only two questions more.

OLIVER: Well?

WALTER: Look at them. Have you a heart hard enough to kill 'em?

OLIVER: I have.

    (WALTER *runs between* OLIVER *and the* CHILDREN.)

WALTER: Why then, have you an arm strong enough to fell me down, you damned dog?

    (WALTER *draws his sword directly.*)

OLIVER: Fell you?

WALTER: Yes, for you must do that before you shall touch a hair of their heads!

OLIVER: Indeed? We'll try that!

    (OLIVER *and* WALTER *fight, the* CHILDREN *hide behind a tree.* OLIVER *gains ground upon* WALTER *and strikes the sword out of his hand. The* GIRL *runs and picks up* WALTER's *sword, gives it to him just as* OLIVER *is aiming to run him through the body.* WALTER *renews the fight and kills* OLIVER *offstage. Re-enter* WALTER *with his sword and hand bloody.*)

WALTER: Damme, I didn't think I had so much pluck in me! There he lies. Come forth, my little tremblers, I am your champion.

CHILDREN: Have you killed Oliver?

WALTER: Dead as a door nail!

BOY: Go kill him again. Such a rascal as he cannot be too dead.

GIRL: Oh, poor Walter, your hand bleeds. Here, I'll bind it up (*tearing off some of her garment and binding up Walter's hand*). Look, I'll kiss it and make it well.

BOY: Shall we return to our uncle's, Walter?

WALTER: Alas, poor dears, you have no home! — Let me consider what's best to be done. I have it: I'll leave them here, return to that rascal their uncle, get the reward and Josephine, and steal something from the buttery. Then we'll go far enough out of the reach of that villain. — I say, dears, I'll go and bring Josephine to you. Will you stay here till I come back?

BOY: We'll do anything that Walter bids us.

WALTER: I'll soon come back. Look, here's a pretty arbour, and here's my cloak to sit down upon, and here are victuals. Now don't stir from this spot, I charge you. Good-bye, I won't be long. — Oh Lord, oh Lord, if doing one

worthy action gives such joy, how happy might the great be who have opportunities of doing them daily!

>   (*Exit* WALTER.)

BOY: Look, sister, what quantities of blackberries and nuts there are in that bush.

GIRL: Let's go pluck them. We can soon find our way back again, you know, and
they are nicer than the beef and manchets Walter left us.

BOY: I should like to live here always, to have nothing to do but to play all day,
catch birds and eat berries!

Song. GIRL.

>   See, brother, see on yonder bough
>   The robin sits! Hark, hear it now!
>   Listen, brother, to the note
>   From pretty Robin Redbreast's throat,
>   Sweetest bird that ever flew,
>   Whistle Robin, loodle loo!
>   Loodle loo, sweet Robin!

>   (*As the curtain drops, the* CHILDREN *retire hand in hand up the
>   stage.*)

ACT II

Scene 1. *An apartment in the Castle.*

>   (*Enter* SIR ROWLAND *followed by a* SERVANT.)

SIR ROWLAND: To speak with me? Is it Oliver or Walter? Heaven forfend any ill
should come to my children.

SERVANT: I never saw this man before, sir. He says his business is urgent.

SIR ROWLAND: Admit him.

>   (*Exit* SERVANT.)

Who can it be?

>   (*Enter* GABRIEL, *drunk.*)

(*aside*) Ha? My brother's servant? Should he be alive? — Gabriel, I am glad to
see you!

GABRIEL: The joy is mutual, your honour. But your honour looks a little pale.
Your countenance hasn't that rosy, healthful appearance mine has!

SIR ROWLAND: Grief, Gabriel.

GABRIEL: True, your honour, grief brings on drinking, and then what is man? Oh,
never drink, your honour, never drink!

SIR ROWLAND: (*aside*) Now to know my fate. — I shall soon meet my brother
where grief cannot come.

GABRIEL: True, you'll meet very soon.

SIR ROWLAND: (*aside*) All's safe, I find! — Where are my brother's sad remains?

GABRIEL: Remains? Oh, he remains but a little way off, your honour.

---

manchet: dialect word for a small loaf or roll usually made of fine wheaten bread.

SIR ROWLAND: Gabriel, this drunken guise little becomes your mournful errand!

GABRIEL: Why, you see, your honour, I was sent before to get everything in readiness, but living on salt provisions at sea gave me such a confounded thirst that I was forced to stop every mile just to moisten my mouth with a quart of ale. So, on my second day's journey, my master overtakes me. 'So,' says he — says he — 'Gabriel,' says he —

SIR ROWLAND: Says? Who says?

GABRIEL: My master, I tell you! 'Gabriel,' says he, 'I discharge you.' But my sweet mistress cried I might stay, 'for,' says she, 'if ever we part with Gabriel, we shall lose the only sober servant we have got'. So my master only gave me a kick and sent me forward again.

SIR ROWLAND: Idiot! Wretch! He's dead!

GABRIEL: Dead, is he? I could show you the mark he made with his foot, and if you call that a blow for a dead man to give, why — However, if you won't credit the mark of his foot, here's the mark of his hand.

(GABRIEL *shows a letter.* SIR ROWLAND *snatches it.*)

SIR ROWLAND: Damnation!

GABRIEL: Damnation? A comical way of expressing joy! 'Your brother's arrived,' says I. 'Damnation,' says he! But I hope your honour has taken care of the children.

SIR ROWLAND: (*poring over the letter*) Aye, aye, they're taken care of.

GABRIEL: If that cursed thirst had not seized me, I would have been here yesterday.

SIR ROWLAND: Oh, had you come but yesterday! — Begone! Leave me, drunkard!

GABRIEL: Yes, your honour, I'll go to the cellar, for I feel a kind of dryness on the palate yet. Your brother and his lady will soon be here, your honour; they were not far behind me, and I have a notion I didn't come here quite straight!

(*Exit* GABRIEL, *reeling.*)

SIR ROWLAND: Confusion! Ruin! Yet, if the hand of Heaven has been stretched forth to save the innocent, if the children live! —

(*Enter* WALTER *with caution.*)

(*to* WALTER) Say quick!

(WALTER *advances, draws the sword and shows it bloody.*)

It is concluded. Where's Oliver?

WALTER: Gone, Heaven knows whither. I have fulfilled my oath. Just mention the reward, your honour, the prize of angels, your honour, Josephine, your honour, the —

SIR ROWLAND: Wretch! Murderer! Avoid me! Take my curses; such be ever the reward of villainy!

WALTER: (*aside*) So say I — But, your honour, consider. I killed —

SIR ROWLAND: Slave, dare but to name the foul act and, by hell, thou shalt be rewarded! A halter-villain! Go from the haunts of men and devour thy heart in misery and contempt!

---

halter-villain: i.e. a gallows-bird; someone who deserves to be hanged.

WALTER: I should be a devil of a fool to do that. Make a companion of my conscience. Does your honour find yours so pleasant a one?

SIR ROWLAND: Leave me fellow, or — (*putting his hand on his sword*)

WALTER: I go. I'm gone, sir. — (*apart*) Heigho! (*putting his hand on his heart*) What would he give to do this! Now to steal something from the buttery, endeavour to find Josephine, and away again to the children. Aye, fret and fume! They say villains inflict misery on their fellow creatures, but I think they can make none so miserable as they make themselves!

    (*Exit* WALTER.)

SIR ROWLAND: Lost beyond hope! How shall I act? How? How? But on. My purpose was my brother's family should meet in Heaven, and it shall be accomplished. I'll chaunt my coffers and to some thriftless rascals throw down the dazzling ore, and, while their senses are misled by the damning dear delusion, I'll lead them to destroy this hated brother. Fortune continue dull and blind! Now for happiness — or perdition!

    (*Exit* SIR ROWLAND.)

    Scene 2. *A wood.*

    (*Enter the* CHILDREN *at the top of the stage, hand in hand, picking berries &c.*)

BOY: (*supporting the* GIRL) How do you do, sister?

GIRL: Very tired and very hungry. I could eat some of the meat Walter left us.

BOY: I wish we hadn't left the place. Let's try to find it.

GIRL: I can't. Indeed, I can't. I'm so sleepy, and the wood turns round. But brother, as we may sleep a long time, look, I'll put my Mamma's picture here (*kissing the picture, giving it to her brother to kiss and then putting it in her bosom*), for Josephine told me if I were sick and should sleep a long while I should go where my Mamma is. So, she'll know us by the picture.

    (*A storm begins. Thunder. The* CHILDREN *appear frightened and cling together.*)

BOY: Are you frightened, sister?

GIRL: (*trembling*) No, not much.

BOY: Look, yonder's a place to hide us. I'm sure the thunder can't shoot us there. Come, sister.

GIRL: I can't walk. Indeed, I can't. I'm so sick. — Don't cry, brother.

BOY: I don't cry.

    (*Thunder*)

Do try to walk a bit. There, see, I'll help you. Very well! Very well!

    (*Thunder and rain. Exit* BOY, *supporting the* GIRL.)

---

chaunt my coffers: a colloquial usage of *chaunt* (chant) particular to this period was in connection with fraudulent selling, especially of horses. In this context, Sir Rowland, by chaunting his coffers, is intending to trade under false pretences by not declaring, at the time when he scatters coins under the noses of 'some thriftless rascals', that the money is intended as an advance payment to them for an act of murder. Hence the full force of Sir Rowland's reference to 'the damning dear delusion'.

Scene 3. *Another part of the wood. A cloak laid out on the stage.*

(*Enter* WALTER *with a basket.*)

WALTER: Zounds, what a peppering storm! Sweet souls, how glad they'll be to see me. The cunning rogues have got under the cloak and, I dare say, have got fast asleep.

(WALTER *sets down the basket, withdraws the cloak and starts at not seeing the* CHILDREN. *Runs to the front of the stage and looks about.*)

What a damned villain I am! Gone? Murder! Murder!

(WALTER *runs up the stage and pauses.*)

Oh, they have hid themselves to frighten me! I see you, I see you! You may as well come out, I see you! (*pausing*) They're gone! They're gone! Can I ever sleep again? — Ha! The print of a foot!

(*Exit* WALTER *pursuing the footprints. Re-enters greatly alarmed at not finding the* CHILDREN.)

What the devil do I stand here for? I'll roar myself dumb, I'll — Hollo! Hollo!

(*Exit* WALTER *running.*)

Scene 4. *A road on the outskirts of the wood.*

(*Enter* SIR ROWLAND, *masked, and two* RUFFIANS, *armed. A whistle is heard.*)

SIR ROWLAND: Look-out?

1st RUFFIAN: The travellers have gained the hill and are dismounted.

SIR ROWLAND: 'Tis well. Behind that thicket wait their approach. Be firm. (*giving* 1st RUFFIAN *a purse*) Here's encouragement. This way, this way!

(SIR ROWLAND *and the* RUFFIANS *retire. Enter* LORD ALFORD, LADY HELEN *and two* SERVANTS.)

LORD ALFORD: Thou art weary, Helen.

LADY HELEN: In truth, most sadly. But let us on.

LORD ALFORD: No, here rest awhile. This place is most dear to my remembrance. When my good falcon urged on his quarry to this forest's verge, reclined beneath this aged oak I first saw thee, my Helen.

LADY HELEN: Ah, those times, my Alford! What were then our hopes and fears! The remembrance is strong within me still.

Song. LADY HELEN.

Mark the true test of passion where a lover is nigh,
  Its hue is the rose, its language a sigh,
But where doubts interfere and no lover is nigh,
  Its hue is a lily, its language a sigh.

But look, my lord, this avenue displays your Castle's stubborn turrets. The western tower contains our lovely children. Oh, how sweetly fancy, passing the bounds of vision, pictures to me my babes, at great Nature's bidding, stretching forth their little hands to clasp a mother! The thought is rapture! On, on, my dear lord! You never saw the youngest. Indeed, he's most like you, the image of my Alford. — Pardon these foolish tears; they are a mother's joy.

SERVANT: (*looking out*) Master, defend yourself!
           (LORD ALFORD *puts* LADY HELEN *behind him. The* RUFFIANS
           rush on LORD ALFORD *and his* SERVANTS. *One of the*
           RUFFIANS, *with* SIR ROWLAND, *attacks* LORD ALFORD, *the*
           other RUFFIAN *attacks the* SERVANTS *and beats them off. Then*
           enter WALTER *from the wood.*)
WALTER: What, two to one?
           (WALTER *attacks* SIR ROWLAND, *wounds him, and drives the*
           RUFFIANS *off.* LORD ALFORD *retires into the wood with* LADY
           HELEN. *Re-enter the two* RUFFIANS *who go to support* SIR
           ROWLAND.)
1st RUFFIAN: Are you hurt, sir?
SIR ROWLAND: Heed not that. Have you succeeded?
1st RUFFIAN: No, sir. The travellers escaped in the wood.
SIR ROWLAND: Providence, I thank thee!
1st RUFFIAN: Shall we pursue them?
SIR ROWLAND: No, on your souls, forbear! Convey me to the Castle.
1st RUFFIAN: Shall I fly for assistance?
SIR ROWLAND: No, I'll none. Do as I ordered you.
           (*Exit* SIR ROWLAND, *the* RUFFIANS *supporting him. Enter*
           WALTER.)
WALTER: What the devil does all this mean? Where are the people I've been
    fighting for? Where are the people I've been fighting with? — I'm pretty sure
    I've killed one of them. Damme, now my hand's in, I suppose I shall be killing
    a man every day! But these poor children — no finding them. I'm almost mad.
    Night coming on, too. — Ha? Another ruffian? I'll soon do his business!
           (WALTER *runs off, as if in pursuit.*)

        Scene 5. *A wood. Moonlight. Lamps down. On a bank of flowers the*
        CHILDREN *are discovered, seemingly dead, folded in each other's arms,*
        with leaves strewed over them.

           (*Enter* LORD ALFORD *and* LADY HELEN *from the top of the*
           stage, LORD ALFORD *supporting his wife.*)
LORD ALFORD: Courage, my Helen.
LADY HELEN: I'm wondrous faint.
LORD ALFORD: Droop not, my love, we are safe. Here we'll remain tonight.
LADY HELEN: 'Twas most strange. Spoil was not their aim, but blood. A
    thousand fears press on me. The vizored ruffian had an air methought of —
LORD ALFORD: Dearest Helen, calm thy troubled mind. Rest on that verdant
    bank.

Lamps down (s.d.): the remote control of stage lighting did not become possible until the
advent of gas. In 1794, the Haymarket stage was lit by over forty independent oil lamps, and
the difficulties presented by their rapid and simultaneous adjustment should indicate the
theatrical significance attaching to a direction such as 'lamps down'.

(LADY HELEN *reclines on the bank opposite to where the*
CHILDREN *are.*)

My servants, ere this, have gained the Castle. I'm sure my brother's anxious
care will find us ere the morning.

Song. LORD ALFORD.

When first to Helen's lute
    I sung, as she played to me,
How came this then to shoot
    A thrilling sense all thro' me?
        Oh, 'twas love! 'Twas love!
In my eyes it glistened;
    'Twould inspire a brute
To sing, if Helen listen'd.
        Oh, my love! My love!

Why call I with delight
    This ditty's plaintive numbers?
To wrap my fair in night
    And sooth my Helen's slumbers.
        Oh, 'tis love! 'Tis love!
Lullaby, my dearest,
    Care from thee take flight,
And peace thy heart be nearest,
        Oh, my love! My love!

She sleeps. I'll forth and, under covert of the friendly shade, descry if danger
be aloft.

(LORD ALFORD *advances to where the* CHILDREN *are and, seeing*
them, starts.)

Heavenly powers, what's here? Two infants, and cold e'en to death! Poor
wretched babes. Poor wretched parents, what pangs must rend their hearts!
How shall I thank thee, Heaven, for giving mine a brother's fostering care.

(LORD ALFORD *takes the* GIRL *in his arms.*)

Cold! Cold and breathless! — Hold, life seems newly ebbed!

(LORD ALFORD, *putting his hand on the* GIRL's *chest, pulls out*
the picture, comes forward and, looking on it, exclaims:)

Merciful powers, my own children!

(LORD ALFORD *holds the picture tremblingly in his hand.* LADY
HELEN, *alarmed, awakes.*)

LADY HELEN: My Alford!

(LADY HELEN *advances and snatches the picture from* LORD
ALFORD. *Looking on it, she shrieks, falls on the bank, and*
embraces the BOY.)

My child! My child! My darling boy! Dead?

(*Tearing off some of her garments,* LADY HELEN *wraps them*
round the BOY *then takes him in her arms.* LORD ALFORD *takes*
up the GIRL.)

LORD ALFORD: How chillingly cold! Brother? Barbarian!
LADY HELEN: Hush!
> (LADY HELEN *feels for pulsation from the* BOY *and describes by her manner she perceives it. With great anxiety she turns her eyes on the* BOY, *whose own eyes, after a short pause, open.*)
> Oh God, he lives! (*clasping her hands*) He lives, my husband! – Oh, I fear to ask, how is my girl?
LORD ALFORD: She will recover.
LADY HELEN: How came they here? – But let's away.
LORD ALFORD: At the eastern extremity of this forest stands a humble cottage. There we'll hasten. Thy feeble arms cannot sustain –
LADY HELEN: Away, away! Under my own disasters I might droop, but a mother's fears have Amazonian strength. Away, my lord!
> (*Exeunt* LORD ALFORD *and* LADY HELEN *bearing the* CHILDREN.)

Scene 6. *The inside of* WALTER's *house, door open.*
> (*Enter* JOSEPHINE, WINIFRED *and a female* SERVANT.
> WINIFRED *and* SERVANT *go to a table on which are placed wooden trenchers, a roast fowl, knives and forks, &c. Lamps up.*)
WINIFRED: (*speaking as she enters*) I thought so. Well, and so?
JOSEPHINE: And so, goody, a servant came to the Castle and Sir Rowland ordered him to be confined in the Dark Tower. And, do you know, old King says it is a servant of Lord Alford's.
WINIFRED: I thought so. Well, and so?
JOSEPHINE: Why, goody, then Sir Rowland went out disguised, with four men, and in the confusion I slipped out. But, goody, where's Walter?
WINIFRED: Ah, Heaven knows whether we shall see the dear boy again!
JOSEPHINE: Oh dear, you frighten me! Why, goody?
WINIFRED: Why? Do you know I saw a spider crawl up the side of the chimney? And the horseshoe was last night taken off the door!
WALTER: (*without*) Hallo!
> (*Enter* WALTER, *at door, and shuts it after him.*)
JOSEPHINE: Here *is* Walter.
WINIFRED: I thought so.
> (WALTER *is in extreme dejection and looks pale. Takes a chair, brings it forward, and sits down.*)
WINIFRED: Why, child, what's the matter? Have you seen a ghost? Sit cross-legged my dear boy.
WALTER: There, will that please you?
> (JOSEPHINE *taps* WALTER *on the shoulder. He jumps up, alarmed.*)
> Ah, Josephine, is it you?

Brother? Barbarian!: Morton's first attempt at finding an appropriate epithet for Lord Alford to use against his villainous brother led him, unfortunately, to 'onanist' (*vide* Genesis xxxviii 9). The censorious eye of the Reader of Plays retrieved the young playwright from this enthusiastic leap into professional suicide by expurgating the reference from the licensing manuscript.

JOSEPHINE: Well, Walter, where did you leave the children?

WALTER: Under a tree, and told them to stay there till I —

JOSEPHINE: Under a tree? Oh, in the gentleman's garden —

WALTER: No, no! (*recollecting himself*) Yes, yes, to be sure. Where else should I leave them? In a wood where they might be starved?

JOSEPHINE: No, that, I'm sure, you would not.

WALTER: I never was afraid of goblins, but tonight I thought every cow a ghost and took old Fowler for the devil.

WINIFRED: Aye, aye, old Tab did not scratch under her left ear for nothing. A sure sign that somebody will be hanged.

WALTER: Damn old Tab!

JOSEPHINE: Fie, Walter, you have been drinking.

WALTER: (*aside*) My own tears, then.

WINIFRED: But come, here's a capon for your supper.

WALTER: Oh, if the dear children had that capon!

JOSEPHINE: Lord, Walter, why they have plenty!

WALTER: Plenty, have they? (*recovering*) To be sure they have. (*peevishly*) I know that as well as you, Josephine.

JOSEPHINE: (*weeping*) Had I known how cross you would have been, I would not have come.

WALTER: I beg your pardon, Josephine. Don't cry, my girl. I'm almost mad.

      (WALTER *sits down on the side of the table, knocking over the salt.*)

WINIFRED: Oh, he's spilt the salt! (*throwing some over her shoulder*) And I see there's a winding-sheet in the candle.

WALTER: Damn it, mother, don't frighten me so! Josephine, my dear girl, sing me a song. I can't eat. I'm not well.

JOSEPHINE: I'll sing you what I bought of the old blind pedlar who passed by this morning. It's called 'The Norfolk Tragedy', showing how the ghost of a murdered babe —

WALTER: No, don't sing that!

WINIFRED: Yes, yes, sing it, Josephine.

Song. JOSEPHINE.

        A yeoman of no mean degree,
        For thirst of gain and lucre, he
        A pretty babe did murder straight
        By reason of its large estate.

        To vex him to his heart's content
        To him the murdered babe was sent;
        Full blue appeared the candle flame
        And a knocking at the window came.

      (*At the end of this verse a knock without at the window is heard. All start up from the table,* WALTER *extremely alarmed.*)

JOSEPHINE: Walter, why do you tremble? Are you frightened?

WALTER: Me? Frightened? Bless your soul! Nonsense! Go on.

JOSEPHINE: His conscience sorely smited him
And made him tremble, every limb;
With that, the ghost began to roar
And straightway bursted ope the door.

(*A knocking at the door is heard without. They all start as the door is burst open.* WINIFRED *and* JOSEPHINE *hide themselves.* WALTER *remains near the table, fearing to look towards the door.*)

WALTER: Mother! Mother! Mother! Don't run away!

(*Enter* LORD ALFORD *and* LADY HELEN *and the two* CHILDREN. *The* CHILDREN *run to* WALTER.)

CHILDREN: Walter! Walter!

(WALTER *eyes the* CHILDREN *with fear and by degrees recovers himself from the idea of their being ghosts.*)

WALTER: What, alive? Oh Lord! Oh Lord! Oh Lord!

(WALTER *falls on his knees, hugs the* CHILDREN, *laughs, cries, and shows the most extravagant signs of joy.*)

What, my honoured lord and lady, too? Oh, 'tis too much! Josephine, come here. Down on your knees.

LADY HELEN: My faithful girl, explain these wonders.

JOSEPHINE: Indeed, my lady, I cannot. Walter can.

WALTER: Me? I know nothing. Yes, I know everything. You see, my lord, your brother – (*to* CHILDREN) ah, you little rogues to run away! – and so, my lord, your brother sent – (*looking again at the* CHILDREN *and laughing*) – and I, my lord, I, I – It does not signify! I cannot tell you now! (*going to the* CHILDREN *and kissing them*)

GIRL: I'm very hungry.

WALTER: Hungry, are you?

(WALTER *snatches up the* CHILDREN *and carries them to the table.*)

APATHY: (*without*) Let none pass!

(*Enter* APATHY *with* CONSTABLES.)

Let none pass! Seize that murderer of innocents! Bring him away!

(WALTER *laughs.*)

Do you laugh, you murderer?

WALTER: Laugh? (*showing the* CHILDREN) Look there.

APATHY: Bless my soul, there they are at supper! A capon, I declare! Very pretty eating!

(APATHY *is about to eat when he sees* LORD ALFORD *and* LADY HELEN.)

Oh, my lord, your brother is dying! He has confessed he employed Oliver and Walter here to murder your children.

WALTER: True. And I killed Oliver.

LORD ALFORD: My gallant fellow!

APATHY: He then planned your destruction –

LORD ALFORD: Accursed ambition! Wretched brother!

APATHY: And went out with armed ruffians to attack you.

LADY HELEN: But Heaven sent an unknown friend to save us. Walter, could'st thou but find him.

WALTER: (*with modest hesitation*) Why, my lady, I could find him, I believe.

LADY HELEN: Sure, that look — Walter, you protected us!

WALTER: Why, I believe I did.

LADY HELEN: My preserver! (*taking* WALTER*'s hand*)

LORD ALFORD: My friend! (*also taking* WALTER*'s hand*)

WALTER: Dear master, sweet dear lady, don't kill me with kindness! I can't bear it; I'm too happy, Could ill-gotten wealth do this?

LADY HELEN: Name some reward.

WALTER: A treasure!

LORD ALFORD: If India can produce it, it is yours.

WALTER: My lord, you need not go so far. There's the treasure I want. Give me my little Josephine and I am happy.

LADY HELEN: (*to* JOSEPHINE) My dearest girl, receive from my hand your faithful Walter. It shall be my anxious care to reward his virtues.

WALTER: Madam, I'll serve you with my latest breath. But I trust the children in the wood will tonight find better friends than poor Walter the carpenter.

Finale. WALTER, JOSEPHINE, LORD ALFORD *and* LADY HELEN.

WALTER: Have I saved this girl and boy?
                   Is't so understood, sirs?
                   May I hollo now for joy?
                       Are we out of the wood, sirs?
                       Have I saved, &c.

LORD ALFORD: Providence has smiled on me,
                   Happy I as may be,
                   A father here, at either knee
                       A rosy dimpled baby.
                       Have we saved, &c.

JOSEPHINE: Now my Walter I shall wed,
                   Gay my heart, and light, sirs —

WALTER: And I, my girl, have made a bed
                       To fit us right and tight, sirs!
                       Have we saved, &c.

LADY HELEN: Fullest mine of mother's bliss,
                   Fuller nought can make it,
                   Since all tonight who witness this
                       Seem kindly to partake it.
                       Have we saved, &c.

BLUE BEARD, or FEMALE CURIOSITY!

A dramatic romance in two acts by George Colman the Younger

First performed at the Theatre Royal, Drury Lane, Tuesday, 16 January 1798, with the following cast:

ABOMELIQUE	Mr Palmer
IBRAHIM	Mr Suett
SELIM	Mr Kelly
SHACABAC	Mr Bannister junior
HASSAN	Mr Hollingsworth
SPAHIS	Mr Sedgwick, Mr Bannister senior, Mr Dignum, Mr Wathen, Mr Trueman and Mr Maddocks
JANIZARIES	Mr Danby, Mr Wentworth, Mr Brown, Mr Tett, Mr Denman, Mr Atkins, Mr Phillimore, Mr Fisher, Mr Meyers, Mr Peck, Mr Bardoleau, Mr Walker, Mr Cook, Mr J. Fisher, Mr Dibble and Mr Simpson
MALE PEASANTS	Mr Grimaldi, Mr Gregson, Mr Gallot, Mr Aylmer, Mr Potts, Mr Willoughby and Mr Evans
MALE SLAVES	Mr Roffey, Mr Thompson, Mr Whitmell, Mr Wells, Mr Male, Mr Garman, Mr W. Banks and Mr Nicolini
FATIMA	Mrs Crouch
IRENE	Miss De Camp
BEDA	Mrs Bland
FEMALE PEASANTS	Mmes Arne, Roffey, Wentworth, Jackson, Maddocks and Menage
FEMALE SLAVES	Mmes Brooker, Daniels, Brigg, Haskey, Illingham, Byrne, Willis and Vining
PRINCIPAL DANCER	Mlle Parisot

Music by Michael Kelly

181

VI   Title page of the first edition of Michael Kelly's musical score for *Blue Beard* (1798) showing the arrival of the Spahis at the climax of the quartetto in act II, scene 6

ACT I

Scene 1. *A Turkish village. A romantic, mountainous country beyond it.*
SELIM *is discovered under* FATIMA's *window, to which a ladder of silken
ropes is fastened. It is dawn.*

Duet. SELIM *and* FATIMA.

SELIM: Twilight glimmers o'er the steep,
    Fatima! Fatima! Wakest thou, dear?
    Grey-eyed morn begins to peep,
    Fatima! Fatima! Selim's here!
    Here are true love's cords attaching
    To your window. List! List!
    (FATIMA *opens the window.*)
FATIMA: Dearest Selim, I've been watching;
    Yes, I see the silken twist.
SELIM: Down, down, down, down, down,
    Down the ladder gently trip!
    Pit-a-pat, pit-a-pat, haste thee, dear!
FATIMA: Oh, I'm sure my foot will slip!
    (FATIMA *has one foot out of the window.*)
SELIM: Fatima!
FATIMA: Well, Selim?
SELIM: Do not fear!
    (FATIMA *gets upon the ladder.* FATIMA *and* SELIM *keep time in
    singing to her steps as she descends. Towards the end of the last line
    she reaches the ground and they embrace.*)
BOTH: Pit-a-pat, pit-a-pat, pit-a-pat,
    Pit-a-pat, pit-a-pat, pat, pat, pat.

    (*As they embrace,* IBRAHIM *puts his head out from the door of the
    house.*)
IBRAHIM: Ah, traitress, have I caught you? (*coming forward*) Attempt to run
  away with a man? And not only with a man, but a trooper? One of the
  Spahis? Wicked Fatima! Much as Mahomet's brood must have increased, there
  isn't one turtle in all our prophet's pigeon house that wouldn't be ready to
  pick at you. In! In! And repent! (*pushing* FATIMA *into the house*)
SELIM: Hear, me Ibrahim!

One of the Spahis: corruption of *sipahi*, the Turkish name for a cavalryman serving in the
feudal armies of the First Ottoman Empire.
Mahomet's brood: since the prophet had ten wives and between ten to fifteen concubines, the
arithmetical odds upon 'Mahomet's brood' increasing at an impressive speed were high. Fatima,
however, was his favourite daughter.
turtle: i.e. turtle-dove, a bird often used to symbolise conjugal loyalty. The reference here is to
the naturally faithful ostracising an unnatural and disloyal individual. Mahomet is said to have
had a dove which fed on wheat placed in his ear.

IBRAHIM: I won't hear you, as I'm a Musselman!

SELIM: Credit me to suppose that —

IBRAHIM: I won't credit anything, as I'm a True Believer!

SELIM: Did not you promise her to me in marriage?

IBRAHIM: Um — why, I did say something — like getting a licence from the Cadi.

SELIM: And what has made you break your word?

IBRAHIM: A better bridegroom for my daughter!

SELIM: Why better than I?

IBRAHIM: He's richer. You have your merits, but he's a Bashaw — with three tails! tails!

SELIM: Does that make him more deserving?

IBRAHIM: To be sure it does, all the world over. Throw riches and power into the scale and simple merit soon kicks the beam. — Now to cut the matter short: you're a very pretty trooper, so troop off! For Abomelique, the great Abomelique, comes this day to carry my daughter to his magnificent Castle and to espouse her.

SELIM: Abomelique? The pest of all the neighbouring country?

IBRAHIM: Yes, he's by far the best of all the neighbouring country —

SELIM: Who deals, as all around declare, in spells and magic.

IBRAHIM: Aye, you can't say of him, as they do of many great folks, that he's no conjuror.

SELIM: And you think this man calculated to make a good husband to Fatima?

IBRAHIM: Positively.

SELIM: Better than I?

IBRAHIM: Um — comparatively.

SELIM: And you now look upon me with contempt?

IBRAHIM: Superlatively! I do, by the Temple of Mecca!

SELIM: Now, by my injuries, old man — But I curb my just resentment; you are the father of my Fatima. But for my rival —

IBRAHIM: He is able enough to maintain his own cause.

SELIM: Oh, he shall rue the day when, serpent-like, he stung me! Yes, Abomelique, spite of thy wealth and power, thy mystic spells and hellish incantations, a soldier's vengeance shall pursue thee!

Quartetto. SELIM, IBRAHIM, FATIMA *and* IRENE.

SELIM:　　　　　　　Ruthless tyrant, dread my force!
　　　　　　　　　　A soldier's sabre hangs o'er thee!

---

as I'm a Musselman: ultimately derived from the Persian *musulmān* (Moslem), this phonetic equivalent was in English usage since the sixteenth century, Colman's failure to contrive any pun on its meaning suggesting that, to late eighteenth-century audiences, 'musselman' was a correct term of reference.

Cadi: or *kadi*, a provincial Turkish judge.

Bashaw with three tails: a corrupt form of *paşa* and with a largely fictional allusion to Turkish life, the English word denoting a vicious tyrant, the Turkish a provincial governor. Horsetails indicated rank and were carried in the shield.

merit soon kicks the beam: i.e. is outweighed (see above *Inkle and Yarico*, p. 83).

> Thou soon shall fall a headless corse
> Who now would'st tear my love from me.

IBRAHIM: How prettily, now, he rails,
But 'tisn't so easily done as said
To smite a Bashaw and cut off the head
Of a man who has got three tails.

(FATIMA *and* IRENE *come from the house and kneel to* IBRAHIM.)

FATIMA *and* IRENE: Turn, turn, my father! Turn thee hither!
A daughter would thy pity move.

IRENE: Why doom the opening rose to wither?

FATIMA *and* IRENE: Why blight the early bud of love?

ALL: Oh, how teasing! Oh, how vexing!

$$\text{Are the fears which } \left\{ \begin{array}{l} \text{daughters} \\ \text{lovers} \\ \text{fathers} \end{array} \right\} \text{ prove.}$$

How distressing! How perplexing!
Are the cares that wait on love.

FATIMA *and* IRENE: Hear me! Hear me!

IBRAHIM: I'll not hear thee!

FATIMA *and* IRENE: Can you now our suit refuse?
Cheer me! You alone can cheer me.
'Tis a wretched daughter sues.

IBRAHIM: 'Tis a silly daughter sues.

ALL: Oh, how teasing! Oh, how vexing! &c.

IRENE: Dear! How can you think of marrying my sister to this Bashaw?

IBRAHIM: And pray, good mistress Irene, with all the submission of a dutiful father, may I crave to know your objections?

IRENE: Why, in the first place, then, father, he has a blue beard!

IBRAHIM: And who, in the name of all the devils, made you a judge of beards?

IRENE: Well, I do think it was sent as a punishment to him on account of all his unfortunate wives.

IBRAHIM: Ha! Now, under favour, I do think that a man's wives are punishment enough in themselves! Praised be the wholesome law of Mahomet that stinted a Turk to only four at a time!

IRENE: The Bashaw had never more than one at a time, and 'tis whispered that he beheaded the poor souls one after another, for, in spite of his power, there's no preventing talking.

IBRAHIM: That's true, indeed! And if cutting off women's heads won't prevent

---

corse: corpse.
the wholesome law of Mahomet: 'And if ye fear that ye shall not act with equity towards orphans of the female sex, take in marriage of such other women as please you, two, or three, or four, and not more.' *The Koran* (Chapter 4), Bath 1795, p. 92.

talking, I know of no method likely to prosper! But I'll make *you* silent, mistress, depend on't. No more of this prate!

IRENE: I have done, father.

IBRAHIM: Prepare to take up your abode with your sister — at the Castle.

IRENE: Oh, I am very glad I am to be with her! Are not you, Fatima?

FATIMA: I am, indeed, Irene. A loved sister's presence will be a consolation to me in my miseries.

IBRAHIM: Perhaps I may contrive to go with you, too. If I could bring it about, I should dwell there in all the respect due to a relation of the mighty Abomelique. Let me once get footing in old Three-Tails's Castle and I'll tickle up the slaves for a great man's father-in-law, I'll warrant me! Hark, I hear him on the march over the mountain! And here are all our neighbours, pouring out of their houses to see the procession.

> (*The sun rises gradually. A march is heard at a great distance.* ABOMELIQUE *and a magnificent train appear at the top of the mountain. They descend through a winding path. Sometimes they are lost to the sight to mark the irregularities of the road. The music grows stronger as they approach. At length,* ABOMELIQUE's *train range themselves on each side of the stage and sing the chorus as* ABOMELIQUE *marches down through their ranks. The villagers come from their houses.*)

Grand Chorus. ABOMELIQUE'S TRAIN.

Mark his approach with thunder! Strike on the trembling spheres!
    With martial crash
    The cymbals clash,
        'Tis the Bashaw appears!

War in his eyeball glistens! Slave of his lip is law!
    Our life and death
    Hang on his breath,
        Hail to the great Bashaw!

ABOMELIQUE: Now, Ibrahim, I come to claim my bride, the lovely Fatima, to take this village rose from the obscure and lowly shade and place her in a warmer soil, where the full sun of wealth shall shine upon her and add a richer glow to the sweet blush of beauty.

IBRAHIM: Most puissant Bashaw, I am proud that any twig of mine is thought worthy of a place in your shrubbery. Irene, as you desired, shall go with Fatima as companion. For myself, mighty sir, I am a tough stick, somewhat dry and a little too old, perhaps, to be moved. But, to say the truth, since you are going to take off my suckers, if I were to be transplanted along with them, I think I should thrive.

ABOMELIQUE: It shall be ordered so.

Slave of his lip: i.e. slaver, saliva; possibly not ideal as a legislative base, but potent as a gothic image.

IBRAHIM: Shall it? Then, if I don't make shift to flourish, cut me down and make fire-wood of me!

ABOMELIQUE: Be satisfied, you shall along with us. There shall not be one countenance on which my power, and this day's festival, does not impress a smile.

SELIM: That's false, by Mahomet!

ABOMELIQUE: How now? Who dares utter that?

IBRAHIM: Hush! (*stopping* SELIM's *mouth*) He's nobody. Only a poor, mad trooper — you may know he's a trooper by his swearing! — beneath your mighty notice.

ABOMELIQUE: What prompts him to this boldness?

SELIM: Injury. You have basely wronged me!

ABOMELIQUE: Rash fool! Know my power and respect it!

SELIM: When power is respected, its basis must be justice. 'Tis then an edifice that gives the humble shelter, and they reverence it. But 'tis a hated, shallow fabric that rears itself upon oppression. The breath of the discontented swells into a gale around it, till it totters.

ABOMELIQUE: Speak. How are you aggrieved?

FATIMA: Let me inform him.

IBRAHIM: Oh plague, hold your tongue! A woman always makes bad worse.

ABOMELIQUE: Proceed, sweet Fatima.

FATIMA: I was poor — and happy — for my wishes were lowly as my state. Content and peace dwelt in our cottage, nor were these smiling inmates ruffled when love stole in and found a shelter in my bosom. My father placed my hand in this young soldier's and taught me that our fortunes should soon be united. Poor Selim's soul spoke in his eyes, and mine replied — for true love's eyes are eloquent — that, through my life, I wished no other protector than a brave youth whose lot, being humble like my own, the more endeared him to me. Our hopes and joys were ripening daily. You came, and all are blighted!

    (FATIMA *falls into* SELIM's *arms.*)

ABOMELIQUE: Tear them asunder! Insulted? And by a slave that —

    (SELIM *offers to draw and is restrained by* ABOMELIQUE's *attendants.*)

Thou art beneath my notice! You, Fatima, must to the Castle. (*to the attendants*) Prepare the palanquin! (*to* FATIMA) We are advanced too far, lady. We cannot now recede.

    (*A magnificent palanquin is brought in, drawn by black slaves.*)

Grand Chorus. ABOMELIQUE'S TRAIN.

    Advance!
    See us the bride attending!

---

palanquin: an elaborate, enclosed, oriental sedan-chair, with sides formed of louvred panels, carried by a team of up to six men.

Echo shall now the chaunt prolong,
Torn with a lusty Turkish song,
While the Star of the World is ascending.
(ABOMELIQUE *leads* FATIMA *towards the palanquin.*)
Hark to the drum!
Come, comrades, come!
Time will not brook delaying —
(ABOMELIQUE *forces* FATIMA, *struggling, into the palanquin.*)
See, she resists! Her struggles note!

FATIMA
and SELIM } Oh, give me { him / her } on whom I dote!

(ABOMELIQUE *draws his sabre. All the* SLAVES *draw.*)
Sabres are gleaming round the throat
Of beauty disobeying.

(*Exeunt, hurrying off* FATIMA. IRENE *is seated with her in the palanquin.*)

Scene 2. *A hall in* ABOMELIQUE*'s Castle.*

(*Enter* BEDA *with a guitar.*)

BEDA: Where can he be loitering so long? Why, Shacabac, poor melancholy fool, he's in some dark corner of the Castle, now, moping and sighing, as usual. This is the hour he should come to take his daily lesson with me on the guitar. Music is the only thing that makes him merry. Why, Shacabac!
(*Enter* SHACABAC *with a guitar.*)

SHACABAC: Here I am, Beda.

BEDA: Why, where have you been all this time, Shacabac?

SHACABAC: Getting all in readiness for the Bashaw's return with his intended bride. They say she's very handsome. (*half-aside*) Poor soul, I pity her.

BEDA: Pity a woman because she's handsome? Pray, then, keep out of my way, for I don't like to be pitied!

SHACABAC: Did I say pity? Oh no, I didn't intend that. Heigho!

BEDA: Now what can you be sighing for?

SHACABAC: That wasn't sighing. I'm like our old, blind camel: a little short-winded, that's all.

BEDA: I'm sure, Shacabac, you ought to be the happiest creature in the Castle. The Bashaw loads you with his favours.

SHACABAC: Oh, very heavily indeed! I don't dispute that!

BEDA: You are his chief attendant, and he honours you with more employment than all the other slaves put together.

SHACABAC: Works me like a mule. It would be ungrateful to deny it.

BEDA: And everybody thinks that he trusts you with all his secrets.

SHACABAC: (*alarmed*) No! Do they think that?

---

chaunt: chant. This variant was obsolete even in Colman's day and has been imported simply as a means of heightening the theatrical impact of the piece.

BEDA: Yes. And, to say truth, you keep them locked up as close –

SHACABAC: (*starting*) Locked up? How? Where should I keep them locked up?

BEDA: In your breast, to be sure.

SHACABAC: Oh, yes, yes! That is, if he trusts me with any. But to think that a Bashaw would tell his secrets to a slave! Nonsense!

BEDA: Nay, it isn't for nothing he takes you to talk with him, in private, in the Blue Chamber.

SHACABAC: (*very earnestly*) Don't mention that, Beda! Never mention the Blue Chamber again!

BEDA: Why, what harm is there in the Blue Chamber?

SHACABAC: None in the world! But you know I'm full of melancholy fancies, and I never go into that Blue Chamber that I don't feel as if I were tormented with devils.

BEDA: Mercy! What devils, Shacabac?

SHACABAC: (*recovering himself and smiling*) Only blue devils, Beda! Nothing more. – Come, hang sorrow! Let's strike up a tune on the guitar.

BEDA: Aye, that makes you merry at the worst of times.

SHACABAC: That it does, Beda.

Duet. SHACABAC *and* BEDA.

SHACABAC: Yes, Beda, this, Beda, when I melancholy grow,
        This tinking heart-sinking can soon drive away.

BEDA: When hearing sounds cheering then we blythe and jolly grow.
        How do you, while to you, Shacabac, I play?
        Tink, tinka-tinka, tink – the sweet guitar shall cheer you,
        Clink, clinka-clinka, clink – so gaily let us sing!

SHACABAC: Tink, tinka-tinka, tink – a pleasure 'tis to hear you,
        While neatly you sweetly, sweetly touch the string!

BOTH: Tink, tinka-tinka, tink &c.

SHACABAC: Once sighing, sick, dying, sorrow hanging over me,
        Faint, weary, sad, dreary, on the ground I lay.
        There moaning, deep groaning, Beda did discover me.

BEDA: Strains soothing, care smoothing, I began to play.
        Tink, tinka-tinka, tink – the sweet guitar could cheer you,
        Clink, clinka-clinka, clink – so gaily did I sing!

SHACABAC: Tink, tinka-tinka, tink – a pleasure 'twas to hear you,
        While neatly you sweetly, sweetly touched the string!

BOTH: Tink, tinka-tinka, tink &c.

(*A horn is sounded without.*)

SHACABAC: Hark, the horn sounds at the Castle gate! The Bashaw is returned –

BEDA: And brings his bride with him. I long to see her! I must join the rest of the slaves presently. You know, Shacabac, we are all to kneel and cry, 'May she live long and happy!'

---

blue devils: melancholy, the 'blues' (see above *Inkle and Yarico*, p. 90).

SHACABAC: Heaven send she may! Hush! The Bashaw!
  (*Enter* ABOMELIQUE.)
ABOMELIQUE: Oh, you are here!
SHACABAC: To obey your pleasure. Your slave humbly trusts that in preparing for
  our new mistress nothing has been neglected.
ABOMELIQUE: I commend your care. And, while the lovely Fatima is inspecting
  her apartments, I have employment for you. You must attend me.
SHACABAC: Whither, mighty sir?
ABOMELIQUE: To the Blue Chamber.
SHACABAC: The Blue Cha— ? (*dropping the guitar*)
ABOMELIQUE: What ails the driveller?
SHACABAC: No, nothing, nothing. (*half-aside*) That terrible sound sets me a-shivering!
ABOMELIQUE: What say you?
SHACABAC: I say the guitar fell to the ground and I was afraid of its shivering.
ABOMELIQUE: Attend me.
SHACABAC: I follow.
        (*Exit* ABOMELIQUE *followed by* SHACABAC.)
BEDA: Poor Shacabac! What can be the matter with him? Perhaps he has been
  crossed in love. And, now I come to think of it, he must have a mistress some-
  where or he never would be so often alone with me without saying one tender
  thing to me. Ah, love, love! I never shall forget my poor, dear, lost Cassib!

Song. BEDA.

> His sparkling eyes were dark as jet,
>   Chica, chica, chica, cho.
> Can I my comely Turk forget?
>   Oh, never, never, never! No!
> Did he not watch till night did fall
>   And sail in silence on the sea?
> Did he not climb our sea-girt wall
>   To talk so lovingly to me?
> Oh, his sparkling eyes were dark, &c.
>
> His lips were of the coral hue,
>   His teeth of ivory, so white;
> But he was hurried from my view
>   Who gave to me so much delight!
> And why should tender lovers part?
>   Why should fathers cruel be?
> Why bid me banish from my heart
>   A heart so full of love for me?
> Oh, his sparkling eyes were dark, &c.

        (*Exit* BEDA.)

Scene 3. *A blue apartment. A winding staircase on one side. A large door
in the middle of the flat. Over the door, a picture of* ABOMELIQUE

kneeling in amorous supplication to a beautiful woman. Other pictures and devices on subjects of love decorate the apartment.

> (ABOMELIQUE *and* SHACABAC *descend the stair,* SHACABAC *in apparent terror.*)

ABOMELIQUE: You know my purpose?

SHACABAC: I guess it.

ABOMELIQUE: Why do you tremble?

SHACABAC: The air of this apartment chills me, and the business we are going upon isn't the best to inspire courage.

ABOMELIQUE: Fool! When this mysterious portal shall be opened, what hast thou to dread?

SHACABAC: Oh, nothing at all! The inhabitants of the inner apartment might terrify a man of tender nerves, but what are they to me? Only a few flying phantoms, sheeted spectres, skipping skeletons and grinning ghosts at their gambols! And, as to those who had once the honour to be your wives, poor souls, they are harmless enough now, whatever they might have been formerly!

ABOMELIQUE: 'Twas to prevent the harm with which their conduct threatened me that they have suffered. Their crimes were on their heads.

SHACABAC: Then their crimes were as cleanly taken off as scimitar could carry them. That curiosity should cost so much! If all women were to forfeit their heads for being inquisitive, what a number of sweet, pretty, female faces we should lose in the world!

ABOMELIQUE: Such punishment might outrun even Turkish justice. But in me 'tis prudence, self-preservation. You are not ignorant of the prediction?

SHACABAC: That it is your fate to marry, and your life will be endangered by the curiosity of the woman whom you espouse.

ABOMELIQUE: Thou hast the secret. Dare not to breathe it, or —

SHACABAC: Don't look so terrible, then. For, if you scare away my senses, who knows but the secret may pop out along with them.

ABOMELIQUE: Well, I know thou darest not utter it. The mystic ceremonies in which, from mere necessity, I have employed thee, thou weak and unapt agent, bear in them a supernatural force fettering thy tongue in silence.

> (ABOMELIQUE *gives* SHACABAC *a key decorated with jewels.*)

Take the key. Apply it to the door.

SHACABAC. Yes, I — But I was always, from a boy, the merest bungler at a lock that —

ABOMELIQUE: Dastard! Thou know'st how readily 'twill open.

SHACABAC: But must I once more open it? To —

ABOMELIQUE: Be speedy! This talisman must, ere my marriage rites are

---

outrun even Turkish justice: the chauvinistic eighteenth-century English attitude towards Islamic practices, upon which the gothic horror of the play is founded, tended to over-emphasise the concept of retributive justice in Koranic scripture.

solemnised, be placed within the tomb of those whose rashness has laid them cold beneath the icy hand of death.

SHACABAC: Mercy on us! I know not for the icy hand of Death, but if Fear would do me the favour to keep his chilly paws off me, I should be much warmer than I am at present!

ABOMELIQUE: No dallying.

SHACABAC: I obey.

(SHACABAC *puts the key into the lock. The door instantly sinks, with a tremendous crash, and the Blue Chamber appears, streaked with vivid streams of blood. The figures in the picture over the door change their position, and* ABOMELIQUE *is represented in the action of beheading the beauty he was before supplicating. The pictures and devices of love change to subjects of horror and death. The interior apartment – which the sinking of the door discovers – exhibits various tombs in a selpulchral building, in the midst of which ghastly and supernatural forms are seen, some in motion, some fixed. In the centre is a large skeleton seated on a tomb, with a dart in his hand, smiling, and, over his head, in characters of blood, is written: 'THE PUNISHMENT OF CURIOSITY'.*)

ABOMELIQUE: (*pointing to the skeleton*) Thou seest yon fleshless form?

SHACABAC: Oh, yes! And my own flesh crawls whenever I look upon him! (*giving* ABOMELIQUE *the key*)

ABOMELIQUE: Henceforward he must be my destiny. (*addressing the skeleton*) Demon of blood, death's courier, whose sport it is to sound war's clarion, to whet the knife of suicide, to lead the hired murderer to the sleeping babe, and, with a ghastly smile of triumph, to register the slaughtered who prematurely drop in Nature's charnel-house, here, here have I pent thee, a prisoner to my art, here to circumscribe thy general purposes for my particular good. Twelve winters have I kept thee –

SHACABAC: Have you? Allah preserve us! But I must say that, considering the time, he looks so lean that he does his keeper no credit.

ABOMELIQUE: Approach him with respect.

SHACABAC: Who? I? I'd rather keep at a respectful distance!

ABOMELIQUE: Take this talisman.

SHACABAC: 'Tis a dagger!

ABOMELIQUE: 'Tis a charmed one. While it remains beneath the foot of that same ghastly form, I am free from mortal power. Another hand than mine must place it there. Thou must perform the office.

(ABOMELIQUE *gives* SHACABAC *the talisman.*)

SHACABAC: Must I? Well, – I – (*approaching the figure*) Oh, Mahomet! If ever I get away from this gentleman who has jumped out of his skin, I shall jump out of my own for joy!

(SHACABAC *lays the dagger at the foot of the skeleton. It thunders and lightens violently. The inscription over the skeleton's head changes to the following: 'THIS SEPULCHRE SHALL INCLOSE HER WHO MAY ENDANGER THE LIFE OF ABOMELIQUE.' The skeleton raises his arm which holds the dart then lets his arm fall again.* SHACABAC *staggers from the sepulchre into the Blue Chamber and falls on his face, when the door, instantly rising, closes the interior building. The streaks of blood vanish from the walls of the Blue Chamber and* ABOMELIQUE's *picture, with the other pictures and devices, resume their original appearance.*)

ABOMELIQUE: It omens prosperously! 'This sepulchre shall inclose her who may endanger the life of Abomelique.' Her death, then, is the penalty of her rashness. May Fatima be prudent and avoid it. — Rouse thee, dull fool! Thy task is ended. Arise and follow me hence.

SHACABAC: That I will, if my legs have power to carry me. (*getting up*)

ABOMELIQUE: Hark, I hear a foot in yonder gallery! Ascend the stairs with me in silence! Chattering will cost thy life!

SHACABAC: Then I am sure you must pull out my teeth, for they chatter in spite of me.

(ABOMELIQUE *makes a sign to* SHACABAC *to follow.*)

I attend!

(ABOMELIQUE *and* SHACABAC *ascend the staircase and the scene closes.*)

Scene 4. *An apartment in the Castle.*

(*Enter* FATIMA *and* IRENE.)

IRENE: Prithee, dearest sister, take comfort.

FATIMA: Where shall I find it? Torn from the man I love and forced into the arms of one whom I, and all around, detest, where should I look for comfort? My waking thoughts are torments and, since this marriage was proposed, my very dreams have foreboded misery.

Song. FATIMA.

While pensive, I thought on my love,
    The moon on the mountain was bright,
And Philomel, down in the grove,
    Broke sweetly the silence of night.

Oh, I wished that the tear-drop would flow,
    But I felt too much anguish to weep,
Till, worn with the weight of my woe,
    I sunk on my pillow to sleep.

Methought that my love, as I lay,
    His ringlets all clotted with gore,
In the paleness of death seemed to say,
    'Alas, we must never meet more!'

Philomel: the nightingale. In an unpleasant interlude from Greek myth, Philomena was brutally raped by her brother-in-law, Tereus, King of Thrace, who afterwards cut out her tongue. But she succeeded in communicating the crime to her sister, Tereus' wife, whose revenge was suitably vicious. At which point, fortunately, the gods intervened, turning each into an appropriate bird.

> 'Yes, yes, my beloved, we must part!
>   The steel of my rival was true;
> The assassin has struck on that heart
>   Which beat with such fervour for you.'

IRENE: Why, to be sure, 'tis a sad thing to lose Selim. He is a good youth, and we women have, somehow, such a pleasure in looking at a good young man when he happens to be very handsome! Yet the Bashaw, bating his beard, isn't so very ugly, either. Then, you know, he rolls in riches —

FATIMA: He abuses them, Irene. Wealth, when its purpose is perverted, makes the possessor odious. When virtuous men have gold, they purchase their own happiness by making others happy; heap treasure on the vicious, they strengthen their injustice with the sweet means of charity, and turn the poor man's blessing to a curse.

IRENE: Well, now, it's a great pity you happened to love Selim first. Who knows, but the Bashaw may turn out good to us, after all. See what fine clothes he has given us already.

FATIMA: Alas, my sister, these gay trappings communicate no pleasure to an aching heart.

IRENE: I wish they could see us in them in our village, for all that. Then we are to have a fine feast tonight, in honour of your nuptials which are to take place tomorrow.

    (*Enter* SHACABAC.)

SHACABAC: Madam, the Bashaw waits to attend you to the Illuminated Garden.

IRENE: There! The Illuminated Garden! I told you they were getting ready. Come, now, look cheerily and who knows what may happen yet!

FATIMA: (*to* SHACABAC) I attend him. — Come, sister.

    (*Exeunt* FATIMA *and* IRENE.)

SHACABAC: Poor soul! Must she be sacrificed, too, to the Bashaw's cruelty? His savage spirit settles all family disputes with the edge of the scimitar!

Song. SHACABAC.

> A fond husband will, after a conjugal strife,
>   Kiss, forgive, weep, and fall on the neck of his wife.
> But Abomelique's wife other conduct may dread:
>   When he falls on her neck, 'tis to cut off her head!

> How many there are, when a wife plays the fool,
>   Will argue the point with her, calmly and cool;
> The Bashaw, who don't relish debates of this sort,
>   Cuts the woman, as well as the argument, short.

---

the sweet means of charity: 'The industrious Poor are the support of the Community, consequently, every improvement that tends to the preservation of their lives and healths, is of general utility.' *Account of the Lying-In Hospital*, 1791.

But whatever her errors, 'tis mighty unfair
   To cut off her head just as if it were hair;
For this truth is maintained by philosophers still:
   That the hair grows again, but the head never will.

And among all the basest, sure he is most base
   Who can view, then demolish, a woman's sweet face!
Her smiles might the malice of devils disarm
   And the devil take him who would offer her harm!

(*Exit* SHACABAC.)

Scene 5. *A garden, brilliantly and fancifully illuminated. A fountain play-ing in the middle of it. An elevated sofa on one side, under a rich canopy. A large company of* SLAVES *discovered; some dancers, others with musical instruments. They all appear as preparing for an entertainment.* BEDA *is foremost among them.*

(*Enter* IBRAHIM.)

IBRAHIM: That's right, you poor abominable devils who have the happiness to be slaves to my son-in-law, that's right! Thrum your guitars, puff your trumpets and blow your flutes in honour of your new mistress, my daughter. (*to a* SLAVE *with a trumpet*) Come here, you long-winded dog! Tell me who I am!

SLAVE: You are Old Ibrahim.

IBRAHIM: Old Ibrahim! These slaves are remarkably free! I am the father of the lady who is the wife of the man who is the master of you. — What a fine thing it is to be father-in-law to Three-Tails! Oh dear, (*seeing* BEDA) there's a pretty black-eyed girl! — Come here and tell me your name.

BEDA: My name is Beda, so please you.

IBRAHIM: Beda, is it? Why, you little devil, you're an angel!

BEDA: Oh no, sir, I'm only one of the family.

IBRAHIM: Then give me a family kiss.

BEDA: Dear! If the Bashaw should see you!

IBRAHIM: Then he'd say you have a good taste. — Cheer up, little one, I rule the roost here; it shan't go worse with you that I have power and you have charms. It's amazing, when beauty pleads with a great man, how much quicker it rises to promotion than ugly-faced merit!

(*A flourish of music is heard without.*)

Silence! Here comes the great Abomelique, son-in-law to me who am the father to the lady who is to marry the man that is master to you! Stand aside! Be ready! Strain your throats, kick your heels and show obedience!

(ABOMELIQUE *enters with* FATIMA, IRENE *accompanying them.*
ABOMELIQUE *and* FATIMA *seat themselves under the canopy.*)

A Grand Dance.

Chorus. SLAVES.

Lowly we bend in duty,
   Queen of the peaceful bowers!
We bow to the footsteps of beauty

And strew her path with flowers.
The mellow flute is blowing,
    Bounce goes the tambourine,
Sweet harmony is flowing
    To welcome beauty's queen!

ACT II

Scene 1. *A wood. A company of* SPAHIS *discovered in ambush.*
Glee. SPAHIS.

Stand close! Our comrade is not come
    Ere this, he must be hovering near.
Give him a signal we are here
By gently tapping on the drum.
      Rub, dub, dub.

A comrade's wronged; revenge shall work!
Thus, till our project's ripe, we lurk,
    And still, to mark that we are here
    Yet not alarm the distant ear,
With caution, ever and anon,
The drum we gently tap upon.
      Rub, dub, dub.

1st SPAHI: Selim tarries long.
2nd SPAHI: Disappointed love is a heavy luggage, and he who travels with it generally proceeds slowly.
3rd SPAHI: Not when the hope of redress is packed up with his disappointments. And revenge has long spurs to quicken a dead motion! (*to a sullen, rough-looking companion*) Were you ever in love, comrade?
4th SPAHI: (*very gruffly*) I once knew the tender passion.
3rd SPAHI: Were you successful when you adored?
4th SPAHI: Um — why — the chances were against me!
3rd SPAHI: How so?
4th SPAHI: I adored eleven and obtained but five! 'Twas hard for a man who was so constant to 'em.
1st SPAHI: Well, we are all soldiers. War is the mistress I pursue.
2nd SPAHI: You must take pains to keep sight of her, for you have lost one eye in her service already!
1st SPAHI: Wounds of honour, brother, form the warrior's proudest epitaph. My loss, perhaps, may live in story.
4th SPAHI: It must live in a blind-story, then, if it live at all, brother!

---

blind-story: one of the architectural features of a cathedral, comprising a series of blind windows immediately below the clerestory (clear-story). Interestingly enough, references of such staggering obscurity seemed to have hindered neither the reputation of *Blue Beard* as a play nor of Colman as a wit.

3rd SPAHI: Come, no more of this.

1st SPAHI: Nay, let them proceed; they are only in sport. My comrades know that the breath of a few ribald jesters can never wither the laurels a soldier gains in protecting his country. — Look out, here comes Selim!

(*Enter* SELIM.)

2nd SPAHI: Well met! We have been a full hour at our post, here.

SELIM: Your pardon. The entanglements of the wood retarded my progress.

3rd SPAHI: Now, comrade, the time's at hand when we will redress you.

SELIM: I know your zeal. A Spahi never permits a brother's injuries to remain unrevenged.

4th SPAHI: We'll seize upon Blue Beard and dry-shave him with a two-edged scimitar!

SELIM: If it be expedient to attack the Castle, be cautious, friends, in the procedure. My Fatima, else, may fall in the confusion.

2nd SPAHI: Fear not that. We'll crack the walls like a nutshell and extract your mistress, safe and sound, like the kernel!

4th SPAHI: Our horses stand a few paces hence. Let us mount and away!

SELIM: We will, my comrades! We have some distance yet to ride ere we reach the domain of Abomelique. Prepare; I'll follow instantly. Thanks for your aid.

1st SPAHI: Nay, we want no thanks. Men are unworthy of succour in their own time of need who will not be active to relieve the sufferings of their fellows. March, comrades!

(*Exeunt* SPAHIS.)

SELIM: Now, Fortune, smile upon a soldier's honest love struggling to rescue injured virtue from oppression!

Song. SELIM.

> Hear me, O Fortune, hear me,
> > Thy aid oh let me prove!
> Now, in this struggle, cheer me
> > And crown the hopes of love!

> Then Vice no more shall revel:
> > Yes, tyrant, we shall meet!
> A soldier's sword shall level
> > Oppression at my feet!

(*Exit* SELIM.)

Scene 2. *An apartment in* ABOMELIQUE's *Castle.*

(*Enter* ABOMELIQUE, FATIMA *and* SHACABAC.)

ABOMELIQUE: Yes, Fatima, business of import calls me. For a few hours I leave you. Soon as the sun slopes through the azure vault of heaven to kiss the mountain's top and evening's lengthened shadows forerun the dew-drops of the night, then look for my return. Then shall our marriage be accomplished.

FATIMA: Alas, if ever pity —

ABOMELIQUE: No more of this. Off with this maiden coyness and, in my absence, be gay and jocund! This Castle can afford diversion, lady. Here are the keys —

SHACABAC: (*involuntarily interrupting*) What, *all* the keys?

ABOMELIQUE: Peace, slave! — Inspect the rich apartments. These open every door. This slave, here, shall conduct you. But, with them, take this caution —

FATIMA: A caution?

ABOMELIQUE: Yes. This key, sparkling with diamonds, opens a door within the Blue Apartment.

SHACABAC: (*sighing*) Oh!

ABOMELIQUE: That door, and that alone, is sacred. Dare to open it, and the most dreadful punishment that tongue can utter will await you!

    (*Here* SHACABAC *gives* ABOMELIQUE *a look of supplication for* FATIMA *and is repelled by a ferocious frown from his master.*)

It is the sole restraint I ever shall impose. In all else you have ample scope. Merit my indulgence and tremble to abuse it! (*giving the keys*)

FATIMA: I tremble now to hear your words and mark your manner.

SHACABAC: (*aside*) So do I, I'm sure!

FATIMA: If this key be of such import, 'twere best not to trust it to my keeping.

SHACABAC: (*anxiously*) Oh, much the best! (*to* ABOMELIQUE) Pray, take it again! Pray, do!

ABOMELIQUE: Be dumb! — No, Fatima, a wife were unworthy of my love could I not confide in her discretion. Prove I may trust in yours implicitly. — (*to* SHACABAC) Follow me, slave, to the gate, then hasten back to your mistress.

SHACABAC: Yes, I — Pray then, don't stir from here till I come, lady! — (*aside*) If the poor soul should get to the Blue Chamber before I return and —

ABOMELIQUE: Farewell, Fatima! — (*to* SHACABAC) Come on!

    (*Exit* ABOMELIQUE.)

SHACABAC: I come. — Oh!

    (SHACABAC *first looks at* FATIMA, *then at his master, between anxiety for the one and terror of the other. Then exit, after* ABOMELIQUE.)

FATIMA: What can this mean? His ferocious look as he pronounced the solemn charge struck horror through me! The countenance, too, of the trembling slave was marked with mystery.

    (*Enter* IRENE.)

IRENE: So, sister, the Bashaw is going, I hear, till the evening! — What are those keys in your hand?

FATIMA: They open every door within the walls. Abomelique has left them with me that we may wander through the Castle.

IRENE: Well, now, that is very kind of him!

FATIMA: I have no joy now, Irene, in observing the idle glitter and luxury of wealth.

IRENE: Haven't you? But I have. We'll have a rare rummage! I won't leave a single nook nor corner unexamined.

FATIMA: That must not be! There is one room we are forbidden to enter.

IRENE: A forbidden room? Dear, now, I had rather see that room than any other in the Castle! Did the Bashaw forbid us?

FATIMA: He did, and with an emphasis so earnest, a manner so impressive, that he has taught me a fatal consequence would wait on disobedience.

IRENE: Mercy! – How I do long to see that room. Do let me just look at the key!

FATIMA: (*showing* IRENE *the key*) Beware, Irene.

IRENE: Dear, there can be no harm in looking at a key! What, is that it? Well, it is a monstrous fine one, I declare! Dear Fatima, how pretty it would be just to take one peep!

FATIMA: Tempt me not to a breach of faith, Irene. When we betray the confidence reposed in us to gratify our curiosity, a crime is coupled to a failing and we employ a vice to feed a weakness. The door within the Blue Apartment must remain untouched.

IRENE: Well, I have done. But we may see the rest of the rooms, I suppose.

FATIMA: If that can please you, sister, I will accompany you.

IRENE: That's my good, kind Fatima! – (*aside*) If I could but get her by degrees to this Blue Apartment! – Come, we'll go and look over the Castle. I saw some rich dresses in a wardrobe at the end of the gallery that would have suited me nicely in the dance last night!

Song. IRENE.

> Moving to the melody of music's note,
>   Observe the Turkish fair advance;
> Lightly as the gossamer she seems to float
>   Thro' mazes of the dance.
>   Sportive is the measure,
>   Thrilling is the pleasure,
>   While in merry glee the sexes join.
>   Deeper-blushing roses
>   Every cheek discloses,
>   Eyes with lustre shine.
>   Moving to the melody, &c.

> When the lover takes her glowing hand
>   With manly grace and ease,
> Can the dancing female then withstand
>   His gentle squeeze?
>   No, she gives him then so languishing a glance,
>   Grown tender, soft and melting with the dance.
>   Cupid, Cupid, god of hearts,
>   Dancing sharpens all your darts!
>   Moving to the melody, &c.

(*Exeunt* FATIMA *and* IRENE.)

Scene 3. *Another apartment in* ABOMELIQUE's *Castle.*

(*Enter* IBRAHIM *running after* BEDA.)

IBRAHIM: Come here, you little slippery jade, and let me look at you! (*taking hold of* BEDA) Tell me, now, don't you think you are very pretty?

BEDA: I am such as Nature made me, sir.

IBRAHIM: Nature has been very kind to you, hussy. She has given you two black eyes.

BEDA: That wasn't very kind of her, sir!

IBRAHIM: Don't you know I am made major-domo?

BEDA: Yes, the Bashaw has given you the command, it seems, over all the slaves.

IBRAHIM: Then obey me!

BEDA: How, sir?

IBRAHIM: How? Why — show me your teeth!

BEDA: My teeth?

IBRAHIM: Yes, giggle!

(BEDA *laughs*.)

Oh Mahomet, there's ivory! She has a handsomer mouth than an elephant! Where were you born, child?

BEDA: In Constantinople, sir. My poor mother was carried off with the plague there. My father had it at the same time.

IBRAHIM: Did it kill him, then?

BEDA: No, sir. He was very bad with it, but when my mother died —

IBRAHIM: Then your father got rid of his plague!

BEDA: Yes, sir.

IBRAHIM: I don't doubt it! And how came you a slave?

BEDA: Oh, that's a very long story.

IBRAHIM: Don't tell it, then. We've no need of long stories while there's opium in Turkey! — But I'll lighten the load of your bondage.

BEDA: Will you, indeed, sir?

IBRAHIM: Yes. — (*aside*) I am a true Turkish lover and know all the amorous phraseology of our country! — You shall be the nutmeg of my affections, my allspice of delight! When I meet you in the Grove of Nightingales, let not your eyes be disdainful as the stag's. — (*aside*) There! — Now go and tell Mustapha to mend the hole the rat gnawed in my slipper last night in that damned cockloft my son-in-law crams me into by way of a bedroom.

BEDA: Am I to go now, sir?

IBRAHIM: Aye. — Stay! Give me a kiss first! What, are you loath to take it?

BEDA: Oh, sir, we slaves must take anything.

(IBRAHIM *kisses* BEDA.)

IBRAHIM: Adieu, crown of my head!

BEDA: Good-bye, sir. — (*aside*) An old dotard!

(*Exit* BEDA.)

IBRAHIM: My fortune's made! Abomelique marries my daughter tonight and puts me into power because he can't help it. What a fine thing it is to be a great man in office when nobody dares turn him out!

---

major-domo: although strictly referring to the highest official in an Italian or a Spanish ruling household, English usage applied the term loosely to the head servant of any establishment overseas.

cockloft: even less salubrious than a garret, this was a room at the very top of a house, immediately under the apex of the roof.

Song. IBRAHIM.

> Major-domo am I
> Of this great family,
>> My word through the Castle prevails:
> I'm appointed the Head
> That must keep up the dread
>> And the pomp of my son-in-law's tails!

> I strut as fine as any macaw,
> I'll change for down my bed of straw,
> On perquisites I lay my paw,
> I pour wine slyly down my maw,
> I stuff good victuals into my craw.
>> 'Tis a very fine thing to be a father-in-law
>> To a very magnificent three-tailed Bashaw!

> The slaves, black and white,
> Of each sex, own my might;
>> I command full three hundred and ten.
> The females I'll kiss,
> But it won't be amiss
>> To fright them with thumping the men.
>> I strut as fine, &c.

> At the head of affairs,
> Turn me out, then, who dares;
>> Let them prove the Head pilfers and steals.
> No three-tailed Bashaw
> Kicks his father-in-law
>> And makes his Head take to his heels.
>> I strut as fine, &c.

> (*Exit* IBRAHIM.)

Scene 4. *The Blue Apartment.* FATIMA *and* IRENE *are discovered at the top of the staircase.*

FATIMA: I am tired already with the search we have made, Irene.

IRENE: Oh, I could never be tired with such fine things as we have seen! Do, now, just come down the stair and walk through this wing of the building.

FATIMA: Well, I —

IRENE: Aye, now that's a sweet, good-natured sister!

> (FATIMA *and* IRENE *descend the stair.*)

Now here's a pretty room! All furnished with blue, I see.

FATIMA: With blue? 'Tis the very chamber we were cautioned to avoid! Imprudent

---

maw: the throat or gullet, usually with reference to animals or fish with particularly voracious feeding habits.

craw: the stomach, a term carrying the same derisive inflexion as *maw* in the previous line.

girl, whither have you led me? Haste, haste, Irene, and let us leave it instantly!

IRENE: Dear! Where's the hurry? I'm sure 'tis a very pretty room. Besides, 'tis only the *door* in this room which leads to another, you know, that we were bid not to touch.

FATIMA: No matter, 'tis rash to tarry. Our being here may excite suspicion.

IRENE: Suspicion? Why, we have no bad purpose! And even if we were to open the door — and there it stands as if it seemed to invite the very key in your hand to come and unlock it — why, I see no such great crime in the action.

FATIMA: The Bashaw's charge, Irene —

IRENE: Is a very ill-natured one. And should you disobey him, we could keep our own counsel. Then, if nobody knows we have found out his secret, what have we to fear while we continue mute as death?

> (*A voice within is heard to intone,* 'Death!' FATIMA *and* IRENE *look at each other and tremble.*)

FATIMA: Did you hear nothing, Irene?

IRENE: Yes, I — I — I thought I heard something that — Stay! Oh, it must be an echo! These large old buildings are full of them!

FATIMA: It had an awful sound. A tone like that, they say, will sail upon the flagged wing of midnight, crossing the fear-struck traveller upon the desert to give him token of a foul murder.

> (*A deep groan is heard from the interior apartment.*)

Oh, Heaven have mercy! What can this mean?

IRENE: I know not. It seems the accent of distress. If so, it were humanity to succour the wretched soul who breathes it.

FATIMA: Humanity alone, my sister, could induce me to penetrate the mystery this portal, here, incloses.

IRENE: No eye can see us!

Duet. FATIMA *and* IRENE.

> All is hush'd! No footstep falls!
> And silence reigns within the walls!
> The place invites, the door is near,
> The time is apt, the key is here.
> Say, shall we? Yes? Say, shall we? No?
> What is it makes us tremble so?
>
> Mischief is not our intent,
> Then wherefore fear we should repent?
> Say, shall we? Yes! The door is near.
> Say, shall we? Yes! The key is here.

> (*At the end of the duet,* FATIMA *puts the key in the door, which instantly sinks and discovers the interior apartment as at first rep-*

---

the flagged wing of midnight: the image used here is suitably gothic, suggesting the passage of a black bird gliding lazily on large outstretched wings.

> *resented. The inscription over the skeleton's head is now 'THE*
> *PUNISHMENT OF CURIOSITY'. The Blue Chamber undergoes the*
> *same change as in the first instance. FATIMA and IRENE shriek, run*
> *to each other, and hide their heads in each other's bosoms. At this*
> *moment, SHACABAC appears at the top of the staircase, then runs*
> *down hastily. As he descends, the door rises and the chamber*
> *resumes its original appearance.*)

SHACABAC: (*speaking as descending*) Oh, 'tis as I feared! This comes of her not waiting for me. She knows the secret and she dies! Oh, lady, what have you done?

FATIMA: Begone! You knew of this! Your look, when late Abomelique left me,
now is explained. You are an accomplice in this bloody business.

SHACABAC: I?

FATIMA: My death, no doubt, is certain, and, in you, perhaps, I see my
executioner.

SHACABAC: How a man's looks may belie him! This comes, now, of my being
such an ugly dog! I wouldn't hurt a hair on your head to be made a Sultan.

FATIMA: Prove it, then, by saving us.

SHACABAC: How?

IRENE: Conduct us from the Castle.

SHACABAC: Impossible. The outward gates are closely guarded.

FATIMA: Nay, nay, you do not pity us!

SHACABAC: Not pity you? Oh, he must have a hard heart to see a lovely woman
in extremity and not try to soften her distress! — Stay! Perhaps we may
conceal the — Where's the key?

IRENE: It fell upon the ground and —

SHACABAC: The ground? Aye, here. (*taking up the key*) Perhaps we may be able
to — Nay, then, every hope is lost! The key is broke!

FATIMA: All is discovered, then.

IRENE: Certain. Oh, Fatima, would the Bashaw had any humanity within his
breast and that fatal key could unlock it!

SHACABAC: Oh, would he had! I'd stuff the key down his throat as soon as he
came home to get at it!
> (*The horn of the Castle gate is sounded.*)
There! The Bashaw returned, full six hours before his time!

IRENE: Oh, Heaven, what are we to do?

FATIMA: I am reckless of the future. Perhaps 'twere better I should die! 'Twill end
a life which promised nought but misery!

IRENE: Die? (*embracing* FATIMA) Oh, sister!

SHACABAC: Do not weep! Do not weep! I'm almost distracted. Hurry hence!
Come, lady, meet him as if nothing had happened. Collect your spirits,
smooth your looks. This way, now! Oh, if my choking can save your life, my
sorrow for you bids fair to preserve it! Come, lady, come!
> (*Exeunt SHACABAC, FATIMA and IRENE up the staircase.*)

Scene 5. *Another apartment in the Castle.*

> (*Enter SHACABAC, looking behind him as he enters.*)

SHACABAC: I have left them on the top of the stair, that I may avoid observation.

If they get far enough from the Blue Chamber before enquiry is made for them, they may conceal the —

(*Enter* HASSAN. SHACABAC *runs against him.*)

Umph! — Who's that?

HASSAN: Hassan, the black eunuch.

SHACABAC: Whither are you going?

HASSAN: To seek the Lady Fatima — by the Bashaw's order.

SHACABAC: Are you? — (*aside*) If he meets them so near the fatal chamber and mentions it to the Bashaw, they are lost! I must detain him. — I — Hassan — I say, Hassan. How d'ye do, Hassan?

HASSAN: I'm well, thank you, Shacabac.

SHACABAC: Well, are you? Are you sure you are well?

HASSAN: Very well.

SHACABAC: Very well? Very well, I'm glad of it. So am I, thank you, Hassan. That is, I'm tolerable, as the time goes. But you had never the kindness to ask me, *me*, your fellow slave! Pray, now, do ask me! Do! — (*aside*) for that will take up a little time.

HASSAN: Why then, how d'ye do, Shacabac?

SHACABAC: Very ill indeed, Hassan! Only feel my pulse. Count it till it beats just one hundred and twenty. — (*aside*) Twice sixty seconds will delay him about two minutes!

HASSAN: I don't know how to count, Shacabac.

SHACABAC: Don't you? Why not?

HASSAN: I can't read.

SHACABAC: That's a good reason! — (*aside*) I should think, ere this, they are far enough from the Blue Chamber to — A little longer to make all sure. — I have been thinking, Hassan, why you and I should be of different colours.

HASSAN: Fortune has disposed it so. She has made me black and you white. But don't let that mortify you.

SHACABAC: It shan't. But, as you say, Hassan, Fortune will make men of different shades. Fortune's chequered, and she chequers men alternately, black and white, like the squares in the Bashaw's chessboard. When I think how much Fortune is chequered, I think — I think that — (*aside*) I think I have almost kept you long enough for my purpose! — What are the Bashaw's orders to the Lady Fatima?

HASSAN: That she must attend him instantly in the garden.

SHACABAC: In the garden? Was that the command, Hassan?

HASSAN: It was, Shacabac.

SHACABAC: Then I'll tell you what, Hassan, if ever the Master of the Slaves gave you a sound drubbing for staying so long on a message, you'll get one now!

HASSAN: Why have you delayed me, then?

SHACABAC: I? You have delayed *me*. You have a brain for business, Hassan, but whenever you meet anyone in your way you will stop and gabble. That's your fault. Away!

HASSAN: I'll go find her.

(*Exit* HASSAN.)

SHACABAC: And I'll go to the garden to watch her interview with the Bashaw.
    And, weak as my means are, I'll catch at every straw to preserve her!
        (*Exit* SHACABAC.)

    Scene 6. *A garden, in the back of which is part of* ABOMELIQUE's *Castle
    and a drawbridge leading to the Castle gate. A corridor before the apart-
    ments on the first storey with a door beneath it. A turret on the top of the
    building, overlooking the country.*

        (*Enter* ABOMELIQUE *and a* SLAVE.)
ABOMELIQUE: Is Fatima informed I wait her presence here?
SLAVE: Hassan, by your command – She comes.
        (*Enter* FATIMA.)
ABOMELIQUE: Leave us!
        (*Exit* SLAVE.)
FATIMA: (*in apparent confusion*) This speedy return I – I looked not for.
ABOMELIQUE: I had accounts to settle, with traders, merchants from Gallipoli.
    But when worldly business draws men abroad who leave their hearts at home,
    then, Fatima, Love's wings give swiftness to the leaden hours of dull nego-
    tiation, and the mercurial spirit of an enamoured mind consolidates a volume
    ere Commerce, dozing o'er his day-book, can plod a page. How have your
    hours passed in my absence? Have you viewed the Castle?
FATIMA: I have, sir.
ABOMELIQUE: Well, saw you aught worthy your inspection?
FATIMA: Worthy, sir?
ABOMELIQUE: Aye, worthy. There are sights here, perhaps, that common eyes
    ne'er looked upon.
FATIMA: There are, indeed!
ABOMELIQUE: Now, please you, give me back the keys.
FATIMA: They are here. (*delivering them in great agitation*)
ABOMELIQUE: How now? You tremble!
FATIMA: Tremble, sir? Why should I?
ABOMELIQUE: You can best answer that. Sometimes, lady, 'twill betray guilt.
FATIMA: And know you, then, no instance where the guilty do *not* betray them-
    selves by trembling?
ABOMELIQUE: Umph – I comprehend not that. (*sternly*) One key is wanting!
    Where is it?
FATIMA: I have it.
ABOMELIQUE: Give it me!
FATIMA: Be not impatient. 'Tis in my pocket.
ABOMELIQUE: Produce it!
FATIMA: I shall. – But, by mere accident, you see, 'tis broken. (*giving it*)

---

day-book: a daily record, kept in journal form, of business transactions. Eighteenth-century
book-keeping usually registered all classes of transactions in a single, continuous, chronological
sequence.

ABOMELIQUE: Damnation! Lady, this key is charm-fraught, forged in a sulphurous cave within whose blood-besprinkled mouth nothing but Witchcraft enters, to celebrate her frantic revels. This speaks a damning proof against you, and you die!

> (ABOMELIQUE *draws his scimitar and holds it over* FATIMA*'s head. She falls on her knees.*)

FATIMA: Oh, spare me, spare me! I never approached the door but to –

ABOMELIQUE: No protestations! (*going to strike*)

FATIMA: Beseech you, hold! Alas, if I must die, grant me some little time for preparation!

ABOMELIQUE: (*after a short pause*) Well, be it so. (*pointing to an apartment within the corridor*) Yonder's your chamber. Thither instantly. Soon expect me there, then to expiate your crime by death. Before me to the Castle!

> (*Exit* FATIMA *through the door under the corridor,* ABOMELIQUE *following her with his drawn scimitar. Enter* SHACABAC *on the opposite side.*)

SHACABAC: Allah preserve her poor soul! But I fear she goes to certain death. Oh, that I were able to save her! Are there no means to? This hellish Abomelique whips off women's heads as if they were a parcel of buttons! Let me listen.

> (FATIMA *comes from her apartment upon the corridor.*)

Hist! Lady! Lady Fatima!

FATIMA: Oh, get you hence, good fellow! Your anxiety may make you a sharer with me in the Bashaw's resentment.

SHACABAC: Where is he?

FATIMA: I expect him instantly to ascend the stair and execute his dreadful purpose.

SHACABAC: Oh Mahomet, holy prophet, if ever you break a Bashaw's neck over a staircase, now's your time!

FATIMA: Hark, I hear him! – No.

> (IRENE *appears on the top of the turret.*)

IRENE: Sister! Sister Fatima!

FATIMA: Irene, is it you? Oh, sister, fare you well! I die a cruel death!

IRENE: My heart bleeds for you!

SHACABAC: So does mine, I'm sure!

IRENE: Should travellers appear, I'll call to them to succour us.

ABOMELIQUE: (*calling from* FATIMA*'s apartment*) Fatima!

FATIMA: Oh Heaven, he has entered the apartment!

ABOMELIQUE: (*without*) Why, Fatima!

SHACABAC: 'Tis he! (*retiring under the corridor*)

FATIMA: One moment, I beseech you! I have but one poor prayer to offer up to Heaven and then I come. – Is there no help?

Quartetto. ABOMELIQUE, FATIMA, IRENE *and* SHACABAC.

FATIMA: Look from the turret, sister dear,
                 And see if succour be not near!
                 Oh, tell me, what do you descry?

IRENE: Nothing but dreary land and sky.

FATIMA
IRENE   } Alas, alas, then { I / you / she } must die!
SHACABAC

ABOMELIQUE: Prepare!

FATIMA: He calls! Look out again!
          Look out, look out, across the plain!
          Ah me, does nothing meet your eyes?

IRENE: I see a cloud of dust arise.

FATIMA
IRENE   } That cloud of dust a hope supplies!
SHACABAC

ABOMELIQUE: No more delay!

FATIMA: A moment, stay!
          Oh, watch the travellers, my sister dear!

IRENE: I'll wave my handkerchief, 'twill draw them near.

SHACABAC: They'll see it speedily and hurry here.

ABOMELIQUE: Prepare!

IRENE   } I see them galloping, they're spurring on amain!
SHACABAC } Now faster galloping, they skim along the plain!

ABOMELIQUE: No more delay!

FATIMA: A moment, stay!

FATIMA
IRENE   } They come!
SHACABAC

ABOMELIQUE: Prepare!

FATIMA
IRENE   } They'll be too late!
SHACABAC } Now they dismount! They're at the gate!

ABOMELIQUE: Prepare!

> (ABOMELIQUE, *as they finish the quartetto, rushes from the apartment upon the corridor, seizes* FATIMA *and is on the point of beheading her when* SELIM *and his companions, having crossed the drawbridge, sound the horn loudly at the gate.* ABOMELIQUE, *alarmed at the noise, retires hastily, dragging* FATIMA *into the apartment.* SHACABAC *comes out from under the corridor.*)

SHACABAC: (*to* SELIM, *who is on the drawbridge*) You'll get no entrance there!

SELIM: Say, where is Fatima?

SHACABAC: Trembling under the Bashaw's clutches.

SELIM: We force the gate, then.

SHACABAC: 'Tis impossible. Get round to the eastern battlement; we are weakest there! Away, and success attend you!

SELIM: To judge from your conduct, you should be a friend. Say, what are you?

SHACABAC: What every man should be: a friend to virtue in distress wherever I meet it. Away, or you will be too late.

SELIM: Come, comrades, be firm, fight lustily! Quick march!

(SELIM *and his companions hurry from the bridge to quick martial music. Exit* SHACABAC.)

Scene 7. *An apartment in the Castle. Alarums, shouts, &c.*

(*Enter a body of* SLAVES.)

1st SLAVE: We are attacked. Up to the ramparts! Where's Ibrahim, our leader?

2nd SLAVE: He's nowhere to be found.

1st SLAVE: We must begin without him, then. It is the Bashaw's order. — Follow!

(*Exeunt* SLAVES. *Shouts are heard without. Enter* IBRAHIM.)

IBRAHIM: Mercy on me, I quake in my clothes like a cold jelly in a bag! They are battering the Castle to pieces! I am the unluckiest Musselman in all Turkey: here's a building that has stood wind and weather this age, and, the moment I pop my nose into it, it begins tumbling about my ears!

(*A cry of* 'To arms! To arms!' *is heard.*)

To arms? Oh dear, I had much rather to legs, if I knew which way to escape! Now shall I be expected to put myself in the front of the ranks because I am major-domo. But, if I do, I'll give them leave to mince the major-domo for his son-in-law's supper!

(*Alarums are heard. Re-enter* 1st SLAVE.)

Oh Mahomet, what's that?

1st SLAVE: An enemy is on the walls.

IBRAHIM: Then, you cowardly rascal, do you go and knock him into the ditch!

1st SLAVE: We wait for you. You are appointed our leader. There is no discipline without you. We want a head.

IBRAHIM: Do you? So shall I, if I go with you! Get on before, tell 'em to fight like fury, and I'll be with them to reward their valour when it's all over. Run that way; that leads into the action!

1st SLAVE: I will.

(*Exit* 1st SLAVE.)

IBRAHIM: And I'll run this way, that leads out of it!

(*Exit* IBRAHIM *amidst shouts and alarums.*)

Scene 8. *The inside of the sepulchre. The inscription over the skeleton's head is now:* 'THIS SEPULCHRE SHALL INCLOSE HER WHO MAY ENDANGER THE LIFE OF ABOMELIQUE.' *The shouts and alarums continue.*

(*Enter* ABOMELIQUE *with his scimitar drawn, dragging in* FATIMA.)

ABOMELIQUE: On every side it rages! The slaves give way! — You are still in my power. You, sorceress, have led me to the toil. Your death will extricate me. Meet it then here, here in the sepulchre which you have violated!

FATIMA: Nay, take me hence. Let me not perish in this abode of horror!

ABOMELIQUE: Thy prayers are in vain!

(*As* ABOMELIQUE *raises his scimitar to strike, a near attack is heard, and a violent crash in the building. Part of the wall in the back of the sepulchre, towards the roof, is beat down, and* SELIM *appears in the aperture.*)

SELIM: Hold, ruffian, hold thy arm!

FATIMA: Oh, Selim!

ABOMELIQUE: Rash fool, I know thee and thy purpose! Thy presence, now, swells the full tide of my resentment and gives a higher zest to vengeance. Know the decrees of destiny and curse thy wickedness which would counteract it! 'This sepulchre shall inclose her who may endanger the life of Abomelique': this wretch, here, has endangered it, this sepulchre incloses her, and —

SELIM: But not in death, tyrant! Thy hell-born spells promise not that!

ABOMELIQUE: Does my fate juggle with me, then? Hold. No, yon dagger (*pointint to the talisman*) is my safeguard till mortal hands can reach it. Weak boy, despair and see her die!

FATIMA: While Selim lives — so near me, too — my life is precious and I struggle to preserve it.

> (FATIMA *struggles with* ABOMELIQUE, *who attempts to kill her, and, in the struggle, snatches the dagger from the pedestal of the skeleton. The skeleton rises on his feet, lifts his arm which holds the dart and keeps it suspended. At that instant, the entire wall of the sepulchre falls to pieces and admits* SELIM *to the ground. Behind, among fragments of the building, a body of* SPAHIS *is discovered, on foot, with* ABOMELIQUE'S SLAVES *under their sabres in posture of submission. And, farther back, is seen a large Troop of Horse. The neighbouring country terminates the view.* SELIM *advances towards* ABOMELIQUE.)

SELIM: Now, turn thee hither!

ABOMELIQUE: Baffled? I still have mortal means, and thus I use them!

> (SELIM *and* ABOMELIQUE *fight with scimitars. During the combat, enter* IRENE *and* SHACABAC. *After a hard contest,* SELIM *overthrows* ABOMELIQUE *at the foot of the skeleton. The skeleton instantly plunges the dart, which he has held suspended, into the breast of* ABOMELIQUE *and sinks, with him, beneath the earth. A volume of flame arises and the earth closes.* SELIM *and* FATIMA *embrace.*)

SHACABAC: Huzza! If ever the Bashaw was in fit company, he has got into it now!

FATIMA: Oh, Selim!

SELIM: Thus, safe at last, I clasp thee!

IRENE: Joy, joy, my sister! We have conquered!

FATIMA: Where is my father?

SHACABAC: Hid in the dust-hole! When the noise is over, we may chance to get sight of him.

SELIM: All shall be explained. Our marriage, now, my Fatima, may meet his

---

dust-hole: quite literally dug out of the ground, this primitive prototype of sanitary engineering, if used as a means of concealment, would have offered the facility of self-preservation at its most ambiguous.

sanction. (*to* SHACABAC) And you, my honest fellow, must not go unrewarded. Thanks, my honest comrades!

(SPAHIS *and* SLAVES *come forward.*)

We are victors! And, in the countenance, here, of every slave, I see a smile impressed which betokens joy in having lost a tyrant.

SLAVES: Thanks to our deliverer!

SELIM: Come, Fatima, let us away from this rude scene of horror and bless the providence which nerves the arm of virtue to humble vice and oppression!

Chorus. SELIM, FATIMA, IRENE, SHACABAC, SLAVES *and* SPAHIS.

> Monsters of hell and noxious night,
> Howl your songs with wild delight,
> To your gloomy caves descending,
> His career of murder ending!
>> Now the tyrant's spirit flies
>>> Bathed in a flood
>>> Of guilty blood,
>> He dies! He dies!
> How great is the transport, the joy how complete
> When, raised from despair, thus Love's votaries meet!
>> Sweet the delight that lovers prove!
>>> Sweet when, Fortune tired of frowning,
>>> Hymen comes, with pleasure crowning
>> Happy love!

SPEED THE PLOUGH

A comedy in five acts by Thomas Morton

First performed at the Theatre Royal, Covent Garden, Saturday, 8 February 1800, with the following cast:

SIR PHILIP BLANDFORD	Mr Pope
SIR ABEL HANDY	Mr Munden
BOB HANDY	Mr Fawcett
FARMER ASHFIELD	Mr Knight
HENRY	Mr H. Johnston
MORRINGTON	Mr Murray
EVERGREEN	Mr Davenport
GERALD	Mr Waddy
PETER	Mr Atkins
POSTBOY	Mr Abbot
SERVANT	Mr Klanert
MISS BLANDFORD	Mrs H. Johnston
LADY HANDY	Mrs Dibdin
SUSAN ASHFIELD	Miss Murray
DAME ASHFIELD	Mrs Davenport
Countrymen	

Incidental music by John Moorehead

Mrs Davenport

in the Character of Mrs Grundy.

VII   Mrs Davenport as Dame Ashfield in *Speed the Plough*, from the
       portrait by Samuel de Wilde. The error of the original caption writer
       is evidence of the speed with which Morton's 'Mrs Grundy' device
       rooted itself in the popular imagination.

ACT I

Scene 1. *In the foreground a farmhouse. A view of a castle at a distance.*
FARMER ASHFIELD *discovered at a table with his jug and pipe.*

> (*Enter* DAME ASHFIELD *in a riding dress and a basket under her arm.*)

ASHFIELD: Well, dame, welcome whoam. What news does thee bring vrom market?

DAME: What news, husband? What I always told you, that Farmer Grundy's wheat brought five shillings a quarter more than ours did.

ASHFIELD: All the better vor he.

DAME: Ah, the sun seems to shine on purpose for him!

ASHFIELD: Come, come, missus, as thee hast not the grace to thank God for prosperous times, dan't thee grumble when they be unkindly a bit.

DAME: And, I assure you, Dame Grundy's butter was quite the crack of the market.

ASHFIELD: Be quiet, woolye? Aleways ding-dinging Dame Grundy into my ears! — What will Mrs Grundy say? What will Mrs Grundy think? — Can'st thee be quiet, let ur alone, and behave thyzel pratty?

DAME: Certainly I can! — I'll tell thee, Tummas, what she said at church last Sunday —

ASHFIELD: Can'st thee tell what parson said? Noa? Then I'll tell thee. 'A zaid that envy were as foul a weed as grows, and cankers all wholesome plants that be near it. That's what 'a zaid.

DAME: And do you think I envy Mrs Grundy, indeed?

ASHFIELD: Why dan't thee letten her aloane, then? I do verily think, when thee goest to t'other world, the vurst question thee ax 'll be if Mrs Grundy's there! Zoa be quiet and behave pratty, do ye. Has thee brought whoam the Salisbury news?

DAME: No, Tummas, but I have brought a rare wadget of news with me. First and foremost, I saw such a mort of coaches, servants and waggons, all belonging to Sir Abel Handy and all coming to the Castle, and a handsome young man, dressed all in lace, pulled off his hat to me and said, 'Mrs Ashfield, do me the honour of presenting that letter to your husband.' So there he stood, without his hat! Oh, Tummas, had you seen how Mrs Grundy looked!

ASHFIELD: Dom Mrs Grundy! Be quiet and letten I read, woolye? (*reads*) 'My dear Farmer' — (*taking off his hat*) Thank ye, zur. Zame to you, wi' all my heart and soul — 'My dear Farmer' —

DAME: Farmer? Why, you're blind, Tummas. It is, 'My dear Father.' 'Tis from our own dear Susan.

Mrs Grundy: more famous, sadly, than the play in which she fails to appear, Mrs Grundy has entered the English literary tradition as a stock motif for the personification of disapproval. Peter Fryer used her name as the title of his studies in English prudery published in 1963.
wadget: a pocket full to bursting (dialect).
mort: a great number or quantity (dialect).

ASHFIELD: Od's dickens and daizeys, zoo it be, zure enow! 'My dear Feyther, you will be surprised' — Zoo I be, he! he! What pretty writing, bean't it? All as straight as thof it were ploughed — 'Surprized to hear that in a few hours I shall embrace you. This is to acquaint you' — 'This is'? Ah, that be wrong. You know, dame, it should be 'Thix be' — 'To acquaint you that Nelly, who was formerly our servant, has fortunately married Sir Abel Handy-Bart.'

DAME: Handy-Bart? Pugh, bart. stands for baronight, mun!

ASHFIELD: Likely, likely. Drabbit it, only to think of the zwaps and changes of this world!

DAME: Our Nelly married to a great baronet! I wonder, Tummas, what Mrs Grundy will say?

ASHFIELD: Now, woolye be quiet and letten I read. 'And she has proposed bringing me to see you, an offer, I hope, as acceptable to my dear Feyther' —

DAME: 'And Mother' —

ASHFIELD: Bless her, how prettily she do write feyther, dan't she?

DAME: And mother.

ASHFIELD: Ees, but feyther first, though. — 'As acceptable to my dear Feyther and Mother as to their affectionate daughter, Susan Ashfield.' Now, bean't that a pratty letter?

DAME: And, Tummas, is not she a pretty girl?

ASHFIELD: Ees, and as good as she be pratty. Drabbit it, I do feel zoo happy and zoo warm, for all the world like the zun in harvest!

DAME: Oh, Tummas, I shall be so pleased to see her I shan't know whether I stand on my head or my heels.

ASHFIELD: Stand on thy head? Vor sheame o' thyzel, behave pratty, do!

DAME: Nay, I meant no harm. Eh, here comes friend Evergreen, the gard'ner from the Castle. Bless me, what a hurry the old man is in!

(*Enter* EVERGREEN.)

EVERG: Good day, honest Thomas.

ASHFIELD: Zame to you, Measter Evergreen.

EVERG: Have you heard the news?

DAME: Anything about Mrs Grundy?

ASHFIELD: Dame, be quiet, woolye now?

EVERG: No, no. The news is that my master, Sir Philip Blandford, after having been abroad for twenty years, returns this day to the Castle, and that the reason of his coming is to marry his only daughter to the son of Sir Abel Handy, I think they call him.

DAME: As sure as tuppence, that is Nelly's husband!

Od's dickens and daizeys: a compound oath. *Od's dickens* was an exclamation popular throughout the seventeenth and eighteenth centuries, carrying a reference to the devil still current in modern usage as 'what the dickens!' *Daizeys*, however, is less readily decipherable. Hardy, in the first of his *Wessex Tales*, has Fennel the shepherd declare, 'Daze it, what's a cup of mead more or less?', this dialect curse suggesting that *daizey* could be considered as an alternative written form of *Daze 'ee* ('damn you!'), *daizeys* simply being the whole phrase treated as a substantive and then thrust into the plural, i.e. 'damn-yous'.

EVERG: Indeed! Well, Sir Abel and his son will be here immediately, and, farmer, you must attend them.

ASHFIELD: Likely, likely.

EVERG: And, mistress, come and lend us a hand at the Castle, will you? Ah, it is twenty long years since I have seen Sir Philip. Poor gentleman! — bad, bad health — worn almost to the grave, I am told. What a lad do I remember him, till that dreadful — (*checking himself*) But where is Henry? I must see him, must caution him.

> (*A gun is discharged at a distance.*)

That's his gun, I suppose. He is not far then. Poor Henry!

DAME: Poor Henry? I like that, indeed! What, though he may be nobody knows who, there is not a girl in the parish that is not ready to pull caps for him. The Miss Grundys, genteel as they think themselves, would be glad to snap at him. If he were our own, we could not love him better.

EVERG: And he deserves to be loved. Why, he's as handsome as a peach tree in blossom, and his mind is as free from weeds as my favourite carnation bed. But, Thomas, run to the Castle and receive Sir Abel and his son.

ASHFIELD: I wool, I wool. Zo, good day. (*bowing*) Let every man make his bow and behave pratty, that's what I say. Missus, do ye show un Sue's letter, woolye? Do ye letten see how pratty she do write feyther.

> (*Exit ASHFIELD.*)

DAME: Now Tummas is gone, I'll tell you such a story about Mrs Grundy. But come, step in; you must needs be weary, and I am sure a mug of harvest beer, sweetened with a hearty welcome, will refresh you.

> (*Exeunt into the house.*)

Scene 2. *Outside and gate of the Castle.* SERVANTS *cross the stage, laden with different packages.*

ASHFIELD: Drabbit it, the wold Castle 'll be hardly big enow to hold all this lumber.

SIR ABEL: (*without*) Gently, there! Mind how you go, Robin. (*a crash*)

ASHFIELD: Who do come here? 'A do zeem a comical zort of a man. Oh, Abel Handy, I suppoze.

> (*Enter SIR ABEL HANDY, SERVANT following.*)

SIR ABEL: Zounds and fury! You have killed the whole country, you dog, for you have broke the patent medicine chest that was to keep them all alive! Richard, gently! Take care of the grand Archimedian cork-screws! Bless my soul, so much to think of! Such wonderful inventions in conception, in concoction and in completion!

> (*Enter PETER.*)

Well, Peter, is the carriage much broke?

PETER: Smashed all to pieces. I thought as how, sir, your infallible axle-tree would give way.

SIR ABEL: Confound it, it has compelled me to walk so far in the wet that I declare my waterproof shoes are completely soaked through!

> (*Exit PETER.*)

ASHFIELD: (*loud and bluntly*) Zarvent, zur! Zarvent!

SIR ABEL: (*starting*) What's that? Oh, good day. (*aside*) Devil take the fellow!

ASHFIELD: Thankye, sir. Zame to you, with all my heart and zoul.

SIR ABEL: Pray, friend, could you contrive *gently* to inform me where I can find one Farmer Ashfield?

ASHFIELD: Ha! ha! ha! (*laughing loudly*) Excuse my tittering a bit, but your axing mysel vor I be so dommed zilly! (*bowing and laughing*) Ah, you stare at I becas I be bashful and daunted.

SIR ABEL: You are very bashful, to be sure! I declare I'm quite weary.

ASHFIELD: If you'll walk into the Castle, you may sit down, I dare zay.

SIR ABEL: May I, indeed? You are a fellow of extraordinary civility.

ASHFIELD: There's no denying it, zur.

SIR ABEL: No, I'll sit here.

ASHFIELD: What? On the ground? Why, you'll wring your ould withers.

SIR ABEL: On the ground? No, I always carry my seat with me. (*spreads a small camp-chair*) Here I'll sit and examine the surveyor's account of the Castle.

ASHFIELD: Dickens and daizeys, what a gentleman you would be to show at a vair!

SIR ABEL: Silence, fellow, and attend! (*reads*) 'An account of the castle and domain of Sir Philip Blandford, intended to be settled as a marriage-portion on his daughter and the son of Sir Abel Handy. By Frank Flourish, surveyor. Imprimis, the premises command an exquisite view of the Isle of Wight.' Charming! Delightful! I don't see it, though. (*rising*) I'll try with my new glass, my own invention.

> (SIR ABEL *looks through the glass and* ASHFIELD *places himself before him, observing him.*)

Yes, there I caught it! Ah, now I see it plainly! With a common glass I should not be able to see it at all. Eh? No, I don't see it.

> (SIR ABEL *takes the glass from his eye and sees* ASHFIELD *looking in at the other end.*)

Do you?

ASHFIELD: Noa, zur, I doan't. But little zweepy do tell I he can see a bit on't from the top of the chimbley. Zo, an you've a mind to crawl up, you may see un, too! He! he!

SIR ABEL: Thank you. (*aside*) But damn your titter. (*reads*) 'Fish ponds well stocked.' That's a good thing, farmer.

ASHFIELD: Likely, likely. But I doan't think the vishes do thrive much in theas ponds.

SIR ABEL: No? Why?

ASHFIELD: Why, the ponds be always dry i' the zummer, and I be tould that bean't wholesome vor the little vishes.

SIR ABEL: Not very, I believe. Well said, surveyor! — 'A cool summer-house.'

ASHFIELD: Ees, zur, quite cool, by reason the roof be tumbled in.

SIR ABEL: Better and better! — 'The whole capable of the greatest improvement.' Come, that seems true, however. I shall have plenty to do, that's one comfort. I have such contrivances! I'll have a canal run through my kitchen — (*aside*) I must give this rustic some idea of my consequence. — You must know, farmer, you have the honour of conversing with a man who has obtained

patents for tweezers, tooth-picks and tinder-boxes, to a philosopher who has
been consulted on the Wapping docks and the Gravesend tunnel and who has
now in hand two inventions which will render him immortal: the one is converting saw-dust into deal boards and the other is a plan for cleaning rooms
by a steam engine. And, farmer, I mean to give prizes for industry; I'll have a
ploughing match.

ASHFIELD: Will you, zur?

SIR ABEL: Yes, for I consider a healthy young man between the handles of a
plough as one of the noblest illustrations of the prosperity of Britain.

ASHFIELD: Faith and troth, there be some tightish hands in theas parts, I promise
ye.

SIR ABEL: And, farmer, it shall precede the hymeneal festivities.

ASHFIELD: Nan!

SIR ABEL: Blockhead! The ploughing match shall take place as soon as Sir Philip
Blandford and his daughter arrive.

ASHFIELD: Oh, likely, likely.

        (*Enter* SERVANT.)

SERVANT: Sir Abel, I beg to say my master will be here immediately.

SIR ABEL: And, sir, I beg to ask who possesses the happiness of being your
master?

SERVANT: Your son, sir, Mr Robert Handy.

SIR ABEL: Indeed! And where is Bob?

SERVANT: I left him, sir, in the belfry of the church.

SIR ABEL: Where?

SERVANT: In the belfry of the church.

SIR ABEL: In the belfry of the church? What was he doing there?

SERVANT: Why, sir, the *natives* were ringing a peal in honour of our arrival when
my master, finding they knew nothing of the matter, went up to the steeple
to instruct them and ordered me to proceed to the Castle. Give me leave, Sir
Abel, to take this out of your way (*takes the camp-chair*). Sir, I have the
honour —

        (*Exit* SERVANT *with a bow.*)

SIR ABEL: Wonderful! My Bob, you must know, is an astonishing fellow! You

---

patents: the Industrial Revolution, which had been in progress for almost fifty years by the
time *Speed the Plough* came to be written, was clearly a motive force behind the phenomenal
inventiveness which blossomed in England towards the close of the eighteenth century. In
1750, 70 patents were granted; by 1780 the annual total had risen to over 500.

Wapping docks: subsequently known as the London Dock, designed by Daniel Alexander and
built at Wapping, to the east of the Pool of London on the north bank of the Thames. Opened
in 1805 at a cost of £2 million, London Dock, with its tide-free moorings, massive quayside
warehouses and towering security walls, was considered to be one of the wonders of the world.

Gravesend tunnel: another potential wonder of the world, this scheme, in fact, was to have a
different fate. Intended to provide a direct link between Gravesend in Kent and Tilbury in
Essex by means of a tunnel under the Thames, work began in the Spring of 1799, a *Times*
correspondent noting, on 13 May of that year, 'two walls of large diameter' which were nearing
completion. Further progress reports, however, were unforthcoming.

have heard of the *admirable Crichton*, maybe? Bob's of the same kidney! I contrive, he executes; Sir Abel *invenit*, Bob *fecit*. He can do everything, everything!

ASHFIELD: All the better vor he. I zay, zur, as he can turn his hand to everything, pray in what way med he earn his livelihood?

SIR ABEL: Earn his livelihood?

ASHFIELD: Ees, sir, how do he gain his bread?

SIR ABEL: Bread? Oh, he can't earn his bread, bless you! He's a genius.

ASHFIELD: Genius! Drabbit it, I have got a horze o' thic name, but, dom'un, he'll never work, never!

SIR ABEL: Egad, here comes my boy Bob! Eh? No, it is not! No.

    (*Enter* POSTBOY *with a round hat and cane.*)

Why, who the devil are you?

POSTBOY: I am the postboy, your honour, but the gem'man said I did not know how to drive, so he mounted my horse and made me get inside. Here he is.

    (*Enter* HANDY JUNIOR *with a postboy's cap and whip.*)

HANDY JUN: Ah, my old Dad, is that you?

SIR ABEL: Certainly. The only doubt is if that be you!

HANDY JUN: Oh, I was teaching this fellow to drive. Nothing is so horrible as people pretending to do what they are unequal to. (*to the* POSTBOY) Give me my hat. That's the way to use a whip.

POSTBOY: Sir, you know you have broke the horse's knees all to pieces.

HANDY JUN: (*apart*) Hush! There's a guinea.

SIR ABEL: (*to* ASHFIELD) You see, Bob can do everything! But, sir, when you knew I had arrived from Germany, why did you not pay your duty to me in London?

HANDY JUN: Sir, I heard you were but four days married, and I would not interrupt your honey-moon.

SIR ABEL: (*sighing*) Four days! Oh, you might have come!

HANDY JUN: I hear you have taken to your arms a simple rustic, unsophisticated by fashionable follies, a full-blown blossom of nature.

SIR ABEL: Yes.

HANDY JUN: How does it answer?

SIR ABEL: So-so.

HANDY JUN: Any thorns?

SIR ABEL: A few.

HANDY JUN: I must be introduced. Where is she?

SIR ABEL: Not within thirty miles, for I don't hear her!

ASHFIELD: Ha! ha! ha!

HANDY JUN: Who is that?

SIR ABEL: Oh, a pretty-behaved tittering friend of mine.

ASHFIELD: Zarvent, zur. No offence, I do hope. Couldn't help tittering a bit at

---

the admirable Crichton: this was James Crichton, a sixteenth-century Scottish prodigy who, at the age of seventeen, is said to have been fluent in at least twelve languages, to have had a memory capable of the instant recall of anything he had heard or read, and to have been an accomplished fencer and horseman.

Nelly. When she were zarvent maid wi' I, she had a tightish prattle wi' her, that's vor zartain.

HANDY JUN: Oh, so then my honoured Mamma was the servant of this tittering gentleman! I say, father, perhaps she has not lost the tightish prattle he speaks of.

SIR ABEL: My dear boy, come here. (*apart*) Prattle? I say, did you ever live next door to a pewterer's? — that's all. You understand me? Did you ever hear a dozen fire-engines at full gallop? Were you ever at the Stock Exchange on a settling-day or at Billingsgate in the sprat season?

HANDY JUN: Ha! ha!

SIR ABEL: Nay, don't laugh, Bob.

HANDY JUN: Indeed, sir, you think of it too seriously. The storm, I dare say, soon blows over.

SIR ABEL: Soon! You know what a trade wind is, don't you, Bob? Why, she thinks no more of the latter end of her speech than she does of the latter end of her life.

HANDY JUN: Ha! ha!

SIR ABEL: But I won't be laughed at! I'll knock any man down that laughs!

HANDY JUN: I beg your pardon. But how, in the name of Babel, did she wheedle you into matrimony?

SIR ABEL: Why, she dealt with me as the devil deals with a witch: humoured me for a time that I might be her slave forever! I thought I was marrying a notable woman who would have taken care of my gouty leg and eased my head of part of its burden, instead of which —

HANDY JUN: She has added to its burden.

SIR ABEL: You know, my dear boy, my aim is to make myself useful.

HANDY JUN: And her aim, I suppose, is to make it ornamental.

SIR ABEL: Bob, if you can say anything pleasant, I'll trouble you; if not, do what my wife can't: hold your tongue!

HANDY JUN: I'll show you what I can do: I'll amuse you with this native.

SIR ABEL: Do, do! Quiz him! At him, Bob!

HANDY JUN: I say, farmer, you are a set of jolly fellows here, an't you?

ASHFIELD: Ees, zur, deadly jolly, excepting when we be otherwise, and then we bean't.

HANDY JUN: Play at cricket, don't you?

---

settling-day: on the modern London Stock Exchange, this occurs once a fortnight and is the day when the money connected with any transfer of stocks and shares prior to that date actually changes hands, transactions between settling-days being conducted on the basis of notes of contract. In the eighteenth century, since most stock was exchanged and paid for on the spot, settling-days were considered necessary only for the convenience of overseas dealers and took place, in a frenzy of activity, at quarterly intervals.

Billingsgate: until 1981, the site of London's fish market. Sir Abel's allusion might also extend to the shade of meaning captured by *Bailey's English Dictionary* (1736), which lists Billingsgate as a common noun defined bitterly as 'a scolding, impudent slut'. Since the sprat season was both short and lucrative, this would have been a period when the vocal retailing practices of Billingsgate's fishwives reached their zenith.

ASHFIELD: Ees, zur, we Hampshire lads conceat we can bowl a bit, or thereabouts.

HANDY JUN: And cudgel, too, I suppose?

SIR ABEL: At him, Bob!

ASHFIELD: Ees, zur, we sometimes break oon another's heads, by way of being agreeable, and the like o' that.

HANDY JUN: Understand all the guards? (*putting himself in an attitude of cudgelling*)

ASHFIELD: Can't zay I do, zur.

HANDY JUN: What, hit in this way, eh?

> (HANDY JUN. *makes a hit at* ASHFIELD *which he parries, hitting young* HANDY *violently.*)

ASHFIELD: Noa, zur, we do hit thic way.

HANDY JUN: Zounds and fury!

SIR ABEL: Why, Bob, he has broke your head.

HANDY JUN: Yes, he rather hit me. He somehow —

SIR ABEL: He did indeed, Bob.

HANDY JUN: Damn him! The fact is I am out of practice.

ASHFIELD: You need not be, zur; I'll gi' ye a bellyful any day, wi' all my heart and zoul.

HANDY JUN: No, no, thank you, Farmer — What's your name?

ASHFIELD: My name be Tummas Ashfield. (*threatening*) Anything to say against my name?

HANDY JUN: No, no! — (*aside*) Ashfield! Should he be the father of my pretty Susan? — Pray, have you a daughter?

ASHFIELD: Ees, I have. Anything to say against she?

HANDY JUN: No, no, I think her a charming creature.

ASHFIELD: Do ye, faith and troth? Come, that be deadly kind o' ye, however. Do you zee, I were *frightful* she were not agreeable.

HANDY JUN: Oh, she's extremely agreeable to me, I assure you.

ASHFIELD: I vow, it be quite pratty in you to take notice of Sue. I do hope, zur, that breaking your head will break noa squares. She be a-coming down to theas parts wi' Lady — our maid Nelly, as wur — your spouse, zur.

HANDY JUN: (*aside*) The devil she is! That's awkward!

ASHFIELD: I do hope you'll be kind to Sue when she do come, woolye, zur?

HANDY JUN: You may depend on it.

SIR ABEL: I dare say you may. Come, farmer, attend us!

ASHFIELD: Ees, zur, wi' all respect. Gentlemen, pray walk thic way and I'll walk before you.

> (*Exit* ASHFIELD.)

SIR ABEL: Now, that's what he calls behaving pretty!

---

cricket . . . Hampshire: according to hallowed cricketing tradition, the first regularly constituted cricket club is said to have been formed in 1750 at Hambledon, in Hampshire.
cudgelling: probably a reference to the once popular rural sport of singlesticks.
break noa squares: 'breaking squares' was a dialect phrase which denoted the causing of disagreement or disturbance.

HANDY JUN: Susan Ashfield coming here?
SIR ABEL: What, Bob, some intrigue, eh?
HANDY JUN: Oh, fie!
SIR ABEL: Consider, sir, you come here to marry the beautiful and accomplished
Miss Blandford. And consider, on the other hand, you have already got a
slight memorandum of the farmer's agreeable way!
(*Exeunt.*)

Scene 3. *A grove.*

(*Enter* MORRINGTON. *He comes down the stage wrapped in a
greatcoat. Looks about, then at his watch, and whistles, which is
answered. Enter* GERALD.)
MORRINGTON: Here, Gerald! Well, my trusty fellow, is Sir Philip arrived?
GERALD: No, sir, but hourly expected.
MORRINGTON: Tell me, how does the Castle look?
GERALD: Sadly decayed, sir.
MORRINGTON: I hope, Gerald, you were not observed.
GERALD: I fear otherwise, sir. On the skirts of the domain I encountered a strip-
ling with his gun, but I darted into that thicket and so avoided him.
(HENRY *appears in the background in a shooting dress, attentively
observing them.*)
MORRINGTON: Have you gained any intelligence?
GERALD: None. The report that reached us was false; the infant certainly died
with its mother. Hush! Conceal yourself, we are observed! This way!
(MORRINGTON *and* GERALD *retreat,* HENRY *advances.*)
HENRY: Hold! As a friend, one word!
(*Exeunt* MORRINGTON *and* GERALD. HENRY *follows them, and
returns.*)
Again they have escaped me. 'The infant died with its mother'? This agony of
doubt is insupportable.
(*Enter* EVERGREEN.)
EVERG: Henry, well met.
HENRY: Have you seen strangers?
EVERG: No.
HENRY: Two but now have left this place. They spoke of a lost child. My busy
fancy led me to think that I was the object of their search. I pressed forward,
but they avoided me.
EVERG: No, no, it could not be you, for no one on earth knows but myself and —
HENRY: Who? Sir Philip Blandford?
EVERG: I am sworn, you know, my dear boy. I am solemnly sworn to silence.
HENRY: True, my good old friend, and, if the knowledge of who I am can only be
obtained at the price of thy perjury, let me forever remain ignorant, let the
corroding thought still haunt my pillow, cross me at every turn and render me
insensible to the blessings of health and liberty. Yet, in vain do I suppress the
thought: who am I? Why thus abandoned? Perhaps the despised offspring of
guilt — (*seizing* EVERGREEN *violently*) ah, is it so?
EVERG: Henry, do I deserve this?

HENRY: Pardon me, good old man, I'll act more reasonably. I'll deem thy silence mercy.

EVERG: That's wisely said.

HENRY: Yet it is hard to think that the most detested reptile that Nature forms, or man pursues, has, when he gains his den, a parent's pitying breast to shelter in. But I —

EVERG: Come, come, no more of this.

HENRY: Well. — I visited today that young man who was so grievously bruised by the breaking of his team.

EVERG: That was kindly done, Henry.

HENRY: I found him suffering under extreme torture, yet a ray of joy shot from his languid eye, for his medicine was administered from a father's hand; it was a mother's precious tear that dropped upon his wound. Oh, how I envied him!

EVERG: Still on the same subject. I tell thee, if thou art not acknowledged by thy race, why then, become the noble founder of a new one. The most valuable carnations were once seedlings, and the pride of my flower-bed is now a Henry, which, when known, will be envied by every florist in Britain. Come with me to the Castle for the last time.

HENRY: The last time?

EVERG: Aye, boy, for when Sir Philip arrives you must avoid him.

HENRY: Not see him? Where exists the power that shall prevent me?

EVERG: Henry, if you value your own peace of mind, if you value an old man's comfort, avoid the Castle.

HENRY: (*aside*) I must dissemble with this honest creature. — Well, I am content.

EVERG: That's right, that's right, Henry. Be but thou resigned and virtuous, and He who clothes the lily of the field will be a parent to thee.

(*Exeunt.*)

ACT II

Scene 1. *A lodge belonging to the Castle.* DAME ASHFIELD *discovered making lace.*

(*Enter* HANDY JUNIOR.)

HANDY JUN: A singular situation this my old Dad has placed me in: brought here to marry a woman of fashion and beauty, while I have been professing and, I've a notion, feeling the most ardent love for the pretty Susan Ashfield! Propriety says take Miss Blandford, love says take Susan, fashion says take both. But would Susan consent to such an arrangement? And, if she refused, would I consent to part with her? Oh, time enough to put that question when the previous one is disposed of. (*seeing* DAME ASHFIELD) How do you do? How do you do? Making lace, I perceive. Is it a common employment here?

---

breaking of his team: the sense here is of breaking away rather than of breaking up, the animals harnessed to the vehicle having run out of control.

DAME: Oh no, sir, nobody can make it in these parts but myself. Mrs Grundy, indeed, pretends. But, poor woman, she knows no more of it than you do!

HANDY JUN: Than I do? That's vastly well! My dear madam, I passed two months at Mechlin for the express purpose.

DAME: Indeed.

HANDY JUN: You don't do it right. Now, I can do it much better than that. Give me leave and I'll show you the true Mechlin method.

(HANDY JUN. *turns the cushion round, kneels down and begins working.*)

First, you see, so. Then, so.

(*Enter* SIR ABEL *and* MISS BLANDFORD.)

SIR ABEL: I vow, Miss Blandford, fair as I ever thought you, the air of your native land has given additional lustre to your charms. – (*aside*) If my wife looked so! – Ah, but where can Bob be? You must know, miss, my son is a very clever fellow. You won't find him wasting his time in boyish frivolity! No, you will find him – (*seeing him*)

MISS B: Is that your son, sir?

SIR ABEL: (*abashed*) Yes, that's Bob.

MISS B: Pray, sir, is he making lace or is he making love?

SIR ABEL: Curse me if I can tell! (*hitting him with his stick*) Get up, you dog! Don't you see Miss Blandford?

HANDY JUN: (*starting up*) Zounds, how unlucky! (*endeavouring to hide the work*) Ma'am, your most obedient servant. (*throwing it off*) Curse the cushion!

DAME: Oh, he has spoiled my lace!

HANDY JUN: (*aside*) Hush! I'll make you a thousand yards another time. – You see, ma'am, I was explaining to this good woman what – what – need not be explained again. – (*aside*) Admirably handsome, by Heaven!

SIR ABEL: (*aside*) Is not she, Bob?

HANDY JUN: (*to* MISS BLANDFORD) In your journey from the coast, I conclude, you took London in your way? (*to* DAME ASHFIELD) Hush!

MISS B: Oh no, sir, I could not so soon venture into the *beau monde*, a stranger just arrived from Germany.

HANDY JUN: The very reason; the most fashionable introduction possible! But, I perceive, sir, you have here imitated other German importations and only restored to us our native excellence.

MISS B: I assure you, sir, I am eager to seize my birthright: the pure and envied immunities of an English woman.

HANDY JUN: Then I trust, madam, you will be patriot enough to agree with me that, as a nation is poor whose only wealth is importation, the humble native artist may ever hope to obtain from his countrymen those fostering smiles without which genius must sicken and industry decay. But it requires no *valet de place* to conduct you through the purlieus of fashion, for now the way of

---

most fashionable introduction: referring to the sudden vogue for the 'German drama' in London at this time (see Introduction, pp. 47–8).

valet de place: a man employed to act as a guide to visitors, especially at some of the statelier of English homes, where persons of quality might be admitted as tourists.

the world is for everyone to pursue their own way, and following the fashion is differing as much as possible from the rest of your acquaintance.

MISS B: But surely, sir, there is some distinguishing feature by which the votaries of fashion are known?

HANDY JUN: Yes, but that varies extremely. Sometimes fashionable celebrity depends on a high waist, sometimes on a low carriage, sometimes on high play and sometimes on low breeding. Last year it rested solely on green peas!

MISS B: Green peas?

HANDY JUN: Green peas. That lady was the most enchanting who could bring the greatest quantity of green peas to her table at Christmas! The struggle was tremendous! Mrs Rowley Powley had the best of it by five pecks and a half, but, it having been unfortunately proved that at her ball there was room to dance and eat conveniently, that no lady received a black eye and no coachman was killed, the thing was voted decent and comfortable, and scouted accordingly.

MISS B: Is comfort, then, incompatible with fashion?

HANDY JUN: Certainly. Comfort in high life would be as preposterous as a lawyer's bag crammed with truth or his wig decorated with coquelicot ribbons! No, it is not comfort and selection that is sought, but numbers and confusion, so that a fashionable party resembles Smithfield market — only a good one when plentifully stocked — and ladies are reckoned by the score, like sheep, and their husbands by droves, like horned cattle!

MISS B: Ha! ha! And the conversation?

HANDY JUN: Oh, like the assembly: confused, vapid and abundant, as 'How do, ma'am! No accident at the door? He! he!' 'Only my carriage broke to pieces.' 'I hope you had not your pocket picked. Won't you sit down to faro?' 'Have you many tonight?' 'A few — about six hundred.' 'Were you at Lady Overall's?' 'Oh yes, a delicious crowd! And plenty of peas! He! he!' And thus runs the fashionable race.

SIR ABEL: Yes, and a precious run it is: full gallop all the way. First they run on, then their fortune is run through, then bills are run up, then they are run hard, then they've a run of luck, then they run out, and then they run away! But I'll forgive fashion all its follies in consideration of one of its blessed laws.

HANDY JUN: What may that be?

SIR ABEL: That husband and wife must never be seen together!

---

high play: gambling, usually at cards, for enormous stakes.
scouted: openly mocked, by word of mouth rather than in print.
coquelicot: a brilliant orange-red, the colour of the common red poppy.
Smithfield market: graphically described by Dickens in *Oliver Twist*, Smithfield was at this time a *live* meat market, its four-and-a-half acre open-air site to the east of the City of London handling one-and-a-half million head of livestock annually by the end of the eighteenth century. All animals were driven in on the hoof.
the fashionable race: newspaper reports bear up Morton's caricature of *fin-de-siècle* London social life, one observer commenting: 'to such a profusion of expense and extravagance do some of our fashionable females extend their hospitality, that they distribute to their friends all the *un*-necessaries of life. But do they think of those who are in want of the necessaries?' *The Times*, 15 May 1799.

(*Enter* SERVANT.)

SERVANT: Miss Blandford, your father expects you.

MISS B: I hope I shall find him more composed.

HANDY JUN: Is Sir Philip ill?

MISS B: His spirits are extremely depressed and, since we arrived here this morning, his dejection has dreadfully increased.

HANDY JUN: But I hope we shall be able to laugh away despondency.

MISS B: Sir, if you are pleased to consider my esteem as an object worth your possession, I know no way of obtaining it so certain as by your showing every attention to my dear father.

(*As they are going, enter* ASHFIELD.)

ASHFIELD: Dame! Dame! She be come!

DAME: Who? Susan? Our dear Susan?

ASHFIELD: Ees, zo come along! Oh, Sir Abel, Lady Nelly, your spouse, do order you to go to her directly.

HANDY JUN: Order? You mistake —

SIR ABEL: No, he don't. She generally prefers that word.

MISS B: Adieu, Sir Abel!

(*Exeunt* MISS BLANDFORD *and* HANDY JUN.)

SIR ABEL: Oh, if my wife had such a pretty way with her mouth!

DAME: And how does Susan look?

ASHFIELD: That's what I do want to know, zoa come along, woolye, though! Missus, letten us behave pratty! — Zur, if you pleaze, dame and I will let you walk along wi' us.

SIR ABEL: How condescending! Oh, you are a pretty-behaved fellow!

(*Exeunt.*)

Scene 2. *Farmer Ashfield's kitchen.*

(*Enter* LADY HANDY *and* SUSAN.)

SUSAN: My dear home, thrice welcome! What gratitude I feel to your ladyship for this indulgence!

LADY H: That's right, child.

SUSAN: And I am sure you partake my pleasure in again visiting a place where you received every protection and kindness my parents could show you, for, I remember, while you lived with my father —

LADY H: Child, don't put your memory to any fatigue on my account! You may transfer the remembrance of who I was to aid your more perfect recollection of who I am.

SUSAN: Lady Handy!

LADY H: That's right, child. I am not angry.

SUSAN: (*looking out*) How luxuriantly the honeysuckle has grown that I planted. Ah, I see my dear father and mother coming through the garden!

LADY H: Oh, now I shall be caressed to death! But I must endure the shock of their attentions.

(*Enter* ASHFIELD *and* DAME ASHFIELD, *who run to* SUSAN, *with* SIR ABEL.)

ASHFIELD: My dear Susan!

DAME: My sweet child! Give me a kiss!

ASHFIELD: Hald thee! Feyther first, though. Well, I be as mortal glad to zee thee as never war. And how be'st thee? And how do thee like Lunnun town? It be a deadly lively place, I be tuold.

DAME: Is not she a sweet girl?

SIR ABEL: That she is.

LADY H: (*with affected dignity*) Does it occur to anyone present that Lady Handy is in the room?

SIR ABEL: Oh, Lud! — I'm sure, my dear wife, *I* never forget that you are in the room.

ASHFIELD: Drabbit it, I overlooked Lady Nelly, zure enow! But consider, there be zome difference between thee and our own Susan. I be deadly glad to zee thee, however.

DAME: So am I, Lady Handy.

ASHFIELD: Don't ye take it unkind I ha'n't a-buss'd thee yet. Meant no slight, indeed. (*kissing her*)

LADY HANDY: (*aside*) Oh, shocking!

ASHFIELD: No harm I do hope, zur?

SIR ABEL: None at all.

ASHFIELD: But, dash it, Lady Nelly, what do make thee paint thy vace all over wi' red ochre zoo? Be it vor thy spouse to knaw thee? That be the way I do knaw my sheep!

SIR ABEL: The flocks of fashion are all marked so, farmer.

ASHFIELD: Likely. Drabbit it, thee do make a tightish kind of a ladyship, zure enow!

DAME: That you do, my lady! Do you remember the old house?

ASHFIELD: Aye, and all about it, doan't ye, Nelly — my lady?

LADY H: (*in a rustic manner*) Ees, he! he! he! — (*aside*) Oh, I am quite shocked at myself! I must be gone! — Susan, child, prepare a room where I may dress before I proceed to the Castle.

(*Exit* SUSAN. *Enter* HANDY JUN.)

HANDY JUN: I don't see Susan. I say, Dad, is that my Mamma?

SIR ABEL: Yes. Speak to her.

HANDY JUN: (*chucking her under the chin*) A fine girl, upon my soul!

LADY H: Fine girl, indeed! Is this behaviour?

HANDY JUN: Oh, beg pardon, most honoured parent.

(LADY HANDY *curtsies.*)

That's a damned bad curtsey! I can teach you to make a much better curtsey than that!

LADY H: You teach me, that am old enough to — ? Hem!

HANDY JUN: Oh, that toss of the head was very bad indeed! Look at me. — That's the thing!

LADY H: Am I to be insulted? Sir Abel, you know I seldom condescend to talk.

SIR ABEL: Don't say so, my lady; you wrong yourself.

LADY H: But, when I do begin, you know not where it will end.

SIR ABEL: (*aside*) Indeed, I do not!

LADY H: I insist on receiving all possible respect from your son.

HANDY JUN: And you shall have it, my dear girl — madam, I mean!

LADY H: I vow, I am agitated to that degree — Sir Abel, my fan!

SIR ABEL: Yes, my dear. — Bob, look here; a little contrivance of my own. While others carry swords and other suchlike dreadful weapons in their canes, I more gallantly carry a fan.

> (SIR ABEL *removes the head of his cane and draws out a fan which he presents to* LADY HANDY.)

A pretty thought, isn't it?

ASHFIELD: (*to* HANDY JUN.) Some difference between thic stick and mine, bean't there, zur?

HANDY JUN: (*moving away*) Yes, there is. (*to* LADY HANDY) Do you call that fanning yourself? (*taking away the fan*) My dear ma'am, this is the way to manoeuvre a fan.

LADY H: Sir, you shall find I have power enough to make you repent this behaviour, severely repent it! Susan!

> (*Exit* LADY HANDY *followed by* DAME ASHFIELD.)

HANDY JUN: Bravo! Passion becomes her! She does that vastly well!

SIR ABEL: Yes, practice makes perfect.

> (*Enter* SUSAN.)

SUSAN: Did your ladyship call? — Heavens, Mr Handy!

HANDY JUN: Hush, my angel! Be composed!

> (HANDY JUN. *gives* SUSAN *a letter, which is noticed by* ASHFIELD.)

That letter will explain. Lady Handy wishes to see you.

SUSAN: Oh, Robert!

HANDY JUN: At present, my love, no more.

> (*Exit* SUSAN *followed by* ASHFIELD.)

SIR ABEL: What were you saying, sir, to that young woman?

HANDY JUN: Nothing particular, sir. Where is Lady Handy going?

SIR ABEL: To dress.

HANDY JUN: I suppose she has found out the use of money?

SIR ABEL: Yes, I'll do her the justice to say she encourages trade. Why, do you know, Bob, my best coal-pit won't find her in white muslins. Round her neck hangs an hundred acres, at least. My noblest oaks have made wigs for her, my fat oxen have dwindled into Dutch pugs and white mice, my India bonds are transmuted into shawls and otto of roses, and a magnificent mansion has shrunk into a diamond snuff-box.

> (*Enter* COUNTRYMAN.)

COUNTR: Gentlemen, the folks be all got together and the ploughs are ready — and —

SIR ABEL: We are coming.

> (*Exit* COUNTRYMAN.)

HANDY JUN: Ploughs?

---

otto of roses: fashionable corruption of *attar* of roses, a highly perfumed essence imported, at considerable expense, from Persia.

SIR ABEL: Yes, Bob, we are going to have a grand agricultural meeting.

HANDY JUN: Indeed!

SIR ABEL: If I could but find a man able to manage my new-invented *curricle plough*, none of them would have a chance.

HANDY JUN: My dear sir, if there be anything on earth I can do, it is that.

SIR ABEL: What?

HANDY JUN: I rather fancy I can plough better than any man in England.

SIR ABEL: You don't say so? What a clever fellow he is! I say, Bob, if you would —

HANDY JUN: No, I can't condescend.

SIR ABEL: Condescend! Why not? Much more creditable, let me tell you, than galloping a maggot for a thousand, or eating a live cat, or any other fashionable achievement.

HANDY JUN: So it is. Egad, I will! I'll carry off the prize of industry!

SIR ABEL: But should you lose, Bob —

HANDY JUN: I lose? That's vastly well!

SIR ABEL: True, with my *curricle plough* you could hardly fail.

HANDY JUN: With my superior skill, Dad! — Then, I say, how the newspapers will teem with the account!

SIR ABEL: Yes.

HANDY JUN: 'That universal genius, Handy junior, with a plough' —

SIR ABEL: Stop! — 'Invented by that ingenious machinist, Handy senior' —

HANDY JUN: 'Gained the prize against the first husbandmen in Hampshire. Let our Bond Street butterflies emulate the example of Handy junior' —

SIR ABEL: 'And let old city grubs cultivate the field of science, like Handy senior.' Ecod, I am so happy!

LADY H: (*without*) Sir Abel!

SIR ABEL: Ah, there comes a damper!

HANDY JUN: Courage! You have many resources of happiness.

SIR ABEL: Have I? I should be very glad to know them.

HANDY JUN: In the first place, you possess an excellent temper.

SIR ABEL: So much the worse, for if I had a bad one I should be the better able to conquer hers.

HANDY JUN: You enjoy good health.

SIR ABEL: So much the worse, for if I were ill she wouldn't come near me.

---

curricle plough: the curricle was a lightweight, two-wheeled, very fast and very fashionable carriage of the late eighteenth century, hence the humour of the combination of opposites implicit in Sir Abel's design, similar, in the modern idiom, to a formula one steam-roller.

galloping a maggot for a thousand, or eating a live cat: bizarre wagers, stretching into new realms of absurdity the frontiers of human endeavour, reached epidemic proportions during the 1790s. Besides eating contests, which were rife and often fatal, men were prepared to bind themselves to the wheels of hackney coaches (August 1793), chalk every tree in St James's Park in less than forty-five minutes (June 1791) and ride pigs at full gallop from Cornhill to Milk Street (October 1790). In the latter attempt both pig and rider died.

Bond Street butterflies: a reference to the London dandies who became, during the Regency period, more popularly known as Bond Street Loungers. Bond Street was a grand emporium dedicated to servicing the gentlemen of pleasure; London's most fashionable men's outfitters were situated here.

HANDY JUN: Then you are rich.

SIR ABEL: So much the worse, for had I been poor she would not have married me. But, I say, Bob, if you gain the prize, I'll have a patent for my plough.

LADY H: (*without*) Sir Abel, I say!

HANDY JUN: Father, couldn't you get a patent for stopping that sort of noise?

SIR ABEL: If I could, what a sale it would have! No, Bob, a patent has been obtained for the only thing that will silence her.

HANDY JUN: Aye? What's that?

SIR ABEL: (*in a whisper*) A coffin! – Hush, I'm coming, my dear!

HANDY JUN: Ha! ha! ha!

    (*Exeunt.*)

    Scene 3. *A parlour in Farmer Ashfield's house.*

    (*Enter* ASHFIELD *and* DAME ASHFIELD.)

ASHFIELD: I tell ye, I zee'd un gi' Susan a letter, an' I dan't like it a bit.

DAME: Nor I. If shame should come to the poor child! I say, Tummas, what would Mrs Grundy say then?

ASHFIELD: Dom Mrs Grundy! What would my poor wold heart zay? But, I be bound, it be all innocence.

    (*Enter* HENRY.)

DAME: Ah, Henry, we have not seen you at home all day.

ASHFIELD: And I do zomehow fanzie things dan't go zo clever when thee'rt away from farm.

HENRY: My mind has been greatly agitated.

ASHFIELD: Well, won't thee go and zee the ploughing match?

HENRY: Tell me, will not those who obtain prizes be introduced to the Castle?

ASHFIELD: Ees, and feasted in the Great Hall.

HENRY: My good friend, I wish to become a candidate.

ASHFIELD: You, Henry?

HENRY: It is time I exerted the faculties Heaven has bestowed on me. And, though my heavy fate crushes the proud hopes the heart conceives, still let me prove myself worthy of the place Providence has assigned me. – (*aside*) Should I succeed, it will bring me to the presence of that man who, I know not why, seems the dictator of my fate. – (*to them*) Will you furnish me with the means?

ASHFIELD: I will. Thou shalt ha' the best plough in the parish – I wish it were all gould, for thy zake! – and better cattle there can't be noowhere.

HENRY: Thanks, my good friend, my benefactor. I have little time for preparation, so receive my gratitude, and farewell!

    (*Exit* HENRY.)

DAME: A blessing go with thee!

ASHFIELD: I zay, Henry, take Jolly and Smiler and Captain, but dan't ye take thic lazy beast, Genius! – I'll be shot if having vive load an acre on my wheat land could please me more.

DAME: Tummas, here comes Susan, reading the letter.

ASHFIELD: How pale she do look, dan't she.

DAME: Ah, poor thing, if –

ASHFIELD: Hauld thy tongue, woolye?

> (ASHFIELD *and* DAME ASHFIELD *retire. Enter* SUSAN, *reading the letter.*)

SUSAN: Is it possible? Can the man to whom I've given my heart write thus? – 'I am compelled to marry Miss Blandford, but my love for my Susan is unalterable. I hope she will not, for an act of necessity, cease to think with tenderness on her faithful Robert.' – Oh man, ungrateful man, it is from our bosoms alone you derive your power? How cruel, then, to use it in fixing in those bosoms endless sorrow and despair! – 'Still think with tenderness'? – Base, dishonourable insinuation! He might have allowed me to esteem him!

> (SUSAN *locks up the letter in a box on the table. Exit weeping.* ASHFIELD *and* DAME ASHFIELD *come forward.*)

ASHFIELD: Poor thing. What can be the matter? She locked up the letter in thic box and then bursted into tears. (*looking at the box*)

DAME: Yes, Tummas, she locked it in that box, sure enough. (*shaking a bunch of keys that hangs at her side*)

ASHFIELD: What be doing, dame? What be doing?

DAME: (*with affected indifference*) Nothing. I was only touching these keys.

> (ASHFIELD *and* DAME ASHFIELD *look at the box and keys significantly.*)

ASHFIELD: A good, tightish bunch!

DAME: Yes, they are of all sizes.

> (ASHFIELD *and* DAME ASHFIELD *look, as before.*)

ASHFIELD: Indeed – Well, eh? – Dame? – Why dan't ye speak? Thou can'st chatter fast enow zometimes.

DAME: Nay, Tummas, I dare say, if – You know best, but – I think I could find –

ASHFIELD: Well, eh? – You can just try, you knaw – (*greatly agitated*) You can try – Just vor the vun on't – But mind, dan't ye make a noise!

> (DAME ASHFIELD *opens the box.*)

Why, thee hasn't opened it?

DAME: Nay, Tummas, you told me!

ASHFIELD: Did I?

DAME: There's the letter.

ASHFIELD: Well, why do ye gi't to I? I dan't want it, I'm sure!

> (*Taking it,* ASHFIELD *turns it over.* DAME ASHFIELD *eyes it eagerly.* ASHFIELD *is about to open it.*)

She's coming! She's coming!

> (ASHFIELD *conceals the letter. Both tremble violently.*)

No, she's gone into t'other room.

> (ASHFIELD *and* DAME ASHFIELD *hang their heads dejectedly, then look at each other.*)

What mun that feythur an' mother be doing that do blush and tremble at their own dater's coming? (*weeping*) Dang it, has she desarv'd it of us? Did she ever deceive us? Were she not always the most open-hearted, dutifullest, kindest – And thee to goa, like a domm'd spy, and open her box, poor thing!

DAME: Nay, Tummas!

ASHFIELD: You did. I zaw you do it myzel! You look like a thief, now, you doe.

— Hush! — No, dame, here be the letter; I won't read a word on't. Put it where thee vound it and as thee vound it.

DAME: With all my heart!

> (DAME ASHFIELD *returns the letter to the box.* ASHFIELD *embraces her.*)

ASHFIELD: Now I can wi' pleasure hug my wold wife and look my child in the vace again. I'll call her and ax her about it, and, if she dan't speak without disguisement, I'll be bound to be shot. Dame, be the colour of sheame off my face, yet? I never zee'd thee look ugly before. — Susan! My dear Sue, come here a bit, woolye?

> (*Enter* SUSAN.)

SUSAN: Yes, my dear father?

ASHFIELD: Sue, we do wish to give thee a bit of admonishing and parent-like conzultation.

SUSAN: I hope I have ever attended to your admonitions.

ASHFIELD: Ees, bless thee, I do believe thee hast, lamb. But we all want our memories jogg'd a bit, or why else do parson preach us all to sleep every Zunday? Zo thic be the topic: dame and I, Sue, did zee a letter gi'd to thee, and thee bursted into tears and lock'd un up in thic box, and then dame and I, we — that's all.

SUSAN: My dear father, if I concealed the contents of that letter from your knowledge, it was because I did not wish your heart to share in the pain mine feels.

ASHFIELD: (*to his wife*) Dang it, didn't I tell thee zoo?

DAME: Nay, Tummas, did I say otherwise?

SUSAN: Believe me, my dear parents, my heart never gave birth to a thought my tongue feared to utter.

ASHFIELD: There, the very words I zaid!

SUSAN: If you wish to see the letter, I will show it to you. (*searching for the key*)

DAME: Here's a key will open it.

ASHFIELD: (*aside*) Drabbit it, hold thy tongue, thou wold fool! — No, Susan, I'll not zee it; I'll believe my child.

SUSAN: You shall not find your confidence ill-placed. It is true the gentleman declared he loved me. It is equally true that declaration was not unpleasing to me. Alas, it is also true that his letter contains sentiments disgraceful to himself and insulting to me.

ASHFIELD: Drabbit it, if I'd knaw'd that when we were cudgelling a bit, I would ha' lapt my stick about his ribs pretty tightish, I would.

SUSAN: Pray, father, don't you resent his conduct to me.

ASHFIELD: What? Mayn't I leather un a bit?

SUSAN: Oh, no, I've the strongest reasons to the contrary!

ASHFIELD: Well, Sue, I won't. I'll behave as pratty as I always do. But it be time to go to the Green and zee the fine zights. — How I do hate the noise of thic domm'd bunch of keys! — But bless thee, my child. Dan't forget that vartue to a young woman be vor all the world like — like — dang it, I ha' gotten it all in my head, but zomehow I can't talk it! — but — vartue be to a young woman what corn be to a blade o' wheat, do you zee? For while the corn be there it be glorious to the eye and it be called the staff of life, but take that treasure away and what do remain? Why, nought but thic worthless straw that man and beast do tread upon.

(*Exeunt.*)

Scene 4. *An extensive view of a cultivated country. A ploughed field in the centre, in which are seen six different ploughs and horses. At one side, a handsome tent. A number of* COUNTRY PEOPLE *assembled.*

(*Enter* ASHFIELD *and* DAME ASHFIELD.)

ASHFIELD: Make way, make way for the gentry! And, do ye hear, behave pratty, as I do. Dang thee, stond back or I'll knock thee down, I wool!

(*Enter* SIR ABEL *and* MISS BLANDFORD *with* SERVANTS.)

SIR ABEL: It is very kind of you to honour our rustic festivities with your presence.

MISS B: Pray, Sir Abel, where is your son?

SIR ABEL: What? Bob? (*nodding significantly*) Oh, you'll see him presently. Here are the prize medals, and, if you will condescend to present them, I'm sure they'll be worn with additional pleasure. I say, you'll see Bob presently. — Well, farmer, is it all over?

ASHFIELD: Ees, zur. The acres be plough'd and the ground judged, and the young lads be coming down to receive their reward. Heartily welcome, miss, to your native land. Hope you be as pleased to zee we as we be to zee you, and the like o' that. — Mortal beautizome, to be zure! I declare, miss, it do make I quite warm zomehow to look at ye!

(*A shout is heard from without.*)

They be coming. Now, Henry!

SIR ABEL: Now you'll see Bob! Now my dear boy, Bob! Here he comes!

(*Amidst another shout, enter* HENRY *and two young* HUSBANDMEN.)

ASHFIELD: 'Tis he! He has done't! Dang you all, why dan't ye shout? Huzza!

SIR ABEL: Why, zounds, where's Bob? I don't see Bob. Bless me, what has become of Bob and my plough?

(SIR ABEL *retires and takes out his glass.*)

ASHFIELD: Well, Henry, there be the prize, and there be the fine lady that will gi' it thee.

HENRY: Tell me, who is that lovely creature?

ASHFIELD: The dater of Sir Philip Blandford.

HENRY: What exquisite sweetness! Ah, should the father but resemble her, I shall have but little to fear from his severity.

ASHFIELD: Miss, thic be the young man that ha' gotten the goulden prize.

MISS B: This? I always thought ploughmen were coarse, vulgar creatures, but he seems handsome and diffident.

ASHFIELD: Ees, quite pratty-behaved. It were I that teach'd un.

MISS B: What's your name?

HENRY: Henry.

MISS B: And your family?

(HENRY, *in an agony of grief, turns away, strikes his forehead, and leans on the shoulder of* ASHFIELD.)

DAME: (*apart to* MISS BLANDFORD) Madam, I beg pardon, but nobody knows about his parentage, and when it is mentioned, poor boy, he takes on sadly. He has lived at our house ever since we had the farm, and we have had an allowance for him, small enough, to be sure, but — good lad! — he was always welcome to share what we had.

MISS B: I am shocked at my imprudence. (*to* HENRY) Pray, pardon me. I would not insult an enemy, much less one I am inclined to admire (*giving her hand, then withdrawing it*) — to esteem. You shall go to the Castle; my father shall protect you.

HENRY: Generous creature! To merit his esteem is the fondest wish of my heart, to be your slave the proudest aim of my ambition.

MISS B: Receive your merited reward.

(HENRY *kneels.* MISS BLANDFORD *places a medal round his neck, then does the same to the two* HUSBANDMEN.)

SIR ABEL: (*advancing*) I can't see Bob. Pray, sir, do you happen to know what is become of my Bob?

HENRY: Sir?

SIR ABEL: Did not you see a remarkable clever plough and a young man?

HENRY: At the beginning of the contest I observed a gentleman. His horses, I believe, were unruly. But my attention was too much occupied to allow me to notice more.

(*Laughter is heard from without.*)

HANDY JUN: (*without*) How dare you laugh?

SIR ABEL: That's Bob's voice!

(*Amidst more laughter, enter* HANDY JUN. *in a smock-frock, cocked hat, and a piece of plough in his hand.*)

HANDY JUN: Dare to laugh again and I'll knock you down with this! Ugh, how infernally hot!

(HANDY JUN. *walks about,* SIR ABEL *following.*)

SIR ABEL: Why, Bob, where have you been?

HANDY JUN: I don't know where I've been.

SIR ABEL: And what have you got in your hand?

HANDY JUN: What? All I could keep of your nonsensical ricketty plough!

SIR ABEL: Come, come, none of that, sir! Don't abuse my plough to cover your ignorance, sir! Where is it, sir? And where are my famous Leicestershire horses, sir?

HANDY JUN: Where? Ha! ha! ha! I'll tell you as nearly as I can. Ha! ha! What's the name of the next county?

ASHFIELD: It be called Wiltshire, zur.

---

how infernally hot: contact with summer weather was generally disapproved of by the *beau monde* if there was any risk of it leaving its mark. Suntans were especially to be avoided, fair skins amongst the fairer sex being the fashionable ambition, and putting up with the heat was thought to betray a want of breeding.

HANDY JUN: Then, Dad, upon the nicest calculation I am able to make, they are at this moment engaged in the very patriotic act of ploughing Salisbury Plain! Ha! ha! I saw them fairly over that hill, full gallop, with the *curricle plough* at their heels!

ASHFIELD: Ha! ha! A good one! Ha! ha!

HANDY JUN: But never mind, father, you must again set your invention to work, and I my toilet; rather a deranged figure to appear before a lady in.

(*Fiddles strike up.*)

Hey-day! What, are you going to dance?

ASHFIELD: Ees, zur. I suppoze you can sheake a leg a bit?

HANDY JUN: I fancy I can dance every possible step, from the *pas ruse* to the war dance of the Catawbaws.

ASHFIELD: Likely. I do hope, miss, you'll join your honest neighbours. They'll be deadly hurt an you won't gig it a bit wi' un.

MISS B: With all my heart.

SIR ABEL: Bob's an excellent dancer.

MISS B: I dare say he is, sir, but on this occasion I think I ought to dance with the young man who gained the prize; I think it would be most pleasant — most proper, I mean — and I am glad you agree with me. So, sir, if you'll accept my hand —

(HENRY *takes* MISS BLANDFORD's *hand.*)

SIR ABEL: Very pleasantly settled, upon my soul! Bob, won't you dance?

HANDY JUN: I dance? No, I'll look at them. I'll quietly look on.

SIR ABEL: Egad, now, as my wife's away, I'll try to find a pretty girl and make one among them!

ASHFIELD: That's hearty! Come, dame, hang thy rheumatics! Now, lads and lasses, behave pratty and strike up!

(*A dance.* HANDY JUN. *looks on a little and then begins to move his legs, dashes into the midst of the dance and endeavours to imitate everyone opposite him. Then, being exhausted, he leaves the dance, seizes the fiddle and plays till the curtain drops.*)

ACT III

Scene 1. *An apartment in the Castle.* SIR PHILIP BLANDFORD *discovered on a couch, reading.* SERVANTS *attending.*

SIR PHILIP: Is not my daughter yet returned?

SERVANT: No, Sir Philip.

SIR PHILIP: Dispatch a servant to her.

(*Exit* SERVANT, *then immediately re-enter.*)

---

pas ruse: possibly a misspelling of *pas russe*, 'Russian dance'.

war dance of the Catawbaws: the Catawba were an Indian tribe inhabiting an area of North America which, after colonisation, became South Carolina. Although in continual conflict with tribes from the north, they were remarkably friendly towards the English settlers with whom they fought against the British government during the War of Independence.

SERVANT: Sir, the old gardener is below and asks to see you.

SIR PHILIP: (*rising and throwing away the book*) Admit him instantly, and leave me.

> (*Exit* SERVANT. *Enter* EVERGREEN, *who bows, then looking at*
> SIR PHILIP *clasps his hands together and weeps.*)

Does this desolation affect the old man? — Come near me. Time has laid a lenient hand on thee.

EVERG: Oh, my dear master, can twenty years have wrought the change I see?

SIR PHILIP: (*striking his breast*) No, 'tis the canker here that hath withered up my trunk. But are we secure from observation?

EVERG: Yes.

SIR PHILIP: Then tell me, does the boy live?

EVERG: He does, and is as fine a youth —

SIR PHILIP: No comments!

EVERG: We named him —

SIR PHILIP: Be dumb! Let me not hear his name. Has care been taken he may not blast me with his presence?

EVERG: It has. And he cheerfully complied.

SIR PHILIP: Enough! Never speak of him more. Have you removed every dreadful vestige from the fatal chamber?

> (EVERGREEN *hesitates.*)

Oh, speak!

EVERG: My dear master, I confess my want of duty. Alas, I had not courage to go there.

SIR PHILIP: Ah!

EVERG: Nay, forgive me. Wiser than I have felt such terrors. The apartments have been carefully locked up, the keys not a moment from my possession. Here they are.

SIR PHILIP: Then the task remains with me. Dreadful thought! I can well pardon thy fears, old man. Oh, could I wipe from my memory that hour when —

EVERG: Hush! Your daughter.

SIR PHILIP: Leave me. We'll speak anon.

> (*Exit* EVERGREEN. *Enter* MISS BLANDFORD.)

MISS B: Dear father, I came the moment I heard you wished to see me.

SIR PHILIP: My good child, thou art the sole support that props my feeble life. I fear my wish for thy company deprives thee of much pleasure.

MISS B: Oh, no! What pleasure can be equal to that of giving you happiness? Am I not rewarded in seeing your eyes beam with pleasure on me?

SIR PHILIP: 'Tis the pale reflection of the lustre I see sparkling there. My love, did you enjoy the scenes you beheld?

MISS B: Greatly. How strongly they contrast with those we witnessed abroad!

SIR PHILIP: True. Happy country, which, in the midst of direful war, can draw out its rustic train to join the festive dance as securely as if peace again had blessed the world! But, tell me, did your lover gain the prize?

MISS B: Yes, papa.

SIR PHILIP: Few men of his rank —

MISS B: Oh, you mean Mr Handy?

SIR PHILIP: Yes.

MISS B: No, he did not.

SIR PHILIP: Then whom did *you* mean?

MISS B: Did you say lover? I — I mistook. No, a young man called Henry obtained the prize.

SIR PHILIP: And how did Mr Handy succeed?

MISS B: Oh, it was so ridiculous! I will tell, papa, what happened to him.

SIR PHILIP: To Mr Handy?

MISS B: Yes. As soon as the contest was over, Henry presented himself. I was surprised at seeing a young man so handsome and elegant as Henry is. Then I placed the medal round Henry's neck and was told that poor Henry —

SIR PHILIP: Henry? So, my love, this is your account of Mr Robert Handy?

MISS B: Yes, papa — No, papa, he came afterwards, dressed so ridiculously that even Henry could not help smiling.

SIR PHILIP: Henry again!

MISS B: Then we had a dance.

SIR PHILIP: Of course, you danced with your lover.

MISS B: Yes, papa.

SIR PHILIP: How does Mr Handy dance?

MISS B: Oh, he did not dance till —

SIR PHILIP: You danced with your lover?

MISS B: Yes — no, papa! Somebody said — I don't know who — that I ought to dance with Henry because —

SIR PHILIP: Still Henry! Oh, some rustic boy! My dear child, you talk as if you loved this Henry.

MISS B: Oh, no, papa! And I am certain he don't love me.

SIR PHILIP: Indeed?

MISS B: Yes, papa, for when he touched my hand he trembled as if I terrified him, and instead of looking at me as you do, who I am sure love me, when our eyes met he withdrew his and cast them on the ground.

SIR PHILIP: And these are the reasons which make you conclude he does not love you?

MISS B: Yes, papa.

SIR PHILIP: And probably you could adduce proof equally convincing that you don't love him!

MISS B: Oh, yes, quite, for in the dance he sometimes paid attention to other young women, and I was so angry with him! Now, you know, papa, I love you, and I am sure I should not have been angry with you, had you done so.

SIR PHILIP: But one question more: do you think Mr Handy loves you?

MISS B: I have never thought about it, papa.

SIR PHILIP: I am satisfied.

MISS B: Yes, I knew I should convince you!

SIR PHILIP: Oh Love, malign and subtle tyrant, how falsely art thou painted blind! 'Tis thy votaries are so, for what but blindness can prevent their seeing thy poisoned shaft which is forever doomed to rankle in the victim's heart.

MISS B: Oh, now I am certain I am not in love, for I feel no rankling at my heart; I feel the softest, sweetest sensation I ever experienced. But, papa, you must come to the lawn. I don't know why, but today nature seems enchanting: the birds sing more sweetly and the flowers give more perfume!

SIR PHILIP: (*aside*) Such was the day my youthful fancy pictured. How did it close!

MISS B: I promised Henry your protection.

SIR PHILIP: Indeed! That was much! Well, I will see your rustic hero. This infant passion must be crushed. Poor wench, some artless boy has caught thy youthful fancy. — Thy arm, child.

(*Exeunt.*)

Scene 2. *A lawn before the Castle.*

(*Enter* HENRY *and* ASHFIELD.)

ASHFIELD: Well, here thee'rt going to make thy bow to Sir Philip. I zay, if he should take a fancy to thee, thou'lt come to farm and zee us zometimes, wo'tn't, Henry?

HENRY: (*shaking his head*) Tell me, is that Sir Philip Blandford who leans on that lady's arm?

ASHFIELD: I don't know, by reason, d'ye zee, I never zee'd un. Well, good-bye! I declare thee doz look quite grand with thic golden prize about thy neck, vor all the world like the lords in their stars that do come to theas parts to pickle their skins in the zalt zea ocean! Good-b'ye, Henry!

(*Exit* ASHFIELD.)

HENRY: He approaches! Why this agitation? I wish, yet dread, to meet him!

(*Enter* SIR PHILIP *and* MISS BLANDFORD, *attended.*)

MISS B: The joy your tenantry display at seeing you again must be truly grateful to you.

SIR PHILIP: No, my child, for I feel I do not merit it. Alas, I can see no orphans clothed with my beneficence, no anguish assuaged by my care.

MISS B: Then I am sure my dear father wishes to show his kind intentions, so I will begin by placing one under his protection.

(MISS BLANDFORD *goes up the stage and leads down* HENRY.

SIR PHILIP, *on seeing him, starts, then becomes greatly agitated.*)

SIR PHILIP: Ah, do my eyes deceive me? No, it must be him! Such was the face his father wore!

HENRY: Spake you of my father?

SIR PHILIP: His presence brings back recollections which drive me to madness! How came he here? Who have I to curse for this?

MISS B: (*falling on* SIR PHILIP's *neck*) Your daughter!

HENRY: Oh, sir, tell me — on my knees I ask it — do my parents live? Bless me with my father's name and my days shall pass in active gratitude, my nights in prayers for you!

(SIR PHILIP *views* HENRY *with severe contempt.*)

Do not mock my misery! Have you a heart?

---

lords in their stars: reference to the insignia of the English peerage.

SIR PHILIP: Yes, of marble, cold and obdurate to the world, ponderous and painful to myself. Quit my sight forever!

MISS B: Go, Henry, and save me from my father's curse.

HENRY: I obey. Cruel as the command is, I obey it. (*touching the medal*) I shall often look at this and think on the blissful moment when your hand placed it there.

SIR PHILIP: Ah, tear it from his breast!

(SERVANT *advances.*)

HENRY: Sooner take my life! It is the first honour I have earned, and it is no mean one, for it assigns to me the first rank among the sons of industry. This is my claim to the sweet rewards of honest labour. This will give me competence, nay more, enable me to despise your tyranny!

SIR PHILIP: Rash boy, mark! Avoid me and be secure; repeat this intrusion and my vengeance shall pursue thee!

HENRY: I defy its power! You are in England, sir, where the man who bears about him an upright heart bears a charm too potent for tyranny to humble. Can your frown wither up my youthful vigour? No! Can your malediction disturb the slumbers of a quiet conscience? No! Can your breath stifle in my heart the adoration it feels for that pitying angel? Oh, no!

SIR PHILIP: Wretch, you shall be taught the difference between us!

HENRY: I feel it now, proudly feel it! You hate the man that never wronged you; I could love the man that injures me. You meanly triumph o'er a worm; I make a giant tremble.

SIR PHILIP: Take him from my sight! Why am I not obeyed?

MISS B: Henry, if you wish my hate should not accompany my father's, instantly be gone.

HENRY: Oh, pity me!

(*Exit* HENRY. MISS BLANDFORD *looks after him.* SIR PHILIP, *exhausted, leans on his* SERVANTS.)

SIR PHILIP: Supported by my servants? I thought I had a daughter!

MISS B: (*running to him*) Oh, you have, my father! One that loves you better than her life!

SIR PHILIP: (*to* SERVANTS) Leave us.

(*Exit* SERVANTS.)

Emma, if you feel, as I fear you do, love for that youth, mark my words: when the dove woos for its mate the ravenous kite, when nature's fixed antipathies mingle in sweet concord, then, and not till then, hope to be united!

MISS B: Oh Heaven!

SIR PHILIP: Have you not promised me the disposal of your hand?

MISS B: Alas, my father, I didn't then know the difficulty of obedience.

You are in England, sir: the technical term for this type of speech was clap-trap. Although clap-traps were derided as a cheap means of winning applause, both Morton and Colman slipped them into their work wherever they sensed the theatrical need for instant euphoria, 'the praise of Laws, Jack Tars, Innocence, an Englishman's *castellum*, or Liberty' being amongst the most effective, according to Frederick Reynolds (*Reminiscences*, London 1827, II, 227).

SIR PHILIP: Hear, then, the reasons why I demand compliance. You think I hold these rich estates? Alas, the shadow only, not the substance.

MISS B: Explain, my father!

SIR PHILIP: When I left my native country, I left it with a heart lacerated by every wound that the falsehood of others, or my own conscience, could inflict. Hateful to myself, I became the victim of dissipation. I rushed to the gaming-table and soon became the dupe of villains. My ample fortune was lost. I detected one in the act of fraud, and, having brought him to my feet, he confessed a plan had been laid for my ruin, that he was but a humble instrument, for the man who, by his superior genius, stood possessed of all the mortgages and securities I had given was one Morrington.

MISS B: I have heard you name him before. Did you not know this Morrington?

SIR PHILIP: No. He, like his deeds, avoided the light, ever dark, subtle and mysterious. Collecting the scattered remnant of my fortune, I wandered, wretched and desolate, till, in a peaceful village, I first beheld thy mother, humble in birth but exalted in virtue. The morning after our marriage she received a packet containing these words: 'The reward of virtuous love, presented by a repentant villain', and which also contained bills and notes to the high amount of ten thousand pounds.

MISS B: And no name?

SIR PHILIP: None. Nor could I ever guess at the generous donor. I need not tell thee what my heart suffered when death deprived me of her. Thus circumstanced, this good man, Sir Abel Handy, proposed to unite our families by marriage and, in consideration of what he termed the honour of our alliance, agreed to pay off every incumbrance on my estates and settle them as a portion on you and his son. Yet still another wonder remains. When I arrive, I find no claim whatever has been made, either by Morrington or his agents. What am I to think? Can Morrington have perished and with him his large claims to my property? Or does he withhold the blow, to make it fall more heavily?

MISS B: 'Tis very strange! Very mysterious! But my father has not told me what misfortune led him to leave his native country.

SIR PHILIP: (*greatly agitated*) Ha?

MISS B: May I not know it?

SIR PHILIP: Oh, never, never, never!

MISS B: I will not ask it. Be composed. Let me wipe away those drops of anguish from your brow. How cold your cheek is! My father, the evening damps will harm you. Come in. I will be all you wish, indeed I will.

(*Exeunt.*)

Scene 3. *An apartment in the Castle.*

(*Enter* EVERGREEN.)

EVERG: Was ever anything so unlucky? Henry to come to the Castle and meet Sir Philip! He should have consulted me. I shall be blamed, but, thank Heaven, I am innocent.

LADY H: (*without*) I will be treated with respect.

SIR ABEL: (*without*) You shall, my dear.

(*Enter* SIR ABEL *and* LADY HANDY.)

LADY H: But how? But how, Sir Abel? I repeat it.

SIR ABEL: (*aside*) For the fiftieth time.

LADY H: Your son conducts himself with an insolence I won't endure. But you are ruled by him; you have no will of your own.

SIR ABEL: I have not, indeed.

LADY H: How contemptible!

SIR ABEL: Why, my dear, this is the case: I am like the ass in the fable, and, if I am doomed to carry a pack-saddle, it is not much matter who drives me.

LADY H: To yield your power to those the law allows you to govern! −

SIR ABEL: Is very weak indeed!

EVERG: Lady Handy, your very humble servant. I heartily congratulate you, madam, on your marriage with this worthy gentleman. Sir, I give you joy.

SIR ABEL: (*aside*) Not before 'tis wanted.

EVERG: Aye, my lady, this match makes up for the imprudence of your first.

LADY H: Hem!

SIR ABEL: Eh? What? What's that, eh? What do you mean?

EVERG: I mean, sir, that Lady Handy's former husband −

SIR ABEL: Former husband? Why, my dear, I never knew! Eh?

LADY H: (*aside*) A mumbling old blockhead! − Didn't you, Sir Abel? Yes, I was rather married many years ago, but my husband went abroad and died.

SIR ABEL: Died, did he?

EVERG: Yes, sir. He was a servant in the Castle.

SIR ABEL: Indeed. So he died, poor fellow!

LADY H: Yes.

SIR ABEL: What, you are sure he died, are you?

LADY H: Don't you hear?

SIR ABEL: Poor fellow. Neglected, perhaps. Had I known it, he should have had the best advice money could have got.

LADY H: You seem sorry.

SIR ABEL: Why, you would not have me pleased at the death of your husband, would you? A good kind of a man?

EVERG: Yes, a faithful fellow. Rather ruled his wife too severely.

SIR ABEL: Did he? (*apart to* EVERGREEN) Pray, do you happen to recollect his manner? Could you just give a hint of the way he had?

LADY H: Do you want to tyrannise over my poor, tender heart? 'Tis too much!

(LADY HANDY *seems to faint away*.)

EVERG: Bless me, Lady Handy is ill! Salts! Salts!

SIR ABEL: (*producing an essence box*) Here are salts, or aromatic vinegar, or essence of −

EVERG: Any! Any!

SIR ABEL: Bless me, I can't find the key!

---

like the ass in the fable: by Apuleius (second century), adapted and given wider currency by Boccaccio. Lucian, a young man passing through Thessaly, is accidentally metamorphosed into an ass and, in this form, realises just what it is to receive rough treatment at the hands of owners as diverse as a magistrate and a eunuch before finally regaining his human appearance.

EVERG: Pick the lock.

SIR ABEL: It can't be picked; it is a patent lock.

EVERG: Then break it open, sir.

SIR ABEL: It can't be broke open; it is a contrivance of my own. You see, here comes a horizontal bolt which acts upon a spring, therefore –

LADY H: (*starting up*) I may die while you are describing a horizontal bolt! Do you think you shall close your eyes for a week for this?

(*Enter* SIR PHILIP BLANDFORD.)

SIR PHILIP: What has occasioned this disturbance?

LADY H: Ask that gentleman.

SIR ABEL: I am accused –

LADY H: Convicted! Convicted!

SIR ABEL: Well, I will not argue with you about words, because I must bow to your superior practice! But, sir –

SIR PHILIP: Pshaw! – (*apart*) Lady Handy, some of your people were enquiring for you.

LADY H: Thank you, sir. Come, Sir Abel!

(*Exit* LADY HANDY.)

SIR ABEL: Yes, my lady. – (*to* EVERGREEN) I say, couldn't you give me a hint of the way he had?

LADY H: (*without*) Sir Abel!

SIR ABEL: Coming, my soul.

(*Exit* SIR ABEL.)

SIR PHILIP: So, you have well obeyed my orders in keeping this Henry from my presence!

EVERG: I was not to blame, master.

SIR PHILIP: Has Farmer Ashfield left the Castle?

EVERG: No, sir.

SIR PHILIP: Send him hither.

(*Exit* EVERGREEN.)

That boy must be driven far, far from my sight. But where? No matter, the world is large enough!

(*Enter* ASHFIELD.)

Come hither. I believe you hold a farm of mine.

ASHFIELD: Ees, zur, I do, at your zarvice.

SIR PHILIP: I hope a profitable one?

ASHFIELD: Zometimes it be, zur. But thic year it be all t'other way, as 'twur. But I do hope, as our landlords have a tightish big lump of the good, they'll be zo kind-hearted as to take a little bit of the bad.

SIR PHILIP: It is but reasonable. I conclude, then, you are in my debt.

ASHFIELD: Ees, zur, I be, at your zarvice.

SIR PHILIP: How much?

ASHFIELD: I do owe ye a hundred and fifty pounds, at your zarvice.

SIR PHILIP: Which you can't pay.

ASHFIELD: Not a varthing, zur, at your zarvice.

SIR PHILIP: Well, I am willing to give you every indulgence.

ASHFIELD: Be you, zur? That be deadly kind! Dear heart, it will make my auld dame quite young again, and I don't think helping a poor man will do your honour's health any harm, I don't indeed, zur! I had a thought of speaking to your worship about it, but then, thinks I, the gentleman, mayhap, be one of those that do like to do a good turn and not have a word zaid about it. Zo, zur, if you had not mentioned what I owed you, I am zure I never should, should not, indeed, zur.

SIR PHILIP: Nay, I will wholly acquit you of the debt, on condition —

ASHFIELD: Ees, zur?

SIR PHILIP: On condition, I say, that you instantly turn out that boy, that Henry!

ASHFIELD: Turn out Henry? Ha! ha! ha! Excuse my tittering, zur, but you bees making your vun of I, zure.

SIR PHILIP: I am not apt to trifle. Send him instantly from you or take the consequences!

ASHFIELD: Turn out Henry? I do vow I shouldn't knaw how to zet about it, I should not, indeed, zur.

SIR PHILIP: You hear my determination. If you disobey, you know what will follow. I'll leave you to reflect on it.

(*Exit* SIR PHILIP BLANDFORD.)

ASHFIELD: Well, zur, I'll argufy the topic, and then you may wait upon me and I'll tell ye. — (*making the motion of turning out*) I should be deadly awkward at it, vor zartain. However, I'll put the case. — Well, I goes whiztling whoam. Noa, drabbit it, I shouldn't be able to whiztle a bit, I'm zure! Well, I goes whoam and I zees Henry zitting by my wife, mixing up someit to comfort the wold zoul and take away the pain of her rheumatics. Very well. Then Henry places a chair vor I by the vireside and says, 'Varmer, the horses be fed, the sheep be folded, and you have nothing to do but to zit down, smoke your pipe and be happy!' Very well. (*becoming affected*) Then I zays, 'Henry, you be poor and friendless, zo you must turn out of my house directly.' Very well. Then my wife stares at I, reaches her hand towards the vire-place and throws the poker at my head. Very well. Then Henry gives a kind of aguish shake and, getting up, zighs from the bottom of his heart, then, holding up his head like a king, zays, 'Varmer, I have too long been a burden to you. Heaven protect you as you have me. Farewell! I go!' Then I zays, (*with great energy*) 'If thee doez, I'll be domm'd!' — Hollo! You, mister Sir Philip! You may come in!

(*Re-enter* SIR PHILIP BLANDFORD.)

Zur, I have argufied the topic — and it wouldn't be pratty — zo I can't.

SIR PHILIP: Can't? Absurd!

ASHFIELD: Well, zur, there is but another word: I won't.

SIR PHILIP: Indeed!

ASHFIELD: No, zur, I won't. I'd zee myzelf hang'd first, and you, too, zur, I would indeed! (*bowing*)

SIR PHILIP: You refuse, then, to obey?

ASHFIELD: I do, zur, at your zarvice. (*bowing*)

SIR PHILIP: Then the law must take its course!

ASHFIELD: I be zorry for that, too, I be indeed, zur. But if corn wouldn't grow, I couldn't help it; it weren't poisoned by the hand that zow'd it. Thic hand, zur, be as free from guilt as your own.

SIR PHILIP: (*sighing deeply*) Oh!

ASHFIELD: It were never held out to clinch a hard bargain, nor will it turn a good lad out into the wide wicked world because he be poorish a bit. I be zorry if you be offended, zur, quite. But come what wool, I'll never hit thic hand against here but when I be zure that zomeit at inside will jump against it with pleasure. (*bowing*) I do hope you'll repent of all your zins, I do, indeed, zur. And, if you should, I'll come and zee you again as friendly as ever, I wool, indeed, zur.

SIR PHILIP: Your repentance will come too late!

    (*Exit* SIR PHILIP BLANDFORD.)

ASHFIELD: Thank ye, zur. Good morning to you. I do hope I have made myzel agreeable, and zo I'll go whoam.

    (*Exit* ASHFIELD.)

ACT IV

Scene 1. *A room in Ashfield's house.* DAME ASHFIELD *discovered at work with her needle,* HENRY *sitting by her.*

DAME: Come, come, Henry, you'll fret yourself ill, child. If Sir Philip will not be kind to you, you are but where you were.

HENRY: (*rising*) My peace of mind is gone forever; Sir Philip may have cause for hate. Spite of his unkindness to me, my heart seeks to find excuses for him, for, oh, that heart dotes on his lovely daughter!

DAME: (*looking out*) Here comes Tummas home at last. Hey-day, what's the matter with the man? He doesn't seem to know the way into his own house!

    (*Enter* ASHFIELD *musing. He stumbles against a chair.*)

Tummas, my dear Tummas, what's the matter?

ASHFIELD: (*not attending*) It be lucky vor he I bees zo pratty-behaved, or dom if I — (*doubling his fist*)

DAME: Who? What?

ASHFIELD: Nothing at all. Where's Henry?

HENRY: Here, farmer.

ASHFIELD: Thee woultn't leave us, Henry, wou't?

HENRY: Leave you? What, leave you now, when by my exertion I can pay off part of the debt of gratitude I owe you? Oh, no!

ASHFIELD: Nay, it were not vor that I axed, I promise thee. Come, gi' us thy hand on't, then (*shaking hands*). Now I'll tell ye: Zur Philip did zend for I about the money I do owe un and said as how he'd make all straight between us —

DAME: That was kind!

ASHFIELD: Ees, deadly kind. — Make all straight, on condition I did turn Henry out o' my doors.

DAME: What!

HENRY: Where will his hatred cease?

DAME: And what did you say, Tummas?

ASHFIELD: Why, I zivelly told un if it were agreeable to he to behave like a brute, it were agreeable to I to behave like a man.

DAME: That was right. I would have told him a great deal more!

ASHFIELD: Ah, likely! Then 'a zaid I should ha' a bit a laa vor my pains.

HENRY: And do you imagine I will see you suffer on my account? No, I will remove this hated form. (*going*)

ASHFIELD: No, but thee shat'un, thee shat'un, I tell thee! Thee have givun me thy hand on't, and dom'me if thee sha't budge one step out of this house!

Drabbit it, what can he do? He can't send us to jail. Why, I have corn will zell for half the money I do owe un. And ha'n't I cattle and sheep — deadly lean, to be zure? And ha'n't I a thumping zilver watch, almost as big as thy head? And dame, here, a got — how many silk gowns have thee got, dame?

DAME: Three, Tummas, and sell them all! And I'll go to church in a stuff one and let Mrs Grundy turn up her nose as much as she pleases!

HENRY: Oh, my friends, my heart is full. Yet, a day will come when this heart will prove its gratitude.

DAME: That day, Henry, is every day.

ASHFIELD: Dang it, never be down-hearted! I do know, as well as can be, zome good luck will turn up. All the way I comed whoam, I looked to find a purse in the path. But I didn't, though.

(*A knocking at the door is heard.*)

DAME: Ah, here they are, coming to sell, I suppose.

ASHFIELD: Lettun! Lettun zeize and zell! (*striking his breast*) We ha' gotten here what we won't zell and they can't zell.

(*Knocking is heard again.*)

Come in, dang it! Don't ye be shy!

(*Enter* MORRINGTON *and* GERALD.)

HENRY: Ah, the strangers I saw this morning! These are not officers of law.

ASHFIELD: Noa? Walk in, gem'men! Glad to zee ye, wi' all my heart and zoul!

Come, dame, spread a cloth, bring out cold meat and a mug of beer.

GERALD: (*to* MORRINGTON) That is the boy? (MORRINGTON *nods.*)

ASHFIELD: Take a chair, zur.

MORRINGTON: I thank, and admire, your hospitality. — Don't trouble yourself, good woman; I am not inclined to eat.

ASHFIELD: That be the case, here. Today none o'we be auver hungry. Misfortin be apt to stay the stomach confoundedly.

MORRINGTON: Has misfortune reached this humble dwelling?

ASHFIELD: Ees, zur. I do think, vor my part, it do work its way in everywhere.

MORRINGTON: Well, never despair.

ASHFIELD: I never do, zur. It is not my way. When the sun do shine, I never think of voul weather, not I, and when it do begin to rain, I always think that's a zure zign it will give auver.

MORRINGTON: Is that young man your son?

ASHFIELD: No, zur. I do wish he were, wi' all my heart and zoul.

GERALD: (*to* MORRINGTON) Sir, remember —

MORRINGTON: Doubt not my prudence. — Young man, your appearance interests me. How can I serve you?

HENRY: By informing me who are my parents.

MORRINGTON: That I cannot do.

HENRY: Then, by removing from me the hatred of Sir Philip Blandford.

MORRINGTON: Does Sir Philip hate you?

HENRY: With such severity that, even now, he is about to ruin these worthy creatures because they have protected me.

MORRINGTON: Indeed? Misfortune has made him cruel. That should not be.

ASHFIELD: Noa, it should not, indeed, zur.

MORRINGTON: It shall not be.

ASHFIELD: Sha'n't it, zur? But how sha'n't it?

MORRINGTON: I will prevent it.

ASHFIELD: Woolye, faith and troth? Now, dame, did not I zay zome good luck would turn up?

HENRY: Oh, sir, did I hear you rightly? Will you preserve my friends? Will you avert the cruel arm of power and make the virtuous happy? My tears must thank you. (*taking* MORRINGTON*'s hand*)

MORRINGTON: (*disengaging his hand*) Young man, you oppress me. Forbear. I do not merit thanks. Pay your gratitude where you are sure 'tis due, to Heaven. Observe me: here is a bond of Sir Philip Blandford's for £1,000. Do you present it to him and obtain a discharge for the debt of this worthy man. The rest is at your own disposal. — No thanks.

HENRY: But sir, to whom am I thus highly indebted?

MORRINGTON: My name is Morrington. At present that information must suffice.

HENRY: Morrington.

ASHFIELD: (*bowing*) Zur, if I may be zo bold —

MORRINGTON: Nay, friend —

ASHFIELD: Don't be angry, I hadn't thanked you, zur, nor I won't. Only, zur, I were going to ax when you would call again. You shall have my stamp-note vor the money, you shall, indeed, zur. And, in the meantime, I do hope you'll take zomeit in way of remembrance, as 'twere.

DAME: Will your honour put a couple of turkeys in your pocket?

ASHFIELD: Or pop a ham under your arm? Don't ye zay no if it's agreeable.

MORRINGTON: Farewell, good friends. I shall repeat my visit soon.

DAME: The sooner the better.

ASHFIELD: Good-bye to ye, zur. Dame and I wool go to work as merry as crickets. Good-bye, Henry.

DAME: Heaven bless your honour! And I hope you will carry as much joy away with you as you leave behind you, I do indeed.

(*Exeunt* ASHFIELD *and* DAME.)

---

stamp-note: properly applied to the official schedule of the cargo carried by merchant ships, and under eighteenth-century anti-smuggling law it was a serious offence for the master of any ship within four leagues of the British coast not to be able to produce one on demand. Ashfield is applying a familiar term loosely to denote a receipt.

MORRINGTON: Young man, proceed to the Castle and demand an audience of Sir Philip Blandford. In your way thither, I'll instruct you further. Give me your hand.

(*Exeunt* MORRINGTON, *looking steadfastly on* HENRY, GERALD *following.*)

Scene 2. *An apartment in the Castle.* SIR PHILIP BLANDFORD *discovered,* MISS BLANDFORD *reading.*

MISS B: Shall I proceed to the next essay?
SIR PHILIP: What does it treat of?
MISS B: Love and friendship.
SIR PHILIP: A satire?
MISS B: No, father, a eulogy!
SIR PHILIP: Thus do we find in the imaginations of men what we in vain look for in their hearts. Lay it by.

(*A knocking at the door is heard.*)
Come in!

(*Enter* EVERGREEN.)

EVERG: My dear master, I am a petitioner to you.
SIR PHILIP: (*rising*) None possesses a better claim to my favour. Ask, and receive.
EVERG: I thank you, sir. The unhappy Henry —
MISS B: What of him?
SIR PHILIP: Emma, go to your apartment.
MISS B: Poor Henry!

(*Exit* MISS BLANDFORD.)

SIR PHILIP: (*turning to* EVERGREEN *with resentment*) Imprudent man!
EVERG: Nay, be not angry. He is without and entreats to be admitted.
SIR PHILIP: I cannot, will not, again behold him.
EVERG: I am sorry you refuse me, as it compels me to repeat his words. 'If,' he said, 'Sir Philip denies my humble request, tell him I demand to see him.'
SIR PHILIP: Demand to see me? (*sarcastically*) Well, his *high* command shall be obeyed, then. Bid him approach.

(*Exit* EVERGREEN. *Enter* HENRY.)

SIR PHILIP: By what title, sir, do you thus intrude on me?
HENRY: By one of an imperious nature, the title of a creditor.
SIR PHILIP: I *your* debtor!
HENRY: Yes, for you owe me justice. You, perhaps, withhold from me the inestimable treasure of a parent's blessing.
SIR PHILIP: (*impatiently*) To the business that brought you hither.
HENRY: This then: (*producing a bond*) I believe this is your signature.
SIR PHILIP: Ah! — (*recovering himself*) It is.
HENRY: Affixed to a bond of £1,000 which, by assignment, is mine. By virtue of

---

a bond . . . which, by assignment, is mine: the term *bond* referred during this period to the document of debenture by which the financially reckless, of which there was a significant population, could secure instant cash liquidity by signing an agreement transferring from them the ownership of stated property assets, often worth very much more than the money advanced, this, I discharge the debt of your worthy tenant, Ashfield, who, it seems, was guilty of the crime of vindicating the injured and protecting the unfortunate. Now, Sir Philip, the retribution my hate demands is that what remains of this obligation may not be now paid to me, but wait your entire convenience and leisure.

SIR PHILIP: No, that must not be!

HENRY: Oh, sir, why thus oppress an innocent man? Why spurn from you a heart that pants to serve you? No answer? Farewell! (*going*)

SIR PHILIP: Hold! One word before we part. Tell me, — (*aside*) I dread to ask it — how came you possessed of this bond?

HENRY: A stranger, whose kind benevolence stepped in and saved —

SIR PHILIP: His name?

HENRY: Morrington.

SIR PHILIP: Fiend! Tormentor! Has he caught me? You have seen this Morrington —

HENRY: Yes.

SIR PHILIP: Did he speak of me?

HENRY: He did, and of your daughter. 'Conjure him,' said he, 'not to sacrifice the lovely Emma by a marriage her heart revolts at. Tell him the life and fortune of a parent are not his own; he holds them but in trust for his offspring. Bid him reflect that, while his daughter merits the brightest rewards a father can bestow, she is by that father doomed to the harshest fate tyranny can inflict.'

SIR PHILIP: (*with vehemence*) Torture! Did he say who caused this sacrifice?

HENRY: He told me you had been duped of your fortune by sharpers.

SIR PHILIP: Aye, he knows that well. Young man, mark me: this Morrington, whose precepts wear the face of virtue and whose practice seems benevolence, was the chief of the hellish banditti that ruined me.

HENRY: Is it possible?

SIR PHILIP: That bond you hold in your hand was obtained by robbery.

HENRY: Confusion!

SIR PHILIP: Not by the thief who, encountering you as a man, stakes life against life, but by that most cowardly villain who, in the moment when reason sleeps and passion is roused, draws his snares around you and hugs you to your ruin. Then, fattening on the spoil, he insults the victim he has made.

HENRY: On your soul, is Morrington that man?

SIR PHILIP: On my soul, he is.

HENRY: Thus, then, I annihilate the detested act (*tearing the bond*), and thus I tread upon a villain's friendship.

SIR PHILIP: Rash boy, what have you done?

HENRY: An act of justice to Sir Philip Blandford.

SIR PHILIP: For which you claim my thanks?

HENRY: Sir, I am thanked already, (*pointing to his heart*) here. Curse on such wealth! Compared with its possession, poverty is splendour. Fear not for me,

---

on the expiry of a stated period, which was frequently very short. Additionally, bonds could be traded by their owners, hence the practice of *assignment*, or legal transfer.

I shall not fear the piercing cold, for in that man whose heart beats warmly for his fellow creatures, the blood circulates with freedom. My food shall be what few of the pampered sons of greatness can boast of, the luscious bread of independence, and the opiate that brings me sleep will be the recollection of the day passed in innocence.

SIR PHILIP: Noble boy! Oh, Blandford!

HENRY: Ah?

SIR PHILIP: What have I said?

HENRY: You called me Blandford.

SIR PHILIP: 'Twas error! 'Twas madness!

HENRY: Blandford! A thousand hopes and fears rush on my heart. Disclose to me my birth, be it what it may, I am your slave forever; refuse me, you create a foe, firm and implacable as —

SIR PHILIP: Ah, am I threatened? Do not extinguish the spark of pity my breast is warmed with.

HENRY: I will not. Oh, forgive me!

SIR PHILIP: Yes, on one condition: leave me. Ah, someone approaches. Begone, I insist! — I entreat!

HENRY: That word has charmed me. I obey. Sir Philip, you may hate, but you shall respect me.

(*Exit* HENRY. *Enter* HANDY JUN.)

HANDY JUN: At last! Thank Heaven, I have found somebody! But, Sir Philip, were you indulging in soliloquy? You seem agitated.

SIR PHILIP: No, sir. Rather indisposed.

HANDY JUN: Upon my soul, I am devilish glad to find you. Compared with this Castle, the Cretan labyrinth was intelligible, and, unless some Ariadne gives me a clue, I shan't have the pleasure of seeing you above once a week.

SIR PHILIP: I beg your pardon. I have been an inattentive host.

HANDY JUN: Oh, no! But when a house is so devilish large and the party so very small, they ought to keep together, for, to say the truth, though no one on earth feels a warmer regard for Robert Handy than I do, I soon get heartily sick of his company. Whatever he may be to others, he's a cursed bore to me.

SIR PHILIP: Where's your worthy father?

HANDY JUN: As usual, full of contrivances that are impracticable and improvements that are retrograde, forming, altogether, a whimsical instance of the confusion of arrangement, the delay of expedition, the incommodiousness of accommodation, and the infernal trouble of endeavouring to save it. He has now a score or two of workmen about him and intends pulling down some apartments in the east wing of the Castle.

SIR PHILIP: Ah, ruin! — (*calling*) Within, there!

(*Enter a* SERVANT.)

Fly to Sir Abel Handy. Tell him to desist. Order his people, on the peril of their lives, to leave the Castle instantly! Away!

(*Exit* SERVANT.)

HANDY JUN: Sir Philip Blandford, your conduct compels me to be serious.

SIR PHILIP: Oh, forbear! Forbear!

HANDY JUN: Excuse me, sir. An alliance, it seems, is intended between our

families, founded on ambition and interest. I wish it, sir, to be formed on a
nobler basis: ingenuous friendship and mutual confidence. That confidence
being withheld, I must here pause, for I should hesitate in calling that man
father who refuses me the name of a friend.

SIR PHILIP: (*aside*) Ah, how shall I act?

HANDY JUN: Is my demand unreasonable?

SIR PHILIP: Strictly just! But, oh, you know not what you ask. Do you not pity
me?

HANDY JUN: I do.

SIR PHILIP: Why, then, seek to change it into hate?

HANDY JUN: Confidence seldom generates hate, mistrust always.

SIR PHILIP: Most true.

HANDY JUN: I am not impelled by curiosity to ask your friendship; I scorn so
mean a motive. Believe me, the folly and levity of my character proceed
merely from the effervescence of my heart. You will find its substance warm,
steady and sincere.

SIR PHILIP: I believe it, from my soul. Allow me a moment's thought. – (*aside*)
Suspicion is awakened. Does not prudence as well as justice prompt me to
confide in him? Does not my poverty command me? Perhaps I may find a
sympathising friend. The task is dreadful, but it must be so. Perhaps he will
perform the awful task of visiting the chamber and removing every vestige of
guilt. – (*to him*) Yes, you shall hear my story. I will lay before your view the
agony with which this wretched bosom is loaded.

HANDY JUN: I am proud of your confidence and am prepared to receive it.

SIR PHILIP: Not here. Let me lead you to the eastern part of the Castle. My young
friend, mark me: this is no common trust I repose in you, for I place my life
in your hands.

HANDY JUN: And the pledge I give for its security is what alone gives value to my
life: my honour.

(*Exeunt.*)

Scene 3. *A gloomy gallery in the Castle. In the centre, a strongly barred
door. The gallery hung with portraits.* HENRY *discovered examining a
particular portrait which occupies a conspicuous situation in the gallery.*

HENRY: Whenever curiosity has led me to this gallery, that portrait has attracted
my attention. The features are peculiarly interesting. One of the house of
Blandford. Blandford! My name! Perhaps my father! To remain longer
ignorant of my birth, I feel impossible. There is a point when patience ceases
to be a virtue. – Hush, I hear footsteps! Ah, Sir Philip and another in close
conversation! Shall I avoid them? No. Shall I conceal myself and observe
them? Curse on the base suggestion! No!

(*Enter* SIR PHILIP *and* HANDY JUN.)

SIR PHILIP: That chamber contains the mystery.

HENRY: (*aside*) Ah!

SIR PHILIP: (*turning round*) Observe that portrait.

(SIR PHILIP *sees* HENRY *and starts.*)
Who's there?

HANDY JUN: (*to* HENRY) Sir, we wish to be private.

HENRY: My being here, sir, was merely the effect of an accident. I scorn intrusion. (*bowing*) – (*aside*) But the important words are spoken: 'that chamber contains the mystery'.

> (*Exit* HENRY.)

HANDY JUN: Who is that youth?

SIR PHILIP: You there behold his father, my brother. (*weeping*) I've not beheld that face these twenty years. Let me again peruse its lineaments. (*in an agony of grief*) Oh, God, how I loved that man!

HANDY JUN: Be composed.

SIR PHILIP: I will endeavour. Now listen to my story.

HANDY JUN: You rivet my attention.

SIR PHILIP: While we were boys, my father died intestate, so I, as elder born, became the sole possessor of his fortune. But the moment the law gave me power, I divided in equal portions his large possessions, one of which I with joy presented to my brother.

HANDY JUN: It was noble.

SIR PHILIP: At least it was just. – We lived together, sir, as one man. As my life I loved him, and felt no joys but what he shared. Sorrow I knew not.

HANDY JUN: Such love demanded a life of gratitude.

SIR PHILIP: (*with suppressed agony*) You shall now hear, sir, how I was rewarded. Chance placed in my view a young woman of superior personal charms. My heart was captivated. Fortune she possessed not, but mine was ample. She blessed me by consenting to our union, and my brother approved my choice.

HANDY JUN: How enviable your situation!

SIR PHILIP: (*sighing deeply*) Oh! On the evening previous to my intended marriage, with a mind serene as the setting sun whose morning beam was to light me to happiness, I sauntered to a favourite tree, where, lover-like, I had marked the name of my destined bride, and, with every nerve braced to the tone of ecstasy, I was wounding the bark with a deeper impression of the name when – oh, God! –

HANDY JUN: Pray, proceed.

SIR PHILIP: When the loved offspring of my mother, and the woman my soul adored – the only two beings on earth who had wound themselves round my heart by every tie dear to the soul of man – placed themselves before me. I heard him – even now the sound is in my ears and drives me to madness – I heard him breathe vows of love which she answered with burning kisses. He pitied his poor brother, and told her he had prepared a vessel to bear her forever from me. They were about to depart when the burning fever in my heart rushed upon my brain. Picture the young tiger when first his savage nature rouses him to vengeance. The knife was in my gripe. I sprung upon them. With one hand I tore the faithless woman from his damned embrace and, with the other, stabbed my brother to the heart.

HANDY JUN: (*starting with horror, then recovering*) What followed?

SIR PHILIP: At that dreadful moment, my brother's servant appeared, and the vessel that was to waft him to happiness bore away his bleeding body. A few days brought the news that he had died suddenly in France, and all enquiry ceased.

      (SIR PHILIP, *exhausted, falls into* HANDY JUN.*'s arms.*)

HANDY JUN: You are faint; let me lead you from this place. Yet, hold! The wretched woman —

SIR PHILIP: Was secretly conveyed here, even to that chamber. She proved pregnant and, in giving birth to a son, paid the forfeit of her perjury by death.

HANDY JUN: Which son was the youth that left us.

SIR PHILIP: Even so. Tell me, could wretch be born possessed of a more solid title to my hate?

HANDY JUN: Yet, he is innocent.

SIR PHILIP: My task being ended, yours begins.

HANDY JUN: Mine?

SIR PHILIP: Yes. That chamber contains evidence of my shame; the fatal instrument, with other guilty proofs, lies there concealed. Can you wonder I dread to visit the scene of horror? Can you wonder? I implore you, in mercy, to save me from the task. Oh, my friend, enter the chamber, bury in endless night those instruments of blood, and I will kneel and worship you.

HANDY JUN: I will.

SIR PHILIP: (*weeping*) Will you? (*embracing* HANDY JUN.) I am unused to kindness from man, and it affects me. Oh, can you press to your guiltless heart that blood-stained hand?

HANDY JUN: Sir Philip, let men without faults condemn. I must pity you.

      (*Exeunt,* HANDY JUN. *leading* SIR PHILIP.)

ACT V

Scene 1. *A wooded view of the country.*

      (*Enter* SUSAN ASHFIELD, *who looks about with anxiety and then comes forward.*)

SUSAN: I fear my conduct is very imprudent. Has not Mr Handy told me he is engaged to another? But 'tis hard for the heart to forgo, without one struggle, its only hope of happiness, and, conscious of my honour, what have I to fear? Perhaps he may repent of his unkindness to me. At least I'll put his passion to the proof: if he be worthy of my love, happiness is forever mine, if not, I'll tear him from my breast, though from the wound my life's blood should follow. Ah, he comes! I feel I am a coward, and my poor alarmed heart trembles at its approaching trial. Pardon me, female delicacy, if for a moment I seem to pass thy sacred limits.

      (SUSAN ASHFIELD *retires up the stage. Enter* HANDY JUN.)

HANDY JUN: By Heavens, the misfortunes of Sir Philip Blandford weigh so heavily on my spirits that — but confusion to melancholy! I am come here to meet an angel who will, in a moment, drive away the blue devils like mist before the

---

blue devils: melancholy, the 'blues' (see above. *Inkle and Yarico*, p. 90).

sun. Let me again read the dear words (*reading a letter*): 'I confess I love you still.' (*kissing the letter*) – But I dare not believe their truth till her sweet lips confirm it. Ah, she's there! Susan, my angel, a thousand thanks! A life of love can alone repay the joy your letter gave me!

SUSAN: Do you not despise me?

HANDY JUN: No, love you more than ever!

SUSAN: Oh, Robert, this is the very crisis of my fate. From this moment we meet with honour, or we meet no more. If we must part, perhaps, when you lead your happy bride to church, you may stumble over your Susan's grave. Well, be it so.

HANDY JUN: Away with such sombre thoughts!

SUSAN: Tell me my doom. Yet hold! You are wild, impetuous, you do not give your heart fair play, therefore promise me – perhaps 'tis the last favour I shall ask – that, before you determine whether our love shall die or live with honour, you will remain here alone a few moments and that you will give those moments to reflection.

HANDY JUN: I do. I will.

SUSAN: With a throbbing heart, I will wait at a little distance. – (*aside*) May virtuous love and sacred honour direct his thoughts!

(*Exit* SUSAN ASHFIELD.)

HANDY JUN: Yes, I will reflect – that I am the most fortunate fellow in England! She loves me still. What is the consequence? That love will triumph, that she will be mine, mine without the degradation of marriage: love, pride, all gratified. How I shall be envied when I triumphantly pass the circles of fashion! One will cry, 'Who is that angel?' – another, 'Happy fellow!' Then Susan will smile around. Will she smile? Oh, yes, she will be all gaiety, mingle with the votaries of pleasure, and – What? Susan Ashfield the companion of licentious women? Damnation, no! I wrong her! She would not. She would rather shun society. She would be melancholy, melancholy. (*sighing and looking at his watch*) Would the time were over! – Pshaw, I think of it too seriously! – 'Tis false. I do not! Should her virtue yield to love, would not remorse affect her health? Should I not behold that lovely form sicken and decay, perhaps die? Die? Then what am I? A villain loaded with her parents' curses and my own! Let me fly from the dreadful thought. But how fly from it? By placing before my imagination a picture of more honourable lineaments: I make her my wife. Ah, then she would smile on me! There's rapture in the thought! Instead of vice producing decay, I behold virtue emblazoning beauty; instead of Susan on the bed of death, I behold her giving to my hopes a dear pledge of our mutual love: she places it in my arms, down her father's honest face runs a tear, but 'tis a tear of joy. Oh, this will be luxury! Paradise! – (*calling*) Come, Susan! Come, my love, my soul, – my wife!

(*Enter* SUSAN. *She at first hesitates. On hearing the word 'wife', she springs into his arms.*)

SUSAN: Is it possible?

HANDY JUN: Yes, those charms have conquered.

SUSAN: Oh, no, do not so disgrace the victory you have gained! 'Tis your own
virtue that has triumphed.

HANDY JUN: My Susan! How true it is that fools alone are vicious. But let us fly
to my father and obtain his consent. On recollection, that may not be quite
so easy; his arrangements with Sir Philip Blandford are — are — not mine, so
there's an end of that. And Sir Philip, by misfortune, knows how to appreciate happiness. Then, poor Miss Blandford; upon my soul, I feel for her.

SUSAN: (*ironically*) Come, don't make yourself miserable! If my suspicions be
true, she'll not break her heart for your loss.

HANDY JUN: Nay, don't say so. She will be unhappy.

ASHFIELD: (*without*) There he is! Dame, shall I shoot at un?

DAME: (*without*) No!

HANDY JUN: Shoot! What does he mean?

SUSAN: My father's voice!

ASHFIELD: (*without*) Then I'll leather un wi' my stick!

HANDY JUN: Zounds! No! — Come here.

(*Enter* ASHFIELD *and* DAME ASHFIELD.)

ASHFIELD: What do thee do here with my Sue, eh?

HANDY JUN: With your Sue? She's mine, mine by a husband's right!

ASHFIELD: Husband? What, thee Sue's husband?

HANDY JUN: I soon shall be.

ASHFIELD: But how, tho'? What, — faith and troth! — What, like as I married
dame?

HANDY JUN: Yes.

ASHFIELD: What, axed three times?

HANDY JUN: Yes, and from this moment I'll maintain that the real temple of love
is a parish church, Cupid is a chubby curate, his torch is the sexton's lantern,
and the according paean of the spheres is the profound nasal thorough-bass
of the clerk's 'amen'.

ASHFIELD: Huzza! Only to think, now. My blessing go with you, my children!

DAME: And mine!

ASHFIELD: And Heaven's blessing, too. (*to* HANDY JUN.) Ecod, I believe now, as
thy feyther zays, thee can'st do everything!

HANDY JUN: No, for there is one thing I cannot do: injure the innocence of
woman.

ASHFIELD: Drabbit it, I shall walk in the road all day to zee Sue ride by in her
own coach!

SUSAN: You must ride with me, father.

---

axed three times?: Farmer Ashfield is referring not to a brutal courtship ceremony but to the
calling of the banns.

his torch: apart from the ubiquitous bow and arrows, Cupid's cache also included a flaming
torch with which he would raise the fires of passion in unsuspecting hearts.

paean of the spheres: florid variant of the 'music of the spheres' (which, according to
Pythagoras, was produced by the spheres of the heavens moving in harmony). Originally a
hymn to Apollo, *paean* has come to denote any song of thanksgiving.

DAME: I say, Tummas, what will Mrs Grundy say then?

ASHFIELD: I do hope thee will not be asham'd of thy feyther-in-laa, woolye?

HANDY JUN: No, for then I must also be ashamed of myself, which I am resolved not to be again.

(*Enter* SIR ABEL HANDY.)

SIR ABEL: Hey-day, Bob, why an't you gallanting your intended bride? But you are never where you ought to be.

HANDY JUN: Nay, sir, by your own confession I *am* where I ought to be.

SIR ABEL: No, you ought to be at the Castle! Sir Philip is there, and Miss Blandford is there, and Lady Handy is there, and therefore —

HANDY JUN: You are *not* there! In one word, I shall not marry Miss Blandford.

SIR ABEL: Indeed? Who told you so?

HANDY JUN: One who never lies and, therefore, one I am determined to make a friend of: my conscience.

SIR ABEL: But, zounds, sir, what excuse have you?

HANDY JUN: (*taking* SUSAN's *hand*) A very fair one, sir, is not she?

SIR ABEL: Why, yes, sir, I can't deny it. But 'sdeath, sir, this overturns my best plan!

HANDY JUN: No, sir, for a parent's best plan is his son's happiness, and that it will establish. Come, give us your consent. Consider how we admire all your wonderful inventions.

SIR ABEL: No, not my plough, Bob. But 'tis a devilish clever plough.

HANDY JUN: I dare say it is. Come, sir, consent, and perhaps, in our turn, we may invent something that may please you.

SIR ABEL: He! he! he! But hold! What's the use of my consent without my wife's? I dare no more approve without —

(*Enter* GERALD.)

GERALD: Health to the company!

SIR ABEL: The same to you, sir.

HANDY JUN: Who have we here, I wonder?

GERALD: I wish to speak with Sir Abel Handy.

SIR ABEL: I am the person.

GERALD: You are married?

SIR ABEL: Damn it, he sees it in my face! — Yes, I have that happiness.

GERALD: Is it a happiness?

SIR ABEL: To say the truth, — Why do you ask?

GERALD: I want answers, not questions. And, depend on't, 'tis your interest to answer me.

HANDY JUN: An extraordinary fellow this!

GERALD: Would it break your heart to part with her?

SIR ABEL: Who are you, sir, that — ?

GERALD: Answers, I want answers. Would it break your heart, I ask?

SIR ABEL: Why, not absolutely, I hope. Time and philosophy and —

GERALD: I understand. — What sum of money would you give to the man who would dissolve your marriage contract?

HANDY JUN: He means something, sir.

SIR ABEL: Do you think so, Bob?

GERALD: Would you give me a thousand pounds?

SIR ABEL: No!

HANDY JUN: No?

SIR ABEL: No, I would not give one, but I would give five thousand pounds!

GERALD: Generously offered. A bargain. I'll do it.

SIR ABEL: But, an't you deceiving me?

GERALD: What should I gain by that?

SIR ABEL: Tell me your name.

GERALD: Time will tell that.

LADY H: (*without*) Sir Abel, where are you?

GERALD: That's your wife's voice. I know it.

SIR ABEL: So do I!

GERALD: I'll wait without. Cry 'hem' when you want me.

SIR ABEL: Then you need not go far.

> (*Exit* GERALD.)
> I dare not believe it. I shall go out of my wits. And then, if he fail, what a
> pickle I shall be in! Here she is.
> (*Enter* LADY HANDY.)

LADY H: So, sir, have I found you at last?

HANDY JUN: My honoured Mamma, you have just come in time to give your consent to my marriage with my sweet Susan.

LADY H: And do you imagine I will agree to such degradation?

ASHFIELD: Do'e, Lady Nelly, do'e be kind-hearted to the young loviers.
> Remember how I used to let thee zit up all night a-sweethearting.

LADY H: Silence! (*to* SIR ABEL) And have you dared to consent?

SIR ABEL: Oh, no, my lady.

HANDY JUN: Sir, you had better cry 'hem'.

SIR ABEL: I think it's time, Bob. – Hem!

HANDY JUN: Hem!

LADY H: What do you mean by 'hem'?

SIR ABEL: Only, my dear, something troublesome I want to get rid of. – Hem!
> (*Enter* GERALD.)
> There he is! Never was so frightened in all my life.
> (GERALD *advances.* LADY HANDY *shrieks.*)

LADY H: Gerald!

GERALD: Yes.

LADY H: An't you dead, Gerald? Twenty years away and not dead?

GERALD: No, wife.

SIR ABEL: Wife? Did you say wife?

GERALD: Yes.

SIR ABEL: Say it again!

GERALD: She is my wife.

SIR ABEL: Once more!

GERALD: My lawful, wedded wife.

SIR ABEL: (*embracing* GERALD *and the rest*) Oh, my dear fellow! Oh, my dear boy! Oh, my dear girl! (*running to the former* LADY HANDY) No. Yes, now she an't my wife, I will! (*embracing her*) – (*to* GERALD) Well, how will you have the five thousand? Will you have it in cash or in bank notes − or stocks, or India bonds, or lands, or patents, or − ?

GERALD: No, land will do. I wish to kill my own mutton.

SIR ABEL: Sir, you shall kill all the sheep in Hampshire.

GERALD: Sir Abel, you have lost five thousand pounds and, with it, properly managed, an excellent wife, who, though I cannot condescend to take again as mine, you may depend on't, shall never trouble you. (*beckoning to his wife*) Come! This way! − Important events now call on me and prevent my staying longer with this company. Sir Abel, we shall meet soon. (*to his wife*) Nay, come! You know I'm not used to trifle. Come! Come!

> (*The former* LADY HANDY *reluctantly but obediently crosses the stage and runs off.* GERALD *follows.*)

SIR ABEL: (*imitating*) Come! Come! − That's a damned clever fellow! − Joy, joy, my boy! Here, here, your hands; the first use I make of liberty is to give happiness. I wish I had more imitators! (*walking about exultingly*) Well, what will you do? Where will you go? I'll go anywhere you like. Will you go to Bath, or Brighton, or Petersburg, or Jerusalem, or Seringapatem? All the same to me. We single fellows, we rove about; nobody cares about us, we care for nobody!

HANDY JUN: I must to the Castle, father.

SIR ABEL: Have with you, Bob. (*singing*) 'I'll sip every flower, I'll change every hour.' − (*beckoning*) Come! Come!

> (*Exeunt* SIR ABEL, HANDY JUN. *and* SUSAN. SUSAN *kisses her hand to* ASHFIELD *and* DAME ASHFIELD.)

ASHFIELD: Bless her, how nicely she do trip it away with the gentry!

DAME: And then, Tummas, think of the wedding.

ASHFIELD: (*reflecting*) I declare I shall be just the zame as ever. Maybe I may buy a smartish bridle or a zilver backy stopper, or the like o'that.

DAME: (*apart*) And then, when we come out of church, Mrs Grundy will be standing about there −

ASHFIELD: (*apart*) I shall shake hands agreeably wi' all my friends −

DAME: (*apart*) Then I just look at her in this manner −

ASHFIELD: (*apart*) 'How dost, Peter?' 'Ah, Dick, glad to zee thee, wi' all my heart and zoul.' −

> (ASHFIELD *bows towards the centre of the stage.*)

DAME: (*apart*) Then, with a kind of half-curtsey, I shall −

> (DAME ASHFIELD *advances to the centre also. Their heads meet.*)

ASHFIELD: What an wold fool thee be'st, dame! Come along, and behave pratty, do'e!

> (*Exeunt.*)

St Petersburg: adopted by the English *beau monde* as a summer resort in the 1790s, when the Revolutionary War in Europe made access to many of its climatically more favoured capitals potentially hazardous.

Seringapatem: 'Timbuktu'; this obscure place-name had risen to prominence during the preceding year on the tide of controversy which greeted the bloody Siege of Seringapatem, an incident which reopened the British war in India.

Scene 2. *The gallery in the Castle.*

(*Enter* HANDY JUN. *with caution, bearing a light and a large key.*)

HANDY JUN: Now to fulfil my promise with Sir Philip Blandford by entering —
that — chamber — and removing — 'Tis rather awful! Somehow everything is
so cursedly still. What's that? I thought I heard something! No. Why, 'sdeath,
I am not afraid. No, I'm quite su-su-sure of that. Only, everything is so
cursedly hush and —

(*A flash of light and a tremendous explosion takes place.*)

What the devil's that? (*trembling*) I swear I hear someone — lamenting! Who's
there?

(*Enter* SIR ABEL.)

(*trembling*) Father?

SIR ABEL: (*trembling*) Bob?

HANDY JUN: Have you seen anything?

SIR ABEL: Oh, my dear boy!

HANDY JUN: Damn it, don't frighten one!

SIR ABEL: Such an accident! Mercy on us!

HANDY JUN: Speak!

SIR ABEL: I was mixing the ingredients of my grand substitute for gunpowder
when somehow it blew up and set the curtains on fire and —

HANDY JUN: Curtains? Zounds, the room's in a blaze!

SIR ABEL: Don't say so, Bob.

HANDY JUN: What's to be done? Where's your famous preparation for extinguish-
ing flames?

SIR ABEL: It is not mixed.

HANDY JUN: Where's your patent fire-engine?

SIR ABEL: 'Tis on the road.

HANDY JUN: Well, you are never at a loss.

SIR ABEL: Never.

HANDY JUN: What's to be done?

SIR ABEL: I don't know! — I say, Bob, I have it: perhaps it will go out of itself.

HANDY JUN: Go out? It increases every minute! Let us run for assistance. Let us
alarm the family.

(*Exit* HANDY JUN.)

SIR ABEL: Yes. — Dear me! Dear me!

SERVANT: (*without*) Here, John, Thomas! Some villain has set fire to the Castle.
If you catch the rascal, throw him into the flames!

(SIR ABEL *runs off. The alarm-bell rings.*)

Scene 3. *The garden of the Castle. The effects of the fire shown on the
foliage and scenery.*

(*Enter* HENRY, *meeting* EVERGREEN.)

HENRY: The Castle in flames! What occasioned it?

EVERG: Alas, I know not!

HENRY: Are the family in safety?

EVERG: Sir Philip is.

HENRY: And his daughter?

EVERG: Poor lady! I just now beheld her looking with agony from that window!

HENRY: Ah, Emma in danger? — Farewell!

EVERG: (*holding him*) Are you mad? The great staircase is in flames!

HENRY: I care not! Should we meet no more, tell Sir Philip I died for his daughter.

EVERG: Yet reflect —

HENRY: Old man, do not cling to me thus. 'Sdeath, men will encounter peril to ruin a woman, and shall I hesitate when it is to save one?

(*Exit* HENRY.)

EVERG: Brave, generous boy! Heaven preserve thee!

(*Enter* SIR PHILIP BLANDFORD.)

SIR PHILIP: Emma, my child, where art thou?

EVERG: I fear, sir, the Castle will be destroyed.

SIR PHILIP: My child! My child! Where is she? Speak!

EVERG: Alas, she remains in the Castle!

SIR PHILIP: Ah, then I will die with her! (*going*)

EVERG: Hold, dear master! If human power can preserve her, she is safe. The bravest, noblest of men has flown to her assistance.

SIR PHILIP: Heaven reward him with its choicest blessings!

EVERG: 'Tis Henry.

SIR PHILIP: Henry! Heaven will reward him! I will reward him!

EVERG: Then be happy: look, sir!

SIR PHILIP: Ah, dare I trust my eyes?

EVERG: He bears her safe in his arms!

SIR PHILIP: Bountiful Creator, accept my thanks!

(*Enter* HENRY *bearing* MISS BLANDFORD *in his arms.*)

HENRY: There is your daughter.

SIR PHILIP: My child, my Emma, revive!

HENRY: (*apart*) Aye, now to unfold the mystery. The avenue to the eastern wing is still passable, the chamber not yet in flames. The present moment is lost, and all is closed forever. I will be satisfied, or perish!

(*Exit* HENRY.)

MISS B: Am I restored to my dear father's arms?

SIR PHILIP: Yes, only blessing of my life! In future, thy wishes shall be mine, thy happiness my joy.

(*Enter* HANDY JUN. *and* SUSAN ASHFIELD.)

HANDY JUN: My dear friend safe? And the lovely Emma in his arms? Then let the Castle bonfire blaze!

SIR PHILIP: Know you, sir, what caused this alarming accident?

HANDY JUN: Yes, I do. And here comes the unfortunate incendiary.

(*Enter* SIR ABEL, *abashed.*)

SIR ABEL: Sir Philip, I am quite ashamed! I am very sorry indeed, sir! But I'll build you a new house, fireproof, upon a plan of my own.

SIR PHILIP: No apologies, Sir Abel. All the treasure I valued is safe!

(MISS BLANDFORD *talks apart to* SUSAN ASHFIELD.)

(*aside to* HANDY JUN.) My young friend, do you mark? The flames will save the trial I imposed on you. Behold, they already burst from the eastern turret!

Ere this, they must have reached the chamber. That consumed, the secret is with us secure.

MISS B: Oh, father, this unkind man has refused me and given his hand to that sweet girl!

HANDY JUN: I confess 'tis true. Your eyes can only fail to conquer those who are before subdued.

SIR PHILIP: But, Emma, where is your Henry? I wish to be just to him; I wish to thank him.

MISS B: He has withdrawn, to avoid our gratitude.

EVERG: No, I saw him again rush into the Castle —

MISS B: Heaven forbid!

SIR ABEL: I endeavoured to prevent him, but with fury in his eye he exclaimed, 'I will penetrate that chamber or perish in the attempt!' Then, with desperation, he plunged into the flames.

SIR PHILIP: Then all is discovered!

HANDY JUN: (*aside to* SIR PHILIP) Hush! For Heaven's sake, collect yourself!
(MISS BLANDFORD *shrieks.*)

MISS B: Ah, thank Heaven he's safe! My deliverer comes! What frantic wildness in his gestures, what terror in his looks!
(*Enter* HENRY *in great agitation.*)
What urged you, Henry, again to venture in the Castle?

HENRY: Fate! The desperate attempt of a desperate man!

SIR PHILIP: Ah!

HENRY: Yes, the mystery is developed. In vain the massy bars, cemented with their cankerous rust, opposed my entrance, in vain the heated suffocating damps enveloped me, in vain the hungry flames flashed their vengeance round me! What could oppose a man struggling to know his fate? I forced the doors — a firebrand was my guide — and among many evidences of blood and guilt I found — these!
(HENRY *produces a knife and bloody cloth.*)

SIR PHILIP: (*starting with horror, then, with solemnity*) It is accomplished. Just Heaven, I bend to thy decree. Blood must be paid by blood. Henry, that knife, aimed by this fatal hand, murdered thy father.

HENRY: Ah!
(HENRY *raises the knife.* MISS BLANDFORD *places herself between him and her father.*)

MISS B: Henry!
(HENRY *drops his hand.*)
Oh, believe him not! 'Twas madness! I have heard him talk thus wildly in his dreams! We are all friends; none will repeat his words, I am sure none will! My heart will break. Oh, Henry, will you destroy *my* father?

HENRY: Would I were in my grave!
(*Enter* GERALD.)

SIR PHILIP: Ah, Gerald here? How vain concealment! Well, come you to give evidence of my shame?

GERALD: I come to announce one who many years has watched each action of your life.

SIR PHILIP: Who?

GERALD: Morrington.

SIR PHILIP: His name shoots life through me! I shall then behold the man who has so long avoided me —

GERALD: But ever has been near you. He is here.

(*Enter* MORRINGTON, *wrapped up in his cloak.*)

SIR PHILIP: Well, behold your victim in his last stage of human wretchedness! Come you to insult me?

(MORRINGTON *clasps his hands together and hides his face.*)

Ah, can even you pity me? Speak! Still silent, still mysterious? Well, it is no matter. Let me employ what remains of life in thinking of hereafter. (*addressing Heaven*) Oh, my brother, we shall soon meet again. And let me hope that, stripped of those passions which make men devils, I may receive the heavenly balm of thy forgiveness as I, from my inmost soul, do pardon thee.

(MORRINGTON *becomes convulsed with agony and falls into* GERALD'*s arms.*)

Ah, what means that agony? He faints! Give him air!

(*They throw open* MORRINGTON'*s cloak and hat.* SIR PHILIP *starts.*)

Angels of mercy, my sight thickens! Support me!

(HENRY *and* MISS BLANDFORD *support* SIR PHILIP.)

(*advancing*) My brother! 'Tis he! He lives! — Henry, regard not me, support your father!

HENRY: (*running to* MORRINGTON) Ah, my father! He revives!

SIR PHILIP: Hush!

(MORRINGTON *recovers. Seeing his brother, he covers his face with shame, then falls at his feet.*)

MORRINGTON: Crawling in the dust, behold a repentant wretch!

SIR PHILIP: (*indignantly*) My brother Morrington?

MORRINGTON: Turn not away! In mercy hear me!

SIR PHILIP: Speak.

MORRINGTON: After the dreadful hour that parted us, agonised with remorse, I was about to punish home what your arm had left unaccomplished when some angel whispered: 'Punishment is life, not death. Live and atone!'

SIR PHILIP: Oh, go on!

MORRINGTON: I flew to you. I found you surrounded by sharpers. What was to be done? I became Morrington, littered with villains, practised the arts of devils, braved the assassin's steel, possessed myself of your large estates, lived hateful to myself — detested by mankind — to do what? To save an injured brother from destruction and lay his fortune at his feet!

(MORRINGTON *places parchments before* SIR PHILIP.)

SIR PHILIP: Ah, is it possible?

MORRINGTON: Oh, is that atonement? No. By me you first beheld her mother.

---

littered with villains: slept rough. *Littered* is used with particular reference to the bedding down of animals, hence its contemptuous force here.

'Twas I that gave her fortune. — Is that atonement? No. But my Henry has saved that angel's life. Kneel with me, my boy, lift up thy innocent hands with those of thy guilty father, and beg mercy from that injured saint.

(HENRY *kneels with* MORRINGTON.)

SIR PHILIP: Oh God, how infinite are thy mercies! Henry, forgive me. Emma, plead for me. (*joining their hands*) There. There.

HENRY: But, my father —

SIR PHILIP: (*approaching*) Charles!

MORRINGTON: Philip!

SIR PHILIP: Brother, I forgive thee.

MORRINGTON: Then let me die, blessed, most blessed!

SIR PHILIP: No, no! (*striking his breast*) Here, I want thee here! Raise him to my heart!

(*They raise* MORRINGTON. *In the effort to embrace, he falls into their arms exhausted.*)

Again!

(SIR PHILIP *and* MORRINGTON *sink into each other's arms.* HANDY JUN. *comes forward.*)

HANDY JUN: If forgiveness be an attribute which ennobles our nature, may we not hope to find pardon for our errors, *here*?

(*The curtain falls.*)

THE PLAYWRITING CANON OF
GEORGE COLMAN THE YOUNGER
AND THOMAS MORTON

In the lists that follow, material included under the heading of *Description* has been drawn where possible from original title pages and playbills. Entries appear under the heading of *Printing* in cases where authorised publication did not take place in the year of performance. The abbreviation 'n.p.' means 'not printed'.

A. GEORGE COLMAN THE YOUNGER

Title	Description	Place and date of first performance	Printing
The Female Dramatist	musical farce, two acts	Hm 16 Aug. 1782	n.p.
Two to One	comic opera, three acts	Hm 19 Jun. 1784	1795
Turk and No Turk	comic opera, three acts	Hm 9 Jul. 1785	n.p.
Inkle and Yarico	opera, three acts	Hm 4 Aug. 1787	
Ways and Means	comedy, three acts	Hm 10 Jul. 1788	
The Family Party	farce, two acts	Hm 11 Jul. 1789	
The Battle of Hexham	play, three acts	Hm 11 Aug. 1789	1808
The Surrender of Calais	play, three acts	Hm 30 Jul. 1791	1808
Poor Old Haymarket	occasional drama, one act	Hm 15 Jun. 1792	
The Mountaineers	play, three acts	Hm 3 Aug. 1793	1795
New Hay at the Old Market	occasional drama, one act	Hm 9 Jun. 1795	
The Iron Chest	play, three acts	DL 12 Mar. 1796	
My Nightgown and Slippers	entertainment in verse	DL 28 Apr. 1797	
The Heir at Law	comedy, five acts	Hm 15 Jul. 1797	1808
Blue Beard	grand dramatic romance, two acts	DL 16 Jan. 1798	
Blue Devils	farce, one act	CG 24 Apr. 1798	1808
Feudal Times	drama, two acts	DL 19 Jan. 1799	
The Castle of Sorrento (with Henry Heartwell)	comic opera, two acts	Hm 17 Jul. 1799	
The Review	musical farce, two acts	Hm 2 Sep. 1800	1808
The Poor Gentleman	comedy, five acts	CG 11 Feb. 1801	
John Bull	comedy, five acts	CG 5 Mar. 1803	1805
No Prelude!	occasional drama, one act	Hm 16 May 1803	n.p.
Love Laughs at Locksmiths	comic opera, two acts	Hm 25 Jul. 1803	1808
The Gay Deceivers	farce, two acts	Hm 22 Aug. 1804	1808
Who Wants a Guinea?	comedy, five acts	CG 18 Apr. 1805	
We Fly By Night	musical entertainment, two acts	CG 28 Jan. 1806	
The Forty Thieves (with R.B. Sheridan and C. Ward)	grand romantic drama, two acts	DL 8 Apr. 1806	1814
The Africans	play, three acts	Hm 29 Jul. 1808	
X.Y.Z.	farce, two acts	CG 11 Dec. 1810	1820

It Came from Memphis

＃

ROBERT GORDON

Secker & Warburg

LONDON

Published by Secker & Warburg 1995

2 4 6 8 10 9 7 5 3

Copyright © Robert Gordon 1995

Robert Gordon has asserted his right under the Copyright, Designs
and Patents Act 1988 to be identified as the author of this work

Secker & Warburg
Random House, 20 Vauxhall Bridge Road,
London SW1V 2SA

Random House Australia (Pty) Limited
20 Alfred Street, Milsons Point, Sydney,
New South Wales 2061, Australia

Random House New Zealand Limited
18 Poland Road, Glenfield,
Auckland 10, New Zealand

Random House (Pty) Limited
Endulini, 5A Jubilee Road, Parktown 2193, South Africa

The Random House Group Limited Reg. No. 954009
www.randomhouse.co.uk

A CIP catalogue record for this book
is available from the British Library

ISBN 0 436 20145 3

Papers used by Random House are natural, recyclable products
made from wood grown in sustainable forests; the manufacturing processes
conform to the environmental regulations of the country of origin

Printed and bound in Great Britain by Mackays of Chatham

For Tara
With love

We knew what it was like to be born in the old world.
We had everything to consume and nothing to conquer. We had to invent everything ourselves.

Godard
Jean Collet

Contents

⇒⊩⊩⇐

Foreword

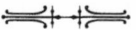

FOR THE LAST TWENTY-FIVE YEARS, MORE OR LESS, I'VE BEEN WALKING
the streets of Memphis looking for ghosts, listening for echoes, trying to construct a vision in my mind of *what it must have been like*. I must admit, I was
egged on by Stanley Booth and Jim Dickinson, I was prompted by the lure of
history (Sun, Stax, and beyond)—but until I met Robert Gordon I was beginning to think I was crazy. And now I'm not so sure which one of us is.

I don't know how old Robert is exactly, and I don't want to suggest in any
way that he has lived other lives or possesses what is called an "old soul"—but
there's something going on here. Because Robert *is* possessed by an imaginative grasp of history, he does see and hear ghostly footsteps, he is a young
writer/filmmaker/musical avant gardist/Memphis literateur living very much
in the present who is no less connected to the past.

I first met Robert not long after I started work on my Elvis Presley biography, when I saw the film he codirected, *All Day & All Night: Memories from
Beale Street Musicians*. Not long after that he started showing me his Memphis. We went to the Antenna Club. We went to the Paradise. We went to
Green's Lounge. And we went to Riverside Park. That was where I disappointed Robert. We went out there because that was where Elvis and his
friends hung out as teenagers. I wanted to find the location of the pavilion area
and Rocky's Lakeside refreshment stand. Searching for someone who might
remember, we went into the golf course clubhouse. Not only did we find
someone who knew: we met Rocky's wife. But I was too embarrassed to reveal myself. I don't think Robert ever looked at me the same way again: *he*
knew that a good reporter doesn't get embarrassed.

Robert proves to be not only a good reporter but a wonderful raconteur and host in the pages of this book. He introduces me to people I have never met and to people I have met many times—to my equal edification. He presents the spirit of Memphis—an equal parts mix of genius, orneriness, and individuation—without trying to bottle it. He tells a good story—he tells *a lot of good stories*—without ever abandoning the free-flowing, Shandyan form of digressive discourse that seems so well-suited to Memphis. He captures Memphis, and he captures an era, for the very reason that he does not seek to capture it. As Randall Lyon says, of a movement that would be rightfully insulted if you called it a movement: "We had what you would call in rhetoric *eroico furore,* poetic furor. We were inspired, we were in a frenzy. . . . I always figured that was the best thing that could happen to you, to be caught up with a group of people with enthusiasm for what they're doing. And how it's received was beyond our consideration."

I've had a lot of fun with Robert. And I expect to have a lot more fun. But I've never had more fun than reading the pages of this remarkable book.

—*Peter Guralnick*

Acknowledgments

I WROTE THIS BOOK WITH THE LOVE OF MY PARENTS, WHOSE DEVOTION and support has shaped my life. With the love of my brother. Of Odessa Redmond, a lifelong inspiration.

Memphis: Jim Dickinson and his family. Dickinson prefers the dark corners, and there he is a light. Mud Boy and the Neutrons: Sid Selvidge, Lee Baker, Jimmy Crosthwait, Dickinson. Belinda Killough, wittiest transcriber in the west. Adam Feibelman, diligent research. Susan Thomas. The history staff, the arts and music staff at the main library. All of those kind enough to let me interview them, especially those whom I did not quote. And everyone else who made the music.

The writers who sparked my interest: Walter Dawson, Stanley Booth, Robert Palmer, Greil Marcus, and writer Peter Guralnick, who was instrumental in bringing this book to light. Dick McDonough, stalwart agent. Betsy Uhrig, encouraging editor. Rob Bowman, Stax man. Stu Abraham.

I imposed this manuscript on several opinionated readers, and I am indebted to each of them: Joy Tremewan, Bobby Caudle Rogers, Joe Purdy and Donna Rifkin, my parents, and RJ Smith, an astute and insightful editor and a good friend.

John Fante, in *Full of Life*, must have been thinking of Tara McAdams when he wrote: "My prose, such as it was, derived from her. For I was always quitting the craft, hating it, despairing, crumpling paper and throwing it across the room. But she could forage through the discarded stuff and come up with things, and I never really knew when I was good, I thought every line I ever wrote was no better than ordinary."

xi

CHAPTER ONE

The Dream of a Common Language

THE ROLLING STONES INTRODUCED ME NOT TO THE BLUES, BUT TO THE bluesmen. The players. On a sweltering Fourth of July, 1975, the summer before I entered ninth grade, they delayed their Memphis performance by placing a wooden stool at stage center and then bringing out a fragile black gentleman with a guitar. The crowd of 50,000 was hot and impatient, but Furry Lewis came up playing medicine shows in the 1920s and he knew more than a little about entertaining. Though solo blues wasn't what a lot of weary rednecks wanted to hear, I'm sure I was not the only new fan he won.

The next time I saw Furry, the crowd numbered less than fifty. At the end of tenth grade, an upperclassman brought him to school during lunch. He sat on a porch near a parking lot and played for a small gathering. A hat was passed. I asked how the performance had been arranged and was given Furry Lewis's phone number. So within two years of being one in 50,000 to see him at a Rolling Stones concert, I did exactly what Mick Jagger and Keith Richards had done—I phoned Furry Lewis. He invited this unknown voice to visit him, accepting my offer to bring whiskey. His brand was Ten High.

Within a year, the phone list on the linen closet door of my parent's house included the names of guitarists, piano players, Sonny "Harmonica" Blake, Saxman Brad's business card, and various schoolmates who liked to kick a soccer ball. These phone numbers were not trophies, though having them made me feel cool. I was a gangly suburban teenager, middle-class, the braces finally removed from my teeth. My neighborhood was like a thousand others across the country. The blues musicians were giving me a geographical and historical grounding in Memphis. Their lives were the product of this particular place.

At the end of eleventh grade, some friends and I pitched in to bring the piano player Mose Vinson for a lunchtime performance. When Mose rests his hand on a table, his fingers look like rows of a furrowed field. I've since learned he was the janitor for Sam Phillips at Sun Records during the pre-Elvis days, and several of his previously unreleased recordings appeared on the *Sun Blues Box*. In 1977, Vinson was a regular at a bar called Birth of the Blues. So were we, getting drunk on Billy Beer and leading parades around the club with salt shakers as castanets. Furry also played there, and the owner booked a handful of other local giants. I think the club was open less than a year.

On the morning when Mose was to perform at school, he arrived late and drunk. Our friend who'd picked him up found him half-lit and had to coax him into the car with more beer. The quick refurbishment during the ride had produced a head rush in the elderly man: He entered the auditorium stagger-ing and slobbering. I think the end-of-lunch bell rang as his hands reached for the first notes. He could not form words. When he tried to speak or sing, he emitted deep-throated moans and grunts. Drool accumulated around his fingers on the keys. I remember the reaction of a squat senior from a rich fam-ily, a kid who could make a difference: He cackled loudly.

I still hate that kid. Perhaps because in his action I saw a part of me, that sense of detachment. In Mose Vinson's talent, I was finding meaning in this particular place on Earth, a meaning that also encompassed this student, the son of a cotton baron. The very cotton bolls which formed Mose Vinson's piano style paid this ingrate's private school tuition. The lack of respect in his life's breath exposed the disparities upon which Memphis is founded. This kid would inherit his father's civic influence and power, and the city would remain divided as his family sees it, not between rich and poor, but between whites and "niggers."

The evil behind that word lives and breathes in Memphis. The city was built on that word. Rock and roll is a response to that word. Rock and roll re-jected the idea of enforced segregation, mixing cultures as it mixed musical genres. On the streets today, the populations mix, but it's a surface politeness, a charming civic trait. Oppression is not unique to Memphis, though it is neatly encapsulated here. It's the sort of environment where great art develops in obscurity. The ideas are strong because, like weeds growing in a concrete sidewalk, they must force themselves through.

Concrete sidewalks have proliferated in these times of urban sprawl. Walk-ing out the front door to a landscape that could be anywhere has taken a new meaning since the microchip met the fiber-optic cable: Walking out the front door is no longer necessary. Today, particulars everywhere are made generalities. There is as much Cajun cooking in a Long Island fast-food joint as there is Americana at Euro-Disney. "Authenticity" is mass-produced.

This age of access, however, has not erased history and cannot completely remove an area's innate characteristics. Natural light in California is conducive to filmmaking, hot peppers grow next to fish ponds in southern Louisiana, cattle and cowboys come from the Midwest because the prairies are there. If aerial photographs could reveal energy the way infrared photographs reveal heat, Memphis would be surrounded by vectors pointing toward it: This is the place.

Memphis was founded on a Mississippi River bluff, safe from the flooding which defines the Delta south of it. Before clothing the world in cotton, the region's fertility fed Native Americans. Sun Studio, the site where black and white cultures merged as rock and roll, stands on what was a river trail heavily traveled by the Chickasaw Indians. Sun's current proprietor has the receipts from T-shirt sales to prove that people will always pass by his door.

Memphis is the capital of the large rural region that surrounds it. You can drive two hundred miles in any direction before hitting another city of size. There are small towns, and smaller ones. The Ozark Mountains are to the west; the distant Appalachian range cascades eastward from across the state, flattening into farmland before finally spilling into Memphis and the river; the Mississippi Delta sprawls south in the shape of a chicken leg, and the conversion of crops to cash has always taken place in Memphis. As a natural crossroads, the city has been influenced by many cultures, but its insulation has deterred European sophistication.

Since its founding in 1819, Memphis has been a place for innovation. Among its contributions are such ubiquitous concepts as the supermarket (Piggly Wiggly, 1916), drive-in restaurants (Fortune's, 1906), motel chains (Holiday Inn, 1952), and efficient overnight package delivery (Federal Express, 1972). Recording music is another part of Memphis's entrepreneurial spirit. The audio recording process was successful here even before the equipment was locally available. Field recordings were made of fife and drum music, work songs, field hollers, and other African expressions that mutated in the Delta. Once facilities in Memphis were available and flourishing, the artists traveled instead of the equipment. Sam Phillips gambled his cozy job recording radio transcriptions of big bands for a shot with an independent label and a new kind of music. He recognized the business of music, and his maverick attitude pointed the industry in a new direction. Oldies radio, alternative rock, and the other stops on today's dial remain a response to or re-working of the ideas he assembled under the aegis of Sun Records.

In Memphis, the studios generated human cultural collisions, not just the inanimate interactions between the listener and circular vinyl spinning at seventy-eight rotations per minute. The initial area recordings were the fiber optics of their time, enabling people to experience another culture without leaving home to do it. Though Delta blues could be imitated by anybody anywhere who heard a recording, Delta bluesmen could be imitated only by those with whom they interacted. They defined regionality, the product of a distinct place.

The blues is a sophisticated music. The Delta musicians created art that was fully realized, that when assessed needs no handicap or critical crutch. As Western scholarship has explored broader horizons, reckoning with the subjectivity and imperialistic attitudes that distorted previous investigations, it has recognized the complexity of expressions once thought "primitive," recognized the traditions and heritage that produced the blues. Unlike other immigrants, when Africans came to these shores they were not permitted to preserve their culture in the new land. Africans underwent a forced transformation. Slave-owners imposed the breakup of families, the mixing of tribes, the acceptance of Christianity. What was produced was something new. Rather than a sterile hybrid, a vibrant, vital culture emerged. Memphis has enslaved this culture; Memphis has nurtured it.

Co-opt, preempt, recycle. I first heard that description of popular culture from a Memphis musician named Jim Dickinson. Like a sponge, pop culture can absorb anything, defying the context of whatever it takes and making it part of the here and now. That's fine and dandy for the pop scene, but it's not necessarily good for what's being absorbed. As pop music, rock and roll has co-opted blues, gospel, and country, preempted the original artists, and recycled their techniques and ideas. We all have a story of learning that our favorite song by the Rolling Stones, Rod Stewart, or Michael Bolton did not originate with these artists; likewise, we know that Arthur "Big Boy" Crudup languished in poverty while Elvis Presley got rich singing his songs. The "original artists" are crassly exploited. Diluted imitators reap fame and fortune, while the preempted musicians receive neither, nor even acknowledgment. The popularizers are not legally obligated to pay homage to their predecessors, only royalties, and often they avoid paying even those. (Chris Strachwiz, of the Arhoolie label, once suggested implementing a Miranda act for musicians, reading bluesmen their rights before they sign a contract.) Even right-minded, moral "disciples" who have tried to remunerate their predecessors have found their way thwarted by thieves calling themselves publishers who wield shady contracts that allow them to divert money. The pop industry, of which the music industry is only a part, is founded on concepts of exploitation and greed. Recycling ideas can be both a tribute and a sham.

Pop culture is novelty-hungry, and the cultural divides between the races are a quick source for new trends. When the rock and roll sound was pioneered by black artists like Ike Turner and Roy Brown, both black and white audiences perceived it as a part of rhythm and blues; there was no novelty in blacks revving up R&B. White imitations of it, however, were freakish. Whites were unable to exactly mimic black music, and their failure created another hybrid. People of all colors gawked. This interaction is really what's being discussed when people ask the question, Can whites play the blues? That phrasing misses the point. What's meant is, What do we call it when whites try to play the blues? As a definition for rock and roll, I suggest: Rock and roll was white rednecks trying to play black music. Their country music background hampered them and they couldn't do it. That's why we don't call what they made rhythm and blues.

In the 1950s, with Elvis as an icon, white audiences were ready for new artists like Little Richard and established artists like Ike Turner, whom they'd previously missed. Though rock and roll now sells everything from hamburgers to presidential candidates, white society did not readily embrace such interracial, intercultural concepts. Segregation was still the law of the land in the 1950s, and anyone who respected black culture was given the same treatment as blacks: second-rate. Only when white eyes witnessed blacks laying down their lives for their country in World War II did some begin perceiving blacks as their allies. That slight opening of the door coincided with a push from the other side. Black witnesses to their brothers' deaths—deaths for a country that enforced apartheid—moved their community to rebel en masse: the Civil Rights movement and desegregation. Despite the passage of laws and the enforcement of various race-mixing programs, this conflict is still being resolved today. Welfare, substandard housing and education, prejudice from the bank's loan desk—the violence that is a response to this covert domination is a testament to the chasm that still runs beneath our society.

This same lack of understanding between the races is responsible for the innovation of rock and roll. Most of the machinery for recording and manufacturing was owned by the whites, and when they got in the studio with blacks, a bridge had to be established. An example of the cultural collision is cited by the aforementioned Jim Dickinson. "There's a box set of Little Richard outtakes that's out on CD, with Earl Palmer on the drums. Brilliant drummer, one of the most influential in early rock music. You hear the first take of, I think, 'Lucille,' and they run it down several times till they get the master. Lee Allen, the sax player, plays the same solo from the first cut to the last. But Earl Palmer starts out playing a shuffle! He's not playing the eighth note thing that became Little Richard's signature. He's shuffling. You also hear these insets of white voices on the talkback, and one of 'em is Cosmo Matassa, the engineer, and the other one must be the producer Art Rupe. And Rupe is saying the most insensitive, typically white things. But those had to be said in order to make the shuffle into what we now know as rock and roll. The racial collision, it has to be there."

The forces of cultural collision struck thrice in the Memphis area, first with the Delta blues, then with Sun, then Stax. These sounds touched the soul of society; unlike passing fads, these sounds have remained with us. By definition, most of popular culture is disposable, but Memphis music has refused to disappear. In electrified civilization, even when stripped of the particular racial and social context in which it was born, what happened in Memphis remains the soundtrack to cultural liberation.

Jim Dickinson has another saying that goes something like this: The best songs don't get recorded, the best recordings don't get released, and the best releases don't get played. It's the antithesis to corporate music mentality, and it also explains why Memphis is so full of treasures. Though no city has had more of a lasting impact on modern culture, Memphis has never been a company town. The forces have all been independent, renegade. Dickinson's maxim defends obscurity by attacking popular culture's drive toward mediocrity. Reaching the most people through the lowest common denominator denigrates individuality, destroys artistry. There is no reason that every song has to be a hit, but there's every reason for the song to *be*.

In 1978, a depressed girl I was dating recommended an album to me, *Big Star 3rd*. It so happened that Big Star was from Memphis, though I was not familiar with them or their first two albums. *3rd* had been recorded in 1974 and languished on a shelf for four years after being roundly rejected by record companies. Everything about it was mysterious. The company that released it was so small that the label on one side of the vinyl listed all of the songs and the label on the other side was a generic design. Yet a major name like Steve Cropper, playing guitar, was printed right there on the back. I recognized a couple of the other musicians, most from Memphis and among them two of my favorites, Lee Baker and Jim Dickinson. The record came with an extensive essay full of references I didn't know. When I played the album, it was un-like any I'd ever heard. One side began with backward-sounding strings and the other with something like cartoon music. Neither opening song sounded like the beginning of an album. Entering *Big Star 3rd* was like entering a movie after it's begun.

Which, in a way, it was. The band leader was Alex Chilton, who had found fame eleven years earlier at sixteen when he first entered a studio and recorded "The Letter" with the Box Tops. By the time of *3rd*'s release, his interest had moved to other types of music. *3rd*, I came to understand, was a response to his career to date: the commercial success and artistic frustration of the Box Tops, the artistic success and commercial frustration of Big Star. It was a record of introspection. The darkness of the music was immediately gripping, and the elusiveness of the lyrics encouraged repeated listenings. It was almost three years—I was by then unhappy in college—before I discovered the lines, "Get me out of here/I hate it here/Get me out of here." That I cannot readily name the song is indicative of the record's lasting beauty.

Since 1978, *Big Star 3rd* has been rereleased at least twice. (As per Dickin-son's maxim, one of the best songs, "Dream Lover," was not included until the second issue.) The band's other two albums, widely praised, poorly dis-tributed, and long out of print, have also been made newly available. The influence of Alex Chilton and this once-overlooked group has become so widespread that Big Star practically defines a category in modern rock. *3rd*'s sound may have been a generation removed from cultural collision, but its re-sult reverberated with the bluesmen: obscurity. The bluesmen did not stop making music after the 1920s and 1930s; they were just no longer recorded. Their material did not hinge on the critical acclaim they may have briefly en-joyed. Rather, it was an extension of their being, and if a record man was will-ing to part with a ten-dollar bill to hear them do their thing, that was fine. But if not, it didn't stop the music. With their lives and not just their words the blues artists had taught those following them to trust their ideas. The audi-ence's response becomes secondary.

The rediscovery of the Delta blues artists began in the later 1950s, shortly after the introduction of Elvis and rock and roll. The first rock and roll audi-ence was also the first blues renaissance audience, and those listening—the witnesses—bore the dual responsibilities of keeping an old tradition alive and of creating a new genre. By the middle and late sixties, the audience had so ex-panded and the industry become so secure that, although less than ten years earlier the witnesses couldn't imagine a career in music, by Chilton's genera-tion it was every kid's dream. The difference, in a word, was the Beatles.

For me, the Beatles were a Saturday morning TV cartoon long before I appreciated their social impact. They allowed my generation to take for granted the possibility of a career in rock music, or even rock music journalism. In my rock and roll youth, the music was losing its social meaning and becoming a service industry, becoming, in fact, a cartoon. But in Memphis, I'd felt the bluesmen's power.

In May 1978, when my sense of place was solidifying and I was distressed that such great music as Furry Lewis and *Big Star 3rd* seemed available only to locals, I was exposed to a Memphis band called Mud Boy and the Neutrons. Rock and roll witnesses all, their music was the missing link between the Rolling Stones and Furry Lewis. Mud Boy and the Neutrons were four white guys who fused the washboard, the electric guitar, and field hollers; where the Stones had come up emulating blues records, Mud Boy emulated bluesmen. I saw them perform as part of a two-day music festival honoring the city's heritage. Gospel music, rock and roll, blues, and country all shared the same stage. Part of the event's pleasure was experiencing the common spirit in such diverse music.

On the main stage, a Delta blues group was winding up their set. Alex Chilton was in the artist's tent and, though not scheduled to play, he'd been inspired. The event was loose enough to allot him some time, and several members of Mud Boy joined him. Their impromptu set, including a menacing run-through of Chilton's hit, "The Letter," introduced punk rock to Memphis at large. Before Mud Boy began, a member of their entourage, Guru Biloxi, came out carrying a spear and ranting, pushing the energy cautiously high. Wound up tight like a heart attack, stomping the stage, Chilton introduced the band—"some very good friends, some very close personal friends of mine," he shouted like a man about to pull the trigger—and Mud Boy kicked into Chuck Berry's "Little Queenie." The four core members were backed by a drummer and a bassist; three dancing girls were part of the act.

"Dancing" doesn't convey what I saw these women do. I was seventeen and so drunk on cheap white wine that I had to put my hand over one eye to rightly hear what was going on. The music was rumbling the way a house shakes when a heavy truck passes out front. There was a sense to the chaos, a sense not so much based on beat or rhythm, though this music was as full of both as any could be without exploding—but a sense based on emotion. The women were dancing, yeah, they were dancing all right. They were fucking the music. They were slithering up and down that rumble. This was fuck music. Not the wet sensuality of Al Green, not the sultry innuendo of the early blues queens. This was the guttural howl of the bump and grind, the madness of urge, the flaunting of that which we've been taught to repress. The power of the blues—the violence, the energy, the sex—was laid bare.

The plug was pulled on their performance. The band was one song into what was the set of a lifetime when the authorities decided it was too much truth for the public good. Johnny Woods and Prince Gabe had each performed that day on the same stage, and they had told the truth. Grandma Dixie Davis would perform parlor piano later on that same stage, and she would tell the truth. Phineas Newborn Jr.—God rest his weary soul—would speak the gospel; B. B. King, Big Sam, Carla Thomas. But only Mud Boy's truth was censored.

The band revolted. There was a shouting match, a sit-in. The audience responded with raised fists. Oho Mick Jagger, oho Johnny Rotten, the real shit went down that day. Rock and roll busted loose from its chains and wasn't a commodity to place between radio commercials or at the top of charts. Music came back to life that afternoon with all the energy of Elvis Presley's 1954 hips.

Mud Boy, for the most part, revels in their obscurity. They continue to perform occasionally, and in 1993, seven years after their first album, twenty-one years after their inception, they even graced their audience with a second record. Naturally, it's on a small French label and difficult to find in the United States. Personally, that no longer frustrates me. If people need to find this band, they will. Those who have continue to come out of the woodwork when they perform, responding to the tribal shamans who call on our behalf to the spirit voice in the woods.

Memphis music is an approach to life, defined by geography, dignified by the bluesmen. This is the big city surrounded by farmland, where snug businessmen gamble on the labor of fieldhands, widening the gap between them, testing the uneasy alliance. Memphis has always been a place where cultures came together to have a wreck: black and white, rural and urban, poor and rich. The music in Memphis is more than a soundtrack to these confrontations. It is the document of it. To misquote W. C. Handy's "Beale Street Blues," If the Mississippi River could talk, a lot of great folks would have to get up and walk.

One summer day, before I entered the twelfth grade, I spotted a couple of tourists downtown. Bullish on Memphis, I stopped to offer assistance. They were French, in town only for the day, and expressed an interest in Memphis music. They got in my car and, though it was my custom to phone first, we showed up unannounced at Furry Lewis's rundown duplex. Over the past year I'd visited him regularly enough that he and his lady friends recognized me when I appeared. This was 1978. Furry was near eighty. I was seventeen. Furry was black. I was white. The French tourists were wearing short pants.

There were hellos all around as we were welcomed, and, once seated, the question was raised about Furry playing the guitar. Seems like he might could play the tourists a song, uh-huh. Did my friends drink whiskey, he wanted to know. I was translating. I didn't speak French, but their schooling hadn't prepared them for Furry's accent. One tourist asked for water, but then decided he'd drink with us. There was a liquor store around the corner, and a friend of Furry's volunteered to get the bottle for us, how much did we want, a half gallon? It was about four in the afternoon. Furry had his guitar out, tuning. Beautiful. I recently found a cassette recording of all this. Thinking only of myself, and only of the present, a half gallon seemed a bit much. Don't try to take *me*. When we sent the friend off for our whiskey with just two dollars, enough for a half pint, barely a swallow all the way around, Furry put the guitar down and said, "The rheumatism got me this afternoon, I can't get myself together."

French wasn't the only language I was learning that day.

CHAPTER TWO

Tell 'Em Phillips Sentcha

WHEN NIGHT SETTLED ON THE TOWN IN 1949, MEMPHIS, DESPITE ITS big-city aspirations, was as quiet as a distant country crossroads. Citizens sighed in the glow of their American happiness. In November of that year, after months of test patterns, Memphis's first TV station began filling homes with the warmth that radiated from their newfangled sets. "We the People," "Circus Animals," and the puppet show "Kukla, Fran & Ollie" all confirmed that life after World War II was good.

In the older medium of radio, 1948 had been a watershed year. In late October, the city's sixth station, WDIA, confronted the audience's lack of interest in yet another place on the dial playing country, pop, and light classical. With nothing to lose but their failing year-old operation, owners Bert Ferguson and John Pepper enlisted Nat. D. Williams, a nationally syndicated black Memphis journalist, as host of a forty-five-minute afternoon show. The response was so overwhelming—including the mandatory bomb threats—that by the summer of 1949 WDIA was the first station in the United States with an entire cast of black disc jockeys.

The impact was enormous. The bullets of World War II had recognized no color, and the movement toward civil rights was fomented by the returning servicemen and their demands for equality. In an era of condoned, organized racism, WDIA became a community bulletin board, a public institution that celebrated instead of insulted 40 percent of Memphis's population.

"I remember when the black ambulances could not haul white people," says Gatemouth Moore, the station's first gospel programmer. "They had a white company, I'll never forget, called Thompson's. I was on my way to the station, and when I come around the curve there was the ambulance from

S. W. Quall's with the door open, and there was a white lady laying in the ditch, bleeding. And they were waiting for Thompson's to come and pick her up. Quall's couldn't pick her up. I guess I waited thirty or forty minutes and still no ambulance. They tell me that the lady died. So I came to WDIA and told the tale. I said, 'Look here.' I said, 'Black folks put their hands in your flour and make your bread, they cook the meat, they clean up your house, and here's this fine aristocratic white lady laying in the ditch bleeding and they won't let black hands pick her up and rush her to the hospital.' And the next week, they changed that law where a black ambulance could pick up anybody. I got that changed on WDIA."

In a few years, WDIA would be the most powerful station in Memphis, but the repercussions of its format were felt immediately. WHBQ, an older station also in financial straits, put economics before apartheid and, when "the mother station of Negroes" went off the air at sunset, WHBQ began broadcasting rhythm and blues. Not ready to hire a black personality, they relied instead on one of their dulcet-toned announcers, Gorden Lawhead. He named the program after a Broadway play, "Red, Hot & Blue," but it was none of those. Lawhead epitomized the white radio announcer of the era, as innocuous as its pop music: Perry Como's "Some Enchanted Evening," Evelyn Knight's "A Little Bird Told Me," Gene Autry's "Rudolph the Red-Nosed Reindeer." Lawhead neither understood nor appreciated R&B.

In the spacious night, after parents retreated to the soothing murmur of TV, the baby boomers found comfort in their radios. They kept the volume low enough not to attract attention, and the light from the dial fought off the total darkness. A little fiddling with the tuner brought in creatures from another dimension. Many of the local stations vanished with the sun, leaving chasms filled by voices from Mexico, from Nashville, from alien places that played alien music. Pop had a certain glide to it, but this music went *thump. Thump thump.* It was mysterious how far the sound traveled to reach beneath the cotton blankets.

The distance was twice what these future witnesses to the birth of rock and roll realized. Many of these records originated right in their own town but had to travel to distant radio stations to achieve their popularity. Howlin' Wolf. Rosco Gordon. Junior Parker's "Mystery Train." The music sounded crazy when border radio stations sent it through the reaches of dark night, but that paled beside the frightening live performances of the artists, just a few miles away and tanked on bad whiskey. Beale Street was in downtown Memphis, and it was the Mississippi Delta's largest plantation. It was where black people could relax, unencumbered by Jim Crow because few whites patronized Beale. Those who did were mostly landlords, and they liked to see a busy place.

Beale Street and the surrounding neighborhood was the mid-South's African-American commerce center, adjacent to downtown Memphis, the white commerce center. Beale was more compact and always hopping. This is the street where the zoot suit was created, where a single amateur night produced Rufus Thomas, B. B. King, Bobby Bland, and Johnny Ace, and the clubs launched Howlin' Wolf, Hank Crawford, and Phineas Newborn Jr. "Wide open" is the term usually applied to Beale.

"Gambling, drinking, policy shaking—had a joint on every corner," recalls pianist Booker T. Laury, born in 1914. "They had a restaurant in the front, you get hot dog and fish sandwiches, and ladies in the next room had a little place set aside for a dance hall. Every crap house had a dance hall. On back a little further, they had a dice table, and the men leave the women up there to be entertained whilst I'd play the blues to 'em. The men would be back there shooting a few craps. Every day, that was the routine. The doors didn't close. They stayed open all night. Changed shifts from twelve to twelve."

The multitude of clubs on Beale Street established the thriving music scene in the city and, by attracting and nurturing regional talent, was directly responsible for both Sun and Stax Records. The core of venues on Beale spawned other clubs around town and also across the river in West Memphis, Arkansas. There was plenty of work for a large number of bands, orchestras, and soloists, and the constant flow of people in and out of Memphis assured an audience. The variety of styles, the opportunities to mix them together, and the plenitude of venues helped forge the groundwork for a musical aura in Memphis that remains to this day.

In 1949, eleven years after Orson Welles's "War of the Worlds" demonstrated the power of radio, another voice from outer space chewed up Memphis and spit it back in its ear. Channeling the same spirit that had lain dormant since tuning Robert Johnson's guitar at the crossroads, a white disc jockey interrupted the satiny WHBQ broadcast. "Dee-gaw!" the radio squawked, and it chilled the parents' bones because they heard something different and knew it meant change. "Dee-gaw!" the detached voice drawled, and the kids leaned closer to the speaker. They heard a jumble of words that was like Captain Midnight's code; you had to listen closely to keep up. Dewey Phillips wasn't coming in for a landing; he was taking off. The excitement was intensified by this alien's proximity. Broadcasting from right downtown "on the magazine—uh, mezzanine floor of the Chisca Hotel," he was no farther away than where a visiting relative or a father's war buddy might stay.

Elvis Presley and Dewey Phillips at WHBQ radio, circa 1956. Photo courtesy of Jim Dickinson.

Thump. Thump thump.

Daddy-O Dewey. He is best known as the first disc jockey to play Elvis Presley, but the legacy of Dewey Phillips is every attempt by a white Memphis kid to play black music, from the first generation of rock and roll right through Stax Records. His listeners learned not to distinguish between races or genres. He demonstrated that the boundaries of "normal" were arbitrary and heralded a freedom that society shunned. Many took heart in the realization that they might be able, like Dewey, to parlay their own particular weirdness, oddity, or eccentricity into a career. Nowhere else in society was such nonconformist thought publicly condoned. It has taken forty years of corporate rock and roll to rebuild the walls Dewey Phillips broke down.

The very fact that Dewey got on the air indicates the force of his character. He was everything that a deejay in 1949 was not. He had been spinning records in the phonograph department at W. T. Grant's in downtown Memphis, howling over the store's intercom and causing a roo-kus. Rocking and rolling. People, including Sam Phillips (no relation), would come in just to listen to his mad ramblings. Management found that Dewey was unmanageable, but they couldn't argue with the crowds he drew. Something about whatever it was he was doing worked.

Grant's was on Main Street, near WHBQ's Gayoso Hotel location. Lawhead and the other radio announcers were more than familiar with Dewey, who accosted them regularly, excited and impassioned, stepping on their feet, sputtering while he pushed whatever new release had caught his ear—and, he was sure, would catch ears all over Memphis if one of the deejays would just play it on the radio. And kindly plug his department at Grant's.

"Dewey was not physically well organized," says Lawhead. But he got what he wanted. It's said that Dewey, yearning for an on-air time slot, visited "Red, Hot & Blue" one night and, pained by how wrong Lawhead was for the job, stepped into the hallway beyond the announcer's sight and set a trash can aflame. When Lawhead ran for the fire extinguisher, Dewey jumped for the microphone. Callers phoned in their support. Lawhead says, "I thought, God, has radio come to this? That we're putting this character Dewey on the air?"

Dewey began with forty-five minutes but soon commanded a three-hour show, five nights a week: "Red, Hot & Blue," nine to midnight. Dewey always called it "The hottest thing in the country."

"As screwed up as he eventually got," says Jim Dickinson, a fan he inspired who would record with the Rolling Stones, "Dewey never lost that warm and almost loving feeling when he was broadcasting. You could tell it was an act of communion between you and this crazy guy. And that he was really playing this music for a purpose, unlike all the other insincere bastards at the time."

For a decade, Dewey reigned supreme. And supremely insane. According to Charles Raiteri, who has chronicled much of Dewey's life and produced an album of his radio shows (*Red Hot & Blue,* Zu-Zazz Records), "B. B. King called him Daddy-O. The [Howlin'] Wolf, to whom all whites were suspect, called him 'brother.' He played draw poker with Johnny Ace. And he and Ike Turner shared co-billing as talent scouts for Sam's Sun label." That's heavy company in an era when sipping cool water from the wrong fountain on a hot day could cost a life.

By 1954, when Sam Phillips cut his acetate of Elvis's first single, "That's All Right," Dewey was the natural test market. He had the kids' ears. The story goes that Dewey screened it over the air and liked it so much that, while playing it continuously—yakking his patter over it all the while—he phoned Elvis's mom to find the kid and bring him to the station. When he showed up, Dewey told him not to cuss. He surreptitiously opened the mike but acted like he was spinning a record, and casually began asking questions. Dewey first established what high school Elvis went to, indirectly informing the listeners that no matter how black the record sounded, the kid was a honky. When it was all done, Elvis asked, "Aren't you going to interview me, Mr. Phillips?"

A white kid sounding black was perfect for Sam Phillips, and perfect for Dewey Phillips. Sam knew he could market it; Dewey liked the confusion. While other deejays kept to a mellifluous format, he made car wrecks of genres, defying the pop charts by whimsically counting down his own top ten, Dewey's Top Ten, the hottest ten records in Dewey country at this very moment in time—and changing fast.

"It was years before I figured out that this stuff I heard on Dewey Phillips's show wasn't popular music," says Dickinson. "I certainly didn't realize that he was playing things that nobody else played. Like 'Red Hot' by Billy Lee Riley —I didn't realize that wasn't a hit until I moved to Texas for college. He'd play Little Richard, then he'd play Sister Rosetta Tharpe. He'd play a country song, he'd play a rock song, he'd play a blues song. And the mindset I learned, listening to that music, is what has enabled me to make a living in the music business."

Indeed, the Dewey Phillips mindset, a reflection of the city's geographic and economic crossroads, defined Memphis music for years to come. He could propel a song like Carl Perkins's "Blue Suede Shoes" to national attention, where it became the first song to simultaneously top the pop, country, and R&B charts. His fans—and almost without exception every Memphis musician raised during his era was a fan—knew no boundaries. Do not tune in now for three hours of light classical; do not tune in now for three hours of down-home blues; do not tune in now with expectations of any sort whatsoever, because never has so untamed a person had so much broadcast power behind him, and whatever happens in the next three hours will be completely different from whatever you heard last week and bear little or no similarity to what you will hear next week, podnuh podnuh.

"I was fourteen and I didn't realize that Dewey wasn't being heard all over America," says Don Nix, who, along with his schoolmates, would capture the Dewey energy in the Mar-Keys, a band of white boys who broke racial barriers playing the black circuit in 1961. "I thought everybody in every town had a disc jockey that in one night, three hours, you could hear anything you wanted to hear. I wasn't allowed to listen to the radio at night, so my brother and I used to sneak the radio under the covers and listen after my parents went to bed. He played Little Walter and Johnny Ace, but also Frank Sinatra, Nat King Cole, and Jimmy Reed. I listened to WDIA too, but not as much, because WDIA played only one kind of music. Dewey played it all."

Dewey played what he liked, and if he liked something a lot, he'd repeat it. "One of the things I remember most, and anybody who listened to Dewey back then will tell you, was a record called 'Tell Me Why You Like Roosevelt' by a gospel guy named Otis Jackson," says Milton Pond, a longtime Memphis record retailer and collector. "It was never on the charts anywhere, but Dewey made it a hit. He liked it and played it on his program every night." Dewey made hits of such contrasting songs as a gospel number called "Down on My Knees" and an R&B song called "Drunk." When Dewey said, "It's a hit," there was no other authority to defy him.

Future Sun musician Randy Haspel remembers Dewey "unashamedly" playing "Heartbreak Hotel" twenty times in a row. Conversely, as record producer Jim Blake adds, what Dewey didn't like, he wouldn't play. "He'd pick up records in the middle, screech 'em off, 'Well that ain't gonna get it.'"

What got it was that the kids—and anybody else who dared to listen—could hear a real person playing real records and could pick up on the enthusiasm. "Nobody knew what Dewey looked like," continues Blake. "We imagined he looked like a fucking Martian. And when he got his TV show [in late 1956] and we saw him, he *was* a Martian. Dewey was hip, he was beat, he was everything all at the same time."

"I didn't listen to black music but I listened to Dewey," says Roland Janes, the legendary Sun guitarist. "He played all kinds of records, but you listened to Dewey as much for Dewey as for the music he played."

Dewey's patter was integral to his show. An offhand phrase one night would become street lingo the next day. He tumbled through a roster of characters, drawing from popular Red Skelton sketches and from people he met on Beale Street. He'd do poor imitations of Tennessee Ernie Ford or Dizzy Dean and fumble with crude sound effects. He'd whip his head from position to position carrying on conversations with himself, each personality occupying a particular place in his mind. The attentive listener heard a song somewhere between the one-man dialogue. This is a straight transcription, Dewey doing all the voices, and beneath him some poor vocalist fighting to be heard: "Ain't that right, Diz? That's right. Did you call Sam, podnuh? No, I gotta call Sam, Diz. [Screeching:] Hi, Phillips, how you, Phillips? How you getting along, Lucy Mae, what's the matter w'you? I'm looking for my husband, Phillips. I ain't seen your husband. You'd better call Sam."

"Mostly what I remember about Dewey," says Jimmy Crosthwait, Memphis's premier white washboard player, "was his beer ads. Falstaff was a sponsor, and he'd say, 'If you can't drink it, freeze it and eat it. Open up a rib and pour it in.'"

Another of Dewey's sponsors was Poplar Tunes, a record store on Poplar Avenue run by Joe Cuoghi, John Novarese, and Frank Berretta. Poplar Tunes's early success allowed the owners to branch out, with Cuoghi founding Hi Records—home to Ace Cannon and Bill Black, and later, under the guidance of Willie Mitchell, home to Al Green; Novarese established a jukebox agency; and Berretta ran the store, which sold retail to the public, wholesale to jukebox operators, and served smaller stores as a one-stop (the middle man who carried recordings from all the various distributors). Cuoghi was a private man, and much to his chagrin, Dewey loved to say his name on the air, embellishing it as far as he could stretch it. "Go on down to Poplar Tunes, get you a wheelbarrow full of—full of—[searching] full of mad dogs, see Papa Joe-Joe Da-Coogie [dramatic pause before yelling] and tell 'em Phillips sentcha."

Milton Pond thought he was the luckiest guy in the world to get a job at Poplar Tunes when he was seventeen. He remembers: "Dewey had made Joe Cuoghi legendary, a household name. But Joe Cuoghi didn't want to be famous. He used to hate whenever Dewey would come in the store because he created such chaos. Dewey'd get back behind the counter and handle records, put 'em on the turntable, didn't care if there was two or two hundred people in the store. He'd say, 'Joe! Come here, buddy boy, I want you to hear this. Hottest record in the country!' He'd crank the volume up, Joe would say, 'Turn that goddamn shit down, there's customers in here.'" "Every time Dewey was in Poplar Tunes," adds Jim Blake, "Cuoghi's asshole would clamp up and he'd chomp down a little harder on that cigar."

"Dewey was great to introduce you to stuff," says the Memphis painter Charlie Miller, "but he'd talk all through the songs. I really wanted to listen to the music, so I liked WDIA better." Many of Dewey's coworkers never adjusted to his style. "The guys at HBQ didn't like Dewey," Pond says. "He did more by accident than a lot of those guys did on purpose. That's what really bothered them. They said, 'Look at this guy, he's a goddamn lush, he's a pill-head, he doesn't know what the fuck he's doing. Yet people love him and he's breaking records. How can Dewey do this and we can't?'"

Dewey was a pillhead, an addiction resulting from injuries suffered in a couple of car wrecks. But when he was successful, not even pain pills could slow him down. Riding high in late 1956, Dewey began an afternoon simulcast on radio and TV. "Phillips' Pop Shop" aired daily from 3:30 to 4:30, and the soda fountains didn't know what hit them. Instead of hanging out there, the kids raced home after school to catch Dewey's act on TV. It's an exaggeration to say that Dewey's show made Ernie Kovacs, the great madman of early 1950s television, look like a funeral, but not by much. Certainly Dewey was the rock and roll Kovacs, whether sticking his face in the camera or walking behind it to bust the fourth wall. He was unscripted, unplanned, untethered, and completely live on two media. His breakaway times for each were not simultaneous, and he loved the confusion of not knowing which audience he was addressing. He kept his nighttime radio slot throughout his TV reign, and though it would have been easier, "Pop Shop's" director says they never considered losing the radio simulcast. Dewey enjoyed the pay and the power.

"Pop Shop" aired before there was a concept for playing records on television. The national broadcast of "American Bandstand" was not until the late summer of 1957, and making short films to illustrate the music was decades away from popularity. Dewey's show originated from the same studio as local TV wrestling, and it drew from the same muse. The magic of rock and roll was capturing the manic moment, and instead of music television, or phonograph or lip-synch television, Dewey created rock and roll television. He let the music inspire him, and the show's object was to seize the inspiration.

"Dewey Phillips on television was one of the strangest things I think I've ever seen, anywhere," says Memphis recording executive John Fry. "They would make technical operations very apparent, violating the cardinal rule of broadcasting. They spent half the time on television dragging the cameras out in front of the audience. And Dewey would carry on arguments with the guy at the radio station about when they were going to do a commercial. They would argue back and forth on the air. It was crazy."

The dynamics of the TV show grew from Dewey's relationship with his assistant, Harry Fritzius. By the time "Pop Shop" debuted, Dewey was thirty years old, a large man and somewhat avuncular. Fritzius was twenty-four, scrawny and vigorous. Both had instinctual timing and a flair for the absurd, and like many great partnerships, theirs fed on mutual disdain.

Fritzius had moved to Memphis from Blytheville, Arkansas, when he was eighteen to attend the Memphis Academy of Arts. The school had sought him, offering an unsolicited scholarship. At the end of his first term, serious Harry Fritzius won his class prize. Shortly thereafter, his first solo painting exhibition nearly sold out and drew a rave review from Memphis's daily newspaper, the *Commercial Appeal.* Several pieces were mentioned in the article, and one, *The Young Artist,* was reproduced. "*The Young Artist,*" the critic wrote, "was inspired by a quotation which is by the artist himself: 'I wander through deserted rooms where memories hang like tattered things upon the walls . . . and loneliness is always at my side.' Here, the gaunt, ashen-faced young artist sits alone under the dreadful glare of a bare light bulb in a deep brown room which is like a recess in the mind, and on the walls are the curled parchment-like scraps of paper, the tattered memories."

Hardly the makings for a manic sidekick.

Upon completion of his art degree, Fritzius took a job at WHBQ-TV, working as a set designer, floor director, and director. His innate ability and creativity were quickly recognized, making his transition onto "Pop Shop" easy. "We were all on camera," says Durelle Durham, the show's director. "We were required to fill an hour's time with video, and you couldn't just follow Dewey around. So it was up to the studio crew to make stuff up. We'd show engineers working on equipment, people climbing the light grid—anything to have something crazy going on while Dewey spun the record. Harry just carved himself a permanent slot in the show."

Serious Harry loosened up. Concrete would loosen after enough time around Dewey. "Whatever his medium," the *Commercial Appeal* said of Harry's second solo art show, "whatever his approach to his subject, he attacks the matter with complete originality." For "Pop Shop" Harry created "Harry," a character that Durham describes as Fritzius's alter ego, as distant from the artist's nature as could be imagined. "Harry" wore a lecherous trench coat, rubber boots, a hunter's hat, and a mask that was either a cockeyed ape or a caveman. This character did not speak; he just acted, pantomimed, and ran amok. To transform himself, Harry entered the dressing room alone; Durham remembers that he never took off the mask even in front of the studio crew.

"Harry" put Fritzius on Dewey's level. Crew members agree that Harry and Dewey did not particularly get along; Harry felt Dewey was beneath him. "Everybody felt superior to Dewey," says one crew member. "He was a coun-

Dewey Phillips and company. Left to right: Harry Fritzius, Dewey Phillips, rockabilly star Billy Lee Riley, Phillips's associate Claude Cockrell, Sam Phillips. Photo courtesy of Dot Phillips.

try boy that people liked. He had a mass appeal to the kids with his craziness, but you wouldn't ascribe much to his intelligence."

"Dewey played records, and Harry was the creative artist on the show," Durham says. "Harry was quite an actor, quite an improviser. You never knew what he was going to do. We interviewed people like Jane Russell and just every star that came to town would be a guest on Dewey's show. We had a starlet there one day and Harry got a chocolate pie and told her to hit him with it. As she was reaching to slam it into his face, he grabbed her wrist and smashed it right into her face. When he put that mask on, he became an entirely different person. It was just uncanny. He was two personalities, he was truly that personality that he was playacting.

"One of his routines was to open mail on the air. Harry would stomp on these packages, kick them around, and then see what he had smashed. One day there was a round tube, like a Quaker Oats container, that he beat up and then held up to the lens and pulled the top off. It was a hornet's nest with live hornets in it. When the cameramen fled, no one was sure they would come back. It turned out that Harry had sent it to himself. We all had cans of bug spray the rest of the week, and that became a running gag."

Another time, Harry took a cameraman up to the building's roof. At his instruction, the camera tilted up from him to the sky, then back down. Harry was gone. People watching at home figured he had jumped off the roof and someone called the fire department's rescue team. When they showed up un-expectedly, Harry put them on the air.

Harry Fritzius, tech crew, made appearances on the show as floor director. His hair was getting a little longer, he grew sideburns, and there was a glint of craziness in his eye. The edge of live TV, where there is no safety net, appealed to him. He also continued his work on the station's other shows. On Christmas Eve, WHBQ broadcast Christmas carols to the image of Harry Fritzius, not as madcap sidekick but as serious art student, painting a Madonna and child live in the studio. Over the course of a few hours, he created the classic image and when he was done, the station went off the air. Jim Dickinson, an ardent fan of "Pop Shop" ("I never saw anything funnier on television, period"), went to the studio to look at the painting. "On television, it was black and white and looked perfectly straight. At the studio, I saw that the faces were green, the hair was orange, there was purple in it—he'd Harry'ed it. And never cracked a smile."

When "American Bandstand" went national in August of 1957, ABC pressured WHBQ to run it in the after-school time slot occupied by Dewey. The affiliate was forced to defy the network. "Pop Shop" was too popular. The network feed still came into the station, and whenever Dewey or the show's director felt like incorporating a taste of "Bandstand," which had yet to debut in Memphis, they didn't hesitate to cut in. "They'd punch up the picture portion off the network and you'd see these kids in Philadelphia dancing," says John Fry. "You had no idea where it was coming from, and they were dancing to a different song so it was all out of time. For a period of months we'd see this bizarre live television picture coming from somewhere without having any idea what it was. He made pieces of 'American Bandstand' just another ingredient that he put into the stew."

"I remember watching 'Pop Shop' when they cut in Jerry Lee Lewis from 'Bandstand,'" says Dickinson. "I'm sure Dewey felt like he owned the Memphis artists and had the right to show it." He used the same proprietary air to justify stealing a test pressing from Elvis in California. The singer was working on *Jailhouse Rock,* his third film, when he paid Dewey's way to visit him. On the MGM lot, they met the actor Yul Brynner. "You a short mother, ain'tcha?" was Dewey's response, and his ticket home.

While there, he'd pilfered a copy of "Teddy Bear," still weeks away from release. Back on his TV show, Dewey boasted about his connections with the superstar and announced that he was going to give Memphis a preview of ol' Elvis. And, against all rules, he played the unreleased test pressing. Never one for understatement, he repeated the preview the next day, as he no doubt intended to do for as long as he was the only disc jockey with a copy. While Dewey was building up to spinning it, the costumed Harry appeared at his side. Harry looked down at the turntable, looked up at the camera, manipulated a quizzical look on his mask, and as Dewey's face turned to disbelief, Harry put the record in his mouth and chewed it into little pieces.

After six months of resistance, WHBQ finally yielded to ABC. On Monday, January 6, 1958, Dewey's show was rescheduled to midnight and renamed "Night Beat." "American Bandstand" assumed its "rightful" time. After a solid dose of Dewey Phillips, Dick Clark was completely unhip.

On Thursday, January 9, the fourth night of the new show, "Night Beat" was abruptly cancelled. "Harry" had gotten out of hand. The incident involved a life-size cutout of Jayne Mansfield that hung on the set behind the turntables. During the usual mayhem, the costumed Harry stood in front of it, his back to the camera, and, according to the evening *Press Scimitar*, "carried on some foolishness." In popular lore, his actions have become grander, more outrageous and suggestive. Director Durelle Durham was there: "He pinched her on the rear and turned his back to the camera, loosened his belt and his zipper, you've done it yourself a million times, and retucked his shirt. He did it with his back to the camera and that was construed as him unzipping himself and playing with that life-size cutout. But it was pressure from ABC that really put Dewey off the air. We all knew the midnight show didn't have the same spunk that it had in the afternoon. We were a very, very popular program."

An old movie ran the following night in Dewey's time slot. TV manager Bill Grumbles fired Fritzius and was quoted in the *Press Scimitar* as saying, "This has been the most miserable week I've spent in broadcasting." Of Harry, he said, "He is very talented, if he could just discipline his talent."

Fritzius could not be reached for comment, but a friend quoted him: "This is probably the best thing that ever happened to me. I'm twenty-five, and it's time I found something to do with my life." The article continues, "The same friend said: 'He can be a very nice person when he doesn't get into one of his moods, and then he lets the pixie part of him overrule his better judgment.'"

True to his word, Harry Fritzius left Memphis, to pursue his artistic career in New York. A play he wrote, *Summer Is a Game We Play,* won a competition and was performed off-Broadway. Then he moved to the West Coast, taking a minor role in the TV series "The Alaskans," and painting.

In the mid-1980s, when Charles Raiteri, who worked at WHBQ for nearly two decades, was seeking information on Harry for a screenplay about Phillips, everyone he encountered spoke of him with the highest respect. Raiteri says, "When rumors were floating around the station that Harry was gay, he called his crew together one night, took them down the street to the Normal Tea Room [named for the neighborhood], and explained to them all about what being gay was and that he was gay. And they accepted that and that was the end of it.

"Anybody who ever worked with Harry described him as a genius. They never talked about anybody else that way. They all thought he was the most talented guy that ever lived. When I was trying to find him, I heard rumors that he was preaching in Greenville, Mississippi, that he'd fallen off an oil barge, things like that. Eventually, a friend of mine found him in San Francisco. I had seen a review of an art show that Harry had there, and the tone of this review made him sound like a very respected, well-known artist. His stuff was also being shown in Europe. This friend said Harry was living in a great big loft, and he had these gigantic paintings all over the place, lying on the floor, everywhere. And he was drinking vodka from the bottle and smoking continuously, and he would finish a cigarette and flick it across the room on his paintings. At that time, this guy said that he didn't think Harry had much more than a year to live. He soon had a heart attack and died."

After "Pop Shop's" demise, WHBQ's only local music program was a Saturday afternoon show hosted by Wink Martindale, later of "Tic Tac Dough" fame. "Every kid resented Wink Martindale hosting 'Dance Party,'" explains Jim Blake, "because we remembered him on 'Space Patrol.' That had been a kiddie program where you sat in a rocket ship and Wink showed Flash Gordon serials. Seeing him trying to be hip didn't work because we all thought of him in his space suit."

Dewey began falling out of favor with radio when contrivances like structured playlists and rigorous formulas were introduced. Gearing itself more toward selling advertising than pleasing listeners, radio needed predictability and fixity. Dewey was nothing if not spontaneous, and unfortunately he became nothing.

In 1958, when Phillips and WHBQ parted ways, it was as if his life became untethered. He began drifting until a new station, WHHM, hired him, hoping to capitalize on his name. They went bankrupt. More drifting, a divorce, more drinking and pill-popping. He settled for a time at a small station outside Memphis. The fans who had been raised on him and now had their own bands hired Dewey to spin records during the breaks at their dances. He enjoyed a small burst of popularity, developing a new audience while reacquainting himself with his original one.

Dewey's injury, or the pills he took, prevented him from driving. "I used to drive him home from the dances," says Randy Haspel. "Those of us who knew the Dewey Phillips story from childhood were eager to overlook his faults. We ignored the ever-present bottle and the vials of pain pills. We assumed it was part of the Phillips persona and we were too naive to recognize his abuses."

Haspel remembers the night Dewey called and invited him and three friends out to meet Elvis. "We thought he was joking but he insisted he was not. Dewey had said, 'Say something, Elvis,' and a voice sounding remarkably like Presley's came over the telephone and invited us to Graceland that very night. We picked ·Dewey up at his house, thanking him profusely for this tremendous favor. We stopped four miles before Graceland at the Manhattan Club, a very popular night club featuring Willie Mitchell and the Four Kings as the house band. Dewey informed us that Elvis was sending someone down in a Cadillac to meet us and we would follow him up to the house. In the meantime, Dewey waited inside the bar. We were not yet seventeen, so we waited in the car. Nearly two hours passed. No Cadillac ever stopped for us. We decided to go home. Dewey would not hear of it and in a final act of bravado, he jumped into the front seat and told us to drive to Graceland. Events happened mercifully fast after that as we anxiously watched Dewey confront the gatekeeper outside the Presley mansion. I remember angry words, defiant gestures, and Dewey, red-faced and frustrated from the ordeal, stalking rapidly away from the driveway with his loping stride.

"He rode back with us silently, humiliated in front of a bunch of idolizing kids when all he really wanted was a ride to the Manhattan Club."

Roland Janes says, "One night someone was supposed to pick him up down at my studio. It was bitter cold then, like ten degrees outside, and Dewey said, 'Well I'm gonna wait out here for a minute and they'll be by to get me.' I thought he was gone. Thirty minutes later when I locked up and was going home, I walked out and he was still standing out there. He knew what was happening. I said, 'Dewey, what the hell you still doing out here?' and he said, 'Oh, so-and-so's gonna be by here in a minute to get me.' I said, 'Nah, he ain't coming, c'mon get in the car.' There was a lot of that happening during the later days."

"He lived with his mother," says Milton Pond. "He had this prescription that he could get filled anytime he needed it. He'd come by Pop Tunes late, when I was working till nine, and ask me if I could give him a ride. I'd say, 'Sure, no problem,' because I thought that was nice. He was one of my heroes. He would always say, 'Now don't forget, Elvis'—by then he was calling everyone Elvis—'on my way home, I've got to stop by Doc's.' He'd stop at Doc Russell's pharmacy and run in and get his prescription filled."

On September 28, 1968, the forty-two-year-old Dewey went to sleep at his mother's house and didn't wake up. He'd helped bring musicians into this world, and musicians helped carry him out. Among his pallbearers were the songwriter Dickie Lee, Sam Phillips, and Sam's two sons, Knox and Jerry.

CHAPTER THREE

The World's Most Perfectly Formed Midget Wrestler

THE LINK BETWEEN THE EXPANSE OF THE DELTA AND THE TWELVE-INCH pieces of black vinyl that recorded the region's sound is peanuts. Beale Street would have drawn Howlin' Wolf from the country, stinking like a mule and covered in mud. Wolf might have established himself as a performer in the clubs and enjoyed a career, in Memphis or Chicago, thrilling audiences with his powerful persona. But getting that power onto vinyl, and getting the vinyl to places where people could hear it and buy it, was the direct result of what easy-listening, big-band fan Buster Williams learned as a young man selling peanuts.

In 1949, the same year Dewey Phillips went on the air, Williams established Plastic Products, Inc., an independent, nonaffiliated record pressing plant. Just two weeks earlier, a plant on Long Island, New York, became the first independent. Williams was the second, but in Memphis he was far from the controlling influence of the major labels on either coast. His empire already included jukeboxes (Williams Distributors) and record distribution (Music Sales). The Quonset hut called Plastic Products completed the system that would lay the physical groundwork for the growth of rock and roll.

"What it boils down to," says the soft-spoken Leon "Mack" McLemore, who, as manager of the record distributorship, was at the nucleus of Williams's web, "is that Buster knew that through Music Sales putting the records out on our jukeboxes, we could sell enough to get our money back for pressing costs. We had all these machines and routes spanning three or four states, and if a record was playing in one jukebox, another guy would want that record too and he would have to get it from us. A thousand was easy, and if you sold that thousand, the label got a little money, Buster got paid, and they could press another thousand. He could gamble on a thousand records, extending credit, and not stand the chance of losing any money." Buster helped Sun Records, Chess Records in Chicago, and later Stax in Memphis, and a host of other independent labels across the country come to their ultimate fruition: getting their music on vinyl, their vinyl into customers' hands.

Buster Williams was born January 14, 1909, in the appropriately named city of Enterprise, Mississippi. At age twelve, he began selling peanuts at high-school football games. The business drained his mother, who was slave to the roaster the day of the event. His first innovation was to replace the paper bag with wax paper, sealing a piece of string along the edge to create an easy open pull-tab. It kept the peanuts fresher, allowing his mother to spread the roasting over a couple days. At sixteen, he had made enough money from peanuts to buy a drugstore, and the profit he made from the store's coin-operated machines led him to jukeboxes and Williams Distributors. "There were some snack machines and several slot machines in that drugstore," says Robert Williams, Buster's son. "He saw what kind of profits those slot machines turned, and soon he had a route for those all over this four-state area. Jukeboxes were a natural thing from the coin machines."

Before the prevalence of radio, jukeboxes were the primary outlet for exploiting a record. Wherever there was a jukebox, there was a crowd. And there was money. After the war, Williams could afford to establish the Music Sales record distributorship and pay himself, essentially, to move records from the wholesale side of the warehouse to the retail side. Independent labels could now get their record in a store as well as on a jukebox. "When he opened the record distribution, it wasn't for local artists or local labels," says his son. "At that time there wasn't anything happening in Memphis. He was looking at a much wider geographic scope. He soon had offices in New Orleans, Shreveport, St. Louis, and Chicago and was shipping independent records all over the country."

The majors also controlled the pressing plants. Independents could bring them work, but outside jobs received lower priority than the in-house jobs; subject to the whims of the factory, it was impossible for a small label to plan a release date and maintain stability. Plastic Products was the next logical step for Williams, completing his control of manufacturing, distribution, and the means of consumption.

As Chess Records in Chicago was developing, Leonard Chess realized the importance of dependable pressing. He came to Memphis on a bus—"Dad picked him up at the bus station"—and after a meeting with Williams, left with a credit line and the understanding that when Chess could pay, Plastic

Products would be at the top of the list. According to Robert Williams, there was no contract, nothing written on paper. And that became the standard. As their business relationship blossomed, the Williams family and the Chess family became close-knit. "The first time I ever had gone to Chicago, I must have been eight or nine years old," remembers Robert. "Leonard had his offices in a rough area but, as you can imagine, he was well connected. The cab pulled up and dropped us off in front of the offices there. Dad put the bags out and turned around to pay and when he turned back, the bags were gone. He walked in and told Leonard, and we had the bags back in about twenty minutes. I mean, they were delivered back with apologies."

The Chess account became a bedrock for the pressing plant, which fed Williams's record distribution and jukebox companies. "Once Music Sales got established in retail stores, we did a whale of a business," says McLemore. "One thing that helped establish us was acquiring MGM as a client. People had to have Hank Williams. We'd tell 'em, 'You buy some of my records, I'll sell you some Hank Williams.' Aside from just the popular stuff, the shops had a section on R&B music and country music. We handled all that."

"These record distributors supported a lot of us little guys," says Roland Janes, whose labels included Rolando, Renay, and Rita Records (the original "Mountain of Love"). "They could almost assure us to break even on our little releases, which would give us an opportunity to work the radio stations and try to get some recognition there. Buster Williams helped everybody, and not just locally." On a runaway hit, Mack McLemore says, they could unload fifty thousand singles between their jukeboxes and retail shops.

In 1961, Williams outgrew the Memphis plant and opened a much larger one in Coldwater, Mississippi. By 1973, he had reopened in Memphis and was running both plants with three shifts, twenty-four hours a day to meet the demand. (He also bought into Eastern Manufacturing, a pressing plant in Philadelphia.) When they were at their peak, Plastic Products's southern plants could press over 150,000 singles a day; they also pressed twelve-inch albums. By the early 1970s, the pressing plant's major clients included Stax, Atlantic, Chess, ABC-Paramount, and MGM. "We shipped a lot of records by air back then," says McLemore. "Tonnage-wise, Coldwater was the largest shipper that Delta Airlines had for a couple of years. We'd send a truckload of records into the airport every night. We had ninety-seven presses at one time down there. Sure it's incredible."

One day in 1966, Buster phoned Mack and told him that he was shutting down the Music Sales distributorship. "I was kind of taken by surprise. We had a real steady business going. But Buster owned the company and if he decided he wanted to close, he'd just close." Thirty years later, McLemore still says with pride that he was able to find new distributors for nearly all the labels they handled. He stayed with Williams as manager of the pressing plant operations until the proliferation of cassettes and the demise of the seven-inch forced them out of business. By that time, Williams had become attracted by a larger hustle, establishing himself as a major wildcatter in the oil business. He rigged a motor vehicle with multiple phone lines and rode throughout the South, checking on his sites and making deals. Oil remained his primary interest until his death in 1992. Buster Williams's Quonset hut stands today as a memorial to the era of vinyl records. Surrounded by the industrial decay of twenty years, much of the equipment remains at the ready, the same as it did on its last shift.

In 1949, RCA Records introduced the seven-inch 45-rpm record. It was meant to compete with the 33⅓ long player successfully marketed the previous year by Columbia. But the improved fidelity of the faster record was not enough to overcome the LP's advantage: symphonies only had to be flipped once, while the 45s spread them over several discs. RCA was ceding the race when they realized the smaller discs were perfect for the youth market's three-minute rock and roll song. The parent's disavowal of the format made it even more attractive. Here kids, identify with these—and collect them all.

The growth of seven-inch 45-rpm records was directly related to the era of the Saturday matinees, when sending kids to the movies was cheaper than hiring a baby-sitter and it occupied them all day. (TV was not yet firmly established, and owning a set was a luxury.) The serials, in which a story was continued week after week, established the rhythms of consumption that would later drive the rock and roll habit. Once the witnesses perceived that rock and roll 45s were going to keep coming out, it made it easier to part with that much silver. You had to get the next installment.

Saturdays at the cinema also produced the original rock and roll screen star. In 1953, the year before Elvis was unleashed, Marlon Brando's character in *The Wild One* summarized the philosophy of the burgeoning youth culture. He was asked, "What are you rebelling against?" He answered, "Whattaya got?" But Brando was spilling popcorn in the aisles when cowboy Lash LaRue premiered in 1945.

Unlike the clean and handsome cowboy stars Roy Rogers and Gene Autry, Lash LaRue was not so obviously a good guy. He brazenly wore black, and though he always performed the hero's duties, it was only after he was mistaken for the bad guy. Teen angst, western style! His cape further distinguished him, and this wild one of his day even had the equivalent of a motorcycle: his bullwhip. Lash could shoot a gun with the best of them, but he achieved his individuality with the whip. His rebellious nature defined cool for the future rock and rollers. Writer Stanley Booth says, "I idolized Lash to the point I didn't just want to be like Lash, I became convinced I was Lash!" Mary Lindsay Dickinson, an early participant in the scene and Jim's wife, sums it up, "Roy Rogers was a wimp, Gene Autry could sing, but Lash LaRue was different. As children, millions of my generation looked up to Lash LaRue as a role model. We memorized his movies, never missed him on television, pored over his comic books, worshiped him at state fairs, and tried to be cool—like him."

The depth of his film character was evident in his first movie, *The Song of Old Wyoming*, in 1945. "They let me pick out my wardrobe," LaRue told me in 1993 while in Memphis for a western film convention, "and I selected a black outfit. I liked black. In the picture, I was the bad guy turned good. The picture closed on my gravestone and it said, 'In the worst of us, there is some good.' I think that sums it up."

Lash had a few interesting encounters during his visits to Memphis. When his movie career began winding down, he led an exhibition rodeo at touring fairs. His name was huge on the banners, and the show featured something for everyone: fancy riding, a comedy mule, pretty women, and the whip. (At one time, Sun's only female recording artist, Barbara Pittman, was a member of his troupe.) While performing in Memphis in 1956, the cowboy lived out a modern interpretation of his films. The *Commercial Appeal*'s front-page headline read, "Lash LaRue Arrested at Fair—Charged with Buying Loot." The article begins, "Police cracked the whip on Lash LaRue last night. . . ." In his possession was a stolen adding machine and three hot typewriters. His cohort Fuzzy St. John was also arrested, along with a showgirl who tried to strangle herself with a scarf during her night in jail.

"I was crushed when he got arrested," recalls Wayne Jackson, later a member of the Stax house band. "I couldn't believe he might steal. I didn't know why he wasn't just rich, why he and his guys were stealing typewriters." Jackson's sentiments are uniformly echoed by his peers. Elvis Presley, by then the world's highest-paid recording star, sent word to Lash that he could help. "I told him to let it go," LaRue says about Presley. "I didn't want anybody getting messed up in it because it was a stinking lousy thing."

At the arraignment, Lash followed his old movie scripts and pleaded innocent. He claimed he'd purchased the goods, valued at $1,200, from a salesman for $105. And, from the same script, a four-day trial commenced, at the end of which he was exonerated. But unlike in his old films, many of his fans forgot the hero's ending, for the shock of the headline still weighed heavily. Typically hip, he summed up the incident to me with a quote from Lord Buckley, the British hepster comedian of the 1950s: "The bad jazz a man blows wails long after he's cut out."

But his exploits over the years kept him endeared to the rock and rollers. In the 1960s, he was reportedly pulled over in Hollywood while driving a red convertible and wearing scuba gear. His car was searched and he went down as an early marijuana bust. When he was in Memphis with another convention in the mid-1970s, several local musicians introduced themselves, explained their longtime admiration of him, and invited him to a party. Intrigued by the company, Lash accepted. He drove from the downtown Peabody Hotel to rural East Memphis to pass an evening. Mary Lindsay Dickinson recounts, "Lash drove through the city, from the Mississippi River past Germantown at literally 110 miles per hour. He had just one finger on the steering wheel of the Red Sled, his Cadillac convertible. We passed many cars and several police cars who completely ignored us. It seemed we were invisible to the rest of the world." The party lasted all night, during which time Lash told of his visits to the planet Jupiter. He produced some pot, which he claimed was a gift from God himself. "Lash leaned back for a thoughtful moment," continues Mary Lindsay Dickinson. "We sorcerer's apprentices sat at the master's feet, hostages for the night to the world's most unique cosmic cowboy. When Lash opened his mouth again, he spoke quietly. 'I have no home in the universe. I am hunted by the police of Jupiter and the police of Earth. My great fear is that they will arrest me at the same moment in each place. That would be more pain than I could bear.'"

Lash LaRue had prepared the witnesses for Dewey Phillips, and although they thought they were just having fun, they unwittingly became students of Dewey's ideas on racial equality and social freedom. For years to come, they would realize the depth of his lessons. One of their earliest chances to test their education was at the wrestling matches, held every Monday night at Ellis Auditorium. In the mid-1950s, a new champion came to town and he had a gimmick unlike anyone else's.

Sputnik Monroe arrived in Memphis in 1957, "220 pounds of twisted steel and sex appeal." He had been garnering acclaim in Mobile, Alabama, and the Memphis promoter, whose receipts had been slipping, was looking for a new draw. Wrestling had been like the movies: either you stood for Good or for

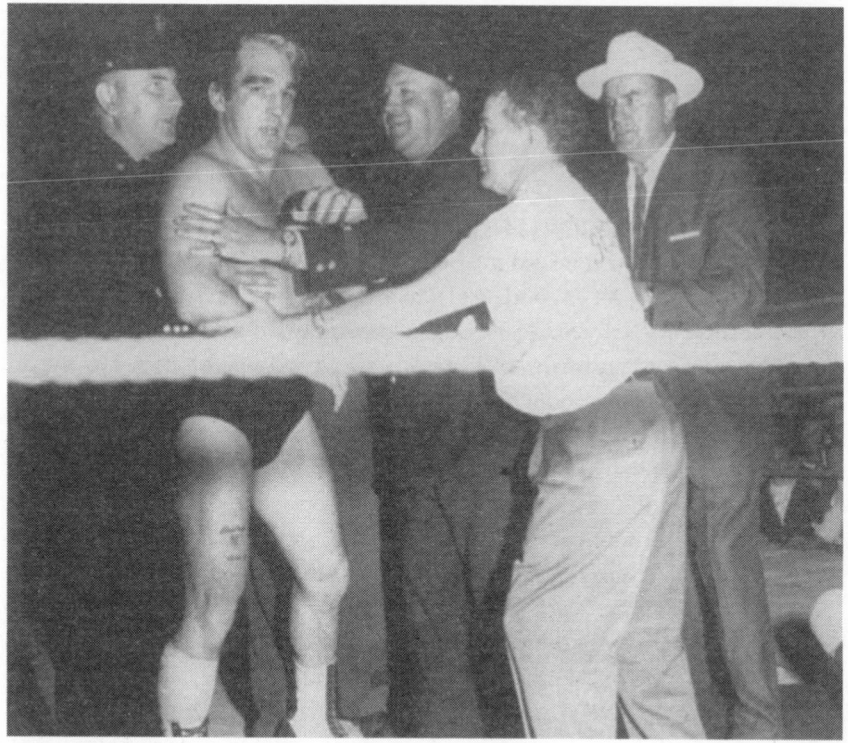

Sputnik Monroe: 220 pounds of twisted steel and sex appeal. Photo courtesy of Sputnik Monroe.

Bad. Monroe wrestled hard and played fair—unless he was losing, and then he cheated. He described himself as neither good nor bad, but "scientifically rough." His philosophy was, "Win if you can, lose if you must, always cheat, and if they take you out, leave tearing down the ring."

Sputnik Monroe, now in his sixties, lives in obscurity in Houston, where he works as a security guard. Although flesh has settled in places where there once was muscle, he is still a large man, powerful and agile. He used to boast, "I'll jump in the air and shit in your hair," and you wouldn't want to dare him today. His ears are cauliflowered, his face battered. As I pass a day with him in an anonymous Houston hotel room, he smokes cigarettes constantly, picking his ears with a toothpick he also chews, speaking in a voice that would carry easily to the cheap seats.

"I started in the carnival athletic show, meeting all comers," he explains. It was nearly half a century ago, when half a sawbuck and plenty of machismo could get you five minutes in the ring with the strong man and a chance at fifty bucks. "Whoever wants to do their thing, however they want to do it," he says. "I had shovel fights, rope fights, pickax-handle fights, wrestled, boxed, one hand tied down, whatever their specialty was. One time I had a guy turn his back to me and hook me by the head, and I realized he'd seen something on TV and wanted to flip me over his back. So I let him flying-mare me. I got up and staggered around, and let him do it to me again. The people cheered and he did it again. And he did it again and he did it again and then he puked and fell over. I never let anybody get out of there a winner."

In addition to brute strength and a consummate understanding of the sport, Monroe had an instinctual mastery of crowd psychology. He would damage or destroy his opponent and then, like the other wrestlers, turn to the audience for approbation and praise. But Sputnik Monroe did not turn to the box seats down front, not to the women in the audience, not to the first balcony, not the second or third balcony. After each triumphant trick, Sputnik Monroe would turn his back on the vanquished, strut like a barnyard rooster, and then fling his hands high above his head, asking for, and receiving, respect and adulation from the black audience, segregated and confined to the crow's nest, the small balcony at the very top of the hall.

"When I arrived in Memphis, I went straight to Beale Street where the blacks hung out and from there straight to jail and got Sugarmon, the black attorney, to defend me in court. They charged me with 'mopery and attempted gawk,' that's an old southern vagrancy thing they made up. I was on Beale Street every night for the first six months. I got arrested three or four times until that didn't work anymore and then the cops left me alone."

Ellis Auditorium, where the big musical acts played and the same site that Elvis had aspired to, began selling out for Monroe's matches. "There got to be a couple thousand blacks outside wanting to get in," he recalls. "So I told the management I'd be cutting out if they don't let my black friends in. I had power because I'm selling out the place, the first guy that ever did it, and they damn sure wanted the revenue."

"I remember one time Sputnik was wrassling in Louisville," says Johnny Dark, now a Memphis sportscaster but then the president and founder of the Sputnik Monroe Fan Club. "In the dressing room, this little black lady came up to Sputnik, she had tears in her eyes, she said, 'You don't remember me, you never met me, but I used to live in Memphis when they made us sit upstairs in those buzzard seats.' She said, 'You're the one who got them to change that.' That was the first time I saw Sputnik with tears in his eyes."

Jim Dickinson was another Sputnik Monroe worshiper. "The way they would cut off the black audience," he explains, "they had a guy counting the white door and a guy on the black door. And they knew how many blacks the section could hold. Sputnik paid the guy who counted the blacks to say a low number every time he was asked, so when the boss said, 'How many have you let in?' he would say 'Twenty-five,' or whatever, and there was five hundred people up there. Finally the audience got so big and so heavily black that they had to integrate the seating. That really is how integration in Memphis started. There's no other single event that integrated the audience other than the wrassling matches and Sputnik paying the guy to lie."

Jim Crow laws were outlawed in 1954, though little practical effect was felt until Lyndon Johnson pushed the Civil Rights Act through Congress in 1964. In Memphis, as in much of the nation, another decade would pass before there was any attempt at mass integration, usually in the form of busing. During the 1968 Memphis sanitation workers' strike, during which the Reverend Martin Luther King Jr. was assassinated, the placards carried by African-American workers had to convey the simplest of ideas to the noncomprehending white majority: "I Am a Man."

"You're talking about separate water fountains, you're talking about back of the bus," says Jim Blake, who managed wrestler Jerry Lawler in the 1970s, and whose Barbarian Records recorded several heroes of the ring. "I went through my whole twelve years at school having never been able to share an experience with a black, and I was starting to resent this, because I was also listening to radio and Dewey Phillips and hearing all these great black records and realizing that these were some talented artists, this was another culture. Where at first we'd gone to the matches hoping to see Sputnik get beat, we started to realize that he was pretty fucking cool. He had his audience and he never played down to 'em, never talked down to 'em. He became a role model."

Monroe had size, demeanor, sharp costumes, good looks, and certainly a boisterous attitude—even without a trademark he would have been conspicuous. Early in his career, however, he was beaned with a wooden chair. After a large splinter was removed from the top of his forehead, a patch of hair grew in white. Perusing Memphis high-school yearbooks from the late 1950s, it's easy to spot the members of his fan club: Johnny Dark is one among many who sported a white streak in his annual photograph. In an early promotional photo, Jerry Lawler, whose career began well after Sputnik's and continues today, also has a white streak.

"If you would have had some kind of election about who was the best-known face in Memphis at the time—Sputnik, Elvis, and the mayor—Sputnik would have been real close to Elvis," says Johnny Dark.

By bonding himself to the tension of the era, Sputnik Monroe became a hero to the rebellious white youth culture. Wrestling, like rock and roll, thrives on the energy of a crowd gone wild. By the end of the decade, Sputnik had become friendly with Sam Phillips, and a hero to Sam's children, Knox and Jerry. Jerry Phillips was an athletic twelve-year-old, short, and wanted to get into wrestling. Sputnik was a muscle of a man who could have flossed his teeth with Phillips, but the young lad's desire gave him an idea. It might not make a main event, but midget wrestling had its possibilities.

That Phillips wasn't a midget was no problem. For Sputnik, in fact, it was the heart of his scheme. Professional wrestling is about frustrating the audience. In a humid and smelly high-school gymnasium, there's a great potential for aggression, the thick smoke and bad lighting, the rage to be roused among a mob of tanked-up hicks—"Get him outta there! He ain't a midget!"—the frustration waiting to be vented by an audience who spent their days and weeks and years beating the earth, powerless against the elements—"That one don't belong! He ain't no midget!" A fan that felt a blue vein popping out of his neck would return week after week.

Sputnik was friendly with Fabulous Frankie Thumb, a midget proper, and he put both Jerry and Frankie in training. He enjoyed Jerry's attitude and soon was taking him to bars and nightclubs, sticking a cigar in Jerry's mouth and lifting him onto the bar. "The bartender would say, 'How old is that guy?'" says Phillips. "And Sputnik would say, 'He's twenty-one. He's with me.' Who's going to argue with Sputnik Monroe? Anywhere that he went, he was king."

They set up matches on the circuit all around Memphis. More midgets came forward and soon there were tag teams. Phillips remembers, "The announcer would say, 'He doesn't have short legs, his arms appear normal . . . ,' that sort of talk. If I had been twenty-five and the size of a midget, it might have been believable, but I was obviously a kid, twelve or thirteen. They'd have me walk through the crowd, chewing a big cigar, taunting the people. Sputnik had taught me pretty good how to pull my pants down and tell 'em to kiss my ass. The audience knew I wasn't real and I just made 'em madder."

The act's run came to a close after a couple short years in a small Arkansas town around 1962. An angry fan—a deranged wrestling zealot caught up in the believability of that which was so plainly not believable—pulled a knife and tried to stab Sputnik Monroe's most inspired idea. Phillips's parents said he could no longer continue the act: "DeLayne" Phillips, the World's Most Perfectly Formed Midget Wrestler, was officially retired.

While white parents hated Sputnik Monroe, the kids loved him. "There was a group of wealthy white kids that dug me because I was a rebel. I'm saying what they wanted to say, only they were too young or inexperienced or afraid to say it. You have a black maid raising your kids and she's talking about me all the time, so I may not be in the front living room, but I'm going in the back door of your goddamn house, feeding your kids on Monday morning and sending 'em to school. And meeting the bus when they come home. Pretty powerful thing."

He created a life of altruism built on self-promotion. After he integrated Ellis Auditorium, his power within the black community caused fear among the city fathers. While the black leaders were arguing about how to protest the segregation of an automobile show, Sputnik called the sponsoring dealership and threatened to open a car lot in a black neighborhood; that night's evening news announced the changed admission policy.

"Another time I give away a hundred watermelons—Sputnik melons; they had white stripes—and announced that I was gonna run for sheriff. People thought prostitution and incest would flourish, 'motherfucker' would be a household word. I could have run for mayor and made it. I could have blackmailed the city. I could have done any goddamn thing I wanted. I was general of a little black army."

By 1960, Monroe couldn't get any bigger in the mid-South. But wrestling has always been regional, and outside his region, his stature was not the same. He wanted to go national. "Before I left Memphis, I read in the paper where Gene Barry was coming to the Mid-South Fair and I went out there to hit him in the nose for copying the way I dress. I was born and raised in Dodge City, Kansas, which is the cowboy town of the world. Gene Barry was the star on 'Bat Masterson' and dressed like I dressed, with a homburg and a vest. I figured if I jerked him off a horse and hit him in the nose for dressing Dodge City –style, I'd get a national reputation."

Police protection kept Gene Barry's nose from its appointment with Sputnik's fist, though the wrestler did manage to pick a fight that night with a cowboy and make the front page of the local paper, his picture as large as President Eisenhower's. After that, Monroe tried becoming a national star by racing from territory to territory, but the lapses between appearances were too great; he couldn't rally the support.

Despite his popularity with the fans, Monroe had always frightened promoters; in wrestling, he was known as a "charger," someone who could lose control. He was not averse to hurting his opponent, nor to being hurt himself. This attitude gave promoters ulcers, because they usually had the same card booked the whole week in different cities and couldn't afford an injury.

The 1960s found Monroe traveling again, but his magic didn't click in other territories. He made occasional returns to Memphis to boost his morale. A bitter divorce led to a drinking binge that spilled from months to years. In the early 1970s, Sputnik wore the Junior Heavyweight Champion belt, and in 1972 he was back in Memphis and back on top.

Randy Haspel, whose band the Radiants was one of the first post-Beatles Memphis bands and one of the last recording acts on the original Sun Records, remembers an encounter with Sputnik in the early 1970s. "I was sitting around Phillips Studio with Skip Owsley, this black conga drummer from my band, and Sputnik came in. He wasn't as active in wrestling as he had been, and he was saying, 'I don't know what to do anymore. I used to be able to tell 'em their wimmin were trash, or I'd shake my ass and them broads would flip out and the guys would want to fight. I can't get these people to hate me like they used to!' This was during the hippie heyday, and we said, 'What people hate now are longhairs. If you talked about love, Sputnik, they'd probably hate you.' Skip, the black guy, said, 'You need to find you a black wrestler and tag team with him.' So two weeks later Sputnik appears on TV with Norvell Austin, and he's dyed a blond streak in *his* hair. They're beating up some designated opponents, and they tied up one guy's arms in the ropes. Sputnik goes over to the corner and gets a bucket and pours it over this guy's head. It's a bucket of black paint. And then Sputnik and Norvell go over to the announcer and Sputnik says, 'Black is beautiful!' and Norvell says, 'White is beautiful!' and Sputnik held up his arm with Norvell's and he said, 'Black and white together is beautiful.' Next time I saw Sputnik he's real excited and says, 'They hate me again!'"

The interracial tag team thrived in Memphis. After he cloned the white streak, the younger Norvell was rumored to be Sputnik's son. The partnership lasted three years, and Norvell, who had never traveled, celebrated his twenty-first birthday while on a Japanese tour. Monroe was left untethered when Austin decided to go solo. After a car wreck in 1978, he recuperated in a Texas beer joint, holding court from a recliner on stacked Coke crates. Since his heyday, he has owned bars and restaurants, sold turquoise jewelry, had a wrecker service, a transmission shop, and taught at wrestling schools. Recently, he has considered becoming a stand-up comedian. Monroe has not wrestled since 1991, but wouldn't decline a challenge. His son Bubba "The Brawler" Monroe, Texas All-Pro in Houston, has been under his father's tutelage for nearly a decade, slowly climbing professional wrestling's ranks.

Despite its popularity, the World Wrestling Federation gets little respect from Sputnik. "My business is dead," he says. "There are no tough guys left in wrestling." The sport of body manipulation has been replaced by acrobats on steroids. Tumbling makes for good television, but the science of the sport has been replaced by pantomime and buffoonery. "Wherever I put your head, your body's gonna follow. Wrestling amounts to one thing: A fulcrum and a lever. Long enough lever, big enough fulcrum, you can pick anything up."

Whatever else the WWF has done, it has finally made wrestling the popular means of expression it had the potential to be back in the 1950s. It may be bereft of the social value it toyed with then, but it is able to tour like rock bands, packing arenas and selling T-shirts and videos. Albums, even.

Sputnik was in the right place to be a societal influence, but his timing was a little late. American youth culture in the 1950s was a whitewall tire about to blow. Major cities were too self-reflexive for this explosion—it needed innocence. A place where racial tensions had been so deeply repressed that society was about to choke on its own sweetness, where urban civilization could obstruct all hopes but a short drive out of town declared the possibilities endless: It was Memphis and it was rock and roll. But had Sputnik Monroe come along a few years earlier, we might all be products of sex, drugs, and wrestling. He is the guy who did not become Elvis Presley.

Sputnik Monroe summed up the common attitude between wrestling and rock and roll one afternoon in a Memphis studio. This was 1972, at the Sam Phillips Recording Service. Jim Dickinson was just off the road from a tour, and he was showing off his recently purchased red, white, and blue leather boots, emblazoned with tricolored double eagles. These were boots that would turn heads, boots that would be the envy of any self-respecting biker and most corrupt sheriff's deputies. If you didn't speak the language in a foreign country, these boots could get you around.

Sputnik Monroe was singularly unimpressed. Perhaps because his feet were not breaking them in. Perhaps because the boots were the center of attention and not him. The reasoning only obscures the facts: These double eagle boots were so engrossing that none of the admirers noticed his eyes impatiently darting.

The compliments continued.

"Those goddamn boots ain't shit," Sputnik finally growled. And before anyone could beg to differ, he continued. "I know a place in Mexico where you can get boots with big dicks and balls on 'em."

The laughter in the room died like a suffocated fire as everyone became aware of Sputnik's smoldering anger. People got fidgety trying to think of what to say. The obvious dawned on someone: "What're you gonna do with boots that have big dicks and balls on 'em, Sputnik?"

Sputnik, like anyone whose work involves shouting over the din of an outraged crowd, has always had a gruff voice, but maybe he plucked a little extra coarseness for this answer. A real attitude is one you can feel. "You go into a bar with them on," he said, "You can get into a fight in fifteen minutes."

CHAPTER FOUR

Nothing Ever Happens But the Impossible

THIS WHITE KID WALKS OUT THE SIDE DOOR OF THE TALL BUILDING where his dad works. It's Saturday, a short day in the office, mostly knocking around. The sunlight is the bluish cool of the coming autumn, and it's not that you can feel winter in the breeze, but the oppression of summer is noticeably absent.

A band is playing, people are dancing. His dad stops and so does the kid. They're in an alleyway, Whiskey Chute, where jugheads used to roll barrels up to town from the riverboats below. Whiskey is in the air. Whiskey is in the white couple dancing, and whiskey is in the black people making the music. Oh it's definitely music, the guitar, the violin confirm that, even though the other instruments seem to be kitchen utensils. A laundry basin. A washboard. A comb. A jug.

The dad nudges the kid once, then a second time, a little harder. The boy nods without hearing. His father puts his big palm on the nine-year-old's shoulder, summoning a little more strength than he thought would be needed, urging his son along. The music follows them as they walk out of the alley into the brighter street. There's summer, felt it then. Dad starts to say something, but he sees the kid's mind is still back at the jug band. Oh well, let him chew on that anyway.

The kid looks over his shoulder when he knows it's too late to see the musicians. "Come on down to my house, honey, ain't nobody home but me." His nine-year-old mind tingles, something juicy in the way the man sang that line. The way the male dancer gestured to the woman every time that chorus came around, and what she did in response. And how often that chorus came around!

41

Blame it on the war. The increase in industrial jobs for the 1940s military effort drew country people to the city, and after the war, letters back home kept them coming. Escaping from the treadmill of sharecropping, out from the shadow of the Man and into the robotic grind of the factory, Delta residents came to Memphis, moved on to St. Louis, and finally to Chicago. They peeled off along the way, finding work, finding family, finding friends. Swept into the urban migration, musicians dotted the exodus. Woe to the bass fiddle player and his baggage. Better learn the violin, buddy.

The city had something that was less common in the country: electricity. And that made the city a louder place. In the rural juke joints, a lone acoustic guitarist or a band of all acoustic instruments could propel a crowd to madness, but in the city, the musicians adapted their style to eardrums attacked by the clanging of streetcars and the pounding of factories and warehouses. They plugged in. But the acoustic instrument retained its place. Instead of waiting for the audience to find the club, street musicians went outside and found a crowd. "Music for sale," they'd hawk between songs and watch their bucket or their hat fill with change. And if a cop came, they could quickly pack up and walk away. Or run.

The jug band, in its forced union of diverse instruments and in the absolute singularity of the jug's sound, is Memphis music incarnate. It creates a sound where there should be none, from instruments intended for other purposes. The kitchen may be for cooking, but if we rattle around it long enough, we'll surely have a hit. Or at least a good time. The handmade equipment—a guitar from a cigar box, a harmonica from a corncob—conveyed an egalitarianism that reached into the guts of their audience, making anyone who clapped along, danced, or even nodded a part of their act. The jug band repertoire was a combination of past and present, culling from the African heritage of the banjo and guitar, the European heritage of the violin, and the contemporary life built around the washboard and the jug. Their music was sophisticated. Just as Harry Fritzius created art from an ordinary mask, the jug blowers made keeping the beat into a touch of God. Their *whoomp whoomp* filled with character, the musicians themselves responding to the audience's disbelief of their capabilities, pushing their talents to new heights while their heads floated off their shoulders from hyperventilating.

Unlike its contents, the sound of a jug cannot be distilled. The joy created from such a household item is at the heart of Memphis music, the simplicity, the eccentricity, the soul. In that peculiar thump is heard the city's avowal to buck trends, to respond, Weird? Damn right it's weird, and if I keep at it a few minutes longer, it may get weirder still. With the Delta blues, Sun, and Stax,

Memphis's peculiar mix became not only popular but also permanent. Pop culture is a continuing shift of trends, disposing of the new with the newer, but *whoomp whoomp,* Memphis won't go away. Three times it hit, which is more than Detroit, more than Cleveland, more than New Orleans, and the music bookies are wagering it'll come from here again. Sometime. And when it does, this bettor says it'll be some variation of its roots.

For the Memphis kids in the 1950s who would witness the creation of rock and roll—unlike for, say, Mick Jagger—pure Delta blues wasn't necessarily a detached voice on a piece of black vinyl. Like the kid in Whiskey Chute—an epiphanic incident in Jim Dickinson's life—the mystery of that sound was not "Where?" but "How?" How do these people next to me on the street do something that I've never seen done by any of my parent's friends or anyone I know? Such cultural collisions are the foundation of Memphis's artistic contribution. "I was struck like it was lightning," Dickinson says of his jug band encounter. "I'd never heard anything like it. I'd heard some Dixieland music on the radio, a little boogie woogie, but that's all that was accessible to a white kid."

Jim Dickinson lives deep in the Mississippi woods today, where he smokes spider webs, eats rattlesnake meat, and fends off voodoo spirits with possum tails. Or so those who don't know him would have you believe. In the contemporary music industry, Dickinson is Memphis Whitey, the beast from underground who travels in the corporate world, the man with corporate connections who prefers the underworld. His head is attuned to the sounds of tomorrow, rooted in the ways of yesterday. He learned music and attitude from bluesmen who created the recorded genre, and that—the music *and* the attitude—scares lots of people. His career parallels rock and roll's evolution from a business to an industry. Like the members of another early Memphis band, the Mar-Keys, he learned the ropes as the ropes were being strung. He's seen every kind of deal. His resume includes playing piano with the Rolling Stones ("Wild Horses"), Arlo Guthrie ("City of New Orleans"), and bluesman Sleepy John Estes; he's produced several Ry Cooder albums and recorded ten soundtracks with him (*Paris, Texas* being one); he's produced the Replacements, Alex Chilton, Toots Hibbert, and pianist Grandma Dixie Davis; and he's written songs that have been sung by Bob Dylan, the Flying Burrito Brothers, and Albert King. While geniuses of the subbasement seek him out, record company executives cringe when his name is mentioned. He has a saying they abhor: "Hits are in baseball, singles are in bars, and your royalty lives in a castle in Europe."

Dickinson's Mississippi home is mostly surrounded by trees; the one clearing reveals a model suburban house across the hill—exactly what he was trying to escape. But paradox vibrates to his core. He's purchased a converted barn that's even more remote, an old structure filled with ghosts. When it's out-fitted with recording equipment, he'll have the ultimate treehouse, moving further from people to attract people. Jim Dickinson oozes funkiness. His beefy appearance carries a biker air, his clothes hail to hippiedom, and his voice, rattling through Delta silt, says backwoods. Gentle by nature, he is a family man. With his wife of thirty years he has raised two rock and roll kids; his ma lives with them. Through fad, fashion, and fallow, Jim Dickinson has plied his trade, taking the accolades with the admonishments. In a business that thrives on the disposable, he continues to make lasting contributions.

In 1950s Memphis, the continuing Civil War was also a war of civility. In the same arena where black and white cultures were roiling, there was a meet-ing of urban and rural. The city's old money had been established on the sweat of the fieldhand's brow, and the conversion of farmland into suburbia created a random mash between country and country club. There were few four-lane roads in the city, none in suburbia. Pecan orchards were coming down as fast as cotton fields were being paved, brick homes going up in their place. Re-stricted neighborhoods all—whites only, with the rural blacks being squeezed into pockets close enough to work in the new neighborhoods' new homes. But the hypocrisy of segregation was laid bare in the very houses where racism was inculcated. The cook in the kitchen was a Sputnik Monroe fan; the man in the yard was an R&B devotee.

"Everybody learned it from the yard man," says Dickinson. "Alex Tiel taught me everything he thought was important to teach a nine-year-old white boy. How to shoot craps, how to throw a knife underhanded—the im-portant life lessons. When it came to something he didn't know, he would run in an expert. He wasn't a musician, but he sang as he worked, unaccompanied, and when he realized I was interested in the music, he brought in a man who taught me this technique that I learned to play from.

"This unknown man told me all music is made up out of 'codes,' and I thought he meant a secret code like Captain Midnight, which I was way off into. But he meant 'chords.' And he says, 'This is how you make a code. You take a note, any note on the piano, and you go three up and four down, like in poker, and that's a code.' Now I know it's a major triad, and the thumb always lands on the key signature note. It works anywhere on the keyboard. And when he showed me that I thought, Alright, this is a system, I can do that. And rock and roll came along soon enough that there was a reason it all started to make sense."

Jim Dickinson with his first guitar, 1956. Photo by Mr. J. B. Dickinson.

Despite its growing presence, black music was still considered trash music by white society. But across town, the white engineer Sam Phillips had begun recording black artists in 1950. He had come to Memphis from the cotton fields of Alabama and was familiar with the sounds that the city slickers gawked at. When he got tired of recording the smooth big bands playing sophisticated music for sophisticated people, he set up a small recording studio and sought what invigorated him. His rural background gave him the insight to tell Howlin' Wolf to play what he played at home, not the more accessible style he'd adapted for the white folks. Phillips knew the audience could conform to the music. His peers asked him why he'd spend time with people who smelled like mules. But the recordings he made—Wolf, Ike Turner, B. B. King—found their way to younger white ears, whose failed attempts at imitating them created rock and roll. The scorn of their parents and friends could not keep them away from this music, and, as is evident in this excerpt of an argument between Sam Phillips and Jerry Lee Lewis right before they recorded "Great Balls of Fire," not even God could deter these men and women who were drawn to what mainstream society referred to as the music of the Devil.

"Jerry. Jerry," Sam Phillips says, calming a man who is haunted by the fire and brimstone of his childhood and the whiskey of his adulthood. "If you think that you can't, can't do good if you're a rock and roll exponent—"

"You can do good, Mr. Phillips, don't get me wrong—"

"Now wait, wait, listen. When I say do good—"

"You can have a kind heart!"

"I don't mean, I don't mean just—"

"You can help people!" Lewis is now chanting a refrain.

"You can save souls!" responds Phillips.

"No! No! No! No!"

"Yes!"

"How can the Devil save souls? What are you talking about?"

The rockabilly artists like Elvis and Jerry Lee brought the strange things that happened on WHBQ at night brazenly into the daylight. To the rest of the world, Elvis in 1954 was as alien as Dewey Phillips's skronk had been coming out of the night sky five years earlier. In his hometown, however, Elvis was not such a freak. Ducktails were becoming common, along with sideburns and a change in the style of dress. What had been latent in the relationship between black and white music was becoming overt; each was groping for the other.

Rock and roll's audience is now a huge chunk of society. But in the music's puberty, the kids and the sound were just becoming acquainted, seeking out each other. Don Nix, a member of the Mar-Keys, remembers, "Nobody had ever heard rock and roll. In Memphis, we were the first ones hearing it, but we didn't realize it. We thought everybody was doing this. Elvis Presley played at our high school, Messick, and it's the first time that I'd ever heard girls scream at anybody they weren't mad at. That was really new to us—screaming! Yeah, I'll have some of that. We formed a garage band, and the fun of it was trying to play whatever was the new record. You never expected to play it like that, but it was fun to try!"

"At our high-school talent show," says Dickinson, "my band and another played rock and roll. The song the other band played had three chords, and the guitarist only knew two, so whenever the third chord came along, he just stopped." The other group had a better drummer and two ducktailed, pretty-boy singers. By the end of the show, Dickinson's band had the singers and the drummer, and everyone else had day jobs.

They became the Regents ('Regency' was the telephone exchange in West Memphis). This was 1957. The core of the group was guitarist Rick Ireland, later an engineer at Memphis's Ardent Studios; a country guitarist named Stanley Neal who had only disdain for rock and roll, except when it paid; the drummer was "Steady" Eddie Tauber. The two singers later enjoyed careers at Stax; Ronnie Stoots toured as the vocalist with the Mar-Keys, and Charles Heinz recorded one of the label's first singles, which won him a career-making offer from ABC. Rejecting it, Heinz cleared the path for Stax's alliance with Atlantic, established with their next release, Rufus and Carla Thomas's "Cause I Love You."

"I didn't take music seriously in high school because the possibility didn't seem to be in the world," says Dickinson, now eligible for a thirty-year pin. "I graduated from high school in 1960, and my band played blues, and we played it like white boys, because that's what we were. And that doesn't seem like a very big deal now, but before the Rolling Stones did it, it wasn't a popular concept. It wasn't okay to play this black music, and it was constantly an issue.

"Other bands in Memphis played some Dixieland, some jazz, various types of more acceptable music, and they might touch one or two rock songs. I maintain that the Regents were the first East Memphis band that played all rock, exclusively rock. And in the whole city, the only other one was what be-came the Mar-Keys, with Packy Axton and Charlie Freeman. There was noth-ing acceptable about the music that we played, and there was certainly no way to make it into a career."

"I had developed quite a collection of Sun 45s," says Rick Ireland. "I had the Sonny Burgess stuff, the Warren Smith stuff. All of Billy Riley's stuff. Roland Janes, who played guitar on most of that, was my idol. Dickinson and I met at high school and were both listening to the same thing. First thing we ever learned as a band was 'My Bucket's Got a Hole in It,' Sonny Burgess's version. And we did 'Red Headed Woman' and all that stuff. Stanley, our other guitarist, didn't approve of that but he would play it. He was country-oriented but we made a nice pair, because he was aggressive and would play a lot of solos, and I had jazz tendencies and would lay back and comp things in the background.

"Stanley got me involved with Roy Cash, Johnny Cash's nephew. We played way out in the country and sometimes on this little radio station called KWAM. I had this real nice fifty-foot black extension cord and after we started playing the radio gigs, it disappeared. We used to say they were running KWAM off the extension cord. One day we went to somebody's house to re-hearse and there were a couple other guys there who played in some country band. I had a Fender Esquire, a flat solid body guitar, and when one of them saw it, he drawled, 'He got one uh them there biscuit board guitars.' I thought, Ooh shit, what kind of company am I in now?"

The innocence in Memphis nurtured wild times. Adults saw a city of beau-tiful trees and churches on every corner. Kids saw everything else. They were dancing in the donut shops, kissing at the drive-ins, and fighting wherever they could. A well-tipped carhop could always produce a round of beers, which directly increased the chances of a rumble between preening teens in the parking lot. One particular gang of toughs were all related, the Tillers. In high school, they'd been football stars and always in the newspapers. Later, they made headlines with their criminal exploits. Jimmy Crosthwait, a puppeteer and the washboard player in Mud Boy and the Neutrons, "saw Mike Tiller carve his initials into a guy's chest, not deep, but with a broken bottle, and it was disgusting. It was then that I noticed that his initials were 'emp— ty.'" Another member of Mud Boy, Lee Baker, saw justice served at a teen hangout called Clearpool. "Some guys jumped Mike Tiller and he was wandering around the parking lot with his damn ear in his hand, trying to hold it on. 'Hey man, take me to the hospital.' We did, we took him up to the emergency room and dumped him. I said, 'I ain't going in there with you because I don't want to get involved in this shit.' That was the fighting days."

White adults wanted to believe that their kids' interest in black music was a passing phase. In their day, they had gone to many of the same clubs their children now frequented, enjoying the black big bands that played for white audiences. But when Dickinson's parents found a photograph taken at the Flamingo Room of their son standing between Ike and Tina Turner, they didn't have to know who they were to know what they meant.

West Memphis, Arkansas, is directly across the Mississippi River from Memphis, Tennessee. There is no natural bluff on the Arkansas side, so every spring the farmlands flood. Some springs, water spreads throughout the whole town. Farmers in Arkansas and Missouri conduct their trade there, where the tallest buildings are storage silos, while across the river, people in skyscrapers bet on their labor as commodities like gamblers at the track. Law was a whimsy in West Memphis. The drinking age had more to do with inches than years; if you were tall enough to hand your fifty cents to the bouncer, you were old enough to get in the club. My father moved to Memphis in 1955, and his uncle drove him around town to get a feel for the city. When they had crossed the Mississippi River and were in West Memphis, Uncle Meyer said, "Let me tell you about where you are. In West Memphis, with three hundred dollars, you could phone the sheriff at any time of the day or night and he would meet you, the mayor in tow. You could shoot the mayor dead in cold blood, and for another three hundred, the sheriff would dispose of the body." (When my mother moved to Memphis from New York later in the decade, she left the city's only Chinese restaurant in tears; the waiters weren't Asian and there was white bread on every table.)

Trumpeter Wayne Jackson, white, was born and raised in West Memphis, an unlikely candidate for house musician at Stax or sideman for Jimmy Buffet and Peter Gabriel. "West Memphis is where everybody came to party back then," he says. "I think they could serve liquor. Or they did, anyway. The sailors from Millington [Naval Base, outside Memphis] would all come across the bridge. And Eighth Street, in quote Colored Town unquote, was big gambling. So all the black people would come over there too, gambling along Eighth, Ninth and Tenth Streets in little old dives and honky tonks, drinking rotgut whiskey. West Memphis was wild. They had cockfights under the bridge. And they had a game called Coon on the Log. They would catch a raccoon and put it on a log in the river, just a few yards off the shore, and everybody brings their coon dogs. And they bet on the dogs to see how long it'll take to get that coon off that log. One dog at a time. A lot of times the coon'd win, stay on the log and just wear the dog out."

"I guess West Memphis had laxer cops or something," says Jimmy Crosthwait. "Everybody over there was drunk, and they had to get from there to here, and nobody ever seemed to stop 'em. West Memphis was where Memphians could get extra wild. They always had the girlie flicks over there, where you could see a little tits and ass, or something weird—nudist colony footage."

Everything was looser across the river. The thriving film scene there was the result of Memphis's puritanical Board of Censors. Dominated by Lloyd T. Binford from 1928 to 1955, its authority was so mighty and its parameters of decency so fastidious that the industry lingo adopted the term "Binfordized" when films were heavily censored or banned. As a child, Binford had witnessed a train robbery, and he seized the opportunity to ban all films on the topic; he forbade anything portraying blacks on a social level equal to whites. Charlie Chaplin was considered a communist sympathizer, his work "inimical to the public welfare"; *Rebel Without a Cause, The Wild One,* and, for its lack of biblical accuracy, Cecil B. DeMille's *King of Kings* were all banned.

But it was the music, especially in a trio of clubs, that drew most people across the bridge. In 1959, at the age of sixteen, guitarist Rick Ireland was asked by the owner of the Cotton Club to join the house band on weekends. That he was not legally of age concerned no one. "These were pretty rowdy crowds. Real heavy-duty, sure-enough, truck-driving rednecks—beer-drinking, truck-driving rednecks. Danny's Club was down the street, and they had chicken wire around the bandstand so the beer bottles wouldn't hit the musicians when the fights broke out. At the Cotton Club, they were a little more subtle. We were open to the flying bottles, but if any of the drunks wandered up to the bandstand, they were in for a little surprise. I was playing my guitar and I noticed this piece of bare copper wire that went all around the two-foot-high

railing on the bandstand. I just followed this thing off to the dressing room by
the stage, and the damn thing was plugged into the wall and in series with a
hundred-watt lightbulb. If somebody ran into that, they were going to get
quite a jolt."

The Plantation Inn (the PI) was the most popular club of the three. The
other two featured country music by white bands, but this one had black
bands playing for white audiences. The PI had been established a generation
before and had catered to the parents of the teenagers who regularly puked in
its parking lot. Throughout the 1950s, the house bands there had evolved from
the big swing sound of the Phineas Newborn Sr. Family Band, featuring
Phineas Jr. on piano and Calvin on guitar, to the smoother sounds of Willie
Mitchell and the Four Kings, and later to an even smaller combo led by Ben
Branch, who would become a prominent civil rights activist. Whatever happened in the audience, the bands never played less than world-class music.
Patrons were allowed to bring in hard liquor, which was not officially sold on
the premises; if one actually had to step outside for the transaction, it was
never farther than the parking lot.

The Plantation Inn was a family place, in the way a swing joint run by the
Addams Family might be. Morris Berger owned it with his son Louis Jack.
There were many places a person could get wild and drunk, but with blacks
and whites in the same room, even if separated by the proscenium, the PI provided a peek behind the wall erected by society. Its spirit was summed up on a
neon sign hanging near the stage that bore the name of a radio show once
hosted by the senior Berger and broadcast from the club: HAVING FUN WITH
MORRIS.

"The Plantation Inn looked like a big two-story house," says Wayne Jackson. "There was a doorway about a quarter of the way down the building, and
you went in there, and the bar part and the restaurant part was on the left and
the bouncer—he was a boxer—you'd go past him and his little desk, and then
you'd turn right into the main, big room. There was a sunken dance floor over
about sixty percent of the area, and the tables were on the sides. The bandstand was on the far end, where Willie Mitchell and the Four Kings would do
their thing. And at the back of the bandstand was a big fan that sucked out the
smoke and hats and ladies' wigs. Before I began sneaking in there, I'd stand
back behind that fan and listen to the band. So, Willie Mitchell always
sounded like, '*Wwwuuhhhwuuuhhhhwuhhh*.'"

"The Plantation Inn was like a roadhouse out of a movie," says Jim Dickinson. "The bouncer's name was Raymond Vega, big ol' nasty guy. Wore a
cast on his arm but his arm wasn't broken. It was for hitting people. Some-

times he'd have a cane, and I remember thinking, What's this crippled guy going to do?"

The Phineas Newborn Orchestra was the house band from 1948, the year Calvin began high school, through 1950, when Calvin and Phineas Jr. went on the road with Ike Turner and Jackie Brenston behind what is often considered the first rock and roll record, "Rocket 88."

"My brother had been in high school a year and I was a grade behind him," says Calvin Newborn, who became renowned for his showmanship. "It was pretty tough even for teenagers, going to school every day and playing from nine until two at night, but I enjoyed it because my pockets stayed full and I was able to buy nice clothes. I wore the best shoes. There was a shoe shop right on the corner where I lived, and every morning when I turned the corner going toward the railroad track that I crossed to get to school, I would get my stumps shined."

The Newborn Orchestra employed sophisticated, schooled musicians who could read charts. Phineas Sr. prided himself on his diversity, maintaining a repertoire that included pop hits and smooth jazz standards. His stage shows competed with those of other bands—and with regular Sunday church services—for the most exciting. "It was hard for me to stay still and play," Calvin says. "I was used to doing like them basketball stars, flying. I would get about six feet in the air playing the guitar. As a matter of fact, I used to think I could fly. I felt like I could make myself as light as I wanted to. Even today I dream that I'm walking down the street and spread my arms and just take off and fly." Mama Rose Newborn, Phineas Sr.'s wife, remembers, "They played 'Tennessee Waltz' at the Plantation Inn every night, and the Plantation Inn wife and husband would come out and dance to it." The band was allowed to dip into bebop but, according to Calvin, only late in the night. The generation that followed Phineas Newborn Sr.—Willie Mitchell, Gene "Bowlegs" Miller, Ben Branch—played with more of a bebop edge. Their bands were smaller, shifting to beat-heavy rhythm and blues. "When I was a kid, I loved to dance," says Charlie Miller, a painter. "At the PI, if there were several people on the dance floor that were really dancing well, the band would keep going. There were times when the bands and dancers were jamming. That was beautiful, it really was."

The peculiar spectrum of a Memphis audience created a challenge for the musicians. Fred Ford, a black saxophone player who toured with B. B. King and Johnny Otis, explains, "When I was coming up in the forties, you had to play everything. Even the radio stations played some of everything. You heard Goodman, you heard Basie, you heard Artie Shaw, you heard Lunceford,

Charlie Miller and date at the Plantation Inn. Photo courtesy of
Charlie Miller.

Ellington, Guy Lombardo. You heard country music. Different tastes from
different people. You couldn't go in and play blues all night or jazz all night or
ballads and love songs all night. You had to be very talented, and you had to
have an open mind."

The Stax sound grew out of the bands that played to white audiences at the
Plantation Inn and to black audiences in Memphis clubs. "Ben Branch was the
leader of the band when I started going," says Dickinson. "He eventually went
to Chicago and got into Operation Bread Basket. He was standing next to
King when he was assassinated. He recorded at Stax when it was still Satellite.
These guys played jazz and they sat down when they played. They had music
stands. I don't know what was on them—racing forms, maybe. But they were
the band that everybody copied—the two horns, the thin sound. As compared
to five horns. Or seven horns. Two horns isn't even a chord, it's just an interval. Two horns takes the keyboard to make a chord. It's less sophisticated. The
horn parts by nature become more percussive and Memphis-y. Packy [Axton,
a saxophonist in the Mar-Keys] learned to play from Gilbert Caples. That's
where the whole Stax sound comes from. It's Ben Branch's band, pure and
simple. The idea of light horns is, I think, the Memphis sound phenomenon."

"If Ben Branch was an influence, Bowlegs Miller was too," says Stax saxman
Andrew Love, who was not allowed in the PI audience because of the color of
his skin. "The Stax sound was just a Memphis sound, and Ben and Bowlegs

were two of the most popular bandleaders in town." In Memphis proper, there were black clubs up and down Beale Street and scattered through the north and south sides of the city. When he was in the tenth grade, Love was tapped by Bowlegs Miller, and he had to regularly sneak out to make the gig; his father, a preacher, forbade this devil's music. Love remembers that his mother bought his horn when his father was out of town. They came out of the store and a big long car drove past, on the side of which was painted "B. B. King." Someone yelled out the car window, "You'll be sorry!"

"Memphis was a music town, and some of the best musicians in the country lived here and played here for five dollars a night, ten, twelve, fifteen dollars on up through the years," says Love. "Fred Ford would come by and borrow my saxophone and before he'd leave, he'd play some licks for me. 'Can you do this?' I learned from Fred Ford, I learned from Robert Talley—I had some of the best teachers around. When I was about fifteen, a big band asked me to try out. They told me I wasn't ready and to come back when I got a little more control of my horn. I was just happy to be there trying, high-school kid."

Underage Memphis teens both black and white were sneaking into clubs across town from each other to hear the same bands, who played differently—like different dialects of the same language—to each audience. That intensified the cultural divide the players had to cross when they finally met. The Stax sound is the result of a post-bebop generation coming together from semi-quarantined cultures to imitate different versions of the same bands.

The West Memphis scene careened along pleasantly for several years. "We thought we were sneaky for going to the other end of West Memphis," says Wayne Jackson, "which was really just down the street from where our parents lived. The deal was that Louis Jack knew all our parents—hell, West Memphis was five thousand people back then. Everybody knew everybody. And they knew where we were the whole time. Louis Jack would say, 'Well, I won't let 'em drink but a little, and I'll run 'em off before it gets too late.' We thought we was hiding. Mamma and Daddy'd say, 'Well, Louis Jack'll run them out about ten-thirty,' and sure enough, we'd be home about eleven."

"Hanging out in nightclubs really wasn't that big a deal," says Charlie Miller, the painter who loved to dance. "After the prom, you were expected to go to the PI. That was a tradition, everybody did it. Our parents knew we were drinking, but it wasn't that big a deal to them. When I was a kid, if you were drinking, it was uncool to look like you were drunk. Anybody that got out of hand was looked down on."

On the morning of February 20, 1960, one girl did not come home on time. One boy had gotten out of hand. "Memphis Girl Found Slain—Police

Say Boy Admits It" screamed the *Memphis Press Scimitar* headline. The sub-headline, "Stamp on Hand," rang the death knell for both fourteen-year-old Carol Feathers and the West Memphis club scene.

By her mother's account in the newspaper, the ninth-grader frequented the Cotton Club. "Carol was a wonderful dancer. Many times they'd clear the floor at the Cotton Club or wherever she was just to let her dance. She could just dance and dance, and she would always stay until the place closed at 4:30 in the morning, just dancing." On the night of the nineteenth, she ran into an old beau, former high-school basketball star Jerry Blankenship, seventeen, married, with a wife nine months pregnant. Blankenship drove her to an abandoned dog track in West Memphis that had become a lover's lane. She refused his advances, a fight ensued, and after hitting her with a tree branch, Blankenship, according to the newspaper, "left the dying girl, her life-blood dripping onto an old crime magazine in the dump."

Heavy emphasis was placed on the Cotton Club stamp still on Blankenship's hand at the time of his arrest. The papers reported that from behind the Crittenden County jail's bars, he shouted, "All this would never have happened if I hadn't been allowed to go in the Cotton Club and drink. They ought to padlock that place and burn it to the ground."

Blankenship got his wish. By the twenty-fourth of the month, the Cotton Club and Danny's were padlocked. The court order cited "public disturbances, unlawful drinking of beer by minors, quarrels, affrays and general breaches of the peace." The Plantation Inn was allowed to remain open. But the Bergers knew they had an image problem and soon built a new establishment, Pancho's, serving Mexican food. It included a club in the back, the El Toro, which became the hip spot in the mid-sixties. During the interim, the PI enacted a change and minors, as the sign had always declared, were not allowed in.

In consideration of their motto, "Having Fun with Morris," a large speaker was attached to the exterior of the building so the music could be heard in the parking lot. "I remember going over there," says Dickinson, "laying in the back of the car listening to the band through that metal horn, just drunk as a dog."

The venues for the young white bands in the late fifties were rental halls, churches, and YMCAs, with sponsors ranging from high-school fraternities to more stoic church groups. The bar scene was not open to them, nor was it appealing. Few of the older white bands were getting lost in the new music.

"We were playing with girl dancers and did the things that should have been impossible for a sixteen-, seventeen-year-old kid from East Memphis to do," says Dickinson. "The two other vocalists sang the slick pretty stuff, and I sang the ugly stuff. We played Muddy Waters and Howlin' Wolf and what I thought then was Chicago music, which was in reality all mid-South music. We used to save blues for the end of the night to run off the crowds, and I remember real well when they began staying, when they started requesting Jimmy Reed songs and stuff like that. I thought, Well obviously something is changing here."

"Jimmy Reed was somebody who had a bunch of harps and was real big and had a big old blue shiny suit and was insane," says Lee Baker, Mud Boy's guitarist. "He had a hit record, 'Baby, What You Want Me to Do?' on the white peoples' chart. They played dances and stuff here, like at Clearpool, and people would go out to hear Jimmy Reed. It was conducive to having a good time because you could hear that stuff, get drunk on your hidden bottle, and just go nuts. Which we did. Hank Ballard, Bo Diddley, Jimmy Reed—that was pop music. Not pop music as such, but talk about hit records and Jimmy Reed was right up there."

"The first records I ever bought were Jimmy Reed and Nat King Cole," says Don Nix of the Mar-Keys. "I thought that Jimmy Reed lived in New York in a penthouse. People that made records, especially records that I loved—I thought that everybody on the Jimmy Reed album was rich. I knew that Nat King Cole was on TV and he was a big star and real wealthy, so I just assumed that Jimmy Reed was, with a chauffeur and all that. And four years later I'm playing with Jimmy Reed! And he's bringing in the equipment out of an old Mercury station wagon, doesn't even have a roadie! It was a shock to me even then."

By the late 1950s, if your guitarist knew a Chuck Berry song, your band could get a gig. (Berry's first hit, "Maybelline," was released in July of 1955.) Most bands had incorporated rock and roll into their repertoire, but few had dispensed with the other, more acceptable styles. Only the Regents and the Mar-Keys devoted themselves to what was termed by others "nigger music," and the anger these two bands heard from those who used the term let them know they were reaching people. Jim King and the Crowns tried. King was a guitarist, and the few years he had on the Regents and the Mar-Keys put a showbiz taint to his act. His vocalist was Jerry McGill, who couldn't shake the full, almost operatic quality of his voice. McGill, however, was a wild man, and his devotion to the rock and roll life earned him a stint as Waylon Jennings's road manager and later a felony conviction for which he went on the lam. Sam Phillips recognized his talent and released one record on McGill and the Topcoats in 1959.

The fraternity parties that did not go to the wild white bands usually went to an older black guitarist named Thomas Pinkston. He maintained a staid society gig at the Tennessee Club, where many of these kids' parents were members. His playing was lively and full of character. But he'd been raised on Beale Street and was not easy to impress. His combos could perform on sandbars and in barns, places where there was no electricity and the focus of the party was a raging bonfire. Pinkston's business card read "World's Finest Negro Hawaiian Guitarist."

While they were dabbling in studios trying to record something good enough to release, the Regents, because nobody else would, began backing a clean-cut local singer named Kimball Coburn. In his favor, he had a regional hit with the apropos title "Cute," and he only joined them for two songs per set. But when he'd lay on his back and wiggle his legs in the air, the Regents didn't feel like the nasty rockers they knew themselves to be.

When Bill Black and Scotty Moore quit Elvis in 1958, their first gig back in Memphis was the going-away party for Wink Martindale's move to California. On the bill were local stars Thomas "Tragedy" Wayne, Warren Smith, and Kimball Coburn, the Regents' recompense. Bill Black was an extremely affable, warm-hearted, and humorous guy who loved and appreciated the craziness in rock and roll, and who died from a brain tumor at thirty-nine in 1965. In the small circle of Memphis rock, he'd previously admired Dickinson's band, noting that they played without a bassist. That night, he volunteered to join them onstage.

Each band did two sets. The Regents opened, with Dickinson singing "Send Me Some Loving." "Warren Smith came up to me after the set and complimented me on that song. It was the first time anybody heavy had said something to me like that. Blew me away." Then Ronnie Angel did two songs, then Kimball did his two. "Bill never asked what key we were in or nothing, just fell right in," remembers Dickinson. "We got to 'Cute,' which had four or five chords in it, which at the time was a hell of a lot of chords. Especially for one song. I was playing an upright piano and there was a microphone stuck up by the strings. Bill leaned down by me and says real loud over the music, right into the microphone, 'What's the name of this song?' He's playing along, and his question is broadcast over the P.A. I say, 'Cute.' He then announces, 'Never heard it.'

"Second set, Bill came on the stage with us again and by that time we were all reasonably lit. It was a little looser. Kimball's big song in the second set was 'Booby Obby Pretty Baby,' which I can still barely bring myself to say. He starts playing this song, and Bill starts shaking his head. He leans down to me, he had to see the microphone there, and he yells, 'Dickinson! What are you doing playing with this fruit?'"

The Regents did make a few early recordings, though they were so dissatisfied with what would have been their first single that they had to threaten to sue to keep it from coming out. How bad could it have been that the band would have canceled their debut? "The session was something our vocalist Stoots put together, and it was humiliating. It was called 'Education Blues,' and the producer's girlfriend had written it. We recorded it because we were children doing what we were told." With no pause in his conversation, Dickinson begins to recite lyrics from three decades back: "Brand-new pair of blue jeans/brand-new white buck shoes/I'm ready for the school days/Got the education blues."

Before Dickinson went to Texas for theater school in 1960, breaking up the Regents, they were invited out to the nascent Stax studio, then called Satellite and located outside of Memphis in Brunswick, Tennessee. "That was the first night I ever met Packy Axton," says Dickinson. "He was cooking hamburgers in the ice cream stand out front and running back and trying to make the music stick to the tape. You could see the needle move but you couldn't hear anything. Boy, that was so primitive, it's hard to imagine how primitive that was. Ampex mono machines. Ricky [Ireland] got on the telephone and called one of the local equipment suppliers. He comes back and says it wasn't working because the tape wasn't lubricated. [Laughs.] And Packy, I thought, We're out here in Brunswick with this fucking teenage wino! That's one of the reasons that I always thought that Stax shit was funny. I know where it came from."

CHAPTER FIVE

Kicks and Spins
and All the Flips

AS THE 1950S BECAME THE 1960S, MORE AND MORE WHITE KIDS WERE
digging black music. Once Stax Records began moving in that direction, the
route from a barn behind a burger stand in Brunswick, Tennessee, to the top
of the pop charts with Otis Redding's "Dock of the Bay" proved astonishingly
direct. However, soul music was not their original intention. The label began
as a hobby for a banker, Jim Stewart, who also played fiddle in a country band.
His sister, Estelle Axton, was a bank teller and thought her brother could carry
a nice little tune. The combination of their last names gave the label its title,
but its direction came from elsewhere.

Estelle Axton had a son, Packy, who was kind of big and a little bit clumsy.
He had a goofy grin and he loved, positively *loved,* having a good time. While
the story of Stax falls like dominoes, it was the oft-overlooked Packy who
tipped the first one. Jim Stewart the fiddle player wasn't considering a career in
black music, Estelle Axton the bank teller sure wasn't, and Steve Cropper, who
was, would never have been around the place had it not been for Packy.

"As it turned out, Packy was a really, really good saxophone player and
played on a lot of hit records and produced a lot of good records over at Stax,"
says Steve Cropper, revealing the happy ending to a story with dubious beginnings: "Charles Axton, we called him Packy, came to me one day in school, we
were going to Messick, and he said, 'Hey, I hear you guys got a band and I'd
like to be in your band.' And I go, 'Well, we're really not looking for anybody,
we're pretty happy with what we've got.' And he said, 'Well, I play horn,' and
I go, 'Well, I don't think we want horns.' It was two guitarists, bass, and
drums, and we were perfectly happy with that. We'd been playing a lot of sock
hops and dances and stuff like that. I asked Packy, 'How long have you been

playing?' and he said, 'I've been taking lessons for about three months,' and I'm going, yeah, okay, great. Somewhere in the conversation he mentioned something about his mother or his uncle having a recording studio, and I went, 'Oh, really?' And it sort of ended with, 'Can you be at rehearsal this coming Saturday?'"

Cropper played with a band called the Royal Spades. Born of youthful exuberance, the name now seems offensive; they say they were named for the suit in cards, like other soul groups. They had a regular gig near the naval base, where the rowdiness of the sailors was second to theirs. The quartet included Donald "Duck" Dunn, who would join Cropper in Booker T. and the MGs, the Stax rhythm section; Charlie Freeman, the envy of most other guitarists in town; and Terry Johnson, the youngest member, playing drums. Before expanding their lineup and releasing their first record, "Last Night," which propelled them to national fame, they changed their name to the Marquis, then changing the spelling so there'd be no doubt about how to pronounce it: The Mar-Keys. They became one of the earliest, if not the first white band accepted on the black music circuit.

"In about 1957, I was taking lessons from the Memphis Symphony and playing in a jug band," says Johnson, the drummer. "What we played wasn't even music, but a shopping center hired us anyway. Steve Cropper lived near there, happened to walk by, and then asked me if I wanted to play with him and Charlie. They told Duck to buy a bass and taught him. We did Chuck Berry songs, Bo Diddley, that kind of stuff. People hired us to play for free beer—I was thirteen so free beer was great—but after we played seven or eight tunes, we'd quit because we didn't know any more."

With the addition of Packy, the band moved their rehearsals to the Brunswick storehouse that his uncle, Jim Stewart, was converting to a one-track studio. Soon they added countrified keyboardist Jerry Lee "Smoochy" Smith; vocalist Ronnie "Angel" Stoots, who'd begun his professional career with Dickinson in the Regents; and perhaps the most important addition, an expanded horn section with Don Nix on baritone sax and occasional vocals and Wayne Jackson on trumpet. Several of the musicians in this lineup had already garnered professional experience. Cropper had written an instrumental, "Flea Circus," when he was fifteen that Bill Justis recorded in 1958, and, concurrent with playing in Brunswick, Cropper was also doing some session work at the Sam Phillips Recording Service. Roland Janes remembers, "We worked three or four sessions there with Jerry Lee Lewis where we had Scotty [Moore] playing the rhythm guitar, Steve Cropper played baritone guitar, and I played lead."

Wayne Jackson, the last member to join the band, regularly sneaked out his bedroom window to establish himself as "the West Memphis Flash," playing trumpet with acts ranging from the Arkansas All State Symphony to "this guy who billed himself as 'Jim Climer, Ninety Pounds of Rock and Roll.'" Married in the eleventh grade, he played the Memphis Rodeo and was paid enough to cover the expenses for his daughter's birth that year. "It dawned on me, I had just made more money in ten days than I had made all year."

His first gig in a nightclub prepared him for the excitement that would later surround the Mar-Keys. "Beale Street wasn't a place I went, but on North Main Street in Memphis there was the Copacabana. I went there with a drummer who was deaf, and I know I was fifteen because I'd just bought my car. I knew 'Cherry Pink and Apple Blossom White,' 'Stardust,' a few songs. We sat in and on the first song a fight broke out. I'd parked that Plymouth near the club. I never will forget, man, the bottles were flying and people were fighting, and I got my trumpet in my case and me and the drummer were headed for the door—Errol Flynn escaping! We were hanging on the wall trying to get by and some pregnant woman was there on crutches. She stood up and said to some guy, 'You can't hit me, I'm pregnant.' And she lowered the boom on him and broke her crutches all to pieces."

Jackson met a couple members of the Royal Spades when they heard him rehearsing with another band. They were looking for a trumpet to finish their lineup, and since the gig came with a recording studio, Jackson accepted. While the kids practiced, Jim Stewart learned to operate the equipment he'd bought when his sister had mortgaged her house. She kept her job at the bank, and she opened the Satellite Dairy, an ice cream stand, to bring in spare change. With the same eye toward cash flow, she began retailing records, discovering that it was a quick way to judge market trends. When Stax was finally established in a former Memphis movie house, they sold records from the theater's old lobby.

In Brunswick, Stewart made several unremarkable attempts at country music and white pop. The first effort released, however, featured the Veltones, a black vocal group. The reason these country gentlemen got involved with soul stirrers is the reason Packy Axton is responsible for Stax's direction: He was taking saxophone lessons from the Veltones' Gilbert Caples, whom he'd admired from the floor of the Plantation Inn. He brought these black musicians to the studio. "Like Dewey Phillips, Packy saw race in a more enlightened way than was typical in Memphis in the late fifties and early sixties," says Jim Dickinson, who later worked with him. "The community aspect of Stax in the ghetto, I don't think any of that would have happened but for the influence of Packy Axton, and Packy's friendship with Bongo Johnny Keyes, his black conga-playing partner who worked at the lobby record store. I don't think there's any doubt about what Packy brought 'em. Jim Stewart was upstairs in the office behind the dragon door. Packy was in the street. And that's where the music is."

"The music was changing in those days," says Andrew Love, the black saxophone player with the Memphis Horns. "The big-band players were getting older and the music in the clubs was rhythm and blues, the kind of music rock and roll came from. They used to call me Andrew 'Honky Tonk' Love, because of how I could play that Bill Doggett tune. I loved jazz then, still do, but I got married at an early age, and jazz players didn't make that many gigs and didn't make as much money around here. I had a young family so I had to stick to the more commercial side, the money-making kind of music."

"The Mar-Keys were the first white band in Memphis to have horn players," says Don Nix, who now lives near Nashville but bypassed all those recording studios to return to Memphis in 1993 to make *Back to the Well,* his first album in over a decade. "We used to sneak over to West Memphis to the Plantation Inn and those places where all the black bands played. And all the black bands had horns. So while everybody else was playing Elvis Presley songs with two guitars and a bass or whatever, we had baritone, tenor, and trumpet, and we played all rhythm and blues music, which no other white band at that time was doing in Memphis. Or anywhere that I knew of. It was just there, and nobody was playing it."

"We were fourteen, fifteen, sixteen years old," says Terry Johnson, the Mar-Keys' drummer, "playing serious black music. Other bands would play 'Walk, Don't Run,' basic guitar rock and roll. We would be out there playing 'You Can't Sit Down' by Phil Upchurch. We'd go over to the Plantation Inn and buy liquor. And they'd let us in over at Currie's Tropicana [off Beale Street], or we'd spend all night sitting on the curb down on Beale Street by the old Handy Club and listen to Evelyn Young play saxophone, sitting there with beer that some black guy had gone around the corner and bought for us. Sometimes we'd encounter 'What are you doing here?' and 'Get out of our neighborhood,' but when we said, 'Hey, we play, we want to learn this stuff,' everybody was real congenial. Particularly because Charlie Freeman was always with us. Charlie was the guy who really had this vision of white guys playing black music. And he had a silver tongue and he was slick and he could talk us into anywhere and get us anything we wanted anywhere we wanted. Charlie could always pull it off."

Painter Charlie Miller was a student at Tech High when Don Nix was sent

there. Tech was the city school's safety net, where the delinquents were transferred and taught a trade before they flunked out completely. "I went out to
listen to the rehearsals in a garage behind somebody's house," says the soft-
spoken Miller. "That was funny, I knew they were pretty heavy back then.
That guy Charlie Freeman, he was a fantastic damn guitar player. When he
played, you knew that was world-class. You knew right away. The others in
the band all respected the hell out of him. They knew how good he was."

But back then, a great guitar player did not a popular band make. Don Nix
was the loaded gun, the one who could draw the crowds because he was so
completely entertaining to watch. "Nix had a hell of a personality," says Miller.
"He could do the same thing anyone else would do, but it would be entertaining. Back when all the guys wanted to be Marlon Brando, Don wasn't quite
like that. Don was a real skinny guy, and everybody had their shirt sleeves
pegged to show off their muscles. Don's would be pegged, and his arms looked
like string hanging out. I remember one time at one of those dances, he walked
up to some real big guy, tapped the guy on the shoulder like you do to cut in,
and when the guy turned around, Don grabbed *him* and started dancing. And
then he could convince the guy not to beat him up."

The Royal Spades continued to play sock hops and teen dances, while
Charlie Freeman's style was winning attention from other musicians in town.
Well-known white bands began hiring him. Between his talent and Duck
Dunn's amiability and eagerness, they wound up integrating the Memphis
club scene.

In 1960 or early 1961, shortly after the Feathers murder, the Penthouse was
an ailing Memphis nightspot. Its owners approached an enterprising young
man named Herbie O'Mell. A go-getter, O'Mell had struck up a friendship
with Dewey Phillips in the 1950s, so that when he sponsored dances at the
Chisca Hotel, the disc jockey would encourage all the kids to come down.
O'Mell was a dancing fiend. He'd won a local twist contest, gone on to Dallas
and then the Palladium in New York where he was declared the national twist
champion. The afternoon gigs were his favorite. "Every club had tea dances,"
he says. "Every day. Boy, you'd go out there at one or two o'clock in the afternoon and I mean every married woman in town would be there, and salesmen
from all over. At five o'clock you'd better be out of the way of the door or
you'd get trampled, those women trying to beat their husbands home."

O'Mell's face is still boyish. Beneath a full head of curly hair, his prominent
eyes shine blue. Several of his clubs have been integral in shaping Memphis
music, and he remains active in the entertainment business, most recently
sought by the casinos now proliferating in the Mississippi Delta. "First thing,

I went upstairs to the Penthouse and saw they had a band in white tux coats with music stands. I closed the place down. I went out to Thomas Street and raided Club Currie's where the Largos and Ben Branch were playing. Ben was the tenor player, Floyd Newman on bari sax, Mickey Collins, he's a great piano player, and Big Bell was playing drums. Then I got Duck Dunn and Charlie Freeman to come up there, and that became the first integrated band in Memphis. It wasn't that I ran out and got Duck and Charlie and said, 'Here are the two guys.' Ben was talking to 'em and I was talking to 'em and they wanted to play."

Isaac Hayes, who sang with Ben Branch for five dollars a night, confirms the event. "It was a big deal. Everybody was talking about it. 'Wow, man! They got a white dude playing with a black band!' That was something in those days."

O'Mell consciously remodeled the Penthouse after the black joints on Beale. He was older than the upcoming crop of musicians; he'd been in the same grade as Elvis, though he attended a different school. As an entrepreneur, he had established friendships with Sunbeam Mitchell and various other Beale Street club owners, gaining access where other whites would not or could not go. "I was really influenced by the black scene in the late fifties. They'd let me in the Hippodrome on Beale and make me stand behind the bar, but I got to watch those acts. I was seeing Joe Henderson and Faye Adams and Evelyn Young and Bill Harvey and the guys that passed through his band. I could get into Club Handy or the Flamingo Hotel, and there wouldn't be three white people in the whole place." O'Mell was taken to the Village Vanguard in North Memphis, a jazz bar; to a no-name place at Third and McLemore beneath the viaduct that promoted younger black bands—like the Impalas, featuring the Hodges brothers before they settled in at Hi Records; to the Rosewood, to Melvin Malunda's and Melvin Bonds's clubs.

Willie Mitchell and O'Mell had become fast friends early on, and the two would go out late nights for chorus girls. "Willie would play until four or five in the morning at the Manhattan Club, and we'd leave there and go over to a place called Tony's and order a plate of what was called 'chorus girls.' They'd take a can of sardines and line 'em up real pretty on a big oval platter, and we'd eat those while listening to Ironin' Board Sam play."

When the Penthouse reopened, it was rocking. It was the first place whites could go within the city limits for the West Memphis experience. "I'll never forget the first two weeks," says O'Mell. "Big Bell made Duck turn around backward to the audience and face him. He said, 'I'll show you how to play that.' I think Duck got a lot of influence there playing with them. It was really a great band." Little Willie John was a regular at the club, living at Herbie's house for a few months. Despite the excitement, the Penthouse lasted less than a year. Though the place stayed jammed, the cops, in an effort to discourage the interracial mixing, regularly stopped cars leaving the club, harassing the patrons. O'Mell says squarely, "The people stopped coming because they got worn out by the Memphis Police Department."

While gigging at the Penthouse, Freeman and Dunn were still rehearsing in the studio with the Royal Spades. Things were heating up there too. They'd moved from the country to an old movie theater that producer Chips Moman found. Mr. Stewart, in Wayne Jackson's words, "was trying to plug the walls into the walls and figure out what a microphone was," and the Mar-Keys were developing a basic riff into a basic song, one that would become a national hit they could call their own, even if no one can say with certainty who is playing what on the record. "'Last Night' was the biggest mistake that ever happened," says Terry Johnson, whose version of events turns magnetic tape to Scotch tape, but it conveys the aptitude—and attitude—of the people trying to plug the walls into the walls. "'Last Night' has eighty-six splices on it," says Johnson, "and probably twenty to twenty-five musicians. I would seriously doubt if any musician played through the whole tune. Probably the most famous person on that record was Steve Cropper, who is a real premier guitarist. People don't even realize there was no guitar on the song! Cropper's alternating keyboards with Smoochy. Some drummer named Curtis Green, I know for sure, did the final roll on it because he could never explain to any of the rest of us how in the hell he did it. And what it was was another mistake. He entered half a beat early, and somehow tripped himself up and caught himself, and it came out to this great, terrific roll. He got his money after the tune was cut and we never saw him again."

Like Ray Charles's "What'd I Say," released two years earlier, "Last Night" is so simple it's almost silly. But it's that very baseness that makes it transcend time, makes it as exciting today as when first released in 1961. "What'd I Say," with its sultry call and response, captured the crest of America's sexual freedom. "Last Night" is about ass-shaking. It's an instrumental with a bunch of horns and a cheesy organ. It says, We know we're not much, folks, but watch this. And then, like a jug band, from nothing they make something great. You can feel the fun in the song, the teenage freedom, the glee of kids achieving something they thought beyond their capabilities. The song takes us with them as they sneak out to West Memphis when they're supposed to be dreaming in the comfort of their parents' heated homes. It plays just behind the beat, most obviously when the musicians break, and suddenly we feel like we're fall-

The Mar-Keys, before and after the innocence.

On tour, August 1961. Left to right: Wayne Jackson, Packy Axton, Steve Cropper, Don Nix, Ronnie Stoots, Terry Johnson, Duck Dunn. Photo courtesy of Memphis Music Hall of Fame.

Mar-Keys promo shot taken in spring 1965 for a Ray Brown–organized European tour that didn't happen. Left to right: Don Nix, Duck Dunn, Terry Johnson, Wayne Jackson, Packy Axton, Steve Cropper. Photo courtesy of the Memphis Music and Blues Museum.

ing forward. The horns, just barely in time, come in and save us. That sort of tension was common in clubs in Memphis and across the river, but the way people outside responded was another hint to locals that things happened differently here. "Last Night" was not Ray Charles, not Motown, not slick, and people all over the country loved it.

And no one had any reason to suspect it was white kids.

The guys who had recently been trying to shoo Packy Axton in the high-school hall were suddenly national sensations. They'd found their voice in R&B, bypassing jazz. Dewey Phillips liked the song because the break gave him a natural place to talk. After the record got hot regionally, Atlantic gave it national distribution. The Memphis company learned of another Satellite label already established, and "Last Night" was reissued as the first release under the new Stax name. The record's staged growth allowed Terry Johnson to finish high school before touring. Between its recording and release, Charlie Freeman had taken a job with a "mickey" band, a road act that wore fancy clothes and played Mickey Mouse music; he was traveling when he heard the song on the radio and did not rejoin the Mar-Keys until Steve Cropper quit during the first tour. For the tour, the Mar-Keys brought Ronnie Stoots as vocalist, Carla Thomas came along to promote her hit, "Gee Whiz," and Mrs. Axton boarded the van as the watchful chaperone. The kids wasted no time in grossing out the women, who quickly fled. With the van to themselves, the road as their home, and stardom as their oyster, they toured for months at a stretch, returning to Memphis long enough to throw wild rock and roll parties, and then setting out for another string of one-nighters.

"We were making a hundred dollars a night in 1961!" remembers Jackson. "Plus royalties—in our teens! People were taking our pictures!"

"What you had was eight guys between the ages of eighteen and twenty who just wanted to get out on the road and play and party their butts off. And that's exactly what we did, and that's probably why the band was a one-hit wonder," says Johnson. "It's hard to be a two-hit wonder when you leave the Dick Clark show and he's waving at you and everybody in the band is shooting him the bird. Typically, they don't play your second record when you do that. It's hard to be more than a one-hit wonder when you've got a twenty-one-day tour of Texas booked and the whole band takes off to Mexico. Mexico was great. We'd come into this little town, all of us in a line like in the Westerns, and the Mexican kids in the street would start going, 'Ba-dup ba-dup,' imitating the horn riff from the song. After three weeks we phoned home and said, 'We've lost our bus because we sold it.' All we wanted to do was have fun, chase women, drink beer. We did it the storybook way. And I think everybody at the time knew we had to live this experience now, and the hell with the consequences. We wanted this memory burning in our brains when we're sitting in the old folks' home, incontinent."

"We were just dumb teenagers that had never been out of Memphis," says Don Nix, "and it's a wonder we're still alive. Ninety percent of the places we were booked were black clubs, and in 1961 it wasn't really cool for either the white side of town or the black side of town to have white teenage boys playing in a black club."

On the road, their gigs were divided between package shows, sometimes with R&B stars, sometimes with country acts, and roadside joints where the Mar-Keys were the evening's entertainment. On the package shows, segregation forced them to stay in separate hotels, eat in separate restaurants, and ride a separate bus. "We were booked at the Regal Theatre, which is in South Chicago, all black neighborhood, all black audience—all black," says Don Nix. "There were nine other acts on the show. LaVern Baker was the headline. Everybody else would go on and do their record at the time, one or two songs, and come off. But LaVern got to do thirty minutes. We came on before her because 'Last Night' was number one and the flip side, 'The Night Before,' was number two in Chicago. So we got to come on next to last.

"The announcer said, 'The Mar-Keys!' and there was a lot of applause as the curtain opened and then everybody in the audience just kind of set and looked. And you could hear throughout the audience: 'White boys!' Oh lord! But the Mar-Keys were a really good band at that time. And we destroyed that audience. So much so that LaVern Baker came out and they didn't really want to hear her. The next day we were the headliners and she was on before us. Which didn't sit very well with LaVern Baker and she cussed us the whole time. We did a whole week there. But on the last night they had a big party for everybody and she bought us a bottle of champagne."

"Don knew more about entertaining than we did," says Wayne Jackson. "He knew to be crazy and do really wild stuff. If the audience threw stuff at us for fun, he'd throw it back and start a riot. We didn't know to do that. So we thought he was nuts. But everybody liked Don, he's got a personality. He looks right and he's funny and he just would be there—that was his talent. Like Steve Cropper had a talent for being in the right place in the right circumstances, Don Nix had his own kind of talent."

Cropper remembers another kind of show. "We played up in Kentucky, a place called the Cherry Club. It was back in the woods, and we had to drive way down this dirt road, took us forever to find it. We got the van to the back door and started unloading these instruments, and there's not too many people around. It comes pretty close to about show time, and now these people are wondering what are all these white guys doing. And we had noticed, when we were setting up the bandstand, there were holes in the wall that looked like they could have been made by a shotgun or something. It turned out to be a pretty rowdy place. And of course they started drinking early. When we got up to play, they all but booed us off the stage. And I never will forget a big lady grabbed a butcher knife out of the kitchen, she had heard us warming up during the day, and she jumped up on stage and says, 'You're gonna listen to these guys!' because she knew that it was us who had the hit record. And so we started right off with 'Last Night' and the minute we hit it, six bars in, everybody's jumping, dancing, and at the end of the night they didn't want us to go.

"People always had doubts about us being the same guys who had 'Last Night,'" he continues. "We'd go out there and open up with Ray Charles's 'Sticks and Stones' or something like that and do all these steps and flips we'd learned from West Memphis, and people were just amazed. Duck and I would flank the horns so you had two guitar players with three horns in the middle—a five-man front—and we're doing kicks and spins and all the flips and all that, people couldn't believe it. We put on a pretty impressive show, and that kept us in big demand everywhere. We drew from what we'd seen in clubs as teenagers, and I think that's what made the Mar-Keys go over. We had to perform that, that service, do more than just play our instruments."

There was an ongoing power struggle in the band and that tension kept their gigs crisp and tight. Packy, drunk and whining, complained that since he'd brought the Mar-Keys into the studio, the band should be his. Cropper, unflappable, stern, the responsible one, wasn't about to yield the reigns to an alcoholic. He did finally quit, though Packy was not the victor. Cropper returned to the studio, his first love, and while touring one day in some indistinguishable city, Packy saw his band's second album for sale.

"Last Night" had made it clear that the Mar-Keys were never assured of playing on their own records. However, they had at least been there; they had even, by default, become the studio's first house band, backing William Bell ("You Don't Miss Your Water") and touring behind him regionally. But as the studio drew more players, and as the Mar-Keys kept to the road, they had less and less involvement with songs issued under that name. (By the mid-sixties, a Mar-Keys performance was simply Booker T. and the MGs with horns.) "The second album was on the market, and it sure as hell couldn't have been us," says Johnson, "because we were gone the whole time. But Floyd Newman and those guys put a nice one together."

"I don't know if Charlie Freeman ever got to play on the Mar-Keys records," says Wayne Jackson. "Terry wasn't a studio-quality drummer—at least that's what they said. And Packy could not get along with Jim Stewart, so he wasn't allowed. Don Nix can't play a lick and he would do strange things, so he was out. He eventually learned to play enough to keep his job. And Smoochy was such a redneck, he didn't fit in the studio scene, although he continued to bring in little songs and things. Duck, Steve, and I were the only ones who actually made it into the studio."

"I didn't play on any sessions after a certain point," says Nix. "Not after they got good musicians to play. I was the only thing they had for a while. I could do it on the road, but for records it was clear they'd rather use other guys. I understood that. Eventually, I was producing, and that's all I ever wanted to do. I wanted to write and to put records together in the studio."

The touring Mar-Keys milked "Last Night" for nearly two years. Stax continued to release follow-up songs ("Banana Juice" is one of my favorites), then a second album. In 1963, they played a cold three-week stint in St. Paul, Minnesota. They had no record on the charts, the gigs were paying less, Freeman and Packy were getting crazier, and it was starting to snow. "We all hated each other's guts from being cooped up on that damn bus for so long," says Johnson, adding that they all remain friends—those who survived. "We quit, and it looked like rats leaving a sinking ship. We'd run the bus into the ocean when we were in South Carolina so we were taking our own personal cars. It was eight guys saying, 'It's over, it's over.' And driving south at this incredible speed just to end it.

"Of the people in the band, the live ones ended up staying in music. When we used to say, 'What do you want to be when you grow up,' none of us ever thought you could make your living as a musician. Packy's ambition, he would tell us, 'Be an alcoholic, that's what I'm gonna be.' And he set out on that trail. And Charlie [Freeman] was set on that path very early. Packy died at age thirty with acute cirrhosis of the liver. And a drug overdose at thirty-one killed Charlie Freeman. They could have been plumbers or fishermen or God knows what, and they'd still be dead."

Theirs were two of the earliest rock and roll deaths, warning flags to others. Axton and Freeman had never suspected their destinies could really be so free. The life their talents brought them made every day an unreality, and indulgences had been part of the unreal game since they'd been invited to play. Two polar opposites taken down in their primes: Packy too big for his body, constantly pushing toward the front; Freeman introspective, a master of subtlety and taste.

Steve Cropper became an essential member of the Stax organization, writing songs, producing, and playing guitar. Don Nix would produce many albums for the company, working hard to bring a white hit to the label—Stax wanted what Sun had gotten—and in the 1970s he worked closely with Leon Russell and Joe Cocker. Terry Johnson became a clinical psychologist, and vocalist Ronnie Stoots a graphic artist; both keep a hand in music. Smoochy Smith has run several nightclubs and always remained a player, currently joining several Sun veterans playing rockabilly in the Sun Rhythm Section. Duck Dunn and Wayne Jackson were integral at Stax, and they also remained active in Memphis clubs until 1967, when they realized they could make a living without wearing themselves out every night. "It takes a long time, when you're good and people love to hear you play, to get the hard-on down," says Jackson. "To get where you don't want to play all the time. I still love it, and if it were fun music, I'd still be out there playing all the time. I moved back to West Memphis after the Mar-Keys, and on the way home from my gig, I'd pass the El Toro, which had been the Plantation Inn. Sitting in on someone's last set was just strictly to do it, to go play something different. Not for pay. A lot of times there would be just enough amphetamine left in me at two A.M. to want to stop in there and play until three, chase some leg around the room and drink enough booze to go home, try and get some sleep and do it again."

CHAPTER SIX

I Know You Can Play, But Can You Dance?

IF YOU BUMPED INTO MUD BOY'S GUITARIST LEE BAKER IN A DARK alley, you wouldn't know whether to run for safety or loan him a quarter. He is burly like a mountain man, bearded, a hippie's ponytail. There's a sense of Harley-Davidson all about him. The biker's road, in fact, is one he might have ridden. In high school, his attitude got him regularly kicked out of his parents' house. He ran with similar company, passing many teen nights with his child-hood friend Mike Alexander, a bassist, talking themselves to sleep in their cars, home away from home, parked in the East High parking lot.

Instead of focusing his renegade spirit on a six-gun, Lee Baker chose six strings, developing his artistic side and creating a singular guitar style that he learned as the country blues masters' chosen one. Baker can hang a note like age, or gnarl his strings like a watch spring uncoiling, resurrecting Furry Lewis and Mississippi Fred McDowell every time he runs a bottleneck slide down the neck of his National steel guitar. He plays with a spaciousness and respect for silence that usually requires more decades than he's yet lived. Barely fifty, he was a malleable teenager when he met the people who created the blues, who invited him to accompany them because he didn't impose his style on theirs. "I can imitate Elmore James and stuff like everybody does," he drawls in his gentle Arkansas accent, "but that's not what Furry did. Furry taught me how to lay back—a whole lot. I think the reason that Furry tolerated me play-ing with him is that I didn't get in the way. I played with him, I played with Bukka White, Gus Cannon, Sleepy John. I'd just play rhythm and listen to what they were doing."

Furry Lewis could express the world with a single note. He taught Baker to respect the space that surrounds that note, before and after, using the silences to create three notes from one. Baker lives with his family in a hundred-year-old cabin on a lake in Arkansas, where the soil and the sky continue to teach his kids what Furry first showed them. From his screened-in porch in late summer, the cicadas may be nearly as loud as the urban noise half an hour away, but they're much more eloquent. He's got roosters, chickens, and dogs in his yard, a big blue tractor, rusting cars. Fields of cotton have become fields of soybeans, reflecting the steadiness of time, the unchanging nature of change.

A few years younger than Dickinson and the Mar-Keys, Baker was an early beneficiary to the small tears they made in society's fabric. When age differences would hold less importance than character, he became a peer. White pop attracted Baker not at all, and after he found a collection of regional blues recorded by folklorist Alan Lomax and released in 1960, he sought his way to the other side. (The series was repackaged in 1993 and rereleased by Atlantic as a four-CD box entitled *Sounds of the South*.) "I came to the blues through people like B. B. King, Freddie King, and Albert Collins," Baker says. "Wes Montgomery. I loved James Burton. I used to watch him on the 'Ricky Nelson Show,' and there was just something about the way he played, real lyrical, a country player. That solo that he does on 'Fools Rush In' is one of my all-time favorite rock and roll guitar solos. But when I first heard Fred McDowell, I said, 'My God, what is this guy doing? Just one guy and he's doing all this?' I said, 'It's only got one chord!' I didn't know how to do it, but I heard it and I started trying to play bottleneck a little bit, just sliding up and down the guitar."

Within a short time of hearing Fred McDowell on record, Baker would find himself at the bluesman's side, and invited to return. His training as an accompanist began when he and Mike Alexander, jamming on teenage ideas while the sun rose over their steering wheels, heard about a gig in the pit band at a talent show. The W. C. Handy Theater was in Memphis's first black subdivision, Orange Mound. They went together to audition. "The group was called the Ultrasonics, about six pieces," says Baker, "and me and Mike were the only white guys. Our job was to back anybody that came in. They wouldn't know what key they were in, they wouldn't know the name of the song, they didn't know nothing. We'd have to feel our way along, and it was cool for me, exactly what I wanted to do, exactly where I wanted to be.

"They had a tenor player named George, and George was nuts, certified crazy. He was a serious jazz player and had played with a bunch of people, but no telling what he took. He'd say he was playing the molecules; not the notes, but the molecules. Our band had a gig backing Rufus Thomas, and George wound up turning everybody out of the club, going crazy, had the band running, breaking bottles and saying, 'All rags must go.' He was so far out that I don't even know what to say about him. The piano player was named Tommy Lemmons. He'd been on the road with the Five Royales or somebody and he was good, but he was a juicehead, it was interfering with his thing. We rehearsed at the promoter's house, right next door to the theater. They called him Longheaded Joe, and he had a brother or something that was also crazy, and we'd go in there and this guy would come in, he had one weird eye, we'd play the old Mar-Keys song 'Bo Time,' and he'd start dancing, just going nuts. We'd be in this guy's living room rehearsing and he'd just yell, 'Hey, "Bo Time!"' and boy, he'd be dancing, it was great."

The pit band's job was not to play the song properly, but play it as the talent thought it went. By the time he met the bluesmen, Baker knew to shape his playing around theirs, to change chords when they changed, whether it was the twelfth bar, the tenth, or the fiftieth. The structure of the blues is the structure of a story being told, and everybody tells theirs differently. It's possible Baker could have enjoyed a career as a successful session guitarist. He has maintained a heartfelt devotion to the instrument and was developing a rare versatility while in his teens. But his music, his life, was irrevocably changed. Rather than generic excellence and commercial potential, he was transformed by the time these bluesmen accorded him. He made himself a willing disciple and, as if from outside his body, he watched everything he'd learned become bent out of shape, rearranged, until today he slips his bottleneck slide over his pinky, and his own parents wouldn't recognize him.

Country blues has always been another name for obscurity, the remarkable sales of the 1990 Robert Johnson box set notwithstanding. So instead of taking calls for session work, Baker spends his time on a tractor, every turn of the blade, every field cut adding a little more frustration for him to release the next time he plays for his small audience. When he tears loose live, he is the definition of rocking, pitching his upper body forth and back in a tight motion, his feet firmly planted to keep him from whirling away like the Tasmanian Devil. He cradles the guitar in his body, his head turned slightly away, safe in case of explosion, soloing, soloing, a spiral staircase with floor on top of floor. Baker has tasted the mass appeal of fame, and the flavor still lingers. In high school, his band was one of the busiest in the region; in the early seventies, he led Moloch, a pioneer blues-rock band, releasing an album on Stax and playing gigs with the MC5 and Iggy Pop. He reached a new audience later that decade through the Alex Chilton records *Big Star 3rd* and *Like Flies on Sherbert*. Baker's hands grip the tractor's steering wheel tighter, his knuckles turning white as some kid in France grooves to his sound on the Mud Boy album

Negro Streets at Dawn, as British kids try to imitate him in their gothic blues creations, as he thinks about talking to people with his guitar. Whether two or two hundred show up to see him play, he speaks the same language.

While training to become unknown at the Handy Theater, Baker was simultaneously playing in the local vacuum left by the Mar-Keys' move to national prominence. "The Blazers had four horns and three rhythm pieces and a vocalist," he says of his high-school band. "Our setup with the horns and all was probably due to the Mar-Keys. Our attitude was, we play the Booker T. shit as good as Booker T. and the MGs. We knew we were bad. We played all the Memphis stuff plus we did lots of James Brown, because we had the horns. We worked up about half the *James Brown Live at the Apollo,* man, we did it with all the kicks and everything. Plus, in those days, you had to play standards, 'Three Coins in the Fountain,' you had to be able to play a lead-out for a prom and stuff like that. And we were playing real hard blues and rock and roll. I respected the early R&B players because they were really jazz cats and could play the funky stuff. I always wanted to be more than just somebody that could play three chords. Shit, in high school, man, I never asked my parents for any money. We could have worked every night if we'd wanted to."

Until the arrival of the Beatles in 1964, the model for success throughout the mid-South remained the West Memphis scene. A whole generation who never experienced the PI spoke of it with reverence, emulating bands who, at their best, were an imitation of the real thing. While a great number of bands like the Blazers were drawing from the rowdiness across the river, clean-cut Tommy Burk and the Counts drew from the slick vocal groups there. They studied harmonies, tuned more to Dion and the Belmonts and other "American Bandstand" acts. They packed their shows with loafer-wearing, bow-headed clean-cut fans. Managers were afraid of the wilder bands, but an entertainment attorney in Memphis named Seymour Rosenberg heard the Counts at his daughter's dance, or maybe he just saw the crowd response, and he offered them a recording contract.

Rosenberg played trumpet, managed Charlie Rich, and cofounded American Recording Studio with Chips Moman, which would later be home to the Box Tops and attract Neil Diamond, Dionne Warwick, Wilson Pickett, and Elvis. The Counts were one of the earliest acts to record there. Though their debt was not to the Mar-Keys, it was to West Memphis. They hit with their second single, copping the Spaniels' arrangement of "Stormy Weather" that they heard at the PI. Area bands couldn't perform without people requesting the Counts' version of "Stormy Weather."

Like the Blazers, the LeSabres were a horn band, and Randy Haspel, who watched from the audience until forming the Radiants, remembers the competition between the two types. "The rivalry between the LeSabres and the Counts was like the Mods and the Rockers in England," he says. "This was like the battle between the greasers and the Ivy Leaguers. The Counts dressed real sharp in blazers that had their own crest on the breast pocket. The LeSabres came straight out of the Elvis, Billy Lee Riley kind of thing. Where the Counts were very disciplined and did steps and had a lot of vocal harmony, the LeSabres were like leather boys with greasy hair, no discipline, smoking on stage and running around."

The Counts won the battles of the bands, but they lost the war in music. Their generic character offered nothing new or unusual, and they could never break beyond a regional following. After the demise of the LeSabres in 1962, their guitarist, Laddie Hutcherson, began touring the region with a racially mixed band, playing colleges. "I met a guy who took me to Beale Street," he says, "took me to see Bowlegs Miller at the Flamingo Room. I topped two flights of stairs and when I opened that door I thought I'd gone to heaven. I thought, This is where my heart is. I started going down regularly and never dreamed I'd get to sit in with the band. But soon enough, Bowlegs told me to bring my guitar and it would alternate between me and Teenie Hodges and Lee Baker."

Bowlegs Miller recorded for Hi Records, where Teenie Hodges and his brothers backed Willie Mitchell and, later, created a classic sound behind Al Green. Bowlegs Miller scouted talent for Hi, discovering Ann Peebles ("I Can't Stand the Rain") among others. Though he never achieved national fame, his patience with younger musicians assured his influence. "Bowlegs didn't try to tell me what to play," explains Hutcherson, "he just showed me when. Andrew Love had a lot to do with it too. I was on a riser behind the horn section, and I didn't know a lot of the songs. The first time Bowlegs counted a tune off, Spencer Wiggins was coming onstage to sing and I had an intro to play. Bowlegs said, 'Five flats,' and started counting. I didn't know five flats from one flat. I said, 'Andrew! What's this?' He said, 'D flat.' So I went into the intro, but when Spencer started singing, I kept on playing. Bowlegs turned around and it just took one look from him for me to learn not to ever, ever step on the singer."

A couple years earlier, it would have been impossible for Hutcherson or Baker to share the stage with black musicians. But two years before President Johnson would sign the Civil Rights Act in 1964, and two years after Charlie Freeman and Duck Dunn had joined Ben Branch's band, Memphis stages and audiences were casually breaking down barriers. Often, it was high schoolers who were blazing the way. If making music was still not something they could do with their lives, it was becoming apparent it might last past graduation. A local booking agency, National Artists Attractions, had a constant demand for talent. Established when the Sun Records explosion had settled and the artists were gigging to earn a living, it drew many clients from Stax. Walking through their office door was stepping from society's strictures into a musician's world. "You could go there and you weren't weird," says Lee Baker. The proprietor was an avuncular man named Ray Brown who'd been a prominent white disc jockey on WMPS, WHBQ's rival station. "Mother Brown's Round Mound of Sound." Brown and his partner were relaxed and friendly, running an office where blacks and whites were equals, and everyone could talk shop and tell Jerry Lee Lewis stories. National Artists Attractions needed its own neon sign: "Having Fun with Ray."

"I got my first gig with Charlie Rich because I was hanging out at Ray Brown's," Baker says. "I was a junior in high school and they needed a guitar player, so I went. Charlie hadn't heard me. Ray said, 'You need to be at this club in Waynesboro, Tennessee, by seven o'clock. Meet Charlie at the motel at five.' First time I ever shook hands with him, his manager gave me some pills. We played several gigs together. The principal would let me and Mike Alexander out at noon Friday so we could go to Columbus, Mississippi, or different places. There are so many opportunities to self-destruct in music, and when you're real young or if you're of an addictive personality or something like that, you can get really messed up. All the people that are still afloat have come through it. Charlie's one of them. One time they carried us offa this base somewhere, and we wound up at this country club in the middle of the night out in God knows where. Charlie's just drunk as a dog and he started playing all this jazz. Really good, intricate piano stuff, and he could hardly hold his drink in hand. I remember thinking, Well, I'm running in pretty fast company here."

Ray Brown's was like a day labor office, where the phone might ring and suddenly there would be a call for guitarists. Each weekend, various versions of the Mar-Keys were put together, two, three, four bands with the same name heading out to different locations. When Lee Baker played a Mar-Keys gig in Texarkana with no horns, the crowd got unruly. "Me and Charlie Freeman worked out the horn stuff on guitars," he says, "so they recognized 'Philly Dog' and all that. And after people had a few drinks, it didn't make any difference." Dickinson and Freeman, who would end up playing together behind Aretha Franklin in an Atlantic Records house band, first played together as Mar-Keys; Laddie Hutcherson's popular sixties group the Guilloteens met the same way. And when the Mar-Keys name wore out, Ray Brown would hurl a *Billboard* at the musicians on his sofa and say, "Who do you want to be this weekend?"

Laddie Hutcherson, late of the LeSabres, went out one weekend as Ronnie and the Daytonas, the Tennessee band that had a surf hit with "Little GTO." When the promoter asked why the van had Mar-Keys written all over it, they said their own vehicle broke down so they'd borrowed their friends'. The second day out, they stopped in a music store and were swarmed by kids. "We're signing autographs as the Daytonas, big-timing it," says Hutcherson. "One of my friends comes up to me, batting his nose, nervous, and says, 'C'mon! We've got to get out of here, right now!' I said, 'Hey man, sign some autographs.' He told me to turn around, and there was a row of records going all the way across the wall, Ronnie and the Daytonas, big photograph of them, and not any of us in the picture."

National Artists Attractions booked stars in distant cities and booked the funkiest places at the smallest crossroads. "There was a gig that was famous among musicians," says Baker. "I played it with Booker T. Jones, a honky-tonk in the middle of nowhere, the Big Apple in Birdsong, Arkansas. For fifty cents you could get a fucking barbecue as big as two hands. I mean it was good! The club had a wood stove and all the country people came, and they didn't want to hear nothing but blues and funky. No jazz, nothing cool, just down-home gut-bucket blues. Whites came in there too, because in the country there's not—there's a distinction but they still socialize. There was usually white men, and not usually white women. They had quarts of beer and they had gambling, people coming to listen to music and play cards."

Original Mar-Keys Wayne Jackson and Duck Dunn were getting studio work at Stax, but they continued to gig regularly in the Memphis area. "You have to remember that I had had a hit record immediately," says Jackson. "Before that, it was just Willie Mitchell through the fan in the parking lot. When I got back from touring, I had to learn showbiz. Club owners wouldn't let you stand still. They'd say, 'I know you can play, but can you dance?' And you better say, 'Yeah, I can dance. And tell jokes too.' You had to make people want to come back and pay that $2.50 again. There's an art to that and I learned it from Robert Talley, a black keyboard player and bandleader who was older than us. I worked with him for two and a half years. Every afternoon he'd come in and teach me a song to do that night. I probably knew three hundred old ballads with beautiful melodies before that was over. Robert taught Duck to hear outside of just R&B. He'd tell you what notes to play coming from that change to this change. And he played so well, it was easy to learn from him. He was a great teacher."

Robert Talley still performs in Memphis clubs, seventy-four years old, retired from the post office, and agile on the piano. He cowrote the Mar-Keys' follow-up hit, "The Morning After." His current groups don't share his breadth of knowledge, and his passion for the old days is almost tangible. "These guys today," he says, "I have to play their program. They don't know the stuff I know. At the Rebel Room, with Wayne and Duck Dunn, Terry Johnson on drums, they kicked my ass. Wayne could hear grass growing, he's a talented guy. We did all kinds of songs and kept that audience dancing. We had a versatile band. When they went to Stax, they were good, but even then it didn't have the fire, the innovation, like we had on that bandstand."

On the folk scene, there was growing personal interaction between the artists, white and black, young and old. By the mid-1960s, these encounters were popular enough to be called a national blues revival. Experiences such as Charlie Miller's were not uncommon for a mid-1950s afternoon. "There was a black guy who had a little shack in one of the alleys in a white neighborhood. My school friends and I would go there, he'd buy wine for us, we'd buy him a bottle too, and he'd sit around and play blues. I can still see that in my mind, the wood-burning stove with this guy sitting there playing, little kids taking a drink of that white port wine and getting just sick as a dog."

No longer schoolkids by the early 1960s, the witnesses discovered languishing Delta greats not only alive, but living around the corner. "Having seen the jug band in Whiskey Chute, I knew these men were out there somewhere," says Dickinson. "Until the Samuel Charters book [*Country Blues*, 1959], there didn't seem any possibility of contacting them. In the summer of 1960, a friend and I followed the trail that Charters left to Gus Cannon, who was the first one I actually met. He was the yardman for an anthropology professor. Gus had told this family that he used to make records and he had been on RCA and they'd say, 'Yeah Gus, sure, cut the grass.' When we met him, he was bending down over the lawnmower, he had this big Russian rabbit hat on, like Davy Crockett. He lived on the property, back over a garage, and he took us up into his room, and on the wall he had a certificate for sales from 'Walk Right In,' for which of course he didn't get any money. And he had a copy of the record that Charters had made for Folkways, but he had no record player. That was a real good introduction to the blues."

In this first encounter with an original bluesman, Dickinson and his cohort were anxious to learn about Cannon's roots. They were taken aback when, after asking where he learned his material, he answered, "From the radio." "It really surprised me," recalls Dickinson. "That's where I had learned stuff. These guys had learned it exactly the way we had. What else were they exposed to except each other?"

By the early 1960s, rock and roll had become formulaic enough to send Jimmy Crosthwait, washboardist in Mud Boy and the Neutrons, searching for new directions. "I was playing in rock and roll bands but I was listening more to jazz. Mose Allison, Charlie Mingus. And by the time of 'Last Night,' I'm just about to be caught up in the folk thing. Dylan comes along around sixty-two and makes folk really kind of fun and not so college preppy as Peter, Paul and Mary or the Kingston Trio and all that creepy shit. There were also all of the Alan Lomax folk albums. Around 1961, when I was about sixteen, I found a little black joint downtown that served greasy hamburgers called the Cotton Row Inn. It had a great jukebox with Art Blakey and the Jazz Messengers, Coltrane, some really good stuff. I loved the atmosphere and I was a beatnik going into the only place I could find where there was jazz on the jukebox. All of this blurs for me with the sit-ins that were happening, black people going into white joints. That's a real confrontational thing, and so I would feel a little leery going into the Cotton Row, but I did, and it worked."

There was a record store on Beale Street called Home of the Blues that attracted blacks and whites. A peculiar white man named Ruben Cherry ran the place. "Lots of people didn't like Ruben," recalls Milton Pond, who frequented every record counter in the city when he wasn't behind the one at Poplar Tunes. "They thought he was pushy and obnoxious. If two people were in there, he made it feel like a crowd. The main thing I remember about him, up by the cash register, he had a nickel glued on the glass counter. He'd wait and wait for somebody to try to pick it up, and when it wouldn't move he'd get the biggest kick out of that."

"I went into Ruben's once with Stanley Booth," says Dickinson. "Ruben kept this rubber rattlesnake behind the counter which he used to scare off would-be stickup men. It was rubber and when he held it, it really looked real. Stanley says, 'Hey Ruben, where's your rubber snake?' As an answer, not as telling a story but just answering the question, Ruben says, 'That goddamn Elvis Presley, he came in here and stole my rubber snake and ran down Beale Street shaking it.' The thing that tickled me so much was that his anger at Elvis was real. He'd known Elvis before he'd started recording. Elvis bought records from Ruben Cherry same as everyone else."

Established in 1949, Home of the Blues had the flavor of old Beale. Wood floors, worn counters, old record racks. Cherry had been president of the local Variety Club, an entertainers organization, and on the walls were casual shots,

enlarged, of him and Jackie Wilson, Marilyn Monroe, and James Brown. He
made a living off the records Poplar Tunes stopped carrying, buying one of
everything released, figuring eventually somebody would come looking for
what didn't sell right away.

Cherry's Beale Street location made him accessible to the numerous musicians who wandered by, and eventually he started his own label. He recorded
black artists like the Five Royales and Willie Mitchell, and white artists like
Billy Lee Riley, as well as lesser-known talents he encountered. Through store
talk, Cherry heard about the Regents' blues numbers and Dickinson's own
scraggly solo renderings. Home from his first year of college during the summer of 1961, Dickinson brought him a tape he'd recorded with his old band,
and Cherry signed him to his label. He called Dickinson "Little Muddy," after
Muddy Waters, and brought in Bowlegs Miller to produce him at Scotty
Moore's Fernwood studio. Cherry liked to play the tape for his black customers and have them guess who was singing; Dickinson thought he sounded
like Ricky Nelson. A popular disc jockey named Hunky Dory became his
manager. "One of my early career problems," says Dickinson, "was calling
WLOK and not knowing whether to ask for Hunky or Mr. Dory." In 1962,
Cherry sold his masters to Vee-Jay, which never issued the tape and went
bankrupt in 1965. Ruben Cherry died at fifty-three in 1976, after twenty-seven
years in the business.

The folk scene introduced the beatnik coffeehouse, which became the
venue for the return of Furry Lewis, Bukka White, Joe Callicott, Nathan
Beauregard, and the other blues pioneers who had been neglected since the
Depression. Initially, coffeehouses in Memphis were stuffy affairs with a
forced seriousness and no tingle whatsoever. The Cottage was the first to explore the territory, opening sometime in 1960, closing the next year. The
house band, a trio from Wisconsin, was so unhip that even the leader's goatee
looked square. When the Pastime opened the next year, it managed to last a
little longer, but its atmosphere never got past encouraging people to quietly
sip their coffee and stare at the piano. The coffeehouse scene was slow in
finding an audience. "The Cottage barely attracted anyone because everybody
was afraid of it," says Mary Lindsay Dickinson. "They were afraid weirdos
would get them. But I already knew I was a weirdo and I was ready to go and
be and have some fun. I was a little young for it but I found it nonetheless."
One of her earliest memories of the place is Jimmy Crosthwait sitting in the
corner beating bongo drums and reading poetry up on the stage. "I met
Jimmy through our involvement in the bohemian circles in Memphis," says
Lee Baker. "Crosthwait was real intense, and he would do things like go sit

under the piano at a crowded party. Good-looking women would bring him drinks."

A bit too devilish to be the brooding artiste, Jimmy Crosthwait has always scouted the future, returning to coax others along. He embraced the coffee-houses precisely because their intention was not defined. Undermining pre-conceptions is his specialty, a part of his being. Slight of frame, he has always worn his hair long, easily mistaken for a hippie. But even a brief conversation with Crosthwait reveals the trickster just below the surface, his eye for the chaos at the core of truth. In a crowded diner known for the electric train that runs around its wall, he'll call you over and say, "Watch this," slipping a melted butter pat onto the tracks. Crosthwait rejected society's conventions before he was of driving age, driving to the Arkansas side of the river and rent-ing a sharecropper's shack. Surrounded by driftwood and river oddities, he shifted from painting to collage, stepping into puppetry through a rejection of kinetic art: "All those people are making one statement that says, 'Here we are in a big mechanical world, see how the big machines work.'" Puppets, instead, are sculptures that happen to move: "Humans provide the movement. The flesh against the spirit is pretty much my entire theme. Finite mortals oppos-ing infinite space, all the mystery is right there in how those two things can exist together and how at one point they become one thing in God."

Crosthwait was entering high school when he first met Dickinson, who was finishing. Their friendship and collaboration broadened in the summer of 1962, when Dickinson returned to Memphis after two years in the drama pro-gram at Baylor College in Texas. Still not considering a career in music, Dickinson enrolled in theater at Memphis State, picking up gigs as a Mar-Key. A small private women's college asked him to direct *The Glass Menagerie*. The night before the show, as he and a cohort were building the set, they discov-ered a shared desire for an alternative performance space. With a third friend they founded the Market Theater in the fall of 1963. "We were leaving the Market Theater one day," remembers Mary Lindsay Dickinson, "when we en-countered Jimmy driving by. He couldn't have been a legal driver. He waved to Jim, we pulled over, and he came up talking about building a guitar out of wood. Shortly after that Crosthwait worked up his act playing garbage cans, and he became a great presence at all the hootenannies."

The Market Theater was an ambitious name for a room in a farmer's mar-ket with holes in the wall. They built a small stage at one end and bought used theater seats from a church. The size of the venue made microphones moot. Jimmy Crosthwait painted the men's bathroom and the women's room was given to Joe McConico, who as Hilton McConico became an acclaimed fashion designer and set designer (*Diva, Confidentially Yours*). "The mayor had decreed Market Theater Day and he was coming, we had Miss Memphis in a swimsuit, we had champagne for celebrities," remembers Dickinson. "A couple days before the opening, this fireman guy with a clipboard shows up, says the city has a hundred-something codes to meet and this place doesn't meet any of 'em. Whenever we had a problem, my partner Phil had an uncle. The phone rings the next day and a voice says, 'How many chairs you got?' We said fifty. He said, 'Take one out and it's not a theater,' and he hung up. So we seated forty-nine, and he told us to charge a dollar or less at the door to avoid entertainment tax. We were on our way."

Their way was brilliantly lit for two and a half months. They performed plays six nights a week, with folk music Sunday afternoons and, because of its popularity, sometimes on weekends. But the space became claustrophobic, with one show rehearsing afternoons while another was staged evenings. As the chilly weather moved in, the charming breaches in the wall became foreboding. Perhaps the most significant contribution of the Market Theater was the new folk audience it revealed. The Hammer Singers at the Pastime wallowed in the commercially appealing style of the Kingston Trio, eviscerating the social significance of the music that the Market Theater hailed—work songs, scruffy blues covers, and obscure Lomax finds. Dylan had released his first album the previous year, and though sappy versions of his material quickly made national hits, the witnesses were reverberating with his renegade spirit, finding in folk and blues more than just a source for pop music.

Sid Selvidge established himself as a folk presence around Memphis in the early 1960s. With Dickinson, Baker, and Crosthwait, he rounds out the core of Mud Boy and the Neutrons. Selvidge brings to the group a voice as pure and sweet as a Delta songbird, with as much range as the expansive sky. It's been said he can perform in a stampede of elephants, which I take to mean that his voice can accompany the marauding beasts, or it could still them. His trademark is a falsetto, which he leaps into for emphasis, not unlike Little Richard's whoops, but less manic. Selvidge only plays acoustic guitar; his piano burned down in a 1970s bar fire and he took it as an omen. Back then, when I began seeing him regularly, he was a folk punk of sorts. The Sex Pistols and the Ramones had just come out, and Selvidge's Peabody label was preparing to record Alex Chilton's *Like Flies on Sherbert*. Sid was playing every weekend at a downtown bar, the business district like civic tooth decay. There'd be a single chair onstage, a single light. He'd appear out of the darkness, walk down the aisle, sit at either the guitar or the piano and not bother to remove his leather jacket. Then, without the slightest acknowledgment to the audience, he'd begin forty-five minutes of beautiful playing. Sometimes when he sipped his water, he'd sneak a look, but usually he furrowed his eyebrows to block us out. When the minute hand made three quarters of a circle, he was up, disappearing again into the darkness.

In 1961, Selvidge came from the Delta to Memphis for college. "I was down in Greenville listening to WLAC [a powerful radio station out of Nashville], disc jockeys like John R., the Hossman, Gene Nobles, and they were playing Muddy Waters and Little Walter and Jimmy Reed. Visiting Memphis, we were driving right up through the Delta, and I didn't know that was where the music came from. I was playing guitar at the time, and my grandfather would come in and say, 'I saw some guy out on a place and he would put a knife between his little finger and this finger. Can you do that?' But that didn't make any sense to me—why would somebody want to play a guitar like that? I would try and the knife would fall between my fingers and the strings were in standard key and it sounded terrible. And being a little dumb white boy I didn't have sense enough to go ten miles away from the house and say, 'Show me how to do that.' My mother hired a guy to teach me how to play a guitar, and he had grown up in the same waifs home as Louis Armstrong in New Orleans. So he played a syncopated thing that didn't have anything to do with what Muddy and them were playing. When I said I wanted to learn folk music, he taught me 'O Spinning Wheel in the Parlor' and I thought I was getting somewhere."

Once in the city, Selvidge befriended Horace Hull and was "catapulted out of the dorm scene." Hull, like Selvidge, had a beautiful voice, and in addition to playing a gut-stringed guitar, he also played banjo. They wore blue work shirts and sang "Cumberland Gap" and other excerpts from the Lomax songbook. Like Selvidge, Hull had incredible vocal range; because of his classical background, his harmonies were Bachian, based on counterpoint. Hull came from a monied family, and he was never able to resolve the conflict of being a folkie from a privileged background. Despite his obvious talent, he'd been taught that folk was not serious music. Hull sometimes snuck to a church pipe organ where he could play Bach fugues in solitude. He and Selvidge were hired by Dickinson in 1963, along with several other young white folk singers, for what was billed as the First Annual Memphis Folk Festival, which was really a concert to help pay off the debts left by the Market Theater. Held at the Overton Park Band Shell—a comfortable outdoor amphitheater built with a utilitarian grace by the WPA in 1936, nestled amongst trees in the open air, lined with rows of wood-backed benches—the show laid a cornerstone for the four Memphis Country Blues Festivals that were to come later in the decade, events that brought the Delta bluesmen and the Memphis hippies into the international spotlight.

Though he'd been unable to meet bluesmen in the Delta, in Memphis Selvidge began sharing the stage with them. Folklorists like Sam Charters and Paul Oliver had begun rediscovering the first recorded blues artists in the late fifties, the witnesses had begun meeting them by the early sixties, but nobody thought to hire them until 1963, when the gregarious Charlie Brown was managing a coffeehouse called the Oso. Brown was born Charles Elmore in Sardis, Mississippi, but his head was round and his face flat, so that he actually looked like the comic strip character. He realized it wouldn't kill the bluesmen to give them gigs; playing, in fact, was their thing. Their rediscovery by a new generation was still draped in awe, academia, and anthropology. Brown simply removed them from the display shelves. Gus Cannon and Furry Lewis became regular performers alongside Selvidge at the Oso, joined by lesser-known artists like the moaning jug band vocalist Van Zula Hunt. They didn't fool with no "Puff the Magic Dragon."

It was 1966 before the ideas of hiring the bluesmen to play and of renting the Overton Park Band Shell would merge, creating the first of the Country Blues Festivals. Memphis's proximity to the Delta meant a concentration of talent in a native setting that was beyond the ken of places up North. The Newport Folk Festival was hosted by Theodore Bikel, and *Newsweek* termed the audience "Milk Drinkers." The games were as different as wiffle ball and pro baseball.

Nashville, where ideas are imported for fun and profit, decided to capitalize on the folk movement in the early 1960s. Former Sun Records horn man and producer Bill Justis, who had achieved national fame in 1957 with the saxophone-based "Raunchy," had moved to Nashville and was cranking out easy instrumental remakes of pop hits for Smash Records. Middle America, housewives, drunk drivers—everybody was scooping them up, which amused Justis greatly. He was a most unusual character. His disdain for rock and roll was only slightly weaker than his desire to profit by it, a fact he did not hide. Sun historian Colin Escott writes in his *Good Rockin' Tonight* (1991) that on session reels Justis can be heard saying, "Let's get real bad now so we can sell some records. Instant crapsville, girls. Here we go. . . ." In his band was a former schoolmate of Dickinson's whose mother had sent him newspaper clippings about the First Annual Memphis Folk Festival. One afternoon in 1963, Dickinson's phone rang and it was his friend saying hold for Bill Justis, who came on the line and offered him a major label recording contract. With it came a priceless music biz lesson.

The very name of the record they made is meaningless: *Dixieland Folkstyle*. The first word implies New Orleans, the second implies Appalachia, vocal harmonies, and an entirely different kind of banjo playing. The studio was filled with entirely too many musicians to replicate any sort of folkiness. And the only hint of the Crescent City, buried among schmaltz like "Green Green" and "Michael Row the Boat Ashore," is the New Orleans standard "St. James Infirmary." But Justis lumped the two buzzwords together, assuring those not in the know that this record was in the know—and then he smiled as sales racked up.

The lesson came on the middle of the third day of sessions. "We're working real tight Nashville union sessions, three hours by the clock, when these big loading doors that had never been touched come flying open in the middle of a cut and in comes this big fat redneck all dressed in black. He had a long, greasy ducktail and mirror sunglasses, and he's talking real fast and loud. I thought, Boy this guy is history. They're gonna throw his ass outta here.

"But everybody stops. He walks up to one of the women who's singing with the Anita Kerr singers and he's saying, 'Baby, I'm sorry, but I just couldn't make up my mind, so I bought both of 'em.' The whole session goes out to the parking lot to look at these two new Jaguars. He was handing out these Sherman Cheroots, brown cigarettes, and from one end of the cigarette to the other, every little microspace occupied, he had his name in gold letters: Shelby S. Singleton. Finally I see my buddy the trumpet player, who had written all the arrangements (not that I could read 'em). I said, 'Who *is* this guy, man?' And he says, 'He's the producer.' And I said, 'The producer! Well what is Justis?' 'Justis is the arranger.' 'Then what are you?' And he says, 'Oh, I'm the copyist.' It was like lightning again. I thought, Somewhere in this is a place for me."

Not long after returning home, Dickinson opened his mail and found a contract. He thought it seemed a little after the fact, so he phoned Justis. "He said, 'Don't you want to make a record?' and I said, 'Didn't we just make a record?' and he said, 'No. You. Don't *you* want to make a record?' I said yeah, sure and he says sign the contract and send it back."

Justis had Sun beams dancing in his head. The lack of polish in Dickinson's voice fit perfectly his notion of rock and roll. Hitsville. Promptly, he sent Dickinson a tape of a Shel Silverstein song, "The Unicorn," and booked time at the Sam Phillips Recording Service, the studio built after Elvis's contract was sold to RCA. Justis produced this session as Shelby Singleton had done the last, *in absentia*. Dickinson noted the pattern. (Actually, the idea of producing *in absentia* is not terribly far-fetched. Consider: The producer is the one who determines the recording's direction. That is, the band performs *for* the producer, playing to satisfy him or her. Some producers are such a presence that they need not even be in attendance for the artist to direct their performance toward them.)

"The Unicorn" was a Nashville folk song, as opposed to a folk song by the folks. Dickinson presumed that the folks might be ready for the folks and pretended he didn't receive Justis's tape. He reached back to his Whiskey Chute encounter, to the abandon of the players whose instruments were portable enough to run from the cops. "Come on down to my house, honey," the singer had encouraged, "there ain't nobody home but me." He formed a jug band, the New Beale Street Sheiks. Crosthwait, a drummer, was dispatched to buy a washboard. They had a guitar-playing friend smart enough to figure out the tub bass. Dickinson pulled the harmonica from his neck rack and replaced it with a kazoo. "We got one gig the night before the recording session and people loved it," Dickinson recalls. "So we went to Phillips the next day. Crosthwait had real long hair, and this was pre-Beatles. He had a rag around his neck and we all looked wretched and they didn't want to let us in. Bill Black and Scotty Moore were there. Scotty Moore never did trust me. But Bill Black thought it was funny. He said, 'No, that's Dickinson, that's his thing, let 'em in.'

"My whole deal was just not to cut 'The Unicorn.' Bill started calling people, telling them to come down. I would see him at the window, talking on the phone and pointing at us. He really thought it was funny. Crosthwait was playing the washboard for the second day in his life. We may have had two microphones on the session, but I think it was one. We cut four songs as a demo. When we were done, Scotty Moore wouldn't even let me have the tape. He said, 'I'm going to Nashville tomorrow and I'm going to take this to Justis.'

"I didn't hear anything for a couple weeks so I finally called Justis and I said, 'What'd you think of the tape?' and he says 'The tape! The tape is great! But what's making that noise on there?' I says, 'It's a zinc tub bass, it's just a tub and a rope.' He shouts, 'It's a rope! A rope! I went all over Nashville trying to e.q. [sonically adjust] a rope!'

"Then he says, 'The record will be out Thursday.' I said, 'But that's the demo, Bill,' and he says, 'No, no man, you could never do it that bad again.' We were talking on the same level and I paused, because I wanted him to hear me. 'Bill,' I said, 'you have no concept of how bad I could do it.'"

"You'll Do It All the Time" backed with "Down and Out," by the New Beale Street Sheiks, was released on Thursday, February 6, 1964. *Billboard* called it "a contagious, hard-driving, pulsating, folk-blueser . . . , appealing,

The New Beale Street Sheiks. Left to right: Sid Selvidge, Jim Dickinson, Bill Newport, Jim Vinson, Jimmy Crosthwait. Photo by Steve Jensen.

nostalgic." The powerful John R. on WLAC, whose R&B radio shows had influenced all these musicians when they were kids, played it. Chet Atkins called Justis, tried to buy the record. Then three days later, on Sunday, February 9, 1964, the Beatles appeared for the first time on "The Ed Sullivan Show." And the American record industry ground to a halt.

CHAPTER SEVEN

What's What

WHEN THE BEATLES HIT, KIDS IN MEMPHIS WERE CAUGHT IN THE dueling forces of the new pop sound and the continuing success of Memphis R&B. The first single on Stax in January of 1964 was Rufus Thomas's "Can Your Monkey Do the Dog." Otis Redding was beginning his ascent, Booker T. and the MGs were preparing to release "Soul Dressing," and the upstart Goldwax label would introduce James Carr that year and begin competing with Hi Records for Memphis soul's second notch. Rockabilly still cast its shadow: Elvis was deep into Hollywood schlock, but 1964 would see the release of one of his best movie singles, "Viva Las Vegas." Charlie Rich moved to Smash that year and would soon enjoy "Mohair Sam."

"The first night I ever heard 'I Want to Hold Your Hand,' I heard it on the radio coming home from a Blazers gig at Ole Miss," says Lee Baker. "I said, 'Goddamn, that's it. You watch.' I'd been hearing about it and I said, 'That's the future knocking at our door, we're gonna have to change,' and we did. Horn bands bit the dust. And everybody had horns. The Beatles were replacing the Mar-Keys, rhythm and blues. The LeSabres had horns, the Blazers, even Mississippi bands—everybody, because it was more versatile. And all that changed."

While the Beatles brought a new excitement to pop music, some people saw them as a distraction. Dan Penn, an Alabama songwriter who would come to Memphis in 1966 and produce the Box Tops, sums up the feelings of those who entered the Memphis coffeehouse era with their mind more on black music than British white. "Tommy Roe was working at Fame studios in Muscle Shoals, and he'd been going to England. He came in one night and he had this test pressing in his hand, and he said, 'Boys, I've got something right here that's going to change the world.' And we said, 'Put it on, put it on!' He puts it on and here it comes, 'I Want to Hold Your Hand.' 'Wha'd y'all think?' Everybody in that room cared for nothing but R&B and nobody said much. And he said, 'What do you think?' And I said, 'Man, if that's gon' change the world, I don't know whether I want to live in it or not.'"

Once the Beatles were in the world and every kid everywhere wanted to be a rock star, guitar manufacturers were in heaven. The trend had begun with Elvis, but by the time of the Fab Four, music stores were increasing exponentially. The 1967 "Report on Amateur Instrumental Music in the United States," a survey of the American Music Conference, indicated that between 1956 and 1966, the number of guitar players rose from 2.6 million to 10 million. Memphis musicians from the witnesses on through bands like Randy and the Radiants and the Box Tops had followed the same pattern when purchasing their instruments. They started with the pawnshops on Beale, which sold equipment cheap.

There were up to fourteen pawnshops within two blocks on Beale, but Lou Rafael at Nathan Novick's Sales Store dominated the music trade. The child of European immigrants, Rafael was born in Brooklyn in 1910 and came to Memphis during World War II. He retired in 1983 after forty-one years on Beale. When we spoke at his home, he was wearing brownish plaid pants, brightly striped suspenders, a peach-colored shirt, a powder-blue baseball cap, and had on terry-cloth house shoes. His face is long, his features large, and a cigar is a permanent fixture beneath his pencil-thin mustache. When I greeted him, my stomach dropped as it had the day in the early 1970s when I'd entered Novick's and asked the cost of a harmonica; he'd pulled the cigar out of his mouth to answer and the sight of that mangled, sopped butt has never left my mind.

You married? You're not? What do you do for aggravation?

When tour buses stopped on Beale in the 1970s and all that was left in the crumbling remains was a few pawnshops, the guides would direct their patrons into Novick's, introducing them to Lou Rafael. "They tell me I sold Elvis his first guitar," he says in a husky voice that still has a trace of a Brooklyn accent. "Whether it's true or not, I don't remember. They've talked me into it. People from all over the world have taken their pictures with me because they were so happy I sold Elvis his first guitar." In the old days, it'd never have happened. Lou would have charged them for the privilege. "I remember one time a man stepped out of a taxicab, he was looking for a pawnshop. What did he have? He wanted to pawn his artificial leg. Sometimes people wanted to pawn their teeth, but I wouldn't take teeth unless there was gold in it."

Lou used to brag that he bought his instruments by the boxcar load, but some kids swore he was buying the trade-ins from other stores around the city. Novick's was dark and filled with whirligigs, but musical instruments and band gear were prominently displayed in the window. An elderly, yellowish man sat in a chair by the door. He wore a hat and usually had his hands folded atop his cane, a place for his chin to rest. When a kid would stop to admire the bait, the shill would laugh softly, wisely, and tell the kid to go on in, look around, gots lots more inside, purrrty ones. The mark entered innocently, and like a scene from the Marx Brothers, Lou pounced. "Ya got any money!" he demanded. "How much! Don't look around, com'ere, whaddaya want!" The shill's soft laugh was distant, like the rattling of a jailer's keychain.

Ever hear about the man who went in the restaurant and ordered a hot bowl of soup? It sits on the table for fifteen minutes. The waiter comes by and says, "I see you're not eating the soup, is anything wrong?" The customer says, "You taste it." The waiter says, "Okay," looks around, then says, "Where's the spoon?"

"My father was a salesman's salesman," says Jim Dickinson, "and he took me to every pawnshop on Beale before we bought my first guitar at Novick's. It was pressed masonite covered in wallpaper. The other guys couldn't sell my father. Lou could." Dickinson went with a friend to buy his second guitar from Lou. His third he bought alone. "We're haggling about it and it's getting very tense. We get it down to seventy-five cents between us, and neither one of us is going to give it up. I walk out the store, down Mulberry to where my car is parked, Lou is chasing me out the door with a cigar in his mouth. 'Come back here you motherfucker,' he's screaming, and I'm cussing him out. Finally he said something that made me turn around, and he had me. He said, 'I wouldn't fuck you.' I turned and said, 'You fucked me twice.' Quick as a flash he said, 'I wouldn't fuck you three times.' We walked back to the store and he agreed to take the seventy-five cents off the price. But when he's writing out the bill, I see that he adds seventy-five cents on the bottom. 'What's that, Lou?' 'I'll itemize,' he says. He writes, 'Fifty cents, Tennessee State guitar tax.' 'What's the quarter?' 'My cigar.'

"I had traded in my beat-up old Gibson, got twenty bucks credit for it. When we were done, I asked him what he was going to do with it. He said, 'I wouldn't sell it for less than a hundred fifty.' I said he'd never get that much money for a beat-up guitar like that. He smiled confidently, said he was going to cover it in shoe polish."

A boy is working for his father. Says, "Pop, I want to go to college and become a lawyer or doctor. I don't want to work in the factory anymore." His father says, "Do you know what's what?" "No." "Go back to work." Six months later the kid comes back and says he wants to go to college. The father is fed up, figures he'll fix him up on a date with his secretary. They go out dining and dancing, have a good time, and they wind up at her apartment. She says, "Sit down, make yourself comfortable." She returns in a negligee and he takes a look at her and says, "What's that?" She says, "What's what?" He says, "If I knew what's what, I'd be in college."

From the pawnshops, most kids went to Boyden's Melody Music, where Jack Boyden would usually forget that he had extended credit to the kids, but his mother would remember. "After you bought your first set from Lou on Beale, you bought better stuff from Jack Boyden," says David Fleischman, the "Flash" in Flash and the Board of Directors. "My organist got his B3 at Melody Music. Everybody bought equipment there. The guy extended credit —to kids! Most kids paid him back, I guess. We did, we were fortunate, we always worked."

Sid Selvidge remembers that Melody Music became a place to meet other musicians because the Boydens would let anybody play any instrument in the store. "People were just hanging out over there and playing guitars. Any new guitar that came in, you could test it out, so you were always gonna run into somebody over there. I'd see Baker, Dickinson, Charlie Freeman, Sid Manker —any guitar player or aspiring guitar player in there."

"I probably met Dickinson at Melody Music," says Lee Baker. "The store was right across the street from a coffeehouse called the Bitter Lemon and we were all running around there."

"Jack's mother made that store a success," continues Selvidge. "If you needed strings and they cost fifteen cents and you didn't have it, he'd put it on account. But come the first of the month, you'd get a call, 'Hello, this is Miss Boyden, you need to bring that fifteen cents in.' Mrs. Boyden died and Jack kept going out on the golf course, and the business went out the window."

As guitars proliferated, so did the demand for music lessons. But these kids didn't want to learn theory; they didn't want notes or scales. They wanted to rock and roll. There weren't as many places to take lessons as there were to buy instruments, but a surprising number of successful artists went to the same teacher: Lynn Vernon, Memphis's secret guitar hero. A jazz player, Vernon led a three-piece combo for twenty years on a morning TV show. He also had a studio where he gave lessons.

Roland Janes, in whose hands a guitar was defenseless, describes Lynn Vernon as a great. Though he was already well established, Janes asked Vernon to give him lessons. "He asked me, 'What do you want me to teach you?' I said basically to read chord charts and maybe read music. After my first lesson, I got real busy and never was able to come back. I always kind of regretted that. He could play a lot of styles, but he was more big band–oriented. Chords, lotta chords. And kind of a jazzy feel." Rick Ireland, the guitarist for the Regents, purchased his guitar from Boyden's and took lessons at the store from a preacher. Later, a singing cop taught him. But when he got proficient and could read music, he asked Lynn Vernon to take him to the next level. "He said, 'Show me what you know,'" says Ireland. "I played him something and he says, 'Well, I'll tell ya, I'm kind of jammed up right now, why don't you start teaching for me?' I was still in my teens, and I had his overflow and some students of my own.

"Unlike most jazz players, Lynn played a Stratocaster. He loved that Stratocaster. I started hanging out with Sid Manker as a result of my association with Lynn. Sid was one of the heavier jazz guitar players around here. He played with Justis, and I used to go with them down to the Press Club. It was a jazz place, heavy duty, across the street from the newspaper, upstairs. Every Saturday night, if there were touring bands, those musicians would come down there. The bands were black and white. This was heavy bebop. I saw my first heroin withdrawal there, a bass player. I was hearing all this really heavy jazz when I was still a teenager, still playing with the Regents.

"I met Charlie Freeman through Lynn because he was one of Lynn's students. Lynn got us a job backing this gal that used to sing on the WREC morning show, Dolly Holiday. Charlie and I were supposed to play for her, and we drove somewhere down in South Memphis to her home and sat out there in the dark waiting for her. The rehearsal never came off but Charlie and I got real friendly." After his lessons, Freeman would share what he'd learned from Vernon with the other guitarist in his band, Steve Cropper.

"Mr. Vernon's studio was upstairs from a girls' ballet studio," says Randy Haspel. "You had a guitar case as big as you were and you had to walk through fifty little girls in tutus to get to the stairs. And when I was walking down, Larry Raspberry [later of the Gentrys and then the Highsteppers] was walking up. Mr. Vernon taught me, he taught Raspberry, he taught Bob Simon [The Radiants], he taught Bobby Manuel [Stax session man; "Disco Duck" cowriter]. We all ended up playing the same style Fender Stratocaster because Mr. Vernon had one. You'd say, 'Mr. Vernon, please can I play your guitar?' He'd make you wash your hands and only after that would he let you play his guitar."

When the Beatles turned the world upside down in February 1964, older bands like Tommy Burk and the Counts and Lee Baker's Blazers were already in college, somewhat set in their ways. "I graduated from East High in 1963," says Lee Baker. "The Blazers stayed together because all of us went to Memphis State. And when the Beatles came in, we kept the band together but nobody wanted to hear horns anymore. The sax player started playing electric piano, and one of the other horn players picked up guitar, and the rest got percussion instruments and harmonicas and stuff like that. We were able to keep that big band together, but we had to revamp and start doing more rhythm section stuff."

For a younger band like Randy and the Radiants, adapting to the Beatles was easy. "The British Invasion made us popular because we were just coming into our own," says Haspel. "I was about sixteen years old. My partner, Bob Simon, had been writing songs since we were kids. When the Beatles hit, we already had a band that was up and working. The next time we had rehearsal, people started to assume their roles. Mike Gardner loosened his trap cymbals and started making those kinda slashing motions that Ringo would make. I learned how to rock back and forth like John Lennon. And we started to get really popular. We had been strictly rhythm and blues until the British Invasion, and then we became sort of a hybrid rhythm and blues/Memphis/British pop. It affected the Gentrys in the same way."

The changing of the guard became apparent at the battles of the bands. Even after the Beatles, when most rules changed, these showdowns remained an indefatigable defense of one's honor. Bands set up at either end of the venue—gym, social hall, hotel ballroom, someone's basement—a box before them. Each ticket buyer would vote by placing their torn stub in one of the boxes. The winning band returned the following week to take on new competition.

Haspel remembers that, until the Beatles, everything his band did was because of the Counts. "They did the Five Royales song list, and when we were little kids coming along, we did the same list. We had heard about the West Memphis scene and we knew we were doing their songs, but we were too young to have experienced it. The Counts began booking themselves with this black band called the Avantis that had sung at the Plantation Inn. When they did that, I figured I should do the same. But this is when high schools were still segregated. I was at Christian Brothers High, and Christian Brothers College was integrated. I was a sophomore and Eddie Harrison was a sophomore in college. He had a vocal group called the Premieres—Eddie, his sister, and three other guys. We started booking out with them. For seventy bucks, you could get the Radiants, for a hundred twenty you could get the Radiants with the Premieres. Eddie said he'd never been to East Memphis before he came to my house to rehearse." Integrating the teen bands was the work of Sputnik

Monroe's shadow; the kids achieved nonchalantly what society seemed unable to forcefully wrestle.

Haspel continues: "The Radiants had beat the Devilles [who later became the Box Tops], the Scepters, the Gentrys—everybody. We'd keep winning and be asked back, and that was how we started getting popular. So the final week they brought in Tommy Burk and the Counts. And I mean we went into rehearsals like this was the biggest show of our life. A gymnasium in East Memphis in the middle of summer, must have been a hundred-six degrees in there, packed out with six hundred kids to watch this battle between us and the Counts.

"And we whupped 'em."

By then, the Radiants also had demographics in their favor. While the Counts were at Memphis State, each of the six Radiants attended a different high school, making them eligible for six times as many gigs. As well, two of the members were Baptists, two were Catholics, and two were Jewish. They had entrance to every venue in town. "Music was everywhere," says one person on the scene. "It had become accepted. By then, department stores would have a promotion of English-looking Beatle clothes and they would hire a band to play."

The success of Randy and the Radiants came to the attention of Sam Phillips. By the time of their showdown with the Counts, several disc jockeys had appeared wanting to manage them. The Radiants selected Johnny Dark, a founding member of the Sputnik Monroe Fan Club. "He was even-tempered and good-humored and closer to our age than the other disc jockeys," says Randy Haspel. "At the time, we didn't know that he was friends with Knox and Jerry Phillips. Knox was going to Southwestern, a local college, where he was a Sigma Chi, so we started playing for Knox's frat parties and they loved us. Between Johnny and Knox, they agreed to have us audition for Mr. Phillips. I think Mr. Phillips had an eye for this new thing that was happening with the teens and he saw that with Bob Simon we had original material. And he was also looking to get Knox into producing. We were the first band that Knox produced. We cut our teeth on each other."

Though they were being courted by the label that inaugurated rock and roll, the Radiants had mixed feelings. "I think the only person left on Sun by 1964 was Jerry Lee, and he was just as dead as dead could get. We knew the label had seen its better days. Still and all, to be ushered in by Mr. Phillips, it was something. First time we met him, he'd just come in off the lake somewhere, he had on a yachtsman's hat, and sat in there and he was just as charming as he could be. After that it was, 'We'll do anything for you, Mr. Phillips.'

But our initial response had been much more hesitant. We had big plans, and we didn't know if Sun would be right for us.

"Our first sessions were produced by Mr. Phillips. We had a handful of original songs but we were warming up with songs from our set list. We started 'Mountain High' by Dick and Deedee and all of a sudden Mr. Phillips comes out of the studio and goes, 'What're ya doin'?' We'd grown up on the Elvis legend, you know, and we said, 'Well, we're just warming up, Mr. Phillips.' 'Keep playing it, I like this.' And we're looking at each other like, What's he trying to pull? And I said, 'Mr. Phillips, this was a hit record just a year or two ago. He said, 'I like it, I want you to do it.' So that became our first record."

Through Johnny Dark, the Radiants became close with Sputnik Monroe. "We were playing a dance at Clearpool, a Memphis teen hangout, and the owner and his two bad-boy bouncers flicked on the lights kind of early and chased everybody out," remembers Haspel. "They were drinking and wanted to get out of there. We were taking our time packing up, and words were exchanged. I got beat up and my drummer did too. It went to court and they were all fined for assault and battery, disorderly conduct. When it came time for us to play Clearpool again—we were the most popular band in town so they couldn't keep us out—we had a security guard go with us. He had a gun. These bouncers had been arrested and they wanted revenge. And made no real secret about that fact.

"The stage was set up right across from the concession stand, and the owner with his two greaser bouncers are standing directly across the room from us, and I mean they're staring daggers. They didn't look like they were too afraid of our security guard. Right in the middle of the set, the Clearpool door swings open and our manager Johnny Dark walks in, and right behind him, here comes Sputnik, and he's doing that strut. The whole party just went crazy. 'SPUTNIK! YAY!!!!!' and that kind of stuff. I was just dumbfounded and I said, 'Well folks, Sputnik Monroe is here at our gig. Sputnik, would you like to say a few words?' He got up there and everybody was clapping and cheering for him and he was flexing his muscles and everything. He waited for the applause to die down, then he pointed his finger across at the concession stand and he says, 'I want everybody to know one thing,' and he paused and pointed his thumb back at us, 'These boys are Sputnik's boys, and if you mess with them, you're messing with Sputnik!' And he left.

"The next morning, like Sunday at nine A.M., I get a call from the owner of Clearpool telling me he's apologizing, he's so sorry that the incident ever happened, he fired those guys that punched me out and he hoped that it would never happen again, we were welcome back at his club anytime we wanted to come. We kind of became Sputnik's boys after that."

A year later, in November 1965, Sun released a Radiants follow-up. "We had a Bob Simon song called 'Truth from My Eyes.' Bob was a year younger than I was, and I'd talked him into letting me sing lead. I think he might regret that to this day. We begged Mr. Phillips to make 'Truth' the A-side. He had a song called 'My Way of Thinking,' written by Donna Weiss, a Memphian who later wrote 'Bette Davis Eyes.' Her song was an absolute rip-off of the Kinks, opening with the same guitar as 'You Really Got Me,' and we thought it was embarrassing. Soon 'Truth from My Eyes' was number one on WMPS every night and we kept getting in the papers, sixteen years old, seventeen years old. But WHBQ wouldn't play it. So I went down there and talked to the program director, and he said, 'We won't play a Sun record.' I was incensed. 'Wha'da'ya mean you won't play a Sun record? What are you talking about!' So I told Knox, who fired off an angry letter to WHBQ." As "Truth" was winding down its two-month crest on WMPS, the song got a whole new life on WHBQ.

In the half-decade since Cropper, the Mar-Keys, Dickinson, and the other witnesses had finished high school, the music world had turned upside down. As teenagers, a *career* in music had been something they didn't even dream about. But for the next generation, the horizon was boundless. The difference is illustrated by Haspel's conclusion of the Radiants' story: "Sam put Jud [Phillips] Sr. on the road for us, a tentative tour of the South was planned, and we were riding high. Somehow it seemed at the last minute things collapsed and the plans fell through. I tell you what, it's tough to think you peaked when you were seventeen years old. I don't think I've ever had a more thrilling time in my life. It seemed like everything was possible, anything could happen, and it was all gonna come true."

Roland Janes, whose patience and wisdom make him the answer to a Zen koan, produced a song in 1964 that merged the city's traditional roots with the new musical energy of the Beatles. "Scratchy" was a collaboration between Janes and Travis Wammack, a teenage hotshot guitarist he'd found in a Memphis redneck suburb. Mostly instrumental and full of youthful energy, the song has a vocal break that sets it apart, epitomizing the Memphis philosophy of bucking trends: It's played backward. The music is built around a ticking of the drumsticks, with the guitar sounding like Wile E. Coyote escaping the Roadrunner. They come to a natural break where the ear expects a vocal refrain—and this incredibly maniacal-sounding backward voice comes out.

Nothing anyone could say forward could be as rebellious or as much fun. The song was a direct result of Janes's studio philosophy: "When I had Rita Records with Billy Riley, we were spending quite a bit of money on studio time. Your best time in the studio is your experimental time, and when you're renting studios, experimental time gets kind of expensive. So I finally got myself together and opened Sonic in 1961. My main purpose was to use it as an experimental laboratory, so to speak. To work with young groups and talented singers and writers. And I got into custom recording for other people to support the studio."

The abundance of music in the city had resulted in the proliferation of recording studios. And of music industry wanna-bes. The success of Sun in the latter 1950s and Stax throughout the 1960s encouraged barbers, auto mechanics, hardware salesmen, and just about anyone else to try their hand at some aspect of the music business. They all could have learned something from Roland Janes. Janes maintained his one-man operation through 1974, refusing lucrative tour offers from Jerry Lee Lewis (for whom he'd played guitar), and turning down other studios and artists interested in his services. He stayed busy, attracting young bands who thought they might be the next Mar-Keys, or the next Beatles. Or the next Travis Wammack. Sonic was a tight little space with a great sounding control room. Janes recorded mono and mixed as the band played, "defensive mixing—try to keep the bad stuff out and the good stuff in." He was limited to seven microphones. "I learned an awful lot watching the master, Sam Phillips," says Janes. "He was a great, great engineer."

Janes remembers his first encounter with the long arm of the Beatles. "I was recording a group at Sonic one day, and a kid came in that had just been to England and he had this weird-looking little hairdo. The musicians were sitting in the control room with me, peeking around at him and laughing and making little remarks about that weird hair. Well, turned out that he had a Beatle hairdo and we just hadn't seen one yet."

Janes's background may have been country and rockabilly, and the Beatles may have been the contemporary rage, and Stax was surely happening across town—but in all recording studios, the game was the same. "I didn't really envision anything other than trying to get a hit, and I really didn't care if it was rockabilly or pure country or rock and roll or what have you. And with the other studios, it was the same. In the early years, nobody could beat my rates. Then they started doing spec sessions and their rates were lower than mine. Basically, you had Sonic and then you had all the four-track studios. Ardent was starting up, Phillips was here. Fernwood, Hi, Satellite. But we all kind of had our own thing going, and we really didn't look at it as a competitive thing. In a way, I guess I was the alternative music of the day."

American Sound Studio began on a scale similar to Sonic but would become a major force in pop music. Between November 1967 and January 1971, American landed 120 records on the *Billboard* charts. One week, over a quarter of *Billboard*'s top 100 not only came from the same studio but featured the same core band backing a variety of artists—black, white, male, female. While the soul labels produced soul hit upon soul hit, American recorded an astonishing diversity of styles: spare soul, funky white stuff, lush pop orchestrations.

American was established by producer Chips Moman on a quiet corner in a black part of North Memphis. The building had a barbershop on one side and a restaurant on the other; when the barbers didn't like the music, they'd stick a loud radio in the common ceiling, making recording impossible. The Phillips family had their record distribution warehouse nearby, and Buster Williams's Plastic Products pressing plant was down the road. Music attorney Seymour Rosenberg had his office around the corner in his family's auto parts shop, and directly across the street from American was Lynn-Lou, a smaller studio run by Bill Black. This corner, by the latter 1970s, would devolve into shells of burned-out buildings, and the studio was bulldozed. But the impact of what happened inside the building affected late 1960s and early 1970s pop radio as much as, if not more than, any other studio in America, including Stax across town and Motown in Detroit.

The secret of American's success was the Memphis air. Or the river. Or the barbecue sauce. In other words: Dewey Phillips. In other words: There was no secret—there is nothing anyone could do to make it happen again. But it happened. It was the happening. Chips Moman was the pivotal figure, producing most of the sessions and responsible for assembling the house band. He'd been around the Memphis scene since Stax's gestation, giving direction to "Last Night" and many of the label's early hits. In late 1961, a dispute at Stax arose over production credit, over influence, over restitution and seniority—a rumble among the egos—and Steve Cropper came to the fore, Packy Axton was pushed to the rear, and Moman stomped out the door. He ultimately got a $3,000 settlement, negotiated by a trumpet-playing attorney who also did work for the defendant. Moman moved into the building, wired a rickety mono recorder, and began creating the machine that would soon dominate pop music. His house rhythm section, unlike the cultural collision at Stax, was a group of musicians raised together and familiar with each other's charms and idiosyncrasies. They simply did what they could do and watched the nation and the world applaud.

The first few years of American are a bit cloudy. Moman was out on a wet one, and though he still had his touch, he'd sold his ownership in the studio for an amount he could calculate in cases of whiskey. Tommy Burk remembers cutting there with his Counts as early as 1962; it was 1964 when Moman regained control, solidifying ownership with a bean farmer from Arkansas named Don Crews. That year, after cutting several hot records, goes a story in Peter Guralnick's *Sweet Soul Music* (1986), a singer friend of Moman's was contacted by a small label in Memphis and told, "We got this producer, man, all we ever have to do is give him a bottle of whiskey and a couple of pills, man, and he'll cut you a fucking record." For the label's next session, Moman charged five thousand dollars, up from his previous rate of twenty. They gladly paid.

At the time, Moman was making spare change shooting pool and painting gas stations. He didn't look much like a record producer when he encountered the group of pretty boys that would change his fortunes. Brandishing the blades of his double-edged life at Berretta's Drive-Inn one night in late 1964, he approached these young men who were dressed like a band. Moman is said to have drawled, "Yew boys got a band? Yew wanner make a record?"

"We had played a free show that night at the Veterans Hospital and were still in our gig clothes," remembers Larry Raspberry of the Gentrys. "Chips didn't look like anything we thought was our genre of music, and frankly we didn't take him seriously. We exchanged phone numbers but I don't think we called him back." Moman, however, pursued them. "We needed to make a tape for 'Talent Party,'" continues Raspberry, "and he told us to come in and do it with him." The Gentrys had been the Gents before the Beatles convinced the septet to add a British flavor to their name. They were a popular group in the city, but not overwhelmingly so. In the talent competition at the Mid-South Fair, they came in third. On the nationally televised "Ted Mack Amateur Hour" (where Memphis's Johnny Burnette and the Rock and Roll Trio had gotten their break a generation before) they won two out of their three appearances.

In 1965 the West Memphis scene was only an afterglow. The Avantis, a black Memphis vocal group that had regularly performed at the PI and followed the scene across the river to Memphis, were performing at venues like Clearpool and various YMCAs, enjoying some small success from their record "Keep on Dancing." The Gentrys had heard the song on the radio and had seen the band perform it live. At the end of a session, at Moman's behest, they cut their own version; Moman shared publishing rights to the song.

Later that year, the Gentrys' version was in the top five on the national charts. Sounding like the senior in high school that he and the others were, Larry Raspberry told a journalist during the band's peak, "Ours is a fast, peppy kind of sound, one that makes you feel good to play. The old slow stuff kind of depresses you. Ours has more drive and makes you want to get out and dance." Raspberry and Randy Haspel had begun a friendly rivalry even before each passed the other on the stairs at Lynn Vernon's guitar lessons. Both had R&B bands before the Beatles, and both rolled with the rock. The Radiants had a hit single first, but when the Gentrys got their turn, they reached new heights.

"Keep on Dancing" was originally a B-side. The band's version was brief, so Moman looped the parts that worked—all twenty-five words—and made it long enough to qualify as a song: two minutes, eight seconds. It sold a million copies. Fan clubs were established across the country. Their national headquarters in Memphis hired two staff members to answer mail. The Gentrys performed a few times on "American Bandstand," appeared on "Hullabaloo" and "Shindig," appeared in the MGM film *The Girl in Daddy's Bikini,* and recorded a national radio commercial for Juicy Fruit gum. "Keep on Dancing" also reconstituted Moman. He traded half his interest in the song to farmer Crews, who had recently bought out the other partners. Moman had half his studio back and was working toward a house band, a concept he'd admired at Stax.

"Mary Lindsay and I were married, and I was living at Memphis State," says Jim Dickinson, talking like 1965 was yesterday. "I was playing an occasional Mar-Key gig, not doing a whole lot. Raspberry called me up, ten o'clock one night, and 'Keep on Dancing' was a hit. He says, 'Dickinson, my record is number fifteen with a bullet.' And I says, 'Good, Larry,' and he says, 'Half my band just quit, both the keyboard players and the girl saxophone player.' I says, 'That's pretty bad, Larry,' and he says, 'Yeah. I gotta go on the road and I can't go on the road without a band. I've gotta turn in an album, and Chips won't start because the band has quit. Will you go on the road with the Gentrys?' I says, 'No, Larry.' He says, 'Will you go to American and tell Chips you will?' I understood exactly where he was and I said, 'Yeah, sure, Larry.' So Mary Lindsay and I went down to the studio, middle of the night. I'd seen Chips around and knew who he was. He says, 'Yew go on the road with the Gentrys?' and I said, 'Yeah, sure.'

"While we were standing around talking, Chips locks the door and doesn't open it 'til noon the next day, at which time we'd made the whole first album and half of the second. Chips says to me, 'Yew too good to go on the road with the Gentrys,' and I says, 'That's what I think!' Suddenly I was back in the music business, Chips paying me $92.50 every other week, maybe. And maybe was part of the deal. My first payment was a counter check on mimeograph paper from Lepanto, Arkansas. It didn't look like it was worth $92.50 but it cashed alright. I stayed for about six months, from the fall of 1965 to the spring of 1966, playing on everything after 'Keep on Dancing' and before Sandy Posey's 'Born a Woman.' When I went to work for John Fry at Ardent, all he did was erase the maybe."

One of the Gentrys who didn't quit was Jimmy Hart. During the band's later days, they backed the king of local wrestling, Jerry Lawler, before the regular—and hallowed—Monday Night Wrestling in Memphis. While Lawler was singing, his then-nemesis Handsome Jimmy Valiant—whose own name had tarnished several pieces of vinyl—broke a guitar over Lawler's head. The incident apparently lit a bulb over Hart's. As of this writing, he is ring manager for wrestling magnate Hulk Hogan; stores sell little plastic dolls in his image.

The Beatles also transformed the LeSabres, the band that Haspel describes as "leather boys with greasy hair, smoking onstage and running around." Their lead singer, Laddie Hutcherson, formed the Guilloteens, a trio with a drummer named Joe Davis and guitarist and vocalist Louis Paul, later a solo artist on Stax. Hutcherson met Davis through Ray Brown's agency, when he sent them out together as Mar-Keys. "We let our hair grow long and we started dressing in English-looking clothes," Hutcherson has said. "We chose a name that sounded European since we were trying to cash in on the Beatles." The Guilloteens got a job as house band at a Memphis teen club, the Roaring Sixties, and soon were drawing a crowd. Despite a lack of original material, their manager, Jerry Williams (later a partner with Steve Cropper in the Trans-Maximus studio and label and also the manager for Paul Revere and the Raiders and, for a time, Neil Diamond), took them to Los Angeles hoping to get a break on the TV show "Shindig." "We bought some Hollywood clothes and auditioned for a gig at the Red Velvet Club," recalls Hutcherson. "We were nervous, but our manager was tight with Elvis, and the club owner thought Jerry could get his daughter a date with him. With those connections, I guess we sounded good."

They shared the Red Velvet stage with bands like the Byrds and the Turtles, also unsigned at the time. "The Righteous Brothers loved Louis's voice and took us under their wing, helped us develop our presentation," says Hutcherson. "They brought Phil Spector in to hear us one night, he was producing them, and Spector asked if we had any original music. We played him 'I Don't Believe,' which was barely a song then. The very next day we were in Spector's living room, which was big as a hotel lobby, and he came down this huge flight of stairs, and I remember he was wearing house shoes that were gorilla's feet." After overhauling the song, he produced a demo in the studio. "It had that wall of sound, Spector sound," says Hutcherson. "It was unbelievable." Williams was in Memphis when the Spector session transpired, and when he found out about it, his ego popped a wheelie. "He worked out a deal with Hanna-Barbera—the cartoon company!" Hutcherson continues. "We had to sign the contract or go home, and we signed. In the back of my mind I'm thinking, Spector's going to be pissed!"

Despite more experience in promoting cartoons than rock and roll records, the company made a minor hit of the recut version in 1966. Part of its distinct sound was Louis Paul playing the electric twelve-string guitar. In Randy Haspel's perception, perhaps because their success began so far from home, they were bigger even than the Gentrys. "This was teenage pop success like no one had ever seen. They were the dream. Los Angeles, long hair, a hit record, 'Shindig.' It was our world." They did a couple tours with Paul Revere and the Raiders, but none of their songs took off after the first one. Hutcherson remains active on the Memphis stage, part of the Funn Brothers duo.

Unlike most of Memphis's other studios, Ardent Recording has survived the bulldozer and the other confrontations from the city and the industry. Ardent was begun by three tenth-graders in 1960, established in John Fry's grandmother's sewing room when she wasn't using it. It's now a multimillion-dollar business and remains among the South's premier studios. Its walls are lined with hit records recorded there—ZZ Top, R.E.M., Leon Russell, the Bar-Kays, Travis Tritt, and its halls are haunted by the spirits of the musicians who've passed through—Big Star, Packy Axton, Isaac Hayes, the Replacements. Ardent's reputation has always been grounded in its equipment; it is a studio that understands new recording products and stays on the crest of technological advances while maintaining an atmosphere of ease and earthiness. It is an approachable place with great sound.

Ardent was initially a response to the only recording studio John Fry had ever been in. He'd begun tinkering with electronics in the late 1950s, especially interested in radio. While a sophomore in high school, he and two friends decided to establish a record label; John King would remain in the music business, but their other friend got out when he went to college. His name was Fred Smith, and he later founded a delivery company called Federal Express.

The Ardent label's first release was on an artist from Jacksonville, Florida. "His name was Freddie Cadell, and he was an acquaintance of the folks who lived across the street from my parents," says Fry. "We didn't know any of the bands around town. Subsequently we found plenty, but in 1960 it was not as common. He was recorded in some strange studio in Jacksonville, very primitive."

"At the Rock House," backed with "Big Fat Mama," didn't leap onto charts anywhere, but it walked the teens through the production and manufacturing process. "The norm at that time," says Fry, "was if somebody could render a fairly complete performance without serious errors in it, and if there was no technical mishap, then you considered the recording completed. And you know," he adds wryly, "maybe we knew more than we realized then. That may have been to the benefit of the music." When technology moved from sixteen tracks to twenty-four, Fry, who recorded and mixed the first two Big Star albums, quit producing records; the process had become too tedious.

Ardent created recording equipment from materials built for radio stations, modified by the young electronics whizzes. The first incarnation of the Ardent label released five singles, achieving some success with a popular band from the nearby University of Mississippi, and establishing themselves with the Shades, a band that included a former member of the Counts, Charlie Hull, and Lee Baker's talent show partner, Mike Alexander. Baker began hanging around the studio with Alexander. "Ardent was all white, but everyone was digging black stuff, really committed," says Baker. "People around town were going, 'Yeah, there's this guy that's got a studio in his house.' It was a big deal."

Then John Fry retired from the recording business. He was seventeen. "My interest was as much in radio as a medium as in the records. I think one of the reasons that we started recording was that we were just a bunch of kids and we sure couldn't get a radio station, so we'd do the next best thing. A little bit after I graduated high school in 1962, a friend of my family's got a grant to build a radio station in Pine Bluff, Arkansas. We thought, Well it's not exactly a major market, but this guy'll let us mess with it some."

"When I came along about 1964, I just talked Fry into reviving the label," says Jim Dickinson. Ardent was still in the sewing room, an enclosed garage across the patio in the back; the control room, along with an office, was in the main house. The talkback between the two was a covered walkway. Fry was cutting radio announcements for Pine Bluff when a friend named Bob Fisher came to him with Lawson and Four More, a group he'd met through the music store where he worked. This group piqued Fry's interest. Fry remembers, "Fisher needed help with Lawson's band and brought in Dickinson. Dickinson had recorded with Bill Justis for a Mercury-affiliated label. He was known to have ideas."

"Fisher calls me up," remembers Dickinson, "and says he's found this band of teenagers, fourteen and fifteen—he was lying because they were a little bit older—and he said they played great. He was lying about that too. He said, 'I need some songs.' I said, 'Sure Fisher, what kind of songs?' He said, 'I want a Kinks song.' 'When do you want it?' 'What about tonight?' I said, 'Come by and pick it up.' So I went in the bedroom and I wrote 'Back for More,' put it down on tape.

"It was done when Fisher came by, but I was already angling to get in at Ardent so I said to Fisher that I wanted to play it for them myself, make sure they understood it. We go over to Fry's house, there's Fry and Lawson and Four More. Little babies. Fisher was going to produce the record, and I just took over. It was the first record I produced. By the time it was done, I had also written and cut the A-side, 'If You Want Me You Can Find Me.' Fry appreciated a cocktail back then and he sat there, stewed, his feet up on the console listening to the playback. 'What do you think of that, John?' He says, 'Best damn sound that ever came over these speakers.' I thought, Yeah, this guy's got something going here."

One of the "little babies" in the band was particularly drawn to the studio experience. He played organ with Lawson's group, but was equally comfortable on guitar and, soon, a variety of instruments. Off the stage he was kind of quiet, but in performance he was as excitable as James Brown and as active as Tarzan. His name was Terry Manning, and he's since coproduced or engineered most every ZZ Top album, several George Thorogood hits, the Staple Singers' "I'll Take You There," was involved in mixing Led Zeppelin's second and third albums, and recorded Furry Lewis playing guitar in his bed with his leg off.

Manning was from El Paso and had played in the Bobby Fuller Four before they had a hit with "I Fought the Law." When he heard "Last Night" on the radio, he made his parents drive him to the nearest department store, where he bought his 45 on the Satellite label and began wearing it out. His father was a preacher who moved frequently, and Terry remembers nagging his parents about Memphis until they consented. A week after he arrived in town, he went to Stax, knocked on the door and said, or thought, "Here I am."

After gaining entrance to Ardent as an artist in the mid-1960s, Manning remained for twenty years. At first he answered the phone a lot, but he was watching everything that had anything to do with running the board and soon was producing. By the late 1960s, Manning was involved in many Stax hits, including albums from the Staple Singers, Booker T. and the MGs, and Isaac

Hayes's breakthrough. By then he was John Fry's main man, and Ardent was Stax's B-studio.

Lawson and Four More—Lost in the Morgue, as one local deejay mispronounced it—didn't do much, but because everyone had had a good time recording it, they got to cut another single. Fry's family was moving from their residence, so in 1966, he moved his business. Once on National Street, Ardent became an alternative rental space in town, featuring high-tech equipment and an open mind, a willingness to work with whatever sort of material was brought in. Black Oak Arkansas cut there when every other studio in town thought they were too loud. Leon Russell, Sam the Sham, and many Stax artists recorded there, too.

"In our early days," says Fry, "Dickinson was the guy who'd had some experience. And he knew some players and had industry connections. As well, he had a healthy skepticism toward the business which probably provided some good guidance for a lot of people. He had an experimental spirit, too, not afraid to try doing stuff a different way. That's a great catalyst to have. He encouraged others not to worry too much in advance about whether it's gonna work out, that will become obvious. Just do it and see what happens.

"Our facility coincided with the rapid upswing of the technology. In 1966, if you had four-track equipment, you had as many tracks as anybody had and more than most. Between 1966 and about 1970, you've gone from four to sixteen tracks and consoles much larger, and gone from almost no outboard equipment to at least some, and to using Dolby noise reduction and equipment that required a fair amount of alignment and attention in order to work right. We wound up mixing a lot of stuff that other people would record because we could apply some technological efforts that seemed to enhance it a little bit. Also, our console was the same make as Stax's, so that somebody from there could feel at home with what we had. We stayed busy."

At the local dances when the bands took a break, a disc jockey kept the music going with records. Dewey Phillips was enjoying a minor comeback, broadcasting from a tiny Millington station, and he began to work the dances. "Dewey's Millington show became very popular with a generation of young people who were just kids when he was making musical history at WHBQ," says Haspel. "This was evidenced by the hundreds of teenagers who would flock to the station-sponsored dances every Saturday night at T. Walker Lewis YMCA in East Memphis. Phillips would play 45s and assault the audiences with his nonstop verbiage. The kids would gather around to hear his jokes and get his autograph."

Radio in the early 1960s was beginning its move toward narrowcasting (epitomized by the strict playlists of today), away from the free-form shows by pioneer tastemakers like Phillips and Rufus Thomas, John R. and Wolfman Jack. Disc jockeys still had some influence on the music they played, and on WHBQ and WMPS, the AM stations catering to the white rock and roll audience, local bands releasing local records were not discounted. The concept of playing a record on TV had advanced a little, but not much. The "American Bandstand" method—a bunch of honkies moving whitely while a band lip-synched—had won out over Dewey and Harry doing whatever crazy things the music inspired them to do. (That same trend, unfortunately, dominates current music videos.) The Wink Martindale Saturday show, "Dance Party," had passed through the hands of a couple other deejays before settling into the twelve-year reign of Dewey's protégé, George Klein, beginning in 1964.

At Humes High, Klein had been class president, and Elvis had been his classmate. His break into radio came in the early fifties when WHBQ hired him to "baby-sit" Dewey Phillips, buffering him from his fans. When Klein began as host, "Dance Party" was still reeling from the station's decree banning the dancers, a move they chose over integrating them. So there was no dancing and it wasn't much of a party. He changed the name to "Talent Party" and put the emphasis on the bands. Touring acts stopped by the station and videotaped interviews and performances; when the show was broadcast on Saturday, Klein made all the acts appear to be live and in the studio. After a couple months, he shanghaied the "Shindig" concept and added a chorus of six dancing girls, the WHBQ-ties; within a couple years, he integrated them.

"Television gave a visibility to pop music that it didn't have before," says John Fry. "All of a sudden you were seeing people interviewed and seeing them perform. It sort of opened up the horizons." Everything about "Talent Party" rode the fine line between amateurism and inspired genius. The sets were models of innovation. One band was surrounded by a few dozen paper plates spray-painted in various colors; the next band performed against a psychedelic backdrop that, on closer inspection, was revealed as the board on which the plates had been painted. Klein's radio popularity brought an audience to the TV show. He also had connections in the business, which assured him the bands' cooperation. His patter was rapid, he favored goofy rhymes, and in the interviews he conducted his nervousness sometimes won out. He once asked B. J. Thomas, "You were fatally stabbed in New York, right?"

But—there was B. J. Thomas, a top star, standing right next to George Klein, and it was live from a studio in the heart of Memphis. With the recording studios in town amassing local talent and attracting international stars, Klein was ideally situated to draw major acts on his local show. Except that so many stars were black and all the guests were white. "We were doing the show for a while," remembers Klein, "and I said to the program director, 'Here we are right in the heart of what's happening—Stax, Hi, American, Sun, all the studios—yet we've never had a black act on the show.' And they saw what was coming, and they said, 'You're exactly right, George, but to open the door can you get a big star?' It just so happened that Fats Domino was coming to town so I went down to Ellis Auditorium, it was about ten-thirty, eleven at night, after Fats's show. And I had told my crew to stay in the studio after the news if possible because I was hoping to get Fats to come down and videotape some songs for 'Talent Party.'

"Fats was a little apprehensive. Black acts hadn't done much TV in the South. After talking about some of his records and talking about Felton Jarvis, a producer of his that I knew, and then the Elvis stories came up—finally when we started communicating really nice I said, 'Fats, you said you don't do local TV, but you'll be the first black guy to ever do the show and it would really open the door for other black entertainers.' He said, 'I'd be the first black guy? C'mon, let's go.' He got in my car, we stopped at a liquor store on the way, got him a little nip. He did four songs and an interview, and in the mid-sixties that just opened the door for integrating the show."

"Talent Party" remained faithful to local talent. Every show featured at least one local act, no matter whether they had a recording contract. If they hadn't recorded, they were directed to Roland Janes's Sonic Studio or, later, to John Fry's Ardent to get their lip-synch tape made. "I charged ten dollars an hour and three dollars for the tape," says Janes. "For thirteen bucks you could come in, set up, and cut four songs and be on television. I got kidded about it a lot. Some of the bands didn't sound so good, but some of 'em sounded really, really good. And for any number of bands, that was their first recording experience. I used to tell them if they didn't behave theirselves I'd spank 'em."

Klein says, "I had acts that couldn't get their records played on local radio, yet I was blowing 'em hard and heavy, breaking 'em on TV like Dick Clark used to break acts on 'American Bandstand.' Sam the Sham with 'Wooly Bully' and the Gentrys with 'Keep on Dancing.' A band could come on 'Talent Party' and reach the entire mid-South area—Tennessee, Arkansas, Mississippi, and some parts of Missouri. It was great exposure for local bands and national acts. After many of these bands got big, they told me they appreciated being able to do a local show before they had to go on the road and do bigger shows. We showed them how the lip-synch situation worked and that when the red light came on they were supposed to sing and the floor directors would point at the camera they were supposed to look at. When they got to Cleveland or Detroit or New York or Los Angeles, they had an idea about what they were doing."

Felton Pilate, now the producer for rap and pop star Hammer, credits "Talent Party" with his start. His band, the Soul Children, formed in California, and after backing up Stax acts during the 1972 WattStax concert in Los Angeles, came to Memphis and changed their name to Con Funk Shun. "'Talent Party' got us a lot of local gigs," he said during the 1992 world tour with Hammer. "But the tape we made at a local recording studio for the television show got us signed to our production company deal, which got us signed to Polygram." Cybill Shepherd is another success story. During the annual Miss Teenage Memphis Pageant, the "Talent Party" fashion coordinator spotted Cybill and said she'd like to send some photos to someone she knew at a modeling agency in New York. "Cybill didn't want to do it," says George, "but we sent 'em up there, the agency accepted her, she became model of the year, and Hollywood came calling."

In 1965, Sam Phillips's younger son Jerry, formerly the World's Most Perfectly Formed Midget Wrestler, was in a band that owed more to his father's earlier work than to the Beatles. Jerry played rhythm guitar in the Jesters and had found a punkish kid with lots of attitude to write songs and play lead. His name was Teddy Paige, though he was born Edward Lapaglio. His values were different than those of the other East Memphis white kids. Dickinson, who recorded with him, remembers that from the stage, he would introduce songs with lines like, "Well, here's another little song you're not gonna like," or, "Something else you've never heard." Randy Haspel remembers Paige's burning electric guitar style as unlike anything he'd yet seen; others cite Paige as the first in the area to say that the Beatles ruined music.

Teddy Paige wrote a song called "Cadillac Man," and it was the last Sun release that had the power of the earlier material. "Cadillac Man" was produced by Jerry's older brother Knox in late 1965, after the Radiants had been through the studio. Like all the Sun sessions from 1958 on—everything after the handclaps and saxophone on "Lonely Weekends"—this one was done at the Phillips studio on Madison. The previous year, Stan Kesler had cut "Wooly Bully" with Sam the Sham there. Knox had not been much interested in music then.

He wore V-neck sweaters with button-down shirts. He married young, he taught Sunday school, he worked hard to keep the madness beneath the surface. Once his divorce transpired, he took more interest in the family heirlooms—music, the studio, mayhem. "I wasn't even in the band," says Dickinson, who played piano and sang the song. "I was under contract to Bill Justis at the time. The Jesters had another singer that Sam hated. The session was supposed to be a demo for Teddy's song, and I was hired to play piano. And then the singer didn't show up, so I read the lyrics off notebook paper while we cut it."

Sam Phillips heard "Cadillac Man" and decided he'd like to release it. Another session was booked, what's called a "smoker," to be filed with the union to explain how the song was recorded. "I really thought we would sit around and smoke," says Dickinson. "But we got there and Sam had a suit and tie on and he was walking around with a clipboard in his hand, writing down microphones and stuff and I got real excited. I'd been around him, but I'd never really met him. That session was the first time I felt the hands of the master. I looked into the black pools of madness in Sam's eyes and I saw the same thing Elvis and Howlin' Wolf saw."

Sam Phillips called Dickinson before the record was pressed, reaching him at Ardent when it was still in Granny's sewing room. "Sam put [his brother] Jud on the phone and Jud was an even more dynamic speaker than Sam is. Jud says to me in this booming voice, 'Boy, you gotta cast your lot.' And I told Jud that my lot was already cast, that I was under contract to Bill Justis. And he said, 'Aw hell, boy, Bill won't mind.' So they put the record out."

CHAPTER EIGHT

The Catfish That Ate Memphis

THE PROBLEM WITH THE BEATLES WAS THEIR MUSIC. AND THEIR FANS. Though their hair was radical for the time, that sort of statement paled next to drinking whiskey with someone who'd invented the blues, the very root of the Beatles' music. Their press conferences were entertaining, and in a few short years they would establish hippie fashion and philosophy, but the Beatles' early career indicates they would have leapt at the opportunity to pass a day or a year in Memphis, keeping the company the witnesses enjoyed. By the time they hit the states in 1964, they were slowly finding their way to the source.

For their first American tour, of all the bands available to open for them, the Beatles selected the Bill Black Combo, a Memphis instrumental group founded by Elvis's former bassist. They'd scored an early hit with the instrumental "Smokie (Part 2)" and had made a career appealing to the same audience that enjoyed Bill Justis albums. The Beatles must have known that Black had retired from the band two years earlier, yet the selection remains puzzling. "*They* requested *us*," remembers guitarist Reggie Young, still sounding amazed. He was the sole original member in the group. In his mid-twenties during the tour, Young was already a studio veteran, playing on Eddie Bond's late-fifties hit "Rockin' Daddy" before joining Black and helping define the Hi Records sound. The Combo traveled with the Beatles, and there is an oft-repeated story about the young George Harrison staring rapt while Young noodled on his guitar, having fun bending strings; Harrison's own sound would develop from the bent-string style. "We all hung out together for the whole thirty days," Young remembers. "We swapped stories and stuff, you know, talking shop. I remember we had some time off in Key West and we got in a little cafe and jammed for two days."

The Bill Black Combo may have been roots music to the Beatles, but the radical route for Memphis witnesses was folk music. Dylan had popularized the tradition, maintaining the elements of social commentary that were the music's guts and that had been diluted by the folk-pop that preceded him. Integration and rejection of the Vietnam war provided two mighty subjects. Folk fans found that the next step forward was two steps back, reaching for their roots and finding the witnesses already there. Greenwich Village contributed the venue—the coffeehouse—and kept the music in the national spotlight. In Memphis, where the attempts of previous coffeehouses had proven arch, the Bitter Lemon, yet more arty than its predecessors, proved innocent enough to work. An accidental sort of place, it was imbued with the character of its owner, John McIntire.

On an otherwise ordinary afternoon in Wellsville, Ohio, in the 1930s, a small propeller plane landed in a cornfield. The kids from town ran to see the contraption. From the cockpit stepped a Chicago millionaire, and he began tinkering with the engine. A lanky high-school senior moved forward. John McIntire watched, along with the other kids, as his eldest brother repaired the plane. John was six. This brother—there were ten siblings—was a gifted artist who had accepted an animator's contract from Disney and was to begin employment upon his high-school graduation, one month away.

Once the engine was working, the millionaire led the kids into town and bought them a meal. He swilled several drinks himself. Like the pied piper, he paraded the entourage back to the cornfield, where he rewarded the eldest McIntire with a ride. The plane took off and all the kids watched as it looped the loop. Then it looped again. Then it came very close to the ground and was out of control. McIntire's brother leapt out. The plane crashed, and both men were killed.

Before that time, John McIntire had dreams of playing professional baseball. Upon his brother's death, he was seized with an artistic drive. He was quickly drawing landscapes, exhibiting an acute understanding of perspective. When he finished high school, he worked two years in a steel mill, attending art classes in Steubenville, Ohio. He was awarded a four-year scholarship to the Cleveland Institute of Art, after which came a scholarship to the prestigious Cranbrook Institute. When the Memphis Academy of Arts solicited a teacher, the other students were busy creating resumes, their eyes on big cities. To McIntire, having a job didn't seem like a bad idea, and creating a resume did.

Within four days of his arrival in Memphis in 1961, he had experienced the city's great social chasm. The Academy, located in the beautifully wooded

Overton Park, lodged him for three nights nearby at the posh Park View hotel. He had room service and amenities he'd never experienced. But those quarters were temporary. He found a small place of his own behind Pappy's restaurant, and his first morning there, he awoke to a child standing over his mattress on the floor, filthy, diapered, sucking his thumb. On his windowsill there was a rooster.

"They used to call me the Warlock at Cranbrook because I had these visions of things that happened," he says. "One time in Memphis I saw a murder. I called the police and they said nothing like that happened. But they hadn't got the news yet. The guy called me back and said, 'How did you know about this?' I said, 'I saw it in my sleep when I lay down, middle of the afternoon.' It just flashed on me, I saw this guy shoot another guy and his girlfriend. It scared me to death."

McIntire had two suitcases when he moved to Memphis, tools and photographs of his work in one, and clothes in the other. But he'd had a dream about coming here and opening a coffeehouse. "I even saw people I was going to know. And when I'd later run into them, we were already friends. Carl Orr, I saw him a long time before I met him. Told people about him. When he came into my class, I stared at this kid. I knew he was different. Carl knows when I'm sick. Randall Lyon's the same way. Randall called me the other day, said, 'What the hell's wrong with you?' I'd been in the hospital and the message was on my machine when I got home. He just felt it."

Sitting with "John Mac" is like sitting with the wind. Part of his presence is his absence. Ideas and thoughts hang all around him, and if they can't be seen, they are definitely felt. His home is not unlike his long-gone coffee shop, every nook and shelf filled with paintings, drawings, and sculptures, or with yard sale remnants and oddities which he collects, ponders, then resells for a small profit or a small loss. Seated in a rocking chair, this virtuoso of chaos remains serene, if forever slightly distracted. He is tall and lanky, his longish hair and bushy mustache making him appear younger than his mid-fifties. He moves with an odd jerkiness, not unlike Charlie Chaplin; watching him sculpt is as pleasing as contemplating his artistic result. His rational side intended to stay just a year in Memphis, but McIntire's visionary side predominates. Twenty-four years of students credit him as a major influence. His imagination has not dimmed with age. He embraces everything and puts his art first.

Overwhelmingly shy, he communicated with his first classes only through the chalkboard. His withdrawn nature was interpreted as an air of mysticism. Combined with his commitment to his sculpting, evident not only in his devotion to his work but in the unusual objects with which he surrounded himself, McIntire became an artistic guru in Memphis. He held yard sales that people attended like church.

The Bitter Lemon Coffee Shop and Gallery was an extension of McIntire's personal life. The scene had begun gathering around him when he moved into an old, big house in semi-disrepair. He'd been in town long enough to recognize a group of artists trying to find each other, and this place at 2166 Madison, near Cooper, had room enough for everybody. There was something magnetic about that corner. It's not a major thoroughfare, though it stays busy, the heart of the large neighborhood known as Midtown. Long before McIntire settled there, a physician named Doc McQueen held musical jams in his home near the corner. Roland Janes met Jack Clement through Doc McQueen in 1955. Chips Moman made contacts there, the rockabilly Burnette Brothers. The corner would draw Crosthwait to stage one of Memphis's first art happenings there, a guerrilla event involving a toilet bowl placed in the intersection and set on fire. The dumbfounded look on the faces of the police and the neighbors indicated its success.

McIntire's house became known as Beatnik Manor. The vision taking shape. At one time, there were fourteen people and fourteen cats living there. Rarely did he have it to himself; he met many of his housemates only after they'd moved in. Bands that played the coffeehouse slept there. Kids who ran away from home stayed there. Allen Ginsberg came through, Marcel Marceau. "I bought an old VW van and everybody used it. Sometimes I wouldn't see it for weeks at a time. Guys would borrow it and leave it, they'd be stoned and wouldn't remember where. We'd have to ride around the streets of Memphis looking for it."

Unlike those around him, McIntire held a regular job. His teaching position came with a studio and materials, and since his art was his life, much of his income was disposable. There was never any shortage of people to help spend it. "I always made sure the refrigerator was full of food," he says. "Randall Lyon would be the head chef, the madam of the house. Madam Randall. He never wore clothes, except a big robe and half the time he didn't wear that. He'd go to the door buck naked. He didn't care."

Randall Lyon, stage center, entrance. A shock of wiry hair, a hulking figure, a walrus mustache. He is the (butter and) eggman. Goo goo goo joob. Coming to Memphis because he outgrew Little Rock, Lyon cavorted with writers and musicians, theatrical talents and film technicians, amalgamating them all into an art of himself, at the fringe of the fringe, articulating the maelstrom while increasing the chaos. A self-described obscurantist, he reaches his widest audience through influencing others, keeping himself in the shadows.

"The Bitter Lemon was the place that drew us together," he says. "If it wasn't
for McIntire being in touch with artists, poets, musicians, there wouldn't have
been a center for the situation, and he was it. The Bitter Lemon was a good
place to play. Fred McDowell was playing there, Furry was playing there, the
Allman Joys played there, Lee [Baker] played there, Don Nix, Charlie Freeman, Dickinson—that was the hub. And you could eat there too. John had a
place to work, a place to crash, a place for food."

"It's hard to describe what kind of figure John was," says Robert Palmer,
an aggressive clarinetist who began regularly trekking from Little Rock to
Memphis in 1965; also a writer, he would become the chief pop music critic at
the *New York Times* from 1976 to 1986, publishing his book *Deep Blues* in 1981.
He and Lyon are lifelong friends. "McIntire was like a nonguru. He was a bit
older than everybody else, he had this great visual eye and this unlimited tolerance for weirdness. I think of his place over there on Madison, Beatnik Manor,
as having been like a salon for artists. There was always somebody doing
something interesting around there. Always. And there were always interesting and very weird people staying there. From night to night it was hard to tell
who lived there and who didn't. The epicenter of the Memphis beatnik scene
was for sure with McIntire. And it wasn't really anything he said or did, although he could be incredibly charming and funny. His attitude was like, I'm
doing art, here's my house, you know, rock on."

Once in the Manor, McIntire met an older art student who shared his desire to open a coffeehouse and was grounded enough to handle the business
affairs. During the summer of 1964, they found a place at a perfect distance
from the house: far enough to keep crowds from spilling into the Manor at
closing, but near enough that residents too drunk or high could walk from
one to the other—and arrive feeling refreshed. The aura of the coffeehouse
came from McIntire's predilection for secondhand merchandise and old junk.
"The Bitter Lemon looked like a beatnik place," says Sid Selvidge, "where the
Oso had been kind of a hootenanny place." "The Bitter Lemon was the personification of the sixties in Memphis," says Lee Baker. "During the folk
times, it was folkie, and when the psychedelic thing began, McIntire painted
the whole inside psychedelic."

"The setup was simple," says McIntire. "If you drank coffee it was okay,
but the other drinks were horrible, sweet and syrupy. We had an ice machine,
and a stupid little pizza oven. We'd buy these commercial pizza crusts and put
ketchup on 'em. It was making lots of money in the beginning, and people
tried to buy that place from me. Everybody said not to sell it. We only had
fifteen hundred dollars or so in it, that was everything. And we were offered

John McIntire. "It's hard to describe what kind of figure John was." Photo courtesy of John McIntire.

ten thousand. People who owned other coffeehouses were calling us from around the country. They wanted to franchise us. We went to Little Rock to look at places to open. Bitter Lemons across the United States."

"The Bitter Lemon was an innocent place that seemed to have all kinds of dark foreboding about it," says Jimmy Crosthwait, an early champion of the scene. "Coffeehouses at the time were dens of something or other. But looking back at it, it was like an ice cream parlor. It had ice cream parlor chairs and little ice cream parlor tables and it really tried to sell different kinds of espresso and shit. But there were poetry readings a little bit and maybe every now and then somebody would want to play chess."

"The rich kids would come and spend money in there," says Marcia Hare, who collected their tips as an underage waitress. "They'd get drinks with names like Passion Fruit and they were real expensive—today it would be a five-dollar drink—and it was nothing but orange juice with an umbrella in it. We'd wait on the customers before the band played, but once they began, you didn't want to interrupt. So we'd just dig the music and then go out in the parking lot and get high with the band. We were real into drugs. I guess the crowd inside was spiking their drinks."

Marcia Hare speaks with a purr in her voice, her coquettishness deflecting her commanding presence. Though she has experienced a very full forty-something years, she looks much younger. "Are you going to put me in your book?" she asks. "Ah'm a tittie dancer." I know who you are, I tell Marcia, and tell her the first time I saw her was when the plug was pulled on Mud Boy at the downtown festival. She has since retired from the stage, invested her earnings in a modest house. She came to Memphis in 1964 at the age of fifteen, moving from Maryland. Her sister and she hated the dullness of the city when they first arrived, but by the following year, when Marcia was in the eleventh grade, she found herself in the midst of a unique experience. "My dad was an English professor, and when we moved down here, first thing he did was take us to the civil rights marches and demonstrations. That didn't make us real popular at school. We were Yankees and nigger lovers too." At her father's instruction, Marcia and her sister volunteered their after-school time at a local theater, helping build sets. "He wanted us to do something educational, to keep us out of trouble," she says with a laugh. "But that's where we got into acid."

Theaters, at the time, were Memphis's bastions of alternative culture, and a couple of older hippies were soon taking these girls around the scene. Daughters of a French mother who was working in a New Orleans strip joint, they were wise without study. "We were desperate for a nightclub," Hare says. "In

D.C., we'd been going to Georgetown, and here there wasn't a thing to do. The Bitter Lemon was somewhere to go to hear music. My older sister got a job there right away, because we always thought if you liked to be somewhere, it's better to work there than pay to go in. Through the theater, we met Bill Barth, and Barth was the big hash dealer and that's how I finally got to meet some blues artists. Blues was the one thing we liked about Memphis when we first moved here." The sisters alternated nights waiting tables.

Marcia's sister began dating a playwright, and Marcia ran with a lighting tech who had come from San Francisco. His California roommate had been a man named Kenneth Owsley, aka Oz, aka the Bear. Owsley was a chemist, then brewing huge batches of LSD, which was not yet illegal. The hallucinogen had been invented in 1938 by the Swiss scientist Albert Hoffman, and was, by the 1960s, being manufactured for behavior-modification programs such as the treatment of alcoholism. Scientists and medical students in San Francisco were using it, as Randall Lyon says, "to figure out a way, through equations, to describe a thunderstorm." Owsley had impressed Hoffman with his knowledge of organic chemistry and become the premier brewmeister. When his roommate moved to Memphis, he found, of course, Beatnik Manor. The LSD connection between Memphis and San Francisco could not have been more direct.

"Little Rock and Memphis had LSD in sixty-five," says Lyon. "Really radical, unmeasured LSD was coming here way before other places. We were a very aggressive bohemian community, and there was this very direct connection between San Francisco and Memphis. We took enormous amounts. Huge doses. Well over five thousand micrograms. Major doses. Everyone was transformed by that experience. That happened before the blues festivals, though they were one of the next things to happen. The LSD gave us the inspiration, the understanding that we could do that. We had no fear."

In San Francisco, the acid scene was still a year away from going public, and two from going national. The three-day "acid test" at the Longshoreman's Hall featuring Ken Kesey's Merry Pranksters was in January 1966. The Fillmore would be opening in a few months. *Surrealistic Pillow*, the first Jefferson Airplane record with Grace Slick, came out in 1967. In Memphis in 1965, acid became an underpinning of the scene. "You could take the stuff and go out and run around and not be paranoid because it wasn't illegal," says Robert Palmer. "I remember in 1965 and 1966 we occasionally used to trip with a few people and go out marching through the streets wearing things like oriental rugs and banging cymbals and playing flutes and chanting and yelling, and cops would drive by and say, 'Oh it's a bunch of drunk college kids, ha ha.'"

The acid arrived in Memphis on powdered chalk. A teaspoon of it made forty gelatin caps of 250 micrograms per dose. One shipment got mixed up, and the dosages were much higher. Marcia Hare remembers they were going to meet for a big trip in Little Rock at a friend of Randall's. "We'd gotten this Owsley package and divvied it up in Memphis, licking our fingers all the while. We were already kind of rubbery when we left and halfway down there, everybody was losing it. Carl Orr had a death grip on the steering wheel and was driving real fast. He wiped out four or five markers on the side of the road. When we got to Little Rock, we realized we didn't have an address and we had no idea where we were going. Little Rock was a tiny town then, and it was about midnight. We could hear Bob Dylan music, and I swear we followed the Bob Dylan music and it led us to the right apartment."

Though some were getting heavily into substances, and surely there was abuse, debauchery was not the goal. The heavier doses of the drug did not send one, like today, out into nightclubs to party, but rather on inward, contemplative trips. "It was exceedingly peaceful," says Lyon. "Peaceful contemplation is what it was about. And McIntire and his coffeehouse provided this surrealistic world, removed from the world because it was all secondhand. He had eighteenth-century actors' robes in there, mannequin heads and weird models and strange glass bead curtains and oriental rugs from Goodwill. We got oriental rugs from the fucking Goodwill! McIntire set the imaginal criteria, helped give a certain visual language. John still does that. But when we were all tripping, the high end of the LSD experience was focused on that world of his."

"I only tried it several times, but it'd put you out for three days," says McIntire. "It was real pure. You'd have a fantastic trip and didn't worry about going crazy. Sometimes I'd ride a bicycle all over Memphis, take photographs. Later on, people started putting speed into it. I never did that anymore. I had goals, I wanted to be an artist and I didn't want to let that get in my road. A lot of these kids were younger, had no goals and went looking for heavier stuff and killed themselves. We lost a lot of friends. Blew their brains out. Usually they had money, and their families abandoned 'em. It was terrible when somebody would be missing and you'd find out they were dead."

While the acid community was relatively small and tight, more and more people from the mainstream began sharing their other interests—folk music, artistic inquiry, and the bluesmen. The Beatles had shaken the moss from society, and when the sun shone in, people began seeing beyond the four square walls that defined them. The Bitter Lemon became a hub for persons involved in a process that consumes society today: deculturalization. Beatniks in the

1950s got out of society's box, recognizing black culture as an alternative to their own. They revered the jazzmen, bop, music made by breaking the rules, and they tried to create a society from the same concept. Their inroads paved the way for the early sixties, when integration began taking hold, and in 1962, when Bob Dylan burst on the scene, the glint of marginalia available to the baby boomers kaleidoscoped, telescoped, landed in their lap and began to touch the mainstream populace.

As an impetus for deculturalization, the Vietnam war cannot be underestimated. Though full-scale rebellion against it was still several years away, by the middle of the decade soldiers were returning from their mandatory service, exposed to new lifestyles. Between 1962 and 1964, Randall Lyon had served two years in intelligence. His clearance was so high that he is nervous talking about it today. "Information was my job. I learned how to do propaganda, psychological warfare, and I went overseas and it changed my whole scene. What can you do when you find out that the government is your enemy? Don't worry about the Vietnamese, worry about your government." Goo goo goo joob.

Segregation remained the law of the land, right down to separate water fountains, but youth culture began embracing the dignity and elegance of those not like themselves. The friendships that the witnesses in Memphis developed with blues artists—older men and women, often rural, always poor, ostracized and rejected by "society"—became living proof that there was life beyond what they'd been raised to see.

The attention given the blues artists was an indication that the walls fortifying "high" art were crumbling. Through the Second World War, photography had been dominated by grandiose drama, Ansel Adams lugging his camera to extreme mountaintops to get shots of what the average person would not ordinarily see. But by the 1950s, artists like Lee Friedlander and Robert Frank were expanding ideas tendered by Henri Cartier-Bresson and his candid photographs capturing Parisian daily life. The Delta-born and raised William Eggleston began making strides in photography at the same time the blues renaissance was happening around him. Moved by the images of Cartier-Bresson, Walker Evans, and Friedlander, he turned his eye toward the details of the world about him—the South at first, but place was not integral to his idea. Eggleston discovered life, passion and color—glorious color—in what others dismissed as mundane. "I got serious about being a photographer in the late fifties," Eggleston says. "During the sixties I had an intense period of development, and by the late sixties I had formed a way of working that was thought out, that is still continuing. It's evolved, but it's identifiable as the same person's work.

"I was interested in taking pictures to make photographs more than to record social things or events. I wanted to make a picture that could stand on its own, regardless of what it was a picture of. I was interested in the photograph and still am. I personally don't have much use for pictures of things that are not images with integrity to them. I have never been a bit interested in the fact that this was a picture of a blues musician or a street corner or something." The public's warming to Eggleston's approach was nearly simultaneous with the realization that the yardman could be a major American artist. It would be the mid-1970s before Eggleston achieved his breakthrough, for the first time elevating color photography from the pages of advertising to the walls of the Museum of Modern Art in New York. In the meantime, he found welcome company among his musician friends, and no lack of challenging work near his Memphis home.

"I can see it in the 1965 sunlight," says Lyon. "Bill and his wife Rosa pulling up to Beatnik Manor in the Lincoln. Bill would step out, always wearing a real severe suit, it was like he was the fucking count. Voluptuous and corrupt, a striking image. It was unreal, what an aura. He stopped traffic. It was absolutely fucking Tennessee Williams beautiful."

"I knew everyone around Beatnik Manor," says Eggleston, "was close friends, but that was not a place I spent much time. For one thing, I had my own house, with a laboratory, with music, with lots of things I needed that were not at Beatnik Manor. I was less of a strolling minstrel than some of those people." Eggleston was older than these beatniks, and in the earlier 1960s he'd been part of a different artists group. "Sidney and Mary Chilton, Alex's parents, were some of my closest friends. They were two of the most important people in Memphis from that time, the Kennedy era. Mary held what you might call a salon, and things happened in the house. People would come there, it was an art gallery. They had a piano and Sidney would play, I would play, different people would come in. He was very good at analyzing how jazz things were constructed. I don't know who else would have fostered what they did. That's the kind of environment Alex grew up in. A different generation than Beatnik Manor, but just as intense."

The sixties brought the issues of racism, sexism, and class inequities to the fore of America's political agenda. A faction of contemporary society believes that change in focus destroyed cultural standards. The disintegration of public schools and the lack of "cultural literacy," the importance now accorded popular culture, even rampant violence, are attributed to the social movements of the sixties. This belief ignores or denies the value of non-European contributions. For our purposes, it denigrates the blues artists and the expression in their music. But since they began recording in the early years of the twentieth century, society has refused to let go of their work. From swing bands to the Rolling Stones to the rhythms of machine-influenced music, the bluesmen's ideas are continually reexamined. The Next Big Thing has always been a reworking of the union between the African tradition and the Western folk tradition. The sixties did not deny the beauties of our classical heritage but elevated other formative influences.

"Furry and Bukka White and Freddy McDowell and Joe Callicott and Nathan Beauregard, those people are really the groundwork of it all," says Lyon. "They were this combination of entertainer, performer, and Magus—they had full Magus status. What they were doing and what they were talking about and dealing with in their music—it was compelling. You had to respond to it."

"We would go out in the country and to joints and stuff and run around with those people," says Robert Palmer. "Randall and I used to ride the hour out to Como, Mississippi, and pick up Fred McDowell at the Stuckeys where he pumped gas. It's something none of us ever got over. It was really interesting to hear those people doing their individual stuff and not really changing, not giving a shit about trends or fashion or anything like that. And we were very, very devoted to those people."

The fact that blacks and whites of any age were at each other's homes was a new experience. The whites had attended their twelve years of school during a time when blacks, even if they lived next to the schoolyard, were not allowed to attend the same facilities. "You could go visit Memphis Minnie," continues Lyon. "Across from her bed, she had this picture of her in a white dress with the guitar and it was in a silver frame. She'd had a stroke. You would say, 'Minnie! I really love your music,' and she would start crying. It was heavy. Sleepy John was a saint. He used to have all these little jokes. If he thought there were people around who didn't know him, he would tell them to go turn the light on. Because he was blind. Somebody would go turn the light on and he would have a fit, just howl! He'd need a ride somewhere and we'd drive out to Brownsville to pick him up. He would call up at three o'clock in the morning, ask us questions about his social security, stuff like that. He felt pretty chipper. To his family, he was a meal ticket, which was normal. He was real old then and probably a real hassle, and we only saw the good part."

"First time I went to visit Furry," says Marcia Hare, "Verlene, Furry's girl-friend, was so nice. She said, 'Come in all you children, make yourself comfortable.' You'd sit down and she would go, 'No, make yourself comfortable!' And she'd make you lie down on the couch until she felt that you were comfortable. One day she called me in the kitchen because something I couldn't see was going on in the living room. I found out later Furry was putting his leg on and didn't want anybody to see him do that.

"Furry was a big ham. One time we were over there and Gus Cannon was there. 'Walk Right In' was a big hit where I was from, so for me it was like meeting a star. I really wanted to hear Gus play the song. Furry did not want him to share the limelight. So they sat there like two little children fighting to see who was going to play. Furry literally snatched the guitar out of Gus's hands to prevent it from happening."

Once communication between the cultures had been established, a last confusion had to be overcome. "They had no idea why we were there," says Lyon, "because we weren't much about money or anything. We just wanted to know 'em, they were fascinating people. Finally, they accepted us. Furry became highly socialized, he could work the white folks. He could send the best of 'em to the pawnshop for that guitar." The pawnshop scheme was a way the bluesmen could be sure they got paid to perform. It was obvious that money didn't stick to the scenesters, yet they could also see how anxiously their own company and performance were desired. To perform, they gots to have they axe, and if they pawned it for eighty-five dollars before a gig and had it redeemed by whomever was hiring them, they were assured some minimum payment.

"To go see Bukka White on a gray afternoon like this," says Lyon, and he leans back in the sofa and shuts his eyes, "and you walk in and there's his wife in the kitchen with a little girl, she's heating up curling irons, holding these things on the gas stove for fifteen minutes till they get really hot, and then putting them in her little girl's hair, and this blue smoke would come up, the little girl would twist around and her mamma would say, 'Stop it,' and Bukka would growl, 'I'm gonna put a screw in your eye.'" Lyon is recalling this story from a sprawling loft on Main Street in tiny North Little Rock, surrounded by compact discs. The sound of traffic drifts in the window, mixed with voices conducting business at the post office across the street. The phone has been turned off, the answering machine operating silently if at all, while Randall retreats to a time and place that may be gone but that continues to move him and that still lies at the heart of modern social issues. He is in the company of dead friends, and his being radiates such an energy that I flash on how the subject of this book became such an interest to me, why I was moved to give myself over to the writing of it. As Lyon describes the scene at Bukka White's home, I am transported to gray afternoons at 811 Mosby, where Furry Lewis lived for the years that I knew him. I was an abashed white kid, high school, a product of the last perimeter before urban sprawl. My suburbia still had a locale, and I had a sense of Memphis as a town, of the river as lifeblood.

I was changed by being in the presence of Furry, in a tiny duplex on a crumbling street that was only a block from Poplar Avenue, Memphis's major traffic corridor, but seemed a giant world away. A room that was always real hot, even in winter, with a girlfriend or two attending Furry, with a picture of Muhammad Ali on the wall, and later one of Burt Reynolds. There was a string rigged from the TV to the bed so Furry could turn it on and off without having to attach his wooden leg and get out of bed. My work is a response to the impression made on an impressionable me when, while sipping Ten High bourbon, I asked Furry why he kept a jar lid on his shot glass, and turning his boney body like a ballet, he answered, "My eyes is bad and I can't see well. Don't want a spider to get in it and bite me when I takes a sip."

We had never had that problem at my home.

Deculturalization.

You had to respond to it.

A mutual interest between a few individuals was becoming a community, with the Bitter Lemon at its core. Although Dylan had opened a new world to the mainstream in 1962, and though the Beatles had rocked that new world in 1964, the dominance of the old world remained clear. Society changes slowly. It would be the 1970s before the 1960s rattled mainstream politics, deposing a president and ending a war. Indicative of how far removed this culture remained in the mid-sixties—the extent of their deculturalization—the painter Charlie Miller remembers walking through Overton Park wearing leather sandals and people stopping their cars to accost him, to yell and ask him if he thought he was Jesus. "I remember teachers getting kicked out of Memphis State for wearing Bermuda shorts and sandals," adds McIntire. "Got fired. And growing a mustache. These guys attacked me one night and were going to beat the hell out of me because I had a mustache."

The Bitter Lemon narrowly escaped immediate failure. Upon opening, a contract was signed with a miasmic local girl to sing "from Saturday to Saturday." "Her big song was 'Mariah.' She could bellow it out real loud," says McIntire. "When people realized she was there every night, they stopped coming in. I was sitting there going, Oh man, we're going to close before we even started. Her mother was smiling at us. She was a lawyer and she had this contract. These two guys were sitting in the audience, I didn't know who they were. They were writing and talking to each other real loud, they didn't pay any attention to her. They were going, Yech. But it was the only coffeehouse

at the time. One of 'em picks up a guitar I had lying around and starts strumming it. I said, 'You guys can play?' He said, 'You want us to play something
for you?'

"They started playing and nobody could believe it. They played this open-
tuned blues guitar, and they went into their act. John Fahey was Blind Joe
Death. He wore dark glasses and Bill Barth would lead him up onstage and
they would play together. When he was done, he took his glasses off and sat
back in the audience. This girl got real jealous because they were stealing her
show. They looked at the contract, it said 'from Saturday to Saturday' on
there. We said to her mom, 'Okay, we'll just let her sing on Saturday.'

"She quit."

Bill Barth had moved to Memphis from Queens, New York, in search of
old bluesmen. John Fahey, who had written his doctoral thesis on the late
Charlie Patton, one of the most popular Delta songsters, joined him in the
hunt. Both were accomplished musicians, annually releasing recordings on
Fahey's label, Takoma. Fahey's style drew from the acoustic bluesmen whom
he admired, employing a percussive finger-picking. Barth began applying
Fahey's style to the electric guitar, drawing out the moods like a raga. "Barth
sat around and played guitar a lot," says Lyon, "which was real nice, it was exceptional. He hates my guts now, but I spent many an afternoon on the front
porch with him playing for six hours." Barth and Fahey made national news
when they located Skip James, a singular Delta artist.

They came from the Northeast, where the Delta blues was shrouded in
mystery and novelty, and was still rare, making it a valuable commodity. The
idea of Furry Lewis or Gus Cannon regularly playing a coffeehouse gave them
palpitations. Nick Perls, their friend at home, had established the Yazoo label
and was reissuing country blues from the 1920s and 1930s. Fahey and Barth
had a sense of what blues, bluesmen, old 78 rpm discs, and such related affairs
were commercially worth. "Nick's reissues sometimes didn't have the words
right," says Lyon, "because they didn't know what a jack and a doney was, for
instance [male and female donkey, literally; slang for a man and woman]. So
when Charlie Patton sang that song, they didn't know what he was talking
about. We were southerners and had been around the rural scene, so at least
we could understand them when they sang. Here's Nick, his father was
Alexander Calder's agent and he'd grown up in the whole 1950s Abstract
Expressionist movement, and he still couldn't understand this shit." Conversely, while this group of locals understood not to take the blues players and
their culture for granted, they needed the outsiders to indicate that careers beyond the Bitter Lemon were possible. Though their goals were compatible,

their styles were different. The northerners thought the southerners lazy; the southerners thought the northerners pushy and coarse.

"Barth was about business all day long," says William Eggleston. "The local people had a resentment against those we called carpetbaggers. I think that's just human nature, though. To most southerners, and it's not founded on any kind of logic, people like them come off as sleazy. But that's been going on for ages down here because unlike, say, New York, the South is not an international place. In New York you're accustomed to seeing foreigners, and you're just not in this part of the country.

"I didn't go out of my way to spend time with Barth, probably because we were from such different backgrounds. I kept hearing people like him going on and on about how fantastic this or that discovery was, and it wasn't a bit unusual to me. When I was growing up in the Delta, the places you would hear this music were 'juk' houses out in the country, which were all black. I had known about these people always, had always heard them, thought they were great."

A tenuous trust developed. "Barth would have been able to discuss modal tuning or whatever with any of these musicians," says Lyon, "but he was a Yankee from Flushing Meadows, and there wasn't any language between them except the music. Barth and Fahey didn't dig that southern thing, they didn't know the ropes. I always thought Barth hated that part the most. He needed some southerners along to mitigate some of that stuff. The old guys didn't know who we were, but they understood we were from around there. But they *really* didn't know who he was."

Rediscovering blues players, as the successes continued to mount, became quite the thing to do. People from all over the country converged on Memphis, heading south with a sense of conquest: Come! There are more figures to liberate! It was a different outlook than what seemed appropriate to the young local whites, who were aware that despite their proximity, they too were invading a culture. Their attitude was more about getting to know the players as humans and not statistics, "feel" and not trophies; the locals were interested in who they were, not what they were. Nonetheless, these people were accomplishing things that the locals had not done; Barth, for example, did not *have* to be here helping blues artists get work. Fahey's Takoma Records documented and released albums by them. (One of their cohorts, David Evans, remained in Memphis and continues to actively record obscure local blues for the High Water label.)

"When Barth and Fahey came down, they would try to go to the black parts of Mississippi and get the black people to talk to 'em," remembers Lyon.

"And that's difficult, because you don't know what order you are impinging on down there. At that time, black people were very suspicious of white people going far enough into the scene to find someone like Fred McDowell or Bukka White. They were protected, they were special resources that the people needed. They didn't like that invasion and the white folks just hated that feeling, so you had to forget all that and grin and bear it and go down there and confront it. It was really something. And Barth and Fahey were the first people I knew that were interested in getting next to that and had the courage to do it. But they sure as fuck brought us along. They were living in our house and we went with them, because we were interested too."

As recently as 1994, I have been eyewitness to intelligent film crews who come down from the Northeast to Memphis and the Mississippi Delta and ask for the crown jewels in exchange for nothing. I have heard forethinking people devoted to breaking down cultural barriers say, We are PBS, we are the taxpayers' dollars, we are going into the homes of sixty million people and this is the kind of publicity that you cannot buy. For more than thirty-five years, for almost four decades, the most and the least well-intentioned carpetbaggers have come to Memphis musicians and asked them for the one thing that over the years of plunder and abuse they have retained as exclusively their own: their presence. In exchange for the one thing that Sleepy John Estes had that put food on his table, the one thing that Gus Cannon could do that would make anyone think of Gus Cannon, the one thing that the Center for Southern Folklore has got when they can't even get a lease on their building—they offer an intangible. I have sat at a table with an internationally respected and admired filmmaker whose work has affected the social conscience of America and heard him say, not with braggadocio in his voice but certainly with a sense of pride, In my twenty-five years of filmmaking, pause, in which I have interviewed—and here there was a long pause, some sort of calculation—ten thousand people—ten thousand interviews, he says—in my twenty-five years of filmmaking in which I have interviewed ten thousand people, I have never paid for an interview.

It's not that he's expecting me to admire this. We've previously agreed on our difference of opinion. But he is exasperated, truly exasperated that he would come to Memphis from New York to earn his living, giving the people of America and the people of the world a documentary in which his interviewees' statements might get a minute of air time, might get two separate forty-five-second bites, and though his documentary won't be what he wants it to be unless these artists give him their presence, these people won't give away to him the only thing they have.

He's shocked—in *Nashville* he didn't have to pay those even more famous artists—but after tiring negotiations, professing still not to understand, he relents. They will pay these people in fifty-dollar lumps for the only thing they have. And then the filmmaker, the producer (who will reneg on the payments after shooting), and the crew get up from the table where they've eaten on the taxpayers' money and I watch them go to one of the three most expensive hotels in the city. On the taxpayers' money.

In 1964. In 1994.

In 1965, while canvassing for old 78s, Bill Barth encountered a family who said, "We ain't got no 78s, but would you want to buy this chere guitar?" Barth strummed the instrument, sat down and played a few songs, thanked the family, and left without the guitar. When he returned a few days later, Nathan Beauregard, born a century earlier into a world as different as another planet, was strumming and singing in the living room. Beauregard had played all his life till he no longer felt like it, had never recorded. When Barth helped inaugurate the Memphis Country Blues Festivals the next year, Beauregard became a featured performer.

Born blind, Beauregard was led around by his nephew, a man in his seventies. His withered skin hugged his high cheekbones, the lids over his sunken eyes always shut so that he resembled a mummy. He'd begun developing his repertoire when Lincoln was still president. He was booked on the blues festival not for what he'd known and lived through but for what he still did; that is, the show was about the music, and he could still play. He'd have his amp set up slightly in front of him and to the left so he could lean over and work feedback into his sound. "Nathan Beauregard was cosmic, like some Tibetan monk." Randall's eyes are once again closed. "He lived behind the B'nai B'rith home in Midtown, in a row of shack houses for maids. Barth discovered him in one of those houses. To be walking past somewhere and hear someone moaning, 'Bumble bee, bumble bee, got a stinger long as my right arm.' And the guy's a hundred and something years old. . . ." And alive and well in an old friend's mind.

These musicians were not just artists but spiritual leaders, spiritual beings whose very presence was an education to people raised during segregation. "I'd had my washboard since recording with Dickinson," says Crosthwait, "and now I was getting to see from whence it all came. By 1968, there's Nathan Beauregard, Reverend Robert Wilkins, Bukka [White]—and I got to play with all these guys. In a way, we have become the white boys that inherited this thing, like it or not. Lee [Baker] is definitely the best blues guitarist from

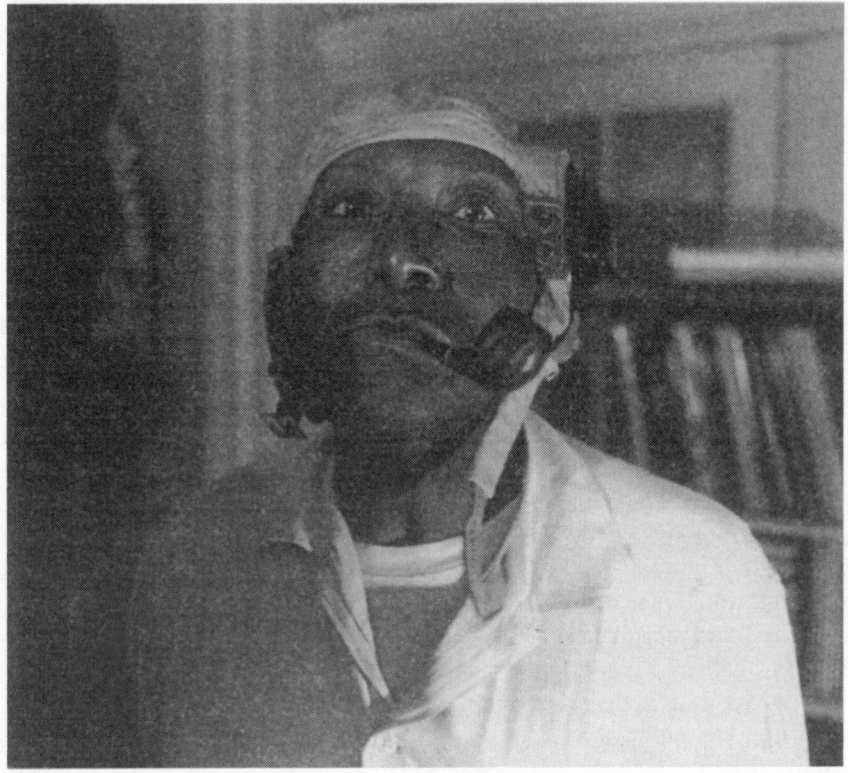

Nathan Beauregard's nephew Marvin: Communication between the cultures is established.
Photo courtesy of John McIntire.

that school. Hell, he played with Furry for so many years, he has inherited that mantle. I don't know of any blues guitar player in the world, from Eric Clapton on, who can say they came by it as honestly as Baker."

"I played in standard tuning and Furry taught me to tune the guitar to the chords," says Baker. "That's how they were making it sound so good with the slide, because it didn't sound good to me just to play the open string. Then I read more about the style off the records. Some of the first electric slide I played was at the Bitter Lemon. Me and Dickinson would do 'Little Red Rooster' there, the Stones had a thing on it and so I would play that. It was all happening at the same time, I got it from Brian Jones, I was getting it from Furry and everywhere."

"The Lemon was the first place where we could play on equal footing with the bluesmen," says Dickinson. "There was a sense of community. We were all there doing the same thing. Some of the stuff that happened there was so cosmically disorganized, it wouldn't have happened anywhere else, even in Memphis, which is typical of the city. There must have been places in Cambridge where the entertainers were on an equal footing, but how could it have been the same? We were all home. On nights when Furry played a third or fourth set, he did it because he wanted to. If there ever was a white place that was like a honky-tonk, it was the Lemon."

The Lemon continued to book Baker, Crosthwait, Dickinson, Selvidge, Don Nix, and a variety of local and out-of-town folk and blues talent from the coffeehouse circuit. "I thought my music career, such as it was, was over when I finished high school," says Dickinson. "But then the blues artists were coming out, and once I heard Dylan there was no doubt. This guy's voice was worse than mine. Doing folk, there was no band to hassle with, you didn't have to carry a bunch of amps, didn't have to worry if the bass player was drunk, and you could keep all the money. I saw folk as an opportunity to mix cowboy songs and Rambling Jack Elliot stuff and gospel. Mixing it all up was what I liked. Then when rock and roll came back around, it was easy to me because I'd already done it."

"When Sid would play, it would pack the place," says McIntire. "I'd make big banners to put across the front of the building, TONIGHT FURRY LEWIS AND SID SELVIDGE. The people would line up outside. One show, then you had to get out. If you wanted to stay again, you'd have to pay another two dollars."

"Our rent was $37.50 including utilities, and we ate for ten dollars a week," says Mary Lindsay Dickinson, "and Husband could go down there and make fifty dollars on a weekend night. We were rich!"

The ages and background of the Bitter Lemon's clientele grew more varied. As the occasional "older" person became interested in youth culture, they went to the Lemon for a firsthand look. Kids from affluent East Memphis would have to sneak there. Certain ages were not allowed in; even though they weren't serving alcohol, the club was hassled by the police over curfews and other blue laws.

McIntire did his best to extend a welcome to minorities, both onstage and in the audience, though he often found circumstances beyond his control. A black friend of his named Ford moved down from New York and into Beatnik Manor. While walking in front of the house one day with fellow lodger Lydia Saltzman, white, Ford was promptly accosted by a Memphis cop for the crime of being black. When he responded to the officer's interrogation in his northern accent, the cops pounded him. "Next time I saw him, he was in the hospital," says McIntire. "He didn't know what happened. He developed a black accent, yassuh, yassuh. He had a job here driving a laundry truck and they followed him around one afternoon and gave him six or eight tickets. Run him off because they didn't like his accent."

Teenie Hodges was already accustomed to crossing over when he played the Bitter Lemon. As a young teen, he'd joined his brother Leroy in the Impalas, where their popularity at black dances began attracting the attention of some whites. Willie Mitchell, the popular black bandleader who was a favorite at the Plantation Inn, soon enlisted Teenie, Leroy, and their brother Charles to accompany him. As Mitchell began moving Hi Records toward soul music, he would make the Hodges Brothers his house band; with his direction and their sound, it was no wonder Al Green enjoyed such crossover success. (The Hi Rhythm drummer, Howard Grimes, had integrated the popular white band Flash and the Board of Directors.) When the Bitter Lemon was hot, Teenie formed a group known as JAMF—Jive Ass Mother Fuckers. "After the Flamingo Room, the Handy Club, the Elks Club, and those other [black] places closed, the Bitter Lemon was one of the first places they started back to playing blues," Teenie says. "We were a little funkier than blues, but we had the blues taste to it. We played four or five hours each night, and the audience danced the whole time."

The Manor remained ever happening. It spilled over into the two apartments next door, and then to three more apartments on the corner. The neighbors complained about a nude man sitting on the porch of Beatnik Manor. It was a sculpture made by one of McIntire's students, who had moved in after he grew a beard and his family kicked him out. Since the manor had become a beatnik haven, the cops always lingered near Madison and Cooper, spying and trying to act tough. "The police used to park behind our hedges and watch the house," says McIntire. "This girl in the house, Jo Lynn, felt sorry for them and used to take them coffee. They'd get real mad at her. They'd split and we'd see them down the street watching from another angle with binoculars, so obvious. It was crazy."

From midnight to three, the house would be awash in music. All the radios in all the rooms were tuned to WDIA, where Rufus Thomas would be broadcasting "Hoot and Holler," a show of lowdown blues, playing it real because he knew the music for real, announcing after each solid sender, "And I ain't just clacking my faucets." Carl Orr was a writer and filmmaker who was awake when others slept, pounding the keys on his typewriters, which he regularly smashed or threw into the river. He usually had a 16 mm movie camera on his arm and a vague idea for a film that would solidify once the processing was done, but if you wouldn't mind standing over there and wearing this sculpture on your head, and sashaying into this room—don't look at the camera—it might be just the scene needed. Regular visits from Marcia Hare's mother never ceased to amaze McIntire. La Paulette was a hostess in a Bourbon Street strip bar and would entertain her daughter's friends in the kitchen with lessons on the seduction of married men.

The presence of bands like JAMF and the coterie of blues artists at the Bitter Lemon was the result of Charlie Brown. The early days of "Mariah" and tame folk passed with the entrance of the new manager. "When Charlie Brown moved into the house, it got notorious," McIntire continues. Brown was the former manager of the Oso who, even when his wildness had him banned from the Bitter Lemon, claimed he ran the place. "He stayed for a long time. He and his friends, they'd sit up all night and drink beer, play poker. All night long. I had to go down the street sometimes and call the police on my own house. They wouldn't listen to me. One night, they piled all the furniture in the front yard, lit it and danced around it. They were shooting out the street light with a .22 rifle and that woke me up. The flames were coming up by my window, and I was on the second floor. A couple of them were buck naked. They wanted to make a ritual. I said, "Charlie, get these people out of here." The cops came and ran everybody off, and in about half an hour they were back again.

"I remember one time, I didn't know what marijuana was. I was cutting a big stone in the backyard, in the shade of all these bushes that Charlie had planted. The police heard me, and came down the alley and through the bushes and asked me what I was doing. We talked for a while. I kept seeing Charlie at the house, peeking around the door, up the steps, across the backyard. After they left, he came down and said, 'Are they gone? I thought we'd had it.' I said, 'Why?' He said, 'You don't know what these are?' It was marijuana plants, like twelve feet high. It was a nice shady place to work. I said, 'Gyad Charlie, isn't it illegal?'"

When his own house got too wild for him, McIntire retreated to his studio at the Academy, or to a completely unpretentious restaurant across the street. Burkle's Bakery was run by an elderly couple, employed elderly waitresses, and featured a sixty-five-cent vegetable plate with rolls and corn bread. "We'd hang out there a lot," says Marcia Hare. "But even that got overrun. McIntire lived across the street, and because John was there all the time, people from the art school were there all the time. It got to be the center for the so-called artists and hippies to hang out. They thought that was the place to be because John was there."

"Charlie Brown had a lot of charisma," says Lyon. "People just met back then. You could tell walking down the street if someone was hip or not. If you met 'em and it was cool, it was like you knew them for all their lives. Charlie was one of those people. He ran the Bitter Lemon. He was going wild of course, like everybody else. We were all totally out of control, no grip on reality whatsoever. The Lemon was packed all the time but it never seemed to make any money. Charlie never paid any of the bills and always took all the cash. Poor McIntire. The thing became an albatross. It was open season on fun. Back then shit was cheap. There was always enough cash around for Charlie to have a ball."

"I never saw any cash," says McIntire. "Most of it went to keeping the place alive, and my salary from the Art Academy paid off the rest. I'd go there, that place would be packed, a waiting line outside, and not a penny in the cash register. It might cost a dollar and a half or two dollars cover charge. I'd go, 'C'mon guys, the place is packed, where's the money?' 'I dunno.' They were buying drugs, it went up in smoke. We were robbed a few times, the sound system, the pizza oven, the supplies. We had a cigarette machine in there, a pay phone, I never saw any of that money. But everybody was doing good. Nobody really had a salary, except the singers mostly. They were on a circuit and demanded their money." One time McIntire's mother got sick and he left town for a week. When he returned, all the posters he'd put up on the walls were gone, many of the musical instruments hanging had disappeared, and so had the Tiffany lamps. "The people who worked there stripped my coffeehouse of everything except some of the chairs and tables. They took a great big bronze statue. Don Nix had the lamps in his apartment. About four or five years ago he came by and said he felt he ought to pay me back. I said, Forget it. It was a good experience. I don't miss that stuff now." Some of McIntire's friends may have been abusing his openness, but the relationship was essentially symbiotic. People lived off of him, but he needed them around, collecting their characters the way he collected secondhand miscellany. The clamoring throng fired his vision.

Many of the traveling bands on the coffeehouse circuit were impressed when they got to Memphis and found themselves sharing a stage with an artist they'd assumed either mythical or dead. One trio from New York, the Solip Singers, was so impressed that they stayed. "We were working our way across the country, with the idea of going to San Francisco," says Solip vocalist Nancy Jeffries, now a senior vice president of A&R at Elektra Records. "We didn't have much money and we'd just go to each town, find the coffeehouse, audition for the owner, sleep on someone's floor. Our trek ended in Memphis because all that stuff was happening, the scene with the young people and the older blues artists. They were so accessible. You didn't realize how important it would be to meet them, and once it happened, it's all I wanted to do."

The Solips were considered a radical band at the time, playing a set made up almost entirely of Bob Dylan songs, supplemented by material from Richard and Mimi Farina, Joan Baez, and the like. Though a folk band, they used electric instruments. Their twelve-string acoustic was amplified, and Nancy played the electric bass. "When Dylan plugged in," she says, "we were for it." In 1966 they got a three-month gig shuttling between Little Rock, where they met Bob Palmer, and Fayetteville, where Randall Lyon was hanging out with John Clellon Holmes, the beat writer (*Go* [1952], *The Horn* [1958]). When the work ended, Jeffries returned to Memphis. The Solips evolved to include Palmer as a regular guest and Barth as a regular member. They would all become instrumental in planning the Country Blues Festivals, and their band would evolve into one of the decade's most experimental amalgamations, the Insect Trust.

In the early spring of 1966, Dylan played Ellis Auditorium, previewing his *Blonde on Blonde* material with the Hawks. The Hawks had been around the same roadhouse circuit that the Solips played, that Dickinson and Baker and everyone who'd been out of town had played. The show was an inspiration, not only because it was the loudest thing anyone had yet heard, but because the band was dressed like the audience, had come from the same place, and was not only getting over nationally, but doing it with something original. Everyone in the audience had hope, a sense of a greater community out there, a sense of the power they would wield in the next decade's battle to depose the president and end the Vietnam war.

"We had what you would call in rhetoric *eroico furore*, poetic furor," says Randall Lyon. "We were inspired, we were in a frenzy. We had heroic passion. I always figured that was the best thing that could happen to you, to be caught up with a group of people with heroic enthusiasm for what they're doing. And how it's received was beyond our consideration."

CHAPTER NINE

Smile on the Outlaw Dreamer's Face

A WEEK BEFORE THE FIRST MEMPHIS BLUES FESTIVAL, THE KU KLUX
Klan held a rally at the event site, the Overton Park Band Shell. Racial tension
was high all over the country in the midsummer of 1966. Riots erupted in several cities. A hundred miles from Memphis, in Grenada, Mississippi, blacks
marched silently around the county courthouse, jeered at by a crowd of
whites; highway patrolmen stood guard with shotguns, gas masks swinging
from their hips. This was a time when the black student union at Memphis
State University was commonly referred to as the "coal bin." At the Shell, segregated rest rooms for the employees were still enforced. "It wouldn't do any
good to take the signs down," a black civic employee told the *Tri-State Defender,* "because the white folks would still give you those nasty looks and try
to take it out on you in some way." At the Klan rally, the imperial wizard of
the United Klans of America spoke to 400 masked followers.

More than double that number showed up at the same location a week later
to see the stage shared by black and white, united in music, new social mores
trimmed and burning. That the first Memphis Country Blues Festival even
happened seems a minor miracle, organized as it was by a group of beatnik
blues fans—social outcasts, experimenters in hard chemical substances. However great the differences between them and the bluesmen, their social ostracism was a common bond. An informal organization formed, the Memphis
Country Blues Society, that put together a sound system and made sure the
bluesmen were paid from the gate receipts. For advertising, Charlie Brown
hawking the event at the Bitter Lemon was all they needed.

The only expense in putting on the show was the deposit required on the
venue. Charlie Brown raised that, convincing Dickinson to sign over the sixty-

five-dollar check that he'd been paid by Sam Phillips for his Jesters session. Barth put up a softball of hash for barter.

The media attention made this first festival a grand coming out, both for the country blues performers, who had been largely ignored by the city since the 1920s, and for the hippie scene, which the city had tried not to notice. As well, it was a positive result of the mix of locals and northerners. Such festivals had begun defining youth culture since the latter 1950s, notably the Newport, Rhode Island, events. The success in Memphis created a community that traveled regularly from here to New York, a community built upon the respect the hippies (née beatniks) and bluesmen had for each other.

Over a thousand people paid the price of admission on July 30, 1966, for the first festival. Honors for kicking off the event went to Lee Baker, who introduced the blues and the hippies with a screaming electric band described in a newspaper review as "near cacophony." His set was the brass blast that precedes the entrance of royalty: Memphis, meet your past, meet your future. Someone was wise enough to invite the black disc jockey and community spokesman Nat D. Williams to speak. He'd been the first black voice on WDIA, was a syndicated columnist for black newspapers across the country, and an influential teacher at Booker T. Washington High School (whence came many of the Stax artists). His recitation of Mr. Handy's "Beale Street Blues" was renowned by then, and he did not disappoint the hippie crowd. Others performing were Mississippi John Hurt, Bukka White, Furry Lewis, Nathan Beauregard, Reverend Robert Wilkins, Fred McDowell, Sid Selvidge, Jim Dickinson and band, and the Solip Singers. The show went off without any unusual hitches, the review in the newspaper somehow validated it, and the bluesmen and the hippies were suddenly a presence.

The corporeal spirituality of the blues musicians was as gripping as their music. What they played was unencumbered by progress, as relevant in 1966 as in 1926. They cut through the urban soundtrack, transporting listeners back in time with them. At their feet, confronted by them, one could not help but be moved. They physically embodied the music: leathery and worn, dusty, dry. The repetition in what they played, the hypnotism, was the sonic equivalent of a plowed field, row upon row. This music is the distinct product of an African musical tradition, an American landscape, and an unjust social system; all blues has to be traced through these Delta artists.

Some newcomer fans had yet to learn the nuances of the styles. One wanted to give Furry Lewis a National steel guitar like Bukka White's. Furry was a subtle player who controlled his instrument like a puppet. Wearing a bottleneck slide, he could make a string coo like a loving cup, then make it

Furry Lewis entertaining a crowd, circa 1970. ("Memphis Sound" T-shirt designed by John McIntire.) Photo by Jimmy Godown.

whine for more. He wound his voice and slide around each other, calling up God from the pit of his soul and hoarsely asking for mercy. He picked where Bukka hammered, and when Furry did strum, it was often just to keep a rhythm on open strings while he told a story. Furry came from a medicine show tradition, and a large part of his performance was shenanigans, entertaining the crowd with stories, playing guitar with his elbow, or some other musical clowning. Bukka White played guitar like John Henry drove steel. A burly man, his voice was gruff, his steel guitar percussive and loud. Both Bukka and Furry could plumb uncharted depths of your soul. Bukka's style was immediate; Furry's required a little more attention.

The kindhearted visitor who gave Furry the steel guitar could probably discourse profoundly on the difference between a Petrarchan and a Shakespearean sonnet. But he was unable to comprehend that Furry and Bukka were writing different poems; the force of Bukka's style had captured him, and he thought the power was in the instrument and not the musician. The way others lumped together verse, he lumped blues. In that sense, it's easy to understand the bewilderment on his face when Furry politely refused his gift.

"Furry played at the Lemon, and I could back him up there," says Lee Baker. "But I still played with a pick when we started those Memphis Country Blues Festivals. I'd have an electric band on the show, and Furry would be there and he'd ask me to come up and play with him. Soon enough, I was also playing with Bukka White, Gus Cannon, Sleepy John—anybody that was there that wanted me to play with 'em, I would. By the last festival, *I was* the backup band."

Baker understands the role of accompanist. He knows that less can be much, much more, that he doesn't have to run through his bag o' tricks every time he's on the stage. "Furry could say more in a couple notes than Stevie Ray Vaughan could say in twenty-four bars. One or two little notes, the way he could do the things he could do, that's becoming a lost art. Everybody, myself included, plays a million notes all the time. But it's the nuances, the inside stuff, that I admire in Furry's playing."

Sid Selvidge got his comeuppance at the blues festival, well after he'd come to know many Delta bluesmen. "Furry had to go on and his guitar was out of tune and I said, 'Oh my God, I'm gon' get this thing in tune because Furry's gonna embarrass himself.' And I went over and meticulously tuned Furry's guitar and gave it to him and he got on stage and strummed it and went *wanga-wang-a-wang* and got it back like he wanted it. I realized, Oh, this is African music we're listening to! It doesn't have anything to do with Mozart!"

The first festival spawned four more, ultimately creating a struggle for ownership of the event between the hippies and the city government. In 1966, however, the future was bright and full of color and long hair, racial and class respect. "I can't even imagine how our friendship happened," says Randall Lyon, who began his long association with Bill Eggleston when he asked to borrow his 16 mm movie camera to document the first blues festival. "I was the kind of person Bill despised the most. He hated hippies and there I was, the arch-hippie. But the first time I met him, borrowing the camera, we got into this long discussion about the piano. Bill had this Steinway Grand and he was very interested in how it was tuned. He was experimenting with tuning. He and Rosa [his wife] had just been to Guatemala and Kathmandu. In the early sixties, he hung out with international jet set bohemians like Charles Henri Ford and Ruth Roman and all those people. He had come home to roost about sixty-four, sixty-five, and I had no idea who he was. We just got to be really good friends. Bill and Rosa were deculturalized by wealth and power, and we were deculturalized by putting ourselves in harm's way, so there was a possible meeting." Eggleston, along with Buster Williams's son Robert, made field recordings of the first festival.

Another result of the first blues fest was the emancipation of the Solip Singers from their folk life. "I had been playing professionally since I was about fourteen and I thought I could get through a blues progression pretty good," says Robert Palmer. "But boy, that first time I tried to play with Bukka, I thought, Oh! This is a whole other thing! Barth and I got to jamming with those guys a lot, with Furry and Bukka and so forth, and then later after we moved to New York, Barth and me and Alan Wilson of Canned Heat used to jam together a lot on country blues stuff. And from all that Memphis experience we were into rocking out jams on country blues tunes, playing the structures of the country blues, thirteen and a half bars, the funny modal tunings, the whole bit. The Insect Trust's version of Elmore James's 'Special Rider' grows out of that."

The Insect Trust was the second incarnation of the Solip Singers, solidifying in New York in 1967. The name comes from William S. Burroughs's *Naked Lunch;* though Palmer couldn't have imagined it at the time, in four years he would travel to Morocco with Burroughs; the two maintain a working relationship. The Solip Singers recorded an album for Chips Moman at American Studios; only a single was released. ("Comeback Baby," an Elmore James-ish blues is on one side, and the folk ballad "He Was a Friend of Mine," with recorder solo, is on the other.) "Privately amongst ourselves," says Palmer, "we referred to the American house band as rednecks. Certainly compared to us they were rednecks. Compared to rednecks they were probably somewhat different. They used to tease us about smoking pot."

Palmer talked his way into a job around American, saving a couple hundred dollars with which he made his first trek to the northeast, 1966. "I was a jazz fanatic, and Barth set me up for the most singular introduction to New York. I drove up there with this girl that I knew from Little Rock who knew a piano teacher there. We drove past Manhattan to Stoney Point, and got to this guy's house and I crashed out on the couch. About twelve hours later I woke up and I looked across this big room, it was a pretty house in the woods, and sitting across from me is John Cage. And what's even weirder about it is I knew exactly who he was and had read his stuff. We proceeded to spend the entire day together, walking around, listening to some of his music that hadn't been commercially issued. We speculated on the future of music in really interesting ways, and that was my first day in New York. Then I went down to Manhattan and I had one contact that Barth had given me, Peter Stampfel of the Holy Modal Rounders. I went over to Stampfel's apartment and it's a block from Bleecker and MacDougal, which was like the center of the universe. I'm not there five minutes and the Blues Magoos come in to score speed and then we all go over to the Kettle of Fish and Bob Dylan was in there and Eric Anderson, Dave Van Ronk, all those guys. That was my second day in New York. Not bad! I also went to see this great Cecil Taylor concert at Town Hall, which was the music for *Unit Structures,* probably the best Cecil Taylor album ever. And I was thinking, Wow, I thought it would be really difficult to get anywhere in New York but this is great. So with some reluctance I went back to Little Rock and finished my last year of college."

The Memphis weave in New York was drawn ever tighter the summer of 1967 with the arrival of Jimmy Crosthwait, his wife Linda, and Chris Wimmer. Crosthwait was working toward merging his artistic handiwork with his natural theatricality; the problem with sculptures, he'd come to realize, was that they didn't move. In 1965, he and Linda were trained in Florida as puppeteers and spent a year touring the Southeast, as far north as Virginia, performing mostly in school assemblies. The next year they did the same for a man in Chicago. "We worked for him during the 1966 and 1967 school year and I was getting tired of doing kid shit," says Crosthwait. "I began conceiving this play with immolating monks and melting-face clowns and was steady building it. A friend from Memphis, Chris Wimmer, who I knew from the Market Theater, came up to Chicago and helped me make the puppets and props. I had been thinking I could book this far-out, hip sort of puppet show in coffeehouses, and just when I was realizing I ain't gonna book this damn thing anywhere, his ex-girlfriend calls from New York and says there's an ad in the *Village Voice,* 'Hippie puppet show wanted.'"

Hippie puppet show they got. Crosthwait's work evolved from his affection for the Bauhaus movement's concept of total theater. In the late 1920s, as the Weimar Republic was dismantling this integration of art and science, the Bauhaus proponents were working with the technological advances created during the First World War—the total war—and conceiving of total theater.

"They wanted to artistically integrate the production of artificial arms and legs that came out of that war," says Crosthwait. "They were conceiving of giant robot actors, plastic stages that you watched from below, things that weren't executed until Disneyland came along. It was carnival, ballet, and sporting event." Of the many things Jimmy Crosthwait has undoubtedly been accused of, being athletic is surely not one. However, ballet, music, and carnival atmosphere are integral to his character.

Crosthwait, twenty-two, a New York virgin, and Wimmer, twenty-one, a minimalist in both figure and theory, were looking for St. Mark's Place as the Six-Day War was coming to a close. At the Electric Circus they met the idea man, Michael Gruener, the son of a successful advertising executive in Los Angeles, and the two people running the show, Jerry Brandt and Stan Freedman, who had caused a splash when they threw green cash money from the balcony onto the floor of the New York Stock Exchange. In the city less than a day, Wimmer was standing at the upstairs window of the club in progress, gawking at the passersby on the street below, when he suddenly bellowed to Crosthwait, "Hey! There's Warren Gardener."

Gardener was a beatnik from Maryland who had lived the bohemian life in Paris, running with William Burroughs and Brion Gysin before heeding the boho call from Memphis in the mid-sixties. In Paris, he had appeared in a Burroughs play, *Junk Is No Good Baby,* in which his character shot up onstage. Let's say Gardener was a method actor. Once in Memphis, he quickly established himself as a live wire and became known as Electricity Man; he referred to himself as Chang, The Unavoidable. "It was almost bad to bring up his name," Crosthwait says, "because then he would show up at your door and live with you until you were out of food and nerves and everything else." One surmises he was more fun to watch than encounter. The last Crosthwait had known of him, Gardener had been busted in Memphis and was doing time in the Shelby County Penal Farm. He'd been at an all-night Memphis diner, laying his jive on another patron, trying to sign him to a recording contract. Gardener boasted of a song called "Smile on the Outlaw Dreamer's Face." This particular diner was near one of the few theaters in Memphis, so the clientele tended toward the artsy. With the guy showing interest, Gardener dug into his bag for a contract. Rooting through, he nonchalantly pulled out his pot and set it on the counter. His future Willie Nelson nonchalantly stopped being an undercover cop and arrested him.

Marcia Hare met Gardener when she was volunteering at the theater after school, knew he was trouble, and followed him everywhere. She reads from a newspaper clipping she's saved: "Gardener pretended he was blind and begged for a living. Gardener's address was termed 'at large' by police. He told police he found the marijuana growing wild in a field in Arkansas. The bearded bicycle rider said his latest musical composition was 'Smile on the Outlaw Dreamer's Face.'" Hare doesn't miss a beat as she puts down the clipping and recites from memory: "It's been down to hell and back again/Been up to heaven once or twice/the Devil and the Lord it seems/both hear and heed the same advice. And that's that/though this life's a whitewashed blackboard baby/don't you know it ain't ever gonna erase/the smile of the blues on the outlaw dreamer's face."

Wimmer yelled to the Manhattan street below. "Warren looks up and sure enough it's him," says Crosthwait. "He beats it up the steps, and when he learns what my puppet show is about, he goes, 'God, that's far out.' And Warren went right up to the executive offices of *Time* magazine, barefoot, talking about, 'Hey man, there's this guy's puppet show and you have got to put a melting clown face on the cover.' And they listened to him."

They didn't put Crosthwait on the cover of *Time,* but it was 1967, it was New York, and everybody was listening to anybody about anything. Things were changing very rapidly, and no one knew what strange idea today would be tomorrow's rage. Which must be how the American Coffee Growers Association agreed to put up a large sum of money to sponsor the Electric Circus. The snack area sold only coffee and coffee-based products, no alcohol.

The Electric Circus became the voice of the psychedelic times. It was mixed media, culling from bands and solo performances, from audio recordings and film, from pantomime to small theater, and the show was not restricted to the stage. Many of the staff were out in the audience, and many of the audience became part of the show. Like a multiring circus, "happenings" would occur simultaneously. "Chris Wimmer was very much a body caricature," says Crosthwait. "He was the guy who would do nothing in the midst of everything. When people were dancing he would come into the middle of the room wearing a kind of white popcorn vendor's uniform, he'd take a white sheet and flip it a few times, lie down and lay the sheet over him. Four or five minutes later he'd get up and walk away."

Wimmer remembers, "I was standing along the wall at the opening night of the Electric Circus and when Crosthwait walked past, Allen Ginsberg turned to Timothy Leary and says, 'Keep your eye on that guy, he's nuts.'"

Who do we listen to?

There were three lighting booths projecting liquid formations, film loops, and slides. Someone had purchased a bulk supply of the sheeny material from which ladies' underwear is made, and the walls and ceiling were scalloped with it. Crosthwait was working a light booth one night when Greta Garbo walked in. "She was being shown around and wanted to see how the booth worked. She was probably sixty and seemed more beautiful than when she was young. Her hair was solid white and she had a lot of curiosity and enthusiasm, a radiant beauty." Before the doors opened one evening, Wimmer tossed a Frisbee with a kindly, slightly older gentleman for fifteen minutes before realizing it was the film director Michelangelo Antonioni, who'd recently made *Blow-Up*. Bill Barth traced Fred Astaire's shoe on a piece of paper and got it autographed. "The Rolling Stones walked into my puppet room at one point, wearing their big furry coats. Jerry Brandt says, 'Rolling Stones, puppeteers . . . puppeteers, Rolling Stones.'"

Security at the club was culled from Black Power activists and the local president of Hell's Angels. "This biker would come to the Circus every night," Crosthwait says, "take off his colors and his gun and his dope, put them in his locker and then transform himself into an Emmett Kelly, a clown with a frown, and he'd walk around with a little stuffed dog and go, 'Ruff ruff! Woof woof woof!'"

Crosthwait's puppet show was held in a little room off the balcony, seating about forty people on the astroturf floor. The show was titled *Iom Dod*, was mostly marionettes, and it ran for half an hour. His immolating monk was the first scene, and the finale was a clown's face melting into a puddle. "The name was a combination of diode and debt-owed and ion dode. It was kind of theater of the absurd, a surrealistic black comedy. I had a Viet clown in his little black pajamas and sanpan hat and he's confronted by the elf plane, which was this military presence. For another scene I reproduced the image on a Camel pack and had an artist pushing his cart at the bazaar."

Crosthwait was responsible for two shows a night, but the second was iffy. While his stage was being reset, he played Tinkerbell in the main hall, donning a clown suit and sliding down a wire from the back of the room to the stage, whipping off his hat and showering confetti. His ride stopped just short of a pedestal upon which he was supposed to dismount, and he made a show of inching and slinking ever closer. Just when it seemed he would make it, he'd fall to the floor and the tall and lanky Chris Wimmer would appear, pick up the scraggly elf, and carry him out. When Jimi Hendrix returned to England after establishing himself in America that summer of 1967, he was asked how things were in the States. His reply was that the only thing really happening was a little guy in a yellow coat hanging on a wire at the Electric Circus.

Upon graduating from college in 1967, Robert Palmer did indeed return to New York. It was the Summer of Love and San Francisco was attracting lots of people and even more media attention, but Palmer knew that New York was jazz, and promptly found himself writing about music for *Go* magazine, employed by Robin "Lifestyles of the Rich and Famous" Leech. (Randall Lyon says of Palmer's writing, "Even as a teenager he had the most finely tuned acoustical sensibility and a way of explaining shit to you. When we were listening to early Coltrane, *Ascension,* for example, it was culturally and musically way beyond my accomplishments, but Bob could make me hear that. And when he told you about it, it was so fucking interesting you just wanted to die. By the time he finished it was, oh my God, the most important thing you ever heard.")

"I was reading the *East Village Other* one night about eleven," says Palmer, "and I see an article on the Electric Circus and I see that Crosthwait, Chris Wimmer, and Lydia Saltzman are there. It's nearly midnight and I go running up to the Electric Circus on St. Mark's Place and talk my way in. We have a little reunion and soon Barth and Nancy [Jeffries] show up, having decided to bail out of Memphis—they'd been traveling back and forth for a while, sometimes with Randall. We start playing together again and get a gig at the Electric Circus, where for most of the summer we opened for Sly and the Family Stone—pre–recording contract. Just awesome music." Lydia Saltzman had been a stalwart at Beatnik Manor. At the Circus, she had a side room where she would do body painting. Palmer played a twenty-five-dollar recorder through a fifty-thousand-dollar state-of-the-art sound system while the lights danced ballet with Wimmer. "There were banks of reverb and echo on my little recorder and it was really loud. I would get all these trills going and then I could just play over 'em and it was like it went on and on and on."

The Memphians whipped up enthusiasm amongst the New Yorkers, and when they returned to Memphis midsummer for the Second Annual Country Blues Festival, Gruener, who'd conceived of the Circus, was among those who joined them. His sophisticated bicoastal mind was blown. He began making arrangements to bring regional country blues artists to Manhattan. Bukka White, Furry Lewis, even Muddy Waters. (When Muddy Waters was asked what he thought of the Electric Circus, his response was reported to be, "Blinking blinking jiving jiving shit.")

For the second festival, even though the Shell is an aesthetically pleasing and stimulating environment, the Country Blues Society wanted to make their own imprint on the environment. That job, naturally, was left to McIntire. "I bought all this crepe paper and gigantic bamboo poles, and I put streamers all the way around the Shell. And bags and bags of balloons." The 1967 festival was also distinguished by the stage props someone had found in an old theater.

A seated golden Buddha towered into the open air, shining over Furry Lewis, over centenarian Nathan Beauregard, and over everyone who took to the stage. Placed beneath his countenance were stage prop trees and stage prop bushes and center stage there was even a stage prop outhouse.

Baker once again opened the festival, this time with a band he dubbed Funky Down Home and the Electric Blue Watermelon. It was a mishmash of blacks and whites and styles, blues-based but free enough for Palmer and New Orleans saxophonist Trevor Koehler to have their reedy way. The Watermelon took the stage and kicked up a solid blues racket, full of guitar but with no sign of the guitarist, Mr. Funky. Suddenly the outhouse door kicked open and there's Baker, wearing a mortarboard hat with a springing plastic flower. To wild applause, he mounted a waiting Harley-Davidson, playing guitar, the picture of rock and roll. "People thought our scene was a bunch of folkies," says Baker. "But we played good, over and above all the theatrics."

At the blues festivals, perhaps because they weren't under his own auspices, McIntire was capable of organizational skills he lacked at his own club. "These two girls were supposed to take the money at the gate," he says, "and they were so stoned that people were just walking in, pretending to put money in the box. So I got under the table, and every time money went in the box, I took it out. I had a shoebox and I was stuffing dollars in it, because I knew they lost money at these things all the time. Nobody could pay anybody and it caused a big fight. After everybody was in, I counted $2,200 in there. That was more money than anybody'd ever seen in their life. I sat in the audience, in the middle, with the box in my lap. When it was all over, Barth was running around, 'Who's got the money?!' I opened the box, and he was so surprised! After it was all done, I ended up making twenty bucks. Twenty dollars off feeding all these damn people. They were sleeping at my damn house, on the floors, I had to keep food in the refrigerators, driving 'em around. But I never got mad. I was too crazy. I'd get mad on the spur of the moment and forget about it. The next day I couldn't remember what I was mad about."

The Memphis Country Blues Festival was becoming a workshop, of sorts, for the younger white musicians. They were able to test the fusion of their diverse influences—culling from the distinguished bluesmen as well as the protonihilism of the Velvet Underground and the Fugs. Solid doses of horns were culled from free jazz as much as R&B. "There was a lot of experimentation that went on at the blues festivals," says Palmer, "a lot of cross-fertilization. Charlie Freeman played with us several times, and we did some free jazz jamming at the blues festival one year. And Dickinson was always real open to all these different kinds of music. Not everybody was. I remember Charlie

Charlie Freeman (middle) with Bill Barth (left) and Joe Grey, performing for Buddha at the 1967 Memphis Country Blues Festival. "There was a lot of experimentation that went on at the blues festivals." Photo by Randall Lyon.

Freeman coming up after the Insect Trust was in New York, doing a session for someone and afterward he came out and jammed with us all day, really stretching out. It was just spectacular. And Dickinson told me at one point that his solo album, *Dixie Fried,* with all those different kinds of music on it, was very much influenced by what the Insect Trust was trying to do. I remember when I came back from Morocco the first time with all these recordings of Moroccan music, Dickinson totally got into it, immediately. He and Charlie could always be counted on for that."

The Insect Trust relocated to New York, joined shortly by saxman Koehler and then rounded out by folk guitar guru Luke Faust. Now a five-piece, they were creating country blues–inspired free jazz. Barth was a serious blues guitarist, Luke was fluent in sea chanties and Appalachian banjo, guitar, and fiddle. Trevor was a postbop jazz musician who had toured with Anita O'Day among others, and Nancy had folk roots. Palmer's role was to tie it together by mashing it up.

Word about the hip Memphis scene spread. Many of the expatriates took over an apartment building in Hoboken. Barth referred to it as the Hoboken Power and Light Company. "There were ten apartments, five floors, in this place on 39 Second Street in Hoboken," says Crosthwait. "On the bottom floor was an alcoholic lady with cigarette burns all on her arms. She made me nervous. I kept waiting for her to nod out with a cigarette on her bunk, and we'd all be gone! I'd think, Just keep dropping 'em on your arm, lady, and wake yourself up."

The landlord was a mafioso who protected the motley crew because they actually tried to pay rent. They learned the power of his connections one day after sax great Pharoah Sanders came by to jam. "We just played at peak intensity for hours and hours and hours," says Palmer, "and I'm sure it could be heard for miles. Three screaming saxophones, electric guitars, and drums. It was really loud and we felt like it got real cosmic, transported us." When they picked up the morning paper the next day, a front-page story told about the noise disturbance and a slew of complaints, and then about the resulting bust of the perpetrators. "The bust never happened," says Palmer. "The story was planted. It was one of those neighborhoods full of candy stores that didn't have any candy but had fifteen telephones, and our landlord fit right in."

Randall Lyon lived in the building for a time. "Rambling Jack Elliot used to pay homage to Luke Faust. Luke's father was the Gloucester fisherman. He knew all these New England shanties and stuff. That was a great building in Hoboken. Sam Shepard was the drummer in the Insect Trust for a brief moment. Sam had this performance art troupe called the Group Image Sucks. He came over to Hoboken to rehearse a few times with the band and I think he freaked out when he realized the potential madness. We were way out of control. Dylan was around but everybody was pissed at him because they figured he ripped off Rambling Jack and Luke. There were fistfights over that, people got hurt in that discussion. That's what was so cool about those people: they didn't accept any authority about anything. They questioned every manifestation of authority. Even if one of their friends became hugely successful, it was still like, Wellll."

Joe Callicott was one of the first bluesmen that Gruener brought to New York, booking him at the swanky 150th anniversary of a large museum. "I was at McIntire's house, had been going to Memphis State for about a month, and these two friends came by and asked if anybody wanted to go to New York," remembers Marcia Hare. "I said, 'Yeah, I do.' I got my coat and my purse and I went to New York. So much for that semester of college. We drove the whole way with Joe in the backseat and the three of us up front. My friends would stop at these gas stations and go in the rest room together, and Joe Callicott would say to me, 'They're fucking.' But they were really shooting Desoxyn. By the time we got to New York, they were in a big fight and Joe was uncomfortable being so far from home. I bumped into Warren Gardener and went off with him." The museum party featured a 1920s orchestra on the main floor, playing in tuxes beneath potted palms. There was a disco on another floor, and then there was Joe Callicott, accompanied by Barth, Luke Faust, and Palmer. Callicott had never been far from his Hernando, Mississippi, home and wore his farm clothes and work boots. In the elevator, he was introduced to Mayor John Lindsay. They stayed in New York less than a week, but Joe Callicott couldn't get home fast enough. Driving back, when they stopped in Nashville to rest for the night, he snuck off. "In the morning," says Hare, "people said they'd seen an old black man hitchhiking on the highway."

Later that summer at a *Go* magazine party, the Insect Trust met Steve Duboff, who became their producer/manager. Without telling them he'd written a couple of Cowsills hits, he asked the band what they thought of the songs. Palmer remembers, "We told him that we thought they were shit, absolute shit. He liked that because he thought they were shit too. He had been involved in bubblegum, and now he was ready for some weirdness, and we gave it to him." They released two albums, one on Capitol in early 1968 and the other on Atco in 1970. Their material included songs learned from Joe Callicott and writings by Thomas Pynchon set to music; they sent a tape of the latter to the author in Mexico for permission. Duboff arranged for their live bookings to be handled by Bill Graham, which resulted in this very strange quintet opening shows for Santana, Frank Zappa, Pink Floyd, and, in Baltimore, the Doors in front of 50,000 people.

Barth was showing signs of strain and left the band during their arena days. Ed Finney, a Memphis jazz guitarist who'd studied with the Regents' Rick Ireland, replaced him. After several more incarnations, including a funk rhythm section, the fabric holding together the Insect Trust became too sheer. Their final dates were a response to the growing slickness of the arena rock around them, with the band abandoning all structure. Palmer went on to collaborate with Lenny Kaye in an early noise band. "Through that last Insect Trust noise thing, then through doing it with Lenny Kaye," muses Palmer, "I feel like our influence filters into a certain wing of the New York punk scene. Lenny took a lot of our thing with him to Patti Smith. When they were first living together in New York, they used to call me all the time and they'd be listening to our tapes. She said that 'Radio Ethiopia' was especially inspired by that music." Joe Callicott probably never heard Patti Smith, but if she was driving to Memphis, he'd have gladly caught a ride.

After seven months, spanning the summer of 1967, Crosthwait had had enough of New York City. What had been "tweed and tennis shoes psychedelia" with the Jefferson Airplane became, he says, "a Quicksilver Messenger Service Methadone nightmare—black and crystal and sharp-edged and New York brittle." Sly Stone had helped launch the Circus, but when he came back six months later, he was "a completely different arrogant son of a bitch."

The Bitter Lemon rode the artistic crest until 1968, when the sublime turned to the loud. When "happenings" were in, McIntire had no qualms about accommodating such events. "We'd close up to get the place fixed up. They were freaky things. One time you had to get on your hands and knees and crawl through boxes to come in the door. Inside, you'd come into a plastic box, and the seats were all pushed together, touching. You could only get twenty-five or thirty people in this plastic box. They had to sit on their haunches, grown people. And the box got real hot. The stage had a guy and a girl sitting on it, staring at each other, with an alarm clock between them, and a spotlight overhead. The alarm would go off and they would do something, stand up or maybe say two words, and then reset the clock and set it down. The people in the box were going nuts, they got to where they couldn't handle it any longer, screaming. They tore the box apart and ripped the whole interior to pieces and ran out the door. It was a real good happening, worked just like it was planned."

The demise and ultimate death of the Bitter Lemon was neither a great ball of fire nor an anguished last gasp. It seemed to mark the changing waves of fashion, the point where one trend was retreating as another came crashing on, and in the collision it simply sank. "The very same people who put me down for being a Yankee and a nigger lover and for the way that I looked were coming out looking exactly like I looked," says Marcia Hare. "I remember Randall and I freaking out that these people who didn't have any sensitivity or any culture or any artistic sense, that they could grow their hair long. It didn't work though, because the fat rednecks still had the goofy look on their face. But we didn't want to get busted for looking weird when everybody suddenly knew what pot was, so we started trying to look straight and let the rednecks wear their hair long."

At the Lemon, bookings became infrequent for traditional blues, for Memphis jazz artists like Fred Ford and Phineas Newborn Jr. The audience was no longer into listening, but rather being loud, a post-Beatles syndrome of yelling and screaming. Roland Robinson, a black bass player in Eddie Floyd's band, remembers stopping by the club because he'd always seen a line going out the door. "I was back from a tour, living at Willie Mitchell's house," he says. "Me and Hubbie, Willie's son, would just get in a convertible and drive around and find a good time. We walked in that place, looked at each other and said, 'Wrong!' And got back in our car and left." It seemed like a long time back when Harry Belafonte had stopped by, or when McIntire had lunched with Marcel Marceau. McIntire became less interested in his own club. And more frightened. The new patrons, attracted by the dissonance that had become hip, did not encourage the tortoise from his shell.

Manager Charlie Brown got more and more crazed. Herpetology, for example. Though it scared most others, it thrilled him. Bluesmen, tending toward frailty, found no pleasure when he excitedly entered the club, piqued everyone's curiosity with a burlap bag, then emptied it on the floor, producing a mass of rattlesnakes untangling themselves. The last to get anywhere safe was Furry Lewis, or Gus Cannon, or whichever elder bluesman was nearby.

Drugs got out of the hands of artists and into the audience. People were milling peyote buttons in the supermarket coffee grinder. Police began raiding the Bitter Lemon so often that McIntire would return to his home and find them there, searching with intimidation instead of a warrant. The vice squad finally chased Charlie Brown to Miami. "It had been a lot of fun but at the end it was scary. I'd sneak in late at night after I knew it was closed, I'd scrub the tables and sweep the floors, wash it all up. I finally just gave the coffeehouse to Herman, the cook. Sign these papers, it's all yours. We went to a lawyer, Herman agreed to pay the $239 in back taxes. But he never reopened it and I ended up having to pay the $239. The government took my paycheck from school. We all sort of left, went to our different places, like an Indian thing, very religious."

On April 4, 1968, Dr. Martin Luther King Jr. was assassinated in Memphis at the culmination of the sanitation worker's strike. Musicians had achieved a racial unity at the start of the decade, but as the strife before the rioting laid bare for the world to see, the Memphis government was not interested in treating all human beings like human beings: full-time pay still left the sanitation workers qualified for welfare. Rioting after the assassination was intense. King's death left Isaac Hayes creatively blocked for a year. Stax saxman Andrew Love remembers, "There was a helicopter flying overhead when the riots were going on, and I actually had thoughts of getting my gun and shooting at it. It *actually* crossed my mind."

"It certainly wasn't any intention of anyone at Stax for all the civil rights implication that came out of soul music," says Terry Johnson, another original Mar-Key. "The music made that happen because people who had never known each other, who had never gone to lunch together, who had never eaten a barbecue together, who had never done the things that typically people do—all of a sudden were doing those things." Stax music appealed to blacks and whites because it came from blacks and whites.

Don Nix, the former Mar-Key who became a songwriter and producer for Stax, remembers going to the studio even when he didn't have work. "You could go and there was Johnnie Taylor and Rufus Thomas and Eddie Floyd—just hanging out! I went there every day not to do anything but hang out with everybody. It was a big family kind of deal. I was at Stax Records the night Martin Luther King was assassinated. When they came to the door and told us he'd been shot, Duck Dunn and I started to go out to our cars. Isaac Hayes came and said, 'No, I'll drive you out,' because it was doubtful whether we would have made it. When I got home and turned on the TV, I found out he had died. It never was the same after that. Maybe for another year I would go over there and hang out, but I think for the black people in that neighborhood, that was the final straw. I knew it was never going to be the same when they were burning everything down and Stax hired a bunch of guys to go up on the roof with machine guns. I saw that and I said, 'Boy! This is it. This is it, folks, it's never going to go back to the way it was.'"

It took the crack of a rifle shot and the silence of Dr. Martin Luther King's last breath for the city fathers to hear the call of their fellow men, but for the blues festivals to get their attention, it only took the ringing of the cash register. The 1968 show was recorded by Seymour Stein—now president of Sire Records—and released as an album by London Records. That outside affirmation of the event's importance was all the city needed to begin interfering, and by the following year, they had just about rurnt it. The city sponsored an indoor stage simultaneous with the 1969 event. Their featured Memphis country blues performer was Johnny Winter, not from Memphis, not a country blues player. In their list of preferred artists, he was behind, as Stanley Booth has written, "such noted blues artists as Louis Armstrong and Marguerite Piazza."

At the Shell, downtown had interfered with down-home. The 1969 festival was expanded from one day to three. Friday afternoon was given over to a rehearsal for the benefit of a public television documentary crew. They were from the "Sounds of the Summer" series, hosted by Steve Allen, who would edit himself in later. Another documentary crew came down from Maryland, their multicamera shoot documenting the multi-cameras present. The album from that year was made in a recording studio. Perhaps the festival's encapsulation is the Bar-Kays' performance. All but two members of the band had been killed in the December 1967 plane crash with Otis Redding. Reformed around the survivors, they came out in full funk garb, and despite the sweat that ran into their eyes, they got down. It was not country blues, but it was part of the extended family. Then Rufus Thomas came out, the funkiest man alive, and the Bar-Kays backed him. And when it was all over and all the water in the Mississippi didn't seem like enough to quench their thirst nor cool them down, a meekish representative from the educational network explained that there had been a mishap in the filming and asked if they would mind performing their sets again. "They did the whole thing over in that fucking heat," says Palmer. "I don't know how they did it, but I remember being real impressed by the fact that they could get through it, and that it was as good as the first one."

The blues show regulars were there: Sleepy John Estes, Mississippi Fred McDowell, Bukka White, Furry Lewis. Johnny Woods made his Memphis debut, playing harmonica like he was from a part of Mississippi not yet discovered. "Johnny Woods had gone from the tractor to the stage," says Marcia Hare. "I was a teenager checking on their dressing rooms, getting 'em what they needed, a bottle of whiskey, pack of cigarettes, trying to make 'em feel comfortable. I asked Johnny Woods if he needed something, and oh boy, did he ever. He took my hand and put it right on his crotch. I went running out of there. I knew then to stick with Furry because he would never do anything like that." Lee Baker's band Moloch performed, and the Insect Trust played. Dickinson, Freeman, and a rhythm section had recently backed Albert Collins for

*Steve Holt, of the Bar-Kays, at the Memphis Country Blues Festival,
circa 1969. Seated in the background (second from left) is Charlie Freeman; seated on right (in hat) is Jim Dickinson. Photo by Marcia Hare.*

his album *Trash Talkin';* that grouping was to become the Dixie Flyers, and
they performed at the festival as the Soldiers of the Cross. Yet and still, when
the Fourth Annual Country Blues Festival is discussed, one need not be very
astute or familiar with the artists to hear the disgust that quickly creeps into
the conversation, and invariably the first thing mentioned is that Johnny Winter played. Real loud.

The city won, and lost. Their bad vibe, combined with Barth's unfortunate
bust (headline: "Drugs Trip Up Blues Promoter") left the Memphis Country
Blues Society crippled. A group of Memphis hippies tried to rescue the tradition the following year, but it proved to be a blues festival coda. New nonprofit organizations began to appear, along with degreed academics. Those
who had done the work but not graduated from the proper institution found
themselves ostracized. "Like twenty years' worth of work wasn't enough, you
had to be at Yale or something," spits Randall Lyon. "They had the meeting of
the National Endowment for the Arts Folklore Panel in Memphis. Bess

Furry Lewis and Marcia Hare at the corner of Madison and Cooper. Photo courtesy of Marcia Hare.

Lomax Hawes was there, a lot of bigwigs. They were setting themselves up to adjudicate who the folk was and what folk wisdom was, circumscribing these artists with this academic folderol, this ethno babble. Several of us stood up and said, 'Folklore is a racist concept and it should not be used.' It was objectionable to everyone present. They disbanded the panel. I videotaped it. I mean, you listen to Skip James and have someone tell you about the folklore of his songs. It was very obnoxious. We figured there was not enough known about class and gender and racism in general for any comment to be made about the aesthetics or the meaning of the music. Other than participating in it and enjoying it as a human fucking being. It was an obscure moment, but still, there ain't no fucking Folklore Panel no more."

CHAPTER TEN

Magic Time

WHEN THE BOX TOPS' "THE LETTER" BECAME THE HIT OF THE SUMMER of 1967, no one was more surprised than the two band members who had quit the group before its release. They would have laughed aloud if told that their average Memphis teen band was going to have such an impact. "The Letter" became one of the biggest-selling records of 1967. Truly, everything was possible, anything could happen, and it all did come true.

"The Letter" introduced a sixteen-year-old vocalist who remains a presence on the pop music scene to this day, a presence so strong that his absences are influential. That session marked the first time that singer Alex Chilton had ever worked in a recording studio, though he'd been surrounded by music all his life. Dan Penn, the song's producer, had also surrounded himself with music, but "The Letter" was his first time in charge of a serious session. And, judging by the other two numbers on the demo tape submitted by Wayne Carson Thompson, the song's young writer, he'd gotten lucky while still learning his craft.

Such left-field success proves there's no formula for a hit, and it keeps the music industry ticking. It's a gambler's business, a game of numbers and chance, and when the long shot wins, it reminds everyone that the fix is never 100 percent. The bread and butter of the industry is the little guy who gambles on being the big guy. All the unheard records are necessary to support those few that actually find their audience. And the occasional jackpot keeps the gifted and the giving returning for one more try.

The only element of "The Letter" with any experience behind it was the American Sound Studio, and one member of the house band describes it as "pretty primitive, one set of earphones, and they were army surplus." Another says it was "really a barely rigged situation." Dan Penn came along while they were still trying to plug the walls into the walls. A singer, songwriter, and future producer, Penn had led several popular white R&B bands around the Muscle Shoals area. "The night Chips invited me to come up and check out his operation," Penn recalls, "his whole studio consisted of an A7 and an RCA board that wasn't very big. They were recording the Gentrys. Rick [Hall] had better equipment [at Fame in Muscle Shoals] than Chips had, but I was hearing something in that record Chips was making. That old studio was really moaning. Chips had done everything wrong, seemed like, to have made it all come out right."

Dan Penn speaks in the softest of southern drawls. His manner today is sagelike, a fine whiskey that's mellowed in a fine oak barrel. The wiry speed demon of yore has been replaced by a ruminator with a trucker's gut. The fire still crackles when his poker face breaks and the corners of a grin creep around the toothpick he chews. Once the wildest of them all, or perhaps the most driven, Penn today spends time in his garden and working on vintage cars. "Me and Moman had tried to produce some records together, but I was having a pretty hard time getting my ideas across," he continues. "So I told him, 'I want to produce a record and I want to do it my way and I want to do it by myself.' It didn't make no difference to me who it was, I would have cut anybody. I was twenty-six years old, I was ready to cut a hit. Chips was big friends with this disc jockey, Roy Mack, and Chips had him bring this band in. Their little singer was acting kind of smart-aleck, so I told Roy to bring me another singer and I'd cut 'em. I handed them a tape with some Wayne Carson songs on there, and told them to pick anything they wanted from this tape, but make sure that we do 'The Letter.' They came Saturday morning about ten o'clock and they had Alex Chilton with them, who I'd never seen. We started running 'The Letter' down, and he sounded pretty good. I coached him a little, not much, told him to say 'aer-o-plane,' told him to get a little gruff, and I didn't have to say anything else to him, he was hooking 'em, a natural singer."

This natural singer, at sixteen, had come up a natural artist. His father's day gig was commercial lighting, but in the glow of the moon he was a well-respected jazz pianist. His mother ran an art gallery out of their expansive home, and the couple's fondness for hospitality created a salon atmosphere in the early 1960s, attracting artists like William Eggleston and Burton Callicott. In a time when hi-fi was rare and record collections more so, Sidney Chilton, Alex's father, had both. From his bedroom at night, Alex heard the sounds of blues and jazz, Chet Baker, Ray Charles, Dave Brubeck, Mingus wafting up the stairs and into his bedroom.

"Taking it all the way back," Chilton says, "the first record that I really became aware of was 'Youngblood' by the Coasters, backed with 'Searchin'.' My oldest brother had a copy of that, and I remember really well him hanging around the house with his girlfriend one night while my parents were out, playing that record over and over again. I was maybe five or six. My dad was into jazz, and I got into his Glenn Miller records first. I became a big fan of Chet Baker singing in about 1957 or 1958, and that was when I first really wanted to sing. He first inspired me. When I got to be eleven or twelve, the start of the sixties, I began listening to the radio. 'Johnny Angel,' the Ronettes' 'Be My Baby,' the Orlons' 'Don't Hang Up.' George Klein was playing a lot of great stuff. I was aware of Elvis and Jerry Lee, I was given a copy of 'Great Balls of Fire' for my seventh or eighth birthday, but I really wasn't much of a fan of all that. And by 1959, Elvis was syrup and Jerry Lee was pretty much gone, and the rockabilly thing was sort of over. I didn't get really caught up in the rock scene until the Beatles came along.

"A lot of times I'd come home from school and my dad and some of his friends would be jamming out. In the 1930s, he'd been traveling around as a musician and when he got married and they started having kids, he blew off music. But I was the youngest kid, and he started playing more music as I was growing up. By the time I was ten, man! It was party time around my parents' house. We moved from the suburbs into the city and I remember countless nights of going to sleep with, like, sixteen jazz musicians playing downstairs."

In 1966, when Alex entered the tenth grade at Central High in Memphis, blacks and whites in the same school district were finally being allowed to attend school together. (Anticipating the possibility of mixed dancing, the school had not held a prom since 1964.) A band of recently graduated Central High students was going through a personnel change and enlisted John Evans, who could double on guitar and organ and who owned his own equipment. "They asked me if I knew anybody who could quote unquote sing like a nigger," Evans recalls in 1994. "In private conversation at the time, that was the word that was used. I was talking to another friend and he told me that at the Central High talent show, there was a kid who sang 'Sunny' by Bobby Hebb, and all the girls really loved him.

"Alex showed up at our practice and he was—" Evans pauses, groping for the right word, "—different. For one, he was a good bit younger than us. We were all about nineteen years old, Alex had just turned sixteen. He had come from a different background. He was wearing—and we would never have worn this, you've got to understand—blue jeans with holes in the knees. We'd have thrown them away if they had holes in the knees. He was wearing a black

T-shirt. I'd never *seen* a black T-shirt, I thought they only came in white. He was wearing a woolen dress scarf like you wear with an overcoat, wrapped around his neck like Bob Dylan, and a blue jean jacket. To all of us, sitting there in our MacGregor and Gant shirts and permanent press cuff trousers with penny loafers, Alex was unusual looking. But he could sing, and sing soulfully."

That was January. In March, Roy Mack, the influential disc jockey who was the band's manager, suggested they return to Moman's studio. At that time, disc jockeys were the interface between the public and the recording industry. If you wanted to book a band but didn't know where to turn, or if you were a band but didn't know where to record, you'd phone the guy who played the hip tunes. Disc jockeys and studios necessarily had friendly relationships: One made records, the other made hits.

The Box Tops' John Evans remembers picking up the Wayne Carson demo tape the night before the session. "We were going to rehearse and we played the tape. I think there were three songs. I've forgotten one altogether. There was one called either 'White Velvet Cat' or 'Pink Velvet Cat' that was unbelievably bad. Imagine a small-town, country-influenced songwriter growing up on fifties rock and roll and trying to write a sophisticated slightly jazzy song based on his experiences at a local bar seeing a beautiful woman. On the other hand, imagine him trying to do something like the Everly Brothers, and that's 'The Letter.'" The band played the demo at their rehearsal on Friday night. Everyone laughed at the cat song and agreed that "The Letter" was the one to cut. While Evans began mapping out the chord changes, one member went off to meet his girlfriend, Alex and another went to get some beer; rehearsal never really happened.

They showed up at the studio promptly at ten A.M. on Saturday. Evans describes Dan Penn's arrival, shortly behind them: "He was wearing this polyester narrow-brimmed fishing hat thing on the back of his head, a white T-shirt with a pack of Lucky Strikes rolled up in his sleeve. We didn't see Lucky Strikes much back then, so that put him on another planet. Also, we came from schools where you had to wear the right brands of the right clothing. One thing you would never do was wear false copies of madras handwoven cloth from India. But he was wearing madras that looked like it came from Kmart. He was wearing Bermuda shorts down to his kneecaps, and sports socks with different colored stripes. And hi-top tennis shoes. He walked in, drawling his talk, and Danny, our drummer, in his white button-down-collar Gant shirt, said, 'Where's Chips?' Dan said, 'Chips won't be in today, I'll be cutting you.' We get introduced all around, and Danny, who'd cut with Chips before, is behind this guy's back, rolling his eyes like, What's wrong with this weirdo and why did Chips do this to us?"

John Evans says "The Letter" took over thirty takes. Penn was twisting knobs and making studio adjustments for about half of them. The rest were working up the song, getting a single take all the way through, the lead vocals becoming slightly more gruff each time. The studio was still moaning: Penn had to get on his knees to change the routing of cords through a patch bay; Chilton remembers that the recording console had big dials for faders, like a radio station board. There was no Leslie speaker for the Hammond organ; instead it was miked from the little built-in cone near the player's feet, giving it a funeral parlor sound. By three o'clock, the band was done.

Penn: "And we cut 'The Letter.' One Saturday morning. Put some strings and horns on it with Mike Leach and then me and this black fellow that used to hang around the studio, we took the jet sound off a record in the office next to the control room. He put the needle on the acetate for me, and I went inside and I was working to get it in the right spot. We got lucky and I mixed it down, and I thought it was okay. 'I'm a Believer' was happening then, little organ going *chink chink chink.* You'll hear a little of that in 'The Letter.' I thought I got my licks in on a little rock and roll record. I didn't think, 'This is a million seller.' But it was."

"I had been playing with a garage group of my friends," says Chilton. "We never had gigs too much, but a few. I was the singer. My father had bought me a guitar, but playing all those notes at once—it seemed impossible to me. One of the guys that we were fooling around with wanted to be in the talent show at school, so I went along with that. Then this band called me up to audition. They had made some records with Moman before, a vocal version of Floyd Cramer's 'Last Date,' and another version of Thomas Wayne's 'Tragedy,' just horrible stuff. But okay, I'll go to the studio and give it a try. We recorded 'The Letter,' and that was how I fell into doing this."

"This" for Chilton has differed from the "this" of most of his musical peers. His career has been long and fragmented, disjointed by explorations into various facets of music, various facets of his background. It is the exploring that sets him apart. Where other musicians have tried to follow the flow, he has ignored it, charting a personal course. In the Box Tops, he may have been the producer's puppet—"The Box Tops are only marginally my records, I listen to them and I hear Dan Penn, I don't hear me"—but through the experience he learned the workings of the recording studio, he learned to play the guitar and write songs, to tour, to survive: Offered the rape stick by the industry as a kid, he succeeded in dismounting and is no longer its fuck-ee. Chilton's artistic horizons broadened in early 1970 when he departed the Box Tops. After attempting a solo career, he found a venue for some of his new ideas in the band Big Star, a group whose influence and popularity has steadily increased since its demise in 1974. He made soul-searching records in the 1970s, soul music in the 1980s, and in 1994 has gone back to the music etched on his childhood soul when his jazzy parents entertained downstairs. The joy, pride, and pleasure he has found in his vocal control is at the heart of his 1994 solo recording, *Clichés*.

Disc jockey and manager Roy Mack had been waiting for the right product to throw his weight behind. He knew that a hit song is never inherent in the music, that it takes promotion, which takes connections, which takes years. All of which he had invested and accumulated. Once "The Letter" broke in Memphis, Mack called on a deejay friend in Birmingham who made the record a hit there. The band was flown down as stars, all expenses paid, and they headlined a big dance in the armory. The teens swarmed, the girls screamed. The deejay made the dance money, Roy Mack proved the record's worth, and the band got to feel good. They were not paid for their appearance. Shortly after the Second Annual Memphis Country Blues Festival, the Box Tops, almost defunct half a year earlier, had a number-one record and more attention coming. "At that time," says John Evans, "the way distribution and promotion were, hits were often not number one nationwide at the same time. Like the whole time we were on the road, we heard the Doors' 'Light My Fire' in different areas of the country at different times." The Mar-Keys had chased their hit all around the country for years, and that honor now belonged to the Box Tops.

The Box Tops were then playing several Rascals covers, "I've Been Lonely Too Long," an assortment of Stax material, and the odd obscure number, like the Wildweeds' "No Good to Cry." As their popularity rose, the technology was unable to keep up. They played arena-sized venues without benefit of stage monitors, without a sound engineer; the audio, in fact, often went out through the sports announcing equipment. But Roy Mack's push got the label's attention. "Our label, Bell, was later home to the Fifth Dimension and Barry Manilow," says Evans, "but at the time we were on it, it was a black label, which meant connections at black radio stations. As a result, not only did 'The Letter' make number one on the pop charts, we got to number three on the R&B charts. We'd go places they couldn't believe we were white."

Even before their record was released to the public, the Box Tops were introduced to the veneer of pop music. Penn had hooked Larry Uttal of Bell Records when Uttal came to American Studio to hear another of his label's acts, James and Bobby Purify. Uttal ordered a Box Tops B-side, but when the band came to record it, they found it was already in the can. Wearing one of its many guises, the American house band, playing behind Alex Chilton, was now the Box Tops, thank you. "We played on 'The Letter' and later on 'Break My Mind,' that's the only other thing," says Evans. "The American house band backed Alex on everything else. I think Dan had it in his mind that if these guys played on something that became a hit, maybe we should give them another chance. He tried, but when we went in to American to cut, Chips had booked the studio on top of us. Dan got pissed off, wanted to do what he wanted to do, so he went to the nearest phone and called Rick Hall at Fame in Muscle Shoals and we drove down there. We'd met at American at seven P.M., so we got to Muscle Shoals really late at night and recorded till the wee hours. To show you how innocent I was, I remember that was the first night I ever drank coffee to stay up. We recorded a few songs, but 'Break My Mind' was the only thing out of that session to ever appear on an album.

"I felt ripped off not playing on our records. Put yourself in my place. A nineteen-year-old kid, you finally get a chance to do something you've wanted to do for years, it ends up being number one in the country for an entire year. Does it even seem reasonable, much less fair that we can't play on our records? By that time, Roy had us under a contract which we signed, being stupid. Our lawyers didn't know the business. I stayed with the Box Tops for a year, and after touring twenty-five days out of each month, my royalties came to $4,000 and I left the band. It's like someone saw those golden eggs come out of that goose and said, Let's cut that sucker open."

"I was coming back through Memphis around 1990, been down in Alabama, and I thought I'd slip over to the old American and see where we'd made all the records." Dan Penn is speaking. "I drove up there and I couldn't find it. The building was nowhere to be seen. I pulled my car up there and sat for a little bit and said, Yeah, there was the control room and here's where we was when we were putting the jet plane on 'The Letter.' Here's where it all happened. It was real strange to see that place gone—as much as had gone down over there. I felt kind of empty, useless, kind of sad. And glad, too, like, this is where we done it, thank God for this little place. But that's Memphis. I'm sure the Stax people, when they pull up over where their building used to be, they must feel the same way. Memphis scraped them away."

Dan Penn's drawl is soft as cotton, slow and thick as mud. Where he's from in Vernon, Alabama, there must be plenty of people who sound just like him, but somehow the forces of life cloverleafed around Dan Penn, creativity intersecting with calculation, poetry getting tangled up in daily language, the passions of black and white culture twining themselves around a country boy, allowing him to effortlessly capture human emotions. His songs sound as truthful today as they did two and three decades ago: "Do Right Woman," "At the Dark End of the Street," "Cry Like a Baby." He began on the fraternity circuit with the soulful Mark V and later with the grittier Pallbearers, who carried him onto the stage in a casket. His vocals, gruff like Chilton's Box Tops era, can be heard on his 1972 debut album, *Nobody's Fool.* His second album, with the working title *Emmett the Singing Ranger Live in the Woods,* was produced by Jim Dickinson, mixed by Penn, and is now languishing in an unknown corner of the Arista vaults (probably near the same forgotten crevice where the Goldwax masters lay). In 1982, he released a gospel album about which he says, "I wouldn't call it great." In late 1993 I was present while Penn recorded a new album in Muscle Shoals for Sire Records, released in 1994 as *Do Right Man.*

The Alabama musical environment was even more insulated from national trends than Memphis's. The story is told of one musician accidentally walking in on a prominent Muscle Shoals producer while he was using the toilet. This producer, involved in bringing the world many hits of the 1960s and 1970s which remain staples of oldies radio—this producer was discovered not to be seated on the toilet bowl, but rather crouching on it, his feet on the rim and his knees around his chin, the technique acquired in the outhouses of his youth which kept his toes from being nibbled on.

Radio was their link with the outer world. "When I first heard Presley I was as enthralled as anybody," Penn says. "Sun was knockout. But it didn't last all that long, because as soon as he started making those slick movies and those funny little teenybopper records, well I slid away real fast and I never did go back. Here comes Ray Charles and then I don't have to worry about it no more because I know which way I'm going." Ray Charles led to Bobby Bland, whose singing was so gutsy that Penn puts him in a pantheon high and separate. In 1993, when Penn was recording his Sire album, he insisted that the photographer shoot him in front of the Muscle Shoals Sound Studio, where he posed in a hat and sunglasses, a coat thrown over his shoulder in direct imitation of and adulation for Bobby Bland's *Two Steps from the Blues.*

Rock and roll didn't get it for Dan Penn, two steps from Bobby Bland. "Chuck Berry didn't register on my little funk meter. He was cute and he was smart, but he never went to church. I never heard that in his voice. And if I can't hear that in your voice, I don't want to listen very long. It's gotta have that soul. Bobby Bland, Ray Charles, Aretha Franklin—Chuck Berry's over there, '*chinkalinkaching*,' and these guys are *serious*." Penn got serious early on, selling a country song to Conway Twitty, "Is a Bluebird Blue." (Not a man of means, Penn was spotted shortly thereafter eating a large steak. "Man with a Conway Twitty song," he said, "can eat what he wants.") When he was interviewed on the radio, the disc jockey said, "Tell us something about yourself, Dan," and teenaged Dan gave his height and weight. And that was it. He'd told 'em something. He's always stood out as someone who knows a little more, shows a little less, and probably has a really good idea if it can be drawn out of him.

In 1967, in the aftermath of the British Invasion, Penn turned out "At the Dark End of the Street," a collaboration with his friend Chips Moman. Pilled up at a music convention in Nashville, they took a break from a poker game, went to a piano, and hammered out the song in less than an hour, returning to play another hand. "We were always wanting to come up with the best cheatin' song. Ever. Me and Rick Hall began looking for the *best* cheatin' song years before 'Steal Away.' I don't know why 'Dark End' is so great. I guess it's the word 'street.' Everybody's interested in that word. The sounds that we were getting back then was the sounds of the street. And streets change. Now we've got *boomboompa boom chichichi*, and that's what the street is now, but in 1967, 'At the Dark End of the Street,' that had the street. I like to collaborate because two heads are better than one. It's easier to perform the miracle. And the miracle is, Can we jerk it out of the air or can't we. There are all kind of ideas always floating around. Other than Spooner Oldham, I guess Moman would be the closest person I ever come to breathing together with. All writers, music people, they can be playing poker, they can be swimming, whatever they're doing, and all they really want is another great song."

Spooner Oldham is a boyhood friend of Penn's who followed shortly behind him to American Sound from Muscle Shoals. Oldham joined the studio staff as a songwriter and also played keyboards on some sessions; this was when two keyboardists were as common as two guitarists. "Dan was living in Vernon, Alabama, when I met him," says the gentle Oldham, one of the few who speak slower than Penn, whose voice is even thicker and softer. "He'd come up to Muscle Shoals occasionally, to a little piano room over the drugstore. I'd written a couple of songs that I'd never showed anybody. I didn't really know any songwriters and there was no market here at the time. So we decided to get together one night and we wrote three or four songs. I can't attest to the quality or integrity, but we did realize that we could sit down and sing and play our instruments and write together. And we've just continued."

I learned all about Spooner Oldham's keyboard playing while watching him smoke a cigarette. After years of admiring his keyboard playing—on soul records, on Bob Dylan records, Ry Cooder, Neil Young, the Box Tops—I spent a few days with him in Muscle Shoals at Penn's recent recording sessions. Pictures of Spooner show a man impossibly thin. Even when you look right at him, you almost can't see him. He's never at the center of a crowd, nor is he far enough outside the edge to draw attention. He's like a distant star made visible only by looking away; if you look too closely at him—or listen too closely to what he's playing—he vanishes. His visage is beautiful, all lines and texture. When he smiles, it involves his whole face.

Spooner keeps a pack of filter cigarettes along with his nonfilters, for variety. He is always smoking, which does not mean that he always has a lit cigarette. The process is such a part of him that when he is empty-handed, he has just finished one or is preparing to light another. He rubs his hand across some pocket and like magic, a cigarette appears between his fingers. Once there, plenty of time will pass before it meets a match. Spooner coddles his cigarette, holds it now by the filter, now by the tip. If he were to do something as direct as point, he might use it for emphasis. But emphasis from "ol' Spoon," as Penn refers to his lifelong friend, comes not directly but indirectly.

On the final day of Dan Penn's 1993 sessions, they were scheduled to record three songs. The album was a mix of old material and new, and while some of the best had already been cut, Penn had saved three doozies for the end: "Dark End of the Street," "You Left the Water Running," and "Do Right Woman." When the Hammond B3 broke down at the start of the day, everybody poured more coffee and continued to schmooze. The delay may have postponed the actual recording, but rehearsal began when the coffee was brewed, when one player entered the studio and encountered another. The way that southern musicians play together is just an extension of the way they interact; their style is evident when they lace their shoes, when they play ping-pong, when they smoke a cigarette.

That last day, they got "Dark End" after five takes but recorded a dozen of the upbeat "You Left the Water Running." They didn't need that many, but playing the song was fun. Each take told the story differently. It was after ten P.M. when they began running down "Do Right Woman," the song that Aretha Franklin and William Bell and Gram Parsons have made a fundamental part of life. Players wandered on and off the studio floor, greeting old friends who dropped by to say hello, who stuck around once ensnared in the magic. Penn grabbed an acoustic guitar and began trying to remember the song's changes. Bassist David Hood was across the room, apparently in a world of his own though he was actually encoding on paper the chord sequence that

Penn was remembering. Spooner was at the organ, following Dan's lead, playing his part slow and full like blood from a deep wound.

No one told the others when to join in, and in the control room, they didn't need to be told when to roll tape. It all happened as naturally as sunset. Suddenly the first take was done, and there was discussion about what to do differently, and by the third take, they had it. "Should we go listen?" someone asked, and Penn paused, because he knew he had a take he could use, but he knew that if they got up, they'd never come back—magic time would be gone—so he said, "Let's do one more," and they did, and it was as if they had turned the first line of the song into credo, dogma, religion: "Take me to heart/and I'll always love you." No one said it but everyone felt that if they walked outside at that moment, the world would have been a different place; what they'd done in this little room a few feet from the Tennessee River seemed to have affected the course of mankind. And when they finished, drained by the intensity, Penn said, "One more."

There were three guitarists on the floor, three keyboardists, bass, and drums. But the space in the song was so wide that a history, a human, a life could get lost in it. From my seat on the sofa in the control room, I looked out at the dimmed room where the musicians' souls were naked as God and they were no longer breathing air but breathing this song, no longer humans but entirely musicians, part of a tribe whose numbers were no greater than those within earshot at that moment. Dan was singing, Reggie Young was strumming, Jimmy Johnson played guitar with his shoulders. My eyes rested on Spooner, an amp partially blocking my vision. His eyes were closed and his head swayed slightly and it would be obvious to a dead person that he was playing his guts out. I leaned to the right to see him better, and I saw his feet dangling from his stool, not touching the floor, not touching the foot pedals. I followed his legs up to his hands, aware I'm witnessing a master at work, a ballet of the greatest depth and dimension: His hands were neatly folded in his lap. His eyes were still closed, his feet still dangled, his palms were together and fingers interlocked: For Spooner, the notes he plays are so big that the space between them can extend for a whole verse. Another chorus passed before he moved, and when he did, David Briggs at the electric piano with his back to Spooner suddenly laid out. No words were exchanged. Spooner stepped into the song like a ghost, as forceful as when he was sitting out, summoning spirits from the vastness.

Oh, you should see him smoke a cigarette.

Spoon was in his late twenties when he came to Memphis, and he stayed for about three years. He and Penn were side by side when the Box Tops took off. "Weeks and months had rolled past since 'The Letter,' and the record company from New York is calling Dan regularly," remembers Oldham. "They keep asking, 'Where's the follow-up? We need a record yesterday.' After this went on for a while, Dan approached me and said, 'Spooner, people have sent me songs, but I really don't like any. All I know to do is you and I just try to write them a song.' So we went to American one evening and each pulled out our list of dozens of titles and ideas and spent ten minutes on each one and there was nothing, really."

Penn: "So me and Spooner stayed up a couple of nights lookin' for a song for the Box Tops' session. I had already booked the band for Saturday, which was like day after tomorrow, at ten o'clock in the morning. I need the song and I don't have a clue. Spooner don't either, and we're just working ourself into nowheres. So then it comes to tomorrow! And about dawn Saturday, we ended up over at Porky's, a restaurant right across from 827 [American's address]."

Oldham: "So daybreak, we go to this little cafe to eat breakfast and consider what to do next because Dan had booked all the musicians for a ten o'clock session to do our song. And we didn't have one yet. We were really getting tired, and considering the possibility of canceling everything."

Penn: "We were just settin' there with the comin' downs. We'd ordered our little bite to eat and figured we'll just mosey on home and crash, because it's been a long two days and we didn't get nothing. I'm looking at Spooner and he's looking at me, big old empty looks. And finally ol' Spooner just laid his head on the table and said, 'I could cry like a baby.'

"I set there a minute and I said, 'What'd you say, Spooner?' He still had his head down, he said, 'I could just cry like a baby.' And it hit me. I said, 'That's it, Spooner!' Ha! He said, 'That's it?' And it hit him 'bout between the booth and the cash register. Magic time had just got here and it was one hundred percent on. Suddenly the air had changed! Just that fast."

Oldham: "I guess we paid our tab and walking across the street, just shoulder to shoulder talking, we had the first verse of that song written before we got to the door. And we got the instruments, piano and guitar, and I think about an hour and a half later we had finished the song, put it on a little demo tape."

Penn: "Spooner's on the way to the organ, I'm on the way to the board to turn it on, throwin' on a piece of tape, he's got the Hammond whirling. We had been ready to give up, there had been no doubt in my mind that the session would not give that morning, but it did. We stayed in the studio—from that moment I would not leave for nothing. When the band got there, we were fresh as a daisy. The song actually gave us eight hours of sleep and I never felt better in my life."

Oldham: "Alex Chilton came walking in at nine A.M., heard the song and I didn't know what we had at that point. We were just exasperated. Alex listened and he just reached his hand out to me and said, 'Thank you.' That was the first glimpse I had that maybe we'd done something right. And then at ten we recorded it."

Penn: "And it was a hit record."

By the time of the Box Tops, Moman's house band had solidified, about to embark on the legion of hits that keeps them in demand as session players to this day. Their chameleon-like capabilities are evident not only in the variety of artists they backed, but also in the different producers under whom they worked and the changes they could make in their sound from day to day. "That hot rhythm section drew a lot of work to that studio at one time," says guitarist Reggie Young. "We were all from Memphis, everybody'd come up on Dewey Phillips. We knew a lot of different kinds of music."

"It was bizarre," says Tom Dowd, whose credits as engineer and producer, and whose long association with Atlantic Records, make him a walking history of pop. "Jerry Wexler, Arif Mardin, myself, we might show up one day with a Lulu or a Dusty Springfield and a week later come in with Herbie Mann or Wilson Pickett—it didn't matter what artist we came in with. We knew we had an accumulation of musicians who were masters of their instruments, who were gracious and took our bizarre direction easily, who didn't rebel, and we enjoyed their company. They were punctual, they were prompt, and it was a pleasure making records. You'd go in and work five, six, seven hours and come back the next day, first thing you know, you had eight or nine songs done and gee, one more day, and ha ha, the album is done."

During the time his teen band was working with Moman, David "Flash" Fleischman remembers, "I walked in one day and Tom Dowd is talking to Dusty Springfield, who is musically educated. He's talking pianissimo and forte and obbligato, and the next day I walk in and he's in there passing a bucket of chicken around with Joe Tex." Working in Memphis seemed to free the producers from the strictures they felt in New York and Los Angeles, surrounded by the industry. The looseness of Memphis kept them from producing a song to death.

"In terms of the races, the sixties was the culmination of the forties and fifties," says Penn. "There were a lot of white people and black people who had tried to bring the R&B and the white side together. It became a white/

black situation, you had white players and black players together. The mixture, who knows what that does to us, but it does something. There was so much respect. Now we get all these white people in the studios. Everybody respects each other but it's like you ain't bringing anything different to 'em. We're trying to make a painting here, what color did *you* bring? You're orange and he's orange and we need some red, we need something different, and back then black people brought so much to the whole thing. I always related a record to painting a picture. Your speakers, or one speaker back then, you stretch this big old canvas—I see it, I try to physically see it. That cross-color respect was a wonderful thing. It carried a lot of power. We don't seem to have much of that now. Hope we get some more of it."

The Box Tops. King Curtis. The Gentrys. Sandy Posey. Lulu (the Brit with the Flip). Elvis. Joe Tex. Neil Diamond. Bobby Womack. Marilee Rush. Herbie Mann. James and Bobby Purify. Dusty Springfield. The Sweet Inspirations. Wilson Pickett.

Atlantic Records. RCA. Uni. Warner Brothers. Decca. Dial. Scepter. MGM. And the list goes on and on and on, ignoring genres, defying categorization, unified not by a sound but by the solid musicianship of the American house band.

House rhythm sections were not unique to Memphis, just successful there. At Sun, Billy Riley, Roland Janes, and J. M. Van Eaton—the Little Green Men—served as a backing unit. At Stax, the Mar-Keys evolved into Booker T. and the MGs—Booker T. Jones, Steve Cropper, Duck Dunn, and Al Jackson Jr.—who had hits on their own and backed countless sessions for others. When Willie Mitchell took the reins at Hi Records, he formed a house band that pooled not only talents but also genes. Three brothers—Teenie, Charles, and Leroy Hodges—locked into a single silky groove behind Al Green, Ann Peebles, and other R&B stars that Mitchell fanned into fires of smoldering sex.

For years the story about the evolution of the American rhythm section, the 827 Thomas Street Band (the studio's address), has been the easy version: They were formed by Stan Kesler and stolen by Moman. According to both Kesler and Young, that's a simplification. "It wasn't like we had never seen each other," says Young. "We intermingled and we played clubs together. But as a group, we were the merger of the Phillips studio section, formed by Stan, and the Hi rhythm section, sort of the Bill Black Combo, which was me and Bobby Emmons and a bass player named Bobby Stewart, who was later replaced by Mike Leach."

"I worked with all of them before they ever went to American," says Kesler, "though not as a group all together. Gene Chrisman, the first session he ever did he probably did for me. Same with Mike Leach and I know it's true with Bobby Wood. I hired Reggie for sessions, and Bobby Emmons was over there at Hi with him. Tommy Cogbill, I used him a lot. I probably could say that I trained them, played a part in getting them studio-wise. When Chips started recording, he had a bunch of success on the front end and that's where the work was. I didn't feel like that band was stolen."

"Some of the early records that Moman was doing, like the Gentrys, he kind of put us together," continues Reggie Young. "I remember we used to go out of town, me and Cogbill and Moman, like to New York and work for Wexler on different projects and go to Nashville, and then we decided to see if we couldn't get some people to come here and us not travel anymore. And it solidified with Gene Chrisman, me, Bobby Emmons, and Tommy Cogbill. We cut a lot of records with Chips producing. They'd have to come to town to get us, and the Atlantic account was one of them. The band bound ourselves together, and then eventually Bobby Wood came in, and then Mike Leach after Tommy started producing a lot and playing less. After we made a commitment to do that, we didn't work out of that studio. And Moman was a drawing card. He was a good writer and a good producer. And other producers would come in and use us. I guess at one point, my amp stayed in one place—I set in front of the control room window for years. I wouldn't take a million dollars for the experience, but it's like if you've been in the service, you don't ever want to do it again. The education was priceless but the job didn't pay that good."

Sharing in the excitement and success formed a natural bond among the members of the American rhythm section, and between them and Moman. One staff songwriter remembers that whatever toys Chips got into, the whole American group would follow. "They went through radios, motorcycles, ham radios, model airplanes, horse racing a little bit when Chips built a race track. Gene Chrisman's motorcycle fell over on him once and he couldn't get it off him. They had to help. Motorcycles didn't last very long." Like the strongest of such bonds, there was a tension to it too, not surprisingly about money. But the band was on a salary, and so to some extent relieved of the worry that comes with the natural ebb and flow of the music business. Moman bore that burden, which meant that the others also felt it.

"I worked all those years thinking every day I was going to lose my job, thinking this can't last forever," Moman says. "A musician ain't supposed to make money, that was what Daddy told us. He said that's nothing but just a waste of your life. That was a real fear. That's how I picked up the name 'the Front Money Kid.'"

(The friendship and the tension between Moman and the band kept each in check for years. And years. They all got fed up with Memphis before they got fed up with each other, and as a group they moved, yanking their kids out of school, packing their homes, and eventually settling in Nashville, where they wrote and cut "Luckenbach, Texas" for Waylon Jennings, creating country music's outlaw movement and making *Ol' Waylon* the second-ever platinum country album.)

Playing for a house rhythm section allows a musician a better shot at surviving in the music industry without the road-weariness or burnout that comes from playing gig after gig, night upon night, anonymous venue after anonymous hotel room. One gets all the thrills, chills, and spills of being a working musician without having to go on the road, although studio burnout remains a reality. However irregular the hours, you have a chance to see your family. You can have a life. So while they were cutting all those hit records, the stars they backed would get off the airplane and into the hotel room and into the limo and into the studio and back in the hotel room and back on the airplane, nodding hiply at the next star disembarking—while the rhythm section poured coffee at home from the same coffee pot (maybe having awakened in the middle of the afternoon), drove across town by the usual route, and went to work making another hit record in the recording studio.

Like the MGs, the American rhythm section worked up head charts on songs as they learned them, and their interpretations were hits. Unlike the MGs, the American music was not distinctly Memphis music. Artists were flying in from around the globe to get chameleonesque, solid backing on *their* music; had Neil Diamond cut at Stax, the resulting album would have stood out in his oeuvre much more than his *Brother Love's Traveling Salvation Show*, which was cut at American. "Memphis is a groove town. Musicians there don't even know they can play together. To them, it's second nature," says Don Nix. "In Nashville, there's a lot of good players. I have a couple of friends who are drummers from Chicago, and you can get that person to play real well on a session. But you can't get anybody to play with him. You've got to get somebody he doesn't know from Alabama to play guitar and a guy from Oklahoma to play bass, and the day they get in the studio they have nothing to do with each other. Although they play real well, it just don't click in the studio."

"Booker T. and the MGs was a perfect example of racial collision—four men who under normal circumstances would not have known each other, much less worked together in ensemble like they did," says Jim Dickinson. "The American rhythm section are guys who play golf together."

But they played golf together in the right town. "If I go to work now in Nashville, if I'm supposed to be there at ten o'clock, I can figure that at 10:01, we're going to be playing," says guitarist Reggie Young. "They've got three hours, and I get paid for three hours, so there's no messing around. At American, we weren't on a schedule, which means nobody really had to get in a hurry, till whoever was producing was in a mood to do that. We didn't have any really set time that we would come in every day. But when we did, most of the time we didn't know when we was gonna be leaving. It could last all day, all night, and all the next day—whatever. There wasn't a clock running. We'd go in a lot of times, set around, talking, talking, talking, which is cool too, building up to the point where, Okay, let's get serious."

Serious meant different things to different people. During one particularly crazy session, well into several days of amphetamines, they were trying to find the right keyboard sound to go with a song. Spooner, so the story went around Memphis shortly after it happened, finally spoke up: "I know the exact sound we need for this," and he walked out the door. The band waited a little while—they were in the middle of a session—and when Spooner didn't come back, they went ahead without him. It was days later, some say two weeks, when he came walking back in the studio door and he had in his hands a toy xylophone. His fellow players said, "Spooner, man, where've you been?" And he said, "Well, we were cutting that session and I knew just the right sound. I'd seen this in a store window in L.A. somewhere, and I went back out there and walked all over the place till I found it."

The late 1960s was the age of amphetamine. There was a lot happening, and people didn't like to sleep for fear they might miss something. "I guess speed was cheap and easy to buy and it was the drug of choice, although a lot of people were going the other direction with downers," Spooner Oldham says. "Myself, I saw more of everything during that period than I have the rest of my life, combined. There were a lot of days when a lot of work was done without anything, but there was a lot of writing done on amphetamines. Especially when one chose to write in the evening after working all day. Stay up all night, that seemed to be the thing to do at the time."

"People wanted to stay up," says Herbie O'Mell. "I was managing Dan Penn, and he and Wayne Carson Thompson, who'd written 'The Letter,' were over at our studio and had been there for about four days. I was coming and going. I would look in and say 'How y'all doing?' And they would go 'Fine,' and that was it. On the fourth day, I looked in and asked how they were, and Wayne fell off his chair. Dan said, 'Oh no, man, you ain't quitting on me now.' And I mean Wayne was out. Dan picked him up and took Wayne's belt off him and tied him to the back of the chair and he said, 'You're finishing this song with me.'"

O'Mell, who had begun the decade running the Penthouse, where Ben Branch integrated the stage with Duck Dunn and Charlie Freeman, ran the hangout for people with shrunken pupils. It was called TJ's. O'Mell and Chips Moman were longtime friends. They'd done bid'ness together. One of American's clients was the Scepter label. Dionne Warwick and B. J. Thomas were from that roster and had scored hits at the studio. When Scepter sent a struggling pop musician named Ronnie Milsap to American, the jigsaw puzzle that was occupying O'Mell fell into place. He'd been eyeing a failed club in Memphis on property owned by some friends; he and Moman were looking for a project to combine their efforts. With Milsap, they had the missing piece.

TJ's had been established by two out-of-towners from a Vegas-style touring show who played Memphis and liked it. T and J returned, and the club was their attempt to bring Vegas to the locals. "They went bankrupt," says O'Mell. "Everything stayed intact and the property owners said if I wanted to open up, pay a month's rent and go ahead. Everything was there, including the sign." Milsap had been in Georgia and was not happy with the direction his career was taking. "I guaranteed that he would make sixty thousand dollars a year," says O'Mell. "At that time he and his band were making about six hundred dollars a week and were living in Georgia playing the Playboy Club in Atlanta and different places. So he moved here, I became his manager, and I got TJ's going because I needed a place to base Ronnie. Moman signed him to a production agreement and we cut records on him over there. I got him gigs all over the South."

The combination worked all around. TJ's became a success. The inside of the club was nice—it had been built for customers expecting a touch of Vegas—and salesmen became steady clientele, buying dinner for clients without having to spring for the fanciest place in town. The weird musician shit happened after they left, or in a separate area of the club, or too subtly for them to notice—except when their attention was caught by flying bottles or a fistfight. O'Mell's years in the Memphis music world—he'd begun promoting dances during Dewey Phillips's glory days—made musicians feel at ease in his company. Moman and his entourage—Dan and Spooner, and whoever else happened to be in town working with them—they could hang out in TJ's without feeling too public. Stax soon moved its corporate offices upstairs, enhancing the insiderness of the club. People from all the studios—and there seemed to be more every day—caught up with each other there. Songwriters coming down off three days of work could nuzzle a bottle of whiskey while eating a steak. TJ's was the kind of place where Dan Penn could turn around at his table and see a stranger removing a handful of pills from his pocket, reach over and take some. "You don't have any idea what you just took," said the stranger. "Don't matter," said Penn, "I just want to ride along with you."

"I must say I frequented TJ's quite often," says Spooner Oldham. "Dan [Penn] and I would sometimes go there and, when we were gonna be in the songwriting mode, have our pencils and write on napkins, tidbits and ideas, and leave and try to finish 'em. TJ's was quite an environment for creativity, not to actually write a song there, but get information out of the air. It was people talking and listening to music. A lot of times we had been working in the studio the day before or maybe that day and it was like a relaxing chatter kind of environment."

TJ's was established before Memphis sold liquor by the drink. With beer and setups, they were supposed to close at three A.M. But when the music was going good and everybody was having a good time—that is to say, regularly—O'Mell would lock the door at the closing hour and not let people in or out until it was over. "We'd stay until five, six, or seven o'clock in the morning, just listening to the music," says O'Mell. "It got to be the place in town. You just never knew."

O'Mell remembers one night when Charlie Freeman came into TJ's and a guest band, "somebody like Creedence Clearwater," was onstage. "Charlie said, 'I'm gonna play.' I said, 'Charlie, you're too messed up, you can't go up there and play.' He said, 'I'm gonna play.' He went out to his car and he got that blue guitar of his, I'll never forget it. I said, 'No, no.' And he said, 'I'm gonna play!' So he walked over to the stage and he sat on the steps but he didn't bring his cord. He must have played for forty-five minutes, sitting on the steps, not plugged in, just doing everything on every song. He came back and he sat down and he said to me, 'I told you I'd burn 'em.'"

Arena bands knew to come to TJ's after their shows. "I had Three Dog Night and Dionne Warwick and Steve Alaimo and Roy Hamilton and Creedence Clearwater—you never knew," says O'Mell. "Musicians worked late, and whatever time we were open till, we served food. We didn't have a big menu, but you could get a good steak on the grill, a baked potato, a vegetable, whatever we were cooking. We had frozen lobster tails and roast beef, too. Chips would send his people over and the band'd be playing and it was one of those things. Oh yeah, well I'll get up and sing a song. That's how Dionne got up. And Steve Alaimo was there and got up with her. You just never knew. I had everybody in there from Elvis Presley to the Bar-Kays and Brother Dave Gardner."

"The bartender was a guy named J.," says Danny Graflund, who frequented the place, "and he had a connection with all these pharmaceutical salesmen. Back then it was loose. We were buying sealed bottles, thirty to a bottle, of Ambar twos, Desoxyn, ups, downs, whatever you wanted. Thirty bucks a hundred. J. bought a grocery store, bought his parents a farm, he was making more money than Herbie. And Herbie was cooking. J. collected empty half-pint bottles, and he would pour rotgut whiskey in 'em, squirt a little simple syrup on top, and guys would drift in from the Naval base outside town, ask, 'Know where I can buy a bottle?' J. would say, 'Well, liquor store's closed but I can sell you one. Taste it.' He'd act like he was breaking the seal, and the guys would go, 'Oh yeah, that's good.' He'd get five bucks for a dollar thirty cents' worth of hooch."

Ronnie Milsap had everything to do with the success of TJ's. He became a club phenomenon, playing an interpretation of popular hits that continues to awe people today. His full and rich voice was outstanding and nothing like what one expected to hear in a club. Though he later became famous as a country artist, his repertoire then was varied, drawing from contemporary hits in all the genres. The audience shared in his excitement as he mixed diverse sounds, working out his act in public. "Ronnie thought that he was the white Ray Charles," says O'Mell. "He could do anything. Anytime a new song came out and he would do it, you'd swear it was the artist singing. It took me a good while to get him to find his own voice. Way I did it, they were rehearsing one day at the club and I walked in and said, 'I want you to put these three songs in the act,' and I gave him three singles. He said okay and I went home. About an hour later he called me and he said, 'About these three songs?' And I said, 'Yeah, they're not hard songs, you can learn 'em can't you?' And he said, 'But they're girl songs.' I'd given him voices he couldn't copy. I said, 'So sing 'em,' and that was the start of getting him into being Ronnie Milsap."

I am familiar with Milsap's Nashville success, and with his 1971 rock and roll album. It's hard for me to imagine what he sounded like in the clubs, but from the impression he left on everyone who saw him at that time, it's obvious he was powerful. Almost to a person, they mention "MacArthur Park" as part of his set. He apparently packed a punch with the cake left out in the rain. "Windmills of My Mind," "Why Don't We Do It in the Road," "Roll Over Beethoven," he lacked nothing for diversity. His show was taped one night, and a few years later, the recordist played it for him. His drummer said that Milsap, by then a country star, never got over how good he sounded and was willing to pay six figures for a copy of that tape.

When Milsap's guitarist at TJ's was off for two weeks, one of the substitutes was Laddie Hutcherson, from the LeSabres and the Guilloteens. "I'd heard Ronnie and I idolized the guy," says Hutcherson. "I went down there and I said, 'Ronnie, I've heard you a bunch of times but I don't know a lot of the stuff you do.' He said, 'What kind of music do you do?' I said, 'Blues and R&B,' and he just made me feel comfortable and played blues and R&B all night long. Real magic night. He sang 'MacArthur Park,' 'Lay Lady Lay,' whatever he did, he made it his. I heard a Dodge commercial that Ronnie cut in Memphis once. It should have been a hit."

In a time and place where anything could happen, where the impossible seemed always occurring, a new rock star emerged. He didn't sing or play an instrument, he never released an album, but he achieved the stature and renown reserved for youth culture icons. The Rolling Stones' bad boy without the music and with triple the attitude, ruffian Campbell Kensinger was a landmark figure at TJ's and later worked for Chips Moman as a bodyguard. "I'll tell you exactly how I met Campbell," says Herbie O'Mell. "A fight broke out in TJ's one night and some guy was coming at me. This other guy that turned out to be Campbell went and grabbed him and physically restrained him. The guy was struggling with Campbell about a yard from me. Campbell says, 'I need a job, do you need a bouncer?' I looked at him and I looked at the guy he was holding. I said, 'You're hired.' He turned the guy around and threw him out. Campbell was a smart fellow, just a little nuts."

CHAPTER ELEVEN

Extreme
Realizations

IN THE BRIEF AND TROUBLED LIFE OF CAMPBELL KENSINGER, VIETNAM was not the source of his rage. The military just gave it shape. Taking note of Campbell's skill in hand-to-hand combat, the U.S. government transferred him to Hawaii, where he trained Marines to kill human beings with their hands. Later, in Memphis, he would revive the classes for his biker gang.

Campbell's fury began at the age of four and ended in a midtown Memphis apartment in 1975 when he was thirty-two, crawling across the floor toward a shotgun, seven holes in his body from a nine millimeter Luger—five in the torso, one in the neck, one in the head. When the police arrived, long after another human would have expired, Campbell realized he would not live to reach the shotgun, would not live to exact revenge on his killer. But within his reach was a telephone, a chance to make a last statement, and he hurled it at the cop standing in the doorway. He growled, then he died.

This from a man on barbiturates.

In the last years of his life, Campbell Kensinger embodied rock and roll, its darker side. He did not play an instrument, he did not sing, he did not have a band. But the lifestyle that radio had hinted at late in the 1949 night found its most extreme realization in the person of Campbell Kensinger. His theatrics extended the proscenium to the edge of death, and then further yet into the darkness. He was completely unfit for society and somehow at the center of it.

The seething ember of Kensinger's character was his size. He was big and strong, but not stocky enough for professional football. Playing pro had been his father's dream, and he wanted his son to realize it for him. Instead of forcing piano lessons on his four-year-old, the senior Kensinger gave the tot a weight-lifting regimen. In contrast, Campbell's mother maintained a large walk-in dollhouse in the backyard.

Campbell grew up in rural East Memphis. Among his neighbors were Jim Dickinson, Jimmy Crosthwait, and Danny Graflund. "First time I met Campbell," says Graflund, who was tough like his friend, "I was at Jump for Joy, this trampoline place near our neighborhood. One night a guy pulled up on a moped, and in his hand he carried a bullwhip. He had a buzzard tattooed on his right forearm that his dad had taken him to get when he was twelve. All the elements were right there. That was Campbell."

In high school, he became a touted athlete, written up in the newspaper. Headlines: "Kensinger Does It Again." Though he was hardy and strong, he did not have the bulk for college ball, for the pros. The Marines was a way of escaping his father's grip. In Hawaii, Campbell married and had a child. His wife wanted their son to have a Christmas tree and Campbell said no. She bought a little decorative tree and put it on top of the TV. When he came home and saw it, he whipped her with it. Three nights later she came back to him, woke him to say she was home. He said, "I told you never to wake me at night." He pulled a .45 from beneath his pillow, put the clip in it, and she ran out the door. "I was standing in the kitchen door," he told Graflund, "popping sand around her feet as she ran."

The Marines discarded Campbell, found him too far gone to be reconditioned for society. Figuring he'd soon effectuate his own demise, they turned him loose, section eight, mental. His first night back in Memphis, he went to his parents' new house in a posh section of East Memphis. He'd had no recent contact with them. It was snowing. He knocked on the door, his father answered in his pajamas. "Hi, I'm home." His father said, "What do you want here?" Campbell kicked his way in, broke open his father's gun cabinet and then the liquor cabinet. His parents ran next door to a neighbor's house and called the sheriff. Campbell told Danny, "I was sitting there in that entry hall in a chair with a bottle of whiskey and a shotgun across my lap. These two sheriff deputies come walking up the driveway and rang the doorbell. I let the action go on that .12 gauge. *Keewanggg*. All I saw was assholes and elbows."

He held them at bay until morning. Graflund: "That's the kind of thing that hurts. That's Campbell saying, 'This isn't fair. I've done nothing wrong except I didn't play for the Green Bay Packers.'" Campbell remained in Memphis but had no more contact with that family. He would name the biker gang he formed around 1972 the Family Nomads. The Family for short.

"By the time everybody in Memphis got to know him and he was like a rock star, head of what was like the Hell's Angels in Memphis," says Graflund,

"I really think he had been hurt so much in his life that he would not take friends, and he would be violent to keep people away. He wanted people at an actual distance. He didn't fear physical violence, but he thought, If I don't ever let you close to me, you can't know me, you can't hurt me. He had been mentally hurt to the point where he put a shell up. You'd have to have grown up with him to know what kind of warm person he really was."

Campbell began working for Herb O'Mell at TJ's in 1967. After the incident with the angry patron, Campbell and Herbie learned they shared several mutual friends. Herbie needed Campbell's help at the club, and Kensinger was impressed by Herbie's range of acquaintances and by his influence; O'Mell seemed on a first-name basis with every person in the Delta triangle. He was—and remains—a man who makes things happen.

There was a waitress at TJ's whom Campbell, divorced from his first wife, wanted to marry. She had been a Vegas showgirl. Graflund went in one night, they were sitting at a table, and Campbell was showing off. He had given her an engagement ring that had a tiny diamond in it. "He was proud. He was a whole new person. He was shaking peoples' hands. He was happy." Graflund continues, "I was also with Campbell the night she threw the ring down and said, 'That little piece of shit you gave me, motherfucker, I wouldn't marry you.' And she left with another girl to go to another club. We went outside and they were driving off in a little yellow Volkswagen bug. The first thing his fist hit was the windshield. In my mind I flashed, Not the windshield—that's the hard part. But the windshield went smash, and he kept hitting it till it broke through. He beat all the windows out, was beating on the car, screaming, hollering, crying, enraged, hurt. 'How can you do this to me!' He wasn't trying to hurt her, because he could have if he wanted to, he was just beating on the car. It seemed like hours, it may have been seconds. Someone got Herbie, Herbie came out, he hovered behind him, says, 'Campbell. Campbell.' Even Herbie wouldn't try to grab him."

Although Campbell shared a longtime trust with Graflund, and had a tight running buddy in another former serviceman, Frank Strausser, his relationship with Herbie was unique. "We were in TJ's one night and a guy walked in whose ass he'd been wanting to whip," says O'Mell, a man you would not mistake for a fighter. "It was just at closing time and Campbell said to the guy, 'I'm gonna whip your ass.' I said, 'Campbell, you can't do it. He's a customer in my place, and you just can't do it.' He said, 'I'm gonna whip his ass,' and I said, 'I tell you what you gotta do. You gotta whip mine first, you gotta go through me to get to him.' He swelled up and he got mad and he picked up a chair and threw it through my front glass door, turned around and said, 'Fuck you,' and walked out. I went after him and said, 'Campbell, come here!' I said, 'You son of a bitch, you broke this glass. Let me tell you something, I ain't got no glassmaker, you gotta stay here all night until I get a glassmaker. You gotta protect this place for me.' Campbell goes, 'Oh, okay.' And when I put a new glass in, I made him pay me five dollars a week."

Campbell was at another club one night when he saw another person who needed a taste of justice. As if psyching himself up for the pounding, he swelled his chest and raised one arm high, then the other, announcing, "In this hand I have life and in this hand I have death." Then he paused. "I am the awesome anaconda." O'Mell had seen him do this prelude before, but this night he bent down and bit a piece of the vinyl upholstery from off the edge of the bar. "I said, 'Aw man, Campbell,' but it was too late. He bit through the guy's shoe, and the guy started screaming. Campbell got up and whipped this guy unmercifully. When he was laid out, Campbell went up with the heel of his boot and stomped him right in his face. I saw it. Campbell was that way. He just would go off, but not on his friends."

Campbell's crank wound a little tighter when he began running with Frank Strausser. One afternoon in an East Memphis bar, Graflund was drinking beer when two of his pals burst in the door, knowing they'd find him there. He had to come right away if not sooner to a Midtown bar called the Toast because there was a guy just back from Nam who claimed he could beat anybody arm wrestling. Danny was comfortable inside and it was hot outside and he was in no mind just then to tangle with the heat. It took a few more beers to loosen him from his roost, which made it time anyway to move to the part of town where the general population was thinking about getting tight.

The upstart arm wrestler was Frank Strausser, fear incarnate. Strausser got wired on his own adrenaline in the Vietnam jungle, and the chill rush of danger in his lower backbone had become as necessary to him as air in and out of his lungs. Upon his return to Memphis, he'd become a cop, the paramilitary uniform and the weapon a sort of methadone. He was yanked from the dangerous Beale Street assignment after the pimps complained that he'd muscled in on their business. East Memphis responded poorly to his being assigned there, and ultimately he left the force. Unable to depart from the streets, he continues stalking the night as a cab driver. "When I get in that cab at three A.M. and I hit the road and I see the lights, I almost get wet," says Strausser. "I've had little punks in my cab who try to beat me out of a fare. I pull over and I face them in the back, say, 'You're giving me a choice. Either I eat the tab, or I shoot you as you're running away, take twenty bucks from my wallet, and put it in your hands and say you robbed me.'"

The tattooed Strausser may have been a grenade with a loose key, but Graflund had the bulk and pinned his arm that afternoon at the Toast. Three days later they were coming down from a run, bonded for life. Strausser and Kensinger become the pair thereafter, but Danny's place was respected. Drinking at TJ's, the two would rise and Campbell would growl at Danny, "You don't want to come with us, we're going to do a man's drug," and they would stroll off to inject amphetamine into the backs of their hands. "Strausser gets this thing when he's speeding," says Danny. "He grits his teeth and you can hear it across the room. When that would start, everybody knew he's about to flip. He'd be getting to a level where he focused on one thing: Punishment would be meted."

"They weren't the type that'd say, 'Let's go out tonight and whip somebody's ass,'" says O'Mell, "but they'd always find somebody. They would go to these places where all the boosters and criminals would show up at one or two in the morning. Thieves out robbing, stealing, breaking in houses. It'd be funny. You'd see one of Campbell's friends, and you'd be saying, 'Oh man, I saw the greatest-looking pair of shoes or gun or bowling ball,' and I remember one of these guys turned around and said, 'Well, I could get one of those for you. Could you just tell me where that item is stocked?'"

"Campbell and Frank got where they were popping that speed hard," Graflund continues. "They would go out to a place on Jackson, the old Broken Wheel, real late-night bar—off-duty policemen, waitresses, whatever. They would go out there and clean that place out, tear it up. They'd get in a fight, get back-to-back, and just fight their way out. Campbell said to me, 'That Strausser, that Strausser, he's alright. We took care of that place. One guy, I had my finger in his eyeball and I was swinging him around the room, threw him over against the wall. I'd have two of 'em and be backing up a little bit, Frank would finish one, turn around, and help me out.' And that's the way it was—'He's alright.' Each was a nice second to the other. Frank ended up marrying a girl I used to live with who was a waitress. They had a fight, got mad, so she fucked Campbell. Then she told Frank, 'Well your best friend came over here and screwed me.' So Frank went to TJ's and wanted to know if that was true. Campbell said, 'Hell yes it's true. If I was you, I'd beat her ass.'"

The American Studio crowd got to know Campbell through TJ's. Imagining a room that could contain a presence so delicate as that of Spooner Oldham's—a spirit—and also contain Campbell may seem scientifically impossible, but that is the kind of place TJ's was. Things at American had taken their own peculiar twists. After Martin Luther King's assassination, guns became common among many of the studio personnel. "Little handguns, something that could fit in a briefcase," remembers one staff person. "I had a little old pistol, something you'd buy over the counter at the 7-Eleven at the time. It wasn't nothing, but it gave me something sort of secure to have. Really, I didn't think it was that bad then. Today it would be a whole 'nother story." Like the radios and motorcycles that American got into, the guns became something to outdo. "I was at American one night when Chips took everyone outside and opened the trunk of his car," remembers a songwriter at the studio. "It was full of weapons. He was showing everybody, 'Hey, look.' I was like, Get me out of here."

Moman eventually hired Campbell as his de facto bodyguard. His duties were varied; it was his presence that counted. He might drop off the laundry for Chips's wife or pick up burgers for the kids; anytime a star was arriving at the airport, Campbell would be dispatched to pick them up in the Rolls Royce. Though he had a beastly side to him, his graces were such that female stars often succumbed to his charms. When Ronnie Milsap went to Muscle Shoals to record his Dan Penn–produced debut album, Campbell was dispatched with the entourage. Unsure how the locals might react to the hippies, he accompanied the wives to the laundromat. Several of the women began passing the time by embroidering, and Campbell asked for his own needle and thread, neatly sewing a row of tombstones around the bottom of his jeans.

Jim Blake, a former marine whose head shops, underground newspapers, and independent record labels made him the scene's impresario, had a reasonable relationship with Kensinger. "Campbell always told me, 'I'm in my world, you're in your world. I can come into your world and visit, and you can take me around, and you can come into my world and visit and I'll take you around. But don't ever try and be a part of my world because then you're playing by my rules—not yours and not anybody else's.'"

"Campbell and I were very close, sort of like opposites attract," says the wispish Bill Eggleston, photographer. "I don't think many people know, he had exquisitely beautiful handwriting. I first heard of him around the time of *A Clockwork Orange* [1971] and I had a mental picture that proved entirely wrong once I got to know him. We practically didn't even have to talk to each other, we got along perfectly. Sometimes he helped me photograph. He would be aware that I was after some kind of new strange bizarre picture, and it was as if he would impresario it and suddenly the event would fall into place. For both stills and video, he was a great help."

"He understood his dilemma," says Marcia Hare, who came to know him well in the last years of his life. She was running with Eggleston in the early 1970s, indulging in downs, and Campbell was employed to drive them around and keep them from getting rolled when rolling around was all they could do. "We'd all be sitting around somebody's house and Campbell would say, 'Excuse me,' and go out the back door. The guys in the living room would explain that he was going out there to get rid of it. He'd come back in and say, 'Sorry man, I tore the doors off your garage. I just didn't want to hurt anybody here.'" Eggleston says, "He kept saying, 'I'm trying to quit.' He meant killing people."

"He had been in the service and been trained to be bad and mean," continues Hare. "I don't think he had any skills other than killing people, and anything you get in the habit of doing is real hard to quit. I learned that from being a tittie dancer all those years. Hustling customers wasn't what I loved, but I'll still sit here, retired, and imagine myself hustling customers because I got in the habit of doing it. Campbell was good at killing people. It was something he did the best and you miss doing things you do well."

"Campbell was a watershed experience," says Randall Lyon, the scene's arch-hippie. "One day he came to my house. I was strung out, really high on narcotics, and Campbell shows up. Real scary guy. He sits on my bed, opens up a briefcase. On one side there's a .38, on the other side there were some Vietnam poems, and he wanted to read them to me because he said I looked like the Canterville Ghost. From then on, Campbell and I had this wonderful relationship. You could go into a place and sit down with Campbell, or if he came and sat with you, you were not going to be fucked with by anyone. Magnificent guy. Brilliant guy. Violent person. You could never tell what he was going to do. He was totally unpredictable."

"When Campbell entered a room," says scrawny Jimmy Crosthwait, "it quickened your spirit. Everything got a little more immediate. If he was sitting at a booth with me, there was always in the back of my mind the possibility that somebody was going to come in to waste his ass with a shotgun and take the booth with him."

Campbell attended parties that were not for the soft-hearted, not for the humane, and really not for human beings. "The group of people around Campbell had parties," says one acquaintance. "Hellbent for leather. They would show me Polaroids of the violence. It was intense. Way over the edge. It was like confessions of a serial killer." At the same time, he had become such a figure around town that knowing him was considered insiderly and hip. "People he didn't know were always inviting him to their parties," says Graflund, "because knowing him became cool. When he'd show up and be Campbell, they'd go, 'Oh no, I didn't think it was going to be this.'"

Reconciling Campbell the beast with Campbell the poet may be impossible for those of us who never knew him. But insight into his character can be gleaned from his friends, people who have a taste for the grit that produces pearls. They saw a reflection of themselves in Campbell, the balance of good and evil distorted by the indiscriminate flux of society. One can only wonder at how resplendent his beautiful side was—however small—that it could counter the monster.

Moman moved his studio to Atlanta in 1972. Campbell stayed in Memphis. O'Mell was running a beer joint where Rita Coolidge was a waitress and Campbell worked there and in a few other bars before ending up at a hot spot called Trader Dick's that featured Tony Joe White, Larry Raspberry, Keith Sykes, and several other soon-to-be national talents. Liquor by the drink had arrived in Memphis in 1969, and the city was still giddy. "The tops came down, the skirts flew up," says Sid Selvidge. "God, I can remember Jimmy Crosthwait out bouncing around on one leg in the middle of the street telling a policeman he hadn't had too much to drink. It was a nuts time."

"Three guys were hassling me in Trader Dick's one night," remembers Crosthwait, "and one said he had a brother who was a policeman in Dallas. I said, 'As far as I can tell, you guys could be fucking cops.' He turned to me and said, 'We could be.' One guy whipped out a little gun he was carrying, a five-shot .22. I suddenly had a real intuitive glimpse of the connection between the Memphis and the Dallas police departments. It's no small coincidence that Kennedy and King were killed in these two cities and there was sort of police help in both of them."

Quaaludes, a horse pill of a downer, had become the rage in the early 1970s. Hardcore partiers in Memphis were into rubberlegging, getting so fucked up they couldn't walk. Many times in Trader Dick's, more of the clientele was moving about on its hands and knees than on its feet. "Campbell was the bouncer there and expected ladies to be ladies," remembers Marcia Hare. "If there was ever some little chick drunk and showing her ass in Trader's, he'd slap her. I always asked Campbell for permission to get wild in there. I'd say, 'Campbell, can we throw all the ketchup bottles out the window?' He'd say, 'Sure, do anything you want.' And we'd do real juvenile stuff. But with Campbell's permission."

Trader Dick's became more and more like a private party. Campbell wrote a song while working there, and when the mood hit, nightly, he'd jump on stage, take the singer's mike and begin: "I am Trader Dick/I will make you sick/I'll beat you with my stick/You dirty little prick." (He and Jim Dickinson intended to cut the song in the studio as a jingle.) Working the door, he would refuse entrance to those he didn't know, and beat up anyone who

Jimmy Crosthwait. "It was a nuts time." Photo by William Eggleston.

didn't like it. "He might let in a couple suckers," Marcia remembers. "They didn't know the honor of being let in was only temporary. He would beat them up later. It got to be where other big guys would come to Trader's to challenge him. Everybody was getting hurt pretty bad out front. If he'd kept that job, he might have gone to jail a couple times but he might have had a chance to adapt to America. When he quit, he got into the biker stuff."

As leader of a biker gang, he no longer had time for jobs. He converted his home into a military-style bunker, the windows covered with chicken wire to keep out grenades. He told his gang, "Carrying a gun isn't going to do you any good unless you know how to use it," and he acquired a bazooka, anti-tank guns, automatic weaponry, and other heavy arms, organizing field trips to the country where he could militarize the bikers. President Nixon visited Memphis in 1973 and the FBI called Campbell and told him their planned route—different from that which had been publicized—and asked for his word that none of his gang would be in the vicinity. He said, "Fine, I got no problem with Nixon."

"That's where he had gotten to," says Graflund, "on the level of, Are you bad?"

Danny Graflund was with Campbell a couple days before he got killed. They went to the apartment where he would soon meet his fate. Campbell threatened the occupants, also bikers, forcing them to give him a handful of pills. When he had eaten all those, he returned, demanding more. It was three in the morning and James Townsend, thirty-three, told reporters that he awoke with Campbell sitting on his bed and was promptly beaten with brass knuckles. Mack McCollum, twenty-six, heard his roommate's cries, knew it was Campbell, and armed himself in his bedroom. When the door opened, he fired. Campbell was hit in the stomach but he did not fall. He took four more bullets in the torso, one in the neck, and one in the head and, though looped on pills, he still crawled across the floor toward that shotgun, still threw the telephone at the Man, still growled. Then he died.

Shortly thereafter, Mack McCollum was gunned down in the parking lot of Peanuts Pub, where Furry Lewis played every Sunday night. The weapon was a nine millimeter Luger. McCollum took five bullets in the torso, one bullet in the neck, one in the head.

The realization of youth power was in the early 1970s air, newfound dimensions, possibilities. Everybody was pushing their own envelope. The Memphis band Moloch, gripped by the blues, worked toward making the music they loved attractive to fans of Hendrix, of Marshall amps, and of outrageousness.

Named for the Babylonian god of chaos, one appeased by the parental sacrifice of children, Moloch was even more difficult to appease, grappling as they were with a musical form that was half a decade away from being acceptable.

"We formed Moloch before heavy metal and stuff like that," says Lee Baker, whose Blazers had faded with pop music's innocence. "We wanted to be loud, rockin' rock and roll, and offensive—grab peoples' attention. Now we'd probably be called heavy metal, but then we really didn't have anything to go by. There was the Yardbirds, but they were peers of ours. It was all happening at the same time, the growth of hard-edged rock and roll. It had been rhythm and blues and blues, and the Stones and the Beatles, and then it got a little harder and a little faster. A deejay gave us that name and we tried to have, not a real bad-guy image because nobody was bad guys, but we had a lot of fun with it. It was sarcastic. 'Yeah, we eat babies. That's what we do.' And people would actually be shocked when they understood the Biblical connotations of the name. It would gross them out. So that was good."

Baker met the charming Eugene Wilkins at a party at Jimmy Crosthwait's house in 1968. Wilkins had a theater background and had returned to Memphis wanting to merge those leanings with the burgeoning eccentricities in rock and roll. He had a good, strong voice. Baker introduced him to a band he'd been jamming with, led by Tarp Tarrant, who'd been Jerry Lee Lewis's wildman drummer. They were stone hippies. They became the band that ate babies. Moloch maintained a house that was a den of escape, where there was always room to sleep one more. Danny Graflund found refuge there and became the band's roadie. He hauled equipment and threatened shady club owners, his presence intensifying the band's bad-boy image. Moloch's other roadie was Randall Lyon. Though he was also built large, muscle was not his strength. Lyon was a junkie, a cook, and very entertaining. Michael "Busta" Jones, who became Moloch's bassist and lived in the house, describes Lyon as "the band's personal guru." Randall did not hide his homosexuality, which made some of his peers uncomfortable. Lee Baker, however, always made Randall feel welcome. "Lee was another person that would sit around and play guitar a lot," says Lyon. "He had some really cool Fender guitars and that was the most excellent thing, to hear somebody who could really play. When I was the roadie, I came on to Fred Nicholson, the organ player. He was born with this talent and while he was really good, he was just coasting on his natural ability. He did model airplanes. I said to him one night, 'Hey, you're here, I'm here,' and Fred came into existence for about twenty seconds and he fainted. Uh oh. Bam."

"Randall would have been a killer songwriter if he ever wanted to work in

Randall Lyon. "You had to respond to it." Photo by John McIntire.

that sort of frame," says Baker, "but he wouldn't be disciplined enough to make the verses come out like they should." His prowess in the kitchen had been established in the early days of Beatnik Manor, and when he didn't spend the food money on smack, he fed the band well. (When Leon Russell was spending a lot of time in Memphis, a battle of the bands was discussed between him and Moloch. An alternative was devised, though never carried out. Leon's entourage included a black cook named Miss Emily, and plans were made for a bake-off, Miss Emily versus Randall.)

Lyon remembers that he had no idea who Isaac Tigrett was when he woke up one day and found him living in the house. Tigrett, whose father's company invented the Slinky and the Happy Drinking Bird, later reaped his own when he founded a restaurant chain called the Hard Rock Cafe. "I have no idea how he ended up in the house," says Lyon. "It was a really weird mix of people, and we were all trying to do the boogie. One day Tigrett said, 'I want to take everybody out for lunch.' I look out the window and there's a 1935 Rolls Royce. We get in the Rolls and drive to his family's country home in Jackson, Tennessee. Goddamn! I mean I was sleeping in a closet."

(The success of the Hard Rock Cafes puts Tigrett in the ranks of Memphis's finest entrepreneurs. The future's celebrity, he is a rock star businessman, a good old-fashioned dealmaker who has taken the business out of conservative suits and dressed it in collarless Indian shirts, sharkskin, sunglasses, and a beard. His capacity for marketing has reached new heights with his latest venture, the House of Blues. A committed blues fan, he is erecting mock country juke joints in urban environments, promoting blues and blues-influenced music. With its extensive merchandising, the House of Blues offers itself as an image, and urbanites from Los Angeles to Tokyo can buy into the authenticity of Mississippi tarpaper shacks without ever riding on a dirt road. It's clean, it's easy, and it's elite. And it's so ingenious that Harvard University is one of his backers. Tied in to the promotion is an educational program entitled Blues in the Schools, attempting to rescue the blues from the junk heap of overlooked Americana. What Furry Lewis taught remains applicable today, even if someone else is to profit by it.)

Moloch's popularity was increasing at the same time as Don Nix's influence at Stax's Enterprise label. The band was signed by Don, and they recorded at Ardent on National. *Moloch* is a beastly-sounding blues-based swirl, the musicians trying to forge new ground with an old formula cranked up to the biggest and the baddest. The album holds a minor place in blues-rock history as the first recording of what's become a standard, "Going Down," since recorded by Joe Walsh, Freddie King, Jeff Beck, and a number of others. Lee Baker calls Moloch's album "a pretty fair representation of what Nix wanted us to do." Graflund refers to it as "the Don Nix Don Nix Don Nix album." On the back of the jacket and on both sides of the album's label is written, "Producer: Don Nix. Arranger: Don Nix. Engineer: Don Nix."

"We went in the studio with no tunes, not even a whole band," says Baker. The process was begun with drummer Phillip Dale Durham, Fred Nicholson on organ, and Baker. "We'd run the tracks down, I'd go back and play bass, put more guitars on 'em, fool with 'em. And we made up these songs. None of us knew then that publishing was where the money was, so we didn't care whose name went on 'em. 'Going Down' was put together in the studio. Nix had a line, we worked the rest out. All the stuff was done like that. The album was pretty much a blues album but it was heavier. We were influenced by Hendrix, like any player who heard him at the time."

Don Nix tells the story of the album's most famous song: "I didn't write 'Going Down,' I got drunk and made it up. I was living in an old apartment on Poplar Avenue in Memphis, sitting with my foot propped up in an open window in the summertime with no air conditioning. I fell asleep and fell two stories out this window into a garbage bin with paper and things. It didn't kill me, but that's when I wrote the song: 'Got my big feets in the window, got my mind down on the ground.'"

Proud of its musical debts, and hoping to give a break to a friend and influence, Moloch brought harmonica player Johnny Woods into the studio. "Gone Too Long" opens side two with Woods talking in his thick Mississippi accent, something about selling his cows, something about maybe needing to move in with his step-daughter, and then the band pounds out its pre-metal blues. Woods's harp fits in like he'd been born and raised in the band house.

By the time they were recording the album for its 1970 release, Baker had been a professional musician for a decade, and several other members had been through the music industry mill. Where management might have helped, they refused to let an outsider shape them. "I figured I knew as good as anybody else what was good for us," says Baker, "and I wasn't going to march to some other person's thing. We considered ourselves players and we weren't going to do anything to cheapen it." Their 1969 pre-album release tour of the Northeast exemplified that attitude. They were booked outdoors at the New York State Pavilion, built during the World's Fair, on a bill with the MC5 and the Stooges featuring Iggy Pop. The gig had come from a well-connected manager who wanted to sign them.

"There were some groupies hanging around backstage," says Baker, "and this manager wanted us to screw these girls onstage. He said, 'If you screw these girls'—and they weren't even good looking—'If you screw these girls onstage, you'll have it made. The world will be at your feet.' I said, 'I'm sorry, they're already at my feet because I'm up onstage. Fuck it, I ain't screwing that whore.' I was looking at Eugene, and we're going, Wow man. What we wanted to do was to burn the fucking show, and we did. We got three standing ovations in our set. We got respect and that's what we wanted.

"Phillip Dale would do this drum solo where he would walk around the drums playing 'em, play the floor, play the fucking mike stands. And he would testify. He'd hug himself, be moaning the blues, singing and playing the drums, sing to the drum solos. We did that shit up in New York and they didn't believe it. They'd never seen anything like us. I was playing a guitar solo and someone threw a cherry bomb that blew up in my face. I never missed a note, and when I got done with my solo I walked to the microphone and said, 'You're going to have to throw some fucking dynamite up here to fool with my playing.' And they went wild cheering. It was great.

"Then the MC5 came out and they blew up amps. And the Stooges, Iggy jumped into the crowd and they beat the shit out of him! Shoved cigarettes in his mouth and he couldn't breathe. The band never quit playing, just grinding away, feedback, awful. It was a cool show, real powerful-sounding stuff, but it wasn't musical. And we were into playing music."

"I was always interested in crossover music," says black bassist Michael "Busta" Jones, who joined the band after the album was recorded. "There were definite black bands and definite white bands in Memphis. And the Shell was a place where both could play. That's probably where I first saw Moloch. There was the Bar-Kays of course, but there was also groups like the Brothers Unlimited, a Sly Stone kind of thing with nine pieces. I had a trio with Willie Mitchell's son Hubbie, we were called Black Rock, and we were trying to do that crossover thing, going for a white audience. I was real interested in English music and British rock bands, and I thought Moloch was the closest to something like that. That's what made me attracted to 'em. Eugene Wilkins always had this Jagger-like attitude about him, and the whole band had a Stonesy thing.

"There was one time I was over at Lee's house, we'd just finished rehearsing and we were sitting around. Don Nix called and said he was going to bring this dude by. All of a sudden [Rolling Stones bassist] Bill Wyman turned up at the door with Don. Everybody always tried to keep Lee abreast of anybody who was coming to town. Anybody asking about music, Lee's name always came up. Still to this day."

In the game of racial Red Rover, by 1970 each side stood just about where the other once had. There were white kids attentive to the old bluesmen and black kids grooving on rocked-out blues progressions. "Lee had more in common with the Delta blues musicians than I did," says Busta Jones. "I'd sit around with them and get the feel of it, but there wasn't really a bass guitar involved, it was mostly acoustic. He was really trying to pick the vibe, get the feelings from them. I was more into the electric power of the whole thing. I went on from Moloch to play with Albert King, and later Stevie Wonder, Brian Eno, and the Talking Heads. No one ever heard of Moloch, but I always credited it as my background."

Midtown, with its grand old homes, wide streets, and beautiful trees has always been a haven for the hip. It's where the action is in Memphis today, though compared to the 1970s, it's like living on the penal farm. "I'll tell you," says Lee Baker, leaning forward in his chair, "during the seventies, Midtown was loose. You'd be playing—acoustic even, like with Sid—and these girls would come up, start dancing, and they'd take their tops off. Just like that. Regular old everyday girls just going nuts. But it happened, it happened a lot."

The last of the original blues festivals had taken place in 1970. While that event was being planned, those who had conceived of the blues festivals were creating the first Dream Carnival, held in July of that year, exactly when their blues show would have been happening. The visionaries behind these new events were Jimmy Crosthwait and Randall Lyon, though every participant brought their own color. "There wasn't anything going on after the blues festivals," says Lee Baker, "so we invented these things called Dream Carnivals. Basically they were just gigs, but with a light show and lots of craziness." Moloch had released their one and only album in February of 1970. They were stone hippies playing loud, electric blues bent way out of shape and bent way, way back. For the first few Dream Carnivals, Moloch drew the sacrificial audience.

Dream Carnivals were the culmination of Crosthwait's affection for Bauhaus and their concept of total theater. He contributed his Electric Circus experience, and the literary background that had led him to puppetry. Lyon brought the enthusiasm, imagination, and, even, organization. "The Dream Carnivals started because we needed something to do," says Lyon. "Basically, they were happenings in the classic sense. I was exercising my full Jean Genet theatrical privileges. We had a Dance of Death with drag queens, bringing gays into straight venues for the first time. We broke a lot of rules. There would be a music section, then a reading might be acted out. Tav Falco might perform or a concept artist named Dixie Ashley, or I would sing Memphis

Minnie—the different parts making it like a carnival. We knew people would want to see us, we knew they'd want to look."

The economic and social risks involved made close friendship a prerequisite for the organizers, though they were not without dissension. Gays were subject to exactly what the civil rights movement had spent decades trying to eradicate. "Our circle wasn't homophobic, but they didn't understand," says Lyon. "They didn't quite get it. I was in on a lot of conceptual stuff, but because I was the hip queer, I did not enjoy total confidant status. Here we were on the heels of having the blues shows stolen from us. I said, 'Hey, we've been putting on these shows, why don't we do a Dream Carnival?' We had a meeting at Crosthwait's and decided we'd do one, even though we'd have to scam everything. It worked, and we always had enough money to pay for the next one. So as long as I was able to make stuff happen, I was allowed around. But I was always out front about my shit, and their problem was they didn't know what to do with a hip queer. It says a lot about the whole period."

Their ability to create a circus out of nothing was respected by a range of people, many of whom encouraged them by supplying materials. Seventy-year-old Cora Wooten, whose family had helped start Memphis's first TV station, waited on Randall or Eggleston whenever they came into her electrical supply shop. "Every time we asked for something, she gave it to us whether we had the money or not, and sometimes we asked for a lot. We might need a hundred feet of coaxial cable with obscure plugs on the end. Then a camera cable. Then a case of videotapes. She would make it no problem. She really inspired us too, encouraging us from her level."

When Lyon took the first Dream Carnival poster to Diamond Printing, he was angling on the best way to ask for a break. "Stefan Diamond looked up and saw me and this crazy poster and he said, 'Anarchists! My man!' One look, 'Anarchists! My man!' He was ready to print our shit in a second. He had those numbers on his arm [a Holocaust survivor], he was up for anything that would fuck things up." Radio commercials were created at John Fry's Ardent studio, where most of the musicians worked. They were able to indulge themselves, designing bizarre audio spots that would stand out even on the free-form, maturing FM radio band. Later carnivals held in a venue near the studio were actually broadcast live on one of the city's most powerful stations, dead air and all.

Although Jim Dickinson was in Florida with the Dixie Flyers, the Memphis house band exported to Miami by Atlantic Records, and was not a part of the first one, there is an anecdote he tells that captures the Dream Carnival spirit. It's about the French writer Alfred Jarry, who wrote the play *Ubu Roi*.

"It's all where you draw the line between the audience and the performer," Dickinson says. "Jarry went to the opera one night and he was wearing a suit that was actually paper on which he'd drawn a suit. They wouldn't let him take his seat in the loge, and he said, 'I don't think it's fair that they won't seat me, yet they've allowed the first five rows to come in carrying musical instruments.'"

"Dream Carnivals were an event," says Crosthwait. "The city woke up and went *bing bing bing* for a little while." The first carnival drew creatures of the night from the cracks in the sidewalk, but the human sacrifice was the high society who had come to rub shoulders with the underground culture. The audience was packed into a warehouse behind a drug-laced strip of blue jeans stores near Memphis State University, and the summer humidity had seeped deep into everyone's cranium. "The first one was like we'd been discovered," says Danny Graflund. "This warehouse was packed full of people, society matrons alongside the hippies. And I remember seeing Marcia Hare and Bill Barth bump into each other, start talking, and suddenly Marcia slaps Bill. 'You son of a bitch!' she shouts. Bill had a giant drink with crushed ice in his hand and he threw it all over her, spraying everywhere, and these rich people who had come to slum go, 'My God, we must get back to the country club immediately.'"

Music was the core of each event, but films, light shows, and theatricality were essential. Randall Lyon might wear a dress and sing, "In My Girlish Days," or carry a spear and pontificate, perhaps adding interpretive dance. "Some of his routines were really wonderful," remembers Crosthwait. "He would go into the Guru Biloxi: 'Yessir, the Guru Biloxi, that southern swami, the Mississippi mystic, get the Guru Biloxi holy dome and trailer park in Biloxi, Mississippi. I am the Guru Biloxi and for five hunnert dollahs I'll heal yawl and straighten yer entiyuh fambly and get you steel-belted radial tires for life. You can get the poster, book, the painting, and the record.'"

Gustavus Nelson (later transformed into Tav Falco) created various personae, including Tube Man. While Mud Boy scraped the guts out of Buffy St. Marie's "Codine," Tav wrapped himself in clear plastic tubing, writhing on the floor in a bondage scene. When he was the Three-Legged Man, he wore a fez on his head and had an artificial leg coming out of his fly. "He could manipulate that third leg, dancing," says Dickinson, "and with all of the other shit going on on the stage, it really looked like the guy had three legs."

Mary Lindsay Dickinson watched that performance from the audience. "He looked like the man on the wedding cake, only he had three legs. The two men in front of me were seriously discussing it. I heard them say, 'Do you think?' And the other one said, 'No, it can't be. But I don't know.' I had just taken my first Quaalude and I was going, Wow, I'm not believing this."

The Mud Boy dancers declared themselves at the Dream Carnivals. Chris Wimmer had returned from the Electric Circus and New York and made his regular disappearances. A performance troupe called the Big Dixie Brick Company formed. And the events were perfect venues for Campbell Kensinger. "After a while, he got to be a show himself," says Graflund. "He didn't have to do anything, because he was there. Since he didn't play guitar and he didn't sing, he didn't go out of style. He was what he was and he was the only one like him around." Other sites for these events included the Skyway of the Peabody Hotel, a grand room once reserved for the highest society. The hotel had become dilapidated; when they went to see the space and were told the guests would have to ride the service elevator to reach the roof, the room was quickly booked. One was held at Clearpool, the site of so many drunken high-school nights, and later, when John McIntire moved into what had been the Jewish Community Center, the building's gymnasium became a favorite site. That was a circle of sorts, because the old JCC was near the illustrious corner of Madison and Cooper, where Beatnik Manor had been.

With the blues festivals and Dream Carnivals under his belt, Randall Lyon was feeling semiconfident and semicompetent as as a promoter. Sojourning in Fayetteville, Arkansas, he schemed to make a little cash and invited Moloch to play the small town. Marcia Hare and one of her friends were part of the band's touring entourage. "We dropped a bunch of acid the night before," says Marcia, "and we were really ragged when it was show time. Randall gave us a hit of morphine to try to straighten us out for the show. We got just fucked up. We didn't know that Fayetteville didn't have an age restriction, and we were too messed up to notice that the house was packed with a bunch of twelve-year-olds. We had been in New Orleans stripping, so when the band started, we got up dancing, got carried away on the morphine, ended up laying on the floor and doing all that sexual stuff, taking our clothes all the way off, G-string, everything. About that time the parents were coming to pick up the kids, and they freaked all the way out. They seriously wanted us arrested. We had to hide out, sneak off before the sun came up and get back across the state line. Randall couldn't go back to Fayetteville for a long time. They didn't know who we were, but they knew his name." By 1972 Moloch had pushed blues-rock to an edge that was still years ahead of its audience, and the satisfaction they'd felt was now outweighed by the feeling of futility. After several changes in band personnel, Baker quit. Wilkins folded the group shortly thereafter.

The idea of two guitars, of three guitars, or of a guitar army is not foreign, but the band that applies that concept to keyboards is the exception. (The Band comes to mind.) This, despite the fact that organ and piano are much more distinct sounds than lead and rhythm guitars. At Stax, Isaac Hayes often sat in on sessions, playing piano when Booker T. was on the organ. Cropper was often the sole guitarist. At American, Bobby Emmons played organ, Bobby Woods played piano, and Reggie Young was usually the lone axeman. The Dixie Flyers, the Memphis house band hired by Atlantic Records to relocate to their new studio in Miami, also used the formula.

Stan Kesler pieced the Dixie Flyers together during his tenure at the Sounds of Memphis studio. Kesler had produced several hits and was looking for a house band that could help him attract major artists and produce several more. "I got them to where they could cut a good session," says Kesler, "and first thing you know Atlantic comes along and, I won't say 'steals' them, but offers them a deal and they left. I don't blame Atlantic for it, I don't blame anybody really. I think the musicians could have been a little more loyal. But people do what they think they have to do."

It was during Moloch's tenure that Dickinson was contacted by Jerry Wexler about moving a Memphis rhythm section to Miami. Dickinson had begun playing with guitarist Charlie Freeman when they would gig as fake Mar-Keys and they had recorded together in 1965 as the Katmandu Quartet. Kesler had been recording Charlie Freeman together with bassist Tommy McClure and drummer Sammy Creason since 1967; Dickinson had recently joined them on several sessions playing piano, notably with Albert Collins for the *Trash Talkin'* album, which was nominated for a Grammy. That was the record that had caught Wexler's ear. (Tarp Tarrant had subbed for Creason on that recording; with Tarp, the four performed as the Soldiers of the Cross at the 1969 blues festival.) Organist Mike Utley was hired when producer Lelan Rogers, Kenny's brother, used the group and wanted two keyboardists. In 1966 in his native Texas, Rogers cut the debut of Roky Erickson's first band, *The Psychedelic Sounds of: The 13th Floor Elevators*. He was in Memphis to record Hank Ballard updating his old hit with "Thrill on the Hill '69" on the King label and Betty Lavette doing "The Man Who Made a Woman Out of Me." Hiring two keyboardists was nothing to him. And that was what defined the section.

Freeman had recently become unwelcome at Sounds of Memphis, the studio where they were getting the most work. "He'd come in there drunk one morning, been out squirrel hunting," says Dickinson. "He wanted this producer to interrupt his session and play this fifteen-minute thing for his hunting friend that he'd been working on. The producer says they're busy and Charlie goes and gets his gun out of his car, blows a hole in the control room ceiling. Fortunately, the call from Wexler came just about then. I didn't care who he wanted, he could have hired Andr'es Segovia, I had to play with Charlie, that's what was working for me."

They signed with Atlantic for a year. The Dixie Flyers were never a touring band or a live band, only a recording band. The difference is that a recording unit does not have to maintain a consistent performance but rather record a series of peak performances. Peaking was what the Dixie Flyers did best. Their first big session was with Aretha Franklin, whose career with Atlantic had sunk into the doldrums. "She hadn't recorded in almost two years when they got her down there, and they weren't sure she was going to show up but they booked time," says Dickinson. "Wexler and I were out on his boat and we got the call on the radio that Aretha was in. Sammy was off fishing and they had to dredge him up from somewhere. We were a bunch of white boys and she didn't know whether she was going to stay. And after we did the first song, she moved in. She drank Orange Tommys, a prepackaged gin drink. She'd get a whole tray of 'em and line 'em up on the piano. And eat pig's feet.

"It was boogie at its highest level. She had an entourage that was three deep all around her. At the time it was very hip to know a Sam Cooke brother, and she had two who stayed throughout the whole session. She could make fourteen notes, seven notes with each hand. 'Don't Play That Song for Me' won her a Grammy, but 'The Thrill Is Gone (From Yesterday's Kiss)' is the best, because of Charlie. We had just done the track and they were envisioning some kind of Claptonesque guitar solo. Charlie went out and played his little Wes Montgomery stuff and I thought Wexler was gonna die. He said 'Baby, baby, I've had great guitar players before but Charlie Freeman is the only one who can take a real solo.'"

In Miami they recorded fourteen albums in six months, with artists as diverse as Carmen McCrae, Delaney and Bonnie, Jerry Jeff Walker, Sam and Dave, Sam the Sham, Lulu, and Ronnie Hawkins. But the work wasn't steady, coming in exhausting chunks followed by irritating lulls. The band was out of their element in Miami, and tried to mentally, and chemically, take themselves back to Memphis as often as possible. "For a while," Wexler writes in his autobiography, *Rhythm and the Blues* (1993), "the Dixie Flyers were flying high. I didn't know that they were doing everything in the drugstore, but I did know they were some wild motherfuckers. . . . I should've known there never were enough projects to keep a house rhythm section working steadily. My conception—to import and keep a cohesive group—was naive."

When things got slow, Atlantic asked the group to record an album of their own, which they would send them on tour to promote. Nobody wanted to tour. Creason and Utley had done long stints, Mar-Keys style, as the Bill Black Combo and had grown unaccustomed to the chore. Dickinson, throughout all his music biz, would not tour until 1972 with Arlo Guthrie and Ry Cooder. The tension got higher when there was disagreement over the album's direction. Dickinson cut a deal with Wexler converting the remaining six months of his contract to a production agreement for a solo album. It began in Miami and ended in Memphis, and is today a sought-after and hard-to-find rocking giant. "I titled the record *Dixie Fried* as a comment on my physical condition," he says. "I'm still satisfied with 'Wild Bill Jones' and 'Casey Jones' and sort of 'Louise.' 'O How She Dances' works real well on the record, but me and Crosthwait used to do that better in our coffeehouse act on any given night. The actual cut of 'Dixie Fried' is maybe the last thing I did. It's all me and Baker.

"I have John Fry totally to thank for the way that record sounds," Dickinson continues. "It was produced by Tom Dowd, and at one point I took it over myself and overdubbed crazy things on it, as I am wont to do. The final mix sounded terrible. I played it on an acetate for Fry and Fry says, 'Waal, Jim, I think that's the worst tape to disc transfer I've ever heard!' And I was out of budget, had nothing left to count on but my fingers. I went back to Fry, and this was long after I had been associated with Ardent, and I begged for help. I got on my knees and begged, which is something I think is very helpful in the music business. A grown man who'll get on his knees in a restaurant—they might give you a budget. Fry's remix is what exists today, which I think is why it has endured to the extent that it has." "*Dixie Fried* is a record I love," says Jerry Wexler, "a masterpiece. My hopes for it upon release were exiguous, to say the least."

The remaining Dixie Flyers worked on an instrumental album of their own. Atlantic rejected the product and the tapes are apparently lost. Most of the band stayed together, accepting a job backing Rita Coolidge and then Kris Kristofferson. Mike Utley went on to play with Jimmy Buffet, producing several hits and employing old friends and bandmates.

Finishing his record in Memphis, Dickinson was able to get back into the local scene. He sat in at the Dream Carnivals, jamming with Lee Baker. He and Dan Penn had become solid running buddies, and he was producing Penn's second album, *Emmett the Singing Ranger Live in the Woods*. CBS Records sent him to Hollywood to produce Brenda Patterson, a Memphis chanteuse who had achieved notoriety singing with Bob Dylan on "Knocking on Heaven's Door." For those sessions, Dickinson hired noted guitarist Ry Cooder and they struck up a quick friendship; Dickinson was cutting the traditional sort of material that Cooder cherished, and in a tradition-friendly manner. "Cooder had just fired Van Dyke Parks as producer of his second album and was in the middle of record company hell," says Dickinson. "After the first day with Brenda, he asked me to go to lunch. It was a very corny Hollywood thing. I thought he was going to ask me to tour with him. It was my first Hollywood session as a producer, I was green as grass. He was asking me to produce him. I said, 'Shit, of course.' I had heard his first album."

He shared production of Cooder's *Into the Purple Valley* with Lenny Waronker, now president of Warner Brothers Records. Los Angeles was dazzled by the southerner's independence, his dedication to roots, his . . . his . . . good God—*authenticity*. Waronker, then head of A&R, called Jim into his office and presented him with the Warner Brothers artist roster. Pick a name you would like to produce, Dickinson was told. "I looked at this list and I asked, 'Don't these people have producers?' 'Just look at the list,' they said. Dionne Warwick was on there, Little Feat. I told them I wanted to produce Bobby Ray Watson. Bobby Ray was from Mississippi and he wasn't on their list, but he was on mine. When Bobby Ray recorded a demo version of this great song of his, 'Fool for a Cigarette,' Cooder played on it, and then stole it, put it on his own album. I figured if they were giving away money, it should go to Bobby Ray. That's what I call regional morality."

CHAPTER TWELVE

That's Mister Boy to You

MUD BOY AND THE NEUTRONS WAS MORE THAN A REUNION OF SEVERAL core players from the Bitter Lemon. Lee Baker, Jimmy Crosthwait, Jim Dickinson, and Sid Selvidge had all felt their lives change after coming under the influence of the bluesmen. Mud Boy was a way for them to continue that discussion in public, sharing what they'd learned with their friends and fans. Though each musician was also a songwriter, the band preferred to reinterpret classics, because the language was universal and also because none figured they could write a better song than "Ubangi Stomp."

Selvidge and Dickinson had been involved in enough music deals gone sour that they were determined to play for themselves and not for the industry. Each was raising a family, and Selvidge also had an academic career; the rock and roll road looked most unattractive and, fortunately, unnecessary. A band that damned the recording business titillated Crosthwait, but Baker— Baker was caught in the crossed desires of artistic expression and commercial success. "I'd have gone on the road," he says. "Hell, I'd go now, playing to people that would like to hear you."

Baker had defected from Moloch in the spring of 1972, and on Halloween of that year, after a summer of rehearsing, Mud Boy made its debut. This Dream Carnival took its name from a line Dracula murmurs when he hears wolves howling: "Behold the Children of the Night." The band name came, indirectly, from Ry Cooder. Dickinson was with him when Cooder's record company asked, completely inappropriately, if he would open a tour for the horrific showman and teen-anthem star Alice Cooper. Cooder turned to Dickinson and exclaimed, "They've got to know I'm not going on the road with no Mud Boy and the Neutrons!" The name had come out of thin air and

Dickinson asked if he had any intentions for it. "One of the reasons we liked the name," he says, "was when people would ask us who was Mud Boy, our reply was going to be, 'That's Mister Boy to you.'"

The Children of the Night show was broadcast in its entirety on the city's most powerful FM station. The site was the former Jewish Community Center, which was only a block from Ardent's new location. Crosthwait says, "Me and Baker got some funny looks taking the wire from Ardent past people's apartment windows and out in their yards, but we just told 'em it was no big deal, not to worry." The show included Johnny Woods and his wife Verlina, he with painted white circles around his eyes and a pink satin shirt, she with a hat that looked like a shoe, bulbous sunglasses, a cape, and a tallboy beer. Furry was in grand form that night, playing to such a large and lively crowd. He thoroughly enjoyed costumes and way-out events, modern-day equivalents of the medicine show atmosphere. He donned a wig and an old smoking jacket worn inside out, the lining showing tigers and psychedelic imagery. "Furry took me and Bobby Ray Watson and Lee Baker aside," remembers Dickinson, "and he said, 'You know that stuff y'all smoke?' And we said, 'Yeah, Furry, you want a joint?' And he says, 'No, no, no, look at this.' And he opened up his coat and he had a lid of pot in each one of the pockets, just showing us he was hip." When Furry performed, he got deep into his playing and refused to quit. Baker sent Connie and Marcia onto the stage in their full regalia—wearing lots and covering little—and they threw a psychedelic net over Furry, dragging him to a cage that was serving as a jail for the evening.

To announce Mud Boy's debut, Crosthwait re-created his sliding wire act from the Electric Circus, running from the rear roof of the gym to a mud-filled raft on the floor in front of the stage. When the pixie hit the raft, the band was born. They had transformed themselves inside and out, way stoned and fully costumed. Each was dressed like something other than who and what he was. Selvidge painted his face blue-black and red like the mandril baboons he studied. Baker was all black hair and beard, his face painted black, a smear in the night, Negative Man. Dickinson wore a flowing orange cape, pink feathered headdress, a jumpsuit with skeleton bones painted on it, and a goat skull for a codpiece. Crosthwait was the most frightening thing he could imagine, a clown at midnight, after the tent is down and the spotlight out. With the dancing girls joined by Big Dixie, the show spilled off the stage. "It was incendiary," says Dickinson. "It made shit happen. The idea was if we put enough of this in front of people, maybe some of it will rub off on them. Memphis is about making chaos out of seeming order."

Itching to have another go at success, Baker urged them toward a recording contract. Dickinson secured a demo deal with Warner Brothers, and the band entered Ardent with McClure, Jerry Lee's former drummer Tarp Tarrant, and Moloch's second guitarist Jimmy Segerson. "It was the most uptight session I have ever been on in my life," says Selvidge. "It got to be like fingernails against the blackboard. Dickinson says that there's some salvageable material, but if there is, someone's going to have to put the happy vibe on it."

"I got one of the best vocals ever on Selvidge at those sessions," says Dickinson. "Most of the stuff was too arranged, and we weren't about arrangements. But even the stuff that didn't work, we at least got some good rhythm tracks."

The months of rehearsal had eroded the group's essence: Mud Boy makes good music because in the core of its conglomerated soul, none of the players is a musician. Each has been able to eke out a living of some sort from his talent, but it's their dimension that makes them what they are: Selvidge may be able to sing high notes of beauty like fields of autumn cotton, but he's an anthropologist. Dickinson is a musical chemist balancing order and chaos, with the approach of an historian. Baker can unleash heroic guitar riffs because he spends all summer atop a tractor cutting grass. Crosthwait may be the world's best white washboard player, but he plays with the hands of a puppeteer. It's the character and personality of each, the wit and response, that propels them. There are ways they play together that they could not train another person to do. They harmonize in their knowledge of each other. It's not a question of being loose or tight, it's a matter of intimacy, performing that intimacy onstage.

Mud Boy only gigs a few times a year, or every other year. Mostly for benefits, rarely out of Shelby County. Gigging infrequently, they maintain their spontaneity. One could say that Mud Boy is the inheritor of the Memphis Country Blues Festivals. Inheritor, yes, but also the inheritance. They played alongside the old bluesmen, they helped them to their chairs, sat behind them and not beside them. They didn't teach the old bluesmen anything and the old bluesmen didn't teach them much—it was osmosis and not a lecture. The medium was a glass of whiskey. The language was the jelly lid over Furry's shot glass.

> Had a little pig and I fed him lots of cheese
> Got so fat that he couldn't see his knees.
> Best ol' pig that we had on the farm
> Who's gonna buy some more bourbon when all this is gone?

Warner Brothers never heard the demos. The band began its self-reparation through live gigs, and fourteen years after their Halloween debut, they finally felt ready to return to the studio, releasing *Known Felons in Drag* in 1986. Their second release, *Negro Streets at Dawn* (a line from Allen Ginsberg's poem "Howl"), would follow seven years later. Mud Boy is the random element, the universal unknown. There are people in small rooms all over the world, in impersonal cubicles in large offices, in malls, in ghettos, and behind fenced mansions—who thrive on a little chaos, enjoy the occasional taste of 220 volts, live for the beauty of the flaw in the grain.

The Mud Boy shows were a continuation of the Dream Carnivals. Even after they stopped wearing costumes, theatricality remained a part of their show, making room for the dancers, for Randall Lyon, for any other appropriately theatrical event to occur. "Husband is smart," says Mary Lindsay Dickinson. (She says people all over the world may address Jim many different ways, but she's the only one who can refer to him as "husband.") "Mick Jagger has to knock himself out and stay skinny as a rail when he could just hire dancing girls. And Mud Boy never even had to hire them. They just wanted to be part of the show."

"The last costume gig we played was the night before Charlie Freeman's funeral," says Dickinson. "And then we just never did the costumes again. Never even talked about it. Baker was just a smear of black and fuzz by then. He had this kind of wooly vest, and his hair was real long and it would be down in his face when he played. He would pull his hair up and put his finger in his ear and sing 'Laundromat Blues' like a twelve-year-old black girl: really really good."

Guitarist Charlie Freeman's death on January 31, 1973, jolted the Memphis scene. Since declaring his subtlety in a garage in the late 1950s, he had never ceased to astound any who'd heard him, achieving in his short life things which compose the dreams or regrets of most thirty-one-year-olds. His legacy included the Mar-Keys, the integration of Memphis stages, and the Dixie Flyers; he'd also laid the groundwork for Phineas Newborn Jr.'s comeback. Freeman was enormously respected in the industry, though stardom had remained elusive; or he had eluded it. These are people skeptical of fame and its reported benefits, forsaking the financial security for the privacy of a normal life. They've watched too many great talents compromise the quality in their playing that makes it special and become just another name for a fast-food hamburger.

Charlie Freeman died of neglect, say some of his friends, and they hold themselves responsible. It seems more likely that he was his own victim: He'd ignored the asthma that had plagued him since he was a child. At the time of his death, he was in Texas, auditioning for a road gig beside Tommy McClure, trying to earn enough money to pay off old doctor's bills so they would examine his fluid-filled lungs. When he passed out from too much dope, his party acquaintances thought they were helping him by moving him from a chair to a bed. But Charlie always slept face down, and his acquaintances did not know that. When he choked on his own vomit, there was no question that he'd died accidentally. As accidentally as his lifestyle allowed. Did he fall from life or slowly jump? Herb O'Mell remembers a session in California. "This was after the Dixie Flyers, and Charlie was out there with Tommy McClure, Sammy Creason, and Spooner, recording with Rita Coolidge. Someone walked in and began passing a magazine with some [cocaine] lines on it. When it got to Spooner, he took the magazine, said, 'Oh,' and he opened it up to read. I saw Charlie go down on that carpet man, just laid on the floor going after that coke."

Freeman loved to party, but he was also a gentleman. "Even though I was a female and real young, Charlie treated me like a human being, not like a piece of ass," says Marcia Hare. "Charlie was the first southerner to accept me. I can say that Charlie Freeman gave me an early start on self-esteem by treating me like I had something important to say, and he would let me say it."

"Charlie and I were fishing buddies," says Jimmy Crosthwait. "Whenever we went fishing he would always, always, lock his keys in his trunk. After a dozen times I thought it might be a joke. He would open his trunk, put the keys in there, look for tackle boxes, and shut the trunk with the keys in it. Every time! And he'd say, 'It's no problem, it's a GM car,' and any GM car that comes by would have a key that would work his trunk. Charlie was drunk a lot of the time and he was part Indian, had a good sense of humor and just a great touch with the guitar. His death was a shame. He probably would have played with Mud Boy but he had to go on the road to make a living. When Charlie died, my next fishing buddy turned into Lee Baker."

The respect that Freeman had shown to Marcia Hare was also felt by Randall Lyon. The simplicity of going together to the cinema meant as much as the wild times they shared. "He radiated this power," says Lyon. "You could just look at Charlie and go, I'm gonna do something wrong and it's going to be great! He was really funny, very hip, and he knew what was going on. Some people on the scene were way far advanced, like angels. They lived in this alternative universe and every once in a while they would touch down. Charlie was working on something really deep. It was magnificent."

"The one gospel song that I'll never forget was at Charlie Freeman's funeral," says Jimmy Crosthwait. "There's Charlie's casket and the piano is back behind this latticework and flowers, you can hear it but you can't see it. I knew it was Jim [Dickinson] playing. An instrumental. And then another instrumental. And then, just out of nowhere, he sang. The third song was 'When the Lord Gets Ready You Gotta Move.' Man, I mean that was a heartbreaker. It was probably one of the best send-offs one musician has ever been able to give to another. It was completely invisible, just the sound. It was really nice. Really nice. 'When the Lord Gets Ready You Gotta Move.'"

"The last time I saw Charlie," says Dickinson, "he gave me Fred Ford's phone number. And it was a big deal because nobody white had Fred Ford's phone number." Fred Ford had played saxophone with Bill Harvey on Beale Street when the young B. B. King was plucked from amateur night and put in front of Harvey's band. Ford used to say he was the first Black Muslim in Memphis. Whether that's literally true, it exemplifies his militancy, an emotion underscored with bitterness. "Thirty years ago they used to call me Sweet Daddy Goodlow," Ford still says, "but I'm Bitter Father Badlow now." Then he sometimes pauses and sometimes he doesn't, but the conviction in his tone is obvious, and extremely unsettling, as he adds, "I have a right to be bitter." It's not that success eluded him—he chose to dedicate himself to the care of Phineas Newborn Jr., remaining in Memphis, and there established his prominence. And there, perhaps thinking of his career that could have been, he's felt the cold touch of exploitation. Among other shadows he's been in, he was with Johnny Otis's band when they backed Big Mama Thornton on "Hound Dog." That, of course, was before Elvis redid it. But even when his anger was fueled by youth, Ford got along with Charlie Freeman. Together, they were able to initiate the comeback of Newborn, one of Memphis's finest musical statements. Freeman had the record label connections and Ford had the artist's trust. Charlie didn't live to see it, but *Solo Piano*, the comeback album by jazz pianist Phineas Newborn Jr., was nominated for a Grammy Award in 1975.

Ours is not an age of genius. Eccentricities are shunned, blemishes quickly covered. But Newborn was a musical maverick. In 1956, at the age of twenty-four, he startled the world with his debut, *The Piano Artistry of Phineas Newborn, Jr.* His greatest achievement, perhaps, was his dexterity, making his left hand as important to the melody as his right. He sounded like more than one person playing. Memphis had long been aware of the boy prodigy. Not only was he born into a musical family—his father was one of the city's most respected bandleaders through the 1950s—but he was also born into a musical

Phineas Newborn Jr., circa 1945.
Photo courtesy of Calvin Newborn.

era. "My dad told Lionel Hampton about my brother," says Calvin New-born, an accomplished guitarist who recorded with Phineas and with Charles Mingus, among others. "Lionel met Phineas on Beale Street and they played 'How High the Moon' for almost two hours, fast as you could play it. Phineas on one end of the keyboard, Lionel on the other. Count Basie gets the credit for discovering Phineas, but it was Lionel Hampton who put his name out there."

In the early 1960s, living in California, Newborn suffered a mental break-down, was misdiagnosed, and never fully recovered. A decade later, he was in Memphis, still receiving treatment at the Veterans Hospital, looked after by Fred Ford. It took two people to fill Freeman's role in the comeback project. Both the writer Stanley Booth and Jim Dickinson were acquainted with Ford, and each had a long relationship with Atlantic; Dickinson could manage studio time in Memphis. The two approached Ford and discussed the project. He was leery but listened. Before parting, Booth pulled a ticket stub from his pocket to write his phone number on the back. Ford took the piece of paper, turned it over, and quickly said, "What do you know about this?" The stub was from a benefit to help Ford's former associate Bill Harvey get a new leg. The bonds were sealed.

"It was a hard deal to put together," says Dickinson. "Fred was existing very much on the borderline of society, and Junior was in the Veterans Hospital, psychiatric. I was cutting Bobby Ray Watson at Ardent and we would slip Junior in whenever Fred could get him out of the hospital. The first two sessions he didn't talk. He would come in wearing his overcoat and his hat and walk straight to the piano. His hands would be outstretched and he would begin playing before he even sat down. Then his foot would move to the pedals and then he would sit. He would take off his hat but not his overcoat. Just shaking hands with Phineas altered the way I play the piano."

Once at the studio's piano, Newborn seemed a recovered man. At one point, there was a baritone sax lying near the piano. Fred picked it up. "Whoa! The two of them playing together again, it was just unbelievable," says Dickinson. "They played 'Cherokee' and it became a game. Phineas was trying to play it faster than Fred, and Fred is playing it faster than you could think it." Newborn conceived music without boundaries, as *Solo Piano*'s medleys make evident; one blends a John Coltrane song, a selection from a Joan Crawford movie, and the pop hit "Where Is the Love." Another of *Solo Piano*'s highlights is the rendition of W. C. Handy's "Memphis Blues." "Fred was talking to Junior," continues Dickinson, "and he says, 'Ain't you got something country you could play?' Phineas broke into 'Memphis Blues.' That's his idea of country."

After that album, Ford and Dickinson worked together on a project for Cybill Shepherd. Phineas, then living with his mother, was to play piano and Fred insisted it be Jim who picked him up. Though Jim's expertise may have been better utilized in the studio, Ford has his own way of imparting knowledge, instructing. "It was like 5:30 in the afternoon, Junior was just waking up," says Dickinson. "I'm sitting at the kitchen table. Mama Rose is cooking him a fish for breakfast. Junior sits down in his overcoat and hat to eat his fish and he's talking to Mama Rose. She's had to deal with three generations of crazy musicians, so Junior ain't nothing to her. She talks in a rough voice like him and Calvin, it's a family trait. She says, growling, 'What do you want for your birthday?' He says, 'I want a pistol.' She says, 'Well, you ain't getting nothing then.' He says, 'You got a pistol.' She says, 'Yeah, but I don't run down the street going "Yeah, yeah, yeah."' She nailed it." Phineas Newborn Jr. never fully recovered his health. He made occasional recordings and performances. On the morning of May 26, 1989, at six A.M. he was found seated on a chair on the porch of his mother's home, dead from natural causes. He was fifty-seven.

"They had these grandiose ideas," says former Ardent engineer Richard Rosebrough, referring to the Memphis pop band Big Star. "It was all a Beatle thing, a Beatle and a Who thing. John Fry was a pilot and so were a few other people at Ardent. At the time, we were all enamored with this plane, the Lockheed Constellation. The plan was for John to buy one and Big Star would tour around the country playing these gigs. They would have this thing above the stage that would drop gold stars on the band while they played and they would get on the cover of *Rolling Stone* and all retire rich and famous. Alex [Chilton] had been a star, and he was disenchanted with the music business, except that he still liked the star thing. And John Fry loved playing the game. He loved radio, and he loved these great pop records, and he always wanted to record his own and have them be great records like Phil Spector and George Martin. But what he needed was somebody to record. And that was Big Star."

The saga of Big Star has to be one of the most unusual in pop history. Although they were unable to reach an audience during their three-year existence, their popularity has snowballed since they disbanded in 1974. Two decades later, they have become so well known and influential that a reunion concert received national publicity, was released as a CD, and the band was invited to perform on "The Tonight Show." Their stardom is a pop culture phenomenon, a word-of-mouth secret that reached Europe, Asia, and Australia without forsaking its intimacy. Music fans bond around informing each other of this talent that, until recently, has been completely overlooked by the mainstream. Big Star's worldwide popularity is the epitome of Memphis obscurity. Not only does the package come with exceptional, durable, bright pop music, there's also a story with it. Hear that lead singer? That's the same guy—*the same guy, man!*—who was the vocalist in the Box Tops. You know, "The Letter," "Cry Like a Baby?" *You don't know the Box Tops? Oh, man* . . . The tale widens to include the Box Tops' producer, Dan Penn, who wrote so many soul hits, and it includes Jim Dickinson, who produced Big Star's third album.

But the heart of the Big Star story is Alex Chilton. On top of the world at sixteen years old, a brilliant pop individualist by twenty-one, and three years later, 1974, he can't get a finished album released, one now recognized as a masterwork. Spurning the industry ever after, he has made a comeback completely on his own terms, a privilege usually reserved for superstars. Unlike most people in the world, Alex Chilton is able to do what he wants with his life. Chilton is his own person, his own artist. Known for being unknown, his independence and unpredictability trap him in the spotlight.

Chilton had begun singing with the Box Tops when he was sixteen in 1967, dropping out of high school to devote himself to the band. When he quit, he was nineteen, free from the state's grip. The band was in England, about to usher in the 1970s with a tour. When it became apparent that the promoter was not going to hold up to his contractual agreements, the singer opted out. He locked his hotel room door, emerging only to catch a plane home. The tour went on without him, but the band, distinguished as it was only by his voice, was effectively over.

Though he could not yet comfortably perform with a guitar while singing, Chilton had become a competent composer. The third Box Tops album included one of his songs, and the fourth and final, with Chips Moman replacing Penn as producer, featured several Chilton compositions. "I Must Be the Devil" drew from the blues that surrounded him. "Together" achieved an anthemic pop feel, revealing a complexity that would distinguish his best Big Star material; the sadness in his vocal, the solemnity of the organ, and the foreboding tone of the fuzzed guitar all belied the joyousness of his lyric. Before departing for the Box Tops' tour that didn't happen, Chilton had begun a solo recording at Ardent, working with engineer and producer Terry Manning. "Ardent was the only place in town that wasn't already locked up with a bunch of Tin Pan Alley writers and these sterile musicians playing all the sessions," he says. The album was completed after he quit the Box Tops; fifteen years later, he released four of these tracks on the self-compiled (and self-titled) *Alex Chilton's Lost Decade*. "Free Again" apparently anticipated his departure: "I'm free again/ to do what I want again/free again/to sing my songs again." Instead of the deep gravelly voice characteristic of the Box Tops, Chilton sang from the top of his throat, the higher register he would employ on the Big Star material.

Returning from England unemployed but with money in the bank, he moved for a brief time to Manhattan, honing a solo coffeehouse gig. "I was just trying to learn how to play," Chilton says. "I was sick of bands. I figured I was going to learn to accompany myself so that I wouldn't even need 'em. It was the first time I'd ever really tried to sing and play guitar together, and sometimes I'd shake so bad that I just couldn't play the guitar at all." Hanging out in various folk joints, he worked up his chops, even jamming with Roger McGuinn from the Byrds. He maintained a working relationship with Dan Penn, writing and demoing songs for his publishing company.

Chris Bell, from an affluent Memphis family, had a friendly disposition, though a tendency toward depression. He'd begun playing guitar at a young age, developing an intuition for acoustical space. He wanted a guitar to sound three-dimensional, and he developed a universality to his voice, a transparency suitable for mass appeal. When the public was not widely appreciative of his talent, he believed himself born out of time and out of place. His mother was British and he used to say he was homesick for England even though he'd never been there. He felt like he should have been a part of the 1960s British Invasion. His fantasy was for the past, but history proved him ahead of his time.

Chilton had known Chris Bell while growing up in Memphis, seen him playing parties since they were in their early teens. They'd worked together briefly in an early band, the Jinx, named in tribute to the Kinks. Chilton remembers, "There was lots of people who were going to parties and people who were having parties. It was a place for young bands to play in front of people. 'Hold It' was a real popular song when we were kids, an instrumental. The Mar-Keys, Willie Mitchell, a bunch of bands recorded it. It's the classic Memphis break song." At his one guitar lesson, Alex asked Sid Manker, a friend of his father's and the guitarist who'd cowritten "Raunchy" with Bill Justis, to teach him "Hold It." "He knew it, and he taught me the swankiest version. There was stuff in there I didn't use until twenty years later." During Chilton's tenure with the Box Tops, Bell had remained in Memphis, playing in bands and gravitating toward Ardent, which had moved from Granny's sewing room to National Street in 1967.

"John Fry had this studio, Terry Manning was working for him, Chris was a hanger-on over there and I became one too," says Tom Eubanks, who quit the band that became Big Star. Eubanks, Bell, and future Big Star rhythm section Jody Stephens and Andy Hummel called themselves Rock City, capitalizing on the tourist attraction in east Tennessee that advertised on the roofs of barns: SEE ROCK CITY. "It was a good scene. When the studio wasn't in use, Fry would let us cut stuff there. That's the way Stax came up with their good stuff, they'd take people off the street and do spec sessions until they got something. We practiced a lot, played every now and then, mostly for rich folks. Everyone else around town was playing soul music, we were doing stuff from *Tommy*, 'Into the Fire' by Deep Purple. I remember we had this job in Ripley, Tennessee, and we were doing 'Shapes of Things' by the Yardbirds, Chris was playing a semi–hollow body guitar, got down on his knees in front of the amp and started getting all this feedback, it sounded like an airplane going over. These country boys stood there with their mouths open, What is this sonuvabitch doing? Another time, we got booked into this building where Sonny and Cher were also playing, in a different room. Somehow, our gig was for a group of American veterans, all these old people. This mafia guy was running the club, and we were drowning out Sonny and Cher, so he started punching our roadie, telling us to turn it down. These old army jerks hated us, they'd come up front and curse us, razz us, so Chris would run over and turn it back up. Here comes the mafia guy beating up the roadie again. But that was Chris too: If somebody insulted him, he'd pour it on."

Drummer Richard Rosebrough played in several bands with Bell and remembers being introduced to Jimi Hendrix records by him. Bell brought Rosebrough to Ardent, and he too became a member of Uncle John's gang. "In 1969, John Fry got sick of us just hanging around there and he decided to teach us to engineer," remembers Rosebrough. "It was me, Christopher Bell, John King, who ran publicity for Ardent, Charlie Hull, who had recorded at the old studio with the Shades, and a few others. Classes were free and held at the studio at eight A.M. Fry would stand in front of a blackboard and teach us. The rule was you could show up drunk, and you could get drunk with the teacher after class, but you had to show up. I like to say that I've got three mentors: John Fry taught me how to record, Jim Dickinson taught me when to record, and Sam Phillips taught me how to make it interesting."

Chilton had already established a working relationship with Ardent, and during a visit to Memphis from Manhattan he encountered Chris Bell at the studio. Impressed with the material Bell was working on, Chilton shared some of his recent efforts. There was talk of forming a folk duo, but Bell was not interested in moving to Manhattan. Realizing the promise of their collaboration, Chilton returned to Memphis. Bell's Beatle fantasy was now complete with this likeness of the Lennon-McCartney partnership. Fry was their George Martin. "We didn't think about doing live gigs," says Chilton. "We thought we were the Beatles and weren't playing live anymore."

Eubanks remembers getting a call from Bell when Big Star's debut was being finished. They had recorded "My Life Is Right," which he'd written with Bell. "Chris asked me if I would let him take my name off the song and put Alex's on, because he wanted to have them all read 'Bell, Chilton' like 'Lennon, McCartney.'" Eubanks's name remains.

Rosebrough, in the meantime, had begun to engineer in earnest. His recording of the Memphis group Alamo, for whom he also drummed, earned them a deal with Atlantic Records. Ardent hired him full-time as an engineer; his next project was the Moloch album. "I was feeling big, an Atlantic recording artist," says Rosebrough, "and when Chris and Alex asked me to play drums for their band, I turned them down. I'd walk into Ardent and I'd see them recording, and I kind of thought they were a kid band. But there was so much energy in the studio when they were around. They were having kid spats, smashing guitars, passing joints when Fry wasn't looking." The Ardent label was blossoming with the infusion of talent. The studio crew regularly marched across the street to the Sweden Kreme for hamburgers and milkshakes, cooling off at a picnic table and dreaming up schemes like traveling in a Lockheed jet. While strumming guitars on Ardent's front stoop, staring at the milkshake palace across the street, Chilton and Bell discussed their group's name. Sweden Kreme didn't have a ring to it, but the supermarket next to it did, so they copped the grocery chain's moniker: Big Star.

Chilton's contemporary view of Big Star is self-deprecating. He's a quarter of a century older now, more worldly and sophisticated, an active artist hounded by past achievements. "The first couple of tunes I wrote in the Box Tops were very simple and very bluesy and very honest. That was easy. It seems like the first song a lot of people write is a good song. But then the more I learned about music and that kind of stuff, the broader the field became, and the more room for error also. So I was really groping around about songwriting and sometimes succeeding. Only about 1975 did I really became self-confident about my abilities, which coincides with the time when I began to write far fewer pieces of music.

"Most of the Big Star stuff was searching for how to get through two verses without saying anything really stupid. Only a few songs really succeeded at that for me. I think 'In the Street' is maybe the best song that I wrote on the first two albums. There are some other good ones. 'Thirteen' is a pretty fair song, but I think the performance on the record is bad. The singing is the wrong approach, but at that time I didn't really know how to make the vocal sounds in the studio that I wanted. 'When My Baby's Beside Me' is a good song. 'O My Soul' is almost a good song, but it was written by a committee, and it's not really about anything. I like the Big Star records okay. There's a lot of nice guitar playing on them, some nice writing and some melodies, and the production's really good. But it's not like I sit around and think those are a phenomenal achievement in recording history. In terms of how they sound, maybe they are. We were doing things in 1971, 1972, and 1973 that sound really good even now. But that kind of guitar playing is something that's there on a guitar neck, and anybody who reaches a certain level of ability is going to play that same stuff anyway."

The Big Star albums have endured. Though technology has changed as much as the global map, these albums continue to reach people. Their immaculate audio quality is largely a testament to John Fry and Chris Bell, who engineered the recordings. But the fact that people still reverberate to the songs is due to the songwriting of Chilton and Bell, and the meaning they convey through their performances. With Fry's meticulous clarity, their sophisticated arrangements and gentle harmonies become gripping, penetrating. Though they are ostensibly a pop band, there's an underlying menace to Big Star's work. They meld the winsome with the twisted. After the soothing tones of "Thirteen" place us like a baby at the doorstep of the next song, we are met by the punch and kick of the electric guitar that opens "Don't Lie to Me," the band's hardest rocker. The song sounds like a domestic quarrel. In the juxtaposition of these two tracks lies Big Star's soul.

"In the sixties, Dan Penn sheltered Alex, kept him from the funkiness," says Randall Lyon. "Then in the seventies, when I met him, he was doing that Big Star stuff and he was absolutely great. I met him through his old man. His father used to play the piano with Bill [Eggleston]. I started going to the Big Star sessions, the power pop stuff with Chris Bell, and it was almost over my head. It was very deep shit. I didn't really know about power pop, had no idea what it was for. It sounded right, but it was way mainstream compared to Sleepy John Estes."

Big Star's 1972 debut (shortly before Ardent cofounder Fred Smith debuted Federal Express) was also the first release on the reactivated Ardent label. Titled *#1 Record*, it confirms the lofty ambitions surrounding the group. Because Ardent was so intertwined with Stax, a distribution deal was easy. By the early seventies, however, Stax was being bought and sold like a Monopoly board property. A month after *#1 Record* was released, Stax entered into a distribution deal with Columbia Records, and Big Star's debut got lost in the shuffle; too much business going on for either company to tend to the actual albums. John King, heading up Ardent's publicity office, was so well organized that Stax was using him for many of their projects. The rave reviews the album received are evidence of his good work. They made the sting from its total unavailability all the sharper. Everyone had given their best effort on this first outing, and they were beaten by an industry gaffe.

David Bell, in his liner notes to *I Am the Cosmos*, the 1992 collection of his brother's work, remembers, "[In 1972] everything seemed to be falling apart with the band, and my brother had apparently tried to do himself in. . . . He was in terrible pain, feeling that he had put forth his best effort . . . only to see his efforts lost in a distribution deal with a record company whose claim to fame had been a score of black Memphis soul products and who obviously didn't know or care how to distribute a white, Anglo influenced rock group. . . . My brother was near rock bottom."

Bell's sense of self-destruction was clearly in evidence during the mastering of Big Star's first album. He went with John Fry to the session where the master stamper was being cut from the tape. Fry worked with engineer Larry Nix, meticulously riding knobs and dials throughout every song, achieving precise tonal qualities. The process took an entire day. When they were done, Bell watched Nix carve the catalog number into the inner groove. "That's the thing that writes on records?" Bell asked. Bell picked up the tool when Nix was done, observed the sturdiness of its arrowlike tip, then dropped it like a dart headlong onto the master, which it pierced. The entire day's work had to be repeated, a session which Bell was invited not to attend.

At Ardent, the group effort in launching the label and running the studio had become a bit cozy. Many of the integral people had been around since the earliest days, working together to advance the studio from a tenth-grade whim to a nationally respected facility. Rosebrough remembers, "We were all drinking, we're taking downs. I start seeing a psychiatrist. I've got a three o'clock appointment and while I'm waiting, [an Ardent employee] and [another Ardent employee] come out from separate doctors. When I come out, Chris Bell is seated where I was, waiting to go in. It's a mind fuck. It was crazy. A lot of crazy situations, competition to make the record company go, people vying for control."

When Bell quit Big Star, he was apparently suffering from paranoia, delusions, and was actually seeing things. Chilton, Stephens, and Hummel threw in the towel as well. But Ardent's publicist John King had conceived of a rock writer's convention to introduce and promote the Ardent label. He had journalists and dealmakers flying in from all over the country, and Big Star agreed to help him out by performing as a trio without Bell. A review of the performance reveals, "They start their set . . . with 'Feel.' And guess what? You could dance to it! By the fourth number the dance floor was packed, and it stayed that way the rest of the evening." Aha! From the days of his preteen parties on through recent shows performed at tiny Memphis clubs, Chilton puts on a great show when the crowd is dancing. It's the shadow of the Plantation Inn, the entertainer entertaining.

"That convention was John King at his finest," says Rosebrough. "He was gaining control at record stations all over the country, and he planned this big show to promote Ardent Records. By then we'd probably released Cargoe's album and were working on a few other acts. So he reserves Lafayette's Music Room, this hip club at Madison and Cooper. [There's that corner again.] The Big Star set was really chaotic, with guitars feeding back, out of tune, lots of people drunk. Big Star was doing T. Rex songs like 'Baby Strange' and 'Jeepster' as well as their own stuff. When they finished their set, Alex put his guitar down and it started feeding back a little. He walked off stage and was gone. Fry and I were in the sound booth and he turned to me and he says, 'Richard, you know the last thing that a captain does on a sinking ship?' I said, 'What John?' He said, 'He sets his ship for normal operation and then he jumps off.' I reach down and I turn all the microphones back on and adjust this one so that it just starts feeding back. John said, 'It's time to leave.' We walked out of the booth, down the steps, all the way around to the back, and as we were walking down the back hall the whole P.A. was really feeding back big-time. John turns to me and he says, 'I love doing this more than anything.' We walked out of there, got in his car, and left."

The reaction to Big Star's set was so overwhelming that the band was convinced to record a second album. The title, *Radio City,* was once again full of romantic expectation. And the music, again, was completely deserving. For album art, Chilton turned to his lifelong friend Bill Eggleston. "I suggested a picture that was radical at the time," says Eggleston, "that picture of a red ceiling. But I thought that picture would look good anywhere. It was modern art, and I thought it was a good idea for a record cover to have a piece of real art on it." Chilton was twenty-three when it was released.

Between Big Star albums, Ardent had moved to the site it continues to occupy, a bigger, more centrally located studio. The corner of Madison and Cooper, a block away, was then being developed as a center for boutiques and nightclubs. Trader Dick's was next to Ardent. Huey's, another hot spot, was a block the other way. "By then, we all had keys," says Rosebrough. "We were going to Ardent at all hours. We could break things and get away with it. Get drunk, take pills, puke on the floor, piss on the wall. Alex and I had become very close then, hanging out all the time, writing and recording together. Three of those songs ended up on *Radio City.* We'd go in at three A.M., just the two of us, recording and experimenting."

When Chilton and Bell parted, they divvied up their material. Though each had contributed to the songs, they agreed on sole proprietorship after the split. So "O My Soul," though "written by committee," bears only Chilton's name; conversely, Chilton's name does not appear on Bell's album. For *Radio City,* Bell also contributed some background vocals and guitar. "The things we did that were unusual, I had nothing to do with," says Fry. "That was Chris and Alex. We had a variety of guitar sounds that played against each other in a kind of pleasing, almost orchestral way. 'September Gurls' used a mandol-guitar that Alex had, a tiny little thing. That real high, piercing guitar you hear is that. We probably featured drums a lot more prominently than it was fashionable to do at the time. And, to the extent that we could, we used a combination of reverb and delay."

Ardent was English in its orientation, subscribing to British studio magazines and mail-ordering records not available in the States. "If they did it in England, our view was that was the right way to do it," says Fry. "Memphis music is peculiar in that it combines so many seemingly contradictory or conflicting elements. You have all this blues and rhythm-and-blues influence, and so much of the recording that we were doing was with various Stax artists, and other people doing similar things. We added an English sensibility to it."

Before the band toured behind the second album, Andy Hummel left the trio; he is today an engineer for General Dynamics. A replacement was enlisted and they drove to the Northeast, where their equipment was promptly stolen. Chilton was the sole guitarist in the band, now confident in his technique. One of their shows was a radio broadcast on a Long Island station, the bootlegged tape of which circulated widely for many years before being properly released by Rykodisc in 1992. In an interview with the disc jockey, Chilton describes "a hard life, out on the road and all, driving around in station wagons. It just wasn't any fun." The band's defiant attitude toward their contemporaries was also manifested in the interview; topping the charts then were Alvin Lee and Ten Years After, Golden Earring, the Mahavishnu Orchestra; the music was self-indulgent progressive rock—songs that were too long, too meaningless, and too boring. Likening the crisp and clean guitar sounds of Big Star to the Byrds and the Beatles, the WLIR disc jockey asked Chilton if they were "anachronistic."

No, Mr. Deejay, they're not from a different time; they're from Memphis, where pop records are not disposable. "Commerciality has always been important to me," Chilton says. "But I don't really approach things from an angle of what's going to get on the radio. I approach things as what I think is appealing. The first time I was in control of the records I was making, the Big Star records, I made them to be commercial. It's so hard for anybody to play what they like in the record business. Big Star was a band playing what we liked to play. It didn't conform to any other corporate crap. After the rhythm and blues and the jazz stuff, the rock and roll that first really captured me was mid-sixties British pop music, and that was all two and a half or three minutes, really appealing songs. So I've aspired to that same format. That's what I like."

CHAPTER THIRTEEN

Stranded in Canton

THE 1970S SEEMED A LONG WAY AWAY FROM THE 1960S. EVERYONE WAS a rock star. The music that had earned disdain when the witnesses began playing it was now the soundtrack for a generation. And everyone wanted to be part of it. Musicians who weren't cutting hits in Memphis were cutting demos. Studios and production companies were everywhere. Teenage bands with a riff had no trouble finding a place to record, a place to spend their money. Songwriters who were making demos of their material opened their facilities to the public and found themselves as studio managers. Businessmen were financing sessions with themselves as producers. It was the beginning of modern, corporate rock, when record label offices went from manufacturing and promoting records to controlling artistic direction: The corporate record is now defined before it is even recorded.

The good cheer even spread to the classical musicians. Pop hits have long featured string arrangements, and in the seventies Noel Gilbert, who led the Memphis Symphony's string section, began appearing on records that would make Beethoven roll over. "The first time we used Noel and the symphony's string section," says Dane Sullivan, a staff writer from American who opened his own facility with his wife Gala, "we split the session four ways. Instead of us having to pay the whole session, we paid one-fourth. Four people got records, and the string section got a full session. They could have spent the same amount of time and gotten paid four times the amount, although then it probably wouldn't have happened." The strange bedfellows enjoyed each other.

"You could hum something to the symphony players and they'd work with you," says Gala. "We'd say, 'Can you do something like this?' It was a real

Rogues' gallery. Left to right: Lee Baker, Wayne Jackson's profile and glass, Waylon Jennings, Jim Dickinson, Jerry McGill, John David, Dan Penn. Photo by Randall Lyon.

give-and-take situation. They didn't demand charts, and they really enjoyed it." By then, Ardent had purchased a Mellotron, a keyboard instrument popularized by the Moody Blues which had tape loops of string sounds inside it; it was like a synthesizer for string sections. The union, anticipating the wave of the future, feared that a whole string section would be replaced by a single person. At Ardent, the Mellotron was kept in a locked room because officially it didn't exist.

It was a good time to be a songwriter. Not only were pop stars employing house bands; they were actively seeking outside material to record. The staff writers at American were supposed to have at least one song for every artist who booked a session there. "Every week there was at least one album session," says Dane Sullivan. "Joe Tex, B. J. Thomas, Elvis, Neil Diamond, Petula Clark, Dusty Springfield—every week there was somebody of that caliber." ASCAP came to Memphis to cultivate writers, offering sizable advances.

Memphis was seen as a test market for records. The audience here approximated what the industry perceived as normal, tending toward the fickle. "If you got sales in Memphis, you could get attention, the ripple effect," says Sullivan. "And if you needed to, you could do it by buying all your own records, going across the bridge and throwing them in the Mississippi River. You create a sales figure, it gets reported. You'd make 'breakout' in *Billboard*.

That was happening all over. We had somebody cut a record that we controlled, and we heard that the pressing plant was shipping 40,000 copies. But no royalty statement was forthcoming from said person, who's located in Kansas or somewhere. How are you gonna fuck with them?"

In a 1969 falling-out between Chips Moman and his bean-farming partner Don Crews, the latter left American and bought a studio across town called Onyx. American's house band and the writers remember working in both places. Onyx became a laboratory for engineers from which a few hits emerged. T. G. Shepherd, a 1970s country star, was a promotion man with an office there. Coming downstairs one day he heard someone in the control room asking about a singer to demo a song. Shepherd volunteered, recorded "Devil in the Bottle," and had a new career. The laboratory atmosphere at Onyx was created around the tape vault. Many of the early American hits were stored there. "We'd go in and pull the original eight-tracks on the old hits," Dane explains, "and see if we could remix it the same, see if we could remix it different. It was fascinating. Between that and the activity that was happening up on Chelsea in the main place, sitting in on sessions and mix and remix sessions, there was plenty of opportunity to learn all about making a record." The Onyx spirit has continued, the studio now home to Easley Recording, the city's foremost alternative venue.

The Sam Phillips Recording Service was another place to get a music education. "Before Sam sold Sun," says Randy Haspel, "I'd be hanging out at the studio, Teddy Paige, Knox, Jerry, whoever. We'd be bored, and so we'd go up and check the library. Box after box, everything you could imagine was in that library, and it was all catalogued on a Rolodex. We'd go, Howlin' Wolf, hmm, E-64, and we'd pull out this box of tape that Mr. Phillips had cut during the early 1950s—unreleased master tapes sitting in the Phillips studio—we'd put it on the board, sit there and listen. Ike Turner, Rufus Thomas, all these people. And when Mr. Phillips sold the label, the library went with it."

"I learned a great lesson when I hired Dr. John to play on the Brenda Patterson album I produced," says Dickinson. "I sat down with him and he said, 'What instrument would you like me to play?' I said, 'Well Mac, I kinda thought I wanted you to play piano.' He said, 'Well I could play bass.' I said, 'Yeah, I got Chris Ethridge, Mac.' He said, 'I could play guitar.' I said, 'Yeah, I got Ry Cooder, Mac.' He said, "You want me to play piano? Full piano?' I said, 'Yeah, yeah, full piano, Mac.' He said, 'Well you know, that's my thing.' I said, 'Yeah, that's why I want you to play it. I want you to play full piano and do your thing.' He said, 'Well full piano, that costs twice as much.'" Dickinson laughs uproariously. "I said, 'Right on, Mac, it's fine with me.' After that,

I did it to some teabags who came to Memphis to record Johnny Halliday, this French Elvis-type. I did the full piano thing, the whole bit.

"Those Johnny Halliday guys, they learned about cutting a record in Memphis. The first night of their session, I was at Ardent with Alex Chilton, Danny Graflund, and Campbell. They saw Graflund as this bodyguard person, so they thought they could send him out for coffee. Graflund went across the street, got a tray of coffee, filled the cups with Ajax, and took it in there and served it. Before their session was over, they had hired Campbell to do security and they had hired me at double scale to play piano, and it was all basically orchestrated by Graflund. They asked if it was the initiation process. While they were there, they crossed up [a Memphis artist] who was cutting down the hall. He went off on them. He had that producer on the floor at Ardent with this machine gun in his mouth. When he got up, he went out in the parking lot, drove to the airport, and was a star in six weeks. He said, 'Man, I just lived through this, I'm gonna do whatever it takes. I'm through fucking around.'"

Former Mar-Key Packy Axton, no longer encouraged to use Stax, was sharing his healthy cynicism with those around Ardent. He would put together a band in the studio, hum a riff, and create an instrumental in as much time as it took to arrange it. In an afternoon, he could do four or more songs, labeling them according to sound, like "medium tempo garbage" or "fast trash." On one, he insisted on an eight-beat break early in the song, telling the band he'd fill it later. Coming off a bender, he was rooting through tapes for something to sell, and had the engineer put that one on while he went to the vocal booth to fill the hole. A break is common on instrumental records; "Last Night" has one. Packy is dead now and there is no way to know what his intention was, but when the time came for him to say whatever he wanted, Packy took a breath, opened his mouth, and vomited. Puked all over the place, and recorded it perfectly on tape. He named the song "Hung Over" and called the band the Martinis.

Near records. Vanity records. Custom records. With so many hits, it seemed so easy. In Nashville, they call naive, money-spending lesser talents "Tex Nobodies." "That name kind of irks me," says Roland Janes, the veteran from Sun, "because everybody is somebody. Without those people, there wouldn't be any music business. They're the bread and butter, what keeps the industry going while you search for hits. A bunch of smaller studios couldn't survive without them. Those people are spending their good, hard-earned money and you're taking that money, so you have an obligation to give them the utmost respect and do the best job you can. I don't have a name for that except 'Ladies and Gentlemen.' I chuckle sometimes like everybody else, but deep down inside I have only respect."

Others no doubt share the respect, but their chuckling is louder. In a boom situation like Memphis's through the 1960s and into the 1970s, the numbers of these wanna-be records increased exponentially. The grist for such a mill is people way on the outside, or on the periphery of the inside of the business, who do not see that the factors involved in making a hit are so multifarious. With so many aspects under so many different domains, with such a large element of chance, that talent, Q quotient, "it"—whatever you call the quality that makes a record special—that quality alone does not guarantee success. As the radio constantly reminds us, talent is not even a necessary ingredient. It's a hustler's business. Any record could be the next big hit. Conversely, the next big hype could be the next big flop. Janes sums up his point: "There's a lot more money spent in the music business than is made in it. You take away that base, and there would be no music business."

Sam Phillips's original storefront business catered to such a clientele. It allowed him to scout talent, and brought us Elvis Presley. Another producer in Memphis named Red Matthews contracted himself out to singers, recording them in ballrooms of hotels with radio station gear. Roland Janes played for Matthews on a session for one Brother Dave Gardner, resulting in "White Silver Sands," one of Memphis's first million-selling records.

One Memphis entrepreneur devoted his career to near records, establishing several labels and releasing some of the worst records of all time—without a trace of irony. The philosophy of Style Wooten was that he was being asked —paid—to record an act, not to determine how worthy or talented they were. His Pretty Girl series was for female singers and featured a photo of the vocalist near her name on the label. Designer was his gospel label; with the built-in audience at their church, gospel acts made steady clients. Camara, Burch-lo, and a variety of other names—Style Wooten was there to suit your every recording need. He worked regularly at Sonic, where he and Roland could cut six gospel acts in a weekend, four songs on each, from which Wooten could pick A- and B-sides. "Style never tried to cut hit records," says Janes. "He just cut custom records for people that wanted records. For a fee he would see it through from start to finish."

Lee Baker's first studio experience was an encounter with both Style Wooten and Roland Janes. "The Blazers had an original song called 'Hard to Please,'" Baker recalls, a hint of a grin creeping onto his face as the memory comes back to him. "We cut 'Bo Diddley' on the other side. Style was gonna take it to Nashville. God knows where it went. I just remember looking at

Style and going, Man! I didn't know a lot about records and stuff like that but I knew that he wasn't it. I knew that Roland was cool and Roland and I got along. The thing about Style, we knew he was bogus—he radiated that he was bogus—but he used Roland and we knew that's where Travis Wammack was. So we thought we had a shot."

Near records were expedited by the proximity of Buster Williams's pressing plants. After spending thirteen dollars at Sonic, for another hundred or so you could get five hundred records pressed by Williams, then take them across town to his Music Sales Record Distributor, and they might pick up a couple hundred for jukeboxes, radio stations, and one-stops. "It was a great thing back then," says Janes. "Everybody worked together and it was conducive to people getting a break."

All the studios—Sun, Stax, Hi, Fame, Ardent, American—cut the near records alongside the real ones. "Everybody wants to be in show business," says Jim Dickinson. "These guys would hang around Chips or somebody and they'd see a session and they'd wanna do it. Especially when they couldn't figure out what anybody was doing. Parks Matthews, I remember cutting his first session. He'd been hanging out over at Chips's. He came in the door of Ardent on National and he had these two skaggy-looking broads with him, two ice chest coolers, and this big orange plastic bag like a trash bag. And he says to me, 'Jimmy! Jimmy! I'm all ready to cut a session. I got the girls—I got the liquor—I got the beer—' and he held up the orange package and he says, 'And I got the pills!' He figured that was all he needed."

The Fame studio in nearby Muscle Shoals was achieving such success with artists like Wilson Pickett, Aretha Franklin, and Clarence Carter that Rick Hall, who ran the place, opened an eight-track facility in an old grocery store in South Memphis. A disc jockey from Tupelo, Sonny Limbo, was sent to run the place. (Limbo discovered country superstars Alabama and also was listed as cowriter on the schmaltzy hit "Key Largo.") "Everybody was trying to build a studio situation where they could work on a larger scale than just booking time on an hourly basis," says Dane Sullivan. "Fame in Memphis was working on a variety of stuff, a lot of it a departure from what they were doing in Muscle Shoals. Jerry Lee Lewis and Tarp Tarrant were cutting there, Bowlegs Miller, even Piano Red." Liza Minnelli came to Memphis and spent time at Fame, but no recording was made. "There was a woman who cut one of our songs who had these Vegas people as her manager," says Dane Sullivan. "Their guy got into some controversy with someone behind them driving with his bright lights on. The manager got out, told the driver to fuck off, and tried to slug him while the window was down. The driver rolled the car window up on his arm and takes off, dropping the window as they curve in front of a semi. The manager was a puddle of blood on the road after that. A lot of projects at Fame seemed to end that way."

The music writer at one of Memphis's daily papers had a bet with Sonny Limbo. The producer claimed he could cut a hit record on the worst singer the writer could find. "Vicky L.?" says Dane Sullivan, his voice rising in a question mark.

"'Knock Three Times' and 'Muleskinner Blues' were her favorite," answers Gala.

"She had Elvis ducktails and a fifty-seven Chevy."

"What they did was really mean. Vicky was working successfully at some club near the airport when [the journalist] found her," continues Gala. "She had written a couple songs and her parents came with her to the studio. The Chivas Regal was flowing."

"Sonny Limbo literally fell out of his chair, onto the console, laughing hysterically," says Dane. "She's out there singing, thinking she's recording a serious session, and they're cutting in on the talkback, 'Sing it whorey, bitch. Sing it like a slut.' And her parents are back there saying, 'You've got to record her on "All I Want for Christmas" and "Jesus Loves Me."'"

"I resented the exploitation aspect of the music business very deeply for a long time," says Dickinson, "until I started to understand it. A couple years after I'd worked for Jerry Wexler at Atlantic in 1970, he came to Memphis for some event, and we ended up at this party, quite a party. Wexler, Sam Phillips, Betty Hayes—who'd booked bands with Ray Brown—my wife and I and Stanley Booth. Wexler had just produced an Aretha Franklin gospel record, and he was real proud of it. He had also, not that recently, done a Tony Joe White record. Well, after dinner everybody was kinda laid-back and Wexler kept trying to play this Aretha Franklin record, but every time he'd start it, Sam would take it off and put on the Tony Joe White record. Sam kept playing the same cut over and over, 'Got a Thing About Ya, Baby,' which was a hit. And finally Wexler says, 'Sam! Baby! You know, I'm really hurt that you're not listening to my Aretha record, baby!' Jerry plays it again and so one more time Sam gets up and takes the record off, puts on 'Thing About Ya, Baby' and says, 'Goddamn, Jerry, that's so good, it don't sound paid for.'

"I thought, By God, that's it. They can hear the difference. To somebody at the level of Sam Phillips and Jerry Wexler, that's what they get off on. Not paying for it! I believe that during the recording process, for a successful record, everybody's got to get off on something. Compromise is the nature of the beast. But everybody, whoever he is in the project, has got to get off on something. I had always taken it real personally when they didn't pay me. I'd say, Oh, the bastards didn't pay me again. Now I understand this sense of larceny as an element of production."

One Memphis music entrepreneur, Eddie Bond, had been angling for a break since he'd established himself as a rockabilly artist with the classic "Rockin' Daddy." Being on a label wasn't enough for Eddie, who has created several recording companies of his own. In his search for the right "in" he has also been a disc jockey, owned radio stations, and hosted his own TV hokum variety show, appealing to an audience of mouth breathers—a friend to the people who needed a friend.

Then Eddie Bond discovered Buford Pusser. The *Walking Tall* sheriff had a natural flair for the dramatic; before becoming a lawman, he'd been a professional wrestler. His battle against evil in McNairy Country (about two hours from Memphis) resulted in great injury on both sides: It took the lives of Buford's wife, several of his friends, and finally his own—but not before he lived to see his life story become such a hit film that a sequel was made. Pusser himself was being considered for the lead role in part three when he was killed. "Eddie introduced Buford to the world, did TV shows about him, had this awful album of tribute songs, just wretched shit," remembers Dickinson. "I'll never forget seeing them pull into that alley at the old Sounds of Memphis studio, the blue lights and the sirens going off in what looked like a pimpmobile. It wasn't at all like *Walking Tall*. Buford was this big Frankenstein motherfucker. All fucked up down one side of his face. Eddie thought he was made, and everybody told him, Eddie they won't let you keep it, somebody's going to take it away. Sure enough, Bing Crosby Productions made the films. That was back when, in Memphis, if you had a crooked disc jockey, seriously now, a crooked disc jockey, a drunken brain surgeon, and a used car salesman, you were making a record. Those were the elements you needed to record somebody's girlfriend singing some dogshit song and everybody pay their rent."

As the seventies continued, so did the Dream Carnivals, though their spirit changed. Instead of a room full of people attempting a unified out-of-body experience, it became more selfish. Blame it on the Quaaludes. At the 1975 St. Valentine's Day Massacre, held at an old movie theater a block from the first Dream Carnival, it was obvious something had changed. "That was a horrible night," says Baker. "The vibes were real fucked up. That was during the days when people were taking a lot of downs and drinking a lot. This 151-proof rum was real popular that night, and people were just fu-ucked u-up. There's a picture of Graflund and Campbell on the stage, and we're playing and they're talking and Campbell looks like a fucking beast."

Graflund recalls that Campbell had, indeed, been a fucking beast that night. "Campbell came with his girlfriend and a couple more of his buddies, and they were sitting out in the audience. Right behind them was Dewitt Jordan, a respected black painter. Dewitt said something to Campbell's girl-friend, something like, 'Hi.' And that was just the end of that guy. Campbell beat him up and threw him out in the street. And I think Dewitt came back in and Campbell beat him and tossed him out again."

Even stoic, predictable, beautiful, traditional folk music was getting fucked up in 1970s Memphis. An urban juke joint for artists called the Procapé Gardens opened in Midtown in 1974. Located next to the city's most promi-nent gay bar (where the floor show was always on when the folk artists were on break), Procapé was an acoustic nightclub that was the inverse of the early coffeehouses, exploring instead of shunning the side of folk music that was wild and unrestrained. Serving hard liquor no doubt helped.

Never having adapted to the electric guitar, Sid Selvidge remained the pre-eminent folkster. For the club's three-year existence, he played Monday through Wednesday nights. "It was great because the only people out on those nights are the ones who want to be there," Selvidge says. Though he usually began each evening alone onstage, rarely did he end that way. Often— regularly—there were too many guitars to fill the stage, so musicians played along from their seats in the audience. "We had a big Bacchanalian time. Everybody would come and sit in, it was quite a scene. Horace Hull, who I'd begun playing in Memphis with a decade earlier, Lee Baker, Alex Chilton was hanging out there. And you could do anything ridiculous that you wanted and nobody would pull your plug."

His last sentence is a definition of the eternal spirit of Memphis music. At Procapé, Memphis photographer Dan Zarnstorff remembers watching Eggle-ston "try to jump up and click his heels twice. He'd throw his legs straight out, horizontal to the floor, and fall flat on the terra-cotta tile. He bounced himself three times before somebody finally stopped him. You could always go to Procapé and know you were going to see some good friends." Marcia Hare was then hanging out with a wealthy junkie who was trying to die. "Larry would take me to the Procapé when Sid was playing and buy me dinner. It was a folkie scene but I would get up in front of Sid and dance, upstaging him. Larry loved for me to sit beside him and toss beer mugs into the middle of the floor and break 'em. I'd do that every night and it was so much fun." When Larry returned to jail, where he successfully overdosed, Marcia went to the

Sid Selvidge gig at Procapé Gardens. Left to right: Gimmer Nicholson, Larry Davis, Alex Chilton, Horace Hull (back to camera), Sid Selvidge, Lee Baker. Photo by Dan Zarnstorff.

club on her own. "I went in there by myself and started breaking beer mugs and they called the police on me. I said, 'What's the deal, I've been doing this for months.' It turned out that Larry had been paying somebody a hundred bucks before I'd start throwing the glass and this time he wasn't there. Instead of going to jail, I had to agree I would never come back in there. That's when I got my first job in Memphis dancing topless. If I was going to be barred from the Procapé, I might as well be able to go somewhere."

Selvidge remembers Chilton using the Procapé to work though an artistic transition. "Alex was at a juncture," he says. "He'd had a real bad experience with the Big Star stuff and was trying to distance himself from his acceptable past, I felt, because what he would do at the Procapé would chase people off. They didn't understand it. His whole concept was, If I were a thirteen-year-old right now and I were just learning my instrument, how would I play guitar? People don't realize what an accomplished guitar player Alex is, his versatility. He's a consummate guitarist. So from that level of sophistication, he was trying to play without knowing all that he knows. He was trying to play note for note what somebody who doesn't play the guitar would play like. That's a pretty convoluted concept, but that was his idea. And it fits perfectly into rock and roll. This was popular music to him, from where he came at it and got his hits in the first place."

The nights at the Procapé contributed to Chilton's album *Like Flies on Sherbert* and, before that, to Selvidge's *Cold of the Morning*. Both were released on Selvidge's Peabody Records. That label was begun by a patron of the Procapé, and Selvidge's album was to be the first release. After the recording and pressing, but before the release, the patron pulled out of the record business. "I was handed several thousand of my own records, and that's basically how Peabody records began."

Cold of the Morning, released in 1976, captures Selvidge's voice in its pristine glory, the finest silk. It's also a good taste of Memphis in the mid-1970s: It's a folk album and Memphis was a folk town—the legacy of the blues players prevailed. But it's folk with a glass-throwing edge, the abandon of rocking demons playing beautifully and unrestrained; Jim Dickinson produced. Much of the album is simply Sid's voice and guitar—"Boll Weevil" is an a cappella field holler. He proudly interprets George M. Cohan alongside Furry Lewis. On the two songs that feature the acoustic version of Mud Boy (plus tuba), one can almost see the players in the studio, nodding to each other, knowing when to lay out and when to come in. Their work is like an oriental carpet with secret messages stitched into the weave.

Selvidge's earlier LP, a 1970 release on the Stax subsidiary Enterprise, does not fare so well. *Portrait* catches him in fine voice, but the day it is reissued on CD, we must question the industry's sanity. Don Nix produced it at Ardent, and in his defense, it was a very purple period in music, tending toward overproduction. "Children's Suite" epitomizes the sappy, overblown production; the strings are so thick, Lawrence Welk would have hanged himself. The couple of songs unencumbered by the cotton candy hold up well, especially "Amelia Earhart," which is practically the artist unadulterated. "I'm not ashamed of this album," Selvidge says. "I did my part, I sung these as demos, and I was young and hot to get on vinyl and somebody walked in with a deal. I'm still glad it didn't hit big because then I'd be stuck with this type stuff the rest of my life."

Selvidge has continued to play clubs regularly, garnering rave reviews in the *New York Times* for his 1977 dates at Tramps; in 1992 he played Carnegie Hall. He has released two more albums on his own Peabody Records, and in 1993 he released *Twice Told Tales* for the Elektra/Nonesuch American Explorer series. If you were going to buy just one Selvidge album, hunt for 1982's *Waiting for a Train*, and play the title cut over and over. Simply the way the track builds from solo artist to the whole band is worth hearing, but you also get a taste of Selvidge's falsetto howl, Baker's insane slide guitar, and Dickinson's piano beating. The album is less folk-oriented than *Cold of the Morning*; the backing band is essentially an augmented Mud Boy, with Selvidge taking all the leads and Jim Spake blowing some mean saxophone. On "Swanee River Rock," Jim Lancaster plays the rockingest tuba solo I've ever heard north of New Orleans. Selvidge's spare rendering of Fred McDowell's "Trimmed and Burning" will satisfy anyone's folk jones.

The disappointing part of seeing a human being bite the head off a live chicken is the ease with which the chicken's neck disengages from its body. Geeks don't so much "bite" the head off as, with the chicken's head in their mouth, they pull the head and neck loose, kind of like sucking your thumb but yanking it all the way off.

This I learned while watching *Stranded in Canton*, a cinema verité document of Memphis made by Bill Eggleston while the Procapé scene was going strong, while Campbell Kensinger was still alive, when video cameras were decades away from being common. Eggleston purchased one of the earliest Porta-Paks made by Sony. Using his knowledge of electronics and photography, he modified the instrument, enhancing the image clarity, crispness, and value. He wanted to document the moving world around him in the same way he'd approached stills.

Stranded in Canton is the name Eggleston has given the footage he shot during the mid-1970s in Memphis, New Orleans, and the Delta between. A geek grudge match on the street in New Orleans is one of the least exciting events he recorded. "We've got our geek here, and we'll put him against your geek," a not quite sane-looking man announces to the small but accumulating crowd. How the show began is not clear, whether the camera stumbled upon it or whether the show is for the camera's benefit. Once headless, the chicken continues flapping its wings and the first geek smiles a feathery grin. Someone in the crowd yells "Geek power," and the second geek steps forward. He doesn't look cocky or nervous or any different from most of the people gathered around him. He strokes the chicken's beak, calming it, and repeats the act just seen. Somehow his show is better. He gets more applause, and Eggleston brings the camera from the front of the crowd around to the side, up close to geek number two. He removes the soup stock from his mouth, and now the geek is glowing, aware he's the winner. When the camera is upon him, he too smiles, chicken cartilage limp on his lip.

Stranded in Canton is a document of the spirit of Memphis, or more precisely, of Midtown Memphis, 1970s. Using natural light (and later infrared tubes), Eggleston shot unobtrusively in bars, backhouses, fields, cars—day and night. "The electricity generated in a room with Campbell and Eggleston videotaping was amazing," says Mary Lindsay Dickinson. "The maestro at work with what may have been the world's first infrared handicam! People would do absolutely anything to be in the movie." Those few who might have been constrained by the camera assumed it was too dark, or that there was no tape in the camera, or that Eggleston was just looped and playing around. "My idea was to shoot whatever was out there," Eggleston says. "The second I saw the first reel—I believe it was when we were in New Orleans and went back to the Royal Orleans and piped it through a TV and it was beautiful—I was very happy. I knew that it was perfect. I was looking forward to more of it and we just kept doing it."

Eggleston is as dashing a man as God has put on this earth. Always trim and crisply dressed, he is quietly acute. Soft-spoken, the precision in his manner, his expression, and thoughts belies charges that his photographs are mere snapshots. To discuss Memphis, we left the city for an antebellum home he is rehabilitating in northern Mississippi. The appreciation and attention to detail he devotes to the refurbishing of this prized home is the same he gives to his photographs and to his subjects in *Canton*. None is more freaky or more precious; all are prized. "*Stranded in Canton* works because of the way the whole takes came off, unedited," he says. "The way it looked, the way it was paced and things moved. The meter to it. It was as if we were looking at something that had been shot fifty times and this was the best take. And always these were the only takes. It was just that good."

"We knew at the time it was really special," says Randall Lyon, who was Eggleston's assistant on many of the shoots. Lyon had worked in television and understood lighting; he became Professor Reflecto, holding the single light employed when a situation demanded it. Usually, however, Lyon was on the other side of the camera, the subject, his innate theatricality and brilliant spontaneity a driving force of the piece. "We tried to do as best as we could. We knew our technical limits and we tried as much as we could to really kick out the jams."

They succeeded. As the various scenes unfold, whether it's Eggleston's underage black chauffeur sitting on Stanley Booth's lap and chewing bubble gum, a drag queen in New Orleans draping himself in toilet paper to perform for the camera in a tiny bar, or a tanked veterinarian in a seersucker suit (Tony the Tiger looking over his shoulder from a box of Frosted Flakes) doubting a statement made by Campbell Kensinger (to which Campbell responds, "I'm heavier than I am tall"), the viewer is quickly and stealthily brought into this other world.

In one scene, the camera follows Lyon into Jim Dickinson's backhouse studio, where assorted people are sitting around: Dickinson and Mary Lindsay; Jerry McGill, who had recorded for Sun and sang with bands around the time of the Regents; Jim Lancaster, who often accompanied Mud Boy on bass and rock and roll tuba; his wife Jill; Marcia Hare, and a few other people. Some of the musicians are jamming—Dickinson is playing an electric guitar, Lancaster is at the piano. McGill, his gaunt and tapered face resembling a cobra's, takes hold of an acoustic guitar and performs a song. When he's done, Lyon begins spouting a soliloquy, holding a bottle of champagne in his hand. The camera surveys the room, but his words are clear. "This is a dis-ass-trous period in our time. We got to respond to what's going ahn or else we got to hang it up with kinder-goddamn-garten." Dickinson accompanies with apocalyptic feedback from his guitar, and it all becomes too claustrophobic for McGill. The camera whips around at the sound of gunfire—McGill has drawn and fired his pistol. He smashes Lyon's bottle with the barrel and then puts the gun against Lyon's head. The voices that squealed when the bullets caught them off guard have suddenly stilled. The guitar continues, a soundtrack like the Wild West saloon player who knows it's best never to stop. The camera remains focused on the gun, the gun always, because whoever may say whatever, the subject in that room is the gun.

"I'm gonna whip you with this gun barrel," says McGill, whose eyes shine like BB's. "Be nice. Be real nice." Lyon is doubled at the waist, his head, his life, in McGill's hands. Then McGill—he is no longer McGill, he is Pancho Villa, he is Jesse James, he is completely and totally Lash LaRue—turns to the camera, sees that it's pointed right at him (he's still holding the gun), and he says, for the camera's benefit, "I don't care nothing about that." He'll do it for the world to see! In an instant, the pistol is waved, smashing the bottle in Randall's hand, and the following instant, smashing the light. The guitar feedback stops with the sudden darkness and the scene, take one, the only take, is over.

After several years, Eggleston began putting together a similar outfit to shoot in color. The equipment proved unsuited, too heavy to allow the necessary portability. *Canton* came to a natural close. "The piece was finished," he says. "It was like a movie, about ninety minutes, and it doesn't need to be any shorter or any longer or lack or need anything. It became apparent that the material was suited for black and white and when I realized what I had, the piece was done." Shortly thereafter, he was asked to be artist in residence at Harvard University. "I showed *Canton* for a solid year," he says. "Later, I showed it at Yale. The audiences seemed very moved. I see a lot of those people now and they always bring it up. I think some people remarked at the time, 'Are these people acting out something you told them to do?' And I would explain, 'No, I'm just recording them, whatever they're doing. I didn't tell them to do this.'"

Canton, friends, life in Memphis, Tennessee.

While still working on the project, Eggleston showed up one night at the Sam Phillips Recording Service. This was 1974, and the divergent forces in Memphis music came together that night at the post-Elvis studio built by the master around the corner from Sun. It was a session for Jerry McGill, whose Topcoats were the South Memphis hoods that East Memphis mamas were too scared to hire in the late 1950s. McGill's one released recording had been for Phillips some twenty-five years earlier. Jim Dickinson was producing the session, using demo money he'd wrangled from Warner Brothers. Those present included Danny Graflund, Campbell Kensinger, Marcia Hare, and another of the Mud Boy dancers. Impresario Jim Blake was there, as was writer Stanley Booth. Knox Phillips, Sam's elder son, was engineering the session.

Campbell sat in the control room drinking Wild Turkey from a fifth, bouncing the bottom of the glass bottle on the floor as he listened to other people tell stories. "You ever think of 'the Law' as a whole fucking table?" Campbell asked. And he picked up a coffee table, put the corner in his mouth, tried to bite it off.

Stanley Booth, who was practiced in the martial arts, leapt up and karate-chopped a decorative wooden shingle into two pieces. Campbell was extremely unimpressed. Jim Blake remembers Campbell saying, "These guys in my gang walk around with these guns and knives and shit and they think they're tough. I'll show you tough," and Campbell whipped his noonchuks from his back pocket and started hitting himself on the left side of the head. "About twenty times," says Blake, "stuff that would have not only disabled me or you but would have given us permanent brain damage. And he said, 'Now that's tough.' Hey, nobody could disagree with that."

Knox and Dickinson, who was thinly slicing hog tranquilizer for sustenance, kept their attention on McGill, who was overdubbing the vocal. Knox remembers that he was singing "Desperados Waiting for a Train," but Dickinson believes it was a Civil War song called "With Sabres in My Hand." In the latter number, there is a line that upset McGill every time, a line at the end of the song in which the southern narrator refers to "the lost cause." Dickinson leaned to Knox to confirm that he'd frisked McGill before the session.

"Yes," said Knox, "I got his gun."

"Gun?" said Dickinson. "Guns!"

At that moment, Eggleston was entering the studio floor with his video camera, and McGill's fuse lit. It was bad enough admitting defeat on audiotape, but to have it chronicled on video was too much. Eggleston crept closer as the song neared its climax. The camera was on top of McGill when he said the scurrilous words. The goddamn line sung, the final notes fading, McGill reached into his jacket and stated, "Lost cause my ass," unloading his six-gun into the ceiling. Instinctively, everyone in the control room ducked. When Dickinson looked up, Knox was still at the board.

"What're you doing?" he hissed.

Stated the ace engineer, "The gun needs mo' echo."

With the vocal completed and the place turning into a mad scientist's laboratory, Knox was ready to call it a night. He had to be convinced that they'd never have such vibes again, and to do the song justice they'd need to finish under these unique circumstances. (And here, Knox, take another slice of this.) For the sake of the song, he agreed. Graflund had moved to the couch out front. Campbell was beside him, bouncing the whiskey bottle on the floor. "He was getting cranked," says Graflund. "He was thinking, 'I should be in the limelight, not this guy.' It was getting to that time of the evening. I knew someone was fixing to get it." Graflund's nose, perhaps from the tension in the room, suddenly began to bleed, so he went in the bathroom. Campbell returned to the control room, and a few minutes later, the bathroom door where Danny was washing his face whipped open. It was one of the girls. "Her nose is bleeding like a faucet," Graflund continues. "She opens up this door screaming, sees me lift my head, I've got blood running down my face, and she just screams and runs out the building. I go back in and say, 'What the hell happened to her?'"

"Campbell was very sorry about that," says Marcia Hare, who'd witnessed the event. "He didn't know her and said that in the future we were to introduce him to anyone we didn't want punched."

As those things go, the night ended suddenly. Stanley Booth and Graflund got their background vocals down, the sun was coming up, and so everyone rose to leave. With feelings of satisfaction and wonder, they all walked outside together. Greeting them was a line of motorcycles up and down the street. Campbell's bike had been brought for him and was by the door. Graflund says, "He walked out, cranked his up, they all cranked up, *varooom,* and they rode down Madison in formation with the sunrise." The tapes remain unreleased.

In the Memphis tradition of recording freaks, the great seventies label was

Jim Blake's Barbarian Records, begun in 1974 and struggling along the margins of society through the end of the decade. Blake is a hustler from way back—swapping comic books in grade school—and the principles he learned on the playground trained him for the record business. The Memphis home he rents is overflowing with unidentified but interesting-looking boxes and stacks of papers, magazines, and records. Some might call him a pack rat, but Blake is waiting for the world to catch up with him, to share his ideas of what is valuable. "I feel like the android in *Bladerunner*," he says. "If people could see the world like I do . . ." He shakes his head. It was Blake who was negotiating for the rights to *Conan* eight years before it was made. When he was promoting records, he tried to get his boss to hire Sweet Connie, the South's most famous groupie, because he understood that she could get records played. Blake could as easily have been a Hollywood mogul as a renter in need of dental insurance. The breaks fell against him, a throw of the coin, but he is the stuff of empires.

Down the block from the first Dream Carnival and a year earlier, Blake was the buyer at one of the city's most popular record stores. Utilizing his place at the epicenter, he began an underground newspaper. He went on to open his own store, expanding into head shops and enlarging his publication business. Memphis disc jockeys realized he was riding the crest and hired him to program their shows. His work got several jocks promoted to prime time on both coasts. In the mid-seventies, Blake himself entered the big time. "Jerry Lawler had drawn comics for my newspapers," he says, "and in 1974 I began to manage him full-time. My interest was in Lawler as an artist. I had this background in records and radio and retail, and I understood that the reason the earlier wrassling records had failed was because they were the wrasslers being projected as wrasslers, instead of as recording artists. The wrestlers never saw records beyond something else to sell at the matches."

Blake helped Jerry "the King" Lawler give personality to his career. He'd seen Lawler's fans dip their wrestling programs in his blood. He knew their devotion. "My whole trip with Lawler was to get him and Elvis together. If I'd ever had that picture, it would have been all I'd needed to take Lawler to the next level as a recording artist. I tried to get them in the ring together, wrestling versus karate." For Lawler's first recording, Blake chose "Bad News." He knew Dickinson from a creative writing class at Memphis State; Dickinson did not have to be convinced into merging wrestling and rock and roll. His participation lured a mixed bag of musicians, ranging from the members of Mud Boy to Teenie Hodges, hot with the Al Green records. This recording set the model for all subsequent Barbarian sessions. "Lawler was a bad guy at the time," says

Blake. "We sold twenty thousand of those records at the matches all over the territory and never got a moment of radio play. People were buying 'em for a buck and a quarter and breaking 'em right in front of us. Great! I'll sell you another."

Blake's friendship with Campbell Kensinger kept the biker from attacking the wrestler. "Campbell Kensinger watching Jerry Lawler on television talk about how bad he is? When Campbell knew that he could beat the piss out of him? I told Campbell I needed both of them around because I was trying to get the film rights to *Conan the Barbarian*. Not that I saw Lawler as Conan; I wanted Lawler to be the bad guy. Campbell should have been Conan. The audience would have known he was real."

Blake began to record other wrasslers, and Lawler went on to record a version of "Cadillac Man" and, in 1977, "The Ballad of Jerry Lawler." "Jerry was at my house and we were working on a poster for that record. Our theme was 'The king is dead. Long live the king.' We were getting ready to print it when the news came on the radio that Elvis died. I pulled the record off the market, even though it was the number-one request record in the area. I didn't want to make any money off of a dead man, it was bad luck. My heart went out of the project after that. The Elvis thing was so important. I'd worked so long and so hard to get Lawler and Elvis together."

During that time, Blake had earned the respect of the musicians. He kept his Barbarian label alive with their help. "The reason I was able to cut the records that I cut was because I'd done the wrassling thing. Everybody had laughed at me and I proved them all wrong. I made everybody in Memphis want to come to the matches. Everybody wanted the records, everybody wanted to meet Lawler. Nobody could afford to laugh at me anymore, and that's how I was able to do my other Barbarian projects, crazy as they were. I recorded giants and midgets and transsexuals. I even got Tommy Burk to reprise 'Stormy Weather.' I was set up perfect when punk came in."

When the Memphis musicians had no other outlet, Blake was cutting them. The more extreme the better. "Everybody liked doing Blake sessions," says one musician, "because they were such a party. You never knew what was going to happen."

"I engineered a bunch of Blake records," says Richard Rosebrough, a former Ardent employee. "I saw all this insanity going down, people making tapes in these insane ways, and I was getting burnt out on recording. I thought of Blake records as Blake tapes, because they'd never come out. We were in la-la land, doing crazy things."

Blake's extensive archives dwarf Barbarian's relatively small output. He did, however, release an outstanding Lisa Aldridge record, and a Dickinson single, "Rumble," that Robert Palmer named in the *New York Times*'s top ten of 1980. Aldridge's version of the Velvet Underground song "Story of My Life" was mixed by the Regents' Rick Ireland. He inverted the standard mixing rule of turn up the good parts and turn down the bad parts. In an entirely pleasing way, he made the guitar distortion and the mistakes the focus of the song; these unique sounds suddenly come way up in the mix and the listener gets whiplash, turning toward the speaker, checking for dust on the needle, before realizing it's part of the song. "Rumble" was recorded over several years, a reel of multitrack tape thrown on the recorder at the end of a bunch of sessions, filled up and mixed down to another reel, which is then filled up and mixed down, again and again. Wholly singular, it sprawls from track to track, session to session, studio to studio, year to year. Impossible to imitate, impossible to duplicate, it is a hit that no one will hear, a pinnacle of the underground, recognized by the *Times* and completely unavailable. (Copies of the single, with its intricate packaging, now sell for two hundred dollars.) Symphonic in its chaos, it is unlistenable because it barely exists.

In Jim Blake's vault are a hundred hours of recordings featuring the finest Memphis musicians. He has solo recordings of Sid Selvidge and a duet between Alex Chilton and Jerry Lawler. He has recordings by the Klitz, by Mud Boy, an a cappella cowboy on downs. Though there have been several times when the market appeared ready for his treasure trove, to Blake, it was not ready enough. Helping him try to get the records released, I had to establish a rule that when we were discussing the price for these tapes, the words "a million dollars" could not enter the conversation. Memphis obscurity personified, Jim Blake should be the head of Disney; he should have a multimedia empire at his feet, conduct meetings around his Beverly Hills swimming pool. The world may not yet share the android's vision, but his patience and unwavering faith is clear to those who know him, clear to those sweating beside him as they cart boxes around a Memphis warehouse.

CHAPTER FOURTEEN

Thank You Friends

IT'S EASY TO DESCRIBE THE POSITION IN WHICH ALEX CHILTON FOUND himself in 1974, impossible to fully realize. At twenty-four, he was an eight-year veteran of the music business, wearing a purple heart. He'd been on the top of the charts, heard his voice define the sound of the street, been on the receiving end of tens of thousands of screaming girls. The singing was his, the talent was his, but the Box Tops belonged to someone else. When he'd declared himself his own artist with Big Star, his work was universally admired and almost completely unavailable. It was a long way to fall.

The disappointment surrounding Big Star was fresh when Chilton returned to the studio the same year that *Radio City* was released. This next album was to be his second solo effort; at the time of this writing, the first has yet to be fully released. Chilton never titled the 1974 effort, but when it finally became available in 1978, it bore the name *Big Star 3rd;* Jim Dickinson's production notes reveal some other names they were considering: *Sister Lovers,* because Alex and drummer Jody Stephens were dating sisters, and also *Beale Street Green,* from a line in the song "Dream Lover." Chilton reflects, "Maybe I would call it Alex Chilton's something or other," indicating the solo effort that it is. Yielding to the vernacular, I will refer to it as *Big Star 3rd.*

The first two Big Star albums are exuberant cries of youthful energy, not just of pleasure but also of pain. Chilton had pushed pop lyricism into a new decade by sloughing off love in "September Gurls," singing, "I loved you/well, never mind." The third album is also exuberant, though the proportions of pain and pleasure have been reversed. The lyrics are cocky and existential, anguished and depressed. In "Big Black Car" Chilton sings, "Nothing can hurt me/Nothing can touch me/Why should I care/ . . . it ain't gonna last."

After the failure of Big Star's first two albums, *3rd* is a confirmation of every-thing shitty about the music business, a sworn testament by an artist who knows.

"That record, to me, is a very unhappy sort of thing," says John Fry, who engineered the recording. "Basically it's a chronicle of all the stuff that was going on in Alex's life at the time, and the thing is by and large pretty depress-ing. You don't do a record that sounds like that when you are feeling real pos-itive about everything that is going on in your life. If an audience is going to get a record like that, it's bought at the expense of somebody's pain some-where.

There were some strange sessions. A lot of it we were hiring the best play-ers you could get and trying to get them to do the worst they could. And then other sessions you'd get the worst players you could and try to see just what they'd do anyway. You never knew whether you were gonna get more than one or two takes. It's almost like a Fellini record, if there were such a thing. It's the juxtaposition of all these bizarre elements, and I guess it makes something that's interesting and it's probably a good thing that it was done and pre-served. But everybody certainly was not having a good time while they were doing it."

3rd is a landmark rock and roll album. Alex Chilton, like many blues artists, looks into the abyss of personal turmoil, chronicles an agonized love relation-ship, and wallows in the darkness of drink and drugs. The audience shares his anguish through the recording; the experience of disappointment is universal, and *3rd* is driven by monumental disappointment. Unlike many such expres-sions, his is not cacophonous. The album is filled with beautiful melodies and somber tones; Chilton's fine and pure voice is a reed bending with the wind. The mix is pristine, as immaculate as the previous Big Star albums, but more capacious. These songs are not about rock and roll guitar, bass, and drums. They are about texture, contrast, blankets of blackness punctuated by minia-ture holes of light. Instruments are often not recognizable, words are unclear, and the effect is one of suggestion, impressionistic rather than representa-tional.

"I thought what the guy would be doing with his music was something that might lead to his getting a record deal so he could keep on recording," re-calls Fry, who was engineering material he felt he'd be unable to sell. "Though at that particular time in his life, I'm not sure he had any choice about it." Our discussion is nearly two full decades after the fact, yet John Fry is visibly un-comfortable when recalling *3rd*. He fidgets, his lips sort of smack in a grimac-ing way, his conversation is filled with pauses. "I think some of the things that

Jim and Alex did and some of the technical things were good, but the overall unpleasant aspect of it to me is the subject matter and the content of most of the songs—and what you could see Alex was going through in connection with that, and what leads up to getting somebody to do that. Jim has a reputation, well deserved I think, for being able to work with and get results from difficult people. And he does that in a pretty unobtrusive way. His approach is to capture the moment and then if there's something we need to add or fix we can deal with that later in a little saner climate. But if it's gonna be insane anyway, just let it get as insane as it can get and see what happens. And we did a lot of that."

"When I first met Alex," says Dickinson, "he was living in his mamma's house and he had the gold record for 'Cry Like a Baby' on the wall. It was sealed, like in a glass box, and the label had peeled off the record and fallen, kind of laying over in the corner. And that summed up Alex for me then." Dickinson was an established producer by 1974. One of his projects had been Dan Penn's never-released second album, *Emmett the Singing Ranger Live in the Woods*. The two friends had fallen out during the mix, over a financial matter, and Dickinson took the Big Star project partly as revenge on Penn; Penn was aching to cut another hit with Chilton. Charlie Freeman, Dickinson's best friend, had died the previous year. The record contains some of his own anguish.

"*Big Star 3rd*, to me," Dickinson continues, "is a catharsis, a series of very different emotional responses that Alex is having. You don't get to do very many *Big Star 3rds*, they don't come along very often—thank God. You'll notice that Alex hasn't done another one either. I've been accused of indulging Alex, and some of it was indulging, but you listen to that record, you hear it working. Where people had normally told him, 'No, you can't do that,' I told him, 'Sure, we can do that.' I figured out a way to do 'that' each time. When it didn't work, we did something else. We created the guitar sound by solving a problem. Alex had this big ol' Ampeg amp, wretched sounding, and he wanted to play real loud. So we turned it all the way up and put the mike across the room, where it wasn't supposed to be. There's great big hunks of that stuff that just worked great. I think back on what it actually is—it's 1974!—what we did seems heroic. The synthesizer parts on 'You Can't Have Me' are being randomly triggered off Tommy Cathey's bass. We did the same thing to an upright bass player we hired and he thought we were completely insane. He would laugh openly while he was playing.

"'Kanga Roo' was when I got Alex's trust. He came in in the middle of the night with Lisa Aldridge and he recorded it with his voice and the twelve-string acoustic guitar on the same track, just to make it harder. He said to me, 'You want to produce something? Produce this.' So I started stacking stuff on it. I did the strings first with the Mellotron, then I started playing guitars. Pretty soon Alex was out there with me."

Lisa Aldridge was Alex's girlfriend at that time, later to lead an all-girl Memphis group, the Klitz. Dickinson posits that *3rd* is about her. "She was a real part of the process, and knew it. And put up with a whole lot of shit for the project and for Alex."

"Femme Fatale" is included as a nod to Lou Reed, who in 1973 released *Berlin*, a dark album similar in tone to *3rd*. The song includes Steve Cropper's contribution to the album. Richard Rosebrough was engineer when it occurred. "Stax was failing then, Cropper's own studio TMI had already failed, and he was at Ardent working with John Prine," says Rosebrough. "Typical Dickinson, if there's someone hot in the building that's playing, get 'em on your album. Cropper didn't want to do it. He said he'd do it if he had time. I knew how to get things in red fast, so when he walked in and said, 'I have ten minutes,' I had a direct line waiting for him. He would not enter more than two feet in the door. He put his guitar on, we did a few passes and that was it. Cropper was scared of it. He thought this was scary evil shit."

3rd has been released at least three times, each time with a different sequence, usually with a few extra tracks. Changing the order of the songs radically affects the meaning of the album. It's like giving the same film to different editors. The latest, and probably the most broadly known sequence, was released by Rykodisc in 1992. I saw a young clerk in a record store put on this new CD and when the opening track came on—one of the album's few upbeat songs—and he played a brief air guitar, I knew he was discovering an album different from the one I first heard. "Thank You Friends," now the second track, used to close the album and serve as an appreciation to pals who would stand close even through times as difficult as those represented in the album's journey; it's now just a rock song that helps cushion you against the blackness, introduced by the third cut, "Big Black Car." By the time the 1992 version gets into a series of heavy songs—"Nighttime," "Blue Moon," "Take Care"—the lighter, rock and roll mood has been well established.

Nat King Cole's "Nature Boy" features Bill Eggleston on the piano; he'd recently hurt his foot and the sound right before we hear Alex smile is Eggleston's crutch falling to the ground. "Dream Lover" was not included on the original release and proves to be a piece of the album's core. Harmonizing with himself, Chilton shows no boundaries, wrenching his guts—wrenching our guts—twice. He gives himself over to the song completely. The gnarled guitar is Lee Baker, and it creates a strong tension against Alex's spare piano, cushioned by the strings. It's full of drama and tension. "We didn't even arrange that song," says Dickinson. "Alex said, 'I've played it twice, once when I wrote it and then I played it for Lisa and I shouldn't have done that.' He said, 'If I play it one more time, I'm going to be bored with it.' That's the kind of thing I'm sympathetic to, so I said, 'Okay Alex, sing a little bit of it, we'll find some stuff, then we'll do it.' Baker didn't even know the changes, he just started playing. It got to this point after the bridge and Alex just leans over and says, 'Play it for me, guitarist,' and Baker goes into a scratchy kind of funk solo. We overdubbed the Memphis Symphony on it, and it's really pretty good."

Early into the sessions, Chilton hired a bodyguard. Don Nix was recording down the hall and he had one, Jerry Lee Lewis had at least one, and Alex figured it might keep him out of trouble. Danny Graflund was a large guy, a friend of Dickinson's, and he needed a job. His relationship with Campbell Kensinger, who was still a year from being killed, added a sense of danger. "It got to be where we started hanging out," says Graflund. "Alex would ask if I wanted to do something, said he was picking up the tab, and when he'd introduce me as his bodyguard, I didn't care. I was having a good time." The two remain friends. Back then, they were drinking heavily, eating downs, rubber-legging their way through Midtown. Part of their professional relationship involved the freedom for Alex to bring situations to a sharp edge, when Danny would step in to resolve them.

Graflund recalls one night when the two of them, accompanied by a girl who had a tape recorder in her purse, went to hear Jerry Lee Lewis at a Memphis club built on the site where Beatnik Manor had once stood. Selecting a large table up front, Graflund removed the "reserved" sign and they occupied it. When the reserved party arrived and tried to claim the table, words got heated. On the tape, one man drawls to Chilton, "You'd better tell your friend to settle down because there's a bunch of folks here that want to whip his goddamn ass." Chilton taunts, "If you're not some kind of chickenshit, you'll tell him yourself." But the redneck never has a chance to speak. Graflund, pilled up and drunk, has become his alter ego, Other Man. ("I drank, waiting for that 'click.' Then I'd be Other Man. People might say, 'You went into a bar last night and grabbed this guy's wife and jumped on the bar,' and I'd say, 'I didn't even go to that club last night.' 'No no, about midnight you were here.' I'd tell people *I* didn't do that. The click came, and I was Other Man.") Like a monster cartoon, Other Man rises from his seat, knocking over the table, sending glasses flying, beer bottles shattering. People scream as if in a 1950s Japanese horror film. Jerry Lee kicks his band into double-time. *3rd*'s "Whole Lotta Shaking" acknowledges the Killer as a kindred spirit.

"It took so long to do *3rd*," says Dickinson. "And we needed all that time to get what we got." Not only were they cutting under the shadow of the previous Big Star disappointments, but this time the Stax organization was caving in while the sessions were underway. "The test pressing has a Stax master number on it," says Dickinson. "As Stax was obviously going down, and Fry was pulling out with all of his product—they were owing him so much money that he couldn't go any further with them—we just kept on cutting." Fry was becoming more and more disturbed by the sessions. The plug was pulled when, on the third day of a bender, Graflund brought a drinking buddy for his first visit to a recording studio. Slim was not accustomed to the isolation necessary for recording, and three days out, the situation disoriented him. After some time, he was led out to the floor and placed in front of the microphone, given the opportunity to sing or speak his mind. It was a moment to be captured. Tape was rolling. What went on in his mind is impossible to determine, what he saw from the other side of the glass or what was reflected back at him. Whether it was because of the three-day binge behind him or because of the fourth day ahead, Slim, at six feet four, 250 pounds, and weathered like a radial tire, stood in the middle of the production room floor, headphones covering his head, and when given the signal that the world was his, the hulking figure grew silent, paused, breathed in, out, then broke down crying, sobbing with no control, blubbering.

Fry declared the recording process over. Despite the high tension, he agreed to participate in the mixing. "Fry was at the peak of his craft then," says Dickinson. "That's the last project that he did hands on, all the way from cutting it through mixing. That record is a testimonial to the engineering expertise of John Fry. That's why it sounds so good. It still sounds contemporary."

But in 1974, it sounded ahead of its time. Record labels uniformly rejected it. Dickinson remembers sending the tape to Jerry Wexler, for whom he'd worked at Atlantic. Wexler phoned him. "Baby," he said, "that tape you sent me makes me very uncomfortable."

"You knew you weren't turning the radio on and hearing anything like that," says John Fry. Fry and Dickinson shopped the tape on the East Coast and then the West, calling on all of their music biz connections. They were turned away repeatedly. When Fry gave up, Dickinson stayed, banging on doors until he was called home for Campbell Kensinger's funeral. The industry's reaction was enough to drive both Fry and Dickinson from the business. If it couldn't accept an album about the dark side of life, the business wasn't for them. Fry put his studio up for sale and tried to get out of the business; Ardent was bought, the sign was taken down, and the offices changed hands, but the new owners defaulted on their payments. Fry continues to run Ardent to this day; his fourth incarnation of the Ardent label began in 1993, and in 1994, Chilton signed a domestic deal with him.

Chilton says that in 1975, "something clicked in my head and from then on I knew how to write a thing that would please me." Cavalierly dismissing the Big Star albums, he is, I believe, an artist too close to his work. He hears only the "mistakes" in the songs, details invisible to the listener not involved in the creation process. In 1975, he recorded with music critic Jon Tiven at Ardent. Tiven was an early believer in Big Star, writing an extended article about the band in the March 1974 issue of *Fusion*. He developed a relationship with the artist and with Fry and came to Memphis in the fall of 1975. An EP was culled from the sessions for the 1977 release, *Singer Not the Song;* Tiven sold more of the tracks to a German label in 1981, and more again to an American label in 1993. The latter two releases share the name *Bach's Bottom*. Frankly, they're hardly special. The songs sound rushed, Chilton is nearly manic, the chaos distracting instead of invigorating, as on 1979's *Like Flies on Sherbert*. There are only two new compositions on these sessions. "All of the Time" was cowritten with Lisa Aldridge and is pleasant, if the performance is a bit grating.

The other new song is "Take Me Home and Make Me Like It," which came out of Chilton's friendship with his bodyguard. The title was a pickup line Graflund used with women in bars. "I'd be out cruising, being Other Man, and I'd see a chick somewhere. I'd move in and say, 'Hi, I'm Danny,' and then get into this rap, Other Man's version of 'Blue Suede Shoes': 'You can knock me down, drag me in the gutter, you can stomp on me, piss on me, spit in my face but then [through gritted teeth] take me home and make me like it.' Me and Alex used to go to his apartment, he would turn on the tape deck and say, 'I'll play, you sing.' I don't think he thought anything serious was going to come out of me, but he wanted a tape of it because it was good for laughs." Danny was in a bar one night when he was told Alex was at that moment cutting his song at Ardent. "I got some chicks, went down to Ardent, the guard let us in. This one girl looked at the control board and said, 'It's Christmas,' and began changing all the knobs."

Richard Rosebrough picks up the story. "This was the first night of the sessions, and Alex was producing instead of Jon Tiven. Alex would stand in the control room with a microphone and sing, while running the board and playing guitar. We're in red, I'm in the studio playing drums. I saw Graflund come in, go to the back wall and I could see by looking at his back—what is he

Dan Penn (in glasses) and Alex Chilton lounging at the Sam Phillips Recording Service, mid-1970s. Photo by William Eggleston.

doing? I could see this spot on the wall getting larger and larger. He peed on the back wall. I was sort of representing the studio at the session. We finished that song, I put my sticks down, and I threw them out. You can't pee on the wall! I have to use this room tomorrow!"

"When the record came out," says Graflund, "I told Alex it was shitty that I was listed as the fourth writer when Mark James and Dan Penn had begged me for the hook to that song. They were going to give me half-writer's credit for just the hook. Not long back, Alex let me know that he'd amended it and I had first writer's."

Ork Records, one of the nation's first punk labels, asked Chilton to play in New York when they were releasing the Tiven sessions EP. They said they'd help him put a band together up there. "They introduced me to Chris Stamey and Chris and I got along fine and had a good time," says Chilton. "We worked together all that year. That's when I produced his single 'Where the Fun Is.' By the end of that year, I don't think Chris wanted to work with me anymore. Chris had decided I was totally unreasonable. At the time, I was a drunk and I'm sure I wasn't the easiest person to get along with in the world."

Chilton's landmark single "Bangkok" was recorded while he was in New York, with Stamey on maracas and Chilton playing everything else. It's an eerie neo-rockabilly original, a lighthouse in Chilton's 1975–79 fog. During that time, he also recorded some demos for Elektra Records. "I think they're available on that bootleg of my stuff, *Dusted in Memphis*. I like the writing on 'My Rival' and a song called 'A Little Fishy.' I did those and a couple of other straight kind of things for a real record label so they wouldn't think I was totally crazy. I hadn't met anybody who didn't think I was crazy in about five years." Elektra's reaction to Chilton's material: "I never heard any reaction at all. They didn't pick it up, I'll tell you that." "Bangkok" is a turning signal that lead down the path to Chilton's 1979 pseudo-reckless masterpiece, *Like Flies on Sherbert*.

The Cramps were another inspiration for *Flies*, and also for the Memphis band the Panther Burns. Chilton met the Cramps in New York in 1977. They were a band from middle America that had moved to Manhattan and, in the face of the exploding punk rock movement, forged their own sound playing horror movie rockabilly. Such a renegade spirit was a natural attraction for Chilton. "I became a big fan of theirs. I would go see them play any time they were in the city. It turned out they rehearsed in the basement where Lisa Aldridge lived, and one day I went over to her house and there they all were in her apartment. We started talking and I suggested we do some recordings together. I told them I could arrange to do it for free at Ardent in Memphis, and when it's all said and done, they could have the tapes and do whatever they want with them. It's the kind of deal I know I would like. So they said okay, so we did."

The Cramps had built a reputation around a small section of New York and were completely unknown outside of that. Shortly after arriving in Memphis, they performed at a college outdoors. Their regalia was a bit baffling to many of these students who'd never left the Bible Belt. They'd seen movies with bikers, so they had some idea about all the leather, but the flamboyant sexuality, the male singer in high heels and leather underwear; it all added up to something beyond them. "We performed outside," says vocalist Lux Interior, "and all these jocks wanted to kill us. At first they thought it was something from New York where you had to have a program and have read five books to understand what it was. Then they got into it. I put Bryan Gregory on my shoulders and his guitar became unplugged. We ran around the crowd going *'ma-mammamamamamamama'* in their faces, while they're hitting their fists into their palms. We won them over, and they were just going nuts. It ended up with them throwing empty beer cans way up into the air and there was just a shower of empty beer cans coming down on us. It was really cool." The Cramps released their Chilton-produced tracks as singles, compiling them onto *Gravest Hits,* which won them international acclaim. They returned to Memphis two years later to record *Songs the Lord Taught Us,* working at Sam Phillips after Chilton had recorded *Flies* there.

When *Flies* was begun, *3rd* was still on the shelf, where it had been gathering dust for four years. "My life was on the skids," Chilton says, "and *Like Flies on Sherbert* was a summation of that period. I like that record a lot. It's crazy but it's a positive statement about a period in my life that wasn't positive." *Flies* began at Phillips, a collaborative effort between Chilton, Dickinson, Sid Selvidge, and Richard Rosebrough. Dickinson produced, Selvidge provided the record label and manufacturing, Rosebrough was a house engineer at the studio. Knox and Jerry Phillips, in their usual sympathetic and philanthropic way, made their studio accessible. Rosebrough remembers, "Alex was getting into his most degenerate mode. He looked really bad, just degenerate enough to where we all wanted to make a tape."

"When I conceived of doing *Like Flies on Sherbert,*" Chilton says, "I thought Jim and I and maybe one or two other people would record. When I turned up for the session, Jim had his whole band there! But I didn't say anything, I thought we should try it and see how it goes. We started recording and I thought, 'Man, these guys don't know the songs,' and I was trying to teach them and they'd go, 'Yeah, we know the songs,' and then they just go and play the first thing they thought of. So we were rolling the tape and doing this outrageous-sounding stuff, and I thought, 'Man, this must sound terrible,' but when I went in the control room and heard what we'd been doing, it was just incredible sounding. Getting involved with Dickinson opened up a new world for me. Before that I'd been into careful layerings of guitars and voices and harmonies and things like that, and Dickinson showed me how to go into the studio and just create a wild mess and make it sound really crazy and anarchic. That was a growth for me."

"I'd been running a bush hog that afternoon before we started," says Lee Baker. "I caught some barbed wire in the thing and it started whipping around. I felt a sharp pain in my leg, got off the thing, got the wire out of the bush hog, and went back to bush-hogging. My leg was still hurting so I reached down there and I had a fucking piece of barbed wire sticking out my leg. I went to the studio that night and I wasn't really feeling too good. And that was the first night of those sessions. I had no idea what we were going to do. Alex had the tunes and he showed us the songs and we played 'em. It was the same as at the Ardent sessions, except with *Sister Lovers* the tunes were weirder. And they were weirder partly because everyone was taking all them damn downs."

Flies is an epitome of Memphis music—a complete rejection of the industry norm. It embraces what traditional studio set-ups reject. It is sloppy, often indecipherable, and very, very alive. *"Like Flies on Sherbert,"* a critic has written, "painfully confirmed the degradation of a once-major talent." Such a statement reveals a dickhead writer with a bad record collection. On the album we hear Chilton bumping into the microphone and then laughing about it— where else but Memphis goddamn rock and fucking roll are you going to hear that allowed onto the finished product? "Fixing" that would have made a super alive moment into stone dead rock. *Flies* presaged the wave of American punk, foretold the return to roots rock, and once again sent out a trend from Memphis that the city itself couldn't swallow until, like a baby bird, it was given predigested versions.

Like Flies on Sherbert was recorded, mostly, over three nights in 1978, mixed over the year that followed, and released to an unsuspecting world in 1979 by Selvidge's Peabody label. While it takes only three minutes to record a song, Chilton emphasizes that sorting it out takes a lot longer, "especially if you cut things that're really crazy." Only five hundred copies of the album were pressed (a British label subsequently pressed a version), and I suspect that through a series of phone calls over a short time, one could locate the individuals who own four hundred of them. Peabody had little distribution, but Chilton's ardent fans managed to acquire their copies.

It's unlikely anyone could have been prepared for what they heard. Like a classic R&B performance, the album opens with a band track, the leader waiting in the wings. The song features drummer Ross Johnson, a librarian with a capacity for verbiage not dissimilar to Dewey Phillips's, and (then) a capacity for beer not dissimilar to a keg's. He and Chilton establish a hard-driving riff, very immediate sounding, upon which Johnson unleashes an extended "Red, Hot & Blue" tale, informing your jugular that whatever else follows will bear little resemblance to the Alex Chilton you have previously known. This was the culmination of the concept that Selvidge saw Chilton experimenting with at the Procapé: If I were a thirteen-year-old right now and I were just learning my instrument, how would I play guitar?

"On *Flies*," says Dickinson, "Alex told me he had gotten too good to play the kind of guitar that he was interested in. Almost all of the piano on *Flies* is Alex."

The Procapé experiment was a success. "You can't pretend to fall together loosely," Chilton says. "You can't know the song really well and then pretend not to know the song and come out with the same effect. Sometimes the first time through on something, people play in a whole different way than they ever play it again, you can hear them learning and hear them thinking the way they'll never think and approach it again." *Flies*, like *3rd*, captures that elusive quality.

There was another quality they were unable to catch. "Alex was real enamored with the way his voice sounded when he woke up," remembers Rosebrough. "So we made these elaborate plans in advance to have everyone at the studio at eight A.M., get the tape on the machine and be ready, and then somebody banged on Alex's door to wake him up. By the time he got to the studio, he was already awake and he'd lost the voice. So he took a handful of Valium. That didn't work, but we tried."

Though Chilton was an established songwriter by the time of *Flies*, fully half the record is other peoples' songs. A similar balance has continued in his work to the present day. But the songs Chilton covers are usually retrieved from deepest obscurity, abandoned gems that serve to direct his listeners along veins of other treasures. Critics who chastise Chilton for not writing all his own material ignore the value of his song scouting. Among the sources for *Flies* are the Greenbriar Boys' bluegrass, the Long Island vocal group the Belltones, and the Carter Family's interpretation of a slave song. If that's not a likely Dewey Phillips set, I don't know what is.

Chilton has written more than his share of great material, so much that it has ensnared him. People don't want Liz Taylor to grow older, they don't want Elvis to be dead, and they won't let Alex Chilton change. But his interpretations are usually his own, and they serve his listeners as signposts to discovering artists that the rest of the world has missed. "I'd like to write some more great songs," he says. "But I'm tired of people bitching at me about 'How come you only write three songs a year?' Fuck it, how many great songs are people supposed to write in a year? If somebody writes one great song in two years—one great song in a lifetime—that's plenty. People are kind of unrealistic about songwriting these days. It's an expectation that the Beatles sort of are responsible for. And if you ask me, the Beatles were stretching themselves way thin." Indeed, the mountains of compact discs that are released every month are a rehashing of old song ideas. The challenge could be not to write your own version of a song, but to make an established song your own,

"Ubangi Stomp" or otherwise. Reinterpretation is a standard practice in jazz, is common in rhythm and blues, and in pop does not preclude a hit. Chilton's interpretation of Ernest Tubb's "Waltz Across Texas" uses a familiar song to create a new message; Tubb's version does not evoke Bonnie and Clyde.

On the morning of Alex Chilton's twenty-eighth birthday, while *Flies* was being created, the newspaper announced Chris Bell's death. He'd had a single-car accident the night before, his 1977 Triumph striking a utility pole at one-thirty A.M. on December 27, 1978. He'd lived to see the rerelease of the first two Big Star albums earlier that year, and enjoyed some acclaim for "I Am the Cosmos," a single he recorded in 1974 that featured Chilton on background vocals. Rykodisc's 1992 compilation of his work, *I Am the Cosmos,* shows an artist in the throes of overwhelming sadness and despair. "Chris had become a very committed Christian by the time of his death," says John Fry. "Nobody really says anything about that." *Cosmos* makes it clear, especially in such songs as "Better Save Yourself" and "Look Up."

"*Flies* nearly killed us," says Randall Lyon, who videotaped much of the recording sessions. "It was a horrible experience from beginning to end. By that time, everybody had a bunch of attitude. It was not a good moment—it's good on record, yeah, but it's because we were all at each other's throats. The music was so heavy. Chris Bell died while Alex was working on that record, and *Flies* to me, is the end process of the whole Chris Bell/Alex freakout. With Chris is where Alex had begun his own individual journey, and when he died, Alex began a period of transformation. It was a crisis. I remember Alex would come over to my place and listen to Tommy James and the Shondells and to Hank Thompson. He was sussing out different parts of things he was interested in. He had to decide what he was going to do, as an artist. He had already been everything. And he turned into this wonderful person. What a survivor that guy is. His influence, his music has been a real motivator. There's so many important movements to him. But he didn't have it easy, he struggled with that shit. I remember when 'The Letter' came out for the first time on a TV collection of the sixties, it was real embarrassing. It really kicked his ass."

"I saw Alex in the parking lot of Goldsmith's department store when some Big Star reissues were coming out along with some new recordings," remembers Mary Lindsay Dickinson. "He was with his mom in his car. He was laughing that he had four records coming out and his mom was taking him to buy some new clothes."

"If you could imagine, even yet, he's a cultural icon and gotten paid damn little," says Selvidge. "Like Furry Lewis would say, Where's my money? Still to this day, Alex gets very little recognition by the labels. They'll put out stuff that's safe, only after everybody's named songs after him."

Perhaps it was *3rd*'s initial release during the punk explosion that refocused attention on Big Star's first two albums. One response to the late-seventies anger was late-seventies melody, and in Big Star's early work, bands like R.E.M., the dB's, and scores of others found inspiration. Rock history now places Big Star on a plane similar to that of the Velvet Underground; it has been said that anyone who heard the Velvet Underground's early records went on to form a band, and the same may apply to Big Star's influence. "I always am amazed that all this stuff has the interest and the following that it does," says Fry. "It's really pretty obscure in terms of the relative handful of records that were out there for anybody to get ahold of. It really strikes me as curious that everybody keeps coming back around to it. I thought the Big Star records were good, and it was fun to be involved in it, particularly the first two, and it was probably worth doing the third, but I didn't expect people to be asking questions about them twenty years later. Music, to me, is essentially an emotional communication, and it either connects or it doesn't connect. I'm not sure why, but all these people resonate to this Big Star business."

Big Star's enduring popularity is intimately related to its obscurity. Their melodic teen angst is a contradiction that continues to reverberate with listeners. But it's not just their music that keeps people's attention. In Big Star's history fans confront the fear of having something important to say that no one will hear. It's taken twenty years, but Big Star has prevailed. The band's cult status helps listeners realize their lives are not in vain.

In 1985, after a long hiatus, Chilton returned to the studio sober and serious, recording the first of a series of R&B-influenced releases. It was a return to his pre-British Invasion roots. That investigation continued right on into a 1993 Big Star reunion CD. Following that, he did not succumb to the obvious commercial step of releasing a new album of Big Star-ish material; instead, he released a solo acoustic disc of vocal standards associated with singers who draw reflexive groans from college kids because they remind them of their parents. *Clichés* forgoes a band for just his voice and acoustic guitar. Drawing on the songs and artists that wafted up the stairwell from his parents' salon below, Chilton creates a direct and personal statement that draws on his past to discuss his present. Unmitigated by other musicians, this effort yields an unusually strong sense of immediacy.

Parallel to Chilton's solo career, Big Star's reunion has assumed its own life. Despite continuing claims of retiring the idea, the band remains in demand, appearing in England, Japan, major cities in the U.S., and finally a homecoming in Memphis. No longer tentative about the material, Chilton and Jody Stephens, aided by two members of the Posies, deliver the old songs with new life. Guitar solos with a 1990s edge enhance the exuberance of Big Star's material, so that fans who may have anticipated an exhibit from the museum of rock and roll history leave the reunion shows with hopes of a new Big Star album.

Alex Chilton is a musician looking to the future, haunted by a following wrapped up in his past. The weight of his achievements creates a dissatisfaction trap: His fans demand, Why don't you do again what you've already done? In the bigger scheme, he does: Big Star was contrary music in 1972, and *Cliches* is contrary to Big Star's latest popularity. The constant is Chilton's aggressive and defiant attitude, while his music, his art, continues to evolve. In this industry, controlling one's own destiny is as difficult as collecting monies owed. His has been a circuitous route, and the path ahead is unclear and unsure—exciting. Despite the naysayers, despite the soothsayers and the imitators, Alex Chilton charts a personal course.

CHAPTER FIFTEEN

Attempted Gawk

MEMPHIS IS A TOWN OF DONUT SHOPS AND CHURCHES. WHATEVER MU-
sical achievements it claims are achievements in spite of the city, not because of
it. A Beale Street merchant, Abraham Schwab, whose family business has been
in the same location for over one hundred years, has a saying: Memphis has
torn down more history than most other cities even have. During President
Nixon's urban renewal program of the 1970s, when the destruction of Beale
left only a shell of its former self, the city tried to make Mr. Schwab leave his
site, for the benefit of the street. He wouldn't budge. They condemned the
building. Today, if he's not too busy stocking items that sell for ninety-eight
cents each, two for a dollar, he will take you on a tour of his basement, pointing out the sturdiness of his structure, laughing about the city's drastic measures against him. Insensitive destruction is part of what defines Memphis.

The country blues from the Delta, the urban blues that evolved on Beale,
Sun Records, then Stax and American and Ardent—Memphis never embraced
them. Blues were shunned as the music of black people, rock and roll the
music of degenerates. During the 1980s, the civic government made an attempt to support the scene, enticing Chips Moman back to town and building
him a studio a block off Beale. The city learned it had no business in the record
business. The very thing that they're afraid of is the thing that made it work.
"The Memphis sound is something that's produced by a group of social
misfits in a dark room in the middle of the night," says Dickinson. "It's not
committees, it's not bankers, not disc jockeys. Every attempt to organize the
Memphis music community has been a failure, as righteously it should be.
The diametric opposition, the racial collision, the redneck versus the ghetto

black is what it is all about, and it can't be brought together. If it could, there wouldn't be any music."

On October 1, 1978, before Alex Chilton's *Like Flies on Sherbert* had been recorded, Mud Boy and the Neutrons gave a farewell performance in the Orpheum Theater, an old-time movie palace that had escaped death by bulldozer like a Saturday serial's hero. It was not the first time Mud Boy retired, nor was it the last. The Tennessee Waltz, as the show was dubbed (a jab at the Band's farewell appearance, the Last Waltz), was the third event in a trilogy sponsored by Dickinson. These shows commemorated the release of *Beale Street Saturday Night,* an album he produced that demonstrated the fluidity running through the diverse branches of Memphis music. (One of the record's achievements was reaching the unconverted. Paid for by a bank, the album was sold at various society functions, at Schwab's store on Beale, and in the Orpheum lobby. Few were sold through record stores.) The first show was a jazz and blues album release party featuring Furry Lewis, Thomas Pinkston, Phineas Newborn Jr., and Fred Ford. The second was a rockabilly tribute named "Red, Hot & Blue" after Dewey Phillips's radio show. The lineup included Harmonica Frank Floyd, Barbara Pittman (one of Sun's few female artists), and the Mississippi soul man Jerry Saylor. The third show was given to rock and roll. Mud Boy played two sets, acoustic and then with a rhythm section.

Their decision to retire was a response to the plug-pulling incident at the Beale Street Music Festival the previous May. An article previewing the show in the *Commercial Appeal* quoted Lee Baker: "The [festival] promoters treated us like we're second-rate, like we're 'white boys' and we don't know how to play." Sid Selvidge was also quoted: "We've been promoting old black folks for years and never taken a nickel from 'em and at great expense to ourselves. We've been knocking City Hall over the head since nineteen-sixty whatever, and all of a sudden it becomes economically feasible for downtown to rip off 'some old niggers.' It's because it's in their economic self-interest—not because it's in the city's or the artists' self-interest—but in their own, lily-white, East Memphis interest to be downtown, and the old bluesmen can help 'em sell it." Mud Boy was angry.

The Tennessee Waltz was a multimedia affair, a Dream Carnival with respect for the proscenium arch. Bill Eggleston and his two preteen sons were onstage with video cameras, recording the event while displaying it on monitors to the audience. The dancing girls, whose quiver had offended the city in May, were front and center. A few other documentarists appeared, and overall, the musicians were pleased to be outnumbered on the stage. Not long before the music began, a former member of the Big Dixie Brick Company, the Dream Carnival's theater troupe, asked Dickinson if he could perform during the intermission. His name was Gus Nelson, but he was about to birth an alter ego. He was thinking of a country blues revival with a punk aesthetic, and it would become the torchbearer of Memphis music, however dim the coals. He would take the name Tav Falco and call the idea Panther Burn. "He came up to me at the show," recalls Dickinson, "and said softly, as is his way, 'Would it be alright if I sang a song, Jim?' I didn't know he sang. I said sure, I thought he meant with the band. He said, 'That's alright, I'll accompany myself.'" In the audience was Alex Chilton.

"I always like to be set up as an antienvironment to something that's happening," says Tav Falco, applying his theater background to music. "That day there was this rock and roll thing going down, so after Mud Boy's first set I went out onstage and set up a guitar, a big black-and-white television monitor, and a Bell and Howell speaker from a motion-picture projector that I was running the guitar through. And I had an electric chainsaw on a stool and an electric skill saw on another stool." Tav cultivated his Charlie Chaplin looks and donned a tuxedo with fingerless gloves, further separating himself from the swirl of color he followed. "This was my first time singing in public. As soon as I began, all these TV cameras came down front and onstage. I had this strange way of playing guitar—nobody else was playing like that that day, believe me—and I started my treatment of Leadbelly's 'Bourgeois Blues.' There was no band up there, just one person on electric guitar with all this sound reinforcement equipment and it just—*phoom!*—was out there, pretty powerful sounding. I was shocked by it because I had never sung through amplified equipment.

"I got further and further into 'The Bourgeois Blues,' you know, 'Home of the brave, land of the free, I won't be mistreated by no bourgeoisie,' and, 'This is a bourgeois town.' I worked up into the height of this frenzy, when I started blowing this police whistle and laid the guitar between the two stools and got the electric chainsaw and started ripping into this guitar. The sound, man, it was just complete sound. Extremely chaotic. People started screaming in the audience, going crazy. And then I got this skill saw and ripped through that guitar, still plugged into the sound system and it created all these scrunching, tearing, slicing, electronic, bursting, exploding noises. And WHBQ is filming in color and there's all these other cameras and Little Bill's got mine going out to the audience on video and I just kind of collapsed. People had to drag me offstage."

"I saw Tav play 'The Bourgeois Blues' and I was really impressed," says

Chilton. "I was interested in looking at a little country blues then and that was his thing, so I started going over to his house. He had a lot more material, and I began showing him some rockabilly stuff and he was interested in that. We just sort of—I said, 'I know a drummer,' and we called up Ross Johnson and we were a band, bingo." Alex charged Tav with leading the band so he could finally enjoy the role of sideman. His presence sanctioned the Panther Burns. Deriving their sound from Tav's initial performance, the band's name reflected the lore surrounding Panther Burn, Mississippi. This town was menaced by an elusive wild beast that, when finally cornered, was set aflame. Its dying shrieks so horrified the citizens that they named their community for it. The moniker was appropriate for Tav's assembly. He says, "I didn't know music, and Ross didn't. We put Eric Hill, an art academy student, on a very crude synthesizer and he didn't know music. Alex was the only one that had an understanding of musical form, which he thought would make it more interesting. The whole No Wave scene was going on in New York, and the Cramps were playing very fundamental, rudimentary things. Here was an art form I could participate in by just picking up the instrument, like a Kodak Instamatic camera. It was the feeling and aesthetic that mattered, more than musicianship or virtuosity. I didn't feel hindered by my lack of conventional guitar knowledge. I just went into it full tilt."

The Panther Burns sounded like wind howling through the cracks in a house. They took Sonny Burgess's "Red Headed Woman" from the Sun label and made it a Memphis punk statement. Falco sifted through the Moon Records catalog, an early Memphis label inspired by Sun and run by a rocking lady named Cordell Jackson. Their revival of her songs "Dateless Night" and "She's the One That Got It" gave new life to her career. She ultimately found wild popularity in a Budweiser commercial, where her expertise on the Hagstrom guitar smoked Stray Cat Brian Setzer. The Panther Burns booked themselves with Charlie Feathers, introducing rockabilly's most devoted practitioner to another generation of fans. They did the same with north Mississippi bluesman R. L. Burnside. In a fallow period of Memphis recordings, the Panther Burns regularly brought studio work to the finest local talent, including Wayne Jackson and Andrew Love—the Memphis Horns, and trumpeter Ben Cauley who had survived Otis Redding's plane crash. Dickinson and Chilton usually alternated as producers. Nearly fifteen years since its inception, the band lives on, their sound a little more polished, their musicianship on equal footing with their theatricality.

When the Panther Burns formed, Falco was a partner with Randall Lyon in a nonprofit video company. TeleVista Projects, Inc., inspired by Eggleston's work, had released documentaries on rural blues joints, profiles of various bluesmen, and interviews with diverse people—the Memphis painter Carroll Cloar, the former Arkansas governor Orval Faubus (who had defied the federal integration order in Little Rock's Central High School), and with Fred Martin, the first black man elected to public office in Arkansas since Reconstruction. By the eighties, their company was still extant on paper, though most of their equipment had been repossessed. By making the Panther Burns the TeleVista orchestra, they received a grant from the National Endowment for the Arts to participate in a global linkup of videographic slow-scan phone-line transmissions.

Tav secured a spot on a local morning talk show, hosted by the polite, matronly Marge Thrasher. Booked during May, their appearance coincided with the Memphis Cotton Carnival, festivities built around secret societies akin to the krewes of New Orleans' Mardi Gras. Befitting an "orchestra's" appearance, Thrasher arranged to have the King and Queen of Cotton Carnival as guests.

"We built this network with Memphis, New York, Toronto, Victoria, Vancouver, and the San Francisco Museum of Art," says Lyon. "When we were on the show, we were sending the signal through telephone transceivers up to a satellite, down to other television stations. We were using broadcast TV systems, cable TV systems, and phone lines. It was one of the first international cyberspace experiments. Corporations were blown away that artists were going to use this technology. Law enforcement and financial markets were the only ones who'd previously used it. Nobody knew what we were doing. We were over the top. After that, I started talking to people about wanting to play the Mississippi River bridge like a musical instrument."

The episode aired live at nine A.M. This was long before the band's musicianship and theatricality were equals. The fleshed-out group featured a second nonmusician on synthesizer, and a trumpeter who wore a wacky horizontally-striped suit and kept his back to the camera. They were a motley crew. The first song was a treatment of the Burnette Brothers' version of "Train Kept a Rollin'." Instead of caterwauling power-punk, the band was restrained and spare, the tentative search for each next note producing a surreal quality. A melody was evident, though camouflaged. Between shots of the band's performance, Lyon interjected slow-scan stills of the same, black and white, the images being sent to the San Francisco Museum of the Arts, the Franklin Street Furnace in Manhattan, and the other participants in the global linkup. At the first song's conclusion, a befuddled Ms. Thrasher asked questions, to which Tav responded:

Thrasher: This is anti-music, is that right?

Falco: An anti-musical environment. The Panther Burns would like to do one more tune—

T: Wait a minute. Wait a minute. That may be the worst sound I've ever heard come out on television.

F: Thank you very much.

T: That's what you want, I'm assuming?

F: Well, the best of the worst is what we're after.

T: Let me get all this straight. Are you all also part of the federal grant of money?

F: No, we're simply an orchestra to accompany this image-feed. You have a sound track and then you have a picture track. In this case, it's a live situation. So we're the live orchestra accompanying the imagery.

T: If I had realized that I'm not sure I would have wanted you to represent the King and Queen of Cotton Carnival. That would not have been my selection of music. [Cutting off Tav's inquiry] Gustavo, what kind of field are you in, because you are so big on telling me that you are anti-music. I don't understand.

F: I don't think anyone else is playing music like this in Memphis or maybe anywhere else in the world.

T: I don't think they are either.

F: In that case, we're doing something quite different, see, something that is not part of the establishment, not part of our everyday environment. We have to create an anti-environment to make more visible real musicians, people who are early Memphis performers—

T: Why do you have to be anti to do it?

F: Because it's all invisible to us. We can't see what's around us. There are blues people here who don't have exposure, rockabilly artists who don't have any exposure. They don't really exist here, they're part of our environment, we see them every day, yet they're invisible to us. We take them for granted. It takes a group like us to create contrast, to create focus—

T: Do people pay you to play this?

F: Occasionally, but we're not in it for the money. We're in it for something else.

T: Art.

F: I don't know if it's art. It's art damage. It's—

T: You're really very bitter, aren't you?

F: I'm not bitter about anything. I get exhilarated by this kind of music. Highly exhilarated.

T: So if you have an outlet, that brings the exhilaration to the forefront.

F: It brings it to a peak, yes.

T: Do you think there's a market for your kind of music?

F: I'm not sure. We're exploring markets. But I'm not that concerned about the marketplace. I'm more concerned about a stage area, a communications environment in which we can set up and focus on our complete environment which has become invisible to us. So we need other forms of music to be able to see what's genuine and what's authentic in our musical environment.

T: Why don't you introduce the band members to us.

F: [Tav introduces Eric Hill, Vincent Wrenn, Ross Johnson, Rick Ivy, Alex Chilton.] We would like to do one more tune, which is a rock and roll tango—

T: Gustavo, we're not quite ready for it. Okay. We're gonna take another break here on "Straight Talk" and we'll be back in just a moment.

The band was allowed to perform their second song, the tango "Drop Your Mask," after which Thrasher said, "An all-time low this morning on 'Straight Talk.' I want to talk about some of our upcoming guests on the program. . . . "

"Because it's all invisible to us," Falco told her, and it is as true in Memphis today as it was when Robert Johnson was ignored by every white man he passed on the street. A Picasso of the blues, Johnson was made transparent by the cultural mindset. That attitude was clearly evident forty years after Johnson's death when one local TV station tried to rectify the situation, producing and broadcasting a one-hour special seeking "to show the variety of music associated with the city . . . blues, gospel, hard rock, rhythm and blues, pop, country and opera." It preempted the John Davidson show at seven P.M., and the following day, the *Press Scimitar* ran some of the viewer comments noted by the station's switchboard operator:

6:55 When is the "John Davidson Show" on?
7:13 Get that Negro — off the TV!
7:35 You won't have anyone looking by the time Marguerite Piazza comes on the program.
7:47 If I want to watch monkeys, I would go to the zoo.

The establishment proffers donuts and churches, ignoring the hole in their middles. Memphis music had come up from the neighborhoods and the neighbors of these diverse musicians, had reached all corners of the earth, breaking down barriers that seemed impenetrable, and Memphians continued to insult Furry Lewis with the basest comments. That same year, 1976, a Memphis disc jockey watched his Estelle Axton-produced comedy song go to number one on the charts; "Disco Duck" eventually sold over five million copies. Rick Dees recently set a fiscal record when he signed a twenty-five million-dollar radio syndication contract with ABC. "Memphis," says Wayne Jackson, "is a tension that expresses itself in music."

On June 5, 1977, Sleepy John Estes died in poverty, and his funeral arrangements were left to his Memphis friends. Bukka White had passed three months earlier. Gus Cannon died November 16, 1979. He was 104. Furry Lewis died at eighty-eight on September 14, 1981. The pawnshop scheme was turned one more time, and invisible Memphians bought him a headstone.

One cold night in the early eighties, the painter Charlie Miller was leaving a club where he'd been painting the dancers. One of the girls was waiting out front for a ride that hadn't shown up. "I told her I'd give her a ride home. On the way, she asked me if I'd stop at one of the convenience stores we passed. She came out of there, had a box of Faultless Starch and a Coca-Cola. We were riding along and she opened that box up and started scooping it out eating it. I said, 'I didn't know you could eat starch like that.' She said, 'Oh yeah, it'll fill your belly up when you're hungry.'"

While on a world tour with Ry Cooder in 1983, Jim Dickinson reunited with Memphis entrepreneur Isaac Tigrett in London. In the course of the day, Tigrett played a tape by a since-deceased Memphis barrelhouse piano player, Big Sam Clark. "I made some sort of bitter remark," says Dickinson, "and Isaac got furious. 'You have no right to be bitter,' he told me. 'You were fortunate enough to witness the end of something truly great, and intelligent enough to understand some of it.' On the road, alone in a hotel room, I thought about that a lot. He is absolutely right. I'm not bitter anymore. I may remain pissed off, but I'm not bitter."

The questions about contemporary Memphis are, Where are the hits? Where are the stars? And the extremely agitating, Why can't Memphis be like Nashville? To the last, one can only respond that Nashville is a company town and Memphis is for renegades. And the relationship works well for both: Nashville acts come to Memphis studios to "get away with" what the corporate environment won't allow, and plenty of Memphians regularly travel up the pike to hawk their talents in the marketplace. The ways of the two cities are as different as New York and Los Angeles, and the comforts of home are as close as a three-hour drive.

A misconception about Memphis is that all the labels departed. But Memphis has never been a record label town. The companies that thrived here were independents, Sun and Stax being the most prominent; in the industry picture, they were rebels. They created a distinct sound—in the case of Sun, a distinct genre—and then they died. Sun died as Sam Phillips returned to his original interest in radio; after Johnny Cash, Jerry Lee Lewis, Howlin' Wolf, and Elvis, what was left for him to do? Stax died when it lost its focus on Memphis music. When they tried to play the big boys' game, Stax was unmercifully hacked by record distributors, oil conglomerates, and—the unkindest cut of all—a local bank.

Memphis has long been a studio town, a place where a person with an idea can find a laboratory to experiment, to create. Since soul music, the city's impact on the music world has been spare. But the blues remains the essence of each new generation's ideas. While other cities have flared up—Athens, Georgia; Seattle; Miami's bass sound; Chicago's house sound—the coals in Memphis still burn, even if coated in ash. "It really is as simple as the blues, Sun, Stax, bam bam bam," says Dickinson, "the same basic thing happening. Memphis created a product that obviously was art because art is enduring, and entertainment is transitory. This stuff that's been done here, by people who were often completely unaware of what they were doing, has endured. The Delta blues is going to be one of the most significant western contributions to the twentieth century."

In 1978, after more than a decade of not producing, Sam Phillips returned to the studio. John Prine, who has often recorded in Memphis, was cutting *Pink Cadillac* at the Sam Phillips Recording Service, with Knox and Jerry Phillips producing. "We had been trying to get a good take on a song called 'Saigon,'" says Knox Phillips. "It was a great song, but we were doing it at like a hundred eighty miles per hour. I thought that was the right way to do it. I had told Sam, 'Dad, this guy sings so bad, you'll love him.' So Sam comes down to help us with the album."

Richard Rosebrough, who was engineering the session, remembers, "It took doing a session with Sam Phillips to make me believe. There was a magic that that man had in his fingertips. There was a magic in that man's eyes, they saw forty feet right through you. The words that came out of that man's mouth were bizarre. I had heard this story that when he cut the Yardbirds, he told the bass player to whip that guitar like a mule's peter. Well, now I believe he said it. When John Prine was working on this song called 'Saigon,' Sam says, 'I want you to slow down that song and draaaaag it out.' The way he used this word 'drag' was the whole thing. 'I want you to drag it out like you're dragging it across the street so you can find out what's on the other side.' It was all in Sam Phillips's control, totally.

"He decided he was going to put some echo on this guitar that started out the song. They had live echo chambers then. He put one hand on the echo-send and the other on the echo-return and he cranked it up. You could hear the speaker disintegrate. You could see sparks down the hall. Everyone went, 'No. No! NO!' sort of automatically reaching for the board. It turned out to be the most magical song on the album. A couple days later I was making tape copies of the whole album. Alex Chilton was down there, hanging out in the lobby on the sofa. This was around the time of his *Like Flies on Sherbert,* so it was going to take a lot to impress him. He heard a song go by and he yawned. Then another. When 'Saigon' started, he slowly got up, followed that sound into the control room. He said, 'What *is* this?' That was Sam Phillips. Sam knows something we don't know."

I heard Sam Phillips once say, "Producing? I don't know anything about producing records. But if you want to make some rock and roll music, I can reach down and pull it out of your asshole." Yessir. Sir.

Across from a donut shop in Midtown, a decaying building is sandwiched between a VCR repair store and a dentist who seems to rarely have customers. It is one of Memphis's last bastions of independent rock and roll. The Antenna Club has no sign, its shabby facade expressing all that anyone needs to know. In the late 1970s, the venue was known as the Well, a beer-only joint where the clientele brought fifths of cheap whiskey to while away the daylight. The Well attracted some musicians living cheaply in Midtown's run-down elegance, and they began performing on the club's small stage in return for the cover charge; the regulars did not have to pay. When punk became a fashion statement, the older couple who ran the place sold it for more than they thought their dive would ever be worth.

Tav Falco and his Unapproachable Panther Burns honed their chops at the Well. The Panther Burns, with Alex Chilton's assistance, have cut a swath for others to follow, often restraining branches so those trailing won't get whipped in the face. From the beginning, Tav employed the Burnettes, a shifting lineup of nubile femmes, and as that group stabilized, they formed an entity of their own. The all-girl Hellcats released a couple albums on the French New Rose label, the outlet that was un–record company enough to gain Chilton's cooperation in the mid-1980s, and that subsequently developed a healthy relationship with the Memphis talent around him. Lorette Velvette, now a solo artist, took her first plane trip with the Panther Burns—a twenty-hour flight to Australia.

After living in Memphis for several years, my girlfriend referred to the Antenna Club scene as car crash music, likening it to a wreck on the highway that makes you slow down as you pass, gawking. The Panther Burns were gawk music. The theatrical basis of their beginnings made them a sight to see, and their drama was enhanced by the musical barbs lobbed from the one musically proficient member. Describing a show in San Francisco, Tav says, "We cleared the house. It was great." The Hellcats, and their phoenix offspring the

Alluring Strange—the band includes three of the five Hellcats—work the same tension. The first time I saw the latter perform they were utterly thrilling, each note sounding as if it might be their last, the whole mess lurching forward with a defiant bravura. Rehearsal has tightened their sound, but sparks still fly on the stage, and on their debut recording, *Will You Marry Me?*

The Country Rockers are misunderstood as gawk music. Their visual appeal is strong: A trio with an octogenarian dwarf drummer and septuagenarian backwoods vocalist and guitarist. But their music is as pure as roadhouse whiskey. Tying them to the modern Memphis scene, their bassist, manager, and producer, Ron Easley, is a regular member of Alex Chilton's touring band and also of the Panther Burns. (The Country Rockers' two completely unpretentious European albums are available domestically on a single CD, *Free Range Chicken.*) Some audiences have the same trouble with Cordell Jackson. Her speed and dexterity on the guitar warrant a listener's attention, but seeing it come from "the rocking granny" in her frilly southern gowns causes people to stare.

A punkish guitar-blur band called the Grifters embraces the independence of the Memphis spirit. Their series of singles drew courtship from several major labels, leading to proffered contracts of five figures. The band has so far declined all, opting for a tiny local label where they know their integrity will not be compromised, as surely as they know the same assurances from the majors were bullshit. Neighborhood Texture Jam has converted the washboard to a larger piece of corrugated tin, enhancing the industrial aura of their rootsy modern rock.

Several of Mud Boy's kids lead popular Memphis bands now. Dickinson's two back up their old man playing blues, rockabilly, and river silt, but on their own, as DDT, create a blistering rock with funk undertones. Selvidge's youngest leads the city's most popular neo–hip hop band, Big Ass Truck. They use their urban R&B mindset to reexamine soul highlights, incorporating samples with a live deejay. Lee Baker's children are younger, but they have already shared the stage with Dickinson's and Selvidge's, launching Son of Mud Boy, aka Three-Legged Puppy.

With over thirty recording facilities listed in the 1994 phone book, bands have no trouble finding a place to get their ideas on tape. Ardent remains busy, though focused more on big-budget projects. The preeminent alternative studio is Easley Recording, validated early on by Chilton. Their work has become identified as the sound of alternative Memphis, and people buy recordings solely because the Easley name is on them.

Not all of the contemporary acts are garage bands. One of the most exciting performers in Memphis is the Grammy-nominated gospel artist O'Landa Draper and the Associates. In his early thirties, Draper commands his sixty-voice choir like a fine painter works the variegated hues of a color. I knew I'd seen something when he rippled the voices across his choir and the image of flowing water was overwhelming. He can command an audience as proficiently as Rufus Thomas, inspiring hundreds of people at once so that each feels Draper is speaking only to him or her. But gospel music is ostracized from popular music, quarantined on specialty stations. Although Draper and his choir accompanied Billy Joel on the 1994 Grammy Awards, the gateway to the mainstream has yet to swing open.

Memphis still sits isolated in a large rural region. In less than an hour's drive, you can be in communities where electricity and indoor plumbing remain uncommon, where juke joints with wood stoves provide a glimpse of blues roots that all the compact disc reissues in the world could never equal. In the city, there are cinder block clubs in black neighborhoods that hint at the old feel—without trying. Outsiders entering either of these types of places must cross a cultural bridge and overcome a sense of invasiveness not unlike that faced by the hippies and folklorists of the sixties.

"All in all, it's been a great battle." Randall Lyon is not wistful with his summation. He continues to influence people, writing for a biweekly newspaper in Little Rock and dispensing advice to young bands trying to get a toehold on today's modern music world. "The cultural forces that prevail, a totally anti-racist, anti–high art/low art approach to the blues—that comes from Bob Palmer's work. People forget there was a school that said there is no African music in the blues, and Palmer refuted it. If that discussion had to take place, we added something to it. And it was going to take place because the records were coming out. People were missing the point. No, these artists aren't playing guitar in a crude manner. That intonation and pitch was worked out in a context. It made people realize that traditional Western scales weren't really the most exciting work ever done. It made people hear music in a different way. The very presence of the artists was inspiring. They had been through so much more than we had been through. And they had worked it out, as a lifestyle. They were the troubadours of the heroic furor."

Robert Palmer is today spread as wide as his ken. He is shaping a ten-part public television series on the history of rock and roll, has codirected a film called *The World According to John Coltrane*, and was the on-screen narrator for a film titled after his book *Deep Blues*. He also continues to work on a book that isolates and follows the strains in music that later became rock and roll.

"It's really surprising," he says, "that some of the major rock and roll bass parts turn out to be from Africa by way of Afro-Cuban music. There's an amazing amount of Afro-Cuban material in early rock and roll—not influence, but actual patterns, riffs, licks."

In John McIntire's twenty-fifth year at the Memphis College of Art, a new administration came in and he was the first faculty member to be let go. He sculpts full-time now, still sifting through yard sales to make an extra buck.

Roland Janes, the Sun guitarist and Zen riddle, is still at the Sam Phillips Recording Service, sipping soda and eating candy bars. The studio doors continue to open and close behind musicians, and as the sound of the street has changed, he has rolled with it. "A lot of people kid me about rap," he says, "and I kid about it myself, but I've worked with enough of it now that I feel like I understand it, the music and the gimmick. It's a message music, really, the message of the streets, told in its rawest form. When most of the musicians first come in here, they're really trying to impress upon me that they're whatever the image they're trying to be. But after they get to know me awhile and they listen to a few of my corny jokes and we get to communicating, then suddenly they become like everybody else. That's the way I see it."

Randy Haspel regularly stops by the studio, greeting Roland, visiting his old partner Bob Simon, who writes in the B room, pushing aside the very board on which Sam recorded Elvis to make desk space. Haspel nearly had a big break in the 1970s, but his audition for John Hammond at Columbia Records was ill-fated. His success as a songwriter is nothing less than a tribute to Dewey Phillips, the artists who have covered his material ranging from Rufus Thomas to George Jones, from Moe Bandy to the Impressions. He still gets a charge out of playing clubs, where the Radiants make filling the dance floor look easy.

As casinos have proliferated in the upper Mississippi Delta, Herbie O'Mell has found himself in demand. He was host of the first one to open, then was hired away by a larger enterprise. He continues to manage affairs for Jim Dickinson, working with O'Landa Draper's gospel choir as well. "Last year I was out at a place and a guy walked up to me, an older guy, kind of rough-looking, and he said, 'You Herbie O'Mell?' And I said yeah, and he said, 'Ya probably don't remember me, I'm a friend of Campbell's.' I said, 'Oh, okay. Nice to see ya.' And he said, 'If you ever need anything, here's my card.' He gave me his card and it had his name, a phone number and it said, 'When in doubt, knock 'em out.'"

Teenie Hodges continues to write crossover hits, recently landing a song on Bonnie Raitt's 1994 *Longing in Their Hearts;* working with his brothers and cousin Roland Robinson, the Hi Rhythm Section released an independent CD, *Perfect Gentlemen,* in 1994. Willie Mitchell is still in the same studio where he cut all the Al Green hits. His control room has been upgraded, but the studio floor remains untouched, the same 1970s carpets and styles oozing feel all over the place. William Brown stays busy as his chief engineer, and Willie's grown children have a rap act of their own, the M-Team, and produce many others.

David Fleischman never got anything national going with Flash and the Board of Directors, but as a promo man for Atlantic, he worked his way into the national office, heading up campaigns behind Ratt, Twisted Sister, Julian Lennon, Mike and the Mechanics, and Robert Plant. He is currently a vice-president at MCA Records.

All the surviving Mar-Keys are still active in music. Cropper and Dunn, along with Booker T. Jones, released a new Booker T. and the MGs album in 1994, after touring behind Neil Young and Bob Dylan. The Memphis Horns (Wayne Jackson and Andrew Love) released a new album in 1992, produced by Terry Manning, who remains active in the studio. Terry Johnson still makes time to write songs. Smoochy Smith has joined Stan Kesler, Sonny Burgess, and a few other friends, performing rockabilly as the Sun Rhythm Section.

"There's something in Memphis that makes people a little crazy," says Don Nix. "Since I moved out of there, I'm not near as crazy as I was. It's something about that town. As far as the music goes, it's always been there, and it always will be, but it's hard to tap it. It will come around again someday, I'm positive of that. It might take another hundred years or it might be next year, I don't know. But Memphis, it's there. It's just there."

"Memphis is the town where nothing ever happens but the impossible always does," says Danny Graflund, sober since evicting Other Man from his life. "People come here and they either love this place, or they don't understand and can't see what's going on. It's its own little thing, and it's always happening." With a video camera at his side, Graflund is attempting to document that which is so hard to see. He is often hired by Mary Lindsay Dickinson, whose latest project is a collaboration with Ardent on a CD-ROM for kids.

Mose Vinson has sobered up since his performance at my high school. He performs every Saturday on Beale Street at the Center for Southern Folklore, and every Sunday at his church. He was featured in 1992 on the National Public Radio program "BluesStage," and in 1994 he entered the studio with Jim Dickinson and recorded a solo CD.

Jimmy Crosthwait performs his puppet show for school kids, working out of the museum system. Recently, he ran an art gallery in Eads, Tennessee, outside of Memphis, where his shows included photograph exhibitions by Stanley Booth and Jim Dickinson. He gave the gallery to someone else when it started drawing too much attention, and he spent two years designing and building a home of his own, progressing from sculptures that move to a sculpture he can move into.

After releasing his major label debut in 1993 and playing Carnegie Hall, Sid Selvidge accepted a commission to write a folk opera, choosing as his subject the Memphis flood of 1923. Bill Eggleston is the scenic director. Eggleston remains prominent in modern photography. Though some magazines became more fascinated by his lifestyle than his art, his books, including *The Democratic Forest* and *Faulkner's Mississippi*, continue to exemplify exciting modern photography.

Lee Baker still rides a tractor around the lake in Arkansas. He has reunited with several former Moloch members and they play around Memphis as the Agitators. Baker and Selvidge recently played an acoustic gig to an audience of five on Beale Street. During the break, Baker and I talked gardens. He'd been told that ground red pepper would keep raccoons away from his corn, but said it wasn't true. "You know what, though," he said, "you get a radiator leak, put that ground red pepper in your radiator, it'll stop it up. That's good to know, 'cause you never know when your radiator will spring a leak. Might not be near a can of Stop Leak, but they got red pepper all over the country." Five of us heard some amazing guitar that afternoon.

Jim Dickinson remains anathema to the music business, producing several albums a year, some over a matter of days for the smallest companies, others done with more grace and bigger budgets. He insists that most of his projects come to Memphis, removing them from the record company loop and helping them broaden their boundaries. Whenever possible, he spreads the money around, bringing in vocalist William Brown or saxophonist Fred Ford. "As sure as I am that corporate rock is overfeeding the public, I am equally sure that the public will get tired of it," he says. "Sooner or later they get tired of everything, they want the antithesis. Always. And as long as that happens, the corporate mentality has to make a place for people like me. Because they only understand the here and now, they don't understand the possible maybe of the future."

Those energy vectors that drew Robert Johnson and Otis Redding and Johnny Cash and Al Green and all these artists to Memphis, those vectors still point here. Disbelievers say the reason can't be in the earth, can't be a product of the dirt or the river. But they ignore empirical facts: People who come to Memphis notice cultural collisions. Other cities may have similar black and white populations that interact or segregate themselves exactly as Memphis does, but something about this city tunes our antennae to such things. Whether knowing its history we project it, or we are drawn to it by forces we cannot see, race relations, also known as music, is the lifeblood of Memphis. The first song to top the pop, country, and rhythm and blues charts came from Memphis forty years ago, Carl Perkins's "Blue Suede Shoes." Memphis music is a concept, not a sound.

This writing occurs four decades after Elvis made his first recordings with Sam Phillips. Rock and roll is middle-aged and fat. Since the 1970s, when the music passed it adolescence, there has been less risk taking. Despite constant predictions of its death or demise, it still struggles to be the voice of rebellion, sponsored by corporate conglomerates that are the object of the overthrow. But the art of rock and roll has moved from the musicians to the businessmen: it's an art of the bottom line. Mass sales, mass popularity. The globe is the market, billboards dwarfed by the Goodyear blimp, dwarfed by the space shuttle. Having achieved the capacity to sell millions of cheeseburgers and hamburgers dressed all different ways, marketing's next step is to sell tens of millions of one kind of hamburger. Moving hundreds of thousands of pieces of product no longer means success. Upstart artists who score hits are ruined by misguided follow-up expectations; upstart artists who don't score hits are abandoned. The business does not allow time for artistic growth. The cultural relevance of music has been replaced by the cultural relevance of sales.

But all is not lost. The Michael Jackson syndrome does not deny the Jim Dickinson factor. As more of the same is heaped on the public, they will, by nature, become bored by that and demand something else. Memphis lives for the "possible maybe of the future." The city has had its days and years in the spotlight, and in this interim (this long, long interim) between thrusts, there are still many people here making a living through music. The machinery is in place. Recording studios proliferate, bands are signed to big and small labels, writers still get their songs on records.

Not long back on a not very busy afternoon, Dickinson and I were a block off Beale Street at a studio. When the doorbell rang, we walked together to the lobby. Dickinson opened the door and a youngish black kid was there. He said, "Is this building a recording studio?" Yes, it is. "I want to make a record, man," the kid said. Dickinson didn't laugh at the naive ambition of the stranger, and he didn't send him off on a goose chase to other studios. Rather, he gave the young man the studio owner's name and phone number, told him he'd have to get permission from him to come in and work. The stranger left, the ticket in his hand, and may or may not have ever made the necessary call. In the early part of this century, the blues came up from Memphis earth, and later rock and roll and soul music followed. Somewhere in Memphis a kid is channeling that same spirit today, and the world waits for that person to come knocking.

Further Reading, Watching, and Listening

≒┼┾

MANY INSIGHTFUL WRITERS HAVE WRITTEN BOOKS ON MEMPHIS MUSIC and Memphis culture. Peter Guralnick's *Sweet Soul Music* (Harper and Row) is a warm introduction to both, with an emphasis on Memphis soul music. Guralnick profiles several other prominent Memphis musicians in his collections *Feel Like Going Home* and *Lost Highway* (both Harper and Row); his biography of Elvis Presley, *Last Train to Memphis* (Little, Brown), has recently been published. Stanley Booth has long been a part of the Memphis scene, and his collection of essays, *Rythm Oil* (Pantheon), is a personal account of both well-known and obscure Memphis musicians. Booth's writing is elegant, and his Rolling Stones book, *Dance with the Devil* (Random House), should not be missed. Michael Bane's *White Boy Singing the Blues* (Da Capo) can't decide if it wants Memphis obscurity or a broader pop music context, and the vacillation probably works in its favor; don't be fooled by the picture of Elvis on the cover of the reissue.

Greil Marcus's *Mystery Train* (Plume) is a provocative and informative examination of roots music and touches on several Memphis artists. For information about the Sun label, turn to *Good Rockin' Tonight* (St. Martin's) by Colin Escott and Martin Hawkins. Louis Cantor, a former disc jockey on WDIA, has written a history of the station entitled *Wheelin' on Beale* (St. Martin's). A handsome pamphlet entitled "LXFi" bootlegs two extended interviews with Alex Chilton, from which I drew several quotes. For a history of William Eggleston and his work, the essay that accompanies his *Ancient and Modern* is a good overview.

And for a solid introduction to the blues, Robert Palmer's *Deep Blues* (Penguin) is outstanding; it's a rare opportunity to drink champagne with Muddy

Waters. Another good book on southern soul, though hard to find, is Barney Hoskyns's *Say It One Time for the Brokenhearted* (Fontana), which focuses on the relationship between black soul and white country music. Amiri Baraka (LeRoi Jones) presents a more sociological approach to blues history in *Blues People* (and he continues the discussion, emphasizing jazz, in *Black Music*).

The larceny and evil of the business behind the music may be an art different from the songs, but as Fredric Dannen's *Hit Men* (Times Books) proves, it's not boring to read about. *The Death of Rhythm and Blues* (Plume) by Nelson George reveals the appetite of the pop machine and how it can swallow a whole culture. Jerry Wexler's autobiography, *That Rhythm Those Blues* (Knopf), is a less dark peek behind the scenes.

For another feel of Memphis, James Conaway has recently published a memoir, *Memphis Afternoons* (Houghton Mifflin), which is a look at what some other people were doing in Memphis around the 1950s. *Memphis Since Crump* (University of Tennessee Press) by David M. Tucker details how integration came to the city that wouldn't. Joan Beiffus's book, *At the River I Stand* (St. Luke's Press), is an oral history of the sanitation workers' strike during which Martin Luther King Jr. was assassinated.

Some other companions in spirit: everything by Nick Tosches; *Love in Vain* (Da Capo), a screenplay by Alan Greenberg about Robert Johnson; poet Etheridge Knight's *Born of a Woman* (Houghton Mifflin); *Let Us Now Praise Famous Men* (Houghton Mifflin) by James Agee and Walker Evans; *Juke Joint* (University Press of Mississippi), a book of photographs by Birney Imes; *Sermons and Sacred Pictures,* an experimental documentary by Lynne Sachs that uses archival film footage from the Memphis black community as its foundation. (Home use: Center for Southern Folklore [see Chapter One below]; institutions: University of California Extension Center for Media and Independent Learning, 2000 Center Street, 4th Floor, Berkeley, CA 94704.)

The following recordings, audio and video, are sources for more information about the subjects in each chapter. Some that were mentioned in the text are repeated here, but many are not.

CHAPTER ONE

Allow me to toot my own horn. *All Day and All Night* is a documentary I directed and edited that uses Beale Street and Beale Street musicians to discuss points similar to many of those made in this book. Rufus Thomas and B. B. King, among others, tell how the Beale Street community nurtured its artists, and through WDIA, helped them go national. The awards the film has received indicate it is as much fun to watch as it was to make. (Center for Southern Folklore, 130 Beale Street, Memphis, TN 38103; 901-525-3655).

There is a lot of scattered film footage of Furry Lewis around, but the piece that first springs to mind is his performance of "Going to Brownsville" in *Good Morning Blues,* a documentary available on the retail market through Yazoo. Another of their reissues, *Out of the Blacks into the Blues,* is also a good genre overview. For listening to Furry, I'd start with *Fourth & Beale* (Lucky 7/ Rounder), recorded by Terry Manning at Furry's home in 1969 and issued only recently. *Live at the Gaslight* sometimes surfaces in used record bins and it has given me hours of enjoyment. And of course, *Furry Lewis in His Prime* on Yazoo.

Mose Vinson went back in the studio in 1994 and was recorded by Jim Dickinson, under the auspices of the Center for Southern Folklore. That recording is available through the Center. Mose, and several of the other 1970s blues survivors, are represented on *The Devil's Music,* an album released in conjunction with a BBC documentary.

The furor of the blues festival incident when the plug was pulled on Mud Boy is evident on a videotape that has no name and is not very widely circulated. The next best thing is Mud Boy's live album, *Negro Streets at Dawn* (New Rose), and really it's a pretty close second.

CHAPTER TWO

The Dewey Phillips album compiled by Charles Raiteri, *Red, Hot & Blue* (Zu-Zazz), is certainly the best way to get a taste of Dewey's power. A Memphis collector recently discovered three more Dewey Phillips acetates at a yard sale, so if anyone is interested in expanding the album for CD issue, be sure to get those extra tracks. Some excellent transcriptions of WDIA gospel broadcasts are collected on the album *Bless My Bones* (Rounder). And Gatemouth Moore, the former WDIA disc jockey, is the eccentric and very entertaining subject of Louis Guida's documentary film *Saturday Night Sunday Morning.*

CHAPTER THREE

The 1968 Memphis Sanitation Workers' Strike, mentioned in this chapter and several others, is examined in depth in the documentary *At the River I Stand.* Three Memphis filmmakers utilized a cache of film footage from the era. It's available to universities and other institutions through California Newsreel (149 Ninth Street, San Francisco, CA 94103).

CHAPTER FOUR

Jim Dickinson's early singles on the Southtown label pop up periodically; his 1972 solo album, *Dixie Fried* (Atlantic), has yet to be issued on CD; if its time ever arrives, it should include bonus tracks. To get a sense of the jug band music, start with *Frank Stokes' Dream*, a collection on Yazoo that also has some great Furry Lewis tracks; a CD called *Wild About My Loving* (RCA) is also good. The sound of the West Memphis scene is probably lost to memories and beer, though some hints can be heard on the Hi Records box *Hi Times* (Right Stuff/Capitol), which includes tracks from Willie Mitchell, Ben Branch, and Bowlegs Miller. Dickinson's *Beale Street Saturday Night* album—used bins in Memphis only—is a modern interpretation of old Beale; the outtakes to that project are called *Delta Experimental Projects Volume 1* (New Rose); *Volume 2*, with which it's coupled on disc, is some of Dickinson's solo soundtracks (including two films by yours truly, *Southern Dust* and *Down*).

CHAPTER FIVE

No record collection is complete without the Mar-Keys' "Last Night," which is on their debut album, *Last Night* (Atlantic). It's also included on *The Complete Stax Singles 1959–1968* (Atlantic). Much of the Stax catalog is now available on CD through Rhino and Fantasy; some have gotten audiophile treatment from Mobile Fidelity Sound Lab. Some other essential Stax listening, beyond the hits: Booker T. and the MGs, *And Now!;* Isaac Hayes, *Hot Buttered Soul;* William Bell, *Soul of a Bell;* Steve Cropper, *With a Little Help from My Friends;* and a live revue recording, *Funky Broadway*.

If anyone is looking for another theme around which to compile a CD, I suggest collecting the best of Charlie Freeman's guitar artistry. It's scattered from early Memphis studio recordings to the Dixie Flyers and beyond, and would tell a tall tale were it all in one place.

CHAPTER SIX

The Alan Lomax series of field recordings that inspired many of the white artists in this book has been reissued as a CD box entitled *Sounds of the South* (Atlantic).

CHAPTER SEVEN

Many of the Ardent singles are available in used record bins around Memphis. Some of the Jesters tracks are included on the Sun Records box *Into the Sixties*

(Bear Family). There's a great *Talent Party* video compilation waiting to happen; George Klein has several reels of choice material, including many prominent artists. They span a decade or more and give a great feel for the times.

CHAPTER EIGHT

Some good blues revival listening: Sleepy John Estes, *Legend of Sleepy John Estes* (Delmark), and his *I Ain't Gonna Be Worried No More* (Yazoo); Fred McDowell, *Amazing Grace* (Testament/Hightone); Bukka White, *Sky Songs* (Arhoolie). The Everest label has proven very trustworthy. Though their packaging is deceptively chintzy, their recordings are consistently exciting and intimate, and I've found it worth buying anything they've put out.

I never heard the Hodges brothers' JAMF, but I'd guess it couldn't be far from the Bar-Kays' *Soul Finger* (Stax), which was issued around this time.

A lot of blues revival film clips have been issued on video, though they seem to be dominated by spiritless performances in uninspiring college auditoriums. You've got to dig for the bright moments. Vestapol, Yazoo, and Shanachie boast extensive catalogs.

CHAPTER NINE

Though each of the Memphis Country Blues Festivals was recorded, only the latter two ever saw audio release. *The 1968 Memphis Country Blues Festival* (London) is long out of print, but I've heard rumors that it was an early issue on CD when the labels weren't yet sure who the CD audience was. The artists from the 1969 show were recorded in the studio; that double LP, *Memphis Swamp Jam,* has been reissued on two discs as *Mississippi Delta Blues Jam in Memphis* (Arhoolie) and includes bonus tracks. A recent issue of an old Reverend Robert Wilkins session, *Remember Me* (Genes), includes one track recorded at the 1969 show and it is incredible. For more information about these festivals, seek out Stanley Booth's article, "Even the Birds Were Blue" (*Rolling Stone,* April 16, 1970).

CHAPTER TEN

The Box Tops albums and various greatest hits packages are readily available in used bins (a reflection on sales, not quality) and they remain good listening. To get a taste of the blues background that Dan Penn refers to, two good starting places would be Bobby Bland's *Two Steps from the Blues* (Duke/MCA) and Ray Charles's *Genius + Soul = Jazz* (Dunhill). As for Dan Penn's solo albums, buy 'em if you find 'em. It's difficult to pick a place to start for the American Rhythm Section—other than just turning on the radio, oldies or new country—but you certainly will not have heard everything until you've heard Dusty Springfield's *Dusty in Memphis* (Atlantic); the CD does not include Stanley Booth's original liner notes. The Sweet Inspirations' "Sweet Inspiration" is another personal favorite. The Rhino compilation *The Muscle Shoals Sound* presents the diversity of that house section.

CHAPTER ELEVEN

No better place to hear Moloch than on their hard-to-find album. The version of the group that included Busta Jones released one single, "Cocaine Katie," on Booger Records. The Don Nix albums are a mixed lot, tending toward overproduction, but several of them have their moments. The double live LP with the Alabama State Troopers includes a whole side of Furry Lewis; the Shelter and Elektra albums are the better bets. His Albert King album on Stax, *Lovejoy, Illinois,* is great. Ever innovative, Don's recent efforts include the Great Southern Musical Memories calender, which draws from his extensive photograph collection, and a book he's currently finishing; one section of the book will be his stories and memories, the second section will be photographs, and the third will be recipes he has collected from various musical luminaries.

The Dixie Flyers need a CD of their own, but until that happens, seek out their work behind Aretha Franklin, *Spirit in the Dark;* Ronnie Hawkins, *The Hawk;* Delaney and Bonnie, *From Bonnie to Delaney;* and Carmen McRae, *Just a Little Lovin'* (all Atlantic).

CHAPTER TWELVE

If you can find the cassette of Mud Boy's *Known Felons in Drag* on Sid Selvidge's Peabody Records, it comes from a different master and has more punch than the French vinyl. Two entrances to Phineas Newborn's oeuvre: his first release, *The Piano Artistry of Phineas Newborn Jr.* (Atlantic/Fantasy), which also features Calvin Newborn, and Phineas's later *Solo Piano* (Atlantic). Big Star's first two albums have been recently reissued by Fantasy on a single disc and are not hard to find; beware the import version, which looks similar but has deleted a couple tracks. The live document recently issued, *Big Star Live* (Rykodisc), is also recommended; rumors are circulating about a soon-to-be-released live-in-Memphis reunion video.

CHAPTER THIRTEEN

Sid Selvidge's best albums (*Waiting for a Train, The Cold of the Morning*) are on his own label and can be purchased through Pop Tunes in Memphis (Pop Tunes, attention Burge, 308 Poplar, Memphis, TN 38103); a fifteen-dollar money order for each CD would include shipping. There has long been talk of making *Stranded in Canton* a laserdisc, and I believe that a deal has been struck with a Japanese company; the work, however, remains undone. There is also talk of reissuing *William Eggleston's Guide,* and a new book is on the horizon. His sons, by the way, manufacture high-end hi-fi speakers; for a pamphlet write EgglestonWorks, at 378 S. Main Street #1, Memphis, TN 38103. The Barbarian archives need desperately to be reissued, and anyone with a pocketful of money wishing to do so is invited to contact me or Jim Blake. Barbarian's Lisa Aldridge and Jim Dickinson singles have been trading for upwards of two hundred dollars.

CHAPTER FOURTEEN

At press time, plans are being finalized to issue Alex Chilton's *Like Flies on Sherbert* (Peabody) as a CD; bonus tracks will be included, and there is talk of including new photos from the sessions. It is essential listening, as is *Big Star 3rd,* now readily accessible thanks to Rykodisc. Chilton's recordings from the 1980s have been recently made available in America through Razor & Tie. He is also in excellent voice on a Chet Baker tribute called *Medium Cool* (Rough Trade).

CHAPTER FIFTEEN

Okay, here's where I plug my own work again: I compiled (but did not name) a cassette called *A Slice of the South,* culling various southern musics from the audio archives at the Center for Southern Folklore (see Chapter One for the address). The tape ranges from a one-string guitarist to fife and drum music to a fiddler imitating a cat fight. It's good.

For modern rural blues, nothing in the world beats the Fat Possum label (P.O. Box 1923, Oxford, MS 38655). Robert Palmer has taken time off from his writing to produce several of their albums; his two Junior Kimbrough albums, his two R. L. Burnside albums, and his Cedell Davis album are essential to any contemporary blues library. He also produced the soundtrack to *Deep Blues* (Atlantic), a film by Bob Mugge about rural blues in the 1990s. (Mugge's must-see Memphis film is *The Gospel According to Al Green.* Another documentary that I found surprisingly good is *The Search for Robert Johnson*[Sony]. These and other good videos of all sorts are easily available [rental and sales] through Facets, 1-800-331-6197.) The Bullseye Blues label, a Rounder subsidiary, has been releasing some good contemporary Memphis blues, including barrelhouse pianist Booker T. Laury, blues-rocker Little Jimmy King, and new albums from Ann Peebles and Otis Clay.

A new Memphis reissue label is Memphis Archive (P.O. Box 171282, Memphis, TN 38187), which, under the direction of former Canned Heat bassist Richard Hite, has compiled several discs of rare country blues. Another source for older blues is the Smithsonian Institution, which keeps all of its titles in print. Two that you must have are *Leadbelly's Last Sessions, Volume 1* and *Music from the South: Country Brass Bands;* these are both in the Folkways series. Write or call for a catalog: Folkways, 955 L'Enfant Plaza 2600, Smithsonian Institution, Washington, DC 20560; 202-287-3262. The Trumpet label, of Jackson, Mississippi, recorded some amazing blues and gospel and their catalog is being tastefully reissued on CD by Alligator Records. The revitalized Capricorn Records has compiled several noteworthy boxed sets of smaller R&B labels. Many of these blues releases—and plenty more—are available through a great blues catalog published by Stackhouse Records, 232 Sunflower Avenue, Clarksdale, MS 38614.

Essential Tav Falco and the Panther Burns begins with their first EP, *Frenzi 2000* and includes *Sugar Ditch Revisited* and *Shake Rag.* You gotta hear the Cramps' *Gravest Hits* (Illegal), the Hellcats' *Hoodoo Train* (New Rose; the CD includes their first EP), the Alluring Strange's *Will You Marry Me* (Safehouse), Lorette Velvette's *White Birds* (Veracity) and her tentatively titled *Rude Angel* on the same label, and the Country Rockers' *Free Range Chicken* (Telstar). There's a hard-to-find compilation called *Swamp Surfing in Memphis* (Frenzi/ Au Go Go) that documents the diversity of the early 1980s Memphis underground from blues to garage art. A duo called OFB (Our Favorite Band) collected many of Memphis's mid-eighties finest to back them on their one release, *Saturday Nights Sunday Mornings* (New Rose).

My favorite album by the Amazing Rhythm Aces is *Too Stuffed to Jump* (ABC). John Prine's *Pink Cadillac* (Oh Boy) is a must. Reggae meets soul on Toots Hibberts's *Toots in Memphis* (Island).

These contemporary Memphis bands also have fine recordings: Big Ass Truck, DDT, '68 Comeback, Neighborhood Texture Jam, the Grifters. Look for singles on the Sugar Ditch and Loverly labels. The Gories' *I Know You Fine But How You Doin* (New Rose) was produced by Alex Chilton, and the Flat Duo Jets' *Go Go Harlem Baby* (Sky) was produced by Jim Dickinson.

Some other musts: Jerry Lee Lewis, *Old Tyme Country Music* (Sun); Charlie Rich, *Pictures and Paintings* (Sire); Blind Willie Johnson, *The Complete Blind Willie Johnson* (Columbia); O. V Wright, *The Soul of O. V. Wright* (MCA); King Curtis, *The New Scene* (OJC/Fantasy); *Memphis Slim with Guests* (Inner City Records); James Carr, *Dark End of the Street* (Blue Side); Larry Davis, *Funny Stuff* (Rooster) (Robert Cray's soul-blues ain't got nothing on this); Slim Harpo, *Knew the Blues* (Excello); Son House, *Father of the Delta Blues* (Columbia); anything by Howlin' Wolf.

When in Memphis, stop by the Center for Southern Folklore at the corner of Beale and Second, catty-corner from B. B. King's Blues Club. Their gift shop includes most of the High Water contemporary blues label. On Beale, also visit A. Schwabs, a general store from the old days with a great voodoo counter. Your visit to Beale will be enhanced by Richard Raichelson's recent publication, *Beale Street Talks: A Walking Tour Down the Home of the Blues*. There are two excellent music museums in the area. The Beale Street Blues Museum in the Old Daisy Theater traces the music's history from the Civil War to W. C. Handy. The story is continued around the corner (across from the Peabody Hotel) at the Memphis Music Hall of Fame.

There's a lot of information in *The Lowlife Guide to Memphis,* published by the friendly folks at Shangri-La Records. Send three bucks for your copy (1916 Madison, Memphis, TN 38104) and be sure to visit their store. Other good, used Memphis records can be found at Audiomania, All the Music, River Records, and Memphis Comics.

Several years back, the Memphis Brooks Museum of Art curated *Memphis 1948-1958;* the catalog includes an essay by Stanley Booth, an article about William Eggleston accompanied by early black and white photographs, and much more history. It's available in the museum gift shop.

Dine at the Buntyn (save room for the cobbler) and the Fourway Grill (barbeque aficionados are directed to Payne's on Lamar), drink cold beer at the P & H Café, talk to Jackie Smith, who lives on the street across from the National Civil Rights Museum and is protesting their exploitation of Dr. King's dream, and tune in to AM 990 for some serious Baptist preaching, and to WEVL-FM 90 for the best in indigenous music. Spend enough time in Memphis and you will appreciate that, truly, nothing ever happens but the impossible always does.

Index of Names

About the Author

Robert Gordon writes authoritatively about Memphis because he was born and raised there, and because, before writing about the city and its music, he was a dedicated member of the audience. His work has appeared in all major U.S. music publications; his documentary about Memphis blues, *All Day & All Night* was broadcast on American national television, was exhibited at the Museum of Modern Art in New York and won numerous awards.